Intermediate Algebra and Problem Solving

Alan Wise
University of San Diego

Richard Nation
Palomar College

Peter Crampton
Palomar College

Harcourt Brace Jovanovich, Publishers
San Diego   New York   Chicago   Atlanta   Washington, D.C.
London   Sydney   Toronto

Cartoon illustrations from the book *According to Guinness* © 1981 by Sterling Publishing Co., Inc., New York, N.Y.

Copyright © 1985 by Harcourt Brace Jovanovich, Inc.

All rights reserved. No part of this publication may be reproduced or transmitted in any form or by any means, electronic or mechanical, including photocopy, recording, or any information storage and retrieval system, without permission in writing from the publisher.

Although for mechanical reasons all pages of this publication are perforated, only those pages imprinted with an HBJ copyright notice are intended for removal.

Requests for permission to make copies of any part of this work should be mailed to: Permissions, Harcourt Brace Jovanovich, Publishers, Orlando, Florida 32887.

ISBN: 0-15-541505-0

Printed in the United States of America

To our families:
Carol, Denise, Gary, Greg, Chris, and Bob
Barbara and Lisa
Susana and Deirdre

This book is part of a three-volume series of guided worktexts. The series includes:

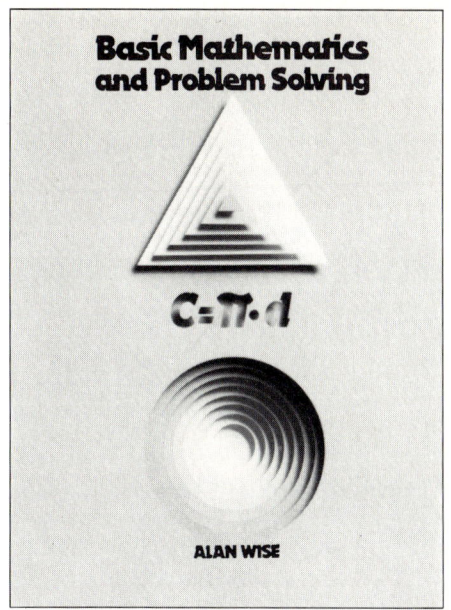

Recommended for a pre-algebra course or for a terminal course in arithmetic.

Topics:

WHOLE NUMBERS AND DECIMALS
Chapter 1   Basic Skills
Chapter 2   Addition and Subtraction
Chapter 3   Multiplication
Chapter 4   Division

FRACTIONS AND MIXED NUMBERS
Chapter 5   Basic Skills
Chapter 6   Multiplication and Division
Chapter 7   Addition and Subtraction

PERCENT AND PROPORTION
Chapter 8   Basic Skills
Chapter 9   Percent
Chapter 10   Proportion

MEASUREMENT AND GEOMETRY
Chapter 11   Measurement
Chapter 12   Powers and Square Roots
Chapter 13   Geometry

INTRODUCTION TO ALGEBRA
Chapter 14   Signed Numbers
Chapter 15   Equations and Rules

Recommended for a first course in algebra or for a diagnostic review of beginning algebra.

Topics:

INTRODUCTION TO ALGEBRA
Chapter 1   Signed Numbers
Chapter 2   Equations and Rules
Chapter 3   Equations and Properties

POLYNOMIALS AND FACTORING
Chapter 4   Exponents
Chapter 5   Polynomials
Chapter 6   Factoring

LINEAR RELATIONS
Chapter 7   Equations and Graphs
Chapter 8   Systems
Chapter 9   Inequalities (Optional)

NONLINEAR RELATIONS
Chapter 10   Rational Expressions
Chapter 11   Radical Expressions
Chapter 12   Quadratic Equations

APPENDIX REVIEW SKILLS
Fractions
Decimals
Percent

Recommended for a second course in algebra or for a pre-college algebra course.

Topics:

REAL NUMBERS AND LINEAR RELATIONS
Chapter 1   Real Numbers and Properties
Chapter 2   Equations and Inequalities
Chapter 3   Equations and Graphs
Chapter 4   Systems

POLYNOMIALS AND FACTORING
Chapter 5   Polynomials
Chapter 6   Factoring

NONLINEAR RELATIONS
Chapter 7   Rational Expressions
Chapter 8   Exponents and Radicals
Chapter 9   Quadratic Equations
Chapter 10   Conics and Systems

FUNCTIONS
Chapter 11   Functions
Chapter 12   Exponential and Logarithmic Functions
Chapter 13   Sequences and Series

Preface

The Worktext Format

Intermediate Algebra and Problem Solving is the third in a series of guided worktexts for college developmental mathematics programs. It provides students at various reading levels with a clear, informal, and nonthreatening means of achieving mastery of fundamental skills. In each lesson, specific easy-to-grasp learning objectives and solved examples form the core of the worktext approach; Make Sure trial problems and ample sets of write-in-text exercises then give students opportunities for both immediate reinforcement and extra practice. A functional pedagogical design highlights this format and creates an attractive, inviting setting for skills development.

Flexible Usage

The books in the Wise series can be used effectively in all teaching formats. In traditional lecture courses, the Make Sure problems can function as blackboard examples. For modified lecture courses, the Make Sures may be assigned as in-class practice. In laboratory and other self-paced approaches, the chapter Reviews will serve as keyed diagnostic tests; in conjunction with the *Student's Guide* and *Computer-Generated Testing System*, the worktext then becomes part of a complete self-study package.

Problem-Solving Instruction

A unique component of this text is a series of twenty Problem Solving Lessons. Complete and self-contained, these lessons teach strategies for solving traditional algebra problems. The authors provide a rich collection of interesting real-world facts and data from a variety of reference sources, with over 500 applied word problems depicting realistic situations that underscore the importance of knowing basic algebra skills.

Each Problem Solving Lesson uses a step-by-step example to illustrate a specific strategy that can then be easily applied to related problems. The Problem Solving Lessons, although a special focus of the Wise series, are nevertheless optional and may be omitted without affecting the main sequence of skills.

Ample Practice

Over 5000 exercises and word problems ensure student mastery of skills. More than 4500 write-in-text exercises are found in Practice, Make Sure, and Review sections, with all exercises referenced to corresponding worked examples. Over 500 word problems underscore the importance of knowing basic algebra skills.

Diagnostic Tools

Chapter Reviews are keyed to worked examples and provide instructors and students with a basis for diagnosis and a tool for checking mastery of skills.

The cumulative Final Review, included at the back of the book, is keyed to worked examples. A placement test for the entire series is included in the *Instructor's Manual with Tests*.

Supplement Package

The following supplements are available:

Instructor's Manual with Tests contains eight different printed tests for each chapter, multiple forms for unit and final examinations, one placement test for the series and helpful teaching notes.

Student's Guide to Make Sure Exercises contains complete step-by-step solutions to all Make Sure exercises.

Answer Book supplies answers to all Practice exercises and word problems.

Computer-Generated Testing System furnishes a limitless variety of practice sheets, quizzes, or examinations (with answers) for use in the classroom or laboratory.

Acknowledgments

Preparing a three-book series for publication requires the effort and skill of many people in addition to the authors. We are grateful to the following people for their many hours spent reading manuscript and for their sensitive and valuable suggestions for its improvement: Helen Joan Dykes, Northern Virginia Community College, and Timothy R. Wilson, Honolulu Community College.

We are also indebted to the many reviewers who have offered their comments and suggestions during the development of this series. In particular, we wish to thank Ruth I. Hoffman, University of Denver; Evelyn M. Neufeld, San Jose State University; James L. Malone, Nassau Community College; Pauline Jenness, William Rainey Harper College; Steve Hinthorne, Central Washington University; Mohammed G. Rajah, Mira Costa College; Jean M. Newton, St. Petersburg Community College; Alice Haygood, Alvin Community College; and Helen Marston, Rutgers University.

At Harcourt Brace Jovanovich, many people gave their time and talent in making this series possible. In particular, we thank: Mary-Ann Courtenaye, Lynn Edwards, Jamie Fidler, Don Fujimoto, Eleanor Garner, Florence Kawahara, Martha Mayall, Pamela Morehouse, and Candace Young.

A special thanks is due Bill Bryden, for initiating the project; Gary Burke for helping us out of many tight spots along the way; John Holland, for his insight and support in creating the organizational format for the series; Cathy Reynolds, for her patience and efforts in designing aesthetically pleasing and functional pages; Audrey Thompson, for her mathematical ability and many extra hours spent in ensuring that the series would meet its deadlines; and Richard Wallis, for representing and supporting our viewpoints at critical times during the project.

Last but certainly not least, we would like to express a very special thank-you to the people who are most responsible for the successful completion of this series: Margie Rogers, for her skillful editing and enthusiastic leadership and support throughout the project, which helped to hold us all together as a team; and our wives for their excellent typing and proofing skills and for their constant support and encouragement.

ALAN WISE
RICHARD NATION
PETER CRAMPTON

Numbered Examples systematically break down each Skill Lesson into manageable learning objectives. Each numbered Example, in turn, is made up of easy-to-grasp sequential steps.

The **three-column format** encourages self-teaching by showing students what to do, how to do it, and why it works.

A **second color** throughout each text highlights important material.

The **Make Sure** write-in-text exercises provide immediate reinforcement of new concepts and are referenced to each numbered Example. Answers usually appear below.

1.3 Compute with Signed Numbers   15

1.3 Compute with Signed Numbers

Addition of two signed numbers is classified as either the addition of two numbers with like signs or the addition of two numbers with opposite signs.

Rules for Adding Two Signed Numbers with Like Signs
1. Find the sum of the absolute values.
2. Write the same like sign on the sum.

Rules for Adding Two Signed Numbers with Opposite Signs
1. Find the difference between the absolute values.
2. Write the sign of the number with the larger absolute value on the sum.

EXAMPLE 1: Add two signed numbers.

Problem ▶	$-13 + (-21) = ?$	-13 and -21 are called *addends*
1. Find the sum (difference) of the absolute values ▶	$-13 + (-21) = ?$ 34	Think: Since the numbers have like signs, you find the sum of their absolute values: $13 + 21 = 34$.
2. Write correct sign ▶	$-13 + (-21) = -34$	Think: Like negative signs mean a negative sum.
Solution ▶	$-13 + (-21) = -34$	

When the addends have like positive signs, the sum will be positive: $13 + (+21) = 34$.

Another Example ▶	$-17 + (+28) = ?$	
1. Find the sum (difference) of the absolute values ▶	$-17 + (+28) = ?$ 11	Think: Since the numbers have opposite signs, you find the difference of their absolute values. larger absolute value ↓ $28 - 17 = 11$
2. Write correct sign ▶	$-17 + (+28) = +11$	Think: The sum of two numbers with opposite signs is positive when the positive addend has the larger absolute value.
Solution ▶	$-17 + (+28) = 11$	

The sum of two signed numbers with opposite signs is negative when the negative addend has the larger absolute value.

Examples ▶ (a) $3 + (-7) = -4$ (b) $-12 + 7 = -5$ (c) $18 + (-25) = -7$

The sum of two opposites is always 0.

Examples ▶ (a) $7 + (-7) = 0$ (b) $-31.7 + 31.7 = 0$ (c) $\frac{3}{8} + (-\frac{3}{8}) = 0$

Make Sure

Add signed numbers.

See Example 1 ▶   1. $-5 + 3$ _____   2. $-3 + (-4)$ _____   3. $12 + (-5)$ _____

MAKE SURE ANSWERS: 1. -2 2. -7 3. 7

---

x

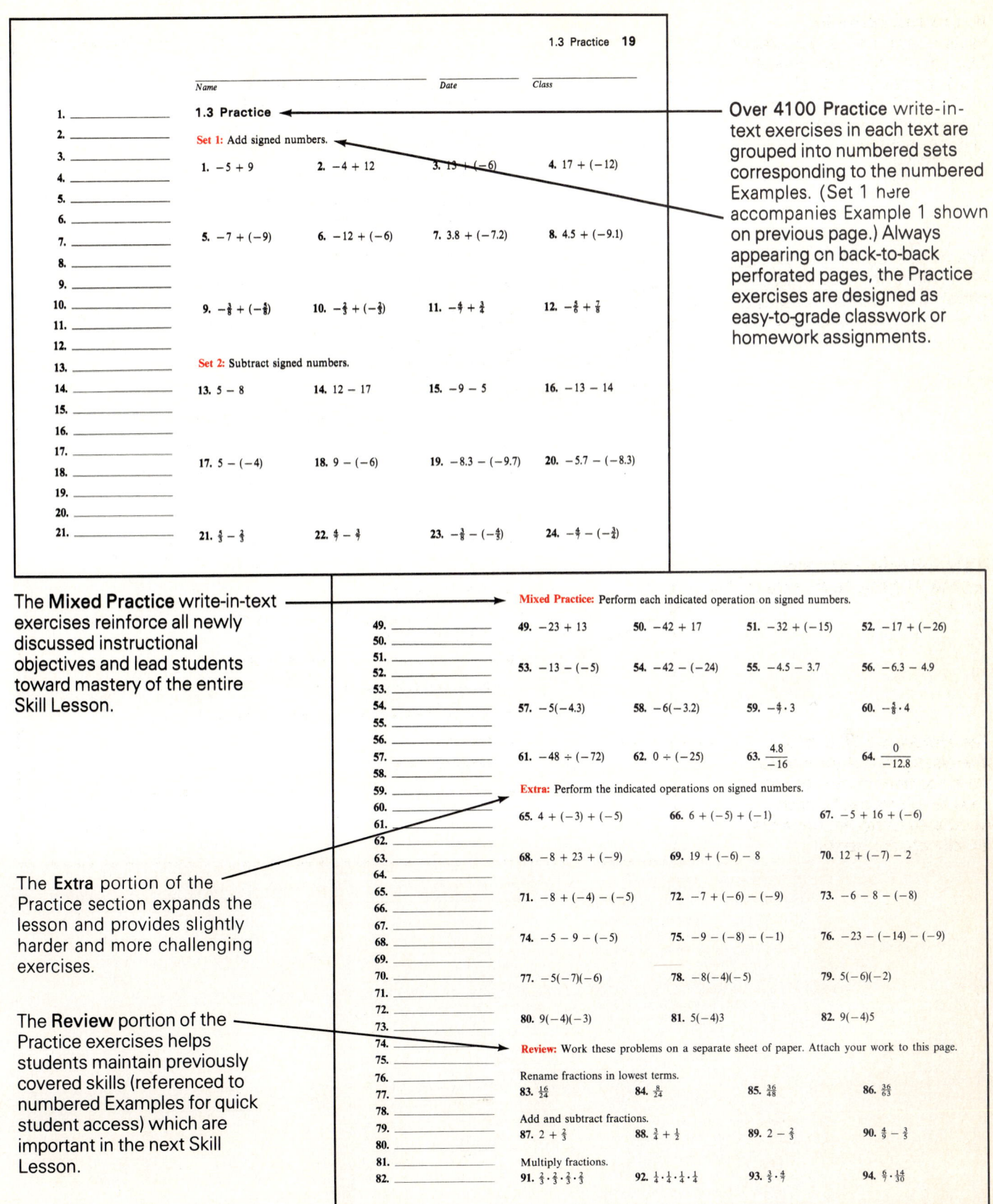

PS 3: Use Linear Relationships 117

Each **Problem Solving Lesson** teaches a specific applications-oriented skill.

Problem Solving 3: Use Linear Relationships

A relationship between two variables that can be described by a linear equation is called a *linear relationship*.

EXAMPLE 1: Solve a problem using a linear relationship given the rate and a data point.

1. Read and identify ▶ A company bus was purchased for $20,000. Assuming the bus depreciates at a constant rate of $4000 per year (straight line depreciation for tax purposes), what is the book value (v) after a time (t) of $3\frac{1}{2}$ years?

Remember, circle the facts and underline the question.

The **step-by-step example** explains a problem solving strategy that can be used for many other types of word problems.

2. Understand ▶ The question asks you to find and evaluate a linear equation in two variables $v = mt + b$ for $t = 3\frac{1}{2}$ given one data point (t, v) of $(0, 20{,}000)$ and the slope (rate) as $m = -\frac{4000}{1}$ (decreases $4000 in 1 year).

3. Decide ▶ To find the equation of a straight line given a point on the line and the slope of the line, you use the **point-slope formula**.

4. Find equation ▶
$v - v_1 = m(t - t_1)$ ◀— point-slope formula with $v = y$ and $t = x$
$v - 20{,}000 = -\frac{4000}{1}(t - 0)$ Think: $(t_1, v_1) = (0, 20{,}000)$
$v - 20{,}000 = -4000t$
$v = 20{,}000 - 4000t$ ◀— proposed equation

5. Evaluate equation ▶
$v = 20{,}000 - 4000(3\frac{1}{2})$ Substitute the given $3\frac{1}{2}$ years(t).
$= 20{,}000 - 14{,}000$
$= 6000$

6. Interpret ▶ $v = 6000$ means that in $3\frac{1}{2}$ years the bus will have a book value of $6000.

Note 1 ▶ In $v = -4000t + 20{,}000$, the negative slope means that the value of the bus is decreasing. That is, the line for $v = -4000t + 20{,}000$ slopes downward. The slope -4000 as a ratio $\frac{-4000}{1}$ describes the rate of decrease. That is, the bus decreases $4000 each 1 year.

Note 2 ▶ The graph of the line for $v = -4000t + 20{,}000$ begins at $t = 0$ and ends at $t = 5$.

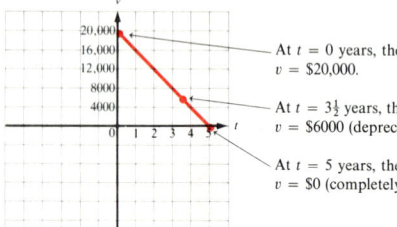

At $t = 0$ years, the book value of the bus is: $v = \$20{,}000$.

At $t = 3\frac{1}{2}$ years, the book value of the bus is: $v = \$6000$ (depreciated $14,000).

At $t = 5$ years, the book value of the bus is: $v = \$0$ (completely depreciated).

1a. _____
b. _____
c. _____

2a. _____
b. _____
c. _____

3a. _____
b. _____
c. _____

Practice: Solve each problem using a linear relationship.

1. A fast-food restaurant finds that daily profit (p) increases by $50 for each increase in sales of 200 hamburgers (h). When 500 hamburgers are sold, the profit is $100. a) Find the linear equation describing this linear relationship. b) What is total profit if 1100 hamburgers are sold in a day? c) How many hamburgers need to be sold to break even?

2. A certain store had a profit (p) of $462 on Monday when 210 customers (c) entered the store. On Tuesday, the number of customers entering the store was 200 more than on Monday, for a profit of $912. Assuming the profit (p) per customer (c) rate is a linear relationship, a) find the linear equation that describes this linear relationship. b) What would the profit be if 404 customers entered the store on Wednesday? c) What is the minimum number of customers needed to enter the store to break even?

3. There is a linear relationship between the temperature (t) in degrees Fahrenheit and the number (n) of times a cricket chirps per minute. A cricket will chirp 72 times per minute at 58°F and 132 times at 73°F. a) Find the linear equation that represents this linear relationship. b) What would the temperature be when

4. There is a linear relationship between the temperature (t) in degrees Celsius and the number (n) of calories needed by a person working. A person working needs 30 more calories per day for each degree that the Celsius temperature drops. An average person working needs 3000 calories per day at 20°C. a) Find the linear

Over 500 **Practice word problems** help students improve their problem solving ability.

xii

70 Chapter 2 Equations and Inequalities

Multiplying both members of an inequality by the same positive term produces an equivalent inequality. Multiplying both members of an inequality by the same negative term and reversing the inequality symbol produces an equivalent inequality. Together these form the *Multiplication Rule for Inequalities*

Multiplication Rule for Inequalities
If $c > 0$ and $a < b$, then $ca < cb$.   If $c < 0$ and $a < b$, then $ca > cb$.
If $c > 0$ and $a > b$, then $ca > cb$.   If $c < 0$ and $a > b$, then $ca < cb$.

Examples ▶ (a) $3 < 5$ therefore $2 \cdot 3 < 2 \cdot 5$ (b) $4 < 7$ therefore $(-2) \cdot 4 > (-2) \cdot 7$.

The Multiplication Rule for Inequalities can help you solve an inequality of the form $Ax < B$.

EXAMPLE 2: Solve an inequality using the Multiplication Rule.

Solve ▶ $-4n < 20$

1. Use Multiplication Rule ▶ $-\tfrac{1}{4} \cdot (-4n) > -\tfrac{1}{4} \cdot 20$ To get n alone in one member, multiply both members by $-\tfrac{1}{4}$. Since the multiplier is negative, reverse the inequality symbol.

2. Simplify ▶ $1n > -\tfrac{1}{4} \cdot 20$
   $n > -5$

Solution ▶ $\{n \mid n > -5\}$

Note ▶ The graph of $\{n \mid n > -5\}$ is: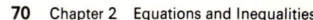
$-7\;-6\;-5\;-4\;-3\;-2\;-1\;\;0\;\;1\;\;2\;\;3$

CAUTION: Be sure to reverse the inequality symbol whenever you multiply both members of an inequality by a negative number.

Screened boxes highlight rules, important concepts, and key definitions for easy student reference.

Numerous **"Caution"** boxes, outlined in color, help students avoid common errors.

"According to Guinness" illustrations and accompanying word problems invite students to sharpen their problem solving skills.

According to Guinness

SMALLEST AND COLDEST PLANET
THE SMALLEST AND COLDEST PLANET, PLUTO (AND ITS PARTNER, CHARON, ANNOUNCED ON JUNE 22, 1978), HAVE AN ESTIMATED SURFACE TEMPERATURE OF -360°F. (100°F. ABOVE ABSOLUTE ZERO). PLUTO'S DIAMETER IS ABOUT 1880 MILES AND IT HAS A MASS ABOUT 1/500 THAT OF THE EARTH.

Use the computation rules for scientific notation to round each answer to the correct number of significant digits. Assume 1880 has 3 significant digits, −360 has 2 significant digits, and 600 has 1 significant digit.

15. _____   15. What is Pluto's surface area? (See Problem 7.)   16. What is Pluto's volume? (See Problem 8.)

Chapter 1 Review 47

Name _____ Date _____ Class _____

Chapter 1 Review

		What to Review if You Have Trouble		
Objectives		Lesson	Example	Page
Find the intersection of two sets ▶	1. $\{-1, 3, 5, 11\} \cap \{-3, -1, 0, 5, 9, 15\} = ?$ _____	1.1	1	3
Find the union of two sets ▶	2. $\{-4, -2, 0, 2, 4\} \cup \{-3, 0, 4, 7\} = ?$ _____	1.1	2	4
Find all subsets of a given set ▶	3. Find all subsets of $\{-5, 2, 3\}$. _____	1.1	3	6
Graph a set of real numbers ▶	4. Graph $\{-5, -0.25, 1, 3.5\}$. _____ $-5\;-4\;-3\;-2\;-1\;\;0\;\;1\;\;2\;\;3\;\;4\;\;5$	1.2	1	9
Find the absolute value of a real number ▶	5. $|-17| = ?$ _____	1.2	2	10

The **Chapter Reviews** help students test their mastery of chapter material and prepare for examinations. Each review problem is keyed to a corresponding numbered Example, providing easy reference to areas needing more study.

xiii

Contents

Selected Symbols (See inside front cover.)

Preface   ix

REAL NUMBERS AND LINEAR RELATIONS

Chapter 1 Real Numbers and Properties   **1**
- 1.1 Use Sets   2
- 1.2 Use the Real-Number Line   9
- 1.3 Compute with Signed Numbers   15
- 1.4 Use Computational Rules   21
- 1.5 Use Properties   31
- 1.6 Use Integral Exponents   37
- Problem Solving 1: Compute Using Scientific Notation   43
- Review   47

Chapter 2 Equations and Inequalities   **49**
- 2.1 Solve First-Degree Equations   50
- 2.2 Solve Equations Containing Like Terms and Parentheses   55
- 2.3 Solve Equations Containing Fractions and Decimals   63
- Problem Solving 2: Solve Problems Using Equations   67
- 2.4 Solve Inequalities   69
- 2.5 Solve Absolute Value Equations and Inequalities   75
- 2.6 Solve Literal Equations   81
- Review   85

Chapter 3 Equations and Graphs   **87**
- 3.1 Use Ordered Pairs   88
- 3.2 Graph Linear Equations   95
- 3.3 Find and Use Slope   101
- 3.4 Find Linear Equations   109
- Problem Solving 3: Use Linear Relationships   117
- 3.5 Graph Linear Inequalities   121
- Review   129

Chapter 4 Systems   **131**
- 4.1 Solve by Graphing   132
- 4.2 Solve Using Algebraic Methods   137
- Problem Solving 4: Solve Mixture Problems   143

4.3	Solve Third-Order Systems	145
	Problem Solving 5: Solve Digit Problems	151
4.4	Evaluate Determinants	153
4.5	Solve Using Cramer's Rule	159
	Review	165

POLYMIALS AND FACTORING

Chapter 5 Polynomials — 167

5.1	Add and Subtract Polynomials	170
	Problem Solving 6: Evaluate Polynomials	175
5.2	Multiply Polynomials	177
5.3	Find Special Products	183
5.4	Divide Polynomials	189
5.5	Use Synthetic Division	197
	Review	203

Chapter 6 Factoring — 205

6.1	Factor Using the GCF	206
6.2	Factor $x^2 + bx + c$	211
6.3	Factor $ax^2 + bx + c$	217
6.4	Factor Special Polynomials	227
6.5	Factor Using the General Strategy	233
6.6	Factor to Solve Equations	239
	Problem Solving 7: Factor to Solve Problems	247
	Review	249

NONLINEAR RELATIONS

Chapter 7 Rational Expressions — 251

7.1	Simplify Rational Expressions	252
7.2	Multiply and Divide Rational Expressions	259
7.3	Add and Subtract Rational Expressions	265
7.4	Simplify Complex Fractions	271
7.5	Solve Equations Containing Rational Expressions	279
	Problem Solving 8: Solve Problems Using Rational Expressions	285
	Problem Solving 9: Solve Formulas Containing Rational Expressions	289
	Review	291

Chapter 8 Exponents and Radicals — 293

8.1	Compute with Rational Exponents	296
8.2	Simplify Radicals	301
8.3	Compute with Radicals	307
8.4	Solve Radical Equations	313
	Problem Solving 10: Solve Formulas Containing Radicals	319
	Problem Solving 11: Evaluate Formulas Containing Radicals	321
8.5	Add and Subtract Complex Numbers	323
8.6	Multiply and Divide Complex Numbers	329
	Review	335

Chapter 9 Quadratic Equations — 337

9.1	Solve Incomplete Quadratic Equations	338
9.2	Solve by Completing the Square	345

9.3	Solve by the Quadratic Formula	351
9.4	Rename and Solve Equations	359
	Problem Solving 12: Solve Problems Using Quadratic Equations	365
9.5	Solve Inequalities	369
	Review	375

Chapter 10 Conics and Systems — **377**

10.1	Graph Parabolas	378
	Problem Solving 13: Find Maximum and Minimum Values	387
10.2	Graph Circles	389
	Problem Solving 14: Use the Pythagorean Theorem	397
10.3	Graph Ellipses	399
10.4	Graph Hyperbolas	405
10.5	Solve Quadratic Systems	412
	Problem Solving 15: Solve Problems Using Quadratic Systems	423
	Review	427

FUNCTIONS

Chapter 11 Functions — **429**

11.1	Find the Domain and Range	432
11.2	Identify Functions	437
11.3	Evaluate Functions	443
11.4	Graph Functions	449
11.5	Find the Inverse of a Function	457
11.6	Solve Variation Problems	465
	Problem Solving 16: Solve Applied Variation Problems	473
	Review	477

Chapter 12 Exponential and Logarithmic Functions — **479**

12.1	Graph Exponential Functions	480
12.2	Graph Logarithmic Functions	485
12.3	Use Properties of Logarithmic Functions	491
12.4	Find Logarithms and Antilogarithms	497
12.5	Solve Exponential and Logarithmic Equations	505
	Problem Solving 17: Solve Problems Using Exponential Formulas	511
	Problem Solving 18: Solve Problems Using Logarithmic Formulas	513
	Review	515

Chapter 13 Sequences and Series — **517**

13.1	Use Sequences and Series	518
13.2	Use Arithmetic Sequences and Series	523
	Problem Solving 19: Solve Problems Using Arithmetic Sequences and Series	529
13.3	Use Geometric Sequences and Series	531
	Problem Solving 20: Solve Problems Using Geometric Sequences and Series	539
13.4	Use the Binomial Expansion	541
	Review	549

Final Review	**551**

Appendix	**A1**
Table 1: Powers and Roots (from 1 to 100)	A-2
Table 2: Common Logarithms (base 10)	A-3

Table 3: Geometry Formulas	A-5
Table 4: Conversion Factors	A-6
Table 5: Exponential Functions e^x and e^{-x}	A-8
Table 6: Natural Logarithms (base e)	A-9
Selected Answers (Answers to odd-numbered exercises and problems and certain Make Sure and Review Exercises.)	A-11

Index I1

Selected Formulas (See inside back cover.)

REAL NUMBERS AND LINEAR RELATIONS

1 Real Numbers and Properties

1.1 Use Sets

1.2 Use the Real-Number Line

1.3 Compute with Signed Numbers

1.4 Use Computational Rules

1.5 Use Properties

1.6 Use Integral Exponents

PS 1: Compute Using Scientific Notation

2　Chapter 1　Real Numbers and Properties

1.1 Use Sets

In algebra it is convenient to classify numbers into groups called sets. To write a set, you can list the *members* or *elements*, separated by commas, inside *braces*.

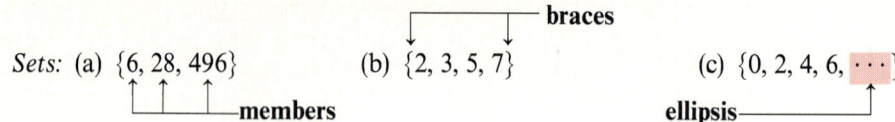

Sets: (a) {6, 28, 496}　　(b) {2, 3, 5, 7}　　(c) {0, 2, 4, 6, ···}

Note ▶ The ellipsis mark (···) in the last set indicates that the list continues on forever.

The method of writing a set by listing the members is called the *roster method*. Capital letters are often used to denote sets. The set of numbers used for counting is denoted by *N*, and is called the set of *natural numbers*.

Natural Numbers: $N = \{1, 2, 3, 4, \cdots\}$

When zero is included with the set of natural numbers, the set is denoted by *W* and is called the set of *whole numbers*.

Whole Numbers: $W = \{0, 1, 2, 3, 4, \cdots\}$

A *number line* can help you visualize sets of numbers. The point that represents the number zero (0) on the number line is called the *origin*.

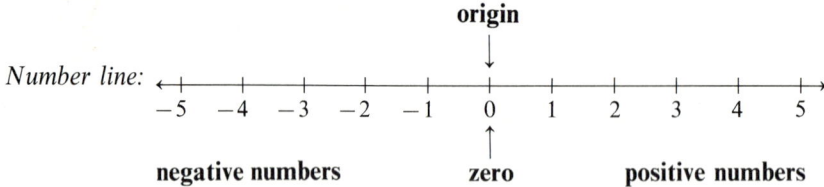

A number is *positive* if it is associated with a point to the right of the origin. A number is *negative* if it is associated with a point to the left of the origin. The number zero (0) is neither negative nor positive. The set of all positive numbers, negative numbers, and zero is sometimes called the *signed numbers*.

To indicate that a number is a negative number, you write a *negative sign* (−) in front of the number. You do not have to write a *positive sign* (+) to indicate a positive number (+3 = 3).

Two numbers with different signs that are the same distance from 0 are called *opposites*, or *additive inverses*.

Example ▶ −3 and 3 are opposites because they are the same distance from 0.

Zero (0) is its own opposite. $0 = -0$.

Read $-a$ as "the opposite of a." Read $-(-a)$ as "the opposite of the opposite of a."

$-(-a) = a$ for every number a.

Examples ▶ (a) $-(-2) = 2$　(b) $-(-(-4)) = -4$

The set of whole numbers and their opposites is called the set of *integers* (*I*).

Integers: $I = \{\cdots, -3, -2, -1, 0, 1, 2, 3, \cdots\}$

1.1 Use Sets

The symbol ∈ is used to indicate set membership.

Examples ▶ (a) $3 \in W$ means 3 is a member of the set of whole numbers.
(b) $0 \notin N$ means 0 is not a member of the set of natural numbers.

> The elements of a set are often denoted by a lowercase letter. If a letter is used to represent a single element of a set, it is called a *constant*. A letter used to represent any typical element of a set is called a *variable*.

Examples ▶ (a) $a \in \{3\}$ means a is a constant because it represents only the number 3.
(b) $x \in N$ means x is a variable because it represents any natural number.

> Two sets are *equal* if both have exactly the same members.

Examples ▶ (a) $\{2, 4, 5, 14\} = \{4, 14, 5, 2\}$ Order is not important when listing members of a set.
(b) $\{3, 7, 10\} = \{3, 3, 7, 10\}$ Repeating a member does not change the set.
(c) $\{4, 5, 20\} \neq \{4, 5\}$ The sets must have exactly the same members to be equal.
         └─ is not equal to

> The set without any members is called the *empty set* or *null set*, and is identified by the symbol ∅, or by braces, { }.

Example ▶ The set of whole numbers between 2 and 3 is the empty set.

> CAUTION: Do not use $\{\emptyset\}$ to represent the empty set.

The set $\{\emptyset\}$ has exactly one member: the empty set.

> The *intersection* of sets A and B, represented by $A \cap B$, is the set of elements in A that are also in B.

EXAMPLE 1: Find the intersection of two sets.

Problem ▶ $\{2, 3, 5, 7, 11\} \cap \{-1, 2, 4, 5, 11, 21\} = ?$

1. Identify common members ▶ $\{\underline{2}, 3, \underline{5}, 7, \underline{11}\}$
$\{-1, \underline{2}, 4, \underline{5}, \underline{11}, 21\}$ The common members are 2, 5, and 11.

2. Write set notation ▶ $\{2, 5, 11\}$ The intersection of two sets is always a set.

Solution ▶ $\{2, 3, 5, 7, 11\} \cap \{-1, 2, 4, 5, 11, 21\} = \{2, 5, 11\}$

Other Examples ▶ (a) $\{1, 2, 3, 4, 5\} \cap \{0, 2, 4, 6, 8\} = \{2, 4\}$
(b) $\{0, 2, 4, 6, 8, \cdots\} \cap \{0, 3, 6, 9, 12, \cdots\} = \{0, 6, 12, 18, 24, \cdots\}$
(c) $\{1, 2, 3\} \cap \{4, 5, 6\} = \emptyset$ These two sets have no common members.
(d) $N \cap W = N$ All natural numbers are also whole numbers.
(e) $A \cap A = A$, where A is any set.
(f) $A \cap \emptyset = \emptyset$, where A is any set.   The ∅ has no elements, so there can be no common elements.

Make Sure

Find the intersection of each pair of sets.

See Example 1 ▶ **1.** $\{2, 4, 6, 8, 10, 12\} \cap \{4, 8, 12, 16\}$ _____   **2.** $\{a, b, c, d\} \cap \{b, d, e, f\}$ _____

MAKE SURE ANSWERS: 1. $\{4, 8, 12\}$ **2.** $\{b, d\}$

4 Chapter 1 Real Numbers and Properties

> The *union* of two sets A and B, represented by $A \cup B$, is the set of elements that are in A or in B, or in both.

EXAMPLE 2: Find the union of two sets.

Problem ▶ $\{1, 4, 7, 11\} \cup \{4, 5, 9, 12, 19\} = ?$

1. List all members ▶ $\{1, 4, 7, 11\}$
$\{4, 5, 9, 12, 19\}$
1, 4, 5, 7, 9, 11, 12, 19   The common 4 is only listed once.

2. Write set notation ▶ $\{1, 4, 5, 7, 9, 11, 12, 19\}$ The union of two sets is always a set.

Solution ▶ $\{1, 4, 7, 11\} \cup \{4, 5, 9, 12, 19\} = \{1, 4, 5, 7, 9, 11, 12, 19\}$

Other Examples ▶
(a) $\{1, 2, 3, 4\} \cup \{-3, 1, 4, 7\} = \{-3, 1, 2, 3, 4, 7\}$
(b) $\{1, 3, 5, 7, \cdots\} \cup \{2, 4, 6, 8, \cdots\} = N$
(c) $A \cup A = A$, where A is any set.
(d) $A \cup \emptyset = A$, where A is any set.   The $\emptyset$ has no elements, so $A \cup \emptyset$ is just A.
(e) $\{0\} \cup N = W$ The set of whole numbers is 0 together with the natural numbers.

Make Sure

Find the union of each pair of sets.

See Example 2 ▶   1. $\{0, 1, 3, 4\} \cup \{3, 4, 5\}$ _____   2. $\{-1, 0, 5\} \cup \{-2, 0, 2\}$ _____

MAKE SURE ANSWERS: 1. $\{0, 1, 3, 4, 5\}$ 2. $\{-2, -1, 0, 2, 5\}$

Every fraction can be written as a decimal.

A decimal that ends in a specific place is called a *terminating decimal*.

Example ▶ 0.75 is a terminating decimal because it ends in the hundredths place.

A decimal that repeats one or more digits without any end is a *repeating decimal*.

Example ▶ $0.8\overline{3} = 0.8333\cdots$ is a repeating decimal because it repeats 3s forever.

Some decimals do not terminate and they do not repeat.

Example ▶ $0.34334333433334\cdots$ is a *nonterminating nonrepeating decimal*. The digits occur in a pattern, but they do not repeat.

Every decimal is either a terminating decimal, a repeating decimal, or a nonterminating nonrepeating decimal.

Any fraction which can be written in the form $\dfrac{m}{n}$, where m and n are integers and $n \neq 0$, can be expressed as either a terminating decimal or a repeating decimal. Also, every terminating or repeating decimal can be expressed as a fraction in the form $\dfrac{m}{n}$ where m and n are integers and $n \neq 0$. This set of numbers is called the *rational numbers*.

> The rational numbers are:
>
> 1. Those numbers that can be written in the fractional form $\dfrac{m}{n}$, where m and n are integers and $n \neq 0$.
> 2. Those numbers which when written as a decimal either terminate or repeat.

Rational Numbers: (a) $\frac{3}{4}$ (b) $\frac{-5}{8}$ (c) 7 (d) -6 (e) $\frac{14}{5}$ (f) 0.5 (g) $1.\overline{4}$

The letter Q is used to denote the set of rational numbers. The letter Q reminds you that all of the rational numbers can be written as a quotient of integers.

At one time all numbers were thought to be rational numbers, but in the sixth century B.C., followers of the Greek mathematician Pythagoras discovered some numbers that are not rational. $\sqrt{2}$ is such a number. That is, $\sqrt{2}$ cannot be written as the quotient of any two integers. Today we say $\sqrt{2}$ is an *irrational number*.

Irrational Numbers: (a) $\sqrt{5}$ (b) $-\sqrt{3}$ (c) $1.31331333133331\cdots$ (d) π

The decimal representation of an irrational number is a nonterminating nonrepeating decimal. The letter $\mathscr{I}$ is used to denote the set of irrational numbers.

> The set formed by the union of the set of rational numbers Q and the set of irrational numbers $\mathscr{I}$ is denoted by $\mathscr{R}$ and is called the set of *real numbers*. $Q \cup \mathscr{I} = \mathscr{R}$

The irrational numbers can be defined in the following three ways.

> The irrational numbers are:
>
> 1. Those numbers that cannot be written in the form $\dfrac{m}{n}$, where m and n are integers.
> 2. Those numbers which when written as a decimal do not terminate and do not repeat.
> 3. Those real numbers that are not rational.

Figure 1.1 illustrates some of these relationships.

Figure 1.1 ▶

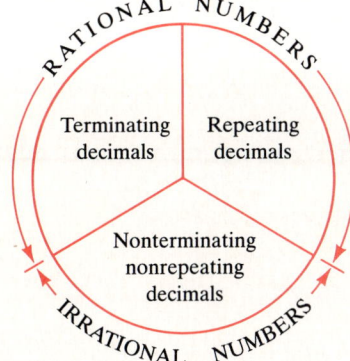

> If all members of a set A are also members of a set B, then A is a *subset* of B. A is a subset of B is written as $A \subseteq B$.

Examples ▶
(a) $\{1, 2\} \subseteq \{0, 1, 2, 3, 4\}$ — The first set is a subset of the second set.
(b) $W \subseteq I$ — The set of whole numbers is a subset of the set of integers.
(c) $I \nsubseteq N$ — The set of integers is not a subset of the set of natural numbers.
(d) $A \subseteq A$ — Every set is a subset of itself.
(e) $\varnothing \subseteq A$ — The empty set is a subset of any set A, because every element of the empty set (there are none) is an element of set A.

6 Chapter 1 Real Numbers and Properties

Listing all the subsets of a set can give you a better understanding of the subset concept.

EXAMPLE 3: List all subsets of a given set.

Problem ▶ List all the subsets of $\{1, 3, 7\}$.

1. List the empty set ▶ $\emptyset$ The empty set is a subset of every set.

2. List sets with one member ▶ $\{1\}, \{3\}, \{7\}$

3. List sets with two members ▶ $\{1, 3\}, \{1, 7\}, \{3, 7\}$

4. List the set itself ▶ $\{1, 3, 7\}$ Every set is a subset of itself.

Solution ▶ The subsets of $\{1, 3, 7\}$ are $\emptyset, \{1\}, \{3\}, \{7\}, \{1, 3\}, \{1, 7\}, \{3, 7\}, \{1, 3, 7\}$.

Make Sure

List all the subsets of each set.

See Example 3 ▶ **1.** $\{a, b\}$ _____   **2.** $\{-1, 0, 1\}$ _____

MAKE SURE ANSWERS: **1.** $\emptyset, \{a\}, \{b\}, \{a, b\}$ **2.** $\emptyset, \{-1\}, \{0\}, \{1\}, \{-1, 0\}, \{-1, 1\}, \{0, 1\}, \{-1, 0, 1\}$

Figure 1.2 shows some of the subset relationships among the sets that have been introduced.

Figure 1.2 ▶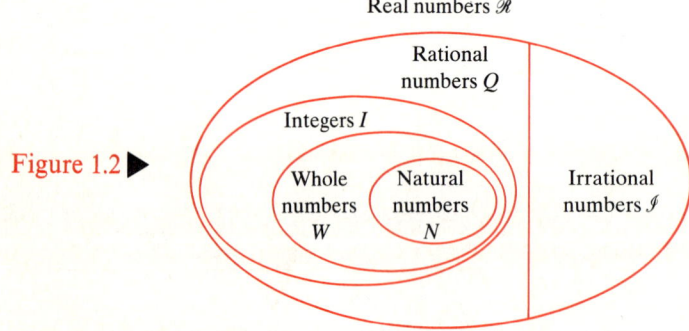

It is important to notice the following relationships:

$N \subseteq W$, $W \subseteq I$, $I \subseteq Q$, $Q \subseteq \mathscr{R}$, $\mathscr{I} \subseteq \mathscr{R}$, $Q \cap \mathscr{I} = \emptyset$,   and $Q \cup \mathscr{I} = \mathscr{R}$.

1.2 Use the Real-Number Line

There is a *one-to-one correspondence* between the real numbers and the points on the real-number line. That is, every real number is represented by a point on the real-number line and every point on the real-number line represents a real number.

> The number associated with a point on the real-number line is called its *coordinate*.

To graph a number, draw a dot on the number line to indicate the location of the point and then label the point with its associated coordinate.

> The dot on the real-number line associated with a number is called its *graph*.

EXAMPLE 1: Graph a set of real numbers on the real-number line.

Problem ▶ Graph $\{-3, -\frac{1}{2}, 1, 2\frac{1}{4}\}$ on the real-number line.

1. Graph each number ▶

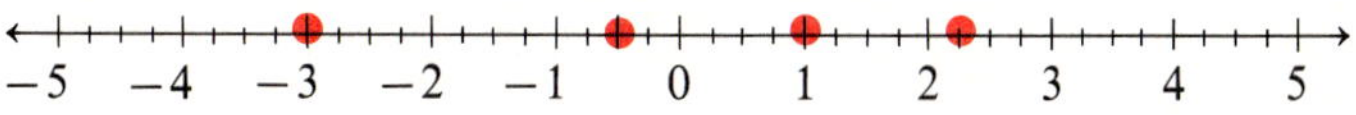

2. Label points ▶

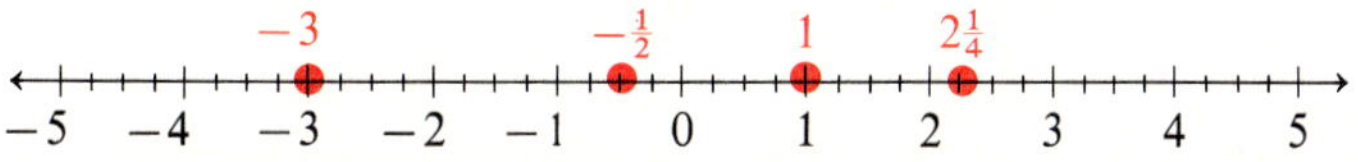

Solution ▶

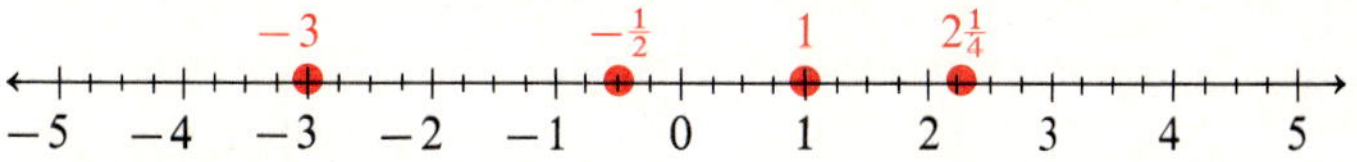

Make Sure

Graph each set of real numbers on the number line.

See Example 1 ▶ 1. $\{-3\frac{1}{2}, -1, 0, 2\frac{1}{2}\}$ 2. $\{-4, -1\frac{1}{2}, 1, 2\frac{1}{2}\}$

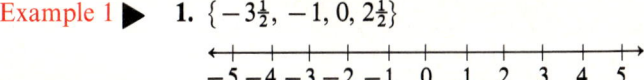   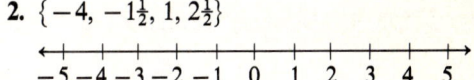

MAKE SURE ANSWERS: See Appendix Selected Answers.

> The relationships $a > b$, $a = b$, and $a < b$ are called *order relations*.

$a > b$ is read "a is greater than b."
$a = b$ is read "a is equal to b."
$a < b$ is read "a is less than b."

$a > b$ means a is to the right of b on the number line.
$a = b$ means a and b are the same point on the number line.
$a < b$ means a is to the left of b on the number line.

> $a > b$ and $b < a$ mean the same.

Example ▶ $-6 > -8$ means -6 is to the right of -8 on the number line.
$-8 < -6$ means -8 is to the left of -6 on the number line.

The symbol ≥ is read "is greater than or equal to." The symbol ≤ is read "is less than or equal to."

> $a \geq b$ means $a > b$ or $a = b$; $a \leq b$ means $a < b$ or $a = b$

Examples ▶ (a) $4 \geq 1$ because $4 > 1$ (b) $3 \geq 3$ because $3 = 3$
(c) $-5 \geq -8$ because $-5 > -8$ (d) $2 \leq 5$ because $2 < 5$
(e) $-4 \leq -4$ because $-4 = -4$ (f) $-2 \leq 7$ because $-2 < 7$

The symbol $\not>$ is read "is not greater than." The symbol $\not<$ is read "is not less than."

> $a \not> b$ means $a \leq b$; $a \not< b$ means $a \geq b$

Examples ▶ (a) $5 \not> 6$ because $5 < 6$ (b) $2 \not> 2$ because $2 = 2$ (c) $-4 \not< -8$ because $-4 > -8$

> The *absolute value* of a number is its distance from zero on the number line. The symbol for absolute value is $|\ |$.

Example ▶ $|-3|$ is read "the absolute value of negative three."

To find the absolute value of a real number, you can use the following definition.

> **Absolute Value**
>
> For any real number a: $|a| = \begin{cases} a \text{ if } a \text{ is positive.} \\ 0 \text{ if } a \text{ is zero.} \\ -a \text{ if } a \text{ is negative.} \end{cases}$

EXAMPLE 2: Find the absolute value of a real number.

Problem ▶ $|-4| = ?$

1. Identify ▶ -4 is a negative number.

2. Write absolute value ▶ $|-4| = -(-4)$ Think: Because -4 is negative: $|-4|$ = the opposite of -4.
                              $= 4$ Recall: $-(-a) = a$

Solution ▶ $|-4| = 4$

Other Examples ▶ (a) $|5| = 5$ (b) $|-15| = 15$ (c) $|0| = 0$ (d) $|-2.3| = -(-2.3) = 2.3$

Make Sure

Find the absolute value of each real number.

See Example 2 ▶ **1.** $|9|$ _____   **2.** $|-3.3|$ _____   **3.** $|0|$ _____

MAKE SURE ANSWERS: 1. 9   2. 3.3   3. 0

1.2 Use the Real-Number Line

> If x is a variable and a is a real number, then: any inequality which can be written in one of the following forms is called a *simple inequality*.
>
> $$x > a, \ x < a, \ x \geq a, \ x \leq a$$
>
> The real number a is called the *boundary value*.

Example ▶ $x \geq 7$ is a simple inequality with a boundary value of 7.

> Any real number that can replace the variable and make a simple inequality a true statement is called a *solution* of the simple inequality. The set of all solutions is called the *solution set*.

Examples ▶ (a) -3 is a solution of $x < 8$ because $-3 < 8$ is a true statement.
(b) 5 is a solution of $x \geq 1$ because $5 \geq 1$ is a true statement.
(c) $4\frac{1}{2}$ is a solution of $x > -2$ because $4\frac{1}{2} > -2$ is a true statement.
(d) 7.1 is not a solution of $x \leq -3$ because $7.1 \leq -3$ is not a true statement.

To graph a simple inequality, you graph its solution set.

EXAMPLE 3: Graph a simple inequality.

Problem ▶ Graph $x > 3$.

1. Locate boundary value ▶ Use an open circle because 3 is not a solution.

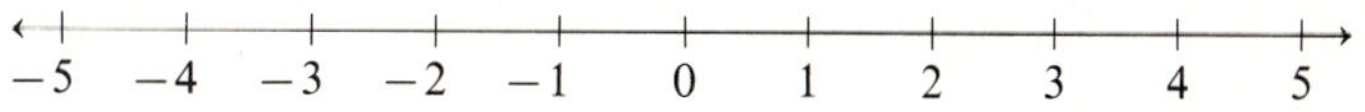

2. Graph solutions ▶ Every number greater than 3 is a solution.

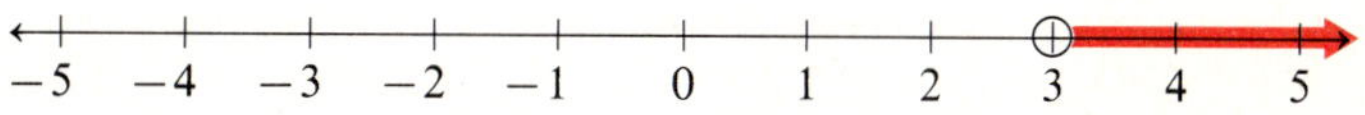

Solution ▶ $x > 3$

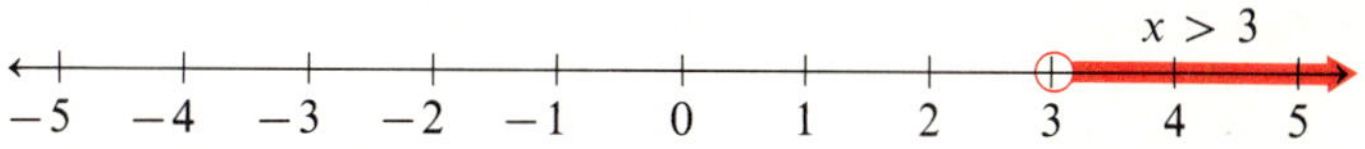

Other Examples ▶ (a) Graph $x \leq -2$.

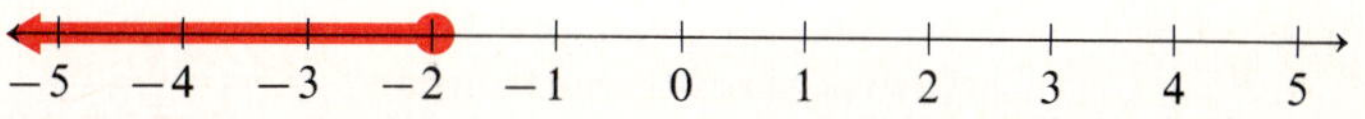

Use a solid circle at -2 because -2 is a solution.

(b) Graph $x \not< 1$.

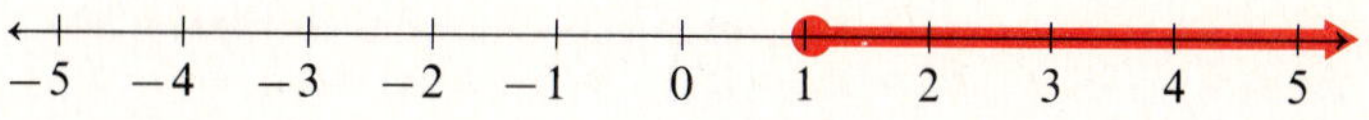

Think: $x \not< 1$ means $x \geq 1$.

Make Sure

Graph a simple inequality.

See Example 3 ▶ 1. $z > -2$ 2. $y \leq 2$

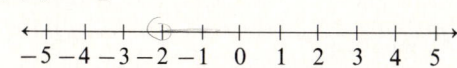
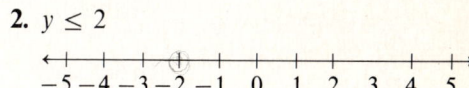

MAKE SURE ANSWERS: See Appendix Selected Answers.

Compound inequalities can be formed by using either *and* or *or* between two simple inequalities.

Examples ▶ (a) $x < 3$ or $x > 5$ (b) $x \leq -2$ or $x > 7$ (c) $x > -1$ and $x < 4$

When graphing compound inequalities, the word *or* will mean to form the union of the solution sets of the simple inequalities. The word *and* will mean to form the intersection of the solution sets of the simple inequalities.

EXAMPLE 4: Graph a compound inequality.

Problem ▶ Graph $x < -1$ or $x \geq 2$.

1. Graph each simple inequality

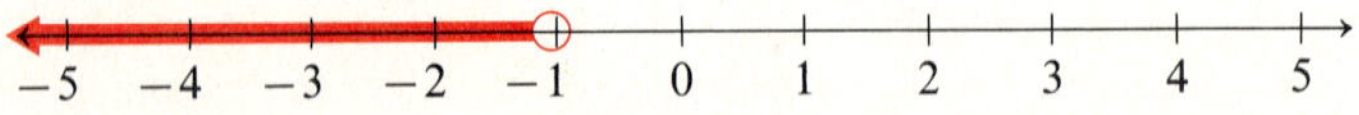

(See Example 3.)

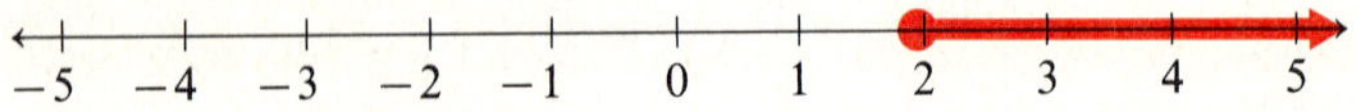

2. Graph the compound inequality

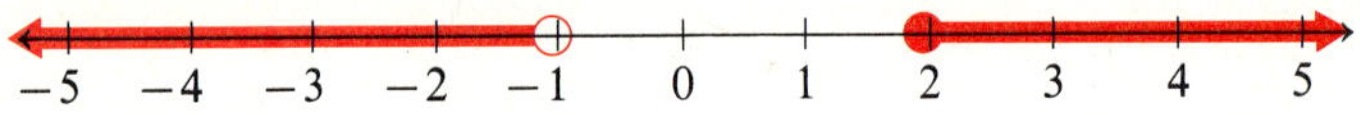

Think: *or* means to form the union of the solution sets of the simple inequalities.

Solution ▶

The compound sentence $x > -2$ and $x < 3$ can be written more compactly as $-2 < x < 3$. The real numbers which satisfy the sentence are both greater than -2 and less than 3.

Example ▶ Graph $-2 < x < 3$.

1. Graph each simple inequality

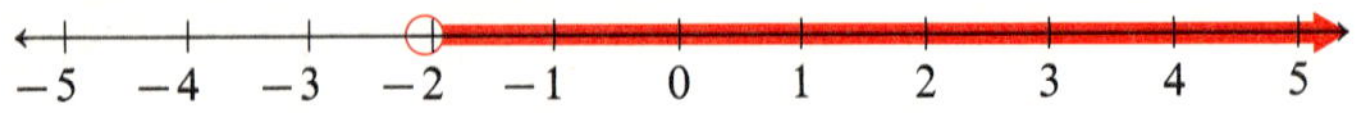

Think: $-2 < x < 3$ means $x > -2$ and $x < 3$.

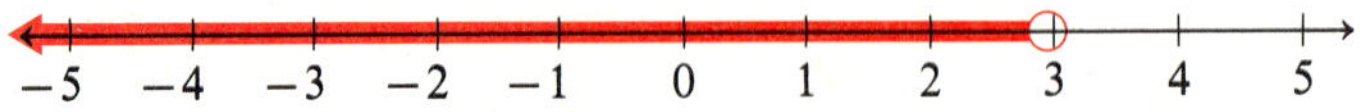

2. Graph the compound inequality

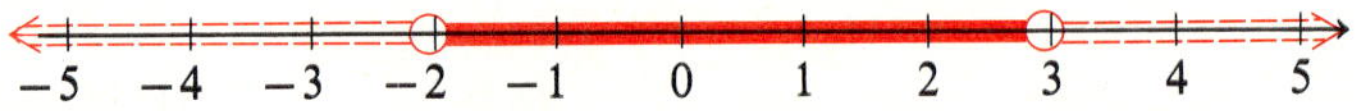

Think: *and* means to form the intersection of the solution sets of the simple inequalities.

Solution ▶

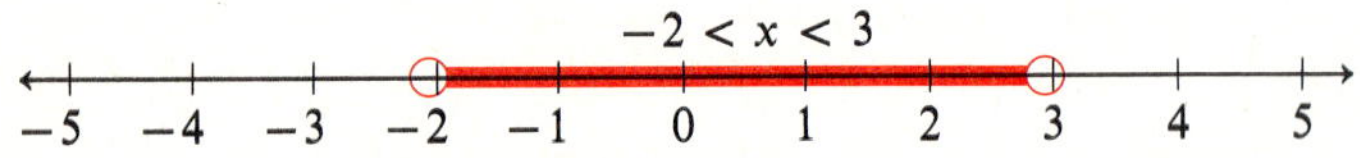

Other Examples ▶ (a) Graph $x > 2$ or $x \leq 4$.

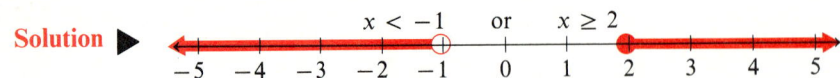

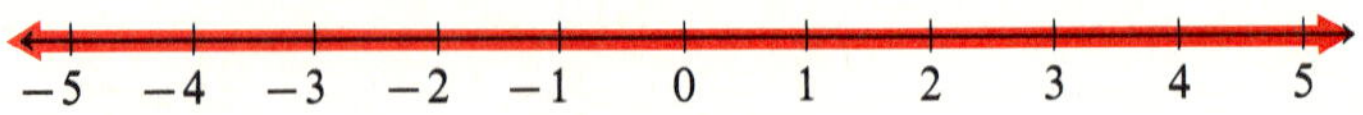

All real numbers make the sentence true.

(b) Graph $x \neq 1$

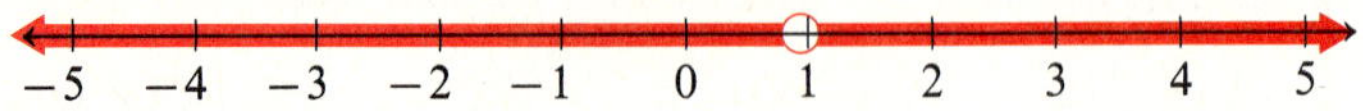

$x \neq 1$ means $x > 1$ or $x < 1$.

Make Sure

Graph each compound inequality.

See Example 4 ▶ 1. $z > 1$ or $z \leq -2$ 2. $-3 < y \leq 2$

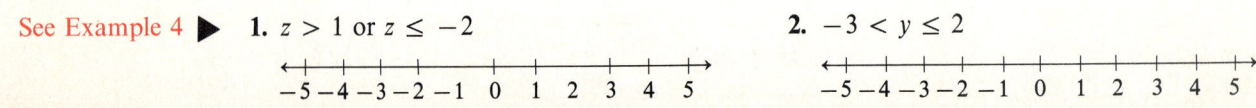

MAKE SURE ANSWERS: See Appendix Selected Answers.

Name　　　　　　　　　　　　　Date　　　　　Class

1.2 Practice

Set 1: Graph each set of real numbers.

1. $\{-4, 0, 1, 4\}$
2. $\{-3, -1, 0, 2\}$
3. $\{-2, 1, 3, 5\}$

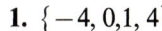

4. $\{-5, -2, 2, 3\}$
5. $\{-2\frac{1}{2}, -1, \frac{1}{2}\}$
6. $\{-4\frac{1}{2}, -1\frac{1}{2}, 3\frac{1}{2}\}$

7. _____
8. _____
9. _____
10. _____
11. _____
12. _____
13. _____
14. _____
15. _____
16. _____
17. _____
18. _____

Set 2: Find the absolute value of each real number.

7. $|7|$
8. $|3|$
9. $|-2|$
10. $|-5|$

11. $|-2\frac{1}{2}|$
12. $|-5\frac{4}{5}|$
13. $|3\frac{1}{3}|$
14. $|6\frac{5}{6}|$

15. $|4.23|$
16. $|5.4|$
17. $|-\sqrt{2}|$
18. $|-\sqrt{5}|$

Set 3: Graph each simple inequality.

19. $z > 1$
20. $y > -2$
21. $x < 4$

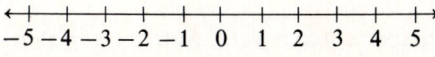 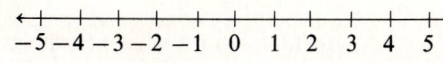 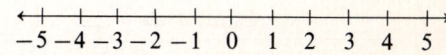

22. $w < -1$
23. $v \geq -2\frac{1}{2}$
24. $u \geq \frac{1}{2}$

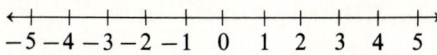 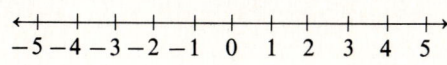 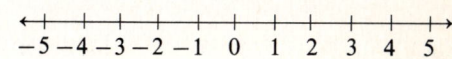

25. $t \leq 3\frac{1}{2}$
26. $s \leq -\frac{1}{3}$
27. $r > 2.3$

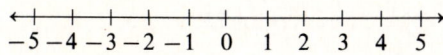 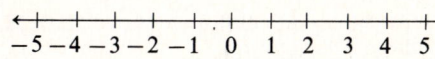 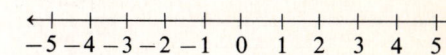

28. $q > -1.3$
29. $p \not> 2$
30. $m \not< -1$

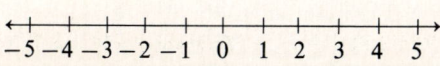 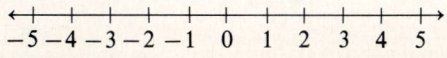 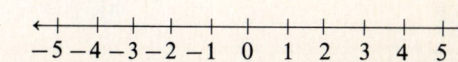

Copyright © 1985 by Harcourt Brace Jovanovich, Inc. All rights reserved.

14   Chapter 1   Real Numbers and Properties

Set 4: Graph each compound inequality.

31. $z > 1$ or $z < -1$

$\longleftrightarrow$
$-5-4-3-2-1\ 0\ 1\ 2\ 3\ 4\ 5$

32. $y > 2$ or $y < -2$

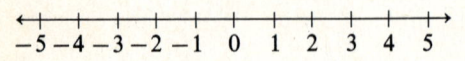

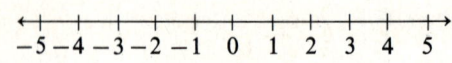

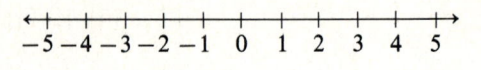

$\longleftrightarrow$
$-5-4-3-2-1\ 0\ 1\ 2\ 3\ 4\ 5$

33. $x \geq 3$ or $x < 1$

$\longleftrightarrow$
$-5-4-3-2-1\ 0\ 1\ 2\ 3\ 4\ 5$

34. $w \geq -1$ or $w < -3$

$\longleftrightarrow$
$-5-4-3-2-1\ 0\ 1\ 2\ 3\ 4\ 5$

35. $v \geq 1\frac{1}{2}$ or $v \leq -1\frac{1}{2}$

$\longleftrightarrow$
$-5-4-3-2-1\ 0\ 1\ 2\ 3\ 4\ 5$

36. $u > -\frac{1}{2}$ or $u \leq -2\frac{1}{2}$

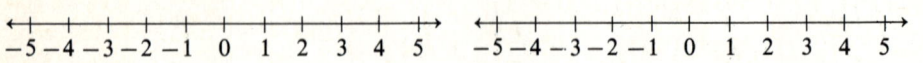

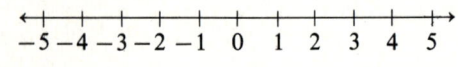

$\longleftrightarrow$
$-5-4-3-2-1\ 0\ 1\ 2\ 3\ 4\ 5$

37. $0 < t < 5$

$\longleftrightarrow$
$-5-4-3-2-1\ 0\ 1\ 2\ 3\ 4\ 5$

38. $-3 < s < 1$

$\longleftrightarrow$
$-5-4-3-2-1\ 0\ 1\ 2\ 3\ 4\ 5$

39. $-4 \leq r \leq -0.5$

$\longleftrightarrow$
$-5-4-3-2-1\ 0\ 1\ 2\ 3\ 4\ 5$

40. $-1.5 \leq q \leq 2.3$

$\longleftrightarrow$
$-5-4-3-2-1\ 0\ 1\ 2\ 3\ 4\ 5$

41. $p \neq 2$

$\longleftrightarrow$
$-5-4-3-2-1\ 0\ 1\ 2\ 3\ 4\ 5$

42. $m \neq -3$

$\longleftrightarrow$
$-5-4-3-2-1\ 0\ 1\ 2\ 3\ 4\ 5$

43. _____

44. _____

45. _____

46. _____

47. _____

48. _____

49. _____

50. _____

51. _____

52. _____

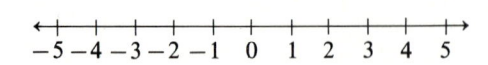

Extra: Determine if each statement is true.

43. $0 > -3$

44. $0 > 2$

45. $-3 > -2$

46. $|-3| \cdot |-2| = |-3(-2)|$

47. $|-5| \cdot |3| = |-5 \cdot 3|$

48. $\dfrac{|-10|}{|-6|} = \left|\dfrac{-10}{-6}\right|$

49. $\left|\dfrac{-8}{6}\right| = \dfrac{|-8|}{|6|}$

50. $|-5| + |-3| = |-5 + (-3)|$

51. $|6| + |-3| = |6 + (-3)|$

52. $|-2 + (-3)| \leq |-2| + |-3|$

Review: Work these problems on a separate sheet of paper. Attach your work to this page.

Add fractions.

53. $\frac{4}{7} + \frac{5}{7}$ 54. $\frac{5}{9} + \frac{7}{9}$ 55. $\frac{2}{3} + \frac{4}{5}$ 56. $\frac{3}{5} + \frac{4}{7}$

57. $\frac{3}{4} + \frac{5}{6}$ 58. $\frac{5}{6} + \frac{4}{9}$ 59. $\frac{11}{12} + \frac{13}{18}$ 60. $\frac{11}{14} + \frac{27}{35}$

Subtract fractions.

61. $\frac{5}{6} - \frac{1}{6}$ 62. $\frac{7}{8} - \frac{3}{8}$ 63. $\frac{4}{5} - \frac{1}{3}$ 64. $\frac{4}{7} - \frac{1}{5}$

65. $2 - \frac{4}{7}$ 66. $3 - \frac{5}{6}$ 67. $3\frac{2}{3} - \frac{5}{6}$ 68. $3\frac{4}{9} - \frac{7}{12}$

Multiply fractions.

69. $\frac{3}{4} \cdot \frac{5}{7}$ 70. $\frac{3}{5} \cdot \frac{4}{7}$ 71. $\frac{4}{5} \cdot \frac{15}{16}$ 72. $\frac{5}{7} \cdot \frac{14}{15}$

73. $\frac{5}{6} \cdot \frac{4}{7}$ 74. $\frac{2}{3} \cdot \frac{6}{7}$ 75. $\frac{3}{4} \cdot 2$ 76. $3 \cdot \frac{4}{5}$

Divide fractions.

77. $\frac{2}{3} \div \frac{3}{5}$ 78. $\frac{4}{7} \div \frac{1}{4}$ 79. $\frac{4}{5} \div \frac{8}{9}$ 80. $\frac{5}{6} \div \frac{10}{18}$

81. $\frac{3}{4} \div 2$ 82. $\frac{5}{6} \div 3$ 83. $5 \div \frac{1}{2}$ 84. $4 \div \frac{2}{3}$

Copyright © 1985 by Harcourt Brace Jovanovich, Inc. All rights reserved.

1.3 Compute with Signed Numbers

Addition of two signed numbers is classified as either the addition of two numbers with like signs or the addition of two numbers with opposite signs.

> **Rules for Adding Two Signed Numbers with Like Signs**
> 1. Find the sum of the absolute values.
> 2. Write the same like sign on the sum.
>
> **Rules for Adding Two Signed Numbers with Opposite Signs**
> 1. Find the difference between the absolute values.
> 2. Write the sign of the number with the larger absolute value on the sum.

EXAMPLE 1: Add two signed numbers.

Problem ▶ $-13 + (-21) = ?$ -13 and -21 are called *addends*.

1. Find the sum (difference) of the absolute values ▶ $-13 + (-21) = ?\ 34$ Think: Since the numbers have like signs, you find the sum of their absolute values: $13 + 21 = 34$.

2. Write correct sign ▶ $-13 + (-21) = -34$ Think: Like negative signs mean a negative sum.

Solution ▶ $-13 + (-21) = -34$

When the addends have like positive signs, the sum will be positive: $13 + (+21) = 34$.

Another Example ▶ $-17 + (+28) = ?$

1. Find the sum (difference) of the absolute values ▶ $-17 + (+28) = ?\ 11$ Think: Since the numbers have opposite signs, you find the difference of their absolute values.
larger absolute value
↓
$28 - 17 = 11$

2. Write correct sign ▶ $-17 + (+28) = +11$ Think: The sum of two numbers with opposite signs is positive when the positive addend has the larger absolute value.

Solution ▶ $-17 + (+28) = 11$

The sum of two signed numbers with opposite signs is negative when the negative addend has the larger absolute value.

Examples ▶ (a) $3 + (-7) = -4$ (b) $-12 + 7 = -5$ (c) $18 + (-25) = -7$

The sum of two opposites is always 0.

Examples ▶ (a) $7 + (-7) = 0$ (b) $-31.7 + 31.7 = 0$ (c) $\frac{3}{8} + (-\frac{3}{8}) = 0$

Make Sure

Add signed numbers.

See Example 1 ▶ **1.** $-5 + 3$ _____   **2.** $-3 + (-4)$ _____   **3.** $12 + (-5)$ _____

MAKE SURE ANSWERS: 1. −2 2. −7 3. 7

16 Chapter 1 Real Numbers and Properties

> If a and b are real numbers, $a - b$ means $a + (-b)$.

You can subtract any two real numbers by using the following rules.

> **Rules for Subtracting Two Signed Numbers**
> 1. Change subtraction to addition.
> 2. Write the opposite of the second addend.
> 3. Follow the rules for adding two signed numbers.

EXAMPLE 2: Subtract two signed numbers.

Problem ▶ $-53 - (+27) = ?$ -53 is called the *minuend*. $+27$ is called the *subtrahend*.

1. Change to addition ▶ $-53 \;\boxed{-}\; (+27) = -53 \;\boxed{+}\; (?)$

2. Write opposite addend ▶ $ = -53 + (\boxed{-27})$ Think: The opposite of $+27$ is -27.

3. Add as before ▶ $ = -80$

Solution ▶ $-53 - (+27) = -80$ -80 is called the *difference*.

Make Sure

Subtract signed numbers.

See Example 2 ▶ **1.** $-5 - 7$ _____   **2.** $-4 - (-7)$ _____   **3.** $8 - (-9)$ _____

MAKE SURE ANSWERS: 1. −12 2. 3 3. 17

The following are equivalent ways to write "r times s":

multiplication symbols   no sign also means multiplication
↓   ↓   ↓   ↓   ↓   ↓
$r \times s \;=\; r \cdot s \;=\; r(s) \;=\; (r)s \;=\; (r)(s) \;=\; rs$

In $r \cdot s = t$, r and s are called *factors*, and t is the *product*.

To multiply two signed numbers, you use the following rules.

> **Rules for Multiplying Two Signed Numbers**
> 1. Find the product of the absolute values of each factor.
> 2. Make the product: (a) positive if the original factors have like signs. (b) negative if the original factors have opposite signs. (c) zero if either original factor is zero (0).

EXAMPLE 3: Multiply two signed numbers.

Problem ▶ $-7 \cdot (-8) = ?$

1. Find the product of the absolute values ▶ $-\boxed{7} \cdot (-\boxed{8}) = ?\; 56$ Think: $7 \cdot 8 = 56$

2. Write correct sign ▶ $\boxed{-}\, 7 \cdot (\boxed{-}\, 8) = +56$ Think: Two like signs mean a positive product.

Solution ▶ $-7 \cdot (-8) = 56$

1.3 Compute with Signed Numbers

When both factors are positive numbers, the product is a positive number: $7 \cdot (8) = 56$. The product of two real numbers with opposite signs is always negative.

Example ▶ $5 \cdot (-4) = ?$

1. Find the product of the absolute values ▶ $5 \cdot (-4) = ?\ 20$ Think: $5 \cdot 4 = 20$

2. Write correct sign ▶ $5 \cdot (-4) = -20$ Think: Two opposite signs mean a negative product.

Solution ▶ $5 \cdot (-4) = -20$

Other Examples ▶   (a) $-9 \cdot 7 = -63$ (b) $8(-6) = -48$ (c) $(-12)(-12) = 144$ (d) $(-3)5 = -15$

Make Sure

Multiply signed numbers.

See Example 3 ▶   **1.** $-3 \cdot 5$ _____   **2.** $6 \cdot (-4)$ _____   **3.** $-5(-8)$ _____

MAKE SURE ANSWERS: 1. −15 2. −24 3. 40

The following are four different ways to write "r divided by s":

(a) Divide r by s.   (b) $r \div s$ (c) $s\overline{)r}$ (d) $\dfrac{r}{s}$

In $r \div s = t$, r is called the *dividend*, s the *divisor*, and t the *quotient*.

Multiplication and division are *inverse operations* because one "undoes" the other.

Example ▶ $3 \cdot 5 = 15$ and $15 \div 5 = 3$

Start with 3 and end with 3.

You can use this inverse relationship to determine the quotient of signed numbers.

Examples ▶   (a) $8 \div 2 = 4$ because $4 \cdot 2 = 8$

(b) $8 \div (-2) = -4$ because $(-4) \cdot (-2) = 8$
(c) $(-8) \div (2) = -4$ because $(-4) \cdot (2) = -8$
(d) $(-8) \div (-2) = 4$ because $4 \cdot (-2) = -8$

Note ▶   The previous examples indicate that the sign of the quotient can be found by using the sign rules for multiplication.

Rules for Dividing Signed Numbers

1. Find the quotient of the absolute values.
2. Write the correct sign on the quotient using the sign rules for multiplication.

Chapter 1 Real Numbers and Properties

EXAMPLE 4: Divide two signed numbers.

Problem ▶ $-143 \div (+11) = ?$

1. Find the quotient of the absolute values ▶ $-\boxed{143} \div (+\boxed{11}) = ?13$ Think: $143 \div 11 = 13$

2. Write correct sign ▶ $\boxed{-}143 \div (\boxed{+}11) = -13$

Solution ▶ $-143 \div (+11) = -13$ Think: Use the sign rules for multiplication.

Other Examples ▶ (a) $24 \div (-8) = -3$ (b) $\dfrac{-42}{-7} = 6$ (c) $\dfrac{51}{-3} = -17$

Division involving the number zero plays a special role in mathematics.

Zero divided by any nonzero number equals zero.

Examples ▶ (a) $0 \div 4 = 0$ because $0 \cdot 4 = 0$

(b) $0 \div (-5) = 0$ because $0 \cdot (-5) = 0$

Division by zero is undefined.

Examples ▶ (a) $4 \div 0$ is undefined. Suppose that $4 \div 0$ equals the real number n.

$4 \div 0 = n$ requires $n \cdot 0 = 4$

Division of a nonzero number a by zero is undefined, because there is no number, which multiplied times zero, equals the nonzero number a.

(b) $0 \div 0$ is also undefined, but for a different reason.

$0 \div 0 = a$ requires $a \cdot 0 = 0$

But $a \cdot 0 = 0$ for every real number a. Division of zero by zero is undefined because the quotient would not be unique.

Other Examples ▶ (a) $\dfrac{0}{7} = 0$ (b) $\dfrac{0}{-3} = 0$ (c) $\dfrac{-4}{0}$ is undefined (d) $\dfrac{0}{0}$ is undefined

Make Sure

Divide signed numbers.

See Example 4 ▶ 1. $-4 \div (-2)$ _____   2. $\dfrac{16}{-4}$ _____   3. $0 \div 7$ _____

MAKE SURE ANSWERS: 1. 2 2. −4 3. 0

1.3 Practice

Name / Date / Class

1. _____
2. _____
3. _____
4. _____
5. _____
6. _____
7. _____
8. _____
9. _____
10. _____
11. _____
12. _____
13. _____
14. _____
15. _____
16. _____
17. _____
18. _____
19. _____
20. _____
21. _____
22. _____
23. _____
24. _____
25. _____
26. _____
27. _____
28. _____
29. _____
30. _____
31. _____
32. _____
33. _____
34. _____
35. _____
36. _____

Set 1: Add signed numbers.

1. $-5 + 9$
2. $-4 + 12$
3. $13 + (-6)$
4. $17 + (-12)$

5. $-7 + (-9)$
6. $-12 + (-6)$
7. $3.8 + (-7.2)$
8. $4.5 + (-9.1)$

9. $-\frac{3}{8} + (-\frac{5}{8})$
10. $-\frac{2}{3} + (-\frac{2}{3})$
11. $-\frac{4}{7} + \frac{3}{4}$
12. $-\frac{5}{6} + \frac{7}{8}$

Set 2: Subtract signed numbers.

13. $5 - 8$
14. $12 - 17$
15. $-9 - 5$
16. $-13 - 14$

17. $5 - (-4)$
18. $9 - (-6)$
19. $-8.3 - (-9.7)$
20. $-5.7 - (-8.3)$

21. $\frac{5}{3} - \frac{2}{3}$
22. $\frac{4}{7} - \frac{3}{7}$
23. $-\frac{3}{8} - (-\frac{4}{5})$
24. $-\frac{4}{7} - (-\frac{3}{4})$

Set 3: Multiply signed numbers.

25. $-3 \cdot 4$
26. $-7 \cdot 8$
27. $5(-6)$
28. $9(-2)$

29. $-5(-4)$
30. $-7(-12)$
31. $2.3(-4.1)$
32. $5.2(-5.5)$

33. $-\frac{4}{5}(-\frac{3}{5})$
34. $-\frac{4}{7}(-\frac{21}{32})$
35. $-\frac{5}{6} \cdot \frac{12}{25}$
36. $-\frac{8}{9} \cdot \frac{15}{16}$

Copyright © 1985 by Harcourt Brace Jovanovich, Inc. All rights reserved.

Chapter 1 Real Numbers and Properties

Set 4: Divide signed numbers.

37. $-6 \div 3$
38. $-16 \div 8$
39. $\dfrac{24}{-6}$
40. $\dfrac{36}{-9}$

41. $\dfrac{-72}{-8}$
42. $\dfrac{-96}{-12}$
43. $\dfrac{-2.4}{-0.6}$
44. $\dfrac{-3.5}{-0.35}$

45. $-\dfrac{2}{3} \div \dfrac{4}{9}$
46. $-\dfrac{4}{5} \div \dfrac{8}{15}$
47. $\dfrac{1}{4} \div \left(-\dfrac{2}{3}\right)$
48. $\dfrac{2}{5} \div \left(-\dfrac{8}{15}\right)$

Mixed Practice: Perform each indicated operation on signed numbers.

49. $-23 + 13$
50. $-42 + 17$
51. $-32 + (-15)$
52. $-17 + (-26)$

53. $-13 - (-5)$
54. $-42 - (-24)$
55. $-4.5 - 3.7$
56. $-6.3 - 4.9$

57. $-5(-4.3)$
58. $-6(-3.2)$
59. $-\dfrac{4}{7} \cdot 3$
60. $-\dfrac{5}{8} \cdot 4$

61. $-48 \div (-72)$
62. $0 \div (-25)$
63. $\dfrac{4.8}{-16}$
64. $\dfrac{0}{-12.8}$

Extra: Perform the indicated operations on signed numbers.

65. $4 + (-3) + (-5)$
66. $6 + (-5) + (-1)$
67. $-5 + 16 + (-6)$

68. $-8 + 23 + (-9)$
69. $19 + (-6) - 8$
70. $12 + (-7) - 2$

71. $-8 + (-4) - (-5)$
72. $-7 + (-6) - (-9)$
73. $-6 - 8 - (-8)$

74. $-5 - 9 - (-5)$
75. $-9 - (-8) - (-1)$
76. $-23 - (-14) - (-9)$

77. $-5(-7)(-6)$
78. $-8(-4)(-5)$
79. $5(-6)(-2)$

80. $9(-4)(-3)$
81. $5(-4)3$
82. $9(-4)5$

Review: Work these problems on a separate sheet of paper. Attach your work to this page.

Rename fractions in lowest terms.

83. $\dfrac{16}{24}$
84. $\dfrac{8}{24}$
85. $\dfrac{36}{48}$
86. $\dfrac{36}{63}$

Add and subtract fractions.

87. $2 + \dfrac{2}{3}$
88. $\dfrac{3}{4} + \dfrac{1}{2}$
89. $2 - \dfrac{2}{3}$
90. $\dfrac{4}{9} - \dfrac{3}{5}$

Multiply fractions.

91. $\dfrac{2}{3} \cdot \dfrac{2}{3} \cdot \dfrac{2}{3} \cdot \dfrac{2}{3}$
92. $\dfrac{1}{4} \cdot \dfrac{1}{4} \cdot \dfrac{1}{4} \cdot \dfrac{1}{4}$
93. $\dfrac{3}{5} \cdot \dfrac{4}{7}$
94. $\dfrac{6}{7} \cdot \dfrac{14}{30}$

1.4 Use Computational Rules

A compact way of writing $4 \cdot 4 \cdot 4$ is 4^3. 4^3 is *exponential notation* for $4 \cdot 4 \cdot 4$.

If n is a natural number greater than one, $a^n = \underbrace{a \cdot a \cdot a \cdots a}_{n \text{ factors}}$.

The a in a^n is called the *base* and the n is called the *exponent*.

Examples ▶ (a) $\overbrace{7 \cdot 7}^{2 \text{ factors}} = 7^{\overset{\text{exponent}}{\underset{\text{base}}{2}}}$ Read 7^2 as "7 to the second power" or "7 squared."

(b) $5 \cdot 5 \cdot 5 = 5^3$ Read 5^3 as "5 to the third power" or "5 cubed."
(c) $3 \cdot 3xxxx = 3^2 x^4$ Read $3^2 x^4$ as "3 squared times x to the fourth power."

For any real number a, $a^1 = a$.

Examples ▶ (a) $4^1 = 4$ (b) $(-3.14)^1 = -3.14$ (c) $(\frac{2}{5})^1 = \frac{2}{5}$

To evaluate a number written in exponential notation, you first write it as a product of repeated factors and then compute the product of the factors.

EXAMPLE 1: Evaluate a number written in exponential notation.

Problem ▶ Evaluate $(-2)^5$.

1. Write as a product ▶ $(-2)^5 = (-2)(-2)(-2)(-2)(-2)$ ⟵ product of repeated factors

2. Multiply ▶ $= 4(-2)(-2)(-2)$
$= (-8)(-2)(-2)$
$= 16(-2)$
$= -32$

Solution ▶ $(-2)^5 = -32$

Other Examples ▶ (a) $(\frac{2}{3})^4 = (\frac{2}{3})(\frac{2}{3})(\frac{2}{3})(\frac{2}{3})$ (b) $(-1)^2 = (-1)(-1)$ (c) $(-1)^3 = (-1)(-1)(-1)$
$= \frac{16}{81}$ $= 1$ $= -1$

Note 1 ▶ An odd power of a negative number is negative.

Note 2 ▶ An even power of a negative number is positive.

Make Sure

Evaluate each number written in exponential notation.

See Example 1 ▶ 1. 3^5 _____   2. $(-3)^4$ _____   3. $(-\frac{3}{4})^3$ _____

MAKE SURE ANSWERS: 1. 243 2. 81 3. $-\frac{27}{64}$

22 Chapter 1 Real Numbers and Properties

You can always write a product of repeated factors in exponential notation.

EXAMPLE 2: Write a product of repeated factors in exponential notation.

Problem ▶ Write $3 \cdot 3 \cdot 3 \cdot 3 \cdot 3 \cdot x \cdot x$ in exponential notation.

1. Write each base ▶

$$\underbrace{3 \cdot 3 \cdot 3 \cdot 3 \cdot 3}_{\text{base}} \cdot \underbrace{x \cdot x}_{\text{base}} \quad 3 \quad x$$

2. Write each exponent ▶

$$\overset{1-2-3-4-5}{\underset{\downarrow\downarrow\downarrow\downarrow\downarrow}{3 \cdot 3 \cdot 3 \cdot 3 \cdot 3}} \cdot \underset{\underset{1-2}{\uparrow\uparrow}}{x \cdot x} \quad 3^5 x^2$$

exponent

Solution ▶ $3 \cdot 3 \cdot 3 \cdot 3 \cdot 3 \cdot x \cdot x = 3^5 x^2$ ⟵ exponential notation

Other Examples ▶ (a) $3.14 \cdot 3.14 = 3.14^2$ (b) $\frac{2}{3} \cdot \frac{2}{3} \cdot \frac{2}{3} = (\frac{2}{3})^3$ (c) $2xxxyy = 2x^3y^2$

(d) $(x + y)(x + y) = (x + y)^2$ (e) $\frac{3xx}{yyzzzz} = \frac{3x^2}{y^2z^4}$ (f) $3aaa + abb = 3a^3 + ab^2$

Make Sure

Write each product of repeated factors in exponential notation.

See Example 2 ▶ 1. $(-3)(-3)(-3)$ _____   2. $(-2)(-2)(-2)a \cdot a \cdot b \cdot b \cdot b$ _____

3. $\dfrac{2 \cdot 2 \cdot x \cdot x}{y \cdot y \cdot y \cdot z \cdot z \cdot z \cdot z}$ _____   4. $(m - n)(m - n)$ _____

MAKE SURE ANSWERS: 1. $(-3)^3$ 2. $(-2)^3 a^2 b^3$ 3. $\dfrac{2^2 x^2}{y^3 z^4}$ 4. $(m - n)^2$

There are several important properties that involve the operation of squaring a real number.

> The square of a real number is never negative: $a^2 \geq 0$.

Examples ▶ (a) $3^2 = 9 > 0$ (b) $(-3)^2 = 9 > 0$ (c) $0^2 = 0$

> The square of the difference $(a - b)$ equals the square of the difference $(b - a)$: $(a - b)^2 = (b - a)^2$.

Example ▶ $(7 - 3)^2 = 4^2 = 16$ and $(3 - 7)^2 = (-4)^2 = 16$

> The square of the absolute value of a number equals the square of the number: $|a|^2 = a^2$.

Example ▶ $|-5|^2 = 5^2 = 25$ and $(-5)^2 = (-5)(-5) = 25$

> If $b^2 = a$, then b is called a *2nd root* or *square root* of a.

Examples ▶ (a) 3 is a square root of 9, because $3^2 = 9$.
(b) -5 is a square root of 25, because $(-5)^2 = 25$.

Every positive real number has a positive square root and a negative square root.

Example ▶ Both 4 and -4 are square roots of 16 because $4^2 = 16$ and $(-4)^2 = 16$.

To write the positive square root of the positive real number a, you use the notation $\sqrt{a}$. The symbol $\sqrt{}$ is called a *radical*.

Example ▶ $\sqrt{16} = 4$

To write the negative square root of the positive real number a, you use the notation $-\sqrt{a}$.

Example ▶ $-\sqrt{16} = -4$

The only square root of zero is zero. $\sqrt{0} = 0$

Negative real numbers do not have a real-number square root because there is no real number b such that b^2 is a negative real number. The square of a real number is never negative.

A number, a variable, or the sum, difference, product, or quotient of numbers and variables is called an *expression*.

When an expression involves more than one operation, *grouping symbols* can be used to indicate which operation is to be performed first. The most common grouping symbols are:

( ) parentheses, [ ] brackets, { } braces, _____ bar.

Examples ▶ (a) $(8 + 4) \cdot 3 = 12 \cdot 3$ (b) $8 + [4 \cdot 3] = 8 + 12$ (c) $\dfrac{8 + 2}{2} = \dfrac{10}{2}$
$= 36$ $= 20$ $\phantom{(c) \dfrac{8+2}{2}\ }= 5$

How should you evaluate $8 + 4 \cdot 3$? The previous examples indicate that you will get a different result depending upon whether you perform the addition first or the multiplication first. To assure a unique result, you use the following rule.

> **Order of Operations Rule**
> 1. Perform all operations inside grouping symbols.
> 2. Evaluate all powers and roots, working in order from left to right.
> 3. Do all multiplications and divisions, working in order from left to right.
> 4. Do all additions and subtractions, working in order from left to right.

EXAMPLE 3: Evaluate using the Order of Operations Rule.

Problem ▶ $5 + \sqrt{36} \div 2 + 5 \cdot 3^2 - 7 = ?$

1. Evaluate powers and roots ▶ $5 + \sqrt{36} \div 2 + 5 \cdot 3^2 - 7 = 5 + 6 \div 2 + 5 \cdot 9 - 7$ Work from left to right.

2. Multiply and divide ▶ $= 5 + 3 + 45 - 7$ Work from left to right.

3. Add and subtract ▶ $= 8 + 45 - 7$ Work from left to right.
$= 53 - 7$
$= 46$

Solution ▶ $5 + \sqrt{36} \div 2 + 5 \cdot 3^2 - 7 = 46$

Chapter 1 Real Numbers and Properties

Other Examples ▶ (a) $14 - 9 + 2 = 5 + 2$ Work from left to right when doing addition and subtraction.
$ = 7$

(b) $24 \div 2^3 \cdot 5 = 24 \div 8 \cdot 5$ Powers and roots are performed first.
$ = 3 \cdot 5$ Work from left to right when doing multiplication and division.
$ = 15$

Make Sure

Evaluate each expression using the Order of Operations Rule.

See Example 3 ▶ **1.** $4 + 16 \div 4$ _____ **2.** $9 \div 3 \cdot 2$ _____ **3.** $\sqrt{25}(-2)^3 - \sqrt{4} \div 4 \cdot \sqrt{36}$ _____

MAKE SURE ANSWERS: 1. 8   2. 6   3. −43

> **CAUTION:** When a grouping symbol appears in a problem, you must first perform operations inside the grouping symbol to *clear the grouping symbol*.

Example ▶ $6 + 10 \cdot (1 + 2^2) = 6 + 10 \cdot (1 + 4)$ Compute using the Order of Operations Rule inside the parentheses.
$ = 6 + 10 \cdot 5$ ⟵ grouping symbol cleared
$ = 6 + 50$
$ = 56$

Grouping symbols inside other grouping symbols are called *nested grouping symbols*.

To remove nested grouping symbols, you work from the inside out.

EXAMPLE 4: Evaluate an expression containing nested grouping symbols.

Problem ▶ $16 - 2\{4 - [5 + (8 - 1)] + 3\} = ?$

1. Compute ▶ $16 - 2\{4 - [5 + (8 - 1)] + 3\} = 16 - 2\{4 - [5 + (7)] + 3\}$
2. Clear ▶ $\phantom{16 - 2\{4 - [5 + (8 - 1)] + 3\}} = 16 - 2\{4 - [5 + 7] + 3\}$
3. Compute ▶ $\phantom{16 - 2\{4 - [5 + (8 - 1)] + 3\}} = 16 - 2\{4 - [12] + 3\}$
4. Clear ▶ $\phantom{16 - 2\{4 - [5 + (8 - 1)] + 3\}} = 16 - 2\{4 - 12 + 3\}$
5. Compute ▶ $\phantom{16 - 2\{4 - [5 + (8 - 1)] + 3\}} = 16 - 2\{-5\}$
$\phantom{16 - 2\{4 - [5 + (8 - 1)] + 3\}} = 16 - (-10)$
$\phantom{16 - 2\{4 - [5 + (8 - 1)] + 3\}} = 26$

Solution ▶ $16 - 2\{4 - [5 + (8 - 1)] + 3\} = 26$

Other Examples ▶ (a) $\sqrt{25} + 3[42 \div (6 - 3 \cdot 2^2)] = \sqrt{25} + 3[42 \div (6 - 3 \cdot 4)]$
$= \sqrt{25} + 3[42 \div (-6)]$
$= \sqrt{25} + 3[-7]$
$= 5 + 3[-7]$
$= 5 + (-21)$
$= -16$

(b) $\dfrac{6 + (-10)}{5 + 4} = \dfrac{[6 + (-10)]}{[5 + 4]}$ The fraction bar is a grouping symbol for both the numerator and the denominator.

$= \dfrac{-4}{9}$

(c) $\sqrt{9 + 16} = \sqrt{(9 + 16)}$ The bar part of the radical symbol is a grouping symbol.

$= \sqrt{25}$
$= 5$

Make Sure

Evaluate expressions containing nested grouping symbols.

See Example 4 ▶ 1. $-5[-3(4 - 7)]$ _____   2. $-3\{-4[-(2 - 3)]\}$ _____

3. $5 - 3\{2 - [(5 - 3) - 2(3 - 7)]\}$ _____

MAKE SURE ANSWERS: 1. −45 2. 12 3. 29

A number, a variable, or the product or quotient of numbers and variables is called a *term*.

Examples ▶ (a) 3 (b) $\dfrac{1}{2}$ (c) −4.5   (d) n (e) x^2 (f) y (g) $2n^3$ (h) $-\dfrac{3}{4}k$ (i) $5xy$

  ⎵⎵⎵⎵⎵⎵⎵⎵⎵⎵⎵⎵   ⎵⎵⎵⎵⎵⎵⎵⎵⎵   ⎵⎵⎵⎵⎵⎵⎵⎵⎵⎵⎵⎵
  number terms   letter terms   general terms

When you write a term, you often omit multiplication symbols.

Examples ▶ (a) $3 \cdot n$ is written as $3n$ (b) $-4 \cdot x \cdot y$ is written as $-4xy$.

The following are all equivalent ways of writing 3 times n:

$3 \cdot n = (3)(n) = (3)n = 3(n) = 3n.$

All the variables in a term form its *literal part*.

Examples ▶ (a) In $3n$, n is the variable and also the literal part.
(b) In $-4xy$, x and y are both variables and xy is the literal part.

The numbers that appear in terms, other than the exponents, are called *numerical coefficients*.

Examples ▶ (a) In $3n$, 3 is the numerical coefficient.   (b) In $-4xy$, -4 is the numerical coefficient.

When there are no numerical coefficients shown, then it is understood that 1 or −1 is the numerical coefficient.

Examples ▶ (a) In x, 1 is the numerical coefficient ($x = 1x$).
(b) In $-y$, -1 is the numerical coefficient ($-y = -1y$).

Addition signs separate the terms of an expression.

Examples ▶ (a) $3n$ ⟵ a single term expression
(b) $5x^2 + 2y$ ⟵ an expression with two terms
(c) $xy^3 + 4yz^2 - 7xz + 3wy$ ⟵ an expression with four terms   Think: $-7xz = +(-7xz)$.

To *evaluate a variable expression*, you substitute given numerical values for the appropriate variables and then perform the indicated operations.

EXAMPLE 5: Evaluate a variable expression for given values of the variables.

Problem ▶ Evaluate $x + 2y - 4x^2y$ for $x = 4$ and $y = -3$.

1. Substitute ▶ $x + 2y - 4x^2y = 4 + 2(-3) - 4(4)^2(-3)$

2. Compute ▶ $\qquad = 4 + 2(-3) - 4(16)(-3)$ Use the Order of Operations Rule.

$\qquad = 4 + (-6) - (-192)$

$\qquad = 190$

Solution ▶ $x + 2y - 4x^2y = 190$ for $x = 4$ and $y = -3$.

To evaluate a variable expression containing grouping symbols, you first substitute the given numerical values for the appropriate variables.

Example ▶ Evaluate $w[3x - (5y - 1)] + x^2y$ for $w = 4$, $x = -3$, and $y = 5$.

1. Substitute ▶ $w[3x - (5y - 1)] + x^2y = 4[3(-3) - (5 \cdot 5 - 1)] + (-3)^2 5$

2. Compute ▶ $\qquad = 4[3(-3) - 24] + (-3)^2 5$ Use the Order of Operations Rule.

$\qquad = 4[-33] + (-3)^2 5$

$\qquad = 4[-33] + 9 \cdot 5$

$\qquad = -132 + 45$

$\qquad = -87$

Solution ▶ $w[3x - (5y - 1)] + x^2y = -87$ for $w = 4$, $x = -3$, and $y = 5$.

Make Sure

Evaluate each variable expression for the given values of the variables.

See Example 5 ▶ 1. $2z + 3y$; $z = 4$, $y = -2$ _____   2. $4w - 3x$; $w = -2$, $x = -5$ _____

3. $\dfrac{r - 2s + t}{2r + s - 3t}$; $r = -1$, $s = 2$, $t = 1$ _____

1.4 Practice

Name _____ Date _____ Class _____

Set 1: Evaluate each number written in exponential notation.

1. 2^4
2. 2^5
3. $(-3)^2$
4. -3^4

5. $(-2)^3$
6. -2^5
7. $(-4)^2$
8. $(-4)^4$

9. $(1.2)^3$
10. $(0.7)^4$
11. $(-\frac{2}{3})^3$
12. $(-\frac{2}{5})^5$

Set 2: Write each product of repeated factors in exponential notation.

13. $5 \cdot 5 \cdot 5$
14. $3 \cdot 3 \cdot 3 \cdot 3 \cdot 3$

15. $(-3)(-3)(-3)(-3)$
16. $(-4)(-4)(-4)(-4)(-4)$

17. $3 \cdot 3 \cdot 3 \cdot 3 \cdot 3 \cdot x \cdot x \cdot x \cdot x$
18. $4 \cdot 4 \cdot 4 \cdot y \cdot y \cdot y \cdot y \cdot y \cdot y$

19. $(-3)(-3)(-3)xxxxy$
20. $-4yyyzzzzz$

21. $(\frac{2}{3})(\frac{2}{3})(\frac{2}{3})(\frac{2}{3})(\frac{2}{3})(\frac{2}{3})www$
22. $(\frac{1}{5})(\frac{1}{5})(\frac{1}{5})(\frac{1}{5})ssst$

23. $(a+b)(a+b)(a+b)$
24. $(c-d)(c-d)(c-d)(c-d)$

Set 3: Evaluate each expression using the Order of Operations Rule.

25. $5 + 36 \div 9$
26. $7 + 49 \div 7$
27. $24 \div 8 \cdot 4$

28. $72 \div 18 \cdot 6$
29. $5 + 3(-2) - (-4)3 - 9$
30. $7 + 4(-5) - (-2)6 - 3$

31. $45 - (-3)^3 + 2^5 - 24 \div 8$
32. $-23 - (-4)^3 + 3^3 - 36 \div 9$

33. $\sqrt{49} \div 7 - (-3)(-5) + 5 \cdot 2^3$
34. $\sqrt{64} \div 4 - (-3)(-4) - 7 \cdot 4^2$

35. $5^3 \cdot 4 - 6 \div (-2)^4 \sqrt{64}$
36. $3^4 \cdot 5 - (-5)^3 \div 35\sqrt{49}$

Set 4: Evaluate expressions containing nested grouping symbols.

37. $5 - [(3 - 4) + 7]$
38. $3 - [(5 - 6) + 4]$
39. $3 + [(3 - 5) + (4 - 2)]$

40. $8 + [(4 - 7) + (5 - 1)]$
41. $(3 - 7)[-5 - (2 - 5)]$
42. $(5 - 9)[-7 - (4 - 9)]$

43. $(-3 + 4)[(-3) + (5 - 3)(-4)]$
44. $[(-5 + 2)(-2) + (7 - 2)](-3)$

45. $2 - \{3 - [(3 - 9) - (4 - 1)] + 4\}$
46. $-5 - \{7 - [(4 - 8) - (9 - 3)] - 5\}$

47. $\sqrt{4}\{-3[(2^2 - \sqrt{9}) - (\sqrt{4} \div \sqrt{25} \cdot 10)]\}$
48. $\sqrt{49}\{3^2 - 2[(12 \div \sqrt{4} \cdot 3) - (\sqrt{16} - 3^2)]\}$

Name _____ Date _____ Class _____

Set 5: Evaluate each variable expression for the given values of the variable.

49. $3z + y$;
    $y = 4, z = -3$

50. $5w + 2x$;
    $w = 3, x = -4$

51. $3u - 4v$;
    $u = -3, v = 3$

52. $4s - 5t$;
    $s = -4, t = 5$

53. $2p - 5q$;
    $p = -3, q = -1$

54. $5m - 2n$;
    $m = -3, n = -4$

55. $x + 2y - 3(z - 2)$;
    $x = -3, y = 2, z = -1$

56. $u + 3v - 2(w - 2)$;
    $u = -3, v = 5, w = -3$

57. $\dfrac{r}{2} - \dfrac{s}{3} + \dfrac{t}{6}$;
    $r = 2, s = -3, t = -2$

58. $\dfrac{p}{2} - \dfrac{q}{4} + \dfrac{r}{8}$;
    $p = 4, q = -3, r = -1$

59. $\dfrac{a(a - b)}{b^2 - \sqrt{c}}$;
    $a = 3, b = -2, c = 36$

60. $\dfrac{d(e - d)}{f^2 - \sqrt{e}}$;
    $d = -2, e = 9, f = 2$

61. $\dfrac{x^2 - y(x - z)}{\sqrt{z} - y^2}$;
    $x = 3, y = -2, z = 9$

62. $\dfrac{u^2 - v(w - u)}{v^2 - \sqrt{w}}$;
    $u = -2, v = -1, w = 4$

Mixed Practice: Evaluate each expression.

63. $4\sqrt{9} \div 6 - 2^3 \div 6\sqrt{9}$

64. $3\sqrt{16} \div 4 - \sqrt{64} \div 8 \cdot 3^2$

65. $3[(9 - 3) - 2(4 - 7)] - (3 - 7)(-4)$

66. $4[(5 - 7) - 3(6 - 1)] - (4 - 1)(-3)$

67. $\sqrt{16} \div 0.4 - 3^2 \div 4\sqrt{64}$

68. $\sqrt{25} \div 0.5 - 5 \div 5^2\sqrt{25}$

69. $4 - 3\{1 - [(5 - 2) - 2(3 - 9)]\}$

70. $5 - 2\{3 - [(8 - 4) - 3(1 - 6)]\}$

71. $-\{5 - [(3 - 8)(4 - 1) - 2(5 - 3)]\}$

72. $-\{3 - [(4 - 3)(5 - 9) - 2(7 - 5)]\}$

73. $3 - 2[24 \div 8 \cdot 4 - 3(2 - 8)] \div 6$

74. $6 - 3[48 \div 16 \cdot 4 - 5(3 - 6)] \div 9$

Review: Work these problems on a separate sheet of paper. Attach your work to this page.

Add signed numbers. (See Lesson 1.3.)
75. $23 + (-42)$ 76. $42 + (-56)$ 77. $32 + (-19)$ 78. $12 + (-9)$

Subtract signed numbers. (See Lesson 1.3.)
79. $-23 - 38$ 80. $-19 - 42$ 81. $-5 - (-34)$ 82. $-12 - (-42)$

Multiply and subtract signed numbers. (See Lesson 1.3.)
83. $3 \cdot 5 - 3 \cdot 4$ 84. $5 \cdot 7 - 5 \cdot 9$ 85. $-4 \cdot 5 - (-4)3$ 86. $-6 \cdot 7 - (-6)4$
87. $3(-4) - 3 \cdot 3$ 88. $5(-6) - 5 \cdot 4$ 89. $-3(-2) - (-3)4$ 90. $-5(-3) - (-5)7$

1.5 Use Properties

When you change the order of real-number addends, the sum does not change. This property of real numbers is called the *commutative property of addition*.

> **Commutative Property of Addition**
> If a and b are real numbers, then: $a + b = b + a$.

Example ▶ $5 + 3 = 3 + 5$ ⟵ different order
$8 = 8$ ⟵ same sum

When you change the order of real-number factors, then the product does not change. This property of real numbers is called the *commutative property of multiplication*.

> **Commutative Property of Multiplication**
> If a and b are real numbers, then: $a \cdot b = b \cdot a$.

Example ▶ $8 \cdot 4 = 4 \cdot 8$ ⟵ different order
$32 = 32$ ⟵ same product

> CAUTION: Real numbers are not commutative with respect to subtraction and division.

Examples ▶ (a) $8 - 3 \neq 3 - 8$ because $5 \neq -5$. (b) $15 \div 3 \neq 3 \div 15$ because $5 \neq \frac{1}{5}$.

When you change the grouping of real-number addends, the sum does not change. This property of real numbers is called the *associative property of addition*.

> **Associative Property of Addition**
> If a, b, and c are real numbers, then: $(a + b) + c = a + (b + c)$.

Example ▶ $(3 + 5) + 7 = 3 + (5 + 7)$ ⟵ different grouping
$8 + 7 = 3 + 12$ Perform operations in parentheses first.
$15 = 15$ ⟵ same sum

When you change the grouping of real-number factors, the product does not change. This property of multiplication is called the *associative property of multiplication*.

> **Associative Property of Multiplication**
> If a, b, and c are real numbers, then: $(a \cdot b) \cdot c = a \cdot (b \cdot c)$.

Example ▶ $(2 \cdot 4) \cdot 7 = 2 \cdot (4 \cdot 7)$ ⟵ different grouping
$8 \cdot 7 = 2 \cdot 28$ Perform operations in parentheses first.
$56 = 56$ ⟵ same product

> CAUTION: Real numbers are not associative with respect to subtraction and division.

Examples ▶ (a) $(9 - 4) - 1 \neq 9 - (4 - 1)$ because $4 \neq 6$.
(b) $[24 \div (-4)] \div 2 \neq 24 \div [(-4) \div 2]$ because $-3 \neq -12$.

32 Chapter 1 Real Numbers and Properties

It is very important to be able to use each of the previous properties and to be able to identify which property has been used to rename an expression.

EXAMPLE 1: Identify the property.

Problem ▶ Which property asserts that $3 + (9 + 5) = (3 + 9) + 5$?

1. Identify changes ▶ $3 + (9 + 5) = (3 + 9) + 5$ ⟵ same order, different grouping

2. Identify operation ▶ $3 \;\boxed{+}\; (9 \;\boxed{+}\; 5) = (3 \;\boxed{+}\; 9) \;\boxed{+}\; 5$ All the operations are addition.

3. Interpret ▶ Changing the grouping of addends is an application of the associative property of addition.

Solution ▶ The associative property of addition states that $3 + (9 + 5) = (3 + 9) + 5$.

Other Examples ▶
(a) $(8 \cdot x)2 = 2(8 \cdot x)$ The commutative property of multiplication.
(b) $(5 \cdot 2)(-3) = 5(2(-3))$ The associative property of multiplication.
(c) $3 + (2 \cdot y) = (2 \cdot y) + 3$ The commutative property of addition.
(d) $(7 + w) + 2 \cdot 3 = 7 + (w + 2 \cdot 3)$ The associative property of addition.

Make Sure

Identify the property that is being illustrated in each problem.

See Example 1 ▶
1. $(5 + 7) + 3 = 5 + (7 + 3)$ _____

2. $(3 - 4)7 = 7(3 - 4)$ _____

3. $(2 + 3)4 = (3 + 2)4$ _____

4. $(3(-2))(-3) = 3((-2)(-3))$ _____

MAKE SURE ANSWERS: 1. Associative Property of Addition. 2. Commutative Property of Multiplication. 3. Commutative Property of Addition. 4. Associative Property of Multiplication.

To multiply a sum by a number, you can first add and then multiply, or you can first multiply and then add.

Examples ▶
(a) $4(5 + 7) = 4 \cdot 12 = 48$ ⟵ same   First add, then multiply.

(b) $4(5 + 7) = 4 \cdot 5 + 4 \cdot 7 = 20 + 28 = 48$ First multiply, then add.

The property of real numbers shown in Example (b) is called the *distributive property of multiplication over addition*.

> **Distributive Property of Multiplication over Addition**
> If a, b, and c are real numbers, then: $a(b + c) = ab + ac$
> and: $(b + c)a = ba + ca$.

Examples ▶
(a) $5(2 + 7) = 5 \cdot 2 + 5 \cdot 7$ (b) $(6 + 1)3 = 6 \cdot 3 + 1 \cdot 3$
$ = 10 + 35$ $ = 18 + 3$
$ = 45$ $ = 21$

To multiply a difference by a number, you can first subtract and then multiply, or you can multiply and then subtract. This property is called the *distributive property of multiplication over subtraction*.

> **Distributive Property of Multiplication over Subtraction**
> If a, b, and c are real numbers, then: $a(b - c) = ab - ac$
> and: $(b - c)a = ba - ca$.

Examples ▶ (a) $7(9 - 5) = 7 \cdot 9 - 7 \cdot 5$ (b) $(8 - 2)3 = 8 \cdot 3 - 2 \cdot 3$
$ = 63 - 35 = 24 - 6$
$ = 28 = 18$

One of the most important uses of the distributive properties is to *clear parentheses*.

EXAMPLE 2: Clear parentheses using the distributive properties.

Problem ▶ Clear parentheses in $4(2y - 3)$.

1. Distribute ▶ $4(2y - 3) = 4 \cdot 2y - 4 \cdot 3$

2. Simplify ▶ $ = 8y - 12$ ⟵ parentheses cleared

Solution ▶ $4(2y - 3) = 8y - 12$

Other Examples ▶ (a) $-7(z - 4) = -7 \cdot z - (-7) \cdot 4$ (b) $(5 - w)6 = 5 \cdot 6 - w \cdot 6$
$ = -7z + 28 = 30 - 6w$

> **SHORTCUT 1.1:** When parentheses are preceded by a positive sign, you can clear the parentheses by just writing what is inside.

Examples ▶ (a) $+(8v - 5) = 8v - 5$ (b) $+(-8 + 9z) = -8 + 9z$

> **SHORTCUT 1.2:** When parentheses are preceded by a negative sign, you can clear the parentheses by writing the opposite of each term inside.

Examples ▶ (a) $-(8v - 5) = -8v + 5$ (b) $-(-2 + 7y) = 2 - 7y$

Make Sure

Remove the grouping symbols using the distributive properties.

See Example 2 ▶ **1.** $3(a + 5)$ _____   **2.** $(2b - 5)(-4)$ _____   **3.** $-(2c - 3d + 4)$ _____

MAKE SURE ANSWERS: 1. $3a + 15$ 2. $-8b + 20$ 3. $-2c + 3d - 4$

Terms that have exactly the same literal part are called *like terms*.

 like terms like terms like terms
Examples ▶ (a) $\boxed{3x} + \boxed{4x}$ (b) $\boxed{xy} - \boxed{3xy} + 5x$ (c) $\boxed{0.5n} + 0.8 \boxed{- n}$

To add like terms, you can use the distributive property.

Example ▶ $3x + 4x = (3 + 4)x = 7x$ The distributive property of multiplication over addition.

Adding or subtracting like terms is called *combining like terms*. The previous example shows that like terms can be combined by adding (or subtracting) their numerical coefficients and then writing the same literal part.

EXAMPLE 3: Simplify an expression with like terms.

Problem ▶ Simplify $3xy + 4w - 5w + xy$.

1. Regroup ▶ $3xy + 4w - 5w + xy = \boxed{(3xy + xy) + (4w - 5w)}$ Do this step mentally.

2. Combine like terms ▶ $\qquad\qquad\qquad\qquad = 4xy + (-w)$

Solution ▶ $3xy + 4w - 5w + xy = 4xy - w$

Make Sure

Simplify each expression by combining like terms.

See Example 3 ▶ **1.** $7a - 3a$ _____ **2.** $b - 9b$ _____ **3.** $3x - 4y + z + 6x + 5y$ _____

MAKE SURE ANSWERS: 1. $4a$ **2.** $-8b$ **3.** $9x + y + z$

The distributive properties can be used to simplify a variable expression containing nested grouping symbols.

EXAMPLE 4: Simplify a variable expression containing nested grouping symbols.

Problem ▶ $-3\{a - 2[-(4a + 2b - 5b) + (5a - b)]\} = ?$

1. Combine like terms ▶ $-3\{a - 2[-(4a + 2b - 5b) + (5a - b)]\} = -3\{a - 2[-(4a - 3b) + (5a - b)]\}$

2. Clear ▶ $\qquad\qquad\qquad\qquad\qquad = -3\{a - 2[-4a + 3b + 5a - b]\}$

3. Combine like terms ▶ $\qquad\qquad\qquad\qquad\qquad = -3\{a - 2[a + 2b]\}$

4. Clear ▶ $\qquad\qquad\qquad\qquad\qquad = -3\{a - 2a - 4b\}$

5. Combine like terms ▶ $\qquad\qquad\qquad\qquad\qquad = -3\{-a - 4b\}$

6. Clear ▶ $\qquad\qquad\qquad\qquad\qquad = 3a + 12b$

Solution ▶ $-3\{a - 2[-(4a + 2b - 5b) + (5a - b)]\} = 3a + 12b$

Make Sure

Simplify variable expressions containing nested grouping symbols.

See Example 4 ▶ **1.** $5[-(a - 2) - 3(2 - a)]$ **2.** $-5\{4 - [3(2a - 3b + 1) + 2(3 - b)]\}$

MAKE SURE ANSWERS: 1. $10a - 20$ **2.** $30a - 55b + 25$

1.5 Practice

Set 1: Identify the property that is being illustrated in each problem.

1. $2 + 5 = 5 + 2$
2. $3 + 7 = 7 + 3$
3. $-3 + 5 = 5 + (-3)$
4. $-3 + (-7) = -7 + (-3)$
5. $4 + (3 + 5) = (3 + 5) + 4$
6. $(7 + 3) + 2 = 2 + (7 + 3)$
7. $4 \cdot 5 + 6 = 6 + 4 \cdot 5$
8. $7 + 2 \cdot 3 = 2 \cdot 3 + 7$
9. $4 \cdot 2 = 2 \cdot 4$
10. $-2(-5) = -5(-2)$
11. $(4 \cdot 3)5 = 4(3 \cdot 5)$
12. $3(7 \cdot 6) = (3 \cdot 7)6$
13. $2(3 + 5) = (3 + 5)2$
14. $(4 + 3)(-2) = -2(4 + 3)$
15. $4(7 + 2) = 4(2 + 7)$
16. $(4 + 3)5 = 5(4 + 3)$
17. $(7 + 5) + 3 = (5 + 7) + 3$
18. $(-3 + (-5)) + 4 = -3 + (-5 + 4)$

Set 2: Remove the grouping symbols using the distributive properties.

19. $5(z + 3)$
20. $7(y + 5)$
21. $(x - 5)(-3)$
22. $(w - 2)(-4)$
23. $-(3v - 5u)$
24. $-(5s - 3t)$
25. $2(2q - 3r + 4s)$
26. $4(3n - 4p + q)$
27. $(3a - b + 2c)(-3)$
28. $(6e - 2f + g)(-4)$
29. $(w - 2x + 3y - 4)5$
30. $(p - 5q + 4r - 3)4$

Set 3: Simplify each expression by combining like terms.

31. $3z + 4z$
32. $5y + 2y$
33. $8x - 3x$
34. $9w - 7w$
35. $v - 5v$
36. $u - 7u$
37. $7t - 5t - 4t + t$
38. $8s - 3s - 9s + s$
39. $5r + 7 - 3r - 11$
40. $8q - 23 - 7q + 9$
41. $3de - df + 5df - 6de + de$
42. $7ab - 4ac - 3ac + 2ab - ac$

36 Chapter 1 Real Numbers and Properties

43. _____

Set 4: Simplify variable expressions containing grouping symbols.

43. $3[a - 2(2 + 5a)]$

44. $-5[2b - 5(3 - 2b)]$

44. _____

45. $4[3(c - 2) - 2(4 - 2c)]$

46. $-2[4(2 - d) - 3(2d - 5)]$

45. _____

46. _____

47. $2x - [-2(x - y) - 3y]$

48. $-2p - [3p - 4(2p - 2q)]$

47. _____

49. $3m - [n + 3(m - n)] - [-2(n - m)]$

50. $2g - [h + 3(h - g)] - [-5(g - h)]$

48. _____

51. $4\{-2a + 3[2 - (a - b)] + 3(b - 2)\}$

52. $5\{-2c - 2[3(c - d) - d] - 2(d - c)\}$

49. _____

53. $3\{-b - [a - 2(b - c) - 5(b - 2c)]\}$

54. $4\{-y - [z - 2(x - y) - 4(2x - 3x)]\}$

50. _____

51. _____

Review: Work these problems on a separate sheet of paper. Attach your work to this page.

Rename each fraction in lowest terms.
55. $\frac{12}{16}$ 56. $\frac{48}{72}$ 57. $-\frac{56}{84}$ 58. $-\frac{26}{65}$

52. _____

Add signed numbers. (See Lesson 1.3.)
59. $3 + (-2)$ 60. $-5 + (-3)$ 61. $-5 + 2$ 62. $-1 + (-1)$

Subtract signed numbers. (See Lesson 1.3.)
63. $1 - (-2)$ 64. $-3 - 4$ 65. $-4 - (-2)$ 66. $-3 - (-5)$

53. _____

Multiply signed numbers. (See Lesson 1.3.)
67. $2(-3)$ 68. $-1(-3)$ 69. $-2(-2)$ 70. $3(-1)$

Evaluate each number written in exponential form. (See Lesson 1.4.)

54. _____

71. 2^3 72. $(-2)^3$ 73. $(-3)^4$ 74. $(-3)^5$

Copyright © 1985 by Harcourt Brace Jovanovich, Inc. All rights reserved.

1.6 Use Integral Exponents

Consider the following lists.

3^4	= 81	→ 81 ÷ 3		10^4	= 10,000	→ 10,000 ÷ 10	
3^3	= 27	←		10^3	= 1000	←	
3^2	= 9			10^2	= 100		
3^1	= 3			10^1	= 10		
3^0	= ?			10^0	= ?		
3^{-1}	= ?			10^{-1}	= ?		
3^{-2}	= ?			10^{-2}	= ?		

Notice that each power divided by its base produces the power in the row below it ($81 \div 3 = 27$). If this pattern is to continue, then each of the following must be as shown.

$3^1 = 3 \longrightarrow 3 \div 3$
$3^0 = 1 \longleftarrow$

$3^{-1} = 1 \div 3 = \dfrac{1}{3} = \dfrac{1}{3^1}$

$3^{-2} = \dfrac{1}{3} \div 3 = \dfrac{1}{3} \cdot \dfrac{1}{3} = \dfrac{1}{3^2}$

$10^1 = 10 \longrightarrow 10 \div 10$
$10^0 = 1 \longleftarrow$

$10^{-1} = 1 \div 10 = \dfrac{1}{10} = \dfrac{1}{10^1}$

$10^{-2} = \dfrac{1}{10} \div 10 = \dfrac{1}{10} \cdot \dfrac{1}{10} = \dfrac{1}{10^2}$

These patterns lead to the following definitions.

> For any nonzero a, $a^0 = 1$.

Examples ▶ (a) $19^0 = 1$ (b) $(-3)^0 = 1$ (c) $(\tfrac{1}{3})^0 = 1$ (d) $(0.4)^0 = 1$

> If $a \neq 0$ and n is any natural number, $a^{-n} = \dfrac{1}{a^n}$ and $\dfrac{1}{a^{-n}} = a^n$.

Examples ▶ (a) $7^{-2} = \dfrac{1}{7^2}$ (b) $\dfrac{1}{3^{-2}} = 3^2$ because $\dfrac{1}{3^{-2}} = 1 \div 3^{-2} = 1 \div \dfrac{1}{3^2} = \dfrac{1}{1} \cdot \dfrac{3^2}{1} = 3^2$.

In general, any nonzero factor in a term may be moved from the numerator to the denominator or from the denominator to the numerator by simply changing the sign of its exponent.

EXAMPLE 1: Write a term without negative exponents.

Problem ▶ Write $\dfrac{3a^{-2}b}{c^{-3}}$ without negative exponents.

1. Identify factors to be moved ▶ $\dfrac{3\;\boxed{a^{-2}}\;b}{\boxed{c^{-3}}}$ Only the bases with negative exponents will be moved.

2. Move bases and change to positive exponents ▶ $\dfrac{3a^{-2}b}{c^{-3}} = \dfrac{3bc^{+3}}{a^{+2}}$ ← without negative exponents

Solution ▶ $\dfrac{3a^{-2}b}{c^{-3}} = \dfrac{3bc^3}{a^2}$

Other Examples (a) $\dfrac{4x^{-3}}{y} = \dfrac{4}{x^3 y}$ (b) $\dfrac{5m^{-2}p}{n^{-4}} = \dfrac{5n^4 p}{m^2}$ (c) $(3z)^{-1} = \dfrac{1}{(3z)^1} = \dfrac{1}{3z}$ (d) $3z^{-1} = \dfrac{3}{z^1} = \dfrac{3}{z}$

Make Sure

Write each term without negative exponents.

See Example 1

1. $\dfrac{2a^2 b^{-2}}{3c^{-1}}$ _____ 2. $\dfrac{3d^{-3} e^{-1}}{4f^2 g^{-2}}$ _____ 3. $\dfrac{2^{-2} p^2 s^{-3}}{3q^{-1} r}$ _____

MAKE SURE ANSWERS: 1. $\dfrac{2a^2 c}{3b^2}$ 2. $\dfrac{4d^3 g^2 e}{3g^2}$ 3. $\dfrac{12rs^3}{p^2 q}$

The associative property of multiplication allows you to rename $a^m \cdot a^n$ where m and n are integers.

Example

$a^2 \cdot a^3 = (a \cdot a) \cdot (a \cdot a \cdot a)$ Write as a product of repeated factors.

$ = (a \cdot a \cdot a \cdot a \cdot a)$ Regroup.

$ = a^5$ Write as exponential notation.

This procedure also applies to cases with exponents that are negative integers.

Examples

(a) $x^4 \cdot x^{-1} = (x \cdot x \cdot x \cdot x) \cdot \dfrac{1}{x}$ Recall: $x^{-1} = \dfrac{1}{x}$

$\phantom{x^4 \cdot x^{-1}} = \left(x \cdot x \cdot x \cdot \cancel{x} \cdot \dfrac{1}{\cancel{x}} \right)$ Regroup and eliminate common factors.

$\phantom{x^4 \cdot x^{-1}} = x^3$

(b) $y^{-2} \cdot y^{-4} = \dfrac{1}{y \cdot y} \cdot \dfrac{1}{y \cdot y \cdot y \cdot y}$

$\phantom{y^{-2} \cdot y^{-4}} = \dfrac{1}{y \cdot y \cdot y \cdot y \cdot y \cdot y}$

$\phantom{y^{-2} \cdot y^{-4}} = \dfrac{1}{y^6}$

$\phantom{y^{-2} \cdot y^{-4}} = y^{-6}$

> **Product Rule for Exponents**
> If m and n are integers, then: $a^m \cdot a^n = a^{m+n}$.

The product rule states that you can multiply factors with like bases in exponential notation by simply adding the exponents.

Examples (a) $5^4 \cdot 5^6 = 5^{4+6} = 5^{10}$ (b) $4^{-3} \cdot 4^8 = 4^{(-3)+8} = 4^5$ (c) $3^{-2} \cdot 3^{-4} = 3^{(-2)+(-4)} = 3^{-6}$

> **CAUTION:** $3^4 \cdot 5^6$ cannot be simplified by the product rule because the bases are different.

1.6 Use Integral Exponents   39

You can use the Product Rule for Exponents to find a Quotient Rule for Exponents.

$$\frac{a^m}{a^n} = a^m \cdot a^{-n} = a^{m+(-n)} = a^{m-n}$$

> **Quotient Rule for Exponents**
> If m and n are integers, then: $\frac{a^m}{a^n} = a^{m-n} (a \neq 0)$.

The quotient rule states that you can divide factors with like bases in exponential notation by simply subtracting the exponents.

Examples ▶ (a) $\frac{3^5}{3^1} = 3^{5-1} = 3^4$ (b) $\frac{5^3}{5^4} = 5^{3-4} = 5^{-1} = \frac{1}{5}$ (c) $\frac{7^2}{7^2} = 7^{2-2} = 7^0 = 1$

> A term is *simplified* if each different base occurs only once and each exponent is positive.

EXAMPLE 2: Simplify a product of terms.

Problem ▶ Simplify $\frac{20x^3y}{z^{-2}} \cdot \frac{y^4 z^{-5}}{4x^{-1}}$.

1. Group like bases ▶ $\frac{20x^3y}{z^{-2}} \cdot \frac{y^4 z^{-5}}{4x^{-1}} = \frac{20}{4} \cdot \frac{x^3}{x^{-1}} \cdot \frac{yy^4}{1} \cdot \frac{z^{-5}}{z^{-2}}$

2. Use product and quotient rules ▶ $= 5 \cdot x^{3-(-1)} \cdot y^{1+4} \cdot z^{-5-(-2)}$
$= 5 \cdot x^4 \cdot y^5 \cdot z^{-3}$

3. Write with only positive exponents ▶ $= \frac{5x^4 y^5}{z^3}$ ⟵ simplified (with only positive exponents)

Solution ▶ $\frac{20x^3y}{z^{-2}} \cdot \frac{y^4 z^{-5}}{4x^{-1}} = \frac{5x^4 y^5}{z^3}$

Other Examples ▶ (a) $(3x^2 y^{-1})(2x^2 y^3) = (3 \cdot 2)(x^2 x^2)(y^{-1} y^3)$
$= 6x^4 y^2$

(b) $\frac{15n^{-2}}{m^3} \cdot \frac{4m^2 n}{p} \cdot \frac{mn^{-1}}{2p^3} = \frac{15 \cdot 4}{2} \cdot \frac{m^2 m}{m^3} \cdot \frac{n^{-2} nn^{-1}}{1} \cdot \frac{1}{pp^3}$
$= 30 m^0 n^{-2} \frac{1}{p^4}$
$= \frac{30}{n^2 p^4}$

Make Sure

Simplify each product of terms.

See Example 2 ▶ 1. $\frac{x^3 y^2}{y^{-1} z^4} \cdot \frac{z^3 x^{-2}}{xy^{-3} z^{-2}}$ 2. $\frac{16a^5 b^3}{25c^4} \cdot \frac{5a^{-3} c^{-2}}{8b^2}$

MAKE SURE ANSWERS: 1. $y^6 z$ 2. $\frac{2a^2 b}{5c^6}$

Expressions of the form $(a^m)^n$ can be rewritten using the definition of exponential notation and the product rule for exponents.

Example ▶ $(3^4)^2$ means $(3^4)(3^4) = 3^8$

The above result could be obtained by simply multiplying the exponents.

> **Power Rule**
> If m and n are integers, then: $(a^m)^n = a^{mn}$.

The Power Rule states that you can find the power of a power by multiplying exponents.

Examples ▶ (a) $(2^4)^3 = 2^{12}$ (b) $(x^{-3})^5 = x^{-15}$ (c) $(y^{-2})^{-4} = y^8$

The power rule can be extended to products and quotients raised to a power.

> **Extended Power Rule**
> $\left(\dfrac{a^p b^q}{c^r}\right)^s = \dfrac{a^{ps} b^{qs}}{c^{rs}}$ where the exponents are integers and the bases are restricted to avoid a zero factor in the denominator.

EXAMPLE 3: Simplify a term raised to a power.

Problem ▶ Simplify $\left[\dfrac{-4a^{-3}b^{-2}}{2ab^{-4}}\right]^{-2}$.

1. Simplify inside brackets ▶ $\left[\dfrac{-4a^{-3}b^{-2}}{2ab^{-4}}\right]^{-2} = [-2a^{-3-1}b^{-2-(-4)}]^{-2}$

$= [-2a^{-4}b^2]^{-2}$

2. Use extended power rule ▶ $= (-2)^{-2} a^8 b^{-4}$

3. Write with only positive exponents ▶ $= \dfrac{a^8}{(-2)^2 b^4}$

$= \dfrac{a^8}{4b^4}$ ← simplified (with only positive exponents)

Solution ▶ $\left[\dfrac{-4a^{-3}b^{-2}}{2ab^{-4}}\right]^{-2} = \dfrac{a^8}{4b^4}$

Make Sure

Simplify each term raised to a power.

See Example 3 ▶ 1. $\left[\dfrac{2x^3 y^{-2}}{3z^{-3}}\right]^3$ _____ 2. $\left[\dfrac{-3a^3 b^{-2}}{2a^{-2} bc^2}\right]^{-4}$ _____

MAKE SURE ANSWERS: 1. $\dfrac{8x^9 z^9}{27 y^6}$ 2. $\dfrac{16 b^{12} c^8}{81 a^{20}}$

1.6 Practice

Set 1: Write each term without negative exponents.

1. $\dfrac{1}{a^{-3}}$
2. $\dfrac{1}{b^{-4}}$
3. c^{-5}
4. d^{-1}
5. $\dfrac{e^{-3}f^{-2}}{g^{-1}h^2}$
6. $\dfrac{m^{-4}n^{-5}}{p^{-1}q^3}$
7. $\dfrac{3r^2s^{-1}}{4t^{-2}u}$
8. $\dfrac{5v^{-1}w^5}{7x^{-3}y}$
9. $\dfrac{6a^{-1}b^2c^{-3}}{7d^{-1}ef^{-4}}$
10. $\dfrac{5g^3h^{-1}k^{-5}}{8mn^{-2}p^{-1}}$
11. $\dfrac{(-3q)^{-2}}{(4r)^{-1}}$
12. $\dfrac{(-3u)^{-3}}{(2v)^{-1}}$

Set 2: Simplify each product of terms.

13. $a^5 \cdot a^4$
14. $b^3 \cdot b^7$
15. $c^6 \cdot \dfrac{1}{c^3}$
16. $d^5 \cdot \dfrac{1}{d^2}$
17. $\dfrac{e^{-3}f^2}{e^2f^{-4}} \cdot \dfrac{e^5f^{-3}}{e^{-1}f}$
18. $\dfrac{g^5h^{-4}}{g^{-1}h^3} \cdot \dfrac{g^{-3}h^4}{gh^{-1}}$
19. $\dfrac{2m^4n^{-4}}{m^{-5}n^2} \cdot \dfrac{3m^{-6}n^3}{8m^3n^5}$
20. $\dfrac{4p^4q^3}{7p^{-7}q^7} \cdot \dfrac{14p^{-3}q^{-4}}{p^5q^{-2}}$
21. $\dfrac{12rst}{13r^2s^{-3}} \cdot \dfrac{2r^{-4}s}{27r^{-5}t^{-3}}$
22. $\dfrac{5u^{-6}v^{-1}}{7v^2w^{-4}} \cdot \dfrac{14uvw}{15u^3v^{-4}}$
23. $\dfrac{7xy^{-1}z^3}{16yz^{-2}} \cdot \dfrac{12x^{-2}yz^{-3}}{14x^{-4}y^{-4}z}$
24. $\dfrac{4a^{-3}b^3c}{6a^2b^{-1}c^2} \cdot \dfrac{9a^5b^{-2}c^{-3}}{10ab^5c^{-5}}$

42 Chapter 1 Real Numbers and Properties

Set 3: Simplify each term raised to a power.

25. $(a^3)^5$

26. $(b^4)^2$

27. $(-3c^3)^4$

28. $(-2d^5)^3$

29. $(3e^4f^{-3})^2$

30. $(2g^5h^{-2})^3$

31. $\left[\dfrac{-2m^{-1}n^{-2}p^3}{3m^2n^{-1}p^{-4}}\right]^4$

32. $\left[\dfrac{-3q^4r^{-1}s^2}{2q^{-1}rs^{-3}}\right]^6$

33. $\left[\dfrac{-2tu^{-1}v^3}{3t^2uv^{-2}}\right]^5$

34. $\left[\dfrac{2w^4x^2y^{-2}}{-3wx^{-3}y^4}\right]^3$

35. $\left[-\dfrac{3a^{-3}b^4c^{-1}}{2a^2b^{-1}c^3}\right]^{-2}$

36. $\left[-\dfrac{2d^3e^2f^{-1}}{3d^{-1}e^3f^2}\right]^{-3}$

Extra: Simplify each problem.

37. $[2a^{-2}(a^3b^{-3})^{-1}]^2$

38. $[4d^2(c^3d^{-1})^{-2}]^3$

39. $[3e^{-1}(e^2f^{-3})^2]^{-3}$

40. $[2g^{-2}(g^3h^{-1})^2]^{-3}$

41. $\dfrac{(3mn^2p^{-1})^{-3}}{(-2m^2np^{-2})^2}$

42. $\dfrac{(2r^{-1}st^2)^{-2}}{(-3r^2s^{-1}t)^3}$

43. $\dfrac{(-2u^3v^2w^{-2})^4}{(-3u^{-2}v^3w^{-1})^{-2}}$

44. $\dfrac{(-xy^2z^3)^{-2}}{(-x^2yz^{-1})^{-3}}$

45. $\dfrac{(-a^{-2}b^{-3}c)^{-3}}{(-2ab^{-1}c^3)^{-1}}$

46. $\dfrac{(-2d^{-2}e^{-1}f^3)^5}{(-3d^{-3}ef^2)^{-3}}$

47. $\left[\dfrac{-3g^2h^{-3}k}{4g^{-1}h^{-4}k^2}\right]^4$

48. $\left[\dfrac{-2m^3n^2p^{-3}}{5m^{-1}n^4p^{-5}}\right]^3$

Review: Work these problems on a separate sheet of paper. Attach your work to this page.

Find the additive inverse of each expression.
49. 3
50. 7
51. -12
52. -15
53. $3a$
54. $5b$
55. $-13c$
56. $-9d$

Simplify each expression that involves additive inverses.
57. $a + 3 - 3$
58. $b + 5 - 5$
59. $c - 7 + 7$
60. $d - 9 + 9$
61. $3e + 2.3 - 2.3$
62. $5f + 4.8 - 4.8$
63. $4g - 4.2 + 4.2$
64. $2k - 5.1 + 5.1$

Find the multiplicative inverse of each number.
65. 4
66. 3
67. -5
68. -9
69. $\tfrac{2}{3}$
70. $\tfrac{4}{5}$
71. $-\tfrac{1}{4}$
72. $-\tfrac{11}{12}$

Simplify expressions that involve multiplicative inverses.
73. $\tfrac{1}{3} \cdot 3a$
74. $\tfrac{1}{5} \cdot 5b$
75. $-\tfrac{1}{8}(-8c)$
76. $-\tfrac{1}{6}(-6d)$
77. $\tfrac{3}{2} \cdot \tfrac{2}{3}e$
78. $\tfrac{4}{7} \cdot \tfrac{7}{4}f$
79. $-\tfrac{2}{5}(-\tfrac{5}{2}g)$
80. $-\tfrac{3}{7}(-\tfrac{7}{3}h)$

Problem Solving 1: Compute Using Scientific Notation

If n is a whole number, then: $10^n = \underbrace{1000\cdots 0}_{n \text{ zeros}}$

and: $10^{-n} = \overbrace{0.000\cdots 01}$.

Examples ▶ (a) $10^0 = 1$ (b) $10^1 = 10$ (c) $10^2 = 100$ (d) $10^{-1} = 0.1$ (e) $10^{-2} = 0.01$

If n is an integer and a is a number between 1 and 10 (including 1 but not 10), then $a \cdot 10^n$ is called *scientific notation*.

Examples ▶ (a) 1×10^0 is scientific notation for the number 1: $1 \times 10^0 = 1 \times 1 = 1$.

(b) 2.4×10^3 is scientific notation for 2400: $2.4 \times 10^3 = 2.4 \times 1000 = 2400$.

(c) 6.25×10^{-4} is scientific notation for 0.000625: $6.25 \times 10^{-4} = 6.25 \times 0.0001 = 0.000625$.

CAUTION: If a is not between 1 and 10, then $a \cdot 10^n$ is not scientific notation.

Example ▶ Neither 24×10^2 nor 0.00625×10^{-1} is scientific notation because neither 24 nor 0.00625 is a number between 1 and 10.

To rename exponential notation such as 24×10^2 and 0.00625×10^{-1} as scientific notation, proceed as shown in the following examples.

Examples ▶ (a) $24 \times 10^2 = (2.4 \times 10^1) \times 10^2 = 2.4 \times (10^1 \times 10^2) = 2.4 \times 10^3$ ⟵ scientific notation

(b) $0.00625 \times 10^{-1} = (0006.25 \times 10^{-3}) \times 10^{-1} = 6.25 \times (10^{-3} \times 10^{-1}) = 6.25 \times 10^{-4}$ ⟵

SHORTCUT 1.3: Scientific notation for 1×10^n is usually written as just 10^n.

Examples ▶ (a) $1 \times 10^1 = 10^1$ or 10 (b) $1 \times 10^2 = 10^2$ (c) $1 \times 10^{25} = 10^{25}$

When $a \cdot 10^n$ is in scientific notation, the digits in a are called *significant digits*.

Example ▶ 6.25×10^{-4} has three significant digits: 6, 2, and 5.

CAUTION: In $a \cdot 10^n$, the last significant digit in a is either a correct or a rounded value.

Example ▶ In 6.25×10^{-4}, 6 and 2 are correct values and 5 is either a correct or a rounded value.

To compute with scientific notation, you use the following rules.

Computation Rules for Scientific Notation

1. Multiply using: $(a \cdot 10^n)(b \cdot 10^m) = (ab)10^{n+m}$.

2. Divide using: $\dfrac{a \cdot 10^n}{b \cdot 10^m} = \dfrac{a}{b} \cdot 10^{n-m} (b \neq 0)$.

3. Find powers using: $(a \cdot 10^n)^m = a^m \cdot 10^{nm}$.

4. Round each result to the fewer number of significant digits in a or b.

44 Chapter 1 Real Numbers and Properties

To find the *attractive force* between two objects, you evaluate the following formula.

$$F = G\frac{Mm}{r^2} \text{ where } \begin{cases} F \text{ is the attractive force measured in } newtons^* \text{ (N)} \\ G \text{ is } Newton's\ Constant: G = 6.67 \times 10^{-11} \\ M \text{ is the mass of one object, measured in kilograms (kg)} \\ m \text{ is the mass of the other object, measured in kilograms (kg)} \\ r \text{ is the distance between the two objects, measured in meters (m)} \end{cases}$$

To evaluate the attractive force formula, you compute using scientific notation.

EXAMPLE: Compute using scientific notation.

1. Read and identify ▶ The Earth's mass is 5.98×10^{24} kg. The Sun's mass is 1.99×10^{30} kg. The distance between the Earth and the Sun is 1.49×10^{11} m. What is the attractive force of the sun on the earth?

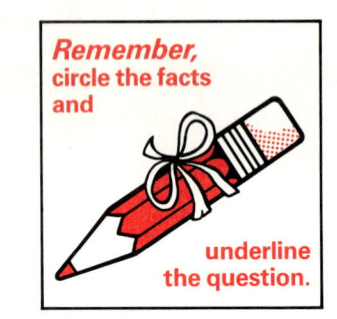

Remember, circle the facts and underline the question.

2. Understand ▶ The question asks you to evaluate the attractive force formula.

3. Decide ▶ To evaluate the attractive force formula, you **compute using scientific notation.**

4. Evaluate ▶ $F = G\dfrac{Mm}{r^2}$ ⟵ attractive force formula

$= (6.67 \times 10^{-11})\dfrac{(5.98 \times 10^{24})(1.99 \times 10^{30})}{(1.49 \times 10^{11})^2}$ Substitute.

$\approx (6.67 \times 10^{-11})\dfrac{(5.98 \times 10^{24})(1.99 \times 10^{30})}{2.22 \times 10^{22}}$ Compute.
Think: $1.49^2 \times (10^{11})^2$

$\approx (6.67 \times 10^{-11})\dfrac{11.9 \times 10^{54}}{2.22 \times 10^{22}}$ Think: $(5.98 \times 1.99) \times 10^{24+30}$

$\approx (6.67 \times 10^{-11})(5.36 \times 10^{32})$ Think: $\dfrac{11.9}{2.22} \times 10^{54-22}$

$\approx 35.8 \times 10^{21}$ Think: $(6.67 \times 5.36) \times 10^{-11+32}$

5. Write scientific notation ▶ $= (3.58 \times 10^1) \times 10^{21}$ Think: $35.8 = 3.58 \times 10^1$

$= 3.58 \times 10^{22}$ ⟵ scientific notation

6. Interpret ▶ 3.58×10^{22} means the attractive force of the Sun on Earth is 3.58×10^{22} N.

Practice: Compute using scientific notation.

1. _____

2. _____

1. The Moon's mass is 7.37×10^{22} kg. The distance between the Earth and Moon is 3.84×10^8 m. Find the attractive force of the Moon on Earth. (Use the Example.)

2. How many times greater is the attractive force of the Sun on Earth than the attractive force of the Moon on Earth? (Use the Example and Problem 1.)

---

*The newton (N) is that force which gives to a mass of 1 kilogram an acceleration of 1 meter per second per second.

PS 1: Compute Using Scientific Notation   45

3. _____

4. _____

5. _____

6. _____

7. _____

8. _____

9. _____

10. _____

3. What is the attractive force between the Sun and Moon during a lunar eclipse? (Hint: During a lunar eclipse, the earth passes between the sun and moon.)

4. What is the attractive force between the Sun and Moon during a solar eclipse? (Hint: During a solar eclipse, the moon passes between the sun and earth.)

5. How many times greater is the Sun's mass than the Earth's mass? (Use the Example.)

6. How many times greater is the Earth's mass than the Moon's mass? (Use the Example and Problem 1.)

7. The average radius of the Earth is 3.96×10^3 miles. What is the Earth's surface area (SA) using: $SA = 4\pi r^2$ and $\pi \approx 3.14$?

8. What is Earth's volume (V) using: $V = \frac{4}{3}\pi r^3$ and $\pi \approx 3.14$? (Use Problem 7.)

9. Our galaxy (the Milky Way) is cylindrical in shape. Its diameter is estimated to be 6×10^5 light-years, with a thickness of 6×10^4 light-years. What is the estimated volume (V) of the Milky Way? (Use $V = \pi r^2 h$ and $\pi \approx 3.14$.)

10. One light-year is 5.88×10^{12} miles. What is the estimated volume of the Milky Way in cubic miles? (See Problem 9.)

Copyright © 1985 by Harcourt Brace Jovanovich, Inc. All rights reserved.

46 Chapter 1 Real Numbers and Properties

11. _____

12. _____

The *density* of an object is the unit rate of mass to volume (density = mass ÷ volume). The density rate measures the weight or mass of an average cubic unit of the given object.

11. Find the density of the Earth with a mass of 5.98×10^{24} kg and a volume of 1.08×10^{21} m³.

12. Find the density of the Moon with a mass of 7.37×10^{22} kg and a volume of 2.21×10^{19} m³.

13. _____

14a. _____

b. _____

13. Find the density of the Sun with a mass of 1.99×10^{30} kg and a volume of 1.41×10^{27} m³.

14. Which is a) the most dense b) the least dense: the Earth, the Moon, or the Sun?

According to Guinness

SMALLEST AND COLDEST PLANET
THE SMALLEST AND COLDEST PLANET, PLUTO (AND ITS PARTNER, CHARON, ANNOUNCED ON JUNE 22, 1978), HAVE AN ESTIMATED SURFACE TEMPERATURE OF −360°F. (100°F. ABOVE ABSOLUTE ZERO). PLUTO'S DIAMETER IS ABOUT 1880 MILES AND IT HAS A MASS ABOUT $\frac{1}{600}$ THAT OF THE EARTH.

Use the computation rules for scientific notation to round each answer to the correct number of significant digits. Assume 1880 has 3 significant digits, −360 has 2 significant digits, and 600 has 1 significant digit.

15. _____

16. _____

17. _____

18. _____

19. _____

20. _____

15. What is Pluto's surface area? (See Problem 7.)

16. What is Pluto's volume? (See Problem 8.)

17. What is Pluto's mass? (Use the Example.)

18. What is Pluto's density? (See Problems 11–17.)

19. What is the Fahrenheit temperature of absolute zero?

20. What is Pluto's surface temperature in degrees Celsius? (Use C = [F − 32] ÷ 1.8.)

Copyright © 1985 by Harcourt Brace Jovanovich, Inc. All rights reserved.

Name _____ Date _____ Class _____

Chapter 1 Review

			What to Review if You Have Trouble		
Objectives			Lesson	Example	Page
Find the intersection of two sets	1. $\{-1, 3, 5, 11\} \cap \{-3, -1, 0, 5, 7, 15\} = ?$	_____	1.1	1	3
Find the union of two sets	2. $\{-4, -2, 0, 2, 4\} \cup \{-3, 0, 4, 7\} = ?$	_____	1.1	2	4
Find all subsets of a given set	3. Find all subsets of $\{-5, 2, 3\}$.	_____	1.1	3	6
Graph a set of real numbers	4. Graph $\{-5, -0.25, 1, 3.5\}$.	_____	1.2	1	9
Find the absolute value of a real number	5. $|-17| = ?$	_____	1.2	2	10
Graph a simple inequality	6. Graph $x < 3$.	_____	1.2	3	11
Graph a compound inequality	7. Graph $x \leq 0$ or $x > 2$.	_____	1.2	4	12
Add signed numbers	8. $14 + (-32) = ?$	_____	1.3	1	15
Subtract signed numbers	9. $15 - (-31) = ?$	_____	1.3	2	16
Multiply signed numbers	10. $-11 \cdot (-5) = ?$	_____	1.3	3	16
Divide signed numbers	11. $\dfrac{-57}{-3} = ?$	_____	1.3	4	18
Evaluate exponential notation	12. Evaluate $(-3)^6$.	_____	1.4	1	21

Copyright © 1985 by Harcourt Brace Jovanovich, Inc. All rights reserved.

48 Chapter 1 Real Numbers and Properties

Write a product of repeated factors in exponential notation ▶	13. Write $2 \cdot 2 \cdot 2 \cdot 2 \cdot y \cdot y \cdot z \cdot z \cdot z$ in exponential notation.	1.4	2	22
Evaluate expressions ▶	14. $6 + 2^3 \div 2 - 4 + \sqrt{16} = ?$	1.4	3	23
	15. $\left\{5 - 3\left[\dfrac{7-(-5)}{4} + 2^3\right]\right\} + 2 = ?$	1.4	4	24
	16. Evaluate $u - 3uv^2 + u^2v$ for $u = 3$ and $v = -1$	1.4	5	26
Identify properties ▶	17. Which property asserts that $5(x + 7) = (x + 7)5$?	1.5	1	32
Clear enclosure symbols ▶	18. Clear parentheses in $-6(3y - 4)$.	1.5	2	33
Simplify expressions ▶	19. Simplify $7x + 9x^2 - 5x + x^2$.	1.5	3	34
	20. Simplify $-\{p + 3[-2(5p + q - 2p) + 5q]\}$.	1.5	4	34
Write a term without negative exponents ▶	21. Write $\dfrac{2y^{-4}z}{2^{-3}x^{-1}y^3}$ without negative exponents.	1.6	1	37
Simplify a product of terms ▶	22. Simplify $\dfrac{36x^{-2}y}{z^{-3}} \cdot \dfrac{3^{-1}y^{-5}}{x^{-2}}$.	1.6	2	39
Simplify a term raised to a power ▶	23. Simplify $\left[\dfrac{-3a^{-2}b^{-4}}{27a^3b^{-1}}\right]^{-2}$.	1.6	3	40
Compute using scientific notation ▶	24. Pluto is the farthest planet in our solar system from the Sun, at 5.91×10^{12} m. The Sun's mass is 1.99×10^{30} kg. Pluto's mass is 1.00×10^{24} kg. Find the attractive force of the Sun on Pluto.	PS 1	—	44

CHAPTER 1 REVIEW ANSWERS: 1. $\{-1, 5\}$ **2.** $\{-4, -3, -2, 0, 2, 4, 7\}$ **3.** $\varnothing$ **4.** $\{-5\}$ **5.** $\{2\}$ **6.** $\{3\}$ **7.** $\{-5, 2\}$, $\{-5, 3\}$, $\{2, 3\}$, $\{-5, 2, 3\}$ **4.** See Appendix Selected Answers **5.** 17 **6.** See Appendix Selected Answers **7.** See Appendix Selected Answers **8.** -18 **9.** 46 **10.** 55 **11.** 19 **12.** 729 **13.** $2^4y^2z^3$ **14.** 10 **15.** -26 **16.** -15 **17.** The commutative property of multiplication **18.** $-18y + 24$ **19.** $2x + 10x^2$ **20.** $17p - 9q$ **21.** $\dfrac{16xz}{y^7}$ **22.** $\dfrac{12z^3}{y^4}$ **23.** $81a^{10}b^6$ **24.** 3.80×10^{18} N

Copyright © 1985 by Harcourt Brace Jovanovich, Inc. All rights reserved.

REAL NUMBERS AND LINEAR RELATIONS

2 Equations and Inequalities

2.1 Solve First-Degree Equations

2.2 Solve Equations Containing Like Terms and Parentheses

2.3 Solve Equations Containing Fractions and Decimals

PS 2: Solve Problems Using Equations

2.4 Solve Inequalities

2.5 Solve Absolute Value Equations and Inequalities

2.6 Solve Literal Equations

2.1 Solve First-Degree Equations

A mathematical sentence with an equality symbol (=) in it is called an *equation*.

Every equation has three parts: left member ⟶ $\boxed{2x + 5}$ $\boxed{=}$ $\boxed{4x - 1}$ ⟵ right member (equality symbol)

> Any number that can replace the variable to make both members of the equation equal is called a *root* or a *solution* of the equation. The set of all solutions is called the *solution set*.

To *solve an equation* means to find all the solutions of the equation. The method of actually finding the solutions depends upon the type of equation to be solved.

> An equation that can be written in the form $Ax + B = C$, where A, B, and C are real numbers and A is not 0, is called a *first-degree equation in one variable*.

Note ▶ In $Ax + B = C$, the letters A, B, and C represent constants, and x represents a variable to the first power.

Examples ▶
(a) $x + 4 = -3$ is a first-degree equation in the variable x.
(b) $6z = 42$ is a first-degree equation in the variable z.
(c) $3y + 7 = -5$ is a first-degree equation in the variable y.

To solve a first-degree equation, you get the variable alone in one member and the constants in the other member. The *Addition Rule for Equations* is sometimes used to achieve this goal.

> **The Addition Rule for Equations**
> If you add the same term to both members of an equation, the solution(s) will not change. For real numbers a, b, and c, if $a = b$ then $a + c = b + c$

To solve an equation of the form $x + B = C$, you add the opposite of B ($-B$) to both members of the equation to get the variable x alone in one member.

EXAMPLE 1: Solve an equation using the Addition Rule.

Solve ▶ $\quad x + 4 = -3$

1. Identify ▶ $\quad x + \boxed{4} = -3$ — Think: The numerical addend with x is 4.

2. Use Addition Rule ▶ $\quad x + 4 + (-4) = -3 + (-4)$ — To get x alone in one member, add the opposite of 4 to both members.

3. Simplify ▶
$\quad x + 0 = -3 + (-4)$ — The steps in the box are often performed mentally.
$\quad x = -3 + (-4)$ — Think: The variable x is alone in one member.
$\quad x = -7$ — Think: -7 is the proposed solution.

4. Check ▶
$\quad x + 4 = -3$ ⟵ original equation
$\quad \dfrac{-7 + 4 \; | \; -3}{-3 \; | \; -3}$
$\quad\quad\quad -3 = -3$

Substitute the proposed solution -7.
Compute.
Compare: $-3 = -3$ means -7 is a solution.

Solution ▶ $\quad x = -7$

2.1 Solve First-Degree Equations

Note ▶ In Step 3 of Example 1, the -7 was referred to as the "proposed solution." The proposed solution may not be correct. It is only a solution if it checks in the original equation.

Another Example ▶ Solve $y - 5 = 3$.

$y - 5 + 5 = 3 + 5$ To get y alone in one member, add the opposite of -5 to both members.

$y = 8$ Check as before.

Make Sure

Solve each equation using the Addition Rule. Check each proposed solution.

See Example 1 ▶   1. $z + 5 = 9$ _____   2. $5 = y - 3$ _____   3. $3 = 7 + x$ _____

MAKE SURE ANSWERS: 1. 4   2. 8   3. −4

To solve some first-degree equations you will need to use the *Multiplication Rule for Equations*.

> **The Multiplication Rule for Equations**
> If you multiply both members of an equation by the same nonzero term, the solution(s) will not change. For real numbers a, b, and c, if $a = b$ then $a\mathbf{c} = b\mathbf{c}$.

To solve an equation of the form $Ax = C$, you multiply both members by the reciprocal of A to get the variable x alone in one member.

EXAMPLE 2: Solve an equation using the Multiplication Rule.

Solve ▶ $6z = 42$

1. Identify ▶ $\boxed{6}z = 42$ Think: The numerical coefficient of z is 6.

2. Use Multiplication Rule ▶ $\dfrac{1}{6} \cdot 6z = \dfrac{1}{6} \cdot 42$ To get z alone in one member, multiply both members by the reciprocal of 6.

3. Simplify ▶ $\left[\; 1z = \dfrac{1}{6} \cdot 42 \;\right.$ The steps in the box are often performed mentally.

$\left.\; z = \dfrac{1}{6} \cdot 42 \;\right]$ Think: The variable z is alone in one member.

$z = 7$ Think: 7 is the proposed solution.

4. Check ▶ $6z = 42$ ⟵ original equation

$6(7) \;|\; 42$ Substitute the proposed solution 7.

$42 \;|\; 42$ Compute.

$42 = 42$ Compare: $42 = 42$ means 7 is a solution.

Solution ▶ $z = 7$

52 Chapter 2 Equations and Inequalities

Make Sure

Solve each equation using the Multiplication Rule. Check each proposed solution.

See Example 2 ▶ 1. $3z = 24$ _____   2. $-35 = -7y$ _____   3. $\frac{3}{5}x = -15$ _____

MAKE SURE ANSWERS: 1. 8 2. 5 3. −25

To solve an equation of the form $Ax + B = C$, you need to use both the Addition Rule and the Multiplication Rule.

EXAMPLE 3: Solve an equation using the rules together.

Solve ▶ $3x + 7 = -5$

1. Use Addition Rule ▶ $3x + 7 + (-7) = -5 + (-7)$ See Example 1.

$3x = -12$

2. Use Multiplication Rule ▶ $\frac{1}{3} \cdot 3x = \frac{1}{3} \cdot (-12)$ See Example 2.

Solution ▶ $x = -4$ Check as before.

Another Example ▶ Solve $\frac{5x}{6} - 8 = 12$.

$\frac{5x}{6} - 8 + 8 = 12 + 8$ The Addition Rule.

$\frac{5x}{6} = 20$

$\frac{5}{6}x = 20$ Think: $\frac{5x}{6}$ means $\frac{5}{6}x$.

$\frac{6}{5} \cdot \frac{5}{6}x = \frac{6}{5} \cdot 20$ The Multiplication Rule.

$x = 24$ Check as before.

Make Sure

Solve each equation using the rules together. Check each proposed solution.

See Example 3 ▶ 1. $3z + 7 = 4$ _____   2. $-6 = 2y - 9$ _____   3. $-1 = 1 - \frac{3x}{2}$ _____

MAKE SURE ANSWERS: 1. −1 2. $\frac{3}{2}$ 3. $\frac{4}{3}$

Name _____ Date _____ Class _____

2.1 Practice: *Check each proposed solution in the original equation*

Set 1: Solve each equation using the Addition Rule.

1. $z + 5 = 7$
2. $y + 3 = 9$
3. $x - 5 = 3$

4. $w - 4 = 3$
5. $-3 = v + 5$
6. $-1 = u + 7$

Set 2: Solve each equation using the Multiplication Rule.

7. $2t = 16$
8. $5s = 20$
9. $-6r = 30$

10. $-9q = 45$
11. $-16 = -2p$
12. $-9 = -m$

Set 3: Solve each equation using the rules together.

13. $2z + 5 = 15$
14. $4y + 3 = 11$
15. $1 = 3x - 7$

16. $4 = 7w - 2$
17. $2v - 5 = -6$
18. $3w - 3 = -1$

1. _____
2. _____
3. _____
4. _____
5. _____
6. _____
7. _____
8. _____
9. _____
10. _____
11. _____
12. _____
13. _____
14. _____
15. _____
16. _____
17. _____
18. _____

2.1 Practice 53

Copyright © 1985 by Harcourt Brace Jovanovich, Inc. All rights reserved.

Mixed Practice: Solve each equation using the rules.

19. $z + 9 = 3$ **20.** $y + 11 = 6$ **21.** $-3 = 4 - x$

22. $-2 = 5 - w$ **23.** $4v = -16$ **24.** $3v = -15$

25. $-14 = -4t$ **26.** $-16 = -6s$ **27.** $2r + 3 = 1$

28. $3q + 5 = 2$ **29.** $-4 = 2 - 3p$ **30.** $-3 = 7 - 5m$

31. $5 - 3z = 5$ **32.** $7 - 4y = 7$ **33.** $3 = 5x + 7$

34. $4 = 7w + 6$ **35.** $4 - 2v = 3$ **36.** $5 - 4u = 1$

Review: Work these problems on a separate sheet of paper. Attach your work to this page.

Remove the grouping symbols using the distributive properties. (See Lesson 1.5, Example 2.)
37. $2(z - 3)$ **38.** $4(y + 3)$ **39.** $(x - 5)(-3)$
40. $(w + 2)(-1)$ **41.** $-2u(3v - t - 25)$ **42.** $-r(q - p + 2m)$

Simplify each expression by combining like terms. (See Lesson 1.5, Example 3.)
43. $2y + 3y$ **44.** $4m + 3m$ **45.** $2p - 5p + p$
46. $7t - 11t + t$ **47.** $4r - 3s - r + 5s$ **48.** $5x - 4y - y - 7x$

2.2 Solve Equations Containing Like Terms and Parentheses

When two or more like terms are all in the same member of an equation, you first combine like terms and then use the rules to solve as in Lesson 2.1.

Example ▶ Solve $5x - 12 - 3x = -7$.

1. Combine like terms ▶
$$(5x - 3x) - 12 = -7$$

$2x - 12 = -7$ ⟵ like terms are combined

2. Solve as before ▶
$$2x - 12 + 12 = -7 + 12$$
$$2x = 5$$
$$\tfrac{1}{2} \cdot 2x = \tfrac{1}{2} \cdot 5$$

Solution ▶ $x = \tfrac{5}{2}$ Check as before.

When like terms are in both members of the equation, you combine like terms in each member, collect the variable terms all in one member, and then solve the equation as in the previous example.

EXAMPLE 1: Solve an equation containing like terms.

Solve ▶ $-y + 4y + 6 = -3y - 4 + y$

1. Combine like terms in both members ▶ $3y + 6 = -2y - 4$

2. Collect variable terms in one member ▶ $3y + 2y + 6 = -2y + 2y - 4$

     Think: To eliminate the $-2y$ term in the right member, add $2y$ to both members.

$5y + 6 = -4$ Think: The variable y only occurs in the left member.

3. Solve as before ▶ $5y + 6 + (-6) = -4 + (-6)$
$$5y = -10$$
$$\tfrac{1}{5} \cdot 5y = \tfrac{1}{5} \cdot (-10)$$
$y = -2$ ⟵ proposed solution

4. Check ▶

$-y + 4y + 6 =$	$-3y - 4 + y$	⟵ original equation
$-(-2) + 4(-2) + 6$	$-3(-2) - 4 + (-2)$	
0	0	⟵ -2 checks

Solution ▶ $y = -2$

Make Sure

Solve equations containing like terms. Check each proposed solution.

See Example 1 ▶   **1.** $5r + 7 - 6r = 5$ _____   **2.** $s - 6 - 4s = 7s - 5 + 23 - 2s$ _____

MAKE SURE ANSWERS: 1. 2, 2. -3

56 Chapter 2 Equations and Inequalities

To solve certain equations containing parentheses, you can use the distributive properties.

EXAMPLE 2: Solve an equation containing parentheses.

Solve ▶ $5(z - 2) = 8z - 2(3z - 7)$

1. Clear parentheses ▶ $\boxed{5z - 5 \cdot 2 = 8z - 2 \cdot 3z - 2(-7)}$ Use a distributive property.

                      $5z - 10 = 8z - 6z + 14$ ⟵ parentheses cleared

2. Solve as before ▶ $5z - 10 = 2z + 14$ Combine like terms.

              $5z + (-2z) - 10 = 2z + (-2z) + 14$ Collect like terms.

                    $3z - 10 = 14$ Combine like terms again.

              $3z + (-10) + 10 = 14 + 10$ The Addition Rule.

                     $3z = 24$

                 $\frac{1}{3} \cdot 3z = \frac{1}{3} \cdot 24$ The Multiplication Rule.

Solution ▶ $z = 8$ Check as before.

Sometimes it is convenient to write a set by stating a characteristic property of its members that they alone possess. The property is written using a variable and *set-builder notation*.

Example ▶ The set-builder notation $\{x \mid x \in N \text{ and } x < 5\}$ is read:

"the set of | all x | such that | x is an element of the natural numbers | and x is less than 5."

You can write some sets by using both set-builder notation and the roster method.

Examples ▶ (a) $\{x \mid x \in N \text{ and } x < 4\} = \{1, 2, 3\}$ (b) $\{x \mid x \in W \text{ and } x < 4\} = \{0, 1, 2, 3\}$

All of the equations considered so far are called *conditional equations* because only certain numbers (not all numbers) make the sentences true.

An equation whose solution set is the set of all real numbers is called an *identity*.

Example ▶ $x + x = 2x$ is an identity because all real numbers make the sentence true. Using set-builder notation, you write the solution set of $x + x = 2x$ as $\{x \mid x \in \mathcal{R}\}$.

Some equations have an empty solution set.

Example ▶ $x + 1 = x$ has no solutions because $x + 1$ is always one greater than x.

> A first-degree equation in one variable that simplifies to a true statement, such as $12 = 12$, is an identity. Its solution set is the set of all real numbers.
>
> A first-degree equation in one variable that simplifies to a false statement, such as $-2 = 3$, has no solutions. Its solution set is the empty set.

Examples ▶ (a) Solve $3(x + 4) = 11x - 2(4x - 6)$.   (b) Solve $5w - 2 = 8w + 3 - 3w$.

             $3x + 12 = 11x - 8x + 12$ $5w - 2 = 5w + 3$

               $12 = 12$ Stop! (true statement) $-2 = 3$ Stop! (false statement)

The solution set is $\{x \mid x \in \mathcal{R}\}$, which means all real numbers are solutions.   The solution set is $\emptyset$, which means there are no solutions.

2.2 Solve Equations Containing Like Terms and Parentheses

Make Sure

Solve equations containing parentheses. Check each proposed solution.

See Example 2 ▶
1. $3(m - 2) = -15$

2. $2(n - 3) = -2(4 - 2n)$

3. $5(v - 1) - 2(v + 2) = -(9 - 3v)$

4. $5u - 2(3 + u) = 5(2u - 1) - 7u$

MAKE SURE ANSWERS: 1. -3 2. 1   3. all real numbers   4. no solutions

---

To solve certain equations containing parentheses, you can use the *zero-product property*.

The zero-product property: If $a \cdot b = 0$, then $a = 0$ or $b = 0$.

EXAMPLE 3: Solve an equation using the zero-product property.

Solve ▶ $(x - 2)(3x + 4) = 0$ Think: $a \cdot b = 0$ with $a = x - 2$ and $b = 3x + 4$.

1. Use zero-product ▶ $x - 2 = 0$ or $3x + 4 = 0$
   property

2. Solve each equation ▶ $x - 2 + 2 = 0 + 2$ or $3x + 4 + (-4) = 0 + (-4)$

                    $x = 2$ or $3x = -4$

                    $x = 2$ or $x = -\frac{4}{3}$

3. Check ▶

$(x - 2)(3x + 4) = 0$		$(x - 2)(3x + 4) = 0$	
$(2 - 2)(3 \cdot 2 + 4)$	0	$(-\frac{4}{3} - 2)(3(-\frac{4}{3}) + 4)$	0
$(0)(10)$	0	$(-\frac{10}{3})(0)$	0
0	0 ⟵ 2 checks	0	0 ⟵ $-\frac{4}{3}$ checks

Solution ▶ $x = 2$ or $-\frac{4}{3}$

Note 1 ▶ When an equation has 2 or more solutions, it is sometimes convenient to list the solutions in a set. The solution set for $(x - 2)(3x + 4) = 0$ is $\{-\frac{4}{3}, 2\}$.

Note 2 ▶ The equation $(x - 2)(3x + 4) = 0$ is not a first-degree equation, but the zero-product property allows you to solve the equation by solving two first-degree equations.

Another Example ▶ Solve $y(3y - 6) = 0$.

$y = 0$ or $3y - 6 = 0$

$y = 0$ or $3y = 6$

$y = 0$ or $y = 2$ Check as before.

> CAUTION: You cannot use the zero-product property unless one member of the equation is zero.

Examples ▶ (a) $w(w + 3) = 1$ does not mean $w = 0$ or $w + 3 = 0$. ← wrong method
(b) $w(w + 3) = 1$ does not mean $w = 1$ or $w + 3 = 1$. ← wrong method

To solve an equation such as $w(w + 3) = 1$, you will need the methods presented in Chapter 9.

Make Sure

Solve each equation using the zero-product property. Check each proposed solution.

See Example 3 ▶
1. $2z(z - 2) = 0$ _____
2. $(y - 3)(y + 4) = 0$ _____

3. $(4x - 1)(2x + 3) = 0$ _____
4. $(3w + 4)(3w + 2) = 0$ _____

MAKE SURE ANSWERS: 1. 0, 2   2. -4, 3   3. $-\frac{3}{2}, \frac{1}{4}$ 4. $-\frac{4}{3}, -\frac{2}{3}$

According to Guinness

TALLEST STRUCTURE
THE TALLEST STRUCTURE IN THE WORLD IS THE RADIO MAST AT KONSTANTYNOW, IN POLAND. IT IS 2,120 FEET, 8 INCHES TALL—MORE THAN FOUR-TENTHS OF A MILE.

BALLOON RECORD
THE HIGHEST ALTITUDE ATTAINED IN A MANNED BALLOON WITH AN OPEN BASKET IS 38,789 FEET BY KINGSWOOD SPROTT JR. OVER LAKELAND, FLORIDA, ON SEPTEMBER 27, 1975.

To the nearest tenth degree, answer questions 5 and 6 using this *altitude/temperature formula*:

$a = 0.16(15 - t)$ where $\begin{cases} a \text{ is the altitude in kilometers (km) above sea level } (a = 0) \text{ but below } \\ 12 \text{ km } (a = 12). \\ t \text{ is the average annual temperature in degrees Celsius (°C).} \end{cases}$

5. _____

6. _____

5. What is the average annual temperature at the highest altitude attained in a manned balloon with an open basket? (Use 1 ft = 0.3048 m.)

6. How much cooler is the average annual temperature atop the world's tallest structure than at ground level in degrees Fahrenheit (°F)? (Use $F = \frac{9}{5}C + 32$.)

2.2 Practice: *Check each proposed solution in the original equation.*

Set 1: Solve equations containing like terms.

1. $3m + 5m = 16$
2. $2n + 7n = 18$
3. $2p + 9p = -33$

4. $q + 5q = -24$
5. $2r - 7r = 20$
6. $5s - 9s = 30$

7. $5 - t + 5t = 3$
8. $7 - u + 7u = 5$
9. $1 = 4 + v - 5v + 7$

10. $6 = 7 - m + 7m - 9$
11. $2.5 = 7n + 0.3 - 3n$
12. $-5 = 1.5 + 1.7p - 0.4p$

13. $5q + 24 = 3q$
14. $6r + 16 = 2r$
15. $4s - 6 = 6s$

16. $3t - 9 = 6t$
17. $3u - 5u = 9 - u$
18. $5v - 9v = 16 - 2v$

19. $6w - 7 - w = 4 - 3w$
20. $7x - 3 - x = 3 - 4x$
21. $3y + 5 + 2y = 5 - 3y + y$

22. $2z + 2 + 3z = 2 - z + z$
23. $1.4m - 1 = 0.7m + 1.1$
24. $2.3 - 3.1n = 5.2n - 30.9$

Set 2: Solve equations containing parentheses.

25. $4(y - 6) = 4$
26. $3(x - 2) = 6$
27. $4(w + 6) = 8w$

28. $3(v + 2) = 6v$
29. $3(u - 6) = 6u$
30. $5(t - 3) = 15t$

31. $4(3s - 3) = -2(1 - s)$
32. $3(2r + 1) = -3(2 - r)$

33. $2(3 - q) - (q + 1) = 7$
34. $3(4m - 1) - 4(m + 1) = 5$

35. $2(n - 1) - (1 - n) = 4(n - 2)$
36. $3(p - 2) - (2 - p) = 3(p - 1)$

Set 3: Solve each equation using the zero-product property.

37. $(q - 3)(q - 2) = 0$
38. $(r - 1)(r - 2) = 0$
39. $(s + 2)(s + 1) = 0$

40. $(t + 3)(t + 4) = 0$
41. $(u - 3)(u + 2) = 0$
42. $(v - 1)(v + 3) = 0$

43. $3w(w + 1) = 0$
44. $3x(x + 3) = 0$
45. $(2y - 3)(3y + 2) = 0$

46. $(4z - 3)(2z - 5) = 0$
47. $(5x - 1)(4x + 1) = 0$
48. $(3y - 1)(5y + 1) = 0$

2.2 Practice

Name _____ Date _____ Class _____

Mixed Practice: Solve each equation.

49. _____

49. $3(z + 6) = 12z$

50. _____

50. $4(y + 5) = 7y$

51. _____

51. $5(3 - 2x) - 5(x + 2) = 0$

52. _____

52. $5(w - 3) - 2(3w + 4) = 0$

53. _____

53. $4(3v - 2) - 3(2 - v) = 10v$

54. _____

54. $2(3 - 2u) - (u - 3) = 3u$

55. _____

55. $5t - 4 + 3t = 6t - 3$

56. _____

56. $7s - 6 - 2s = 3s - 4$

57. _____

57. $4(3r - 1) - 2(3 - r) = 3(1 - r)$

58. _____

58. $5(2m - 1) - 3(2m + 3) = 2(1 - m)$

59. _____

59. $5 = n - 4 + 3n - 8n$

60. _____

60. $6 = p + 3 - 4p - 3p$

61. _____

61. $3 - 4q + q - 5 = 2 + 4q - 1 - 7q$

62. _____

62. $5 - r - 2 + 3r = r - 5 + r + 3$

63. _____

63. $5(2 - 4v) = 10(1 - 2v)$

64. _____

64. $4(1 - 3w) = 2(2 - 6w)$

65. _____

65. $5(3 - x) = 5(2 - x)$

66. _____

66. $3(y - 5) = 3(y - 2)$

Copyright © 1985 by Harcourt Brace Jovanovich, Inc. All rights reserved.

Extra: Solve each equation.

67. $3(z - 4) = -4(3 - 2z)$

68. $-5(y - 3) = 3(5 - 2y)$

69. $2(2x - 1) = 4(x + 2)$

70. $6(w + 1) = -2(1 - 3w)$

71. $6(4v - 1) + 6 = 24v$

72. $5(3u + 3) - 15 = 15u$

73. $2(3 - 8t) = -4(4t - 3)$

74. $6(1 - 4s) = -3(8s - 1)$

75. $2(4r + 2) - 3(r + 1) = -(7 - 3r)$

76. $4(2 - 3q) + 3(2q + 1) = -(5q - 12)$

77. $3(p + 3) - (3 + p) = 4(p + 2)$

78. $5(3n - 2) - 4(2n - 3) = 3(1 - 2n)$

Review: Work these problems on a separate sheet of paper. Attach your work to this page.

Rename each percent as a decimal.
79. 25%
80. 12%
81. 5%
82. 2%
83. $33\frac{1}{3}\%$
84. $66\frac{2}{3}\%$
85. $16\frac{2}{3}\%$
86. 120%

Find the LCD of the given fractions.
87. $\frac{2}{3}, \frac{3}{4}, \frac{5}{6}$
88. $\frac{1}{2}, \frac{3}{4}, \frac{5}{6}$
89. $\frac{2}{3}, \frac{3}{4}, \frac{4}{5}$
90. $\frac{3}{4}, \frac{1}{6}, \frac{2}{9}$

Multiply fractions.
91. $6 \cdot \frac{2}{3}$
92. $12 \cdot \frac{5}{6}$
93. $24 \cdot \frac{3}{8}$
94. $20 \cdot \frac{3}{4}$

Remove the grouping symbols using the distributive properties. (See Lesson 1.5.)
95. $6(\frac{2}{3}x - 4)$
96. $12(\frac{3}{4}y + \frac{2}{3})$
97. $20(\frac{4}{5}z - \frac{3}{4})$
98. $24(\frac{3}{8}w + \frac{5}{6})$

2.3 Solve Equations Containing Fractions and Decimals

To solve an equation containing fractions, you can *clear fractions* by multiplying each term of the equation by the *least common denominator* (*LCD*) of the terms.

EXAMPLE 1: Solve an equation containing fractions.

Solve ▶ $\frac{5}{2}y - 3 = \frac{1}{3}$

1. Find the LCD ▶ The LCD of $\frac{5}{2}$, 3, and $\frac{1}{3}$ is 6.

2. Clear fractions ▶ $6\left(\frac{5}{2}y - 3\right) = 6 \cdot \frac{1}{3}$ Multiply both members by the LCD.

$6 \cdot \frac{5}{2}y - 6 \cdot 3 = 6 \cdot \frac{1}{3}$ Use a distributive property.

$15y - 18 = 2$ ⟵ fractions cleared

3. Solve as before ▶ $15y - 18 + 18 = 2 + 18$

$15y = 20$

$y = \frac{20}{15}$

Solution ▶ $y = \frac{4}{3}$ Check as before.

CAUTION: To clear fractions, you must multiply each term of the equation by the LCD.

Examples ▶ (a) *Wrong Method:* $\frac{5}{2}y - 3 = \frac{1}{3}$ ⟵ equation to be solved

$6 \cdot \frac{5}{2}y - 3 = 6 \cdot \frac{1}{3}$ No! (The -3 must also be multiplied by the LCD, 6.)

(b) *Correct Method:* $\frac{5}{2}y - 3 = \frac{1}{3}$ ⟵ equation to be solved

$6 \cdot \frac{5}{2}y - 6 \cdot 3 = 6 \cdot \frac{1}{3}$ Every term has been multiplied by the LCD, 6.

Make Sure

Solve equations containing fractions. Check each proposed solution.

See Example 1 ▶ 1. $\frac{2}{3}z + \frac{5}{6} = \frac{1}{2}$ _____   2. $\frac{2}{3}(x - 3) = 1 - \frac{3}{4}(1 - 2x)$ _____

MAKE SURE ANSWERS: 1. $-\frac{1}{2}$ 2. $-\frac{27}{10}$

64 Chapter 2 Equations and Inequalities

To solve an equation containing decimals, you can *clear decimals* as shown in Example 2.

EXAMPLE 2: Solve an equation containing decimals.

Solve ▶ $\qquad 1.25z - 0.6 = 0.95$

1. Find the LCD ▶ The LCD of 1.25 ($\frac{125}{100}$), 0.6 ($\frac{6}{10}$), and 0.95 ($\frac{95}{100}$) is 100.

2. Clear decimals ▶
$$100(1.25z - 0.6) = 100 \cdot 0.95 \qquad \text{Multiply both members by the LCD.}$$
$$100 \cdot 1.25z - 100 \cdot 0.6 = 100 \cdot 0.95 \qquad \text{Use a distributive property.}$$
$$125z - 60 = 95 \quad \longleftarrow \text{ decimals cleared}$$

3. Solve as before ▶
$$125z = 155$$
$$z = \frac{155}{125}$$

Solution ▶ $\qquad z = \frac{31}{25}$ or $1\frac{6}{25}$ or 1.24 $\qquad$ Check as before.

Make Sure

Solve equations containing decimals. Check each proposed solution.

See Example 2 ▶ **1.** $0.6a - 0.1 = 0.2a + 1.5$ _____ **2.** $0.4(0.2b - 0.3) = 0.04$ _____

MAKE SURE ANSWERS: 1. 4 2. 2

To solve an equation containing percents, you will need to *clear percents* first.

EXAMPLE 3: Solve an equation containing percents.

Solve ▶ $\qquad 14\%w + 12\%(7000 - w) = 920$

1. Clear percents ▶ $\qquad 0.14w + 0.12(7000 - w) = 920 \qquad$ Rename the percents as decimals.

2. Solve as before ▶
$$0.14w + 840 - 0.12w = 920 \qquad \text{Clear parentheses.}$$
$$0.02w + 840 = 920 \qquad \text{Combine like terms.}$$
$$2w + 84{,}000 = 92{,}000 \qquad \text{Clear decimals.}$$
$$2w = 8000$$

Solution ▶ $\qquad w = 4000 \qquad$ Check as before.

Make Sure

Solve equations containing percents. Check each proposed solution.

See Example 3 ▶ **1.** $25\%a + a = 25$ _____ **2.** $20\%b + 5.5 = 75\%b$ _____

MAKE SURE ANSWERS: 1. 20 2. 10

2.3 Practice 65

Name _____ Date _____ Class _____

1. _____

2.3 Practice: *Check each proposed solution in the original equation.*

Set 1: Solve equations containing fractions.

1. $\dfrac{3}{4}z - \dfrac{1}{4} = \dfrac{3}{4}$ 2. $\dfrac{4}{5}y - \dfrac{1}{5} = \dfrac{3}{5}$ 3. $\dfrac{x+4}{2} + \dfrac{x+1}{4} = 3$

4. $\dfrac{w+3}{6} - \dfrac{w+4}{2} = 2$ 5. $\dfrac{2}{3}(v-4) = 2$ 6. $\dfrac{3}{4}(u-6) = 2$

Set 2: Solve equations containing decimals.

7. $0.8q - 3.2 = 1.6$ 8. $2.3r - 4.7 = 4.5$ 9. $2.3s + 4.7s = 4.9$

10. $5.1m + 2.3m = 2.96$ 11. $0.4(0.2n - 0.3) = 0.01$ 12. $0.8(0.3p - 0.5) = 0.8$

Set 3: Solve equations containing percents.

13. $15\%q = 6$ 14. $30\%r = 9$ 15. $50\%s + s = 12$

16. $75\%t + t = 105$ 17. $20\%u + 25\%u = 18$ 18. $10\%v + 12\tfrac{1}{2}\%v = 36$

Chapter 2 Equations and Inequalities

Mixed Practice: Solve each equation.

19. $\dfrac{1}{4}m + \dfrac{2}{3}m = \dfrac{1}{6}$

20. $\dfrac{1}{3}n + \dfrac{1}{4}n = \dfrac{5}{12}$

21. $\dfrac{3p + 2}{2} - \dfrac{p - 5}{3} = \dfrac{1}{3}$

22. $\dfrac{q - 3}{2} - \dfrac{4q - 1}{6} = \dfrac{2}{3}$

23. $\dfrac{3}{4}(2r - 5) - \dfrac{2}{3}(r - 6) = \dfrac{1}{6}$

24. $\dfrac{3}{4}(s - 2) - \dfrac{3}{5}(2s - 3) = \dfrac{1}{5}$

25. $\dfrac{2}{5} = \dfrac{2t + 3}{6}$

26. $\dfrac{5}{6} = \dfrac{2u - 3}{5}$

27. $\dfrac{v - 3}{3} - \dfrac{v - 2}{2} = \dfrac{4 - v}{4}$

28. $\dfrac{2m - 1}{2} - \dfrac{3m - 1}{3} = \dfrac{4m - 1}{4}$

29. $5 = 0.5(q - 10) + 2$

30. $0.5r = 0.14 + 0.3(16 - r)$

31. $50\%u + 20\%(90 - u) = 30$

32. $20\% + 40\%(25 - v) = 9$

Review: Work these problems on a separate sheet of paper. Attach your work to this page.

Solve each equation using the Addition Rule. (See Lesson 2.1, Example 1.)
33. $z + 5 = 2$
34. $y + 3 = 7$
35. $x - 4 = 3$
36. $w - 7 = 3$

Solve each equation using the Multiplication Rule. (See Lesson 2.1, Example 2.)
37. $3v = 24$
38. $5u = 12$
39. $-4t = 10$
40. $-7s = -12$

Solve each equation using the rules together. (See Lesson 2.1, Example 3.)
41. $3r - 2 = 7$
42. $5q - 7 = 3$
43. $6p + 5 = 17$
44. $7n + 3 = 24$
45. $3w - 4 = 1$
46. $4x - 3 = 2$
47. $5y + 8 = 2$
48. $6z + 9 = 4$

Problem Solving 2: Solve Problems Using Equations

> **How to Solve a Word Problem Using an Equation**
>
> Read the problem very carefully, several times if necessary.
> Identify the facts, the key words, and the question.
> Understand what the unknowns are.
> Decide how to represent the unknowns.
> Draw a picture to help translate when appropriate.
> Translate the question to an equation using key words.
> Solve the equation using the rules and properties.
> Interpret your solution with respect to each represented unknown.
> Check to see if each solution makes sense in the original problem.

EXAMPLE: Solve this word problem using an equation.

1. Read and identify ▶ A 40-foot hose is cut into four pieces. The second piece is twice as long as the first piece. The third piece is half as long as the first piece. The fourth piece is nine times as long as the third piece. How long is each cut piece?

   Remember, identify the facts, the key words, and the question.

2. Understand ▶ The unknowns are { the length of the first piece, the length of the second piece, the length of the third piece, the length of the fourth piece }.

3. Decide ▶
   Let l = the length of the first piece
   then $2l$ = the length of the second piece — because $2l$ is twice l
   and $\frac{1}{2}l$ = the length of the third piece — because $\frac{1}{2}l$ is half l
   and $\frac{9}{2}l$ = the length of the fourth piece — because $\frac{9}{2}l$ is 9 times $\frac{1}{2}l$.

4. Draw a picture ▶

   [Diagram: a hose of total length 40 feet divided into four segments labeled l, $2l$, $\frac{1}{2}l$, and $\frac{9}{2}l$.]

5. Translate ▶ First length plus twice first length plus half first length plus nine times third length is 40.

   $l + 2l + \frac{1}{2}l + \frac{9}{2}l = 40$

6. Solve ▶
   $2(l + 2l + \frac{1}{2}l + \frac{9}{2}l) = 2(40)$ Clear fractions.
   $2l + 4l + l + 9l = 80$
   $16l = 80$ Combine like terms.
   $l = 5$ Use the rules.

7. Interpret ▶
   $l = 5$ means the length of the first piece is 5 feet.
   $2l = 2(5) = 10$ means the length of the second piece is 10 feet.
   $\frac{1}{2}l = \frac{1}{2}(5) = 2\frac{1}{2}$ means the length of the third piece is $2\frac{1}{2}$ feet.
   $\frac{9}{2}l = \frac{9}{2}(5) = 22\frac{1}{2}$ means the length of the fourth piece is $22\frac{1}{2}$ feet.

8. Check ▶ Are the four pieces together 40 feet? Yes: $5 + 10 + 2\frac{1}{2} + 22\frac{1}{2} = 40$.

Copyright © 1985 by Harcourt Brace Jovanovich, Inc. All rights reserved.

68 Chapter 2 Equations and Inequalities

Practice: Solve each word problem using an equation.

1. A 30-foot board is cut into two pieces. The longer of the two pieces is four times as long as the shorter piece. How long is each cut piece?

2. A 40-meter wire is cut into three pieces. The first piece is twice as long as the second piece. The third piece is four times as long as the second piece. How long is each piece?

3. A 117-meter rope is cut into three pieces. The first piece is twice as long as the second piece. The third piece is three times as long as the first. How long is each piece?

4. A 65-meter piece of string is cut into 4 pieces. The first piece is twice as long as the third piece. The second piece is the same length as the third piece. The fourth piece is three times as long as the first piece. How long is each piece?

5. The perimeter of a rectangle is 20 feet. The width is 3 feet shorter than the length. What is the area?

6. The perimeter of a rectangle is 20 meters. The length is 1.5 times the width. What is the area?

7. A standard doubles tennis court is 78 feet long. This is 6 feet longer than twice its width. What is its area?

8. The perimeter of the playing surface of a U.S. football field is 306 yards. Its length is 6 yards shorter than twice its width. What is its area?

9. Alan is twice as old as Richard. Peter is three times as old as Richard. The sum of all their ages is 66. How old is each?

10. Carol is 2 years less than 4 times as old as Ivanhoe. Carol is also 1 year more than 3 times Ivanhoe's age. How old is each?

According to Guinness

WIDEST TREE A FIGURE OF 167 FT. IN CIRCUMFERENCE WAS REPORTED FOR THE POLLARDED EUROPEAN CHESTNUT KNOWN AS THE "TREE OF THE 100 HORSES" ON THE EDGE OF MT. ETNA, SICILY, ITALY, IN 1972.

LARGEST OCEANARIUM THE LARGEST SALT WATER TANK IN THE WORLD IS THAT AT THE MARINELAND OF THE PACIFIC, PALOS VERDES PENINSULA, CALIFORNIA. IT IS 251 ½ FEET IN CIRCUMFERENCE AND 17 FEET DEEP, WITH A CAPACITY OF 640,000 GALLONS. THE TOTAL CAPACITY OF THE WHOLE OCEANARIUM IS 2,500,000 GALLONS.

Use $\pi \approx 3.14$ and a calculator. Round the answer to the nearest tenth when necessary.

11. What is the diameter of the widest tree through the given circumference? (Hint: See Appendix Geometry Formulas.)

12. What is the approximate cross-sectional area of the widest tree through the given circumference? (See Problem 11.)

13. What is the volume of the largest salt water tank in cubic feet?

14. How many gallons are in one cubic foot of water? (See Problem 13.)

15. What is the volume of the largest oceanarium in cubic feet? (See Problem 14.)

16. What percent of the largest oceanarium capacity is the largest salt water tank?

Copyright © 1985 by Harcourt Brace Jovanovich, Inc. All rights reserved.

2.4 Solve Inequalities

Every simple inequality has three parts: left member ⟶ $2x - 5 < 7$ ⟵ right member (order relation)

Note ▶ The order relations are also called *inequality symbols*.

> Any number that can replace the variable to make an inequality true is called a *solution of the inequality*.

Examples ▶
(a) 3 is a solution of $2x - 5 < 7$, because $2 \cdot 3 - 5 = 1$, which is less than 7.
(b) 8 is not a solution of $2x - 5 < 7$, because $2 \cdot 8 - 5 = 11$, which is not less than 7.

> The set of all solutions of an inequality is called its *solution set*. Two inequalities are said to be *equivalent* if they have the same solution set.

Adding the same term to both members of an inequality produces an equivalent inequality with the order preserved. This is known as the *Addition Rule for Inequalities*.

> **Addition Rule for Inequalities**
> If $a < b$, then $a + c < b + c$.
> If $a > b$, then $a + c > b + c$.

To solve an inequality, you need to find its solution set. The Addition Rule for Inequalities can help you solve an inequality of the form $x + B < C$.

EXAMPLE 1: Solve an inequality using the Addition Rule.

Solve ▶ $x + 3 < 14$

1. Use Addition Rule ▶ $x + 3 + (-3) < 14 + (-3)$ To get x alone in one member, add -3 to both members.

2. Simplify ▶
$x + 0 < 14 + (-3)$
$x < 14 + (-3)$

$x < 11$

Solution ▶ $\{x \mid x < 11\}$

Recall ▶ $\{x \mid x < 11\}$ is read "the set of all x such that x is less than 11."

Make Sure

Solve each inequality using the Addition Rule.

See Example 1 ▶ 1. $z - 3 > 5$ _____ 2. $y - 3 < -2$ _____ 3. $4 > 3 - x$ _____

MAKE SURE ANSWERS: 1. $\{z \mid z > 8\}$ 2. $\{y \mid y < 1\}$ 3. $\{x \mid x > -1\}$

Multiplying both members of an inequality by the same positive term produces an equivalent inequality. Multiplying both members of an inequality by the same negative term and reversing the inequality symbol produces an equivalent inequality. Together these form the *Multiplication Rule for Inequalities*.

> **Multiplication Rule for Inequalities**
> If $c > 0$ and $a < b$, then $ca < cb$.   If $c < 0$ and $a < b$, then $ca > cb$.
> If $c > 0$ and $a > b$, then $ca > cb$.   If $c < 0$ and $a > b$, then $ca < cb$.

Examples ▶ (a) $3 < 5$ therefore $2 \cdot 3 < 2 \cdot 5$ (b) $4 < 7$ therefore $(-2) \cdot 4 > (-2) \cdot 7$.

The Multiplication Rule for Inequalities can help you solve an inequality of the form $Ax < B$.

EXAMPLE 2: Solve an inequality using the Multiplication Rule.

Solve ▶ $-4n < 20$

1. Use Multiplication Rule ▶ $-\frac{1}{4} \cdot (-4n) > -\frac{1}{4} \cdot 20$ To get n alone in one member, multiply both members by $-\frac{1}{4}$. Since the multiplier is negative, reverse the inequality symbol.

2. Simplify ▶ $1n > -\frac{1}{4} \cdot 20$

$n > -5$

Solution ▶ $\{n \mid n > -5\}$

Note ▶ The graph of $\{n \mid n > -5\}$ is:

$\leftarrow\!\!+\!\!+\!\!+\!\!\underset{-5}{\oplus}\!\!+\!\!+\!\!+\!\!+\!\!+\!\!+\!\!+\!\!+\!\!\rightarrow$
$-7\ -6\ -5\ -4\ -3\ -2\ -1\ \ 0\ \ 1\ \ 2\ \ 3$

> CAUTION: Be sure to reverse the inequality symbol whenever you multiply both members of an inequality by a negative number.

> Division of both members of an inequality by a negative number also requires that you reverse the inequality symbol.

Example ▶ Solve $-2x > 14$

$\dfrac{-2x}{-2} < \dfrac{14}{-2}$ Think: Dividing both members of an inequality by a negative number results in an equivalent inequality provided you reverse the inequality symbol.

$x < -7$

$\{x \mid x < -7\}$

Make Sure

Solve each inequality using the Multiplication Rule.

See Example 2 ▶   **1.** $5z > 15$ _____   **2.** $2 > -y$ _____   **3.** $-\dfrac{2x}{3} < -4$ _____

MAKE SURE ANSWERS: **1.** $\{z \mid z > 3\}$ **2.** $\{y \mid y > -2\}$ **3.** $\{x \mid x > 6\}$

2.4 Solve Inequalities

To solve an inequality of the form $Ax + B > C$, you need to use the rules together.

EXAMPLE 3: Solve an inequality using the rules together.

Solve ▶ $\qquad -3m + 7 \geq 25$

1. Use Addition Rule ▶ $\quad -3m + 7 + (-7) \geq 25 + (-7)$

$$-3m \geq 18$$

2. Use Multiplication Rule ▶ $\quad -\frac{1}{3}(-3m) \leq -\frac{1}{3} \cdot 18$

Be sure to reverse the inequality symbol when you multiply by a negative number.

$$1m \leq -6$$

Solution ▶ $\{m \mid m \leq -6\}$

Make Sure

Solve each inequality using the rules together.

See Example 3 ▶ **1.** $4z - 3 \leq 5$ _____ **2.** $-3 > 4 - 3y$ _____

3. $4(2 - x) < 2(5 - 2x)$ _____ **4.** $4(2x - 3) \geq 2(4x + 3)$ _____

MAKE SURE ANSWERS: **1.** $\{z \mid z \leq 2\}$ **2.** $\{y \mid y > \frac{7}{3}\}$ **3.** $\{x \mid x \in \mathscr{R}\}$ **4.** $\varnothing$

Compound inequalities can be solved by using the set operations of union and intersection.

Recall ▶ When you use set notation, the word *or* means to form the union of sets, and the word *and* means to form the intersection of sets.

EXAMPLE 4: Solve a compound *or* inequality.

Solve ▶ $\quad 3u - 2 < -8 \quad$ or $\quad 3u - 2 > 13 \quad \longleftarrow$ compound *or* inequality

1. Solve simple inequalities ▶
$\quad 3u - 2 < -8 \quad$ or $\quad 3u - 2 > 13$
$\quad 3u < -6 \quad$ or $\quad 3u > 15$
$\quad u < -2 \quad$ or $\quad u > 5$

2. Form union ▶ $\{u \mid u < -2\} \cup \{u \mid u > 5\}$ $\qquad$ Think: *or* means to form the union of the solution sets.

Solution ▶ $\{u \mid u < -2 \text{ or } u > 5\}$

Note ▶ The graph of $\{u \mid u < -2 \text{ or } u > 5\}$ is:

$\qquad \qquad \qquad \longleftarrow\!\!+\!\!+\!\!+\!\!+\!\!\circ\!\!+\!\!+\!\!+\!\!+\!\!+\!\!+\!\!+\!\!\circ\!\!+\!\!\longrightarrow$
$\qquad \qquad \qquad \quad -4\ -3\ -2\ -1\ \ 0\ \ 1\ \ 2\ \ 3\ \ 4\ \ 5\ \ 6$

Make Sure

Solve each compound *or* inequality.

See Example 4 ▶ 1. $3z - 2 \leq -1$ or $3z - 2 > 4$ 2. $3 - 2y < -2$ or $3 - 2y \geq 5$

MAKE SURE ANSWERS: 1. $\{z | z > 2 \text{ or } z \leq \frac{1}{3}\}$ **2.** $\{y | y \leq -1 \text{ or } y > \frac{5}{2}\}$

EXAMPLE 5: Solve a compound *and* inequality.

Solve ▶ $3w - 7 > -1$ and $3w - 7 \leq 14$ ⟵ compound *and* inequality

1. Solve simple ▶ $3w - 7 > -1$ and $3w - 7 \leq 14$
   inequalities
   $3w > 6$ and $3w \leq 21$
   $w > 2$ and $w \leq 7$

2. Form intersection ▶ $\{w | w > 2\}$ ∩ $\{w | w \leq 7\}$ Think: *and* means to form the intersection of the solution sets.

Solution ▶ $\{w | 2 < w \leq 7\}$

Note ▶ The above example is often solved using its *compact form* $-1 < 3w - 7 \leq 14$.

Example ▶ Solve $-1 < 3w - 7 \leq 14$.
   $-1 + 7 < 3w - 7 + 7 \leq 14 + 7$ Add 7 to all three members.
   $6 < 3w \leq 21$
   $2 < w \leq 7$ Multiply all three members by $\frac{1}{3}$.

$\{w | 2 < w \leq 7\}$

Note ▶ The graph of $\{w | 2 < w \leq 7\}$ is:

Make Sure

Solve each compound *and* inequality.

See Example 5 ▶ 1. $3z + 2 < 8$ and $3z + 2 \geq -1$ 2. $-4 < 3y + 2 \leq 11$

MAKE SURE ANSWERS: 1. $\{z | -1 \leq z < 2\}$ **2.** $\{y | -2 < y \leq 3\}$

2.4 Practice

Set 1: Solve each inequality using the Addition Rule.

1. $z + 4 > 6$
2. $y + 3 > 2$
3. $x - 5 < 2$

4. $w - 7 < 3$
5. $3 \geq 4 - v$
6. $2 \geq 3 - u$

Set 2: Solve each inequality using the Multiplication Rule.

7. $4t < 12$
8. $7s < 14$
9. $-3 \leq 4r$

10. $-7 \leq 3m$
11. $-5 > -\frac{2}{3}n$
12. $-6 > -\frac{4}{5}p$

Set 3: Solve each inequality using the rules together.

13. $2q + 6 > 16$
14. $3r - 2 > -11$
15. $3s + 7 > 4 - 2s$

16. $2t - 3 > 5 - 4t$
17. $2 + \frac{1}{3}u \leq \frac{1}{6}u - \frac{1}{2}u$
18. $4 - \frac{2}{3}v \leq \frac{1}{2}v + \frac{1}{6}v$

Set 4: Solve each compound *or* inequality.

19. $2z + 2 \leq 0$ or $2z + 2 \geq 8$ **20.** $6y + 5 \leq 11$ or $6y + 5 \geq 13$

21. $3 - 6x < 0$ or $3 - 6x > 3$ **22.** $3 - 2w < -3$ or $3 - 2w > 3$

23. $2(v - 5) \leq -6$ or $2(v - 5) > 0$ **24.** $2(2u + 3) \leq 0$ or $3(2u - 3) > 6$

Set 5: Solve each compound *and* inequality.

25. $2z + 3 < 8$ and $2z + 3 > 0$ **26.** $2y - 5 < 7$ and $2y - 5 > -3$

27. $3 < n - 6 < 9$ **28.** $-7 < p - 3 < 0$

29. $3 \leq 3q \leq 9$ **30.** $-7 \leq 5r \leq 5$ **31.** $2 \leq 4s - 6 \leq 6$

32. $3 \leq 3t - 9 \leq 6$ **33.** $2 < 5 - 4u \leq 10$ **34.** $-5 < 3 - 2v \leq 5$

Review: Work these problems on a separate sheet of paper. Attach your work to this page.

Find the absolute value of each real number. (See Lesson 1.2, Example 2.)
35. $|-3|$ **36.** $|\frac{1}{2}|$ **37.** $|-1\frac{2}{3}|$ **38.** $|0|$

Graph each simple inequality. (See Lesson 1.2, Example 3.)
39. $x > -2$ **40.** $x < 3$ **41.** $x > 0$ **42.** $x < -4$

Graph each compound inequality. (See Lesson 1.2, Example 4.)
43. $x > 3$ or $x < -2$ **44.** $x > 1$ or $x < -3$ **45.** $0 < x < 3$ **46.** $-3 \leq x \leq -1$

2.5 Solve Absolute Value Equations and Inequalities

The absolute value of a number is the distance the number is from 0 on the number line.

$|x| = 3$ means "x is 3 units from 0," which is equivalent to $x = 3$ or $x = -3$.

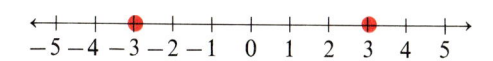

$|x| \geq 3$ means "x is at least 3 units from 0," which is equivalent to $x \leq -3$ or $x \geq 3$.

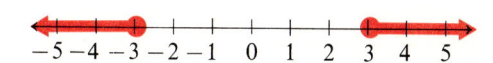

$|x| \leq 3$ means "x is within 3 units of 0," which is equivalent to $x \geq -3$ and $x \leq 3$, or $-3 \leq x \leq 3$

These results can be generalized to variable expressions as follows.

Absolute Value Rules for Equations and Inequalities

For any variable expression E, and any nonnegative number p:

RULE 2.1: $|E| = p$ means $E = p$ $E = -p$.
RULE 2.2: $|E| \geq p$ means $E \leq -p$ or $E \geq p$.
RULE 2.3: $|E| \leq p$ means $E \geq -p$ and $E \leq p$ (compact form: $-p \leq E \leq p$).

To solve an absolute value equation, you use Rule 2.1 of the Absolute Value Rules.

EXAMPLE 1: Solve an absolute value equation.

Solve ▶ $|2x - 1| + 3 = 8$ ⟵ absolute value equation

1. Simplify ▶ $|2x - 1| = 5$ Think: Isolate the absolute value expression in one member.

2. Use Rule 2.1 ▶ $|2x - 1| = 5$ means $2x - 1 = 5$ or $2x - 1 = -5$

3. Solve each equation ▶
$2x = 6$ or $2x = -4$
$x = 3$ or $x = -2$

4. Check ▶

$\|2x - 1\| + 3 = 8$		$\|2x - 1\| + 3 = 8$	
$\|2 \cdot 3 - 1\| + 3$	8	$\|2(-2) - 1\| + 3$	8
$\|5\| + 3$	8	$\|-5\| + 3$	8
$5 + 3$	8	$5 + 3$	8
8	8 ⟵ 3 checks	8	8 ⟵ −2 checks

Solution ▶ $x = -2$ or 3

76 Chapter 2 Equations and Inequalities

If E is an open expression such that $|E| = 0$, then Rule 2.1 states $E = 0$ or $E = -0$, which can be written simply as $E = 0$.

> **Special Case of Rule 2.1:** For any open expression E, $|E| = 0$ means $E = 0$.

Example ▶ Solve $|2 + 6x| = 0$.

1. Use Special Case ▶ $\quad 2 + 6x = 0$

2. Solve ▶ $\quad 6x = -2$

Solution ▶ $\quad x = -\frac{2}{6}$ or $-\frac{1}{3}$ $\quad$ Check as before.

Make Sure

Solve each absolute value equation.

See Example 1 ▶ **1.** $|z + 3| = 6$ _____ **2.** $|2y - 3| = 5$ _____ **3.** $4 = |3 - 2x|$ _____

MAKE SURE ANSWERS: 1. $-9, 3$ 2. $-1, 4$ 3. $-\frac{1}{2}, \frac{7}{2}$

To solve an absolute value inequality of the form $|E| \geq p$, you use Rule 2.2 of the Absolute Value Rules.

EXAMPLE 2: Solve an absolute value inequality of the form $|E| \geq p$.

Solve ▶ $|2x + 3| \geq 11$

1. Use Rule 2.2 ▶ $|2x + 3| \geq 11$ means $2x + 3 \leq -11$ or $2x + 3 \geq 11$

2. Solve inequalities ▶ $\qquad\qquad\qquad\qquad 2x \leq -14 \quad$ or $\quad 2x \geq 8$

$\qquad\qquad\qquad\qquad\qquad x \leq -7 \quad$ or $\quad x \geq 4$

Solution ▶ $\{x \mid x \leq -7 \text{ or } x \geq 4\}$

Note ▶ The graph of $\{x \mid x \leq -7 \text{ or } x \geq 4\}$ is:

$\qquad\qquad\qquad\qquad\qquad\qquad\qquad\qquad -10\ -8\ -6\ -4\ -2\ \ 0\ \ 2\ \ 4\ \ 6$

Rule 2.2 still holds if $\geq$ is replaced by $>$ and $\leq$ is replaced by $<$.

> **Rule 2.2 applied to $|E| > p$:** $|E| > p$ means $E < -p$ or $E > p$.

2.5 Solve Absolute Value Equations and Inequalities

Example ▶ Solve $|x - 5| > 2$.

1. Use Rule 2.2 ▶ $|x - 5| > 2$ means $x - 5 < -2$ or $x - 5 > 2$

2. Solve as before ▶ $x < 3$ or $x > 7$

Solution ▶ $\{x \mid x < 3 \text{ or } x > 7\}$

Note ▶ The graph of $\{x \mid x < 3 \text{ or } x > 7\}$ is:

You may find it necessary to simplify before you apply the rules.

Example ▶ Solve $-6 + |5 + 3x| > 4$.

1. Simplify ▶ $|5 + 3x| > 10$

2. Use Rule 2.2 ▶ $|5 + 3x| > 10$ means $5 + 3x < -10$ or $5 + 3x > 10$.

3. Solve each inequality ▶ $3x < -15$ or $3x > 5$

                                       $x < -5$ or $x > \frac{5}{3}$

Solution ▶ $\{x \mid x < -5 \text{ or } x > \frac{5}{3}\}$

Note ▶ The graph of $\{x \mid x < -5 \text{ or } x > \frac{5}{3}\}$ is:

Make Sure

Solve and graph absolute value inequalities of the form $|E| \geq p$.

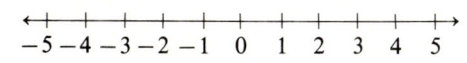

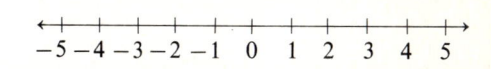

See Example 2 ▶   **1.** $|a - 3| \geq 1$ _____   **2.** $3 < |2b - 5|$ _____

MAKE SURE ANSWERS: See Appendix Selected Answers.

To solve an absolute value inequality of the form $|E| \leq p$, you use Rule 2.3 of the Absolute Value Rules.

EXAMPLE 3: Solve an absolute value inequality of the form $|E| \leq p$.

Solve ▶ $|3x - 4| \leq 8$

1. Use Rule 2.3 ▶ $|3x - 4| \leq 8$ means $-8 \leq 3x - 4 \leq 8$.   Think: Use Rule 2.3 in its compact form.

2. Solve as before ▶ $-4 \leq \;\;\; 3x \;\;\; \leq 12$ Add 4 to all members.

                                  $-\frac{4}{3} \leq \;\;\; x \;\;\; \leq 4$ Multiply all members by $\frac{1}{3}$.

Solution ▶ $\{x \mid -\frac{4}{3} \leq x \leq 4\}$

Note ▶ The graph of $\{x \mid -\frac{4}{3} \leq x \leq 4\}$ is:

78 Chapter 2 Equations and Inequalities

Rule 2.3 still holds if $\leq$ is replaced by $<$ and $\geq$ is replaced by $>$.

> **Rule 2.3 applied to $|E| < p$:** $|E| < p$ means $E > -p$ and $E < p$. (compact form: $-p < E < p$).

Example ▶ Solve $|2x + 7| < 15$.

1. Use Rule 2.3 ▶ $|2x + 7| < 15$ means $-15 < 2x + 7 < 15$

2. Solve as before ▶
$$-22 < 2x < 8$$
$$-11 < x < 4$$

Solution ▶ $\{x \mid -11 < x < 4\}$

Note ▶ The graph of $\{x \mid -11 < x < 4\}$ is:

Another Example ▶ Solve $-3|4 - 2x| > -72$.

1. Simplify ▶ $|4 - 2x| < 24$ Think: The inequality is reversed because each member has been multiplied by $-\frac{1}{3}$.

2. Use Rule 2.3 ▶ $|4 - 2x| < 24$ means $-24 < 4 - 2x < 24$

3. Solve ▶
$$-28 < -2x < 20 \quad \text{Add } -4 \text{ to all members.}$$
$$14 > x > -10 \quad \text{Multiply each member by } -\tfrac{1}{2} \text{ and reverse each inequality symbol.}$$

Solution ▶ $\{x \mid -10 < x < 14\}$

Note ▶ The graph of $\{x \mid -10 < x < 14\}$ is:

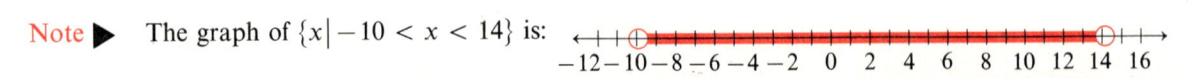

> For all open expressions E and any negative number n, the solution set of $|E| \leq n$ is the empty set because $|E|$ is never negative, and $|E| \geq n$ is the set of all real numbers because every nonnegative number is larger than any negative number.

Examples ▶ (a) The solution set of $|x| \leq -3$ is the empty set. No value of x makes the sentence true.
(b) The solution set of $|x| \geq -3$ is the set of all real numbers. Every value of x makes the sentence true.

Make Sure

Solve and graph absolute value inequalities of the form $|E| \leq p$.

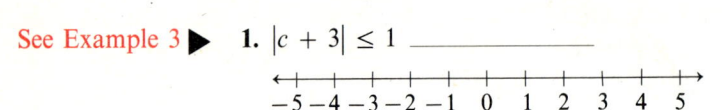

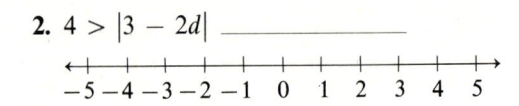

See Example 3 ▶ 1. $|c + 3| \leq 1$ _____

2. $4 > |3 - 2d|$ _____

MAKE SURE ANSWERS: See Appendix Selected Answers.

2.5 Practice

Set 1: Solve each absolute value equation. Check each proposed solution.

1. $|z + 3| = 5$
2. $|y + 2| = 5$
3. $3 = |2 - 3t|$

4. $5 = |4 - 2s|$
5. $|z - 2| + 4 = 9$
6. $|y - 4| + 3 = 11$

Set 2: Solve absolute value inequalities involving $|E| \geq p$. Graph each solution.

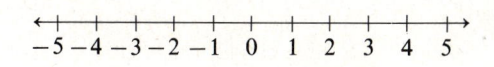

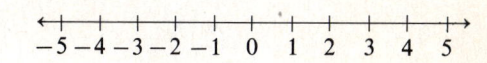

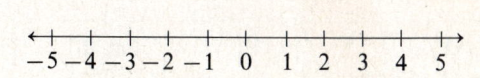

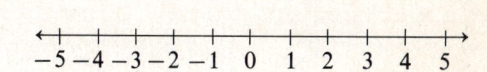

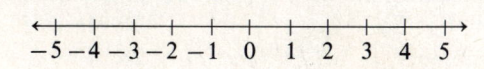

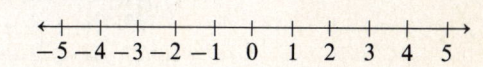

7. $|q - 2| \geq 2$
8. $|r - 1| \geq 3$

9. $|2s + 3| \geq 3$
10. $|3t + 4| \geq 5$

11. $\left|4 - \dfrac{2}{3}u\right| > 1$
12. $\left|3 - \dfrac{3}{4}v\right| > 1$

80 Chapter 2 Equations and Inequalities

13. _____

Set 3: Solve absolute value inequalities involving $|E| \leq p$. Graph each solution.

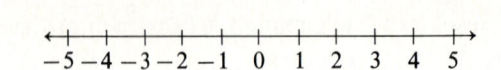

13. $|u + 2| \leq 3$

$\begin{array}{c}\longleftrightarrow\\-5\ -4\ -3\ -2\ -1\ \ 0\ \ 1\ \ 2\ \ 3\ \ 4\ \ 5\end{array}$

14. $|v + 1| \leq 2$

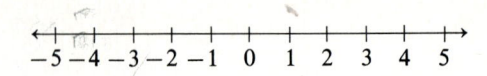

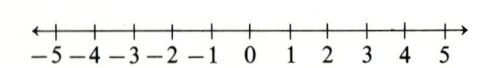

$\begin{array}{c}\longleftrightarrow\\-5\ -4\ -3\ -2\ -1\ \ 0\ \ 1\ \ 2\ \ 3\ \ 4\ \ 5\end{array}$

14. _____

15. $|2w - 3| \leq 3$

$\begin{array}{c}\longleftrightarrow\\-5\ -4\ -3\ -2\ -1\ \ 0\ \ 1\ \ 2\ \ 3\ \ 4\ \ 5\end{array}$

16. $|3x - 4| \leq 5$

$\begin{array}{c}\longleftrightarrow\\-5\ -4\ -3\ -2\ -1\ \ 0\ \ 1\ \ 2\ \ 3\ \ 4\ \ 5\end{array}$

15. _____

16. _____

17. $|3 - 4m| < 4$

$\begin{array}{c}\longleftrightarrow\\-5\ -4\ -3\ -2\ -1\ \ 0\ \ 1\ \ 2\ \ 3\ \ 4\ \ 5\end{array}$

18. $|5 - 2n| < 3$

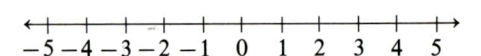

$\begin{array}{c}\longleftrightarrow\\-5\ -4\ -3\ -2\ -1\ \ 0\ \ 1\ \ 2\ \ 3\ \ 4\ \ 5\end{array}$

17. _____

Mixed Practice: Solve each absolute value inequality. Graph each solution.

19. $|z| - 3 \leq -2$

$\begin{array}{c}\longleftrightarrow\\-5\ -4\ -3\ -2\ -1\ \ 0\ \ 1\ \ 2\ \ 3\ \ 4\ \ 5\end{array}$

20. $|y| - 4 \leq -1$

$\begin{array}{c}\longleftrightarrow\\-5\ -4\ -3\ -2\ -1\ \ 0\ \ 1\ \ 2\ \ 3\ \ 4\ \ 5\end{array}$

18. _____

19. _____

21. $-2|t| < -6$

$\begin{array}{c}\longleftrightarrow\\-5\ -4\ -3\ -2\ -1\ \ 0\ \ 1\ \ 2\ \ 3\ \ 4\ \ 5\end{array}$

22. $-3|s| < -6$

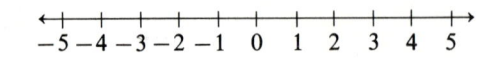

$\begin{array}{c}\longleftrightarrow\\-5\ -4\ -3\ -2\ -1\ \ 0\ \ 1\ \ 2\ \ 3\ \ 4\ \ 5\end{array}$

20. _____

23. $3|p - 1| - 4 \geq 3$

$\begin{array}{c}\longleftrightarrow\\-5\ -4\ -3\ -2\ -1\ \ 0\ \ 1\ \ 2\ \ 3\ \ 4\ \ 5\end{array}$

24. $4|2 - n| - 3 \geq 3$

$\begin{array}{c}\longleftrightarrow\\-5\ -4\ -3\ -2\ -1\ \ 0\ \ 1\ \ 2\ \ 3\ \ 4\ \ 5\end{array}$

21. _____

22. _____

Review: Work these problems on a separate sheet of paper. Attach your work to this page.

Solve equations containing like terms. (See Lesson 2.2, Example 1.)
25. $2a - 4 + 3a = 1 - 3a$
26. $4b - 5 - 2b = 4 - 2b$
27. $5c + 7 = 3 - 2c + 1$
28. $4 - d + 2 = 3d - 1 - 5d$
29. $8 - 7e + 3 = 5 - 3e + e$
30. $8f - 7 + 2f - f = 3 - 4f$

23. _____

Solve equations containing parentheses. (See Lesson 2.2, Example 2.)
31. $3(z - 1) = 5$
32. $2(y + 1) = 7$
33. $2(1 - x) - (x - 3) = 2$
34. $4(2 - w) - (w + 2) = 3$
35. $3(v - 4) - 2(2 - v) = 4(1 - 2v)$
36. $2(u - 3) - 3(2 - 3u) = 5(1 - u)$

24. _____

Copyright © 1985 by Harcourt Brace Jovanovich, Inc. All rights reserved.

2.6 Solve Literal Equations

> An equation in which the constants are represented by letters is called a *literal equation*.

Example ▶ $Ax + By = C$ is a literal equation.

You have *solved a literal equation* for an unknown when you have rewritten it as an equivalent equation with the desired unknown appearing alone as one member of the equation, and when the unknown does not appear in both members.

Example ▶ The literal equation $Ax + By = C$ is solved for C, since C appears alone as one member of the equation and C does not appear in the left member.

To solve a literal equation for a certain unknown, you can use the rules and properties.

> In each of the following examples it is assumed that none of the denominators of the expressions is equal to zero.

EXAMPLE 1: Solve a literal equation for the indicated letter.

Solve for x ▶ $\qquad ax + by = c$

1. Isolate desired term ▶ $ax + by + (-by) = c + (-by)$ Use the Addition Rule.

$\qquad\qquad\qquad\qquad ax = c - by$ Think: The term that involves x is isolated in the left member.

2. Isolate desired letter ▶ $\qquad \dfrac{1}{a} \cdot ax = \dfrac{1}{a}(c - by)$ Use the Multiplication Rule.

$\qquad\qquad\qquad\qquad x = \dfrac{1}{a}(c - by)$ Think: x is alone in the left member and x does not appear in the right member.

3. Check ▶

$ax + by = c$		← original equation
$a\left[\dfrac{1}{a}(c - by)\right] + by$	c	Substitute $x = \dfrac{1}{a}(c - by)$.
$c - by + by$	c	Compute.
c	c	Compare: $c = c$ means $\dfrac{1}{a}(c - by)$ is a solution.

Solution ▶ $x = \dfrac{1}{a}(c - by)$ or $\dfrac{c - by}{a}$

Note ▶ In the previous example, it is assumed that $a \neq 0$.

Make Sure

Solve each literal equation for the indicated letter.

See Example 1 ▶   **1.** $8x + 2y = 6$ for y _____   **2.** $ax - ay = a$ for x _____

MAKE SURE ANSWERS: 1. $y = 3 - 4x$ **2.** $x = 1 + y$

82 Chapter 2 Equations and Inequalities

To solve a literal equation for a letter that appears in more than one term, you isolate all of those terms in one member and then use a distributive property to factor out the desired letter.

EXAMPLE 2: Solve a literal equation containing like terms for the indicated letter.

Solve for y ▶ $-x + dy = x + ey$

1. Isolate ▶ $dy - ey = x + x$ Add x and $-ey$ to both members to isolate the terms that involve the letter y in the left member.

2. Factor ▶ $y(d - e) = 2x$ Factor out y into the left member.

3. Solve as before ▶ $y(d - e) \cdot \dfrac{1}{d - e} = 2x \cdot \dfrac{1}{d - e}$ Multiply both members by $\dfrac{1}{d - e}$, assuming $d - e \neq 0$.

Solution ▶ $y = \dfrac{2x}{d - e}$ Check as before.

Make Sure

Solve literal equations containing like terms for the indicated letter.

See Example 2 ▶ **1.** $3x - 2y = x - 3$ for x _____   **2.** $az - bz = cz + d$ for z _____

MAKE SURE ANSWERS: **1.** $x = y - \dfrac{3}{2}$ **2.** $z = \dfrac{d}{a - b - c}$

EXAMPLE 3: Solve a literal equation containing parentheses for the indicated letter.

Solve for w ▶ $a(w - b) = b(w + a)$

1. Clear parentheses ▶ $aw - ab = bw + ba$ Distribute a in the left member and b in the right member.

2. Solve as before ▶ $aw - bw = ab + ba$ Add $-bw$ and ab to both members to isolate the terms that involve the letter w into the left member.

$w(a - b) = 2ab$ Factor out w in the left member.

$w(a - b) \cdot \dfrac{1}{a - b} = 2ab \cdot \dfrac{1}{a - b}$ Multiply both members by $\dfrac{1}{a - b}$, assuming $a - b \neq 0$.

Solution ▶ $w = \dfrac{2ab}{a - b}$ Check as before.

Make Sure

Solve literal equations containing parentheses for the indicated letter.

See Example 3 ▶ **1.** $a(m + n) = b(m - n)$ for n _____   **2.** $\dfrac{m + n}{m - n} = \dfrac{a}{b}$ for m _____

MAKE SURE ANSWERS: **1.** $n = \dfrac{m(b - a)}{a + b}$ **2.** $m = \dfrac{n(-a - b)}{a - b}$ or $\dfrac{n(a + b)}{a - b}$

2.6 Practice: *Assume all denominators are nonzero.*

Set 1: Solve each literal equation for the indicated letter using the rules.

1. $3y - 2z = 5$ for y
2. $4w - 3x = 7$ for x
3. $az + by = c$ for z
4. $aw + bx = c$ for x
5. $\frac{a}{b}z - ay = a$ for z
6. $\frac{a}{b}z - ay = a$ for y

Set 2: Solve literal equations containing like terms for the indicated letter.

7. $3y + 2z = 5y - 4z + 5$ for y
8. $3w - 5x + 7 = 4w + 2x - 5$ for x
9. $az + ay = 3az$ for z
10. $2aw - ax = aw$ for w
11. $ay - bz = az - by$ for y
12. $aw - bx = bw - ax$ for x

Set 3: Solve literal equations containing parentheses for the indicated letter.

13. $3(z + 2) = 4y$ for z
14. $-(y - 2) = 3z$ for y
15. $a(w - x) = a(x - w)$ for x
16. $x(w - a) = w(a - x)$ for w
17. $\frac{a}{b}(x - a) = x$ for x
18. $a\left(\dfrac{y + b}{b}\right) = y$ for y

Mixed Practice: Solve each literal equation for the indicated letter.

19. $ax + by = cz$ for z
20. $aw - bx = cy$ for w
21. $ax = cy + bx$ for x

22. $ay = c - bw$ for w
23. $a(z + b) = cy$ for z
24. $a(y - b) = cx$ for y

25. $\dfrac{x}{a} = \dfrac{x}{2a}$ for x
26. $\dfrac{x}{a} = \dfrac{x}{a + b}$ for x
27. $a = \dfrac{b}{1 - c}$ for c

28. $d = \dfrac{e}{1 + f}$ for f
29. $a = b(1 - c)$ for c
30. $a = b(c - 1)$ for c

31. $a = \dfrac{b}{cx + c}$ for c
32. $a = \dfrac{b}{cx + c}$ for x
33. $a(x - y) = 1 - y$ for y

34. $a(x - y) = 1 - x$ for x
35. $\dfrac{1}{a} = \dfrac{1}{b} + \dfrac{1}{c}$ for b
36. $\dfrac{1}{a} = \dfrac{1}{b} - \dfrac{1}{c}$ for c

Review: Work these problems on a separate sheet of paper. Attach your work to this page.

Evaluate each expression for the given variable. (See Lesson 1.4.)

37. $2x - 3y$ for $x = 2, y = 1$
38. $4x + 2y$ for $x = -1, y = 3$
39. $3x - 2y$ for $x = -1, y = -3$
40. $5a - 3b$ for $a = -1, b = -2$
41. $a + 3b - c$ for $a = 3, b = -1, c = 0$
42. $2b + c - a$ for $a = 4, b = 1, c = -2$
43. $2pq - 4p$ for $p = 3, q = -1$
44. $5pq - 3q$ for $p = -2, q = -3$

Copyright © 1985 by Harcourt Brace Jovanovich, Inc. All rights reserved.

Name _____ Date _____ Class _____

Chapter 2 Review

			What to Review if You Have Trouble		
Objectives			Lesson	Example	Page
Solve equations using the Addition Rule	1. $y + 5 = -8$	_____	2.1	1	50
Solve equations using the Multiplication Rule	2. $4w = 36$	_____	2.1	2	51
Solve equations using the rules together	3. $2x + 11 = -7$	_____	2.1	3	52
Solve equations containing like terms	4. $-y + 9y + 5 = -2y - 3 + y$	_____	2.2	1	55
Solve equations containing parentheses	5. $4(z - 3) = 7z - 3(2z - 5)$	_____	2.2	2	56
Solve equations using the zero-product property	6. $(x + 5)(4x - 11) = 0$	_____	2.2	3	57
Solve equations containing fractions	7. $\dfrac{3y}{2} - 7 = \dfrac{1}{5}$	_____	2.3	1	63
Solve equations containing decimals	8. $2.25z - 0.8 = 0.75$	_____	2.3	2	64
Solve equations containing percents	9. $13\%w + 9\%(4000 - w) = 480$	_____	2.3	3	64

Chapter 2 Equations and Inequalities

Solve inequalities using the Addition Rule ▶ 10. $x + 8 < 17$ 2.4   1   69

Solve inequalities using the Multiplication Rule ▶ 11. $-6n < 48$ 2.4   2   70

Solve inequalities using the rules together ▶ 12. $-4m + 9 > 24$ 2.4   3   71

Solve compound *or* inequalities ▶ 13. $5u - 3 < -7$ or $5u - 3 > 13$ 2.4   4   71

Solve compound *and* inequalities ▶ 14. $4w - 9 > -3$ and $4w - 9 \leq 11$ 2.4   5   72

Solve absolute value equations ▶ 15. $|3x - 2| = 6$ 2.5   1   75

Solve absolute value inequalities ▶ 16. $|4x + 5| \geq 1$ 2.5   2   76

17. $|2x - 3| \leq 7$ 2.5   3   77

Solve literal equations ▶ 18. Solve $bx - cy = a$ for x.   2.6   1   81

19. Solve $-ax + ey = x - fy$ for y.   2.6   2   82

20. Solve $b(w - k) = a(a - w)$ for w.   2.6   3   82

Solve problems using equations ▶ 21. The perimeter of a rectangle is 34 meters. The width is 7 meters shorter than the length. What is the area?   PS 2   —   67

CHAPTER 2 REVIEW ANSWERS: 1. -13 **2.** 9 **3.** -9 **4.** $-\frac{8}{9}$ **5.** 9 **6.** -5 **7.** $\frac{11}{24}$ **8.** $\frac{31}{45}$ or 0.68 **9.** 3000 **10.** $\{x | x < 9\}$ **11.** $\{n | n > -8\}$ **12.** $\{m | m > -\frac{15}{4}\}$ **13.** $\{u | u < -\frac{2}{5}$ or $u > \frac{16}{5}\}$ **14.** $\{w | \frac{3}{2} \leq w \leq 5\}$ **15.** $-\frac{4}{3}, \frac{8}{3}$ **16.** $\{x | x \leq -\frac{3}{2}$ or $x \geq -1\}$ **17.** $\{x | -2 \leq x \leq 5\}$ **18.** $\frac{a + cy}{b}$ **19.** $\frac{x + ax}{e + f}$ **20.** $\frac{a^2 + bk}{a + b}$ **21.** $60 m^2$

REAL NUMBERS AND LINEAR RELATIONS

3 Equations and Graphs

3.1 Use Ordered Pairs

3.2 Graph Linear Equations

3.3 Find and Use Slope

3.4 Find Linear Equations

PS 3: Use Linear Relationships

3.5 Graph Linear Inequalities

88 Chapter 3 Equations and Graphs

3.1 Use Ordered Pairs

Each point on a number line is associated with a number called its *coordinate*. Each point on a plane is associated with an *ordered pair* (x, y) of numbers called its *coordinates*. The coordinates of a point on a plane are determined by the point's position relative to two perpendicular number lines called *axes*. These axes form a *rectangular coordinate system*.

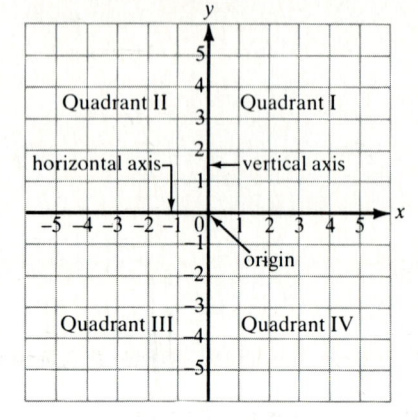

Figure 3.1 ▶

The horizontal axis is called the *x-axis*. The vertical axis is called the *y-axis*. The axes intersect at a point called the *origin*.

Figure 3.1 shows that the axes are labeled so that positive numbers appear to the right of the origin on the *x*-axis and above the origin on the *y*-axis. The axes divide the plane into four regions called *quadrants*, which are numbered counterclockwise as shown.

> The first number in an ordered pair is called the *x-coordinate*, or *abscissa*. The second number is called the *y-coordinate*, or *ordinate*.

To *graph an ordered pair*, you draw a dot at the point associated with the ordered pair.

EXAMPLE 1: Graph an ordered pair.

Problem ▶ Graph $(3, -4)$.

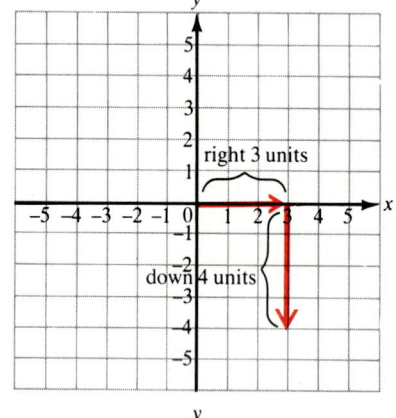

1. Locate point ▶

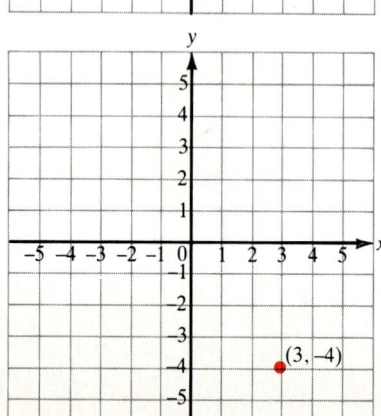

The 3 in $(3, -4)$ means the point is 3 units to the right of the *y*-axis.

The -4 in $(3, -4)$ means the point is 4 units below the *x*-axis.

2. Graph ▶

Solution ▶

Draw a dot and label it with the given coordinates.

3.1 Use Ordered Pairs   89

Graphing an ordered pair is sometimes referred to as plotting the point associated with an ordered pair, or merely as *plotting a point*.

Other Examples ▶ (a) Plot the point for $(-4, -5)$. (b) Plot the point for $(-3, 0)$.

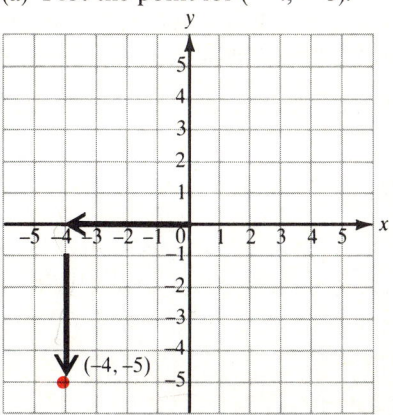

Think: $(-4, -5)$ means move left 4 and move down 5.   Think: $(-3, 0)$ means move left 3 units.

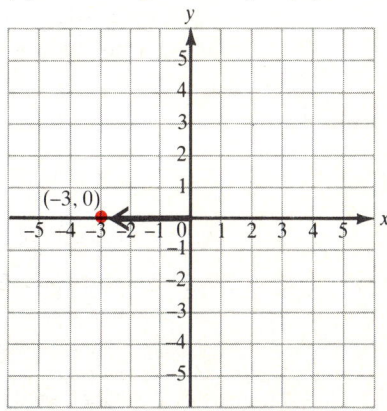

Note 1 ▶ When the *x*-coordinate is 0, the point is on the *y*-axis.

Note 2 ▶ When the *y*-coordinate is 0, the point is on the *x*-axis.

CAUTION: The order of the coordinates is important.

Example ▶ The ordered pair $(-3, 1)$ is not the same as the ordered pair $(1, -3)$. Figure 3.2 shows the graph of the two ordered pairs $(-3, 1)$ and $(1, -3)$.

Figure 3.2 ▶

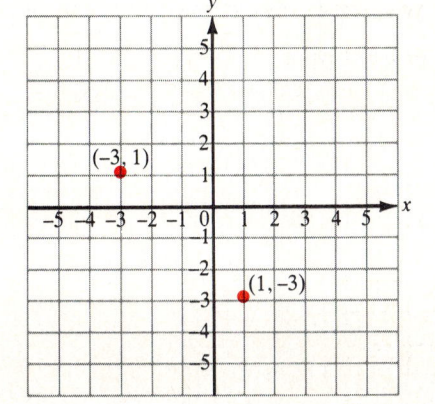

Make Sure

Graph each ordered pair on the given rectangular coordinate system.

See Example 1 ▶ 1. Graph $(4, -1)$.   2. Graph $(-3, 0)$.

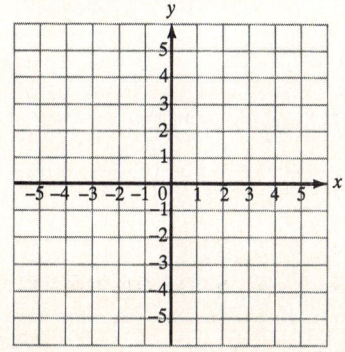

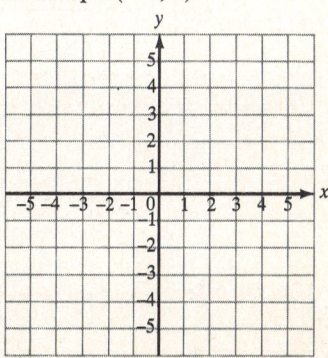

MAKE SURE ANSWERS: See Appendix Selected Answers.

90   Chapter 3   Equations and Graphs

To find the coordinates of a point in a rectangular coordinate system, you project the point onto each of the axes.

EXAMPLE 2: Find the coordinates of a given point.

Problem ▶ Find the coordinates of point P in Figure 3.3.

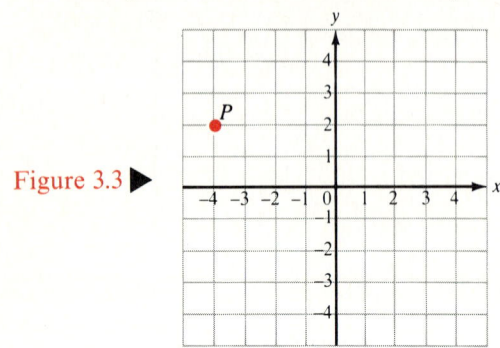

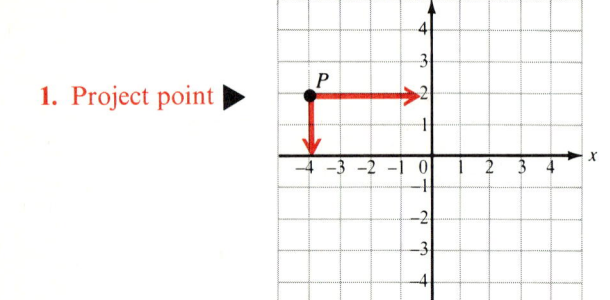

Figure 3.3 ▶

1. Project point ▶ Project the point vertically onto the x-axis, and horizontally onto the y-axis.

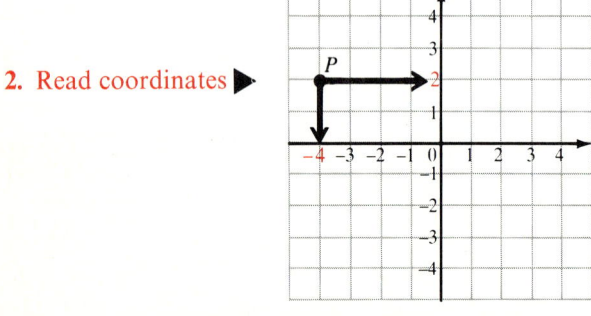

2. Read coordinates ▶ Think: The vertical projection intersects the x-axis at $x = -4$.
The horizontal projection intersects the y-axis at $y = 2$.

Solution ▶ The x-coordinate of P is -4. The y-coordinate of P is 2.

Note ▶ Since the point P is associated with the ordered pair $(-4, 2)$, you can use $(-4, 2)$ to name the point P.

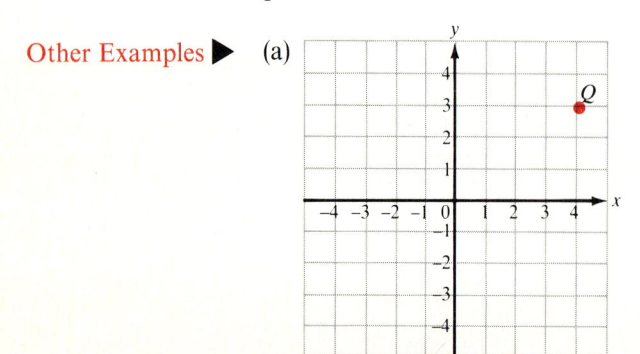

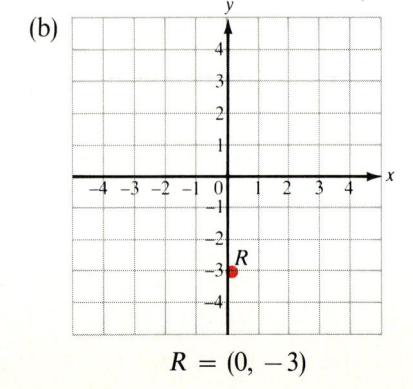

Other Examples ▶ (a)   (b)

$Q = (4, 3)$ $R = (0, -3)$

Make Sure

Find the coordinates of each point in the given rectangular coordinate system.

See Example 2 ▶ 1. [coordinate grid with points A, B, C] $A = (\ ,\ )$
$B = (\ ,\ )$
$C = (\ ,\ )$

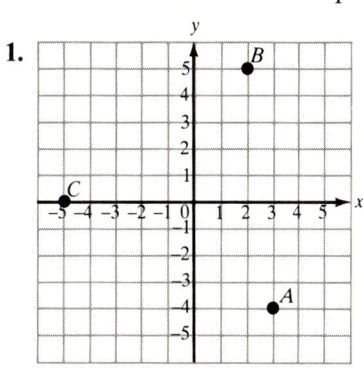

2. [coordinate grid with points D, E, F] $D = (\ ,\ )$
$E = (\ ,\ )$
$F = (\ ,\ )$

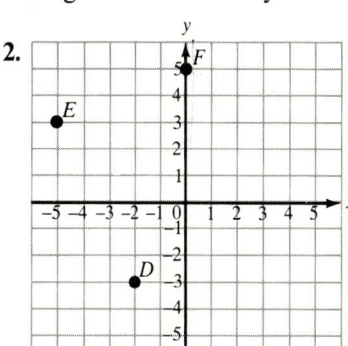

MAKE SURE ANSWERS: 1. $A = (3, -4)$, $B = (2, 5)$, $C = (-5, 0)$. 2. $D = (-2, -3)$, $E = (-5, 3)$, $F = (0, 5)$.

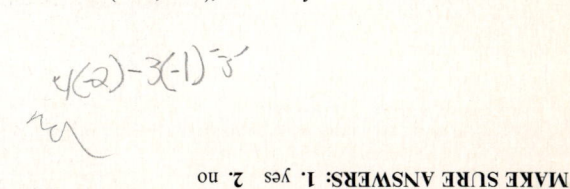

> The ordered pair (m, n) is a *solution of an equation in the two variables x and y* if the equation is a true sentence when m is substituted for x and n for y.

EXAMPLE 3: Determine if an ordered pair is a solution of an equation.

Problem ▶ Which pair, $(-2, -13)$ or $(4, -1)$, is a solution of $-4x + y = -5$?

1. Check first ordered pair ▶

$-4x + y = -5$	← original equation
$-4(-2) + (-13)$	-5	Substitute: $(-2, -13)$ means $x = -2$ and $y = -13$.
$8 + (-13)$	-5	Compute.
-5	-5	Compare: $-5 = -5$ means $(-2, -13)$ is a solution.

2. Check second ordered pair ▶

$-4x + y = -5$	← original equation
$-4(4) + (-1)$	-5	Substitute: $(4, -1)$ means $x = 4$ and $y = -1$.
$-16 + (-1)$	-5	Compute.
-17	-5	Compare: $-17 \neq -5$ means $(4, -1)$ is not a solution.

Solution ▶ $(-2, -13)$ is a solution of $-4x + y = -5$. $(4, -1)$ is not a solution of $-4x + y = -5$.

Make Sure

Is the given ordered pair a solution of the given equation?

See Example 3 ▶ 1. $(-2, 3)$, $4x + 3y = 1$ _____   2. $(-2, -1)$, $4x - 3y = 5$ _____

MAKE SURE ANSWERS: 1. yes 2. no

The ordered pair $(-2, -13)$ is not the only solution of $-4x + y = -5$. You can find other solutions by substituting arbitrary numerical values for one of the variables and solving the resulting equation for the other variable. A table is often used to organize the procedure.

EXAMPLE 4: Find the indicated solutions of a given equation in two variables.

Problem ▶ Find solutions of $-4x + y = -5$ for $x = -1, 0, 1, 2,$ and 5.

1. Solve for y ▶ $-4x + y = -5$ ←—— original equation

$y = 4x - 5$ Add $4x$ to both members.

2. Make a table ▶

x	$4x - 5 = y$
-1	
0	
1	
2	
5	

3. Evaluate ▶

x	$4x - 5 = y$	
-1	$4(-1) - 5 = -9$	When $x = -1, y = -9$.
0	$4(0) - 5 = -5$	When $x = 0, y = -5$.
1	$4(1) - 5 = -1$	When $x = 1, y = -1$.
2	$4(2) - 5 = 3$	When $x = 2, y = 3$.
5	$4(5) - 5 = 15$	When $x = 5, y = 15$.

Solution ▶ $(-1, -9), (0, -5), (1, -1), (2, 3), (5, 15)$ are the respective solutions of $-4x + y = -5$ for $x = -1, 0, 1, 2,$ and 5.

You can also find solutions of an equation in x and y by assigning arbitrary values to y and then solving for the corresponding x values.

Example ▶ Find solutions of $2x - 4y = -6$ for $y = 0, 1,$ and 4.

$2x - 4y = -6$ ←—— original equation

$2x = 4y - 6$

$x = 2y - 3$ ←—— equation solved for x in terms of y

y	$2y - 3 = x$	
0	$2(0) - 3 = -3$	When $y = 0, x = -3$.
1	$2(1) - 3 = -1$	When $y = 1, x = -1$.
4	$2(4) - 3 = 5$	When $y = 4, x = 5$.

$(-3, 0), (-1, 1),$ and $(5, 4)$ are the respective solutions of $2x - 4y = -6$ for $y = 0, 1,$ and 4.

Make Sure

Find each indicated solution for each equation.

See Example 4 ▶ **1.** $3x + 2y = 6$ for $x = 4, 2, 0, -2$ **2.** $4x - 3y = 1$ for $y = 3, 0, -1, -3$

MAKE SURE ANSWERS: 1. $(4, -3), (2, 0), (0, 3), (-2, 6)$ **2.** $(\frac{5}{2}, 3), (\frac{1}{4}, 0), (-\frac{1}{2}, -1), (-2, -3)$

3.1 Practice

Set 1: Graph each ordered pair on the given rectangular coordinate system.

1. $A(3, 4)$, $B(2, -3)$, $C(-1, 4)$, $D(-3, -5)$
   $E(0, 4)$, $F(-3, 0)$, $G(1, 0)$, $H(0, -2)$

2. $S(-1, 3)$, $T(-3, -4)$, $U(4, -3)$, $V(1, 5)$
   $W(3, 0)$, $X(0, -4)$, $Y(-1, 0)$, $Z(0, 3)$

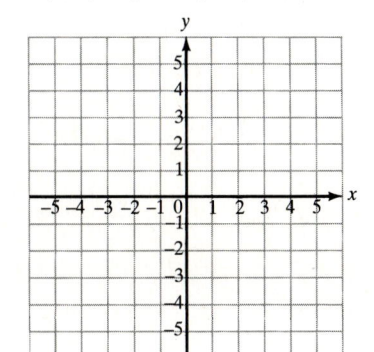

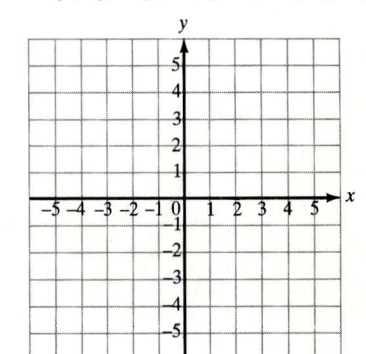

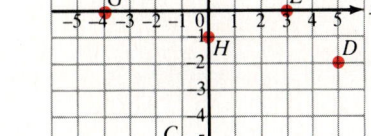

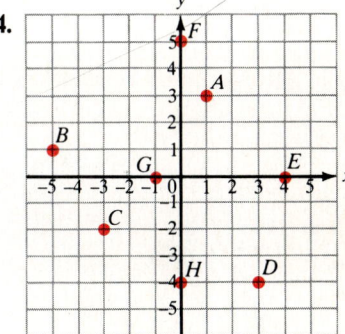

Set 2: Find the coordinates of each point in the rectangular coordinate system.

3.

4.

Set 3: Is the given ordered pair a solution of the given equation?

5. $(2, 3)$, $3x + 2y = 12$

6. $(-1, 3)$, $3x + 4y = 9$

7. $(3, 2)$, $3x + 2y = 12$

8. $(-3, 1)$, $4x - 3y = 9$

9. $(-3, 0)$, $x + 4y = -3$

10. $(3, -2)$, $5x - 2y = 19$

11. $(0, -3)$, $x = 4y - 3$

12. $(-1, -4)$, $y = 7x + 27$

13. $(2.3, 4.1)$, $2x = y + 0.5$

14. $(5.2, 3.4)$, $2x = 3y + 0.2$

15. $(\frac{1}{2}, \frac{1}{3})$, $3x = 4y + \frac{1}{6}$

16. $(\frac{2}{3}, \frac{3}{4})$, $2x = 3y - \frac{5}{4}$

17. _____

18. _____

19. _____

20. _____

21. _____

22. _____

23. _____

24. _____

25. _____

26. _____

27. _____

28. _____

29. _____

30. _____

Set 4: Find each indicated solution for each equation.

17. $2x + y = 7$ for $x = -4, 0, 2$

18. $2x - y = 5$ for $x = -2, 1, 3$

19. $x - 4y = -3$ for $y = -5, 0, 5$

20. $3x - 2y = 6$ for $y = -2, -1, 3$

21. $2x - 3y = 6$ for $x = -3, 0, 3$

22. $2y = 3x + 4$ for $x = -2, 1, 2$

23. $x = y + 4$ for $y = -1.5, 0, 4.3$

24. $x = 5 - y$ for $y = -0.5, 2.3, 4.1$

25. $\frac{1}{2}x + \frac{1}{3}y = 6$ for $x = -4, -2, 4$

26. $\frac{2}{3}x + \frac{3}{4}y = 12$ for $y = -6, 0, 4$

27. $\frac{2}{3}x + 6 = \frac{3}{4}y$ for $y = -3, 3, 4$

28. $\frac{4}{7}x = 10 - \frac{3}{4}y$ for $y = -4, 0, 4$

29. $y = 5 + 3x$ for $x = -2, 1, 3$

30. $y = 4x - 3$ for $x = -5, 0, 5$

Review: Work these problems on a separate sheet of paper. Attach your work to this page.

Evaluate each expression for the given values of the variables. (See Lesson 1.4.)
31. $x - 3$ for $x = 0, 2$.
32. $2x - 5$ for $x = 0, -2$.
33. $3x - 4$ for $x = 0, -1$.
34. $-5x + 4$ for $x = 2, -2$.
35. $\frac{2}{3}x - \frac{1}{6}$ for $x = 0, -3$.
36. $\frac{3}{4}x - \frac{1}{6}$ for $x = 0, -4$.
37. $-\frac{3}{2}x + 4$ for $x = 0, 4$.
38. $-\frac{5}{3}x + 3$ for $x = 0, -6$.
39. $-\frac{4}{3}x - \frac{5}{6}$ for $x = 0, 2$.

Solve each literal equation for y. (See Lesson 2.6.)
40. $3x - 2y = 0$
41. $2x + 3y = 0$
42. $4x - 3y = 6$
43. $4x + 3y = 6$
44. $x - 3y = 4$
45. $x + 3y = 4$
46. $5x - 3y = 4$
47. $5x + 3y = 4$
48. $3x + 2y = 5$

Copyright © 1985 by Harcourt Brace Jovanovich, Inc. All rights reserved.

3.2 Graph Linear Equations

A rectangular coordinate system is also referred to as a *Cartesian coordinate system* in honor of René Descartes (1596–1650). He introduced the idea of coordinates to establish a relationship between equations and *graphs of equations*.

> Equations that can be written in the *standard form* $Ax + By = C$, where A, B, and C are real numbers and A and B are not both 0, are called *linear equations in two variables*.

Examples ▶ (a) $3x - 4y = 7$ is a linear equation.
(b) $5x^2 + 4y + 2 = 0$ is not a linear equation because of the x^2 term.
(c) $\frac{4}{x} + 9y = 1$ is not a linear equation because of the variable in the denominator.

Linear equations in two variables are called linear because they all have graphs that are straight lines. Since every straight line is uniquely determined by any two of its points, you can graph a linear equation by graphing two of its solutions and then drawing a straight line through them. Graphing a third point can serve as a check.

EXAMPLE 1: Graph a linear equation.

Problem ▶ Graph $-3x + y = -2$.

1. Solve for y ▶ $\quad -3x + y = -2 \longleftarrow$ original equation
$\qquad\qquad\qquad y = 3x - 2 \longleftarrow$ equation solved for y in terms of x

2. Find three solutions ▶

x	$3x - 2 = y$	
0	$3(0) - 2 = -2$	$(0, -2)$ is a solution.
1	$3(1) - 2 = 1$	$(1, 1)$ is a solution.
2	$3(2) - 2 = 4$	$(2, 4)$ is a solution.

3. Plot solutions ▶

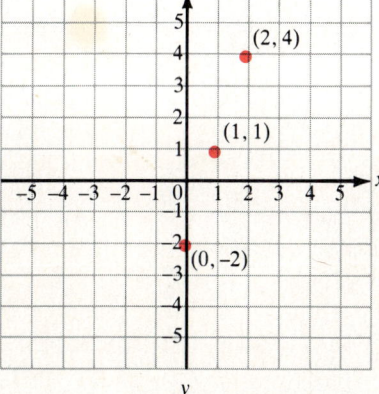

4. Draw graph ▶

   Solution ▶

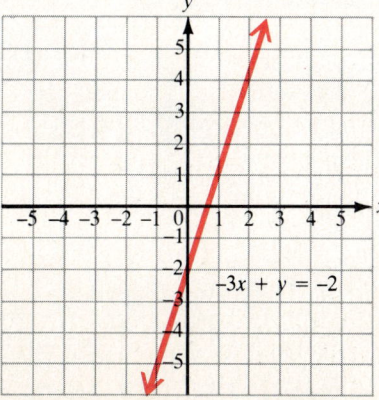

Use a straight edge to draw the line.

The line should pass through all three points.

96 Chapter 3 Equations and Graphs

Make Sure

Graph each linear equation by graphing two of its solutions. Check each graph.

See Example 1 ▶ 1. $2x - 3y = 6$

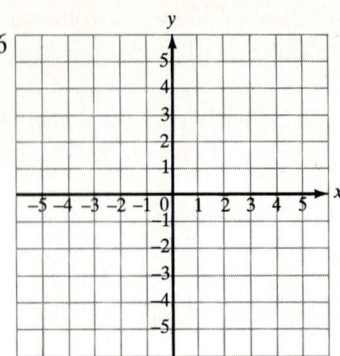

2. $2y = 3x - 4$

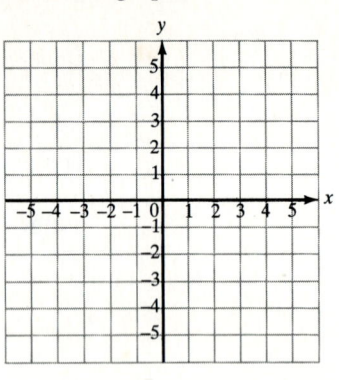

MAKE SURE ANSWERS: See Appendix Selected Answers.

In Chapter 1 you graphed simple equations, such as $x = 3$, as a point on a single number line. It is important to note that you are now graphing equations in two variables and that $x = 3$ is just a short way of writing $1x + 0y = 3$, which is a linear equation whose graph is a straight line that passes through all the points with an x coordinate of 3.

EXAMPLE 2: Graph an equation of the form $x = C$ (or, $y = C$), where C is any real number.

Problem ▶ Graph $x = 3$.

1. Find three solutions ▶

x	y
3	0
3	1
3	4

Think: $x = 3$, regardless of the value of y.

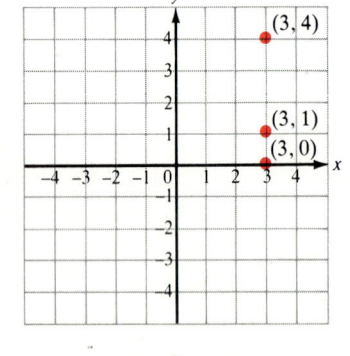

2. Plot the solutions ▶

3. Draw graph ▶

Solution ▶

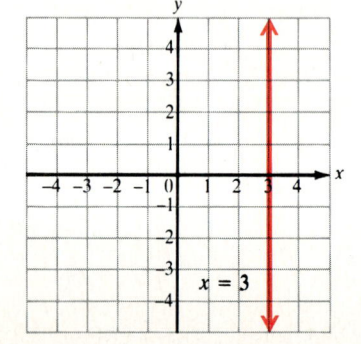

3.2 Graph Linear Equations   97

The graph of every equation of the form $x = C$ is a vertical line which passes through the point $(C, 0)$. The graph of every equation of the form $y = C$ is a horizontal line which passes through the point $(0, C)$.

Example ▶ Graph $y = -1$.

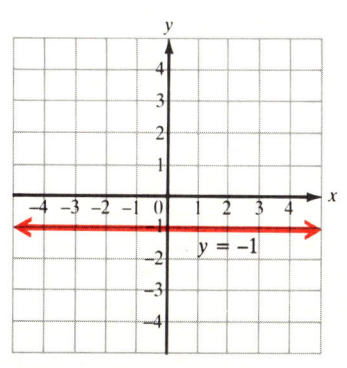

Make Sure

Graph each linear equation of the form $x = C$ or $y = C$.

See Example 2 ▶ **1.** $x = -5$　　　　　　**2.** $y = 4$

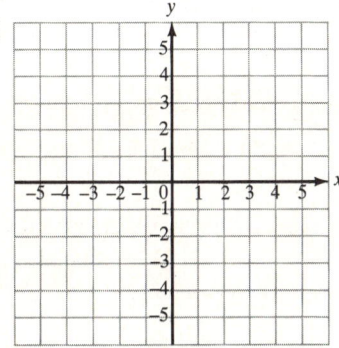

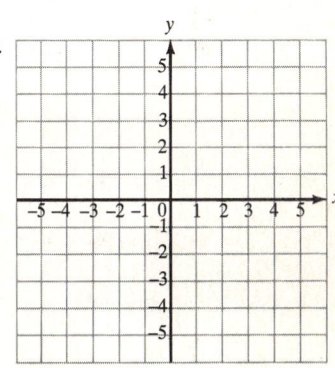

MAKE SURE ANSWERS: See Appendix Selected Answers.

If $A \neq 0$ and $B \neq 0$, then the graph of $Ax + By = C$ will intersect the x-axis at a point called the *x-intercept* and the y-axis at a point called the *y-intercept*. See Figure 3.4.

Figure 3.4 ▶

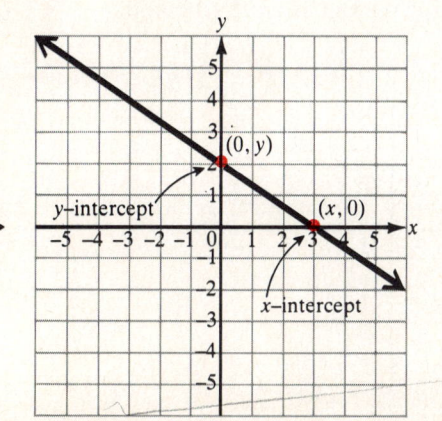

Graphing a line using the x-intercept and the y-intercept is called the *intercept method*.

98 Chapter 3 Equations and Graphs

Note 1 ▶ To find the *x*-intercept, you let $y = 0$ and then solve for *x*.

Note 2 ▶ To find the *y*-intercept, you let $x = 0$ and then solve for *y*.

EXAMPLE 3: Graph a linear equation in two variables using the intercept method.

Problem ▶ Graph $-2x + 3y = 6$ using the intercept method.

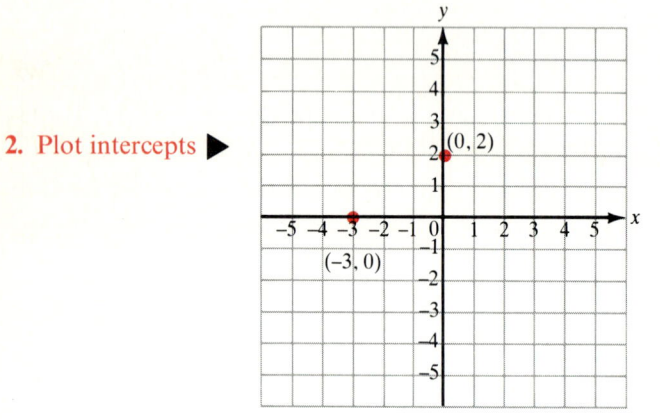

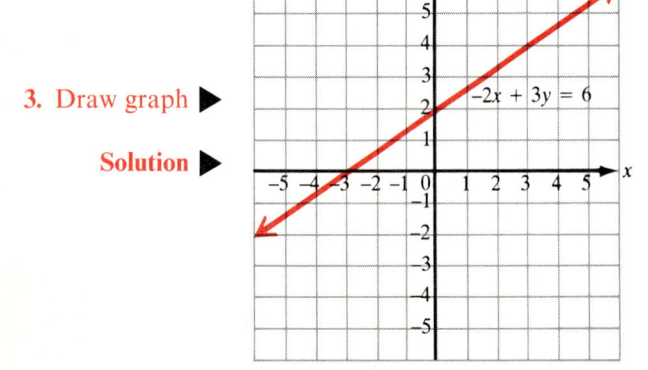

1. Find intercepts ▶

x	y
0	2
−3	0

$-2(0) + 3y = 6$ or $3y = 6$ or $y = 2$ means $(0, 2)$ is the *y*-intercept.

$-2x + 3(0) = 6$ or $-2x = 6$ or $x = -3$ means $(-3, 0)$ is the *x*-intercept.

2. Plot intercepts ▶

3. Draw graph ▶

Solution ▶

Note ▶ To check your solution, just find and plot a third solution to see if it falls on the graph.

Make Sure

Graph each linear equation using the intercept method.

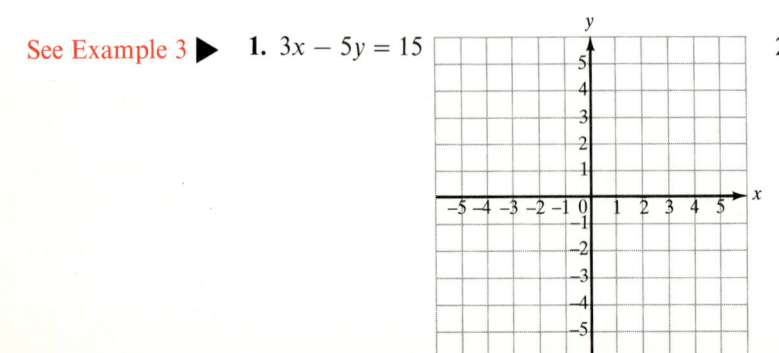

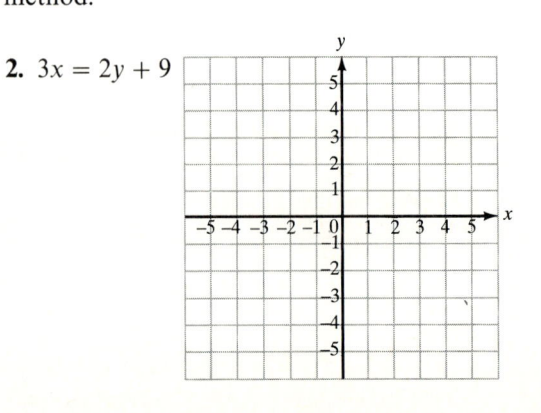

See Example 3 ▶ 1. $3x - 5y = 15$ 2. $3x = 2y + 9$

MAKE SURE ANSWERS: See Appendix Selected Answers.

Name _____ Date _____ Class _____

3.2 Practice: *Check each graph by plotting a third solution.*

Set 1: Graph each linear equation by graphing two of its solutions.

1. $x + 3y = 4$

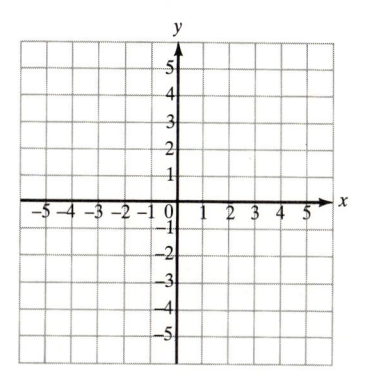

2. $x - 3y = 6$

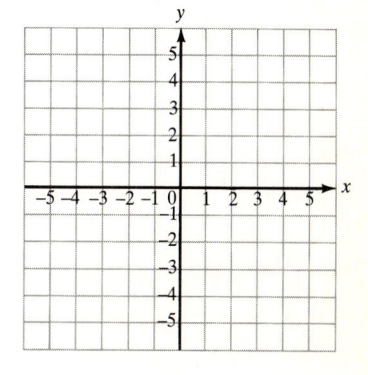

3. $3x - 2y = 12$

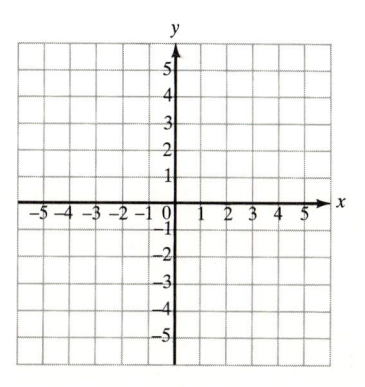

4. $2x + 3y = 6$

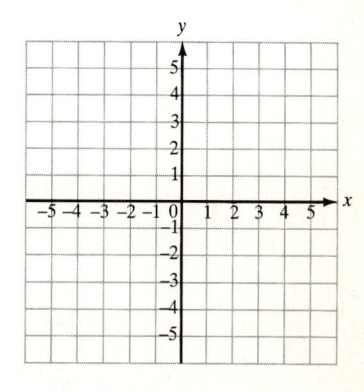

Set 2: Graph each linear equation of the form $x = C$ or $y = C$.

5. $x = 4$

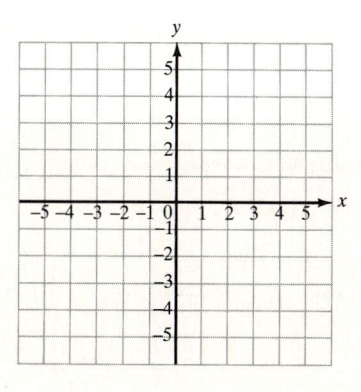

6. $x = -3$

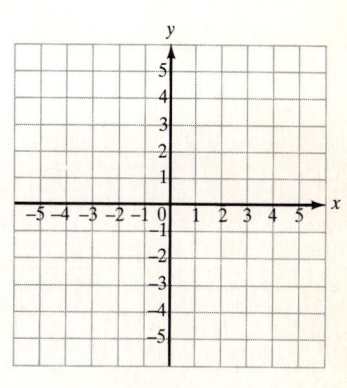

7. $y = -4$

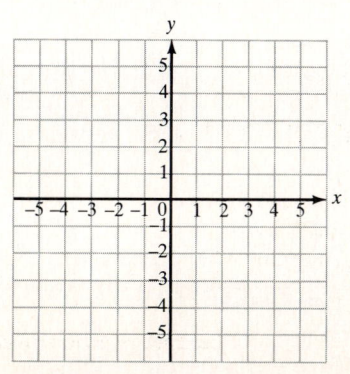

8. $-y = 3$

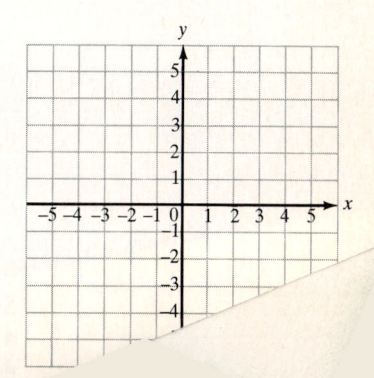

100 Chapter 3 Equations and Graphs

Set 3: Graph each linear equation using the intercept method.

9. $3x + 2y = -6$

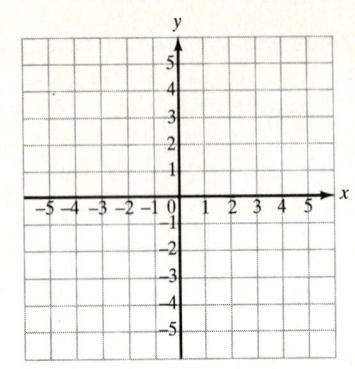

10. $3x + 4y = 12$

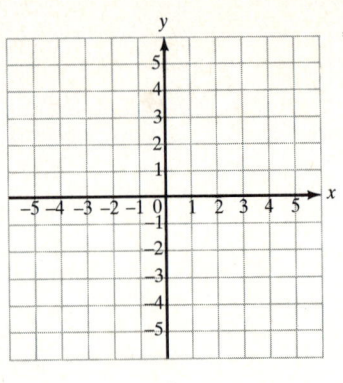

11. $4x - 3y = -24$

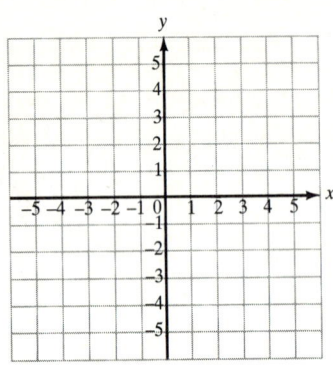

12. $5x - 2y = 10$

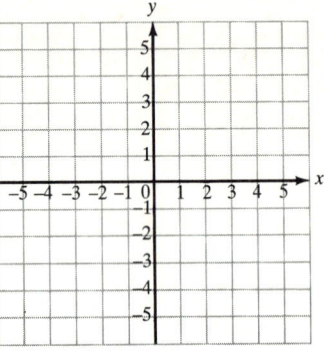

Mixed Practice: Graph each linear equation on graph paper. Attach your work to this page.

13. $4y = 3x - 12$
14. $4x = 5y + 20$
15. $-x = 4$
16. $-x = -2$
17. $2x - 2y = 0$
18. $3x - 4y = 0$
19. $3x = 5y$
20. $4x = 3y$
21. $3x = 5y - 15$
22. $4y = 3x + 12$
23. $\frac{2}{3}x + \frac{3}{4}y = 3$
24. $\frac{4}{5}x - \frac{3}{4}y = 3$

Review: Work these problems on a separate sheet of paper. Attach your work to this page.

Rename each fraction in lowest terms.

25. $\frac{12}{16}$
26. $\frac{-6}{3}$
27. $\frac{-4}{12}$
28. $\frac{6}{-2}$
29. $\frac{8}{-4}$
30. $\frac{-6}{-9}$
31. $\frac{-4}{-6}$
32. $\frac{-12}{-15}$

Evaluate each expression using the Order of Operations Rule. (See Lesson 1.4.)

33. $\frac{5-1}{4-2}$
34. $\frac{3-6}{5-1}$
35. $\frac{6-4}{1-5}$
36. $\frac{-5-3}{4-2}$
37. $\frac{-3-(-4)}{-4-(-3)}$
38. $\frac{-5-(-2)}{-1-(-10)}$
39. $\frac{-3-(-3)}{-4-2}$
40. $\frac{-1-(-3)}{-2-(-2)}$

e each literal equation for y. (See Lesson 2.6.)

- $3 = 2(x - 3)$
42. $y - 4 = -3(x - 4)$
43. $y + 3 = -1(x + 2)$
- $= \frac{2}{3}(x - 3)$
45. $y - 5 = -\frac{3}{5}(x - 1)$
46. $y + 2 = 0(x + 4)$

3.3 Find and Use Slope

Given any two points $P_1(x_1, y_1)$ and $P_2(x_2, y_2)$ on a nonvertical line, you can measure the steepness of the line by computing its *slope*. The symbol for slope is m.

> **Slope Formula**
>
> If a line contains $P_1(x_1, y_1)$ and $P_2(x_2, y_2)$, then its slope is: $m = \dfrac{y_2 - y_1}{x_2 - x_1}$, provided $x_1 \neq x_2$.

EXAMPLE 1: Find the slope of a nonvertical line given the coordinates of two of its points.

Problem ▶ Find the slope of a line that contains $P_1(-4, -1)$ and $P_2(-2, 5)$.

1. Write slope formula ▶ $m = \dfrac{y_2 - y_1}{x_2 - x_1}$

2. Evaluate ▶ $= \dfrac{5 - (-1)}{-2 - (-4)}$ Substitute: $P_1(-4, -1)$ means $x_1 = -4$ and $y_1 = -1$.
$P_2(-2, 5)$ means $x_2 = -2$ and $y_2 = 5$.

$= \dfrac{6}{2}$

$= 3$

Solution ▶ The slope of a line that contains $P_1(-4, -1)$ and $P_2(-2, 5)$ is 3.

The numerator $y_2 - y_1$ of the slope formula is called the *change in y*. It is the measure of vertical change in moving from P_1 to P_2. The denominator $x_2 - x_1$ is called the *change in x*. It is the measure of horizontal change in moving from P_1 to P_2. See Figure 3.5.

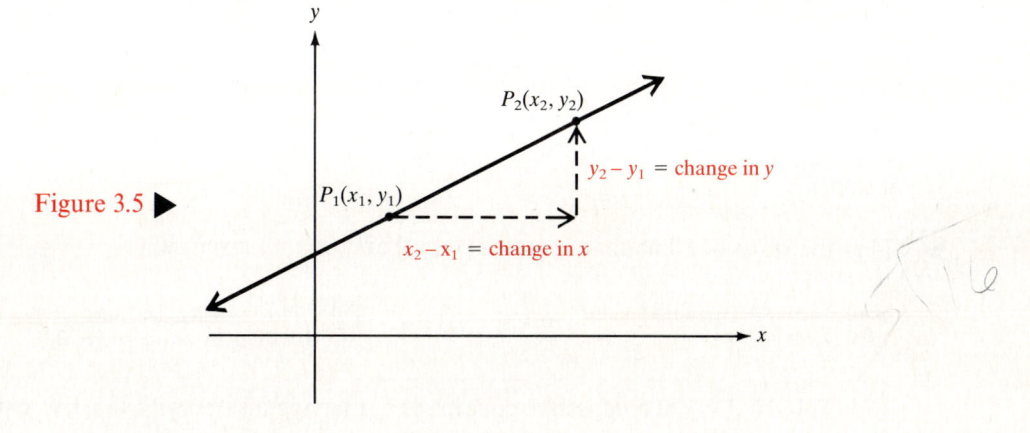

Figure 3.5 ▶

Note ▶ Since $y_2 - y_1$ is called the change in y and $x_2 - x_1$ is called the change in x, the equation $m = \dfrac{y_2 - y_1}{x_2 - x_1}$ is sometimes written as $m = \dfrac{\text{change in } y}{\text{change in } x}$.

The equation $m = \dfrac{y_1 - y_2}{x_1 - x_2}$ can also be used to find slope.

Example ▶ Find the slope of a line that contains $(-3, 1)$ and $(2, -4)$ using both $m = \dfrac{y_2 - y_1}{x_2 - x_1}$ and $m = \dfrac{y_1 - y_2}{x_1 - x_2}$.

Chapter 3 Equations and Graphs

1. Write slope formulas: $m = \dfrac{y_2 - y_1}{x_2 - x_1}$ $\quad\bigg|\quad$ $m = \dfrac{y_1 - y_2}{x_1 - x_2}$

2. Evaluate:
$$= \dfrac{-4 - 1}{2 - (-3)} \quad\bigg|\quad = \dfrac{1 - (-4)}{-3 - 2}$$

$P_1(x_1, y_1) = (-3, 1)$ means $x_1 = -3$ and $y_1 = 1$.
$P_2(x_2, y_2) = (2, -4)$ means $x_2 = 2$ and $y_2 = -4$.

$$= \dfrac{-5}{5} \quad\bigg|\quad = \dfrac{5}{-5}$$

$$= -1 \quad\bigg|\quad = -1$$

Solution ▶ The slope of the line that contains $P_1(-3, 1)$ and $P_2(2, -4)$ is -1.

> **CAUTION:** $m = \dfrac{y_2 - y_1}{x_2 - x_1} = \dfrac{y_1 - y_2}{x_1 - x_2}$ but $m \neq \dfrac{y_2 - y_1}{x_1 - x_2}$ and $m \neq \dfrac{y_1 - y_2}{x_2 - x_1}$.

It is not important which point is labeled P_1 and which point is labeled P_2; however, the order used to subtract x coordinates in the denominator must be the same as the order used to subtract y coordinates in the numerator.

All points on a horizontal line have the same y coordinate. This means that for any two points $P_1(x_1, y_1)$ and $P_2(x_2, y_2)$ on a horizontal line $y_1 = y_2$. Therefore the slope of every horizontal line is 0.

Example ▶ Find the slope of the line containing (x_1, y_1) and (x_2, y_2) given that $y_1 = y_2$ and $x_1 \neq x_2$.

$$m = \dfrac{y_2 - y_1}{x_2 - x_1}$$

$$= \dfrac{0}{x_2 - x_1} \qquad \text{Think: } y_1 = y_2 \text{ means } y_2 - y_1 = 0.$$

$$= 0 \qquad\qquad x_1 \neq x_2 \text{ means } \dfrac{0}{x_2 - x_1} = 0.$$

All points on a vertical line have the same x coordinate. This means that for any two points $P_1(x_1, y_1)$ and $P_2(x_2, y_2)$ on a vertical line $x_1 = x_2$. Therefore any vertical line will not have a slope.

Example ▶ Find the slope of a line that contains (x_1, y_1) and (x_2, y_2) given that $x_1 = x_2$.

$$m = \dfrac{y_2 - y_1}{x_2 - x_1} \text{ provided } x_1 \neq x_2.$$

Think: Since $x_1 = x_2$ the slope formula cannot be used. (Division by zero is undefined)

> **CAUTION:** Do not confuse the concept of not having a slope with the different concept of having a slope of 0.

Recall ▶ (a) A horizontal line has a slope of 0. $\qquad$ (b) A vertical line does not have a slope.

Make Sure

Find the slope of the line that contains each pair of points.

See Example 1 ▶ **1.** $(3, 2)$ and $(1, 5)$ _____ $\qquad$ **2.** $(-3, 2)$ and $(2, -4)$ _____

3.3 Find and Use Slope 103

The following chart illustrates important relationships between the graph of a line and its slope *m*.

Slope	Typical Graph	General Characteristic
Slope *m* = 0.		Graph is a horizontal line.
Slope *m* > 0.		Graph is rising to the right.
Slope *m* < 0.		Graph is falling to the right.
No slope.		Graph is a vertical line.

Figure 3.6 can help you visualize the steepness of lines by comparing various slopes.

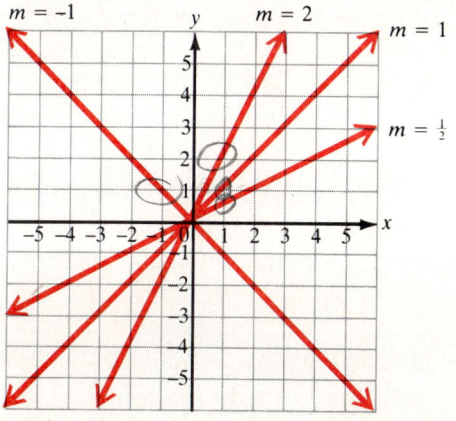

Figure 3.6 ▶

[Graph showing lines through origin with slopes $m = -1$, $m = 2$, $m = 1$, and $m = \frac{1}{2}$]

The graph of any equation of the form $y = mx + b$ will have a *y*-intercept of $(0, b)$. This must be true because any point on the *y*-axis has an *x*-coordinate of 0, and when $x = 0$ $y = mx + b$ simplifies to $y = b$.

Example ▶

[Graph of $y = 3x - 2$ showing line passing through $(0, -2)$]

Observe: The graph of $y = 3x - 2$ intersects the *y*-axis at $(0, -2)$.

Chapter 3 Equations and Graphs

The graph of any equation of the form $y = mx + b$ will have a slope of m. This must be true because a change in x of 1 unit will produce a change in y of m units. Using the concept that slope equals $\dfrac{\text{change in } y}{\text{change in } x}$, you have slope $= \dfrac{m}{1} = m$.

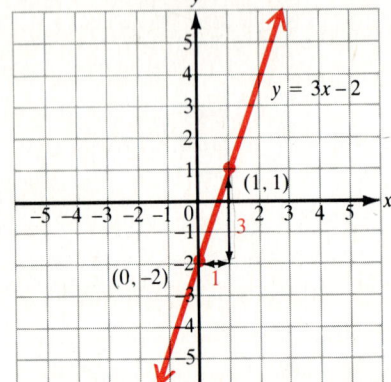

Example ▶

Observe: The graph of $y = 3x - 2$ has a slope of 3.

Let $P_1(x_1, y_1) = (0, -2)$ and $P_2(x_2, y_2) = (1, 1)$, then $m = \dfrac{1 - (-2)}{1 - 0} = 3$.

An equation written in the form $y = mx + b$ is said to be in *slope-intercept form*.

1. A line with a slope of m and a y-intercept of $(0, b)$ will have $y = mx + b$ as an equation.
2. The graph of $y = mx + b$ will be a line with a slope of m and y-intercept $(0, b)$.

EXAMPLE 2: Write the equation of a line in slope-intercept form with the given y-intercept and slope.

Problem ▶ Write the equation of a line in slope-intercept form with y-intercept $(0, 3)$ and slope -2.

1. Write slope-intercept form ▶ $y = mx + b$

2. Substitute ▶ $y = -2x + 3$ Think: A y-intercept of $(0, 3)$ means $b = 3$.
A slope of -2 means $m = -2$

Solution ▶ $y = -2x + 3$ is the slope-intercept form of the equation of the line with y-intercept $(0, 3)$ and a slope of -2.

Make Sure

Write the equation of each line in slope-intercept form given the y-intercept and the slope.

See Example 2 ▶
1. $(0, 2)$, $m = 4$ _____   2. $(0, -3)$, $m = -\frac{3}{4}$ _____

3. $(0, 0)$, $m = 1$ _____   4. $(0, \frac{5}{2})$, $m = -3$ _____

MAKE SURE ANSWERS: 1. $y = 4x + 2$ 2. $y = -\frac{3}{4}x - 3$ 3. $y = 1x + 0$ or $y = x$ 4. $y = -3x + \frac{5}{2}$

3.3 Find and Use Slope

It is often convenient to graph an equation by using its slope-intercept form.

EXAMPLE 3: Graph an equation using the y-intercept and slope.

Problem ▶ Graph $5x - 2y = 8$.

1. Write slope-intercept form ▶

$5x - 2y = 8$ ← original equation

$-2y = -5x + 8$

$y = \dfrac{5}{2}x + (-4)$ ← slope-intercept form

2. Identify m and b ▶ $m = \dfrac{5}{2}$ because $\dfrac{5}{2}$ is the coefficient of x in the above slope-intercept form.
$b = -4$ because -4 is the constant in the above slope-intercept form.

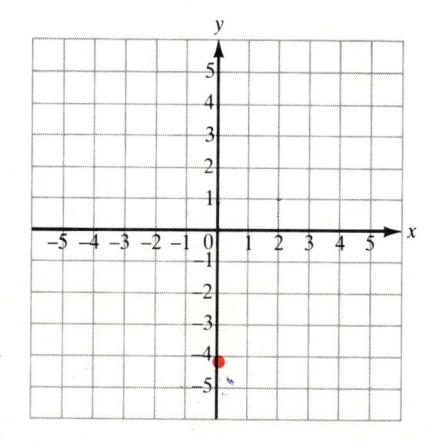

3. Plot (0, b) ▶ Think: Since $b = -4$, the y-intercept is $(0, -4)$.

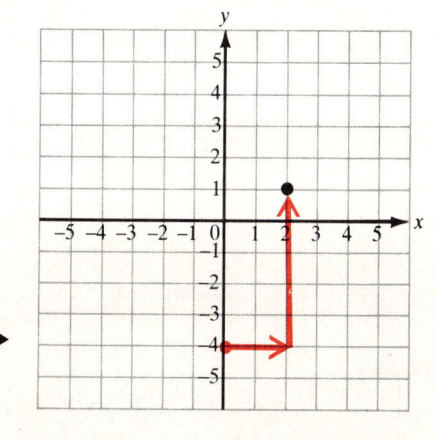

4. Use slope ▶ Think: Start at $(0, -4)$. A slope of $\dfrac{5}{2}$ means that change in x of 2 units followed by a change in y of 5 units will locate another point on the graph.

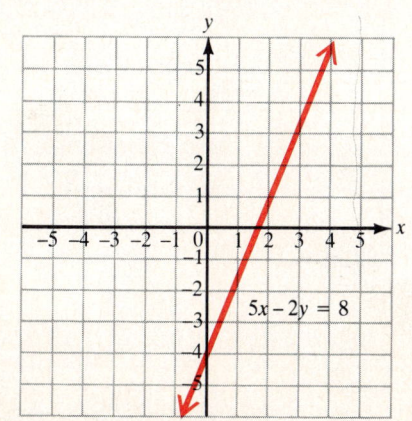

5. Draw line ▶

Solution ▶

106 Chapter 3 Equations and Graphs

Another Example ▶ Graph $y = -\frac{1}{2}x + 1$.

1. Write slope-intercept form ▶ $y = -\frac{1}{2}x + 1$ ⟵ the original equation is in slope-intercept form

2. Identify m and b ▶ $m = -\frac{1}{2}$ and $b = 1$.

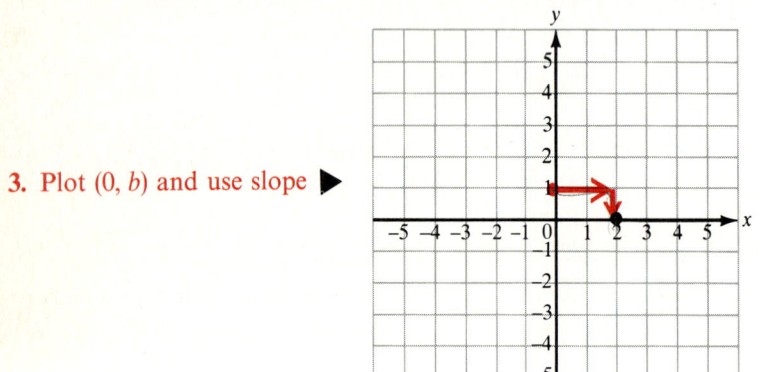

3. Plot (0, b) and use slope ▶

Think: Start at (0, 1). A slope of $-\frac{1}{2}$ means that a change in x of 2 units followed by a change in y of -1 units will locate another point on the graph.

4. Draw line ▶

Solution ▶

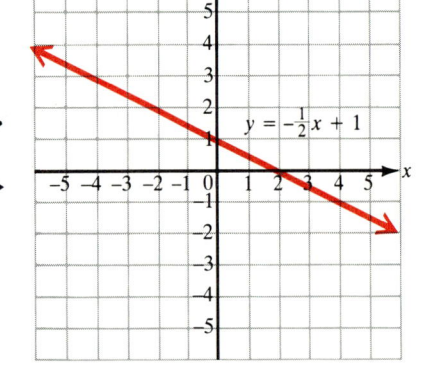

Make Sure

Graph each equation using the *y*-intercept and slope.

See Example 3 ▶
1. $y = \frac{2}{3}x + 4$

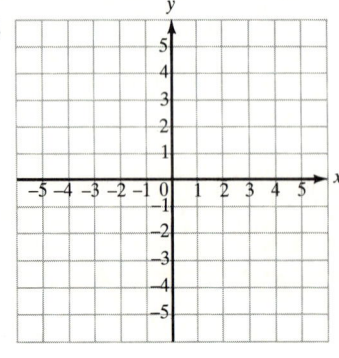

2. $3x + 2y = 6$

MAKE SURE ANSWERS: See Appendix Selected Answers.

3.3 Practice

Set 1: Find the slope of the line that contains each pair of points.

1. (2, 3) and (5, 6)
2. (2, 6) and (4, 7)
3. (3, 5) and (4, 2)

4. (1, 6) and (5, 2)
5. (−3, 2) and (4, 6)
6. (−4, 1) and (2, 5)

7. (−2, 3) and (3, −4)
8. (−1, 5) and (4, −1)
9. (−3, −2) and (−3, −1)

10. (−4, −1) and (−4, −5)
11. (−2, 4) and (3, 4)
12. (−3, 0) and (2, 0)

Set 2: Write the equation of each line in slope-intercept form given the y-intercept and the slope.

13. (0, 2) and $m = 3$
14. (0, 3) and $m = 2$
15. (0, 4) and $m = -4$

16. (0, 1) and $m = -1$
17. (0, −2) and $m = -3$
18. (0, −3) and $m = -2$

19. (0, −2) and $m = 0$
20. (0, 3) and $m = 0$
21. (0, 2) and $m = 1$

22. (0, −1) and $m = 7$
23. (0, 0) and $m = 0$
24. (0, 0) and $m = 4$

25. (0, 4) and $m = \frac{3}{4}$
26. (0, −2) and $m = \frac{2}{3}$
27. (0, 1) and $m = -\frac{5}{3}$

28. (0, −1) and $m = -\frac{7}{4}$
29. (0, $\frac{2}{3}$) and $m = -\frac{4}{7}$
30. (0, $-\frac{2}{3}$) and $m = -\frac{4}{5}$

Chapter 3 Equations and Graphs

Set 3: Graph each equation using y-intercept and slope.

31. $x + y = 7$

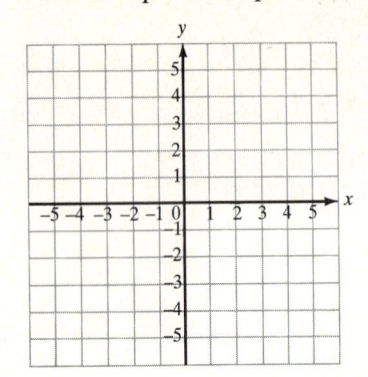

32. $x - y = -5$

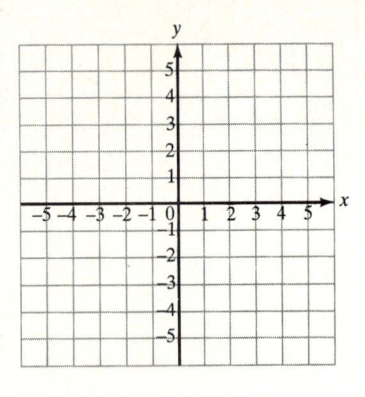

33. $2x + 3y = 9$

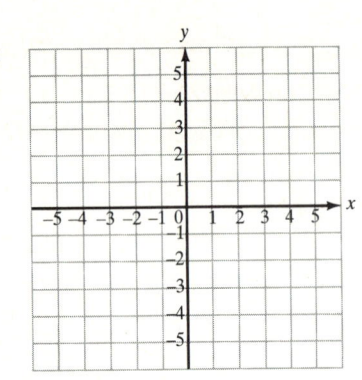

34. $3x - 4y = -15$

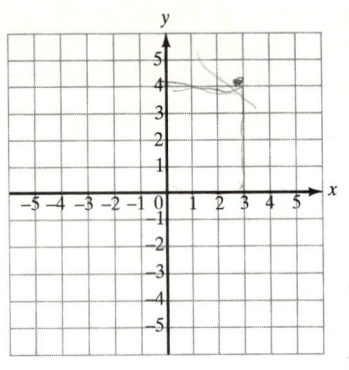

35. $4y - 3x = 2$

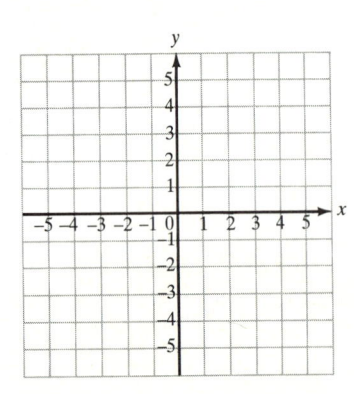

36. $6y - 2x = 3$

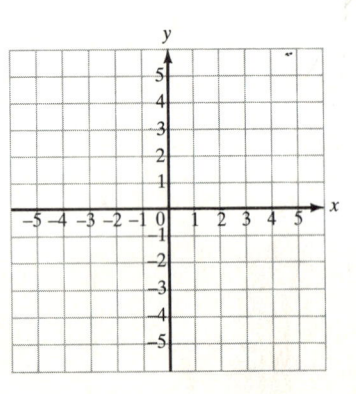

Review: Work these problems on a separate sheet of paper. Attach your work to this page.

Find the multiplicative inverse of each number.
37. 4   **38.** 7   **39.** -3 **40.** -5
41. $\frac{1}{3}$ **42.** $\frac{1}{6}$ **43.** $-\frac{2}{3}$ **44.** $-\frac{5}{7}$

Evaluate each expression using the Order of Operations Rule. (See Lesson 1.4.)

45. $\dfrac{3-4}{4-7}$ **46.** $\dfrac{4-6}{7-3}$ **47.** $\dfrac{7-2}{3-5}$ **48.** $\dfrac{2-(-3)}{3-(-4)}$

49. $\dfrac{-3-5}{-4-2}$ **50.** $\dfrac{-3-(-4)}{-1-(-5)}$ **51.** $\dfrac{-5-(-2)}{-1-(-3)}$ **52.** $\dfrac{-3-(-3)}{3-(-3)}$

Solve each literal equation for the indicated letter. (See Lesson 2.6.)
53. $y - 5 = 2(x - 3)$ for y.   **54.** $y + 5 = -3(x + 2)$ for y.
55. $y + 1 = \frac{2}{3}(x - 3)$ for x.   **56.** $y + 4 = 0(x - 2)$ for y.
57. $y - 3 = 0(x - 0)$ for y.   **58.** $y + 4 = -\frac{5}{3}(x - 4)$ for x.

3.4 Find Linear Equations

The definition of slope can be used to derive a formula for finding an equation of a straight line with a given slope and a given point.

Derive formula ▶ If (x_1, y_1) is any point on a given straight line and (x, y) is any other point on the same line $(x \neq x_1)$, then the slope of the line is given by:

$$\frac{y - y_1}{x - x_1} = m.$$

Since $x \neq x_1$, you can multiply both members of $\frac{y - y_1}{x - x_1} = m$ by $x - x_1$ $(x - x_1 \neq 0)$.

$\frac{y - y_1}{x - x_1}(x - x_1) = m(x - x_1)$ Use the Multiplication Rule.

$\frac{y - y_1}{\cancel{x - x_1}}(\cancel{x - x_1}) = m(x - x_1)$ Divide.

$y - y_1 = m(x - x_1)$ ⟵ derived formula

The derived formula $y - y_1 = m(x - x_1)$ is called the *point-slope formula*.

> **Point-Slope Formula**
>
> If a straight line has slope m and contains a point (x_1, y_1), then an equation of the line is given by the formula: $y - y_1 = m(x - x_1)$.

EXAMPLE 1: Find an equation of a line given the coordinates of a point on the line and the slope of the line.

Problem ▶ Find an equation of the line that contains $P_1(2, 1)$ and has a slope of 3. Write your answer in slope-intercept form.

1. Find an equation ▶ $y - y_1 = m(x - x_1)$ ⟵ point-slope formula

$y - 1 = 3(x - 2)$ Substitute: $P_1(2, 1)$ means $x_1 = 2$ and $y_1 = 1$.

2. Write slope-intercept form ▶ $y - 1 = 3x - 6$

$y = 3x - 5$ ⟵ slope-intercept form

3. Check ▶ Does the line for $y = 3x - 5$ contain the point $(2, 1)$? Yes: $1 = 3 \cdot 2 - 5$.
Does the line for $y = 3x - 5$ have a slope of 3? Yes: $m = 3$.

Solution ▶ $y = 3x - 5$ is the slope-intercept form of the equation of the line that contains the point $(2, 1)$ and has a slope of 3.

Other Examples ▶ (a) Find the equation in slope-intercept form of the straight line that contains $(7, -2)$ and has a slope of $-\frac{5}{4}$.

$y - y_1 = m(x - x_1)$

$y - (-2) = -\frac{5}{4}(x - 7)$

$y + 2 = -\frac{5}{4}x + \frac{35}{4}$

$y = -\frac{5}{4}x + \frac{27}{4}$ Check as before.

(b) Find an equation of the straight line that contains (5, −3) and has a slope of 0.

Method 1: Use Point-Slope Formula.

$y - y_1 = m(x - x_1)$

$y - (-3) = 0(x - 5)$

$y + 3 = 0$

$y = -3$

Method 2: Use $y = $ a constant.

$m = 0$ means the equation is of the form $y = C$.

The line contains (5, −3), therefore $y = -3$.

> **CAUTION:** You cannot use the point-slope formula to find an equation of a vertical line.

Make Sure

Find an equation of each line given the coordinates of one point on the line and its slope. Write each equation in slope-intercept form.

See Example 1 ▶ 1. $(-3, 2)$, $m = \frac{2}{3}$ _____   2. $(1, -2)$, $m = -\frac{4}{3}$ _____

MAKE SURE ANSWERS: 1. $y = \frac{2}{3}x + 4$ 2. $y = -\frac{4}{3}x - \frac{2}{3}$

Given any two different points on a nonvertical straight line, you can find an equation of the line using the definition of slope and the point-slope formula.

EXAMPLE 2: Find an equation of a nonvertical line given two different points on the line.

Problem ▶ Find an equation of the line that contains $P_1(-2, 3)$ and $P_2(6, -1)$. Write your answer in slope-intercept form.

1. Find the slope ▶ $m = \dfrac{y_2 - y_1}{x_2 - x_1}$

$= \dfrac{-1 - 3}{6 - (-2)}$ $P_1(-2, 3)$ means $x_1 = -2$ and $y_1 = 3$.
$P_2(6, -1)$ means $x_2 = 6$ and $y_2 = -1$.

$= \dfrac{-4}{8}$

$= -\dfrac{1}{2}$

2. Solve as before ▶ $y - y_1 = m(x - x_1)$ ⟵ point-slope formula   See Example 1.

$y - 3 = -\dfrac{1}{2}(x - (-2))$

$y - 3 = -\dfrac{1}{2}(x + 2)$

$y - 3 = -\dfrac{1}{2}x - 1$

$y = -\dfrac{1}{2}x + 2$ ⟵ slope-intercept form

3. Check ▶ Does the line for $y = -\frac{1}{2}x + 2$ contain $(-2, 3)$? Yes: $3 = -\frac{1}{2}(-2) + 2$.
Does the line for $y = -\frac{1}{2}x + 2$ contain $(6, -1)$? Yes: $-1 = -\frac{1}{2}(6) + 2$.

Solution ▶ $y = -\frac{1}{2}x + 2$ is the slope-intercept form of the equation of the line that contains $(-2, 3)$ and $(6, -1)$.

Other Examples ▶ (a) Find the equation in slope-intercept form of the straight line that contains $(-3, -6)$ and $(2, 4)$.

$$m = \frac{y_2 - y_1}{x_2 - x_1}$$

$$= \frac{4 - (-6)}{2 - (-3)} \qquad \text{If } (x_1, y_1) = (-3, -6), \text{ then } x_1 = -3 \text{ and } y_1 = -6.$$
$$\text{If } (x_2, y_2) = (2, 4), \text{ then } x_2 = 2 \text{ and } y_2 = 4.$$

$$= 2$$

$$y - y_1 = m(x - x_1) \longleftarrow \text{point-slope formula}$$

$$y - (-6) = 2(x - (-3))$$

$$y + 6 = 2x + 6$$

$$y = 2x \qquad \text{Check as before.}$$

(b) Find an equation of the straight line that contains $(-5, 7)$ and $(3, 7)$.

$$m = \frac{y_2 - y_1}{x_2 - x_1}$$

$$= \frac{7 - 7}{3 - (-5)} \qquad \text{Let } (x_1, y_1) = (-5, 7) \text{ and } (x_2, y_2) = (3, 7).$$

$$= 0$$

$$y - y_1 = m(x - x_1)$$

$$y - 7 = 0(x - (-5))$$

$$y = 7 \qquad \text{Check as before.}$$

Note ▶ In Example (b), you can find the equation $y = 7$ quicker by noting that:
1. $m = 0$, so the line is horizontal with equation $y = $ a constant
   and
2. the y-coordinates of $(-5, 7)$ and $(3, 7)$ are both 7, so the constant is 7: $y = 7$.

Summary ▶ If (x_1, y_1) and (x_2, y_1) are two different points on a line $(x_1 \neq x_2)$, then an equation of the line is: $y = y_1$.

Examples ▶ (a) An equation of the straight line through $(-4, -1)$ and $(18, -1)$ is $y = -1$.
(b) An equation of the straight line through $(3, \frac{1}{2})$ and $(21, \frac{1}{2})$ is $y = \frac{1}{2}$.

The following states a quick way to find an equation of a straight line given two different points with the same x-coordinates.

If (x_1, y_1) and (x_1, y_2) are two different points on a line $(y_1 \neq y_2)$, then an equation of the line is: $y = y_1$.

Examples ▶ (a) An equation of the straight line through $(5, -8)$ and $(5, 17)$ is: $x = 5$.
(b) An equation of the straight line through $(-3, 11)$ and $(-3, 2)$ is: $x = -3$.

Make Sure

Find an equation of each line given the coordinates of two points on the line. Write each equation in slope-intercept form.

See Example 2 ▶ **1.** $(3, -2)$ and $(-2, 4)$ _____   **2.** $(3, -2)$ and $(-2, -2)$ _____

MAKE SURE ANSWERS: 1. $y = -\frac{6}{5}x + \frac{8}{5}$ 2. $y = 0x - 2$ or $y = -2$

Two different lines in a plane either intersect at a single point or they do not intersect at all. Two lines in a plane that do not intersect are said to be *parallel*. It is not necessary to graph two linear equations to determine whether the lines are parallel. Two different (nonvertical) parallel lines have equal slopes. Two different intersecting lines (nonvertical) have different slopes.

Example ▶ Determine whether the graphs of $2x + 3y = 6$ and $x + 2y = 8$ are parallel or intersecting lines.

1. Find slopes ▶

$2x + 3y = 6$ $x + 2y = 8$ ⟵ equations of the lines

$3y = -2x + 6$ $2y = -x + 8$

$y = -\frac{2}{3}x + \frac{6}{3}$ $y = -\frac{1}{2}x + \frac{8}{2}$

$y = \boxed{-\frac{2}{3}}x + 2$ $y = \boxed{-\frac{1}{2}}x + 4$ ⟵ slope-intercept forms

  m_1 m_2

The slopes of the lines are $m_1 = -\frac{2}{3}$ and $m_2 = -\frac{1}{2}$.

2. Compare slopes ▶ $m_1 \neq m_2$

Solution ▶ The graphs of $2x + 3y = 6$ and $x + 2y = 8$ are intersecting lines.

Another Example ▶ Determine whether the graphs of $-3x + y = 4$ and $-6x + 2y = 10$ are parallel or intersecting lines.

1. Find slopes ▶

$-3x + y = 4$ $-6x + 2y = 10$ ⟵ original equations

         $2y = 6x + 10$

$y = \boxed{3}x + \boxed{4}$ $y = \boxed{3}x + \boxed{5}$ ⟵ slope-intercept forms

   m_1 b_1 m_2 b_2

The slopes are $m_1 = 3$ and $m_2 = 3$.

2. Compare slopes ▶ $m_1 = m_2$

Solution ▶ The graphs of $-3x + y = 4$ and $-6x + 2y = 10$ are parallel lines.

If two linear equations have equal slopes and equal y-intercepts, then each equation graphs to be the same line. The lines are said to *coincide*.

Note ▶ The lines in the previous example had equal slopes but different y-intercepts. The lines do not coincide. They are *distinct parallel lines*.

You can now find equations of lines parallel to a given line.

EXAMPLE 3: Find an equation of the line that contains a given point and is parallel to a given line.

Problem ▶ Find an equation of the line that contains $(-4, 2)$ and is parallel to $3x - 4y = 12$. Write your answer in slope-intercept form.

1. Find slope ▶ $3x - 4y = 12$ ⟵ equation of the given line

$-4y = -3x + 12$

$y = \boxed{\frac{3}{4}}x - 3$ ⟵ slope-intercept form of given line
$\quad\quad m$

The slope of the given line is $m = \frac{3}{4}$.
All lines parallel to it will have slope $m = \frac{3}{4}$.

2. Solve as before ▶ $y - y_1 = m(x - x_1)$ ⟵ point-slope formula $\quad\quad$ See Example 1.

$y - 2 = \frac{3}{4}(x - (-4))$ $\quad\quad$ Substitute.

$y - 2 = \frac{3}{4}x + 3$

$y = \frac{3}{4}x + 5$ ⟵ slope-intercept form

3. Check ▶ Does the line for $y = \frac{3}{4}x + 5$ have a slope of $\frac{3}{4}$? Yes: $m = \frac{3}{4}$.

Does the line for $y = \frac{3}{4}x + 5$ contain the point $(-4, 2)$? Yes: $2 = \frac{3}{4}(-4) + 5$.

Solution ▶ $y = \frac{3}{4}x + 5$ is the slope-intercept form of the equation of the line that contains $(-4, 2)$ and is parallel to $3x - 4y = 12$.

Make Sure

Find an equation of the line that passes through a given point and is parallel to a given line. Write each equation in slope-intercept form.

See Example 3 ▶ 1. $(4, -2)$, $6x - 5y = 8$ _____ 2. $(0, -2)$, $y = 4$ _____

MAKE SURE ANSWERS: 1. $y = \frac{6}{5}x - \frac{34}{5}$ 2. $y = 0x - 2$ or $y = -2$

Two lines are *perpendicular* if they intersect at right angles. Perpendicular lines (which are not vertical or horizontal) have slopes that are negative reciprocals of each other. Using algebraic notation this means if two perpendicular lines have slopes m_1 and m_2, then:

$$m_1 = -\frac{1}{m_2}, \quad m_2 = -\frac{1}{m_1}, \quad \text{and} \quad m_1 \cdot m_2 = -1.$$

Note ▶ Horizontal lines have a slope of zero, and zero does not have a reciprocal. Therefore the relationship $m_1 = -\frac{1}{m_2}$ does not hold for horizontal and vertical lines even though they are perpendicular.

EXAMPLE 4: Find an equation of the line that contains a given point and is perpendicular to a given line.

Problem ▶ Find an equation of the line that contains (3, 1) and is perpendicular to $2x - 5y = 8$. Write your answer in slope-intercept form.

1. Find m_1 ▶ $2x - 5y = 8$ ⟵ equation of the given line

$$-5y = -2x + 8$$

$$y = \underbrace{\frac{2}{5}}_{m_1} x - \frac{8}{5} \quad \longleftarrow \text{slope-intercept form}$$

The slope of the given line is $m_1 = \frac{2}{5}$.

2. Find m_2 ▶ $m_2 = -\dfrac{1}{m_1}$

$$= -\frac{1}{\frac{2}{5}}$$

$$= -\frac{5}{2} \qquad \text{Every line perpendicular to } 2x - 5y = 8 \text{ has a slope of } -\tfrac{5}{2}.$$

3. Solve as before ▶ $y - y_1 = m_2(x - x_1)$ ⟵ point-slope formula

$$y - 1 = -\frac{5}{2}(x - 3) \qquad \text{Substitute.}$$

$$y = -\frac{5}{2}x + \frac{17}{2} \qquad \text{Check as before.}$$

Solution ▶ $y = -\dfrac{5}{2}x + \dfrac{17}{2}$ is the slope-intercept form of the equation of the line that contains (3, 1) and is perpendicular to $2x - 5y = 8$.

Make Sure

Find an equation of the line that contains a given point and is perpendicular to a given line. Write each equation in slope-intercept form (if possible).

See Example 4 ▶ 1. $(-3, 2)$, $4x - 3y = 6$ _____   2. $(4, 1)$, $y = -2$ _____

MAKE SURE ANSWERS: 1. $y = -\frac{3}{4}x - \frac{1}{4}$ 2. $x = 4$

Name _____ Date _____ Class _____

3.4 Practice: *Write each equation in slope-intercept form.*

Set 1: Find an equation of each line given the coordinates of one point on the line and its slope.

1. (2, 3) and $m = 2$
2. (4, 3) and $m = 3$
3. (−3, 2) and $m = -1$

4. (−2, −2) and $m = -2$
5. (−1, 0) and $m = \frac{4}{3}$
6. (−3, 0) and $m = -\frac{3}{4}$

Set 2: Find an equation of each line given the coordinates of two points on the line.

7. (3, 2) and (6, 6)
8. (4, 3) and (6, 5)
9. (1, 2) and (4, 0)

10. (2, 5) and (6, 0)
11. (−3, 2) and (3, −1)
12. (−4, 0) and (2, −3)

13. (2, 0) and (0, −4)
14. (−3, 0) and (0, 1)
15. (0, 3) and (−5, 3)

16. (1, −2) and (3, −2)
17. (0, 3) and (0, −2)
18. (−2, 0) and (−2, −3)

1. _____
2. _____
3. _____
4. _____
5. _____
6. _____
7. _____
8. _____
9. _____
10. _____
11. _____
12. _____
13. _____
14. _____
15. _____
16. _____
17. _____
18. _____

116 Chapter 3 Equations and Graphs

19. _____

20. _____

21. _____

22. _____

23. _____

24. _____

25. _____

26. _____

27. _____

28. _____

29. _____

30. _____

31. _____

32. _____

33. _____

34. _____

35. _____

36. _____

Set 3: Find the equation of the line that contains a given point and is parallel to a given line.

19. $(4, 0)$ and $4x - y = 6$
20. $(3, -2)$ and $x - 4y = 6$
21. $(0, 0)$ and $4x + 3y = 12$

22. $(0, -2)$ and $2x + 5y = 9$
23. $(2, -5)$ and $y = 3$
24. $(3, -1)$ and $x = -4$

Set 4: Find the equation of the line that passes through a given point and is perpendicular to a given line.

25. $(0, 4)$ and $2x - y = 4$
26. $(3, 0)$ and $6x - y = 6$
27. $(2, -3)$ and $3x + 2y = 4$

28. $(-3, 4)$ and $2x - 5y = 1$
29. $(-3, -5)$ and $y = 3$
30. $(6, -3)$ and $x = -4$

Extra: Write the equation of the line through:

31. $(4, 0)$ and parallel to the line containing $(1, 0)$ and $(3, -1)$.
32. $(-1, 2)$ and parallel to the line containing $(-2, 1)$ and $(0, 0)$.
33. $(2, 3)$ and perpendicular to the line containing $(2, 0)$ and $(0, 3)$.

34. $(-1, -3)$ and perpendicular to the line containing $(-1, 0)$ and $(0, 3)$.
35. $(2, 3)$ and parallel to the line containing $(0, 3)$ and $(0, -1)$.
36. $(-1, 0)$ and perpendicular to the line containing $(-1, 3)$ and $(2, 3)$.

Review: Work these problems on a separate sheet of paper. Attach your work to this page.

Is the given ordered pair a solution of the given equation? (See Lesson 3.1.)

37. $(0, 0)$, $x + y = 3$
38. $(0, 0)$, $x + y = 0$
39. $(0, 0)$, $x + 3y = 6$
40. $(2, 0)$, $2x + 3y = 4$
41. $(-3, 0)$, $2x - 3y = -6$
42. $(0, -2)$, $3x - 2y = 4$
43. $(2, 3)$, $2x + 3y = 0$
44. $(-2, 3)$, $4x - 3y = 0$
45. $(-1, -3)$, $2x + 3y = 7$

Graph each equation. (See Lesson 3.2.)

46. $x + y = 0$
47. $x - y = 0$
48. $2x - 3y = 6$
49. $4x - 3y = 12$
50. $x - 3y = -3$
51. $3x + y = -3$
52. $y = \frac{2}{3}x$
53. $y = 2$
54. $x = -3$

Copyright © 1985 by Harcourt Brace Jovanovich, Inc. All rights reserved.

Problem Solving 3: Use Linear Relationships

> A relationship between two variables that can be described by a linear equation is called a *linear relationship*.

EXAMPLE 1: Solve a problem using a linear relationship given the rate and a data point.

1. Read and identify ▶ A company bus was purchased for $20,000. Assuming the bus depreciates at a constant rate of $4000 per year (*straight line depreciation* for tax purposes), what is the book value (v) after a time (t) of $3\frac{1}{2}$ years?

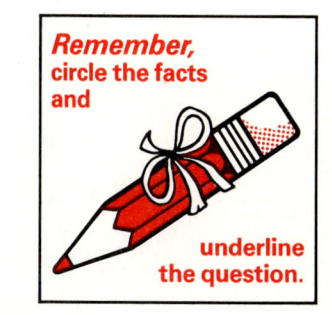

Remember, circle the facts and underline the question.

2. Understand ▶ The question asks you to find and evaluate a linear equation in two variables $v = mt + b$ for $t = 3\frac{1}{2}$ given one data point (t, v) of $(0, 20{,}000)$ and the slope (rate) as $m = -\frac{4000}{1}$ (decreases $4000 in 1 year).

3. Decide ▶ To find the equation of a straight line given a point on the line and the slope of the line, you **use the point-slope formula.**

4. Find equation ▶
$v - v_1 = m(t - t_1)$ ⟵ point-slope formula with $v = y$ and $t = x$
$v - 20{,}000 = -\frac{4000}{1}(t - 0)$ Think: $(t_1, v_1) = (0, 20{,}000)$
$v - 20{,}000 = -4000t$
$v = 20{,}000 - 4000t$ ⟵ proposed equation

5. Evaluate equation ▶
$v = 20{,}000 - 4000(3\frac{1}{2})$ Substitute the given $3\frac{1}{2}$ years(t).
$= 20{,}000 - 14{,}000$
$= 6000$

6. Interpret ▶ $v = 6000$ means that in $3\frac{1}{2}$ years the bus will have a book value of $6000.

Note 1 ▶ In $v = -4000t + 20{,}000$, the negative slope means that the value of the bus is decreasing. That is, the line for $v = -4000t + 20{,}000$ slopes downward. The slope -4000 as a ratio $\frac{-4000}{1}$ describes the rate of decrease. That is, the bus decreases $4000 each 1 year.

Note 2 ▶ The graph of the line for $v = -4000t + 20{,}000$ begins at $t = 0$ and ends at $t = 5$.

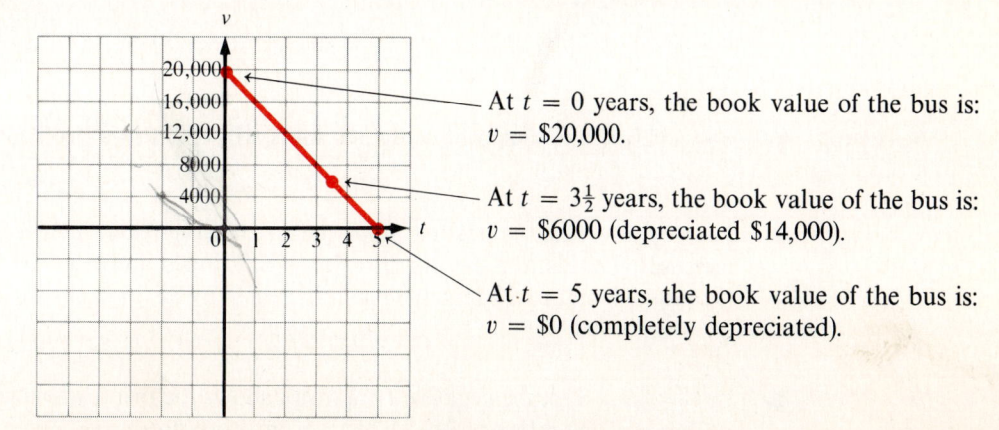

At $t = 0$ years, the book value of the bus is: $v = \$20{,}000$.

At $t = 3\frac{1}{2}$ years, the book value of the bus is: $v = \$6000$ (depreciated $14,000).

At $t = 5$ years, the book value of the bus is: $v = \$0$ (completely depreciated).

EXAMPLE 2: Solve a problem with a linear relationship given two data points.

1. Read and identify ▶ A certain company has a profit of $1000 when 450 units are produced and a profit of $500 when 250 units are produced. Assuming the profit (p) per unit (u) rate is a linear relationship, what is the profit when 500 units are produced?

2. Understand ▶ The question asks you to find and evaluate a linear equation in two variables $p = mu + b$ for $u = 500$ given two data points (u, p) of (450, 1000) and (250, 500).

3. Decide ▶ To find the equation of a straight line given two points on the line, you **use the slope and point-slope formulas.**

4. Find equation ▶ $m = \dfrac{p_1 - p_2}{u_1 - u_2}$ ⟵ slope formula with $x = u$ and $y = p$

$= \dfrac{1000 - 500}{450 - 250}$ 　　　Think: $(u_1, p_1) = (450, 1000)$
　　　　　　　　　　　　　　　　　$(u_2, p_2) = (250, 500)$

$= \dfrac{500}{200}$

$= \tfrac{5}{2}$ ⟵ slope of equation 　　Think: $p = \tfrac{5}{2}u + b$.

$p - p_1 = m(u - u_1)$ ⟵ point-slope formula with $x = u$ and $y = p$

$p - 1000 = \tfrac{5}{2}(u - 450)$ 　　Substitute: $p_1 = 1000$, $m = \tfrac{5}{2}$, and $u_1 = 450$

$p - 1000 = \tfrac{5}{2}u - \tfrac{5}{2}(450)$

$p - 1000 = \tfrac{5}{2}u - 1125$

$p = \tfrac{5}{2}u - 125$ ⟵ proposed equation

5. Evaluate equation ▶ $p = \tfrac{5}{2}(500) - 125$ 　　Substitute the given 500 units (u).

$= 1250 - 125$

$= 1125$

6. Interpret ▶ $p = 1125$ means that producing 500 units will result in a profit of $1125.

Note 1 ▶ In $p = \tfrac{5}{2}u - 125$, the positive slope means that the profit increases when the number of units increases. That is, the line for $p = \tfrac{5}{2}u - 125$ slopes upward. The slope $\tfrac{5}{2}$ as a ratio describes the rate of increase. That is, the profit increases by $5 for each increase in production of 2 units.

Note 2 ▶ In $p = \tfrac{5}{2}u - 125$, the negative y-intercept shows the profit when zero units are produced. That is, there will be a loss of $125 when no units are produced.

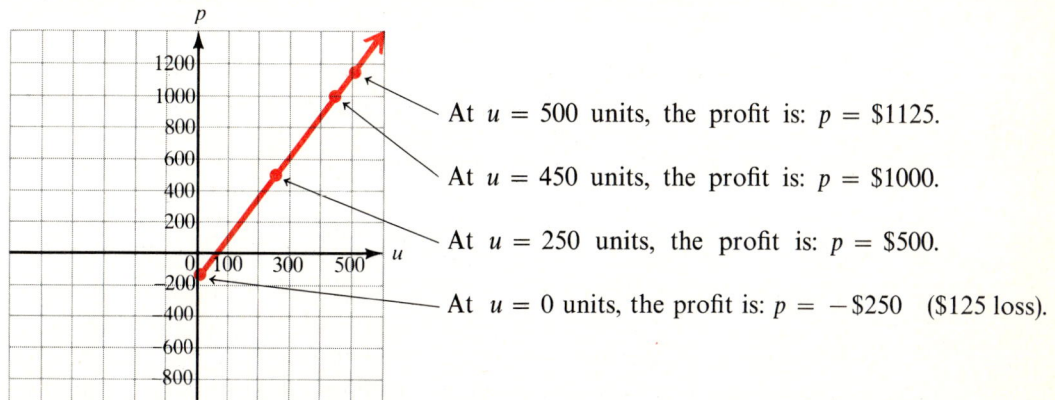

PS 3: Use Linear Relationships 119

Note 3 ▶ The graph of the line for $p = \frac{5}{2}u - 125$ begins at $u = 0$ and continues on for as many units as you wish to consider.

At $u = 500$ units, the profit is: $p = \$1125$.

At $u = 450$ units, the profit is: $p = \$1000$.

At $u = 250$ units, the profit is: $p = \$500$.

At $u = 0$ units, the profit is: $p = -\$250$ (\$125 loss).

Note 4 ▶ The u-intercept is called the *break-even point*. That is, the point at which the profit (or loss) is $0 is the break-even point.

Example ▶ Find the break-even point for the problem in Example 2.

$p = \frac{5}{2}u - 125$ Write the linear equation relating p and u.

$0 = \frac{5}{2}u - 125$ Let $p = 0$ (dollars)

$125 = \frac{5}{2}u$ Solve for u.

$50 = u$ Think: $125 \cdot \frac{2}{5} = 50$ (units)

To break even in Example 2, you need to produce 50 units.

Practice: Solve each problem using a linear relationship.

1. A fast-food restaurant finds that daily profit (p) increases by $50 for each increase in sales of 200 hamburgers (h). When 500 hamburgers are sold, the profit is $100. a) Find the linear equation describing this linear relationship. b) What is total profit if 1100 hamburgers are sold in a day? c) How many hamburgers need to be sold to break even?

2. A certain store had a profit (p) of $462 on Monday when 210 customers (c) entered the store. On Tuesday, the number of customers entering the store was 200 more than on Monday, for a profit of $912. Assuming the profit (p) per customer (c) rate is a linear relationship, a) find the linear equation that describes this linear relationship. b) What would the profit be if 404 customers entered the store on Wednesday? c) What is the minimum number of customers needed to enter the store to break even?

3. There is a linear relationship between the temperature (t) in degrees Fahrenheit and the number (n) of times a cricket chirps per minute. A cricket will chirp 72 times per minute at 58°F and 132 times at 73°F. a) Find the linear equation that represents this linear relationship. b) What would the temperature be when a cricket chirps 100 times per minute? c) How many times per minute does a cricket chirp at 85°F?

4. There is a linear relationship between the temperature (t) in degrees Celsius and the number (n) of calories needed by a person working. A person working needs 30 more calories per day for each degree that the Celsius temperature drops. An average person working needs 3000 calories per day at 20°C. a) Find the linear equation that represents this relationship. b) How many calories per day would a person working need at 40°C? c) At what temperature would a person working need 2000 calories per day?

1a. _____
b. _____
c. _____

2a. _____
b. _____
c. _____

3a. _____
b. _____
c. _____

4a. _____
b. _____
c. _____

Copyright © 1985 by Harcourt Brace Jovanovich, Inc. All rights reserved.

120   Chapter 3   Equations and Graphs

5a. _____

b. _____

c. _____

6a. _____

b. _____

c. _____

7a. _____

b. _____

c. _____

8a. _____

b. _____

c. _____

9a. _____

b. _____

c. _____

10a. _____

b. _____

c. _____

11a. _____

b. _____

c. _____

12a. _____

b. _____

c. _____

5. There is a linear relationship between a person's shoe size number (n) and the length (l) of the person's foot in inches when standing normally. The straight line that represents this linear relationship for a man is parallel to the one for a woman, which is $n = 3l - 22$. A man who is standing normally on a foot that measures 12 inches has a shoe-size number 11. a) Find the linear equation that represents this linear relationship. b) What is the shoe-size number for a man who is standing normally on a foot that measures $10\frac{1}{2}$ inches? c) How long is a man's foot, when standing normally, that requires a size 9 shoe?

6. There is a linear relationship between an average person's ideal weight (w) in pounds and the person's height (h) in inches. The slope of the straight line that represents this linear relationship for the average woman is $\frac{21}{22}$ that of the slope for the average man, which is represented by $w = 5\frac{1}{2}h - 231$. The ideal weight for an average woman whose height is 5 feet 4 inches is 120 pounds. a) Find the linear equation that represents the linear relationship. b) What is the ideal weight for an average woman whose height is 5 feet 2 inches? c) How tall is an average woman whose ideal weight is 100 pounds, to the nearest inch?

7. The first session of the 1st U.S. Congress was in 1789. From then on, a newly elected Congress has had its first session every two years. a) Find the linear equation describing the relationship between the number (n) of the U.S. Congress and the year (y) for the first session of that Congress. b) Find the number of the U.S. Congress that held its first session in 1985. c) In what year was the first session of the 50th U.S. Congress?

8. There is a linear relationship between the temperature (t) at which water boils and altitude (a). At sea level, water boils at 212°F (degrees Fahrenheit). At 2000 feet above sea level water boils at 208°F. a) Find the linear equation that represents this linear relatonship. b) At what Fahrenheit temperature does water boil in mile-high Denver, Colorado, to the nearest whole degree? c) At what altitude does water boil at 200°F?

9. The U.S. population (p) in the year (y) 1960 was 180.7 million. In 1980, the U.S. population was 226.6 million. a) Find a linear equation that approximates this nonlinear relationship. b) Estimate the U.S. population for the year 2000. c) Estimate the year that the U.S. population will reach 300 million.

10. The world population (p) in the year (y) (1980) was about $4\frac{1}{2}$ billion. The average annual percent increase in world population is about 1.6%. a) Find a linear equation that approximates this nonlinear relationship. b) Estimate the world population for the year 2000. c) Estimate the year that the world population reached 2 billion.

11. At age (a) 18, the energy needs of the average man (5 ft 10 in. tall, 154 lbs) are 2800 calories (c) per day. At age 65, his energy needs have decreased to 2200 calories per day. a) Find a linear equation that approximates this nonlinear relationship. b) Estimate the energy needs for an average man at age 45. c) At about what age are the energy needs for the average man about 2000 calories?

12. At age (a) 18, the energy needs for the average woman (5 ft 4 in. tall, 120 lbs) are 2100 calories (c) per day. At age 65, her energy needs have decreased to 1500 calories per day. a) Find a linear equation that approximates this nonlinear relationship. b) Estimate the energy needs for an average woman at age 35. c) At about what age are the energy needs of the average woman about 1800 calories?

Copyright © 1985 by Harcourt Brace Jovanovich, Inc. All rights reserved.

3.5 Graph Linear Inequalities

Every linear equation $Ax + By = C$ graphs to be a line that separates the coordinate plane into two regions called *half-planes*. It is convenient to designate the half-planes as H_1 and H_2 as shown in Figure 3.7.

Figure 3.7 ▶

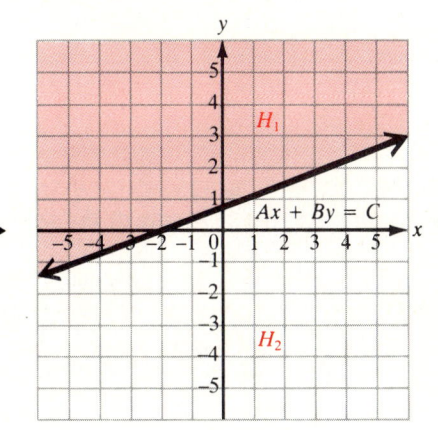

The graph of $Ax + By < C$ is one of the half-planes and the graph of $Ax + By > C$ is the other half-plane. If a point in the half-plane H_1 is a solution of $Ax + By < C$, then every point in H_1 is a solution of $Ax + By < C$. Consequently, you can determine which half-plane, H_1 or H_2, is the solution set of $Ax + By < C$ by testing any point in one of the half-planes. The origin $(0, 0)$ often serves as a convenient test point.

Recall ▶ The word *or* means to form the union, and the symbol $\leq$ means *is less than or is equal to*. Therefore the graph of $Ax + By \leq C$ is formed by the union of the half-plane $Ax + By < C$ and the line $Ax + By = C$.

EXAMPLE 1: Graph an inequality of the form $Ax + By \leq C$ (or, $Ax + By \geq C$).

Problem ▶ Graph $3x + 2y \leq -6$.

1. Graph the line ▶

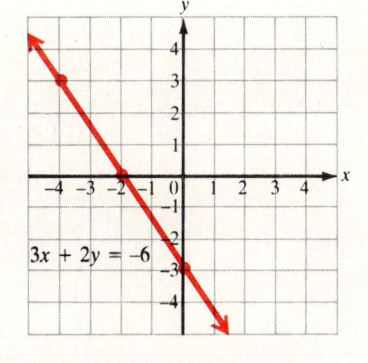

x	y	
0	−3	← y-intercept
−2	0	← x-intercept
−4	3	← check point

2. Test a point ▶

$$\begin{array}{c|c} 3x + 2y & \leq -6 \\ \hline 3(0) + 2(0) & -6 \\ \hline 0 & -6 \\ \hline 0 & -6 \end{array}$$

The origin $(0, 0)$ is not on $3x + 2y = -6$, so it can serve as a test point.

3. Interpret ▶ $0 \not\leq -6$ means $(0, 0)$ is not a solution.

Chapter 3 Equations and Graphs

4. Graph half-plane ▶ Shade the half-plane that does not contain the origin.

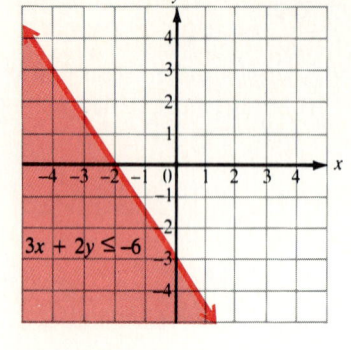

Solution ▶

$3x + 2y \leq -6$

Note ▶ The boundary is a solid line to indicate that it is part of the solution.

Other Examples ▶ (a) Graph $-2x + y \geq 4$. (b) Graph $x - y \leq 0$.

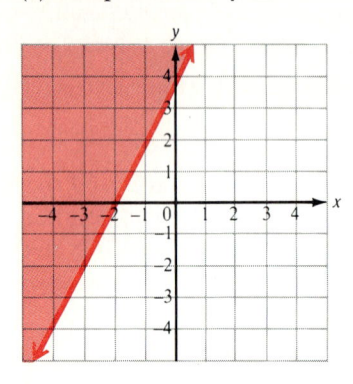

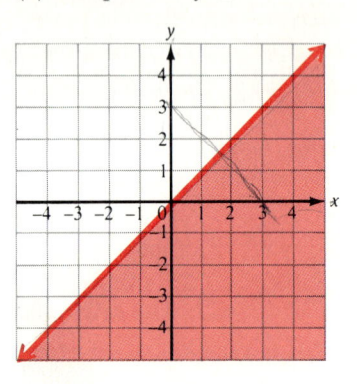

Make Sure

Graph inequalities of the form $Ax + By \leq C$ or $Ax + By \geq C$.

See Example 1 ▶ **1.** $2x + y \leq 4$ **2.** $x - 3y \geq -3$

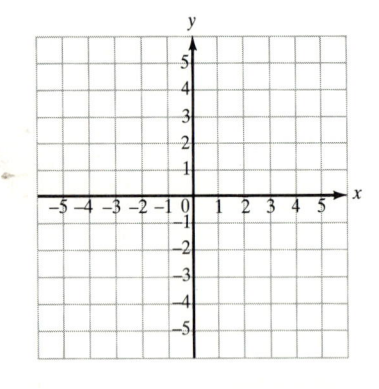

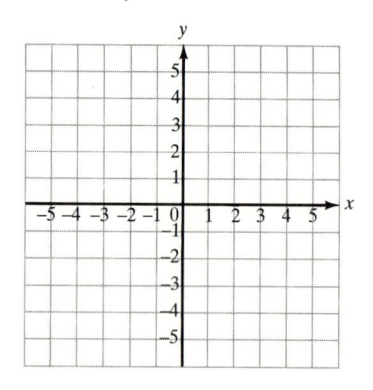

MAKE SURE ANSWERS: See Appendix Selected Answers.

3.5 Graph Linear Inequalities

EXAMPLE 2: Graph an inequality of the form $Ax + By < C$ ($Ax + By > C$).

Problem ▶ Graph $x + 2y < 4$.

1. Graph the line ▶

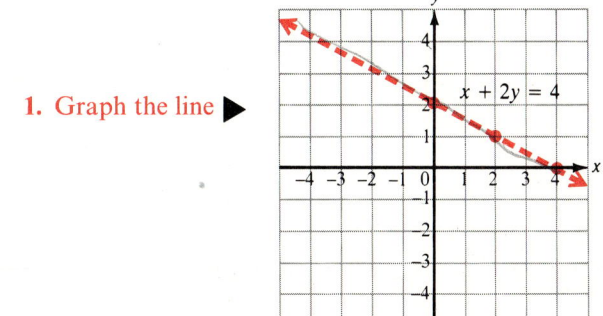

x	y
0	2	← y-intercept
4	0	← x-intercept
2	1	← check point

2. Test a point ▶

$x + 2y < 4$

$0 + 2(0)$	4
0	4
0	4

The origin (0, 0) is not on $x + 2y = 4$, so it can serve as a test point.

3. Interpret ▶ $0 < 4$ means (0, 0) is a solution of $x + 2y < 4$.

4. Graph half-plane ▶

Solution ▶

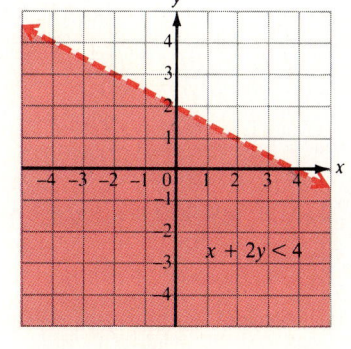

Shade the half-plane that contains the origin.

The graph of $x + 2y = 4$ is a dashed line to indicate that it is not part of the solution.

Make Sure

Graph inequalities of the form $Ax + By < C$ or $Ax + By > C$.

See Example 2 ▶ **1.** $x + 2y < -4$ **2.** $2x - 3y > -6$

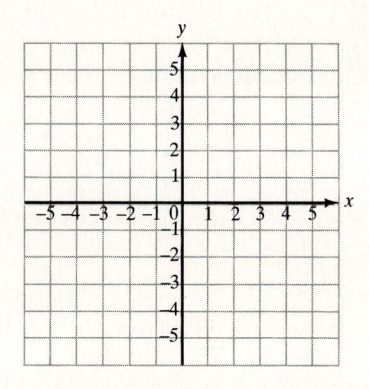

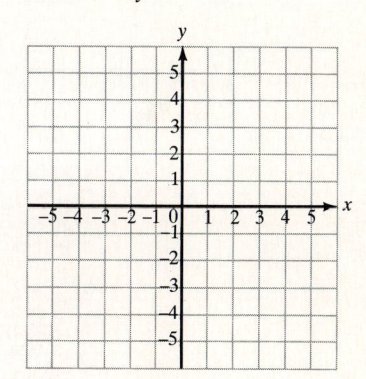

MAKE SURE ANSWERS: See Appendix Selected Answers.

124   Chapter 3   Equations and Graphs

In Chapter 2 you graphed simple inequalities, such as $x > 3$, on a number line. It is important to note that you are now graphing inequalities in two variables and that $x > 3$ is just a compact way of writing $x + 0y > 3$.

EXAMPLE 3: Graph an inequality of the form $x > C$ (or, $y > C$).

Problem ▸ Graph $x > 3$ on a rectangular coordinate system.

1. Draw the line ▸

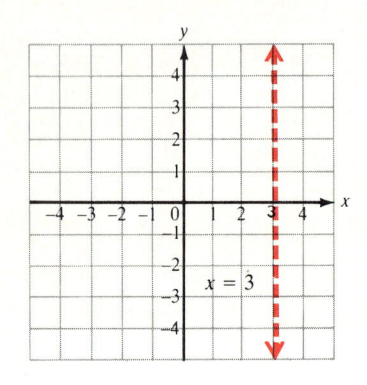

x	y
3	0
3	1
3	2

2. Test a point ▸

$$\frac{x > 3}{0 \mid 3}$$
$$0 \not> 3$$

The origin $(0, 0)$ is not on $x = 3$, so it can serve as a test point.
Compare: $0 \not> 3$ means $(0, 0)$ is not a solution of $x > 3$.

3. Graph half-plane ▸

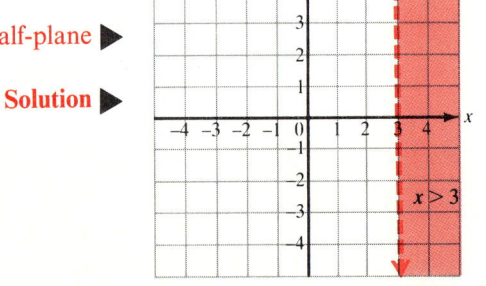

Solution ▸

Shade the half-plane that does not contain the origin.

Recall ▸ The boundary line is a dashed line to indicate that it is not part of the solution.

Another Example ▸ Graph $y > -2$ on a rectangular coordinate system.

$$\frac{y > -2}{0 \mid -2}$$
$$0 > -2$$

The origin $(0, 0)$ is not on $y = -2$, so it can serve as a test point.
Compare: $0 > -2$ means $(0, 0)$ is a solution of $y > -2$.

Shade the half-plane that includes the origin.

Make Sure

Graph inequalities of the form $x > C$ or $y > C$.

See Example 3   **1.** $x > -2$ **2.** $y > 0$

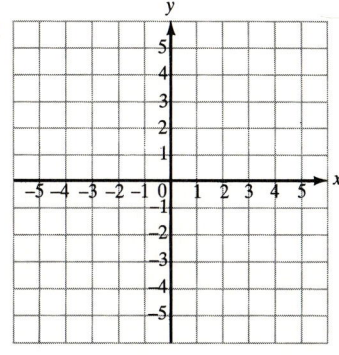

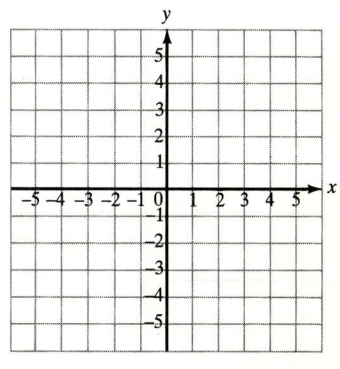

MAKE SURE ANSWERS: See Appendix Selected Answers.

EXAMPLE 4: Graph an inequality of the form $y \geq Ax$ ($y \leq Ax$).

Problem   Graph $y \geq 3x$.

1. Graph the line   Since $y = 3x$ the line has $m = 3$ and $b = 0$.

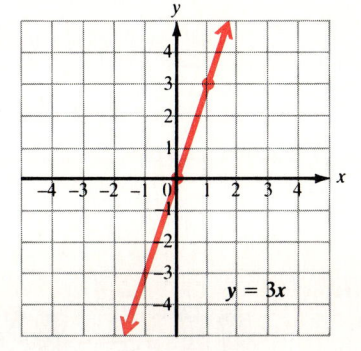

2. Test a point

$$\begin{array}{c|c} y & \geq 3x \\ \hline 0 & 3(2) \\ 0 & \not\geq 6 \end{array}$$

The origin (0, 0) cannot be used as a test point, because it is on the boundary $y = 3x$.
Select a point that is not on the boundary, say (2, 0).
Compare: $0 \not\geq 6$ means (2, 0) is not a solution of $y \geq 3x$.

3. Graph half-plane

Solution   Shade in the half-plane that does not contain (2, 0).

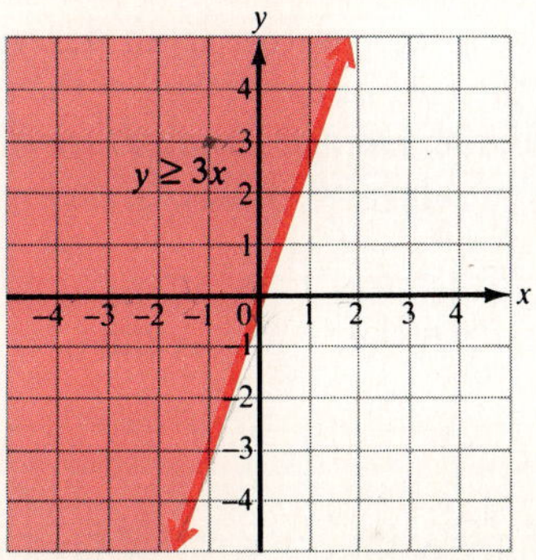

126 Chapter 3 Equations and Graphs

Other Examples ▶ (a) Graph $y \leq -2x$. (b) Graph $y > \frac{3}{2}x$.

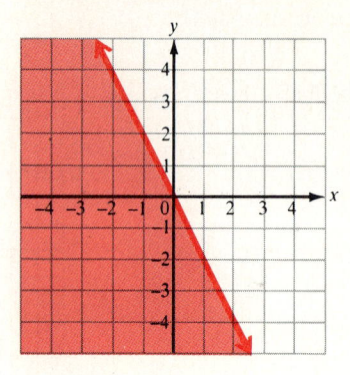

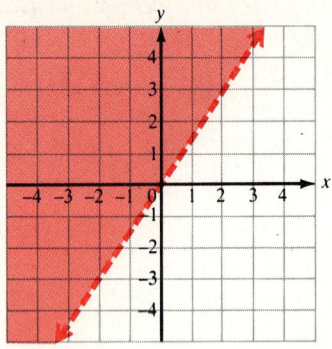

Make Sure

Graph inequalities of the form $y \geq Ax$ or $y \leq Ax$.

See Example 4 ▶ **1.** $y \geq -2x$ **2.** $y \leq \frac{2}{3}x$

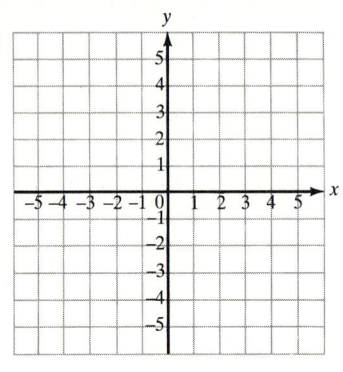

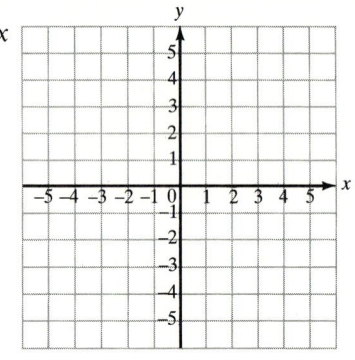

MAKE SURE ANSWERS: See Appendix Selected Answers.

According to Guinness

DEEPEST MINE
BY MAY, 1975, A RECORD DEPTH OF 12,600 FEET HAD BEEN ATTAINED IN THE WESTERN DEEP LEVELS MINE AT CARLETONVILLE, TRANSVAAL, SOUTH AFRICA

There is a linear relationship between temperature (t) in degrees Fahrenheit and depth (d) at 1 to 10 miles below the earth's surface. At one mile below the earth's surface, the temperature is 60°F. (This is also approximately the average temperature on the earth's surface.) For each 100 feet of depth below one mile, but above 10 miles, the temperature rises one degree.

3. Find the linear equation that represents this relationship using 1 mile = 5280 feet. (See PS 3.)

4. What is the temperature of the rock wall at the bottom of the deepest mine?

5. What is the temperature 10 miles below the earth's surface?

6. How far below the earth's surface is the temperature 200°F?

3.5 Practice

Name _____ Date _____ Class _____

3.5 Practice

Set 1: Graph inequalities of the form $Ax + By \leq C$.

1. $x + y \leq 3$

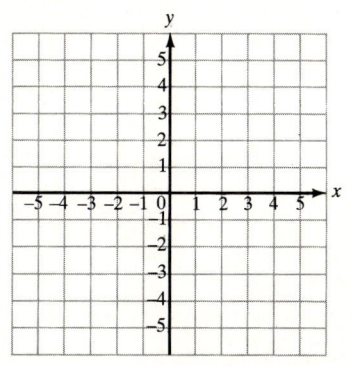

2. $x - y \leq -2$

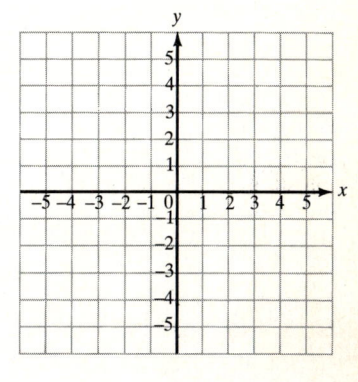

Set 2: Graph inequalities of the form $Ax + By < C$.

3. $5x + y < 5$

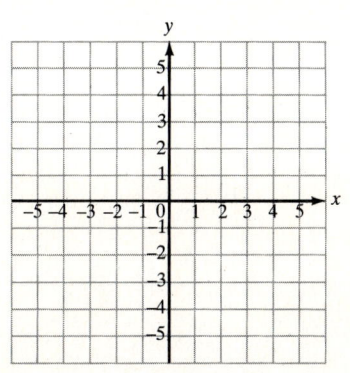

4. $x + 3y < -3$

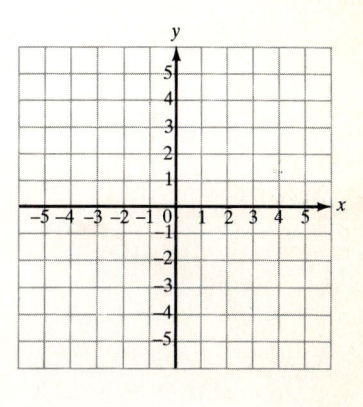

Set 3: Graph inequalities of the form $y > C$.

5. $x > 0$

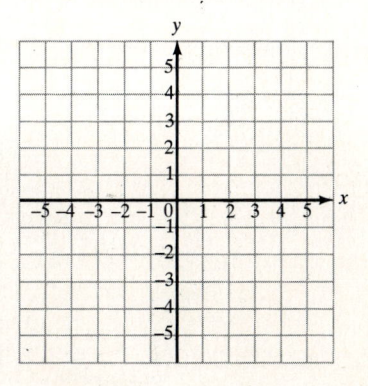

6. $x > 1$

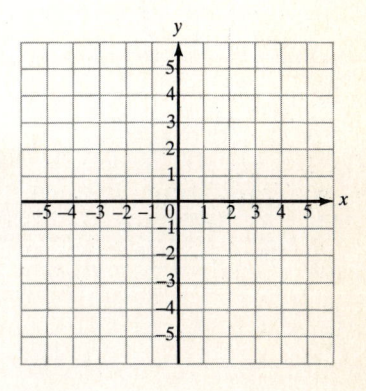

Set 4: Graph inequalities of the form $y \leq Ax$.

7. $y \leq 2x$

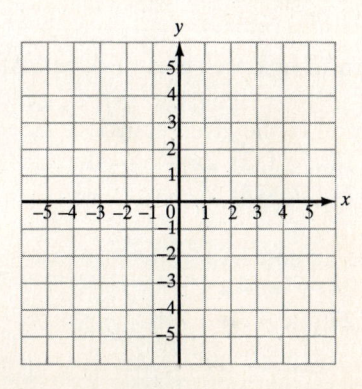

8. $y \leq -4x$

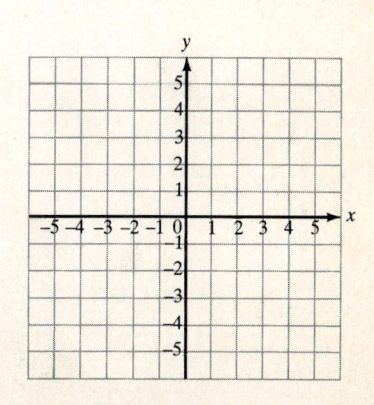

Copyright © 1985 by Harcourt Brace Jovanovich, Inc. All rights reserved.

128 Chapter 3 Equations and Graphs

Mixed Practice: Graph each inequality.

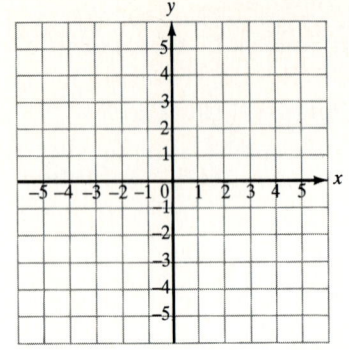

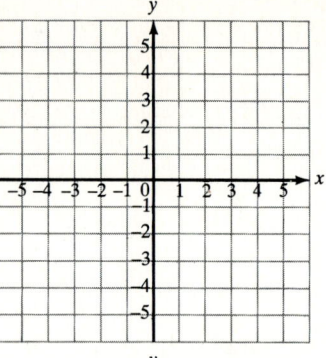

9. $3 \geq y - 3x$

10. $-2 \geq 2y - 4x$

11. $-2 < x - 2y$

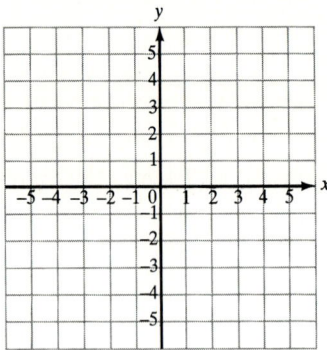

12. $-3 < y - 3x$

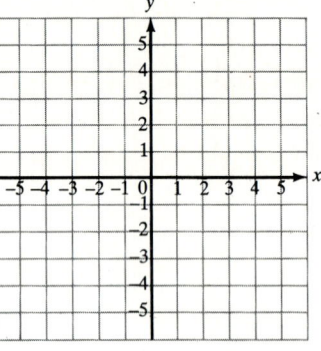

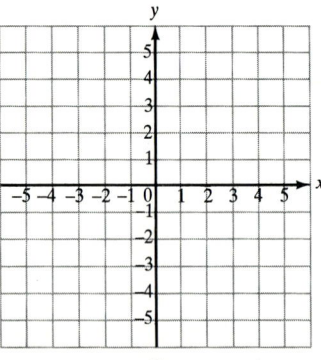

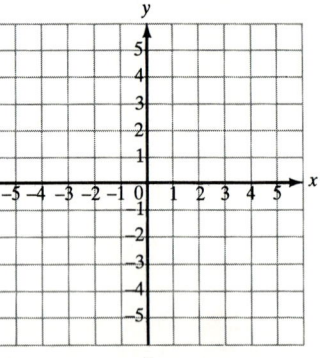

13. $-3 \geq x$

14. $4 \geq y$

15. $2y < 3x$

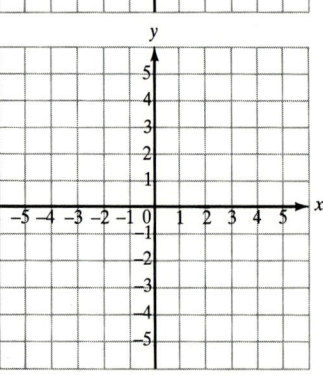

16. $-3y < 2x$

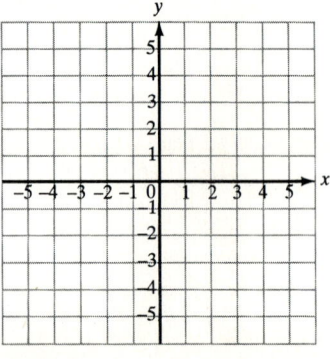

Review: Work these problems on a separate sheet of paper. Attach your work to this page.

Is the given ordered pair a solution of the given equation? (See Lesson 3.2.)
17. $(2, 1)$, $3x - 4y = 2$
18. $(-1, 3)$, $2x + y = 1$
19. $(-2, 1)$, $3x + 5y = 1$
20. $(1, 3)$, $x + y = -4$
21. $(2, -3)$, $x - y = -1$
22. $(-2, -3)$, $2x - 4y = 8$

Graph each linear equation. (See Lesson 3.2.)
23. $x = 3$
24. $y = -4$
25. $3x + 2y = 10$
26. $4x - 3y = 9$
27. $5x + 3y = 18$
28. $2x + 5y = 10$

Copyright © 1985 by Harcourt Brace Jovanovich, Inc. All rights reserved.

Name _____ Date _____ Class _____

Chapter 3 Review

		What to Review if You Have Trouble		
Objectives		**Lesson**	**Example**	**Page**
Graph an ordered pair ▶	1. Graph $(-2, 3)$ on Grid A. _____	3.1	1	88
Find the coordinates of a given point ▶	2. Find the coordinates of point P on Grid A. _____	3.1	2	90
Check proposed ordered pair solutions ▶	3. Which pair, $(2, 3)$ or $(-4, 1)$, is a solution of $-3x + 2y = 14$? _____	3.1	3	91
Find ordered pair solutions ▶	4. Find solutions of $-x + 2y = 10$ for: $x = -1, 0, 1, 2,$ and 4. Put your answers in the table at right.	3.1	4	92
Graph equations by plotting points ▶	5. Graph $2x - y = 3$ on Grid B.	3.2	1	95
	6. Graph $y = -3$ on Grid C.	3.2	2	96
	7. Graph $3x - 2y = 6$ using the intercept method on Grid D.	3.2	3	98
Find the slope of a nonvertical line ▶	8. Find the slope of the line that contains $P_1(-5, 8)$ and $P_2(7, -11)$ _____	3.3	1	101
Write slope-intercept form ▶	9. Write the slope-intercept form of the equation of a line with a y-intercept of $(0, -2)$ and a slope of 5. _____	3.3	2	104
Graph using the slope-intercept method ▶	10. Graph $3x - 2y = 4$ using the slope-intercept method on Grid D.	3.3	3	105

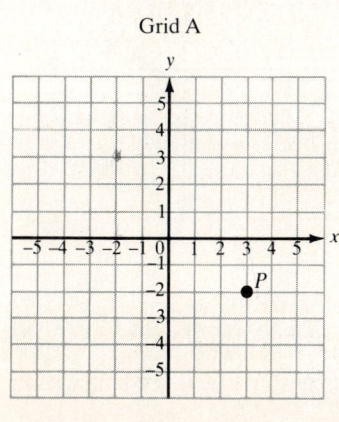

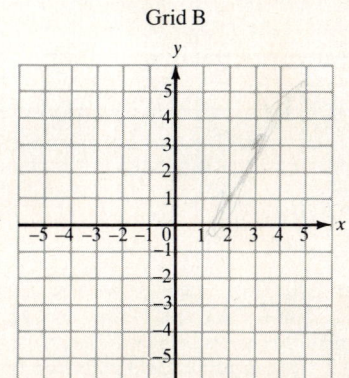

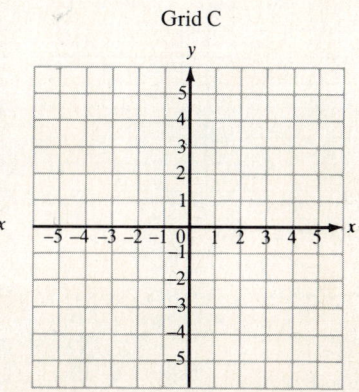

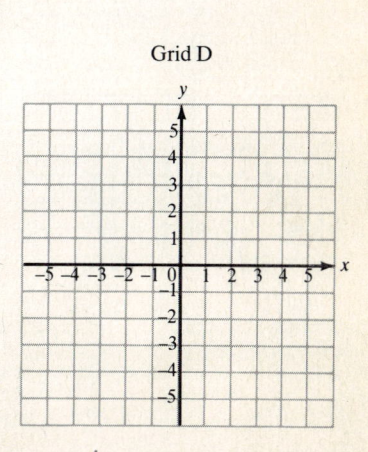

Grid A   Grid B   Grid C   Grid D

Copyright © 1985 by Harcourt Brace Jovanovich, Inc. All rights reserved.

130 Chapter 3 Equations and Graphs

Find Equations ▶

11. Find an equation of the line that contains $P_1(3,4)$ and has a slope of -2. Write your answer in slope-intercept form. _____ 3.4 1 109

12. Find an equation of a line that contains $P_1(-3, 4)$ and $P_2(5, -1)$. Write your answer in slope-intercept form. _____ 3.4 2 110

13. Find an equation of a line that contains $(-3, 2)$ and is parallel to $4x + y = 5$. Write your answer in slope-intercept form. _____ 3.4 3 113

14. Find an equation of a line that contains $(-2, 3)$ and is perpendicular to $3x - 5y = 7$. Write your answer in slope-intercept form. _____ 3.4 4 114

Graph inequalities ▶

15. Graph $2x - 3y \geq -6$ on Grid E. _____ 3.5 1 121

16. Graph $2x - y > 4$ on Grid F. _____ 3.5 2 123

17. Graph $y > -1$ on Grid G. _____ 3.5 3 124

18. Graph $y \leq -2x$ on Grid H. _____ 3.5 4 125

Use linear relations ▶

19. Office equipment was purchased for $10,000 and depreciated, using straight line depreciation, at a rate of $1200 per year. **a.** Find the linear equation describing this linear relationship **b.** What was the book value after $4\frac{1}{2}$ years? **c.** How long did it take the equipment to reach a $1000 *scrap value* (the value at which the equipment is sold for scrap)?
   a. _____
   b. _____
   c. _____
   PS 3 1 117

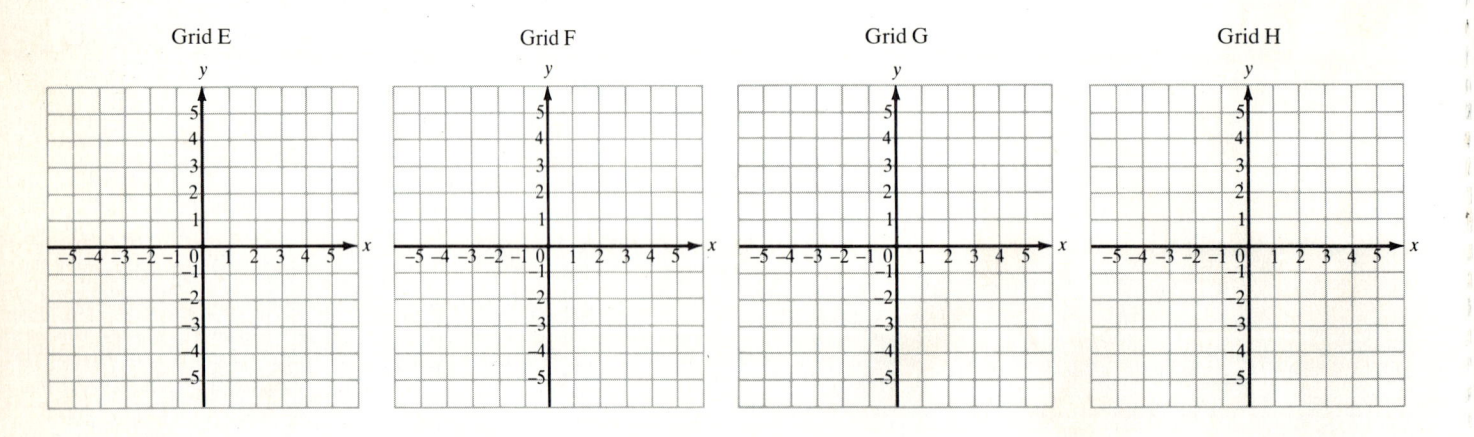

Grid E   Grid F   Grid G   Grid H

CHAPTER 3 REVIEW ANSWERS: 1. See Appendix Selected Answers. **2.** $(3, -2)$ **3.** $(-4, 1)$ **4.** $\{(-1, \frac{2}{9}), (0, 5), (1, \frac{11}{7}), (2, 6), (4, 7)\}$ **5.–7.** See Appendix Selected Answers. **8.** $-\frac{19}{12}$ **9.** $y = 5x - 2$ **10.** See Appendix Selected Answers. **11.** $y = -2x + 10$ **12.** $y = -\frac{5}{8}x + \frac{17}{8}$ **13.** $y = -4x - 10$ **14.** $y = -\frac{3}{5}x - \frac{3}{5}$ **15.–18.** See Appendix Selected Answers. **19a.** $v = 10,000 - 1200t$ **b.** $4600 **c.** $7\frac{1}{2}$ years

REAL NUMBERS AND LINEAR RELATIONS

4 Systems

- 4.1 Solve by Graphing
- 4.2 Solve Using Algebraic Methods
- **PS 4:** Solve Mixture Problems
- 4.3 Solve Third-Order Systems
- **PS 5:** Solve Digit Problems
- 4.4 Evaluate Determinants
- 4.5 Solve Using Cramer's Rule

4.1 Solve by Graphing

Two or more equations considered simultaneously are called a *system of equations*.

Recall ▶ An equation that can be written in the form $Ax + By = C$, where A and B are not both 0, is called a linear equation in two variables.

A pair of linear equations in the same two variables is called a *second-order system*.

Examples ▶ (a) $\begin{cases} x - y = -2 \\ 2x + y = 11 \end{cases}$ is a second-order system. (b) $\begin{cases} 3y = 5 \\ 2x + 7y = 1 \end{cases}$ is a second-order system.

Every ordered pair that is a common solution of both linear equations in a second-order system is called a *solution of the system*.

EXAMPLE 1: Check a proposed ordered pair solution of a second-order system.

Problem ▶ Is $(3, 5)$ a solution of $\begin{cases} x - y = -2 \\ 2x + y = 11 \end{cases}$?

1. Check in each equation ▶

$$\begin{array}{c|c} x - y = -2 \\ \hline 3 - 5 & -2 \\ -2 & -2 \\ -2 & -2 \end{array} \longleftarrow (3, 5) \text{ checks in } x - y = -2$$

$$\begin{array}{c|c} 2x + y = 11 \\ \hline 2(3) + 5 & 11 \\ 6 + 5 & 11 \\ 11 & 11 \end{array} \longleftarrow (3, 5) \text{ checks in } 2x + y = 1.$$

2. Interpret ▶ Since $(3, 5)$ is a solution of each equation in the system, it is a solution of the system.

Solution ▶ Yes: $(3, 5)$ is a solution of $\begin{cases} x - y = -2 \\ 2x + y = 11 \end{cases}$.

> **CAUTION:** An ordered pair is not a solution of a second-order system if it is only a solution of one equation in the system.

Example ▶ $(5, 1)$ is not a solution of $\begin{cases} x - y = -2 \\ 2x + y = 11 \end{cases}$. Think: $(5, 1)$ is a solution of $2x + y = 11$, but it is not a solution of $x - y = -2$.

Make Sure

Is the given ordered pair a solution of the given second-order system?

See Example 1 ▶ 1. $(-2, 3)$, $\begin{cases} 2x + 3y = 5 \\ 5x + 2y = -4 \end{cases}$ _____ 2. $(1, 2)$, $\begin{cases} x + y = 3 \\ 5x + 2y = 12 \end{cases}$ _____

MAKE SURE ANSWERS: 1. yes 2. no

The graph of a second-order system consists of two lines in a plane. Geometrically, a solution of a second-order system is represented by any point which is on both lines.

A system of equations having one or more solutions is called a *consistent* system. A system of equations with no solutions is said to be *inconsistent*.

Consistent systems can be further classified as being *independent* or *dependent*.

A system of equations that has exactly one solution is called an independent system. A system of equations that has more than one solution is called a dependent system.

In an independent second-order system, each equation graphs to be a unique line. The lines intersect at a single point. In a dependent second-order system, both equations graph to be the same line. See Figure 4.1.

The lines intersect at a single point.　　　The lines are parallel.　　　The lines coincide.

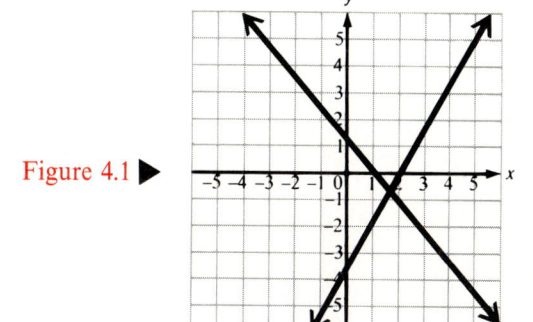

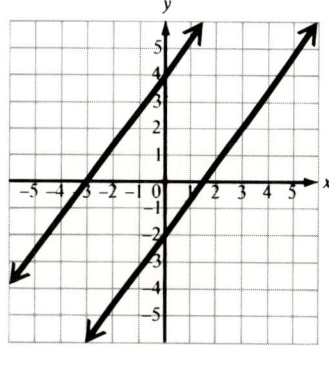

Figure 4.1 ▶

A single solution　　　　**No solution**　　　　**Infinite number of solutions**
Independent System　　　**Inconsistent System**　　**Dependent System**

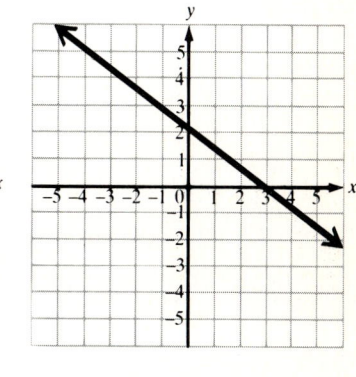

Sometimes you can find a solution of a second-order system by examining its graph.

EXAMPLE 2: Solve a second-order system by graphing.

Problem ▶ Solve $\begin{cases} x + y = 4 \\ -x + y = -2 \end{cases}$ by graphing.

1. Graph each equation ▶

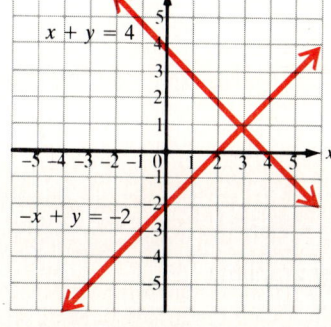

$x + y = 4$　　　$-x + y = -2$

x	y
0	4
4	0
2	2

x	y
0	-2
2	0
4	2

2. Identify intersection ▶ The graphs intersect at (3, 1).

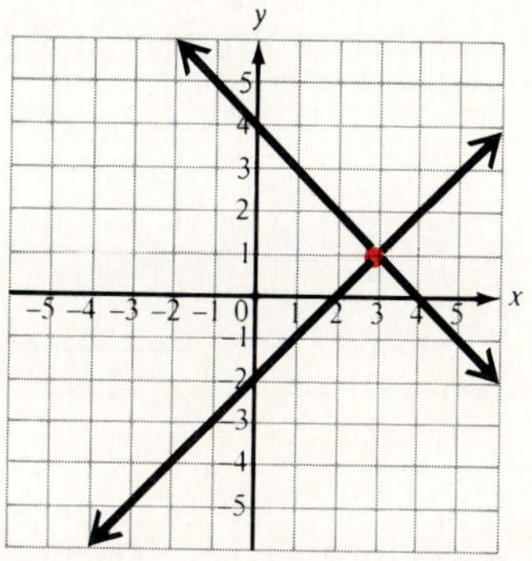

Solution ▶ $\begin{cases} x + y = 4 \\ -x + y = -2 \end{cases}$ has exactly one solution of (3, 1).　　Check as before.

134 Chapter 4 Systems

When solving a system by graphing, you may find that the system is an inconsistent system or a dependent system.

Examples ▶ (a) Solve $\begin{Bmatrix} -\frac{1}{2}x + y = -2 \\ x - 2y = -4 \end{Bmatrix}$ by graphing.

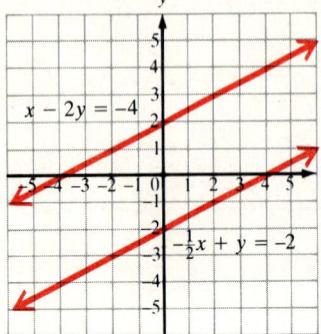

$-\frac{1}{2}x + y = -2$

x	y
0	-2
4	0
2	-1

$x - 2y = -4$

x	y
0	2
-4	0
2	3

Because the lines are parallel, $\begin{Bmatrix} -x + y = -2 \\ 2x - 2y = -6 \end{Bmatrix}$ is an inconsistent system.

(b) Solve $\begin{Bmatrix} -4x + y = 2 \\ 8x - 2y = -4 \end{Bmatrix}$ by graphing.

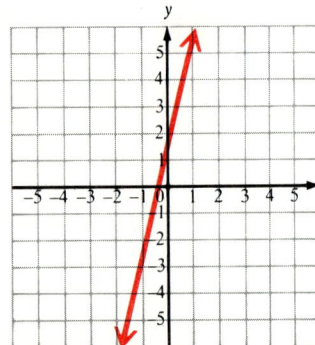

$-4x + y = 2$

x	y
0	2
$-\frac{1}{2}$	0

$8x - 2y = -4$

x	y
0	2
$-\frac{1}{2}$	0

Because the graphs coincide, $\begin{Bmatrix} -4x + y = 2 \\ 8x - 2y = -4 \end{Bmatrix}$ is a dependent system.

Make Sure

Solve each second-order system by graphing.

See Example 2 ▶ 1. $\begin{Bmatrix} 3x + 2y = 5 \\ 2x + 5y = -4 \end{Bmatrix}$ _____

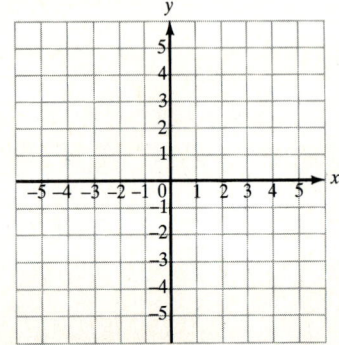

2. $\begin{Bmatrix} 5x + 2y = 12 \\ x + y = 3 \end{Bmatrix}$ _____

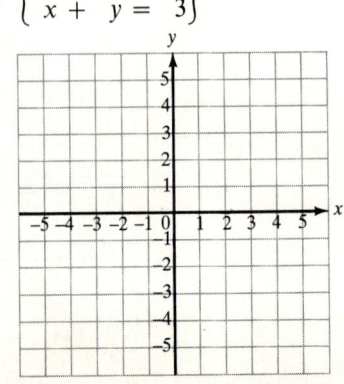

MAKE SURE ANSWERS: See Appendix Selected Answers.

4.1 Practice

Set 1: Is the given ordered pair a solution of the given second-order system?

1. $(2, 1)$, $\begin{cases} 2x - y = 3 \\ 3x + y = 7 \end{cases}$
2. $(3, 5)$, $\begin{cases} 4x + y = 17 \\ 2x - y = 1 \end{cases}$
3. $(0, -3)$, $\begin{cases} 4x - 3y = 8 \\ 5x + 2y = -6 \end{cases}$
4. $(-2, -1)$, $\begin{cases} 5x - 4y = 3 \\ 2x - 4y = 0 \end{cases}$
5. $(\frac{1}{2}, \frac{1}{3})$, $\begin{cases} 6y = 5 - 6x \\ 3y = 2x \end{cases}$
6. $(\frac{2}{3}, \frac{3}{4})$, $\begin{cases} 4y = 5 - 3x \\ 8y = 9x \end{cases}$

Set 2: Solve each second-order system by graphing. If the system has no single solution, indicate whether the system is inconsistent or dependent.

7. $\begin{cases} x + y = 3 \\ x - y = 3 \end{cases}$
8. $\begin{cases} x + y = 3 \\ x - y = -1 \end{cases}$
9. $\begin{cases} 3x + 2y = 4 \\ 4x + 5y = 3 \end{cases}$
10. $\begin{cases} 4x - 3y = 4 \\ 5x - 2y = -2 \end{cases}$
11. $\begin{cases} 4x - y = 4 \\ x + 4y = 1 \end{cases}$
12. $\begin{cases} 2x - 3y = 7 \\ 3x + 2y = 4 \end{cases}$

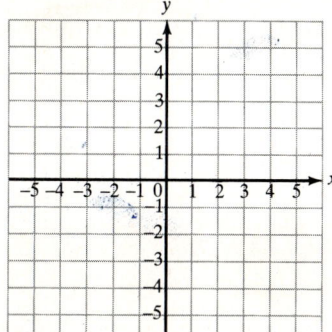

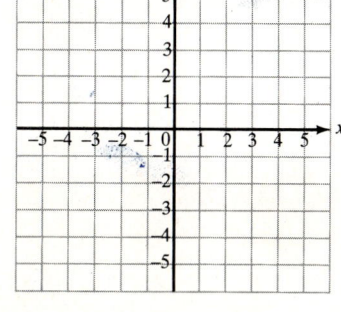

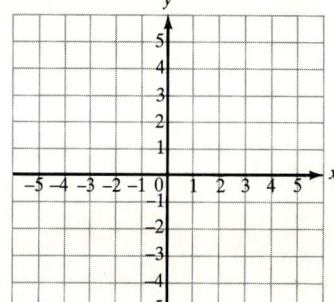

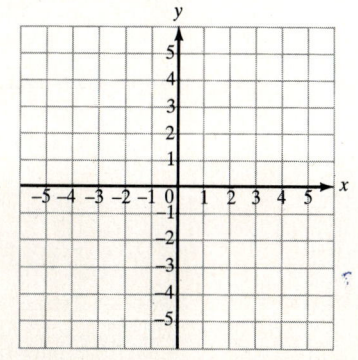

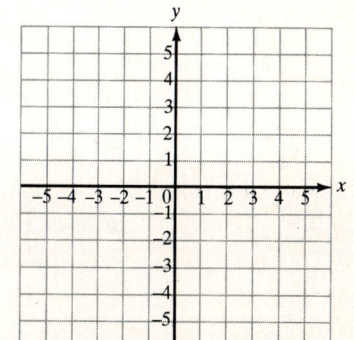

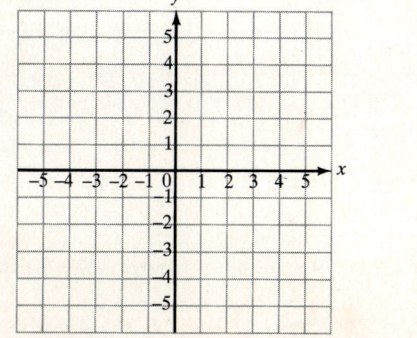

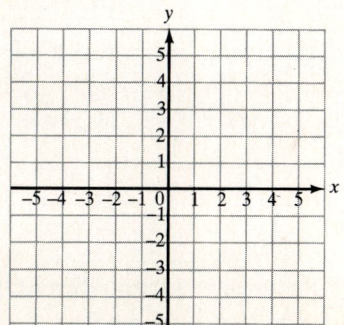

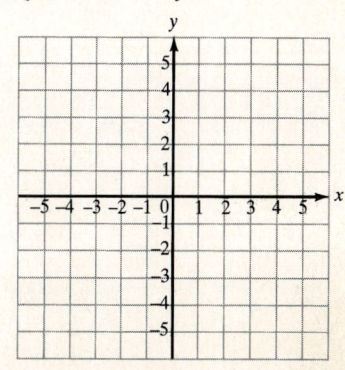

136 Chapter 4 Systems

13. $\begin{cases} 2x = 3y + 1 \\ 4x = 6y + 2 \end{cases}$

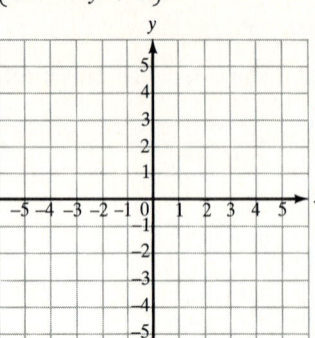

14. $\begin{cases} 3x = y - 2 \\ 6x = 2y - 4 \end{cases}$

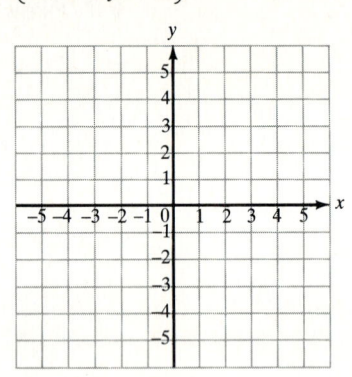

15. $\begin{cases} y = 3x + 1 \\ y = 3x - 1 \end{cases}$

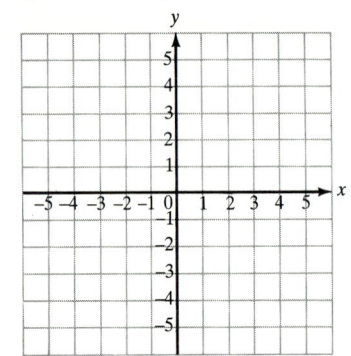

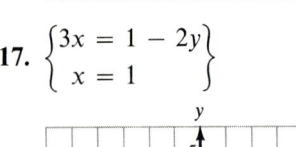

16. $\begin{cases} 2x - 1 = 3y \\ 2x + 2 = 3y \end{cases}$

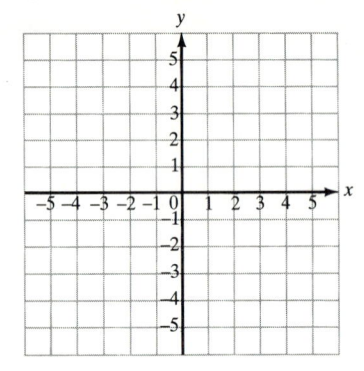

17. $\begin{cases} 3x = 1 - 2y \\ x = 1 \end{cases}$

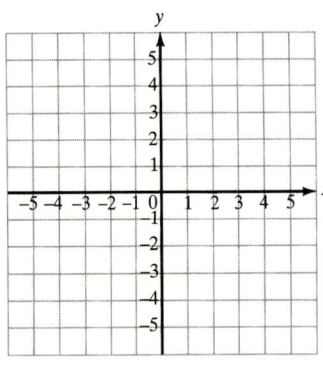

18. $\begin{cases} 3x = 1 + 2y \\ y = 1 \end{cases}$

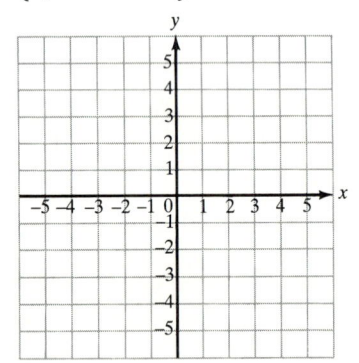

Review: Work these problems on a separate sheet of paper. Attach your work to this page.

Clear parentheses using the distributive properties. (See Lesson 1.5.)

19. $-(2x - 3)$
20. $-(3 - 4x)$
21. $-\left(\dfrac{2x - 3}{4}\right)$
22. $-\left(\dfrac{2 - 5x}{3}\right)$
23. $3\left(\dfrac{4x - 3}{2}\right)$
24. $2\left(\dfrac{4 - 5x}{3}\right)$
25. $-4\left(\dfrac{3 + 6y}{2}\right)$
26. $-3\left(\dfrac{3 - 2x}{-6}\right)$

Solve each literal equation for the indicated letter. (See Lesson 2.6.)

27. $2x - y = 4$ for y
28. $x + 3y = 4$ for x
29. $2x - 3y = 2$ for y
30. $3x - 2y = 3$ for x
31. $3x + 2y = 4$ for x
32. $3x + 2y = 4$ for y

4.2 Solve Using Algebraic Methods

The graphical method of solving a second-order system has several disadvantages. It it time-consuming, and it is not an exact method. The following *substitution method* has the advantage of being an exact algebraic method for solving second-order systems.

> **The Substitution Method for Solving Second-Order Systems**
>
> 1. Solve one of the equations for one of the variables in terms of the other variable.
> 2. Substitute the expression obtained in Step 1 into the other system equation to produce an equation in only one variable.
> 3. Solve the equation obtained in Step 2.
> 4. Substitute the numerical value obtained in Step 3 into the expression obtained in Step 1, and solve for the other variable.
> 5. Check the proposed ordered pair solution in both of the original equations.

EXAMPLE 1: Solve a second-order system by the substitution method.

Problem ▶ Solve $\begin{cases} -4x + 3y = -9 \\ 2x - 6y = 3 \end{cases}$ by the substitution method.

1. Solve one equation ▶ Solve for x first:

$$2x - 6y = 3$$
$$2x = 6y + 3$$
$$x = \frac{6y + 3}{2}$$

OR Solve for y first:

$$2x - 6y = 3$$
$$-6y = -2x + 3$$
$$y = \frac{-2x + 3}{-6}$$

2. Substitute in the other equation ▶

$$-4x + 3y = -9$$
$$-4\left(\frac{6y + 3}{2}\right) + 3y = -9$$

$$-4x + 3y = -9$$
$$-4x + 3\left(\frac{-2x + 3}{-6}\right) = -9$$

3. Solve for one variable ▶

$$-12y - 6 + 3y = -9$$
$$-9y - 6 = -9$$
$$-9y = -3$$
$$y = \frac{1}{3}$$

$$-4x + 1x - \tfrac{3}{2} = -9$$
$$-3x - \tfrac{3}{2} = -9$$
$$-3x = -\tfrac{15}{2}$$
$$x = \frac{5}{2}$$

4. Solve for the other variable ▶

$$x = \frac{6y + 3}{2}$$
$$= \frac{6(\tfrac{1}{3}) + 3}{2}$$
$$= \frac{5}{2}$$

$$y = \frac{-2x + 3}{-6} \longleftarrow \text{equation obtained in Step 1}$$
$$= \frac{-2(\tfrac{5}{2}) + 3}{-6}$$
$$= \frac{1}{3}$$

5. Check in each equation ▶

$$\begin{array}{c|c} -4x + 3y = -9 \\ \hline -4(\tfrac{5}{2}) + 3(\tfrac{1}{3}) & -9 \\ -10 + 1 & -9 \\ -9 & -9 \end{array} \longleftarrow (\tfrac{5}{2}, \tfrac{1}{3}) \text{ checks}$$

$$\begin{array}{c|c} 2x - 6y = 3 \\ \hline 2(\tfrac{5}{2}) - 6(\tfrac{1}{3}) & 3 \\ 5 - 2 & 3 \\ 3 & 3 \end{array} \longleftarrow (\tfrac{5}{2}, \tfrac{1}{3}) \text{ checks}$$

Solution ▶ $(x, y) = \left(\dfrac{5}{2}, \dfrac{1}{3}\right)$

138 Chapter 4 Systems

The previous example had a single solution, so it is an independent system.

> If a second-order system reduces to a true statement, such as $0 = 0$, then the system is dependent. It has infinitely many solutions.
>
> If a second-order system reduces to a false statement, such as $0 = -1$, then the system is inconsistent. It has no solution.

Example ▶ Solve $\begin{cases} x + 2y = 3 \\ 2x + 4y = 5 \end{cases}$ by the substitution method.

1. Solve for a variable ▶ $x + 2y = 3$ ⟵ first equation

$x = -2y + 3$ ⟵ first equation solved for x

2. Substitute ▶ $2x + 4y = 5$ ⟵ second equation

$2(-2y + 3) + 4y = 5$ Substitute $-2y + 3$ for x.

$-4y + 6 + 4y = 5$

$0 = -1$ ⟵ a false statement

Solution ▶ A second-order system that reduces to a false statement is an inconsistent system.

Note ▶ $0 = -1$ is a compact way of writing $0x + 0y = -1$. The left member, $0x + 0y$, always equals 0 regardless of the values of x and y. Since there is no ordered pair (x, y) that is a solution of the equation, the system is inconsistent.

Another Example ▶ Solve $\begin{cases} -2x + y = 3 \\ -4x + 2y = 6 \end{cases}$ by the substitution method.

1. Solve for a variable ▶ $-2x + y = 3$ ⟵ first equation

$y = 2x + 3$ ⟵ first equation solved for y

2. Substitute ▶ $-4x + 2y = 6$ ⟵ second equation

$-4x + 2(2x + 3) = 6$ Substitute $2x + 3$ for y.

$-4x + 4x + 6 = 6$

$0 = 0$ ⟵ a true statement

Solution ▶ A second-order system that reduces to a true statement is a dependent system.

Make Sure

Solve each second-order system by the substitution method. Check each solution.

See Example 1 ▶   **1.** $\begin{cases} x - 2y = 2 \\ 2x + 5y = 13 \end{cases}$ _____   **2.** $\begin{cases} 5x = 7 - 3y \\ 5x = 5 - 3y \end{cases}$ _____

MAKE SURE ANSWERS: 1. (4, 1)   2. no solution (inconsistent)

4.2 Solve Using Algebraic Methods

You can sometimes eliminate a variable from a system of equations by adding equations.

Example ▶ Solve $\begin{cases} x + y = 10 \\ x - y = 2 \end{cases}$.

1. Add ▶
$$\begin{aligned} x + y &= 10 \quad \longleftarrow \text{first equation} \\ x - y &= 2 \quad \longleftarrow \text{second equation} \\ \hline 2x + 0 &= 12 \quad \longleftarrow \text{sum of the equations} \end{aligned}$$

2. Solve ▶
$2x + 0 = 12$
$x = 6$

3. Substitute ▶

$x + y = 10$ ← first equation	OR	$x - y = 2$ ← second equation
$6 + y = 10$ Substitute 6 for x.		$6 - y = 2$ Substitute 6 for x.
$y = 4$		$y = 4$

Solution ▶ $(x, y) = (6, 4)$ Check as before.

If adding equations does not eliminate a variable, then it will be necessary to make use of a process called the *addition method*.

The Addition Method for Solving Second-Order Systems
1. Multiply each equation by the appropriate constants so that the coefficient of one variable is the opposite of the coefficient of the same variable in the second equation.
2. Add the resulting equations.
3. Solve for the remaining variable.
4. Substitute the solution from Step 3 into either one of the original equations and solve for the other variable.
5. Check the proposed ordered pair solution in both of the original equations.

EXAMPLE 2: Solve a second-order system by the addition method.

Problem ▶ Solve $\begin{cases} 3x + 2y = -2 \\ 4x + 5y = 9 \end{cases}$ by the addition method.

1. Use the Multiplication Rule ▶

$\begin{matrix} 4 \\ -3 \end{matrix} \diagdown \begin{matrix} 3x + 2y = -2 \\ 4x + 5y = 9 \end{matrix}$ Think: 4 and -3 are appropriate multipliers since $4 \cdot 3 = 12$ and $(-3) \cdot 4 = -12$.

$\begin{aligned} 4(3x + 2y) &= 4(-2) \longrightarrow 12x + 8y = -8 \\ -3(4x + 5y) &= -3 \cdot 9 \longrightarrow -12x - 15y = -27 \end{aligned}$

2. Add equations ▶
$$\begin{aligned} \cancel{12x} + 8y &= -8 \\ \cancel{-12x} - 15y &= -27 \\ \hline 0 + (-7y) &= -35 \end{aligned}$$
Add the equations to eliminate the x term.

3. Solve for one variable ▶ $y = 5$

4. Substitute and solve for the other variable ▶

$3x + 2y = -2$ ← first equation	OR	$4x + 5y = 9$ ← second equation
$3x + 2(5) = -2$		$4x + 5(5) = 9$
$x = -4$		$x = -4$

Solution ▶ $(x, y) = (-4, 5)$ Check as before.

Note ▶ You could have eliminated the y term first by multiplying the first equation by 5 and the second equation by -2.

140 Chapter 4 Systems

To solve a second-order system that involves fractions, you may find it convenient to first clear fractions.

Example ▶ Solve $\begin{cases} \frac{3}{4}x + \frac{1}{3}y = 2 \\ \frac{2}{3}x - \frac{1}{2}y = 4\frac{1}{6} \end{cases}$ using the addition method.

1. Clear fractions ▶ $12\left(\frac{3}{4}x + \frac{1}{3}y\right) = 12 \cdot 2 \longrightarrow 9x + 4y = 24$ Multiply each equation by the LCD.

$6\left(\frac{2}{3}x - \frac{1}{2}y\right) = 6\left(4\frac{1}{6}\right) \longrightarrow 4x - 3y = 25$

2. Solve as before ▶ $\begin{aligned} 3(9x + 4y) &= 3 \cdot 24 \longrightarrow 27x + 12y = 72 \\ 4(4x - 3y) &= 4 \cdot 25 \longrightarrow 16x - 12y = 100 \end{aligned}$ Add equations.

$\overline{43x + 0 = 172}$

$x = \frac{172}{43}$ Solve for one variable.

$x = 4$

$\frac{3}{4}x + \frac{1}{3}y = 2 \longleftarrow$ equation from the original system

$\frac{3}{4}(4) + \frac{1}{3}y = 2$ Substitute $x = 4$.

$3 + \frac{1}{3}y = 2$ Solve for the other variable.

$\frac{1}{3}y = -1$

$y = -3$

Solution ▶ $(x, y) = (4, -3)$ Check as before.

Make Sure

Solve each second-order system using the addition method. Check each solution.

See Example 2 ▶ 1. $\begin{cases} 2x + 3y = 13 \\ 5x - 3y = 1 \end{cases}$ _____ 2. $\begin{cases} 3x + 2y = 13 \\ x + y = 4 \end{cases}$ _____ 3. $\begin{cases} 2x + 3y = 13 \\ 2x + 3y = 4 \end{cases}$ _____

MAKE SURE ANSWERS: 1. (2, 3) 2. (5, −1) 3. no solution (inconsistent)

4.2 Practice

1. _____

2. _____

3. _____

4. _____

5. _____

6. _____

7. _____

8. _____

9. _____

10. _____

11. _____

12. _____

Set 1: Solve each second-order system using the substitution method. If the system has no single solution, indicate whether the system is inconsistent or dependent. Check each solution.

1. $\begin{cases} x + y = 5 \\ x - y = 3 \end{cases}$

2. $\begin{cases} x - y = 1 \\ x + y = 7 \end{cases}$

3. $\begin{cases} x + y = 5 \\ 2x + 3y = 12 \end{cases}$

4. $\begin{cases} 2x + 3y = 13 \\ x + y = 5 \end{cases}$

5. $\begin{cases} 2x = -3y + 3 \\ 12x = 3y + 4 \end{cases}$

6. $\begin{cases} 3x = 4y + 2 \\ 6x = -8y \end{cases}$

7. $\begin{cases} \frac{3}{4}x + y = 5 \\ x - \frac{1}{2}y = 3 \end{cases}$

8. $\begin{cases} \frac{1}{3}x + \frac{1}{2}y = 1 \\ x - \frac{1}{5}y = 3 \end{cases}$

9. $\begin{cases} x + y = 3 \\ 3x + 3y = 9 \end{cases}$

10. $\begin{cases} 2x - 3y = 6 \\ 6x - 9y = 18 \end{cases}$

11. $\begin{cases} x = y + 6 \\ 2x = 2y + 7 \end{cases}$

12. $\begin{cases} 2x = 3y + 4 \\ 4x = 6y + 6 \end{cases}$

142 Chapter 4 Systems

13. _____

Set 2: Solve each second-order system using the addition method. If the system has no single solution, indicate whether the system is inconsistent or dependent. Check each solution.

13. $\begin{cases} 2x + y = 7 \\ 2x - y = 1 \end{cases}$
14. $\begin{cases} x - 3y = 1 \\ x + 3y = 13 \end{cases}$
15. $\begin{cases} 2x + 3y = 12 \\ 4x - y = 10 \end{cases}$

14. _____

15. _____

16. $\begin{cases} 5x - 2y = 7 \\ 2x + y = 1 \end{cases}$
17. $\begin{cases} 3x = 2y \\ 2x = 7 - y \end{cases}$
18. $\begin{cases} 3y = 2 - 4x \\ 2y = 5 + x \end{cases}$

16. _____

17. _____

19. $\begin{cases} 6 - 3y = 4x \\ -4 + 2y = 3x \end{cases}$
20. $\begin{cases} 19 - 3x = 2y \\ 2x = 5y \end{cases}$
21. $\begin{cases} 3x - 2y = 4 \\ x - \frac{2}{3}y = 4 \end{cases}$

18. _____

19. _____

20. _____

22. $\begin{cases} 2x + 3y = 0 \\ x + \frac{3}{2}y = 1 \end{cases}$
23. $\begin{cases} \frac{2}{5}x - y = -\frac{1}{5} \\ x + \frac{2}{3}y = -\frac{1}{2} \end{cases}$
24. $\begin{cases} \frac{3}{4}x - y = \frac{3}{4} \\ x + \frac{3}{4}y = 1 \end{cases}$

21. _____

22. _____

Review: Work these problems on a separate sheet of paper. Attach your work to this page.

Remove the enclosure symbols using the distributive properties. (See Lesson 1.5, Example 2.)
25. $2(3x - 4y - 2z)$
26. $3(x - 3y + 2z)$
27. $-2(2x - 3y + 4z - 5)$
28. $-4(5x + 3y - 4z + 2)$
29. $-(2x - 3y + z + 4)$
30. $-(5x + 3y - z - 1)$

23. _____

Is the given ordered pair a solution of the given equation? (See Lesson 3.1.)
31. $(1, 1)$, $3x - 2y = 1$
32. $(2, 1)$, $3x + 4y = 10$
33. $(0, 1)$, $3x - 4y = 4$
34. $(-2, 3)$, $4x - 3y = 7$
35. $(-3, 0)$, $3x - 5y = -9$
36. $(-1, -3)$, $3x - 5y = 12$

24. _____

Copyright © 1985 by Harcourt Brace Jovanovich, Inc. All rights reserved.

Problem Solving 4: Solve Mixture Problems

To solve a mixture problem, you first represent the *system unknowns* and then use the system unknowns to write the amount of a given ingredient in each given mixture.

EXAMPLE: Solve this mixture problem.

1. Read and identify ▶ A nurse wants to strengthen 20 cc of a 50%-alcohol mixture to an 80%-alcohol mixture. How much pure alcohol must be added? How much 80%-alcohol mixture will the nurse have then? How much alcohol will be in the 80%-alcohol mixture?

Remember, identify the facts, the key words, and the question.

2. Understand ▶ The unknowns are { the amount of pure alcohol to be added; the amount of 80%-alcohol mixture; the amount of alcohol in the 80%-alcohol mixture }.

3. Decide ▶ Let x = the amount of pure alcohol to be added ← first system unknown
then y = the amount of 80%-alcohol mixture ← second system unknown

4. Make a table ▶

	percent of alcohol	base	amount of alcohol
50%-alcohol mixture	50%	20	50%(20)
added alcohol	100%	x	100%(x)
80%-alcohol mixture	80%	y	80%(y)

5. Translate ▶ amount of 50%-alcohol mixture plus added alcohol equals amount of 80%-alcohol mixture

$$20 + x = y$$

alcohol in 50%-alcohol mixture plus added alcohol equals alcohol in 80%-alcohol mixture

$$50\%(20) + 100\%(x) = 80\%(y)$$

System Equations A: $20 + x = y$ or $x = y - 20$
B: $50\%(20) + 100\%x = 80\%y$

6. Solve as before ▶

$20 + 1x = 1y$
$0.5(20) + 1x = 0.8y$ Clear percents in equation B.

$20 + 1x = 1y$
$-10 - 1x = -0.8y$ Multiply equation B by -1.
$\overline{10 + 0 = 0.2y}$ Add equations A and B.

$10(10) = 10(0.2y)$ Solve for the remaining letter.

$100 = 2y$

$50 = y$

7. Interpret ▶ $y = 50$ means the amount of 80%-alcohol mixture is 50 cc.
$x = y - 20 = 50 - 20 = 30$ means the amount of pure alcohol to be added is 30 cc.
$50\%(20) = 0.5(20) = 10$ means the amount of alcohol in the 50%-alcohol mixture is 10 cc.
$80\%(y) = 0.8(50) = 40$ means the amount of alcohol in the 80%-alcohol mixture is 40 cc.

144　Chapter 4　Systems

8. Check ▶ Does the amount of 80%-alcohol mixture minus the amount of pure alcohol that was added equal the amount of 50%-alcohol mixture?
Yes: 50 cc − 30 cc = 20 cc.
Does the amount of alcohol in the 80%-alcohol mixture minus the amount of pure alcohol that was added equal the amount of alcohol in the 50%-alcohol mixture?
Yes: 40 cc − 30 cc = 10 cc.

Note ▶ It takes 30 cc of pure alcohol to get 50 cc of 80%-alcohol mixture, containing 40 cc of alcohol, from 20 cc of 50%-alcohol mixture.

Practice: Solve each mixture problem.

1. a) How many gallons of pure antifreeze is needed to enrich 4 gallons of a 20%-antifreeze mixture to a 60%-antifreeze mixture? b) How much pure antifreeze will be in the final mixture?

2. A nurse must have 20 ounces of a 25%-peroxide solution. She has 20%- and 40%-peroxide solutions. a) How much of each solution is needed to prepare the correct dosage? b) How much peroxide will be in the final solution?

3. Two types of candy are to be mixed together so that the mixture will sell for a $1/kilogram. One of the candies sells for $1.20/kilogram and the other for 40¢/kilogram. a) How much of each are needed to make 12 kilograms of the mixture? b) What percent of the mixture is the $1.20/kilogram candy? c) What fractional part of the mixture is the 40¢/kilogram candy?

4. Two types of coffee beans are to be mixed so that the blend will sell for $3 per pound. One type of bean costs $2 per pound and the other costs $5 per pound. a) How much of each type is needed to prepare 100 five-pound cans of the coffee blend? b) What percent of the blend is made of the $5-per-pound bean? c) What fractional part of the blend is made of the $2-per-pound bean?

5. A chemist mixes 48 liters of a 4%-saline solution with 40 liters of a 15%-saline solution. a) What percent saline is the mixture? (Hint: Let p = the percent saline in the mixture.) b) What fractional part of the mixture is not saline?

6. A nurse mixes 36 liters of water with 60 liters of an 8%-salt solution. a) What percent salt is the mixture? b) What fractional part of the mixture is water?

7. a) How many gallons of pure acetic acid will a photographer need to strengthen 5 gallons of a 10%-acetic solution to a 20%-acetic mixture? b) How much 20%-acetic mixture will this make?

8. 6 liters of a 40%-glycerin solution and 10 liters of an 80%-glycerin solution are mixed together. a) What percent of glycerin is the mixture? b) How much more of the 40%-glycerin solution would be needed to create a 60%-glycerin solution?

Extra

9. A 20-gallon vat is filled with a 40%-acid solution. a) How much must be drained and replaced with pure acid to increase the acidity to 50%? (Hint: Make a table with 4 rows: original, drained, replaced, and mixture.) b) How much must be drained and replaced with pure water to cut the 50%-acid solution back to the original 40%?

10. A 20-liter radiator is filled with a 20%-antifreeze solution. a) How much must be drained and replaced with pure antifreeze to increase the strength to an 80%-antifreeze mixture? b) How much must be drained and replaced with pure water to bring the 80%-antifreeze mixture back to the original 20%?

1a. _____
 b. _____
2a. _____
   _____
 b. _____
3a. _____
   _____
 b. _____
 c. _____
4a. _____
   _____
 b. _____
 c. _____
5a. _____
 b. _____
6a. _____
 b. _____
7a. _____
 b. _____
8a. _____
 b. _____
9a. _____
 b. _____
10a. _____
  b. _____

Copyright © 1985 by Harcourt Brace Jovanovich, Inc. All rights reserved.

4.3 Solve Third-Order Systems

An equation that can be written in the form $Ax + By + Cz = D$, where A, B, C, and D are real numbers (A, B, and C not all zero), is called a *linear equation in three variables*.

Examples ▶
(a) $3x + 2y - 5z = 8$
(b) $\frac{3}{4}x + \frac{1}{2}z = 4$ ⟵ $\frac{3}{4}x + 0y + \frac{1}{2}z = 4$
(c) $\sqrt{2}x + 0.4y = 0.7$ ⟵ $\sqrt{2}x + 0.4y + 0z = 0.7$

Solutions of a linear equation in three variables are *ordered triples* (x, y, z) of real numbers that make the equation true.

Example ▶ The ordered triple $(3, 4, 1)$ is a solution of $x - y + 2z = 1$ because $3 - 4 + 2 \cdot 1 = 1$.

The graph of an ordered triple is a point in three-dimensional space. The graph of a linear equation in three variables is not a line as might be expected, but rather a *plane* in three-dimensional space.

Any three linear equations in the same three variables are called a *third-order system*.

Example ▶ $\begin{cases} x + 2y - z = -5 \\ -3x + 5y - 2z = 1 \\ 2x + 3y + z = 4 \end{cases}$ is a third-order system.

A *solution of a third-order system* is any ordered triple that is a solution of all three equations in the system.

EXAMPLE 1: Check a proposed ordered triple solution of a third-order system.

Problem ▶ Is $(-2, 1, 5)$ a solution of $\begin{cases} x + 2y - z = -5 \\ -3x + 5y - 2z = 1 \\ 2x + 3y + z = 4 \end{cases}$?

1. Check in each equation

$x + 2y - z = -5$		$-3x + 5y - 2z = 1$		$2x + 3y + z = 4$	
$-2 + 2(1) - (5)$	-5	$-3(-2) + 5 \cdot 1 - 2 \cdot 5$	1	$2(-2) + 3 \cdot 1 + 5$	4
$-2 + 2 - 5$	-5	$6 + 5 - 10$	1	$-4 + 3 + 5$	4
$0 - 5$	-5	$11 - 10$	1	$-1 + 5$	4
-5	-5	1	1	4	4

2. Interpret ▶ Since $(-2, 1, 5)$ is a solution of each equation in the system, it is a solution of the system.

Solution ▶ Yes: $(-2, 1, 5)$ is a solution of $\begin{cases} x + 2y - z = -5 \\ -3x + 5y - 2z = 1 \\ 2x + 3y + z = 4 \end{cases}$.

> **CAUTION:** An ordered triple is not a solution of a third-order system if it is only a solution of one or two equations in the system.

Example ▶ $(1, 2, 3)$ is not a solution of $\begin{cases} A: 2x - 3y + 4z = 8 \\ B: x + y + z = 6 \\ C: x - y - z = 9 \end{cases}$ because $(1, 2, 3)$ is a solution of equations A and B, but it is not a solution of equation C.

146 Chapter 4 Systems

Make Sure

Is the given ordered triple a solution of the given third-order system?

See Example 1 ▶ **1.** $(-1, 2, 1)$, $\begin{cases} 2x + 3y + z = 5 \\ x + y = 1 \\ 3x + 3y - 2z = 1 \end{cases}$ _____ **2.** $(2, 1, 3)$, $\begin{cases} x + y + z = 6 \\ x - y + z = 4 \\ x + y - z = 4 \end{cases}$ _____

MAKE SURE ANSWERS: 1. yes 2. no

Figure 4.2 illustrates some of the ways that the graphs of three planes may be related.

Figure 4.2 ▶

The planes intersect at a single point.
One solution
Consistent and Independent System

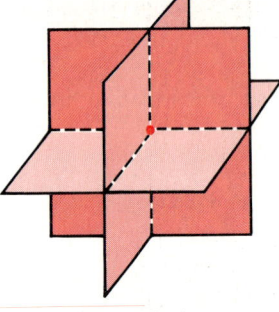

The planes are parallel.
No solution
Inconsistent System

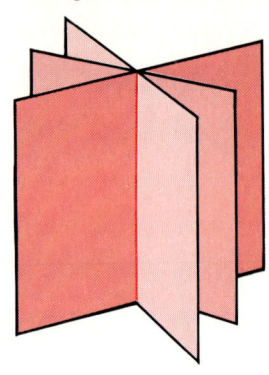

The intersection of the planes is a line.
Infinite number of solutions
Dependent System

It is geneally not convenient to seek a solution of a third-order system by graphing. However, the addition method can be extended to solve third-order systems.

> **The Addition Method for Solving Third-Order Systems**
> 1. Choose two of the equations from the system and eliminate a variable as you did for a second-order system.
> 2. Eliminate the same variable from a different pair of equations.
> 3. Solve the second-order system formed by the equations obtained in Steps 1 and 2.
> 4. Substitute the solution obtained in Step 3 into any of the original system equations and solve for the third variable.

EXAMPLE 2: Solve a third-order system by the addition method.

Problem ▶ Solve $\begin{cases} A: 2x + 4y - z = -5 \\ B: 3x - y + z = 16 \\ C: 4x + 3y - 5z = 2 \end{cases}$ by the addition method.

1. Eliminate a variable ▶
$A: 2x + 4y - z = -5$
$B: 3x - y + z = 16$
$D: \overline{5x + 3y + 0 = 11}$ ⟵ a new equation in only the two variables x and y

2. Eliminate the same variable from a different pair of equations ▶
$B: 5(3x - y + z) = 5 \cdot 16 \longrightarrow 15x - 5y + 5z = 80$
$C: 4x + 3y - 5z = 2 \longrightarrow 4x + 3y - 5z = 2$
$ E: \overline{19x - 2y + 0 = 82}$ ⟵ another new equation in only the variables x and y

4.3 Solve Third-Order Systems 147

3. Solve the resulting second-order system

$D:\ 2(5x + 3y) = 2 \cdot 11 \longrightarrow 10x + 6y = 22$
$E:\ 3(19x - 2y) = 3 \cdot 82 \longrightarrow 57x - 6y = 246$
$\overline{67x + 0 = 268}$ ⟵ one equation in one unknown

$$x = \frac{268}{67}$$

$$x = 4$$

$D:\quad 5x + 3y = 11$
$5(4) + 3y = 11 \qquad$ Substitute: $x = 4$
$20 + 3y = 11$
$3y = -9$
$y = -3$

4. Solve for third variable

$A:\quad 2x + 4y - z = -5$ ⟵ first equation in the original system
$2(4) + 4(-3) - z = -5 \qquad$ Substitute: $x = 4$ and $y = -3$
$8 + (-12) - z = -5$
$-4 - z = -5$
$-z = -1$
$z = 1$

Solution ▶ $(x, y, z) = (4, -3, 1)$ \qquad Check as before.

If one of the equations of a third-order system is missing a variable, then the addition method can be simplified as shown in the next example.

Example ▶ Solve $\begin{cases} A: \quad\ \ -y + z = 25 \\ B: \quad x + y + z = 35 \\ C: -x + 2y + z = 90 \end{cases}$

1. Eliminate a variable

$B:\quad x + y + z = 35$ \qquad Think: Eliminate x since equation A does not have an x term.
$C:\ -x + 2y + z = 90$
$D:\quad\ \ 0 + 3y + 2z = 125$ ⟵ new equation

2. Solve the resulting second-order system

$A:\ 3(-y + z) = 3 \cdot 25 \longrightarrow -3y + 3z = 75$
$D:\quad 3y + 2z = 125 \longrightarrow 3y + 2z = 125$
$\overline{0 + 5z = 200}$
$\phantom{D:\quad 3y + 2z = 125 \longrightarrow 0 + }z = 40$

$A:\ -y + z = 25$
$-y + 40 = 25 \qquad$ Substitute $z = 40$
$-y = -15$
$y = 15$

3. Solve for third variable

$B:\ x + y + z = 35$
$x + 15 + 40 = 35 \qquad$ Substitute $y = 15$ and $z = 40$.
$x = -20$

Solution ▶ $(x, y, z) = (-20, 15, 40)$ \qquad Check as before.

148 Chapter 4 Systems

If the addition method produces a false statement, then the third-order system is inconsistent.

Example ▶ Solve $\begin{cases} A: & x - 2y + 2z = 3 \\ B: & -x - y + 3z = 2 \\ C: & -2x + y + z = 1 \end{cases}$ by the addition method.

1. Eliminate a variable ▶

$A:\quad x - 2y + 2z = 3$
$B:\ -x - y + 3z = 2$
$D:\ \overline{\ 0 - 3y + 5z = 5}$ ⟵ a new equation

2. Eliminate the same variable from a different pair of equations ▶

$A:\ 2(x - 2y + 2z) = 2 \cdot 3 \longrightarrow \quad 2x - 4y + 4z = 6$
$C:\ -2x + y + z = 1 \longrightarrow \ \underline{-2x + y + z = 1}$
$E:\quad 0 - 3y + 5z = 7$ ⟵ another new equation

3. Solve the resulting second-order system ▶

$D:\quad -3y + 5z = 5 \longrightarrow \ -3y + 5z = 5$
$E:\ (-1)(-3y + 5z) = (-1)7 \longrightarrow \ \underline{3y - 5z = -7}$
$0 + 0 = -2$
$0 = -2$ ⟵ a false statement

4. Interpret ▶ A third-order system that reduces to a false statement is an inconsistent system.

Solution ▶ $\begin{cases} x - 2y + 2z = 3 \\ -x - y + 3z = 2 \\ -2x + y + z = 1 \end{cases}$ is an inconsistent system.

> **CAUTION:** If the addition method is used on a third-order system and a true statement such as 0 = 0 is produced, it does not necessarily imply that the third-order system is a dependent system.

Example ▶ $\begin{cases} A: & x + y + z = 4 \\ B: & -x - y - z = -4 \\ C: & -x - y - z = -3 \end{cases}$

$A:\ x + y + z = 4$
$B:\ \underline{-x - y - z = -4}$
$0 + 0 + 0 = 0$
$0 = 0$ ⟵ true

$A:\ x + y + z = 4$
$C:\ \underline{-x - y - z = -3}$
$0 + 0 + 0 = 1$
$0 = 1$ ⟵ false

Observe that for this system the addition of equations A and B has produced the true statement 0 = 0; however, the addition of equations A and C has produced the false statement 0 = 1. In this case, you can see that the system is inconsistent because equation A requires the sum of x, y, and z to equal 4 and the opposite of equation C requires the sum of x, y, and z to equal 3.

Make Sure

Solve each third-order system by the addition method. Check each solution.

See Example 2 ▶

1. $\begin{cases} 3x + 2y + z = 4 \\ 4x - 3y + 2z = 11 \\ x - 5y - 4z = 7 \end{cases}$ _____

2. $\begin{cases} 3x + 2y - z = 8 \\ 4x - 3y + 3z = -18 \\ 2x - y + 2z = -10 \end{cases}$ _____

MAKE SURE ANSWERS: 1. (2, −1, 0) **2.** (0, 2, −4)

4.3 Practice

Set 1: Is the given ordered triple a solution of the given third-order system?

1. $(2, 0, 1)$, $\begin{cases} 2x + 3y - 5z = -1 \\ 3x - 2y - 2z = 4 \\ x + 5y + 3z = 5 \end{cases}$

2. $(1, 2, 1)$, $\begin{cases} 3x - 2y + z = 0 \\ 4x + y - z = 5 \\ x + y + z = 4 \end{cases}$

3. $(-1, 3, 0)$, $\begin{cases} 2x + 2y - z = 4 \\ 3x + y + z = 0 \\ x + 2y - z = 7 \end{cases}$

4. $(0, -2, 3)$, $\begin{cases} x + 2y + 3z = 5 \\ 2x - y + 2z = 8 \\ 3x - 2y - z = 3 \end{cases}$

Set 2: Solve each third-order system using the addition method. If the system has no single solution, indicate whether the system is inconsistent or dependent. Check each solution.

5. $\begin{cases} x + y + z = 6 \\ x - y + z = 2 \\ x + y - z = 4 \end{cases}$

6. $\begin{cases} 3x + 2y + z = 4 \\ 4x - 3y + 2z = 11 \\ x - 5y - 4z = 7 \end{cases}$

7. $\begin{cases} 4x - 3y + z = 4 \\ 2x + 4y = -10 \\ 3y + 4z = 2 \end{cases}$

8. $\begin{cases} x + y + z = 4 \\ 3x - y + 2z = 14 \\ 2x - 2y - z = 6 \end{cases}$

9. $\begin{cases} 2x - y - z = 7 \\ 4x + 2y + 3z = -1 \\ x + 3y - 2z = 8 \end{cases}$

10. $\begin{cases} x - y - z = 5 \\ 2x + y - 2z = 4 \\ 3x + 2y + 3z = 17 \end{cases}$

150   Chapter 4   Systems

11. _____

11. $\begin{cases} 5x + 3y + 2z = 6 \\ x + 2y + 4z = 7 \\ 3x + 4y + z = 1 \end{cases}$ 12. $\begin{cases} x + 2y + 3z = 3 \\ 3x - y - 5z = 9 \\ 5x - 3y + z = 27 \end{cases}$ 13. $\begin{cases} x + 5y + 2z = 18 \\ 3x - 2y - z = -8 \\ 4x - 6y + 5z = 8 \end{cases}$

12. _____

13. _____

14. $\begin{cases} x - y - 3z = 6 \\ x + y = -1 \\ 5x - 4z = 2 \end{cases}$ 15. $\begin{cases} x + y + z = 1 \\ 2x + 3y = 2 \\ 6y + 6z = 1 \end{cases}$ 16. $\begin{cases} 4x - 2y + 6z = 4 \\ 2x - y + 3z = 2 \\ 2x + y - 3z = 10 \end{cases}$

14. _____

15. _____

17. $\begin{cases} 2x - y + 3z = 2 \\ 3x + y + 2z = 4 \\ x - 2y + 3z = 1 \end{cases}$ 18. $\begin{cases} 2x - 3y + z = 2 \\ x + 4y + 5z = 1 \\ 3x + y + 6z = 4 \end{cases}$ 19. $\begin{cases} x + 2y - z = 1 \\ 3x + y + 2z = -7 \\ 2x - y + 3z = -8 \end{cases}$

16. _____

17. _____

18. _____

Review: Work these problems on a separate sheet of paper. Attach your work to this page.

Evaluate each expression using the Order of Operations Rule. (See Lesson 1.4.)
20. $3(2) - 4(1)$ 21. $4(3) - 3(5)$ 22. $2(-3) - 2(4)$
23. $3(-2) - 4(-1)$ 24. $0(-2) - 3(1)$ 25. $-3(0) - (-2)(3)$
26. $-2(-3) - (-2)(0)$ 27. $-1(3) - (-2)(3)$ 28. $-3(-1) - (-1)(-4)$
29. $-3(2) - 3(0) + 2(-1)$ 30. $3(3) - (-2)(0) + (-1)(2)$
31. $2(-3) - 4(-2) + 3(2)$ 32. $-(-1)(-2) + (-3)(2) - 0(1)$
33. $-3(2) + (-2)(3) - (-1)(2)$ 34. $-4(-2) + (-2)(3) - (-1)(-2)$

19. _____

Copyright © 1985 by Harcourt Brace Jovanovich, Inc. All rights reserved.

Problem Solving 5: Solve Digit Problems

EXAMPLE: Solve this digit problem.

1. Read and identify ▶ The sum of the digits of a (3-digit number) is (13). The hundreds digit is (one-half) of the tens digit. When the digits are (reversed) the new number is (99 greater) than the original number. <u>What is the original number?</u>

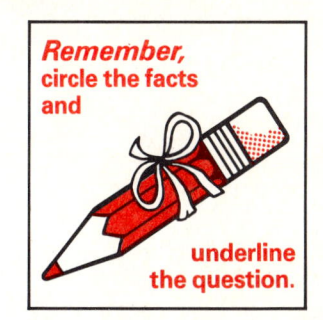

Remember, circle the facts and underline the question.

2. Understand ▶ The unknowns are $\begin{cases} \text{the units digit} \\ \text{the tens digit} \\ \text{the hundreds digit} \\ \text{the expanded notation for the original number} \\ \text{the expanded notation for the new number} \end{cases}$

3. Decide ▶ Let u = the units digit ⟵ first system unknown
then t = the tens digit ⟵ second system unknown
and h = the hundreds digit ⟵ third system unknown

4. Make a table ▶

	hundreds value	tens value	units value	expanded notation
original number	$h(100)$	$t(10)$	$u(1)$	$100h + 10t + u$
new number (digits reversed)	$u(100)$	$t(10)$	$h(1)$	$100u + 10t + h$

5. Translate ▶ The sum of the digits is 13.   The hundreds digit is one-half of the tens digit.

$$u + t + h = 13 \qquad\qquad h = \tfrac{1}{2} \cdot t$$

The new number is 99 greater than the original number.

$$100u + 10t + h = 99 + 100h + 10t + u$$

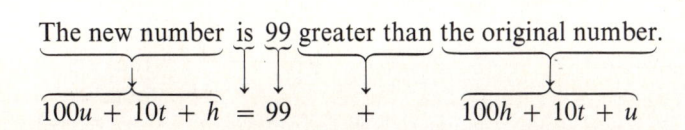

System Equations A: $u + t + h = 13$
B: $h = \tfrac{1}{2}t$
C: $100u + 10t + h = 99 + 100h + 10t + u$

6. Solve by the substitution method ▶

$t = 2h$	Solve equation B for t.
$u + 2h + h = 13$	Substitute equation B into equation A.
D: $u + 3h = 13$	Combine like terms to produce equation D.
$99u - 99h = 99$	Combine like terms in equation C.
$u - h = 1$	Divide equation C by 99.
$-u + h = -1$	Multiply equation C by -1.

$$\begin{array}{r} u + 3h = 13 \\ -u + h = -1 \\ \hline 0 + 4h = 12 \end{array}$$ Add equations D and C.

$h = 3$ Solve for the remaining variable.

152 Chapter 4 Systems

7. Interpret ▶ $h = 3$ means the hundreds digit is 3.
$t = 2h = 2 \cdot 3 = 6$ means the tens digit is 6.
$u + t + h = 13$ or $u = 13 - t - h = 13 - 6 - 3 = 4$ means the units digit is 4.
$100h + 10t + u = 100 \cdot 3 + 10 \cdot 6 + 4 = 364$ means the original number is 364.

8. Check ▶ Is the sum of the digits 13? Yes: $3 + 6 + 4 = 13$.
Is the hundreds digit one-half of the tens digit? Yes: $3 = \frac{1}{2} \cdot 6$.
Is the new number 99 greater than the original number? Yes: $463 = 99 + 364$.

Practice: Solve each digit problem.

1. The sum of the digits of a 2-digit number is 14. The tens digit is 4 more than the units digit. What is the number?

2. The sum of the digits of a 3-digit number is 20. The hundreds digit is 1 less than the units digit. The tens digit is 10 less than the other two digits combined. What is the number?

3. The sum of the digits of a 2-digit number is 10. The tens digit is 4 times the units digit. What is the number?

4. The sum of the digits of a 3-digit number is 18. The units digit is one-third the tens digit. The sum of the tens and units digit is twice the hundreds digit. What is the number?

5. The sum of the digits of a 2-digit number is 14. If the digits are reversed, the new number is 36 less than the original number. What is the original number?

6. The sum of the digits of a 3-digit number is 19. The units digit is one more than the tens digit. If the digits are reversed, the new number is 198 more than the original number. What is the new number?

7. The sum of the digits of a 2-digit number is 11. The units digit is one less than one-third of the tens digit. What is the number?

8. The sum of the digits of a 3-digit number is 10. The hundreds digit is one more than twice the units digit. If the digits are reversed, the new number is 396 less than the original number. What is the original number?

9. The tens digit of a 2-digit number is one-half the units digit. If the number is doubled, it will be 12 more than the reverse of the number. What is the original number?

10. Forty times the sum of the digits of a 3-digit number is 57 more than the reverse of the number. The hundreds digit is 2 less than the units digit. The sum of the hundreds and units digits equals the tens digit. Find the original number.

Extra

11. The sum of the digits of a 2-digit number is 13. The difference of the digits is 3. What two numbers are possible solutions?

12. The sum of the digits of a 3-digit number is 14. The sum and difference of the tens and units digits are 11 and 1, respectively. What are the two possible numbers?

4.4 Evaluate Determinants

The concept of a *determinant* is useful in solving systems of linear equations. A determinant is a square array of numbers written inside vertical bars. It symbolizes the sum or difference of certain products of these numbers.

A *second-order determinant* is a square array of 4 numbers denoted by $\begin{vmatrix} a_1 & b_1 \\ a_2 & b_2 \end{vmatrix}$. The *value of the determinant* $\begin{vmatrix} a_1 & b_1 \\ a_2 & b_2 \end{vmatrix}$ is defined to be the number $a_1b_2 - a_2b_1$.

Some of the terminology associated with second-order determinants is shown below.

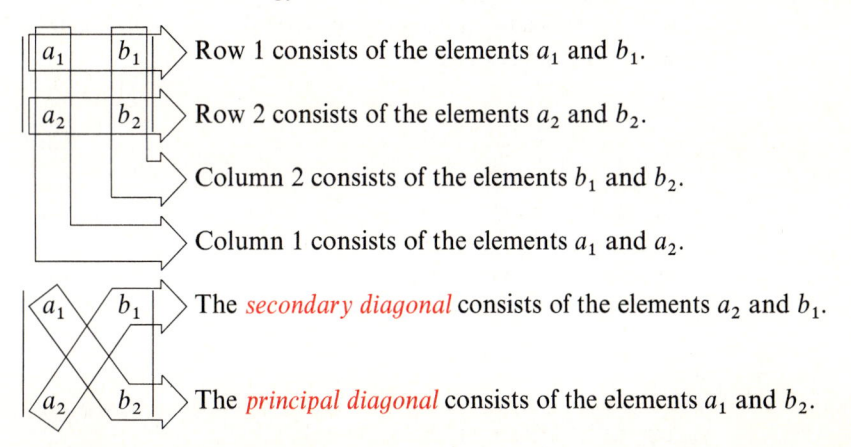

Row 1 consists of the elements a_1 and b_1.

Row 2 consists of the elements a_2 and b_2.

Column 2 consists of the elements b_1 and b_2.

Column 1 consists of the elements a_1 and a_2.

The *secondary diagonal* consists of the elements a_2 and b_1.

The *principal diagonal* consists of the elements a_1 and b_2.

The diagonals are useful in evaluating second-order determinants.

The Diagonal Procedure

You can find the value of a second-order determinant $\begin{vmatrix} a_1 & b_1 \\ a_2 & b_2 \end{vmatrix}$ by using the diagonals.

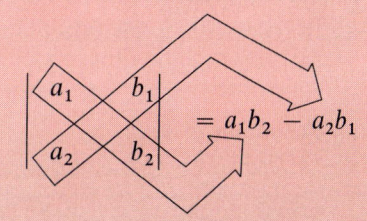

$$\begin{vmatrix} a_1 & b_1 \\ a_2 & b_2 \end{vmatrix} = a_1b_2 - a_2b_1$$

EXAMPLE 1: Evaluate a second-order determinant.

Evaluate ▶ $\begin{vmatrix} 3 & -4 \\ 2 & 9 \end{vmatrix}$

1. Use Diagonal Procedure ▶ $\begin{vmatrix} 3 & -4 \\ 2 & 9 \end{vmatrix} = 3 \cdot 9 - 2(-4)$

2. Simplify ▶ $= 27 - (-8)$

$= 35$

Solution ▶ $\begin{vmatrix} 3 & -4 \\ 2 & 9 \end{vmatrix} = 35$

154 Chapter 4 Systems

Make Sure

Evaluate each second-order determinant.

See Example 1 ▶
1. $\begin{vmatrix} 2 & 3 \\ 1 & 5 \end{vmatrix}$ _____
2. $\begin{vmatrix} 4 & -2 \\ 3 & -1 \end{vmatrix}$ _____
3. $\begin{vmatrix} 0 & -2 \\ -1 & 5 \end{vmatrix}$ _____

MAKE SURE ANSWERS: 1. 7 2. 2 3. −2

Determinants of the third order will be useful in solving third-order systems in Lesson 4.5.

Third-Order Determinant

$$\begin{vmatrix} a_1 & b_1 & c_1 \\ a_2 & b_2 & c_2 \\ a_3 & b_3 & c_3 \end{vmatrix} = a_1b_2c_3 - a_1b_3c_2 + a_2b_3c_1 - a_2b_1c_3 + a_3b_1c_2 - a_3b_2c_1.$$

It is not necessary to memorize the above definition. Applying some algebra to the six terms in the determinant above will allow you to write it as an expression that involves second-order determinants.

$$\begin{vmatrix} a_1 & b_1 & c_1 \\ a_2 & b_2 & c_2 \\ a_3 & b_3 & c_3 \end{vmatrix} = a_1b_2c_3 - a_1b_3c_2 + a_2b_3c_1 - a_2b_1c_3 + a_3b_1c_2 - a_3b_2c_1 \qquad \text{By definition.}$$

$$= a_1 \begin{vmatrix} b_2 & c_2 \\ b_3 & c_3 \end{vmatrix} - a_2 \begin{vmatrix} b_1 & c_1 \\ b_3 & c_3 \end{vmatrix} + a_3 \begin{vmatrix} b_1 & c_1 \\ b_2 & c_2 \end{vmatrix} \qquad \text{The definition written in terms of second-order determinants.}$$

The previous equation is referred to as the expansion of the determinant about its first column. The second-order determinants of A are called *minors*. The minor associated with any element can be obtained by deleting the row and column in which the element occurs.

Examples ▶
(a) The minor of a_1 in $\begin{vmatrix} a_1 & b_1 & c_1 \\ a_2 & b_2 & c_2 \\ a_3 & b_3 & c_3 \end{vmatrix}$ is $\begin{vmatrix} b_2 & c_2 \\ b_3 & c_3 \end{vmatrix}$.

(b) The minor of a_2 in $\begin{vmatrix} a_1 & b_1 & c_1 \\ a_2 & b_2 & c_2 \\ a_3 & b_3 & c_3 \end{vmatrix}$ is $\begin{vmatrix} b_1 & c_1 \\ b_3 & c_3 \end{vmatrix}$.

(c) The minor of 5 in $\begin{vmatrix} 1 & 2 & 3 \\ 4 & 5 & 6 \\ 7 & 8 & 9 \end{vmatrix}$ is $\begin{vmatrix} 1 & 3 \\ 7 & 9 \end{vmatrix} = 1 \cdot 9 - 7 \cdot 3 = 9 - 21 = -12.$

You can expand a third-order determinant about any row or any column. First form the products of each element in a row (column) with its minor. Then use the following *sign array* to determine which products are to be added and which products are to be subtracted.

$\begin{vmatrix} + & - & + \\ - & + & - \\ + & - & + \end{vmatrix}$ ⟵ sign array for third-order determinants

4.4 Evaluate Determinants 155

EXAMPLE 2: Expand a third-order determinant about a given row or column.

Problem ▶ Expand $\begin{vmatrix} 8 & 3 & 4 \\ 2 & 1 & 5 \\ 6 & 7 & -3 \end{vmatrix}$ about its second column.

1. Form product ▶ $\begin{vmatrix} 8 & 3 & 4 \\ 2 & 1 & 5 \\ 6 & 7 & -3 \end{vmatrix} = ?\ 3\begin{vmatrix} 2 & 5 \\ 6 & -3 \end{vmatrix}\ ?\ 1\begin{vmatrix} 8 & 4 \\ 6 & -3 \end{vmatrix}\ ?\ 7\begin{vmatrix} 8 & 4 \\ 2 & 5 \end{vmatrix}$ Form the product of each element in the second column and its minor.

2. Determine each sign ▶ $\begin{vmatrix} + & - & + \\ - & + & - \\ + & - & + \end{vmatrix}$ $-3\begin{vmatrix} 2 & 5 \\ 6 & -3 \end{vmatrix} + 1\begin{vmatrix} 8 & 4 \\ 6 & -3 \end{vmatrix} - 7\begin{vmatrix} 8 & 4 \\ 2 & 5 \end{vmatrix}$ Use the sign array.

Solution ▶ $\begin{vmatrix} 8 & 3 & 4 \\ 2 & 1 & 5 \\ 6 & 7 & -3 \end{vmatrix} = -3\begin{vmatrix} 2 & 5 \\ 6 & -3 \end{vmatrix} + 1\begin{vmatrix} 8 & 4 \\ 6 & -3 \end{vmatrix} - 7\begin{vmatrix} 8 & 4 \\ 2 & 5 \end{vmatrix}$ The determinant expanded about its second column.

Make Sure

Expand each third-order determinant about the given row or column.

See Example 2 ▶ **1.** $\begin{vmatrix} 2 & 3 & 4 \\ 1 & -1 & -2 \\ 5 & -3 & -4 \end{vmatrix}$ Row 2   **2.** $\begin{vmatrix} 3 & 1 & 4 \\ 0 & 2 & -1 \\ 2 & -2 & -3 \end{vmatrix}$ Column 1

MAKE SURE ANSWERS:

1. $-1\begin{vmatrix} 3 & 4 \\ -3 & -4 \end{vmatrix} + (-1)\begin{vmatrix} 2 & 4 \\ 5 & -4 \end{vmatrix} - (-2)\begin{vmatrix} 2 & 3 \\ 5 & -3 \end{vmatrix}$ **2.** $3\begin{vmatrix} 2 & -1 \\ -2 & -3 \end{vmatrix} - (0)\begin{vmatrix} 1 & 4 \\ -2 & -3 \end{vmatrix} + 2\begin{vmatrix} 1 & 4 \\ 2 & -1 \end{vmatrix}$

If all the elements of a third-order determinant are known constants, you can evaluate the determinant by expanding it and simplifying the resulting expression.

EXAMPLE 3: Evaluate a third-order determinant.

Evaluate ▶ $\begin{vmatrix} 3 & -2 & 1 \\ -4 & 2 & -3 \\ -1 & 4 & 5 \end{vmatrix}$ Choose a row or column to expand about.

1. Expand ▶ $\begin{vmatrix} 3 & -2 & 1 \\ -4 & 2 & -3 \\ -1 & 4 & 5 \end{vmatrix} = ?\ 3\begin{vmatrix} 2 & -3 \\ 4 & 5 \end{vmatrix}\ ?\ -4\begin{vmatrix} -2 & 1 \\ 4 & 5 \end{vmatrix}\ ?\ -1\begin{vmatrix} -2 & 1 \\ 2 & -3 \end{vmatrix}$

156 Chapter 4 Systems

2. Determine each sign ▶

$$\begin{array}{ccc} + & - & + \\ - & + & - \\ + & - & + \end{array} \quad +3\begin{vmatrix} 2 & -3 \\ 4 & 5 \end{vmatrix} - (-4)\begin{vmatrix} -2 & 1 \\ 4 & 5 \end{vmatrix} + (-1)\begin{vmatrix} -2 & 1 \\ 2 & -3 \end{vmatrix}$$

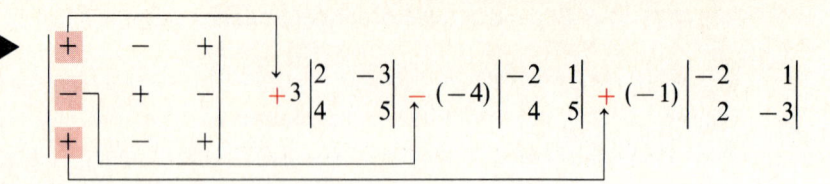

3. Simplify ▶

$$\begin{vmatrix} 3 & -2 & 1 \\ -4 & 2 & -3 \\ -1 & 4 & 5 \end{vmatrix} = +3\begin{vmatrix} 2 & -3 \\ 4 & 5 \end{vmatrix} - (-4)\begin{vmatrix} -2 & 1 \\ 4 & 5 \end{vmatrix} + (-1)\begin{vmatrix} -2 & 1 \\ 2 & -3 \end{vmatrix}$$

$$= 3(10 - (-12)) - (-4)(-10 - 4) + (-1)(6 - 2)$$

Solution ▶ $= 6$

Note ▶ When you evaluate a third-order determinant, you will get the same value regardless of which row or column you expand about. If a third-order determinant contains a zero, then you can evaluate it more easily by expanding it about a row or column that contains the zero.

Example ▶ Evaluate $\begin{vmatrix} 2 & 1 & -3 \\ 4 & -1 & 0 \\ 5 & 3 & -2 \end{vmatrix}$ by expanding about the second row.

1. Expand ▶

$$\begin{vmatrix} 2 & 1 & -3 \\ 4 & -1 & 0 \\ 5 & 3 & -2 \end{vmatrix} = ?\,4\begin{vmatrix} 1 & -3 \\ 3 & -2 \end{vmatrix} ?\,-1\begin{vmatrix} 2 & -3 \\ 5 & -2 \end{vmatrix} ?\,0\begin{vmatrix} 2 & 1 \\ 5 & 3 \end{vmatrix}$$

2. Determine each sign ▶

$$\begin{array}{ccc} + & - & + \\ - & + & - \\ + & - & + \end{array} \quad -4\begin{vmatrix} 1 & -3 \\ 3 & -2 \end{vmatrix} + (-1)\begin{vmatrix} 2 & -3 \\ 5 & -2 \end{vmatrix} - 0\begin{vmatrix} 2 & 1 \\ 5 & 3 \end{vmatrix}$$

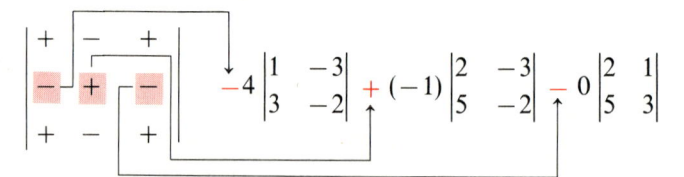

3. Simplify ▶

$$\begin{vmatrix} 2 & 1 & -3 \\ 4 & -1 & 0 \\ 5 & 3 & -2 \end{vmatrix} = -4\begin{vmatrix} 1 & -3 \\ 3 & -2 \end{vmatrix} + (-1)\begin{vmatrix} 2 & -3 \\ 5 & -2 \end{vmatrix} - 0\begin{vmatrix} 2 & 1 \\ 5 & 3 \end{vmatrix}$$

$$= -4((-2) - (-9)) + (-1)((-4) - (-15)) - 0$$

Think: $0 \cdot a = 0$ for any a.

Solution ▶ $= -39$

Make Sure

Evaluate each third-order determinant by the expansion method.

See Example 3 ▶ **1.** $\begin{vmatrix} 4 & 1 & 2 \\ -1 & 3 & -3 \\ 2 & -4 & 3 \end{vmatrix}$ _____ **2.** $\begin{vmatrix} -1 & 0 & 2 \\ 2 & 1 & -3 \\ 0 & -2 & 1 \end{vmatrix}$ _____

MAKE SURE ANSWERS: 1. −19 2. −3

4.4 Practice

Set 1: Evaluate each second-order determinant.

1. $\begin{vmatrix} 2 & 3 \\ 1 & 4 \end{vmatrix}$
2. $\begin{vmatrix} 3 & 2 \\ 2 & 3 \end{vmatrix}$
3. $\begin{vmatrix} 3 & -1 \\ 2 & 2 \end{vmatrix}$

4. $\begin{vmatrix} 4 & 2 \\ -2 & 1 \end{vmatrix}$
5. $\begin{vmatrix} 3 & 0 \\ 4 & 1 \end{vmatrix}$
6. $\begin{vmatrix} 4 & 2 \\ 0 & 1 \end{vmatrix}$

7. $\begin{vmatrix} -3 & 2 \\ 4 & 0 \end{vmatrix}$
8. $\begin{vmatrix} 0 & -1 \\ -4 & -2 \end{vmatrix}$
9. $\begin{vmatrix} -3 & -2 \\ 3 & -1 \end{vmatrix}$

10. $\begin{vmatrix} -1 & 2 \\ -3 & 1 \end{vmatrix}$
11. $\begin{vmatrix} 1 & 0 \\ 0 & 1 \end{vmatrix}$
12. $\begin{vmatrix} 0 & 1 \\ 1 & 0 \end{vmatrix}$

Set 2: Expand each third-order determinant about the given row or column.

13. $\begin{vmatrix} 1 & 2 & 3 \\ -1 & 0 & 2 \\ 2 & 3 & 1 \end{vmatrix}$ Row 1

14. $\begin{vmatrix} -2 & 0 & 1 \\ 4 & 3 & 2 \\ 1 & -1 & -2 \end{vmatrix}$ Row 2

15. $\begin{vmatrix} 2 & -1 & -2 \\ -3 & 4 & 2 \\ 0 & 2 & 3 \end{vmatrix}$ Column 1

16. $\begin{vmatrix} 1 & 0 & 0 \\ 0 & 1 & 0 \\ 2 & -1 & 3 \end{vmatrix}$ Column 2

158 Chapter 4 Systems

17. _____

Set 3: Evaluate each third-order determinant.

17. $\begin{vmatrix} 0 & 2 & 4 \\ 1 & 3 & -2 \\ -1 & 4 & 1 \end{vmatrix}$ 18. $\begin{vmatrix} 1 & 2 & 3 \\ 0 & -1 & 2 \\ -2 & 3 & 1 \end{vmatrix}$ 19. $\begin{vmatrix} -2 & -3 & -1 \\ 4 & 2 & 0 \\ 1 & 2 & 3 \end{vmatrix}$

18. _____

19. _____

20. $\begin{vmatrix} 2 & 1 & 1 \\ 1 & -2 & -2 \\ 3 & 3 & 1 \end{vmatrix}$ 21. $\begin{vmatrix} 2 & 1 & -1 \\ 1 & -2 & 3 \\ 2 & 1 & -2 \end{vmatrix}$ 22. $\begin{vmatrix} 1 & 2 & 2 \\ -2 & 1 & -3 \\ 3 & -2 & 1 \end{vmatrix}$

20. _____

21. _____

23. $\begin{vmatrix} 1 & 2 & 3 \\ -3 & 2 & 1 \\ -2 & -1 & -3 \end{vmatrix}$ 24. $\begin{vmatrix} 0 & 0 & 1 \\ 0 & 1 & 0 \\ 1 & 0 & 0 \end{vmatrix}$ 25. $\begin{vmatrix} 1 & 0 & 0 \\ 0 & 1 & 0 \\ 0 & 0 & 1 \end{vmatrix}$

22. _____

23. _____

Review: Work these problems on a separate sheet of paper. Attach your work to this page.

Divide signed numbers. (See Lesson 1.3.)

26. $\dfrac{-4}{2}$ 27. $\dfrac{-12}{4}$ 28. $\dfrac{24}{-6}$ 29. $\dfrac{48}{-8}$

24. _____

30. $\dfrac{-72}{-12}$ 31. $\dfrac{-96}{-16}$ 32. $\dfrac{-64}{12}$ 33. $\dfrac{56}{-21}$

Evaluate each expression using the Order of Operations Rule. (See Lesson 1.4.)
34. $3(-2) - 0(-3)$
35. $0(-3) - (-3)(3)$
36. $-3(2) - 0(-3)$
37. $0(2) - 3(-2) + (-4)(5)$

25. _____
38. $-3(-2) - (-4)(3) + 0(2)$
39. $-(-2)(3) + (-3)(0) - (-1)(3)$

Copyright © 1985 by Harcourt Brace Jovanovich, Inc. All rights reserved.

4.5 Solve Using Cramer's Rule

Solving a general second-order system by the addition method demonstrates how determinants can be used to solve second-order systems.

Example ▶ Solve $\begin{cases} a_1x + b_1y = c_1 \\ a_2x + b_2y = c_2 \end{cases}$ for x.

1. Eliminate a variable ▶
$$b_2(a_1x + b_1y) = b_2c_1 \longrightarrow a_1b_2x + b_1b_2y = c_1b_2$$
$$-b_1(a_2x + b_2y) = -b_1c_2 \longrightarrow -a_2b_1x - b_1b_2y = -c_2b_1$$
$$\overline{a_1b_2x - a_2b_1x + 0 = c_1b_2 - c_2b_1}$$

2. Solve for x ▶
$$x(a_1b_2 - a_2b_1) = c_1b_2 - c_2b_1$$
$$x = \frac{c_1b_2 - c_2b_1}{a_1b_2 - a_2b_1} \quad \text{Provided } a_1b_2 - a_2b_1 \neq 0$$

3. Write determinant notation ▶ $x = \dfrac{\begin{vmatrix} c_1 & b_1 \\ c_2 & b_2 \end{vmatrix}}{\begin{vmatrix} a_1 & b_1 \\ a_2 & b_2 \end{vmatrix}}$. In a similar manner $y = \dfrac{\begin{vmatrix} a_1 & c_1 \\ a_2 & c_2 \end{vmatrix}}{\begin{vmatrix} a_1 & b_1 \\ a_2 & b_2 \end{vmatrix}}$ provided $a_1b_2 - a_2b_1 \neq 0$

These results are summarized in the following rule, which is named after the Swiss physicist and mathematician Gabriel Cramer (1704–1752).

Cramer's Rule for Second-Order Systems

Given $\begin{cases} a_1x + b_1y = c_1 \\ a_2x + b_2y = c_2 \end{cases}$ and $D = \begin{vmatrix} a_1 & b_1 \\ a_2 & b_2 \end{vmatrix} \neq 0$, $D_x = \begin{vmatrix} c_1 & b_1 \\ c_2 & b_2 \end{vmatrix}$, $D_y = \begin{vmatrix} a_1 & c_1 \\ a_2 & c_2 \end{vmatrix}$, then:

$$x = \frac{D_x}{D} \quad \text{and} \quad y = \frac{D_y}{D}.$$

The determinant D is called the *coefficient determinant*, since it consists of the coefficients of x and y. D_x is a determinant formed from D by replacing the x-coefficients with the constants. D_y is the determinant formed from D by replacing the y-coefficients with the constants.

EXAMPLE 1: Solve a second-order system using Cramer's Rule.

Problem ▶ Solve $\begin{cases} 4x + 5y = 8 \\ 2x + 3y = 3 \end{cases}$ using Cramer's Rule.

1. Compute D ▶ $D = \begin{vmatrix} 4 & 5 \\ 2 & 3 \end{vmatrix} = 12 - 10 = 2$ Observe: $D \neq 0$.

2. Compute D_x ▶ $D_x = \begin{vmatrix} 8 & 5 \\ 3 & 3 \end{vmatrix} = 24 - 15 = 9$ Replace the x-coefficients in D with the constants.

3. Compute D_y ▶ $D_y = \begin{vmatrix} 4 & 8 \\ 2 & 3 \end{vmatrix} = 12 - 16 = -4$ Replace the y-coefficients in D with the constants.

4. Apply Cramer's Rule ▶ $x = \dfrac{D_x}{D} = \dfrac{9}{2}, \quad y = \dfrac{D_y}{D} = \dfrac{-4}{2} = -2$

Solution ▶ $(x, y) = (\tfrac{9}{2}, -2)$

160 Chapter 4 Systems

Cramer's Rule can only be applied if $D \neq 0$. If $D = 0$ and at least one of the determinants D_x or D_y is not zero, the second-order system has no solutions and is inconsistent. If all the determinants D, D_x, and D_y are zero, the second-order system has infinitely many solutions and is dependent.

Example ▶ Solve $\begin{cases} 4x - 2y = 6 \\ 6x - 3y = 1 \end{cases}$ using Cramer's Rule.

1. Compute D ▶ $D = \begin{vmatrix} 4 & -2 \\ 6 & -3 \end{vmatrix} = -12 - (-12) = 0$

2. Compute D_x ▶ $D_x = \begin{vmatrix} 6 & -2 \\ 1 & -3 \end{vmatrix} = -18 - (-2) = -16$

3. Interpret ▶ $D = 0$ and $D_x = -16 \neq 0$ means the system is inconsistent.

Solution ▶ $\begin{cases} 4x - 2y = 6 \\ 6x - 3y = 1 \end{cases}$ is an inconsistent system.

Another Example ▶ Solve $\begin{cases} x - 2y = 5 \\ -2x + 4y = -10 \end{cases}$ using Cramer's Rule.

1. Compute D ▶ $D = \begin{vmatrix} 1 & -2 \\ -2 & 4 \end{vmatrix} = 4 - 4 = 0$

2. Compute D_x ▶ $D_x = \begin{vmatrix} 5 & -2 \\ -10 & 4 \end{vmatrix} = 20 - 20 = 0$

3. Compute D_y ▶ $D_y = \begin{vmatrix} 1 & 5 \\ -2 & -10 \end{vmatrix} = -10 - (-10) = 0$

4. Interpret ▶ $D = D_x = D_y = 0$ means the system is a dependent system.

Solution ▶ $\begin{cases} x - 2y = 5 \\ -2x + 4y = -10 \end{cases}$ is a dependent system.

Make Sure

Solve each second-order system using Cramer's Rule. Check each solution.

See Example 1 ▶
1. $\begin{cases} 4x - y = 3 \\ 2x + 4y = -3 \end{cases}$ _____ 2. $\begin{cases} 6x + 9y = 11 \\ 3x = 4 \end{cases}$ _____

3. $\begin{cases} 2x - 3y = 4 \\ 4x - 6y = 5 \end{cases}$ _____ 4. $\begin{cases} 3y = 2x - 6 \\ y = \frac{2}{3}x - 2 \end{cases}$ _____

MAKE SURE ANSWERS: 1. $(\frac{3}{6}, -1)$ **2.** $(\frac{4}{3}, \frac{3}{3})$ **3.** no solution (inconsistent) **4.** infinite number of solutions (dependent)

4.5 Solve Using Cramer's Rule

Cramer's Rule can be generalized to solve systems of n linear equations in n variables.

Cramer's Rule for Third-Order Systems

Given $\begin{cases} a_1x + b_1y + c_1z = d_1 \\ a_2x + b_2y + c_2z = d_2 \\ a_3x + b_3y + c_3z = d_3 \end{cases}$,

$D = \begin{vmatrix} a_1 & b_1 & c_1 \\ a_2 & b_2 & c_2 \\ a_3 & b_3 & c_3 \end{vmatrix} \neq 0, D_x = \begin{vmatrix} d_1 & b_1 & c_1 \\ d_2 & b_2 & c_2 \\ d_3 & b_3 & c_3 \end{vmatrix}, D_y = \begin{vmatrix} a_1 & d_1 & c_1 \\ a_2 & d_2 & c_2 \\ a_3 & d_3 & c_3 \end{vmatrix}$, and $D_z = \begin{vmatrix} a_1 & b_1 & d_1 \\ a_2 & b_2 & d_2 \\ a_3 & b_3 & d_3 \end{vmatrix}$,

then: $x = \dfrac{D_x}{D}, \quad y = \dfrac{D_y}{D},$ and $z = \dfrac{D_z}{D}$.

EXAMPLE 2: Solve a third-order system using Cramer's Rule.

Problem ▶ Solve $\begin{cases} x - 3y - 2z = 3 \\ 2x + 5y + 2z = -6 \\ x - y + z = 4 \end{cases}$ using Cramer's Rule.

1. Compute D ▶ $D = \begin{vmatrix} 1 & -3 & -2 \\ 2 & 5 & 2 \\ 1 & -1 & 1 \end{vmatrix} = 1\begin{vmatrix} 5 & 2 \\ -1 & 1 \end{vmatrix} - (2)\begin{vmatrix} -3 & -2 \\ -1 & 1 \end{vmatrix} + 1\begin{vmatrix} -3 & -2 \\ 5 & 2 \end{vmatrix}$ Expanding about the first column.

$= 1(5 - (-2)) - (2)(-3 - 2) + 1(-6 - (-10))$

$= 21$

2. Compute D_x ▶ $D_x = \begin{vmatrix} 3 & -3 & -2 \\ -6 & 5 & 2 \\ 4 & -1 & 1 \end{vmatrix} = 3\begin{vmatrix} 5 & 2 \\ -1 & 1 \end{vmatrix} - (-6)\begin{vmatrix} -3 & -2 \\ -1 & 1 \end{vmatrix} + 4\begin{vmatrix} -3 & -2 \\ 5 & 2 \end{vmatrix}$ Expanding about the first column.

$= 3(5 - (-2)) - (-6)(-3 - 2) + 4(-6 - (-10))$

$= 7$

3. Compute D_y ▶ $D_y = \begin{vmatrix} 1 & 3 & -2 \\ 2 & -6 & 2 \\ 1 & 4 & 1 \end{vmatrix} = 1\begin{vmatrix} -6 & 2 \\ 4 & 1 \end{vmatrix} - (2)\begin{vmatrix} 3 & -2 \\ 4 & 1 \end{vmatrix} + 1\begin{vmatrix} 3 & -2 \\ -6 & 2 \end{vmatrix}$

$= 1(-6 - 8) - (2)(3 - (-8)) + 1(6 - (12))$

$= -42$

4. Compute D_z ▶ $D_z = \begin{vmatrix} 1 & -3 & 3 \\ 2 & 5 & -6 \\ 1 & -1 & 4 \end{vmatrix} = 1\begin{vmatrix} 5 & -6 \\ -1 & 4 \end{vmatrix} - (2)\begin{vmatrix} -3 & 3 \\ -1 & 4 \end{vmatrix} + 1\begin{vmatrix} -3 & 3 \\ 5 & -6 \end{vmatrix}$

$= 1(20 - 6) - (2)(-12 - (-3)) + 1(18 - 15)$

$= 35$

5. Apply Cramer's Rule ▶ $x = \dfrac{D_x}{D} = \dfrac{7}{21} = \dfrac{1}{3}, \quad y = \dfrac{D_y}{D} = \dfrac{-42}{21} = -2,$ and $z = \dfrac{D_z}{D} = \dfrac{35}{21} = \dfrac{5}{3}$

Solution ▶ $(x, y, z) = (\tfrac{1}{3}, -2, \tfrac{5}{3})$

Chapter 4 Systems

There are some techniques that you can use to reduce the amount of computation involved in using Cramer's Rule to solve a third-order system.

Example ▶ Solve $\begin{cases} 2x + 4y - z = -3 \\ x - 3y + z = 8 \\ x + 2y = 0 \end{cases}$ using Cramer's Rule.

1. Compute D ▶ $D = \begin{vmatrix} 2 & 4 & -1 \\ 1 & -3 & 1 \\ 1 & 2 & 0 \end{vmatrix} = (-1)\begin{vmatrix} 1 & -3 \\ 1 & 2 \end{vmatrix} - (1)\begin{vmatrix} 2 & 4 \\ 1 & 2 \end{vmatrix}$ Expand about the third column because it has a zero element.

$= (-1)(2 - (-3)) - (1)(4 - 4)$

$= -5$

2. Compute D_x ▶ $D_x = \begin{vmatrix} -3 & 4 & -1 \\ 8 & -3 & 1 \\ 0 & 2 & 0 \end{vmatrix} = -(2)\begin{vmatrix} -3 & -1 \\ 8 & 1 \end{vmatrix}$ Expand about the third row because it contains two zero elements.

$= -(2)(-3 - (-8))$

$= -10$

3. Compute D_y ▶ $D_y = \begin{vmatrix} 2 & -3 & -1 \\ 1 & 8 & 1 \\ 1 & 0 & 0 \end{vmatrix} = 1\begin{vmatrix} -3 & -1 \\ 8 & 1 \end{vmatrix}$ Expand about the third row because of the two zero elements.

$= (-3 - (-8))$

$= 5$

4. Apply Cramer's Rule ▶ $x = \dfrac{D_x}{D} = \dfrac{-10}{-5} = 2, \quad y = \dfrac{D_y}{D} = \dfrac{5}{-5} = -1$

5. Solve for the third variable ▶
$2x + 4y - z = -3$ ⟵ first equation
$2(2) + 4(-1) - z = -3$ Substitute 2 for x and -1 for y.
$4 + (-4) - z = -3$
$-z = -3$
$z = 3$

Solution ▶ $(x, y, z) = (2, -1, 3)$ Check as before.

Make Sure

Solve each third-order system using Cramer's Rule. Check each solution.

See Example 2 ▶ **1.** $\begin{cases} 3x - 4y - 3z = 3 \\ 2x - 3y - 2z = 1 \\ x + y + z = 8 \end{cases}$ _____   **2.** $\begin{cases} x + y + z = 6 \\ x + 2y = 7 \\ x - y - 2z = -1 \end{cases}$ _____

MAKE SURE ANSWERS: 1. (5, 3, 0) **2.** (3, 2, 1)

4.5 Practice: *Check each solution in the original system.*

Set 1: Solve each second-order system using Cramer's Rule.

1. $\begin{cases} 3x + 2y = 13 \\ 2x - 4y = -2 \end{cases}$
2. $\begin{cases} 4x - 3y = -17 \\ x + 5y = 13 \end{cases}$
3. $\begin{cases} x + y = 4 \\ 2x - 3y = 13 \end{cases}$

4. $\begin{cases} 4x - y = 9 \\ 3x + 4y = 2 \end{cases}$
5. $\begin{cases} 2x + 6y = 5 \\ 4x - 3y = 0 \end{cases}$
6. $\begin{cases} 3x + 4y = 5 \\ 9x - 8y = 0 \end{cases}$

7. $\begin{cases} 4x + 9y = 21 \\ 2x - 6y = -7 \end{cases}$
8. $\begin{cases} 6x + 2y = 2 \\ 3x - 6y = -3 \end{cases}$
9. $\begin{cases} 2x - y = 1 \\ 6x - 3y = 3 \end{cases}$

10. $\begin{cases} 5x + 3y = 2 \\ x + 0.6y = 0.4 \end{cases}$
11. $\begin{cases} x - 3y = 4 \\ 3x - 9y = 6 \end{cases}$
12. $\begin{cases} 0.5x - y = 1.75 \\ 2x - 4y = 7 \end{cases}$

164   Chapter 4   Systems

Set 2: Solve each third-order system using Cramer's Rule.

13. $\begin{cases} x + y + z = 6 \\ x + 2y - z = 6 \\ x - y - 2z = -1 \end{cases}$ 14. $\begin{cases} x - y - z = 5 \\ 2x - 4y + 3z = 11 \\ 3x - 2y - 5z = 15 \end{cases}$ 15. $\begin{cases} x = 2y - z + 5 \\ 2x = -3y - z + 2 \\ 0 = y - 2z + 3 \end{cases}$

16. $\begin{cases} 0 = 3y + 2z - 6 \\ 2x = 5y + 2z - 4 \\ 0 = 2y + 3z - 9 \end{cases}$ 17. $\begin{cases} x + y = 1 - 2z \\ 2x + y = 3 - z \\ 2x + 2y = 2 - z \end{cases}$ 18. $\begin{cases} 2x + y = 1 - z \\ x + 2y = z \\ x + y = 1 + z \end{cases}$

19. $\begin{cases} x + y = 1 \\ x + z = 1 \\ y + z = 1 \end{cases}$ 20. $\begin{cases} x - y = -1 \\ x - z = 1 \\ y + z = 0 \end{cases}$ 21. $\begin{cases} x + 2y - z = -3 \\ x + 3y + 2z = 4 \\ 2x + y + z = -3 \end{cases}$

22. $\begin{cases} 5x + 2y - 3z = 27 \\ x - y - 2z = 4 \\ 4x + 5y - z = 27 \end{cases}$ 23. $\begin{cases} x + y = 1 \\ y + z = 1 \\ 2y + z = 2 \end{cases}$ 24. $\begin{cases} x + 2y - 3z = 6 \\ 2x + 4y - 6z = 1 \\ 3x + 6y - 9z = 1 \end{cases}$

Review: Work these problems on a separate sheet of paper. Attach your work to this page.

Remove the grouping symbols using the distributive properties. (See Lesson 1.5, Example 2.)
25. $-(x + 3)$ 26. $-(x - 4)$ 27. $-(x^2 - 3x + 2)$
28. $-(x^2 + 2x - 5)$ 29. $-(3 - 2x - 3x^3)$ 30. $-(4 + 5x - 2x^3)$
31. $-(1 - 3x^4 + 2x)$ 32. $-(5 + 4x^3 - 2x^2 + x)$

Simplify each expression by combining like terms. (See Lesson 1.5, Example 3.)
33. $4x - 3 + 2x + 5$ 34. $5x - 6 - 3x - 4 + 2x$
35. $3x - 5 + 4x + 7 - 5x$ 36. $2x^2 - 3x - 5 - 3x^2 + 4x$
37. $5x - 2x^2 - 3 - 7x + x^2 + 1$ 38. $5x^2 - 7 + 3x - 2 - 7x^2 - 5x$

Name _____ Date _____ Class _____

Chapter 4 Review

What to Review if You Have Trouble

		Lesson	Example	Page

Objectives

Check proposed system solutions ▶ 1. Is $(-1, 3)$ a solution of
$$\begin{cases} 2x - 4y = -14 \\ -5x + 2y = 1 \end{cases}?$$
_____ 4.1 1 132

Solve second-order systems by graphing ▶ 2. Solve $\begin{cases} 2x - y = 5 \\ -x + 3y = 0 \end{cases}$ by graphing.

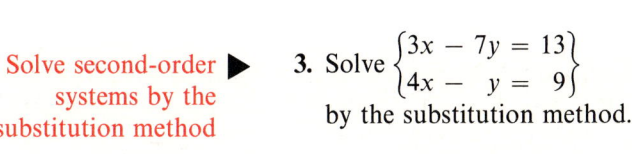

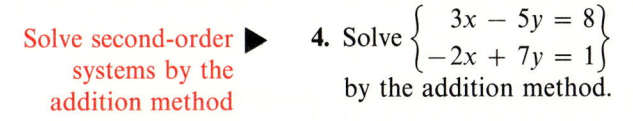

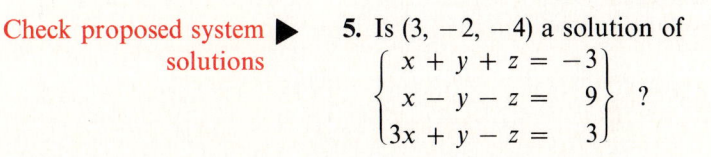

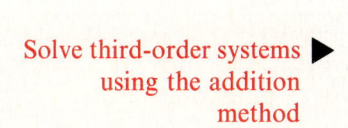

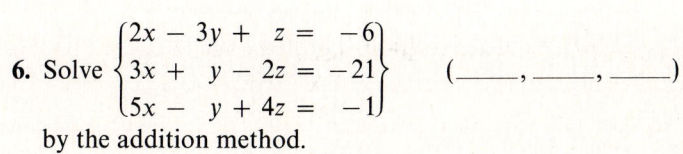

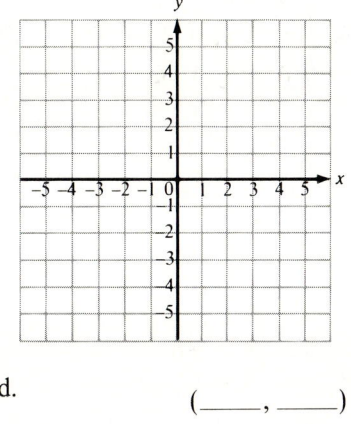

(____, ____)

4.1 2 133

Solve second-order systems by the substitution method ▶ 3. Solve $\begin{cases} 3x - 7y = 13 \\ 4x - y = 9 \end{cases}$ by the substitution method.

(____, ____) 4.2 1 137

Solve second-order systems by the addition method ▶ 4. Solve $\begin{cases} 3x - 5y = 8 \\ -2x + 7y = 1 \end{cases}$ by the addition method.

(____, ____) 4.2 2 139

Check proposed system solutions ▶ 5. Is $(3, -2, -4)$ a solution of
$$\begin{cases} x + y + z = -3 \\ x - y - z = 9 \\ 3x + y - z = 3 \end{cases}?$$
_____ 4.3 1 145

Solve third-order systems using the addition method ▶ 6. Solve $\begin{cases} 2x - 3y + z = -6 \\ 3x + y - 2z = -21 \\ 5x - y + 4z = -1 \end{cases}$ by the addition method.

(____, ____, ____) 4.3 2 146

Copyright © 1985 by Harcourt Brace Jovanovich, Inc. All rights reserved.

166 Chapter 4 Systems

Evaluate second-order determinants ▶	7. Evaluate $\begin{vmatrix} -5 & 7 \\ 3 & -9 \end{vmatrix}$. _____	4.4	1	153
Expand third-order determinants about a given row or column ▶	8. Expand $\begin{vmatrix} 1 & 2 & 4 \\ 5 & 3 & 6 \\ 9 & 7 & 8 \end{vmatrix}$ about the third row.	4.4	2	155
Evaluate third-order determinants ▶	9. Evaluate $\begin{vmatrix} 5 & -2 & 4 \\ -3 & 1 & 6 \\ 0 & 2 & -1 \end{vmatrix}$. _____	4.4	3	155
Solve second-order systems using Cramer's Rule ▶	10. Solve $\begin{cases} 2x + y = -1 \\ -3x - 2y = 5 \end{cases}$. (____, ____)	4.5	1	159
Solve third-order systems using Cramer's Rule ▶	11. Solve $\begin{cases} x + y + z = 0 \\ 2x + 3y + 4z = -4 \\ 3x - 2y + z = -4 \end{cases}$. (____, ____, ____)	4.5	2	161
Solve mixture problems ▶	12. How much water must be evaporated from 120 gallons of a 1%-salt solution to obtain a 3%-salt solution? _____	PS 4	—	143
Solve digit problems ▶	13. The sum of the digits of a 3-digit number is 22. The hundreds digit is one less than twice the tens digit. When the digits are reversed, the new number is 99 less than the original number. What is the original number? _____	PS 5	—	151

CHAPTER 4 REVIEW ANSWERS: 1. See Appendix Selected Answers. **2.** No **3.** (2, −1) **4.** $(\frac{61}{19}, \frac{19}{11})$ **5.** No **6.** (−4, 1, 5) **7.** 24 **8.** $\begin{vmatrix} 2 & 9 \\ 4 & 3 \\ 1 & 6 \end{vmatrix} - 7 \begin{vmatrix} 1 & 4 \\ 5 & 6 \\ 3 & 3 \end{vmatrix} + 8 \begin{vmatrix} 1 & 2 \\ 5 & 3 \end{vmatrix}$ **9.** −83 **10.** (3, −7) **11.** (1, 2, −3) **12.** 80 gallons **13.** 958

POLYNOMIALS AND FACTORING

5 Polynomials

5.1 Add and Subtract Polynomials

PS 6: Evaluate Polynomials

5.2 Multiply Polynomials

5.3 Find Special Products

5.4 Divide Polynomials

5.5 Use Synthetic Division

Introduction to Polynomials

A basic and useful concept occurring often in elementary mathematics is that of the *polynomial*.

> A polynomial in x is either a single term of the form ax^n or an expression that can be written in a form that involves only terms of the form ax^n, where a is any real number and n is a whole number.

Examples ▶ (a) 4 (b) $-3.5x^2$ (c) $3x^2 + 5$ (d) $x^3 - 1$ (e) $4x^2 + 2x - 5$ (f) $3x^3 - x + 1$

polynomials with one term   polynomials with two terms   polynomials with three terms

Note ▶ The constant 4 is a polynomial since: $4 = 4 \cdot 1 = 4x^0$.

Polynomials can be written in terms of any variable.

Examples ▶ (a) $y^2 + 16$ (b) $-8y$ (c) $3w^4 - 5w$ (d) $8w^2 - 3w - 4$ (e) z^5 (f) $z^3 - 1$

polynomials in y polynomials in w polynomials in z

Polynomials can be written in terms of more than one variable.

> A polynomial in x and y is either a single term of the form $ax^n y^m$, or an expression that can be written in a form that involves only terms of the form $ax^n y^m$, where a is any real number and both m and n are whole numbers.

Examples ▶ (a) $2xy$ (b) $3xy^2 - 7y^3$ (c) $3u^2v + 2u^3v$ (d) $4u^2 - 5v$ (e) $w + 2wz$ (f) $4wz + wz^2$

polynomials in x and y polynomials in u and v polynomials in w and z

Not all terms and expressions are polynomials.

Examples ▶ (a) y^{-2} is not a polynomial because -2 is not a whole number.

(b) $\dfrac{3}{x}$ is not a polynomial because $\dfrac{3}{x} = 3x^{-1}$ and -1 is not a whole number.

(c) $5x^{\frac{1}{2}}$ is not a polynomial because $\dfrac{1}{2}$ is not a whole number.

(d) $\dfrac{3}{x+2} + \dfrac{1}{x-5}$ is not a polynomial because it cannot be written as an expression where every term is of the form ax^n where n is a whole number.

A polynomial is said to be simplified if all like terms have been combined.

Examples ▶ (a) $3x + 4x = \underbrace{7x}_{\text{simplified form}}$ (b) $-w^2z + 5wz + 3w^2z = \underbrace{2w^2z + 5wz}_{\text{simplified form}}$

> A simplified polynomial consisting of exactly: one term is called a *monomial*.
> two terms is called a *binomial*.
> three terms is called a *trinomial*.

Example ▶ Classify $3x^2 - 4xy + 3xy + 7y^2$ as a monomial, a binomial, or a trinomial.

1. Simplify ▶ $3x^2 - 4xy + 3xy + 7y^2 = 3x^2 - xy + 7y^2$ Combine like terms.

2. Identify + and − ▶ $ = 3x^2 \boxed{-} xy \boxed{+} 7y^2$ Think: The indicated operations of addition and subtraction determine the beginning of the next term.

1st term   2nd term   3rd term

3. Count terms ▶ $\boxed{3x^2} \boxed{-xy} \boxed{+7y^2}$ Think: A polynomial consisting of three terms is called a trinomial.

Solution ▶ $3x^2 - 4xy + 3xy + 7y^2 = 3x^2 - xy + 7y^2$ is a trinomial.

Note ▶ Polynomials consisting of four or more unlike terms are not generally referred to by a special name.

> In a monomial of the form ax^n, where $a \neq 0$, the whole-number exponent n is called the *degree of the monomial*.

Examples ▶ (a) $4x^3$ is a monomial of degree 3.
(b) $-y$ is a monomial of degree 1 $(-y = -y^1)$.
(c) 5 is a monomial of degree 0 $(5 = 5 \cdot 1 = 5x^0)$.

Note ▶ The monomial 0 is not assigned a degree.

> The *degree of a monomial in more than one variable* is the sum of the exponents.

Examples ▶ (a) $7x^3y^2$ is a monomial of degree 5 $(3 + 2 = 5)$.
(b) $-uv$ is a monomial of degree 2.
(c) $-3.5wz^3$ is a monomial of degree 4.
(d) $-4wx^3y^2z^4$ is a monomial of degree 10 $(1 + 3 + 2 + 4 = 10)$.

> The *degree of a polynomial* is the largest degree of any of its monomial terms (assuming like terms have been combined).

Example ▶ Find the degree of $3x^2y - x^2y^3 + 2x^2y^2 + x^2y^3$.

1. Simplify ▶ $3x^2y - x^2y^3 + 2x^2y^2 + x^2y^3 = 3x^2y + 2x^2y^2$ Combine like terms.

2. Find each degree ▶ $3x^2y$ has degree 3 $(2 + 1 = 3)$.
$2x^2y^2$ has degree 4 $(2 + 2 = 4)$.

3. Compare the degrees ▶ $4 > 3$, so the polynomial is of degree 4.

Solution ▶ $3x^2y - x^2y^3 + 2x^2y^2 + x^2y^3$ is a polynomial of degree 4.

> CAUTION: Be sure to simplify a polynomial before you determine its degree.

Other Examples ▶ (a) $3x^2 + 7x + 1$ is a polynomial of degree 2.
(b) $5x^4$ is a polynomial of degree 4.
(c) 6 is a polynomial of degree 0.
(d) 0 is a polynomial that is not assigned a degree.
(e) $7x^2 - 5xy + 4y^2$ is a polynomial of degree 2.
(f) $x^3 + y^3$ is a polynomial of degree 3.

170   Chapter 5   Polynomials

Polynomials are generally written in descending powers of one of the variables.

Example ▶ Arrange $4xy^3 + 5x^2y^3 - x^3y + 7 - 3xy^3$ in descending powers of x.

1. Simplify ▶ $4xy^3 + 5x^2y^3 - x^3y + 7 - 3xy^3 = xy^3 + 5x^2y^3 - x^3y + 7$ Combine like terms.

2. Identify powers of x ▶ $ = \boxed{x^1}\, y^3 + 5\, \boxed{x^2}\, y^3 - \boxed{x^3}\, y + 7\, \boxed{x^0}$ Think: $7 = 7x^0$

3. Arrange in descending powers of x ▶ $ = x^3y + 5x^2y^3 + xy^3 + 7$

Solution ▶ $4xy^3 + 5x^2y^3 - x^3y + 7 - 3xy^3$ in descending powers of x is $x^3y + 5x^2y^3 + xy^3 + 7$.

It is generally easier to work with polynomials if you first write the polynomials in *standard form*.

> A polynomial is in standard form if:
> 1. the variables are written in alphabetical order in each term;
> 2. the polynomial is simplified; and
> 3. the terms are arranged in descending powers of the first variable.

Example ▶ Write $3z^3w^2 - 7wz + 5 + 2z^2w^3 + zw$ in standard form.

1. Alphabetize the variables ▶ $3z^3w^2 - 7wz + 5 + 2z^2w^3 + zw = 3w^2z^3 - 7wz + 5 + 2w^3z^2 + wz$

2. Simplify ▶ $ = 3w^2z^3 - 6wz + 5 + 2w^3z^2$ Combine like terms.

3. Arrange in descending powers of the first variable ▶ $ = 2w^3z^2 + 3w^2z^3 - 6wz + 5$ Think: $5 = 5w^0$

Solution ▶ $3z^3w^2 - 7wz + 5 + 2z^2w^3 + zw$ in standard form is $2w^3z^2 + 3w^2z^3 - 6wz + 5$.

Other Examples ▶ (a) The standard form of $5 + w$ is $w + 5$.
(b) The standard form of $3x^2 - 7 + 5x$ is $3x^2 + 5x - 7$.
(c) The standard form of $y^2x + x^2y + 4$ is $x^2y + xy^2 + 4$.
(d) The standard form of $-11xy^2 + 5yx^2 + 3x^3 - 7y^3$ is $3x^3 + 5x^2y - 11xy^2 - 7y^3$.

5.1 Add and Subtract Polynomials

You add polynomials by combining like terms.

EXAMPLE 1: Add polynomials.

Problem ▶ $(-2x^2 + 4x - 5) + (7x^2 + 2x - 6) = ?$

1. Group like terms ▶ $(-2x^2 + 4x - 5) + (7x^2 + 2x - 6) = (-2x^2 + 7x^2) + (4x + 2x) + (-5 + (-6))$

2. Combine like terms ▶ $ = 5x^2 + 6x - 11$

Solution ▶ $(-2x^2 + 4x - 5) + (7x^2 + 2x - 6) = 5x^2 + 6x - 11$

5.1 Add and Subtract Polynomials

Sometimes it is convenient to add polynomials by arranging like terms in columns.

Example ▶ Add $4x^3 + 7x - 8$, $6x^3 + 2x^2 - 3x - 5$ and $x^2 - 5x + 3$ in vertical form.

1. Write vertical form ▶
$$\begin{array}{r} 4x^3 + 7x - 8 \\ 6x^3 + 2x^2 - 3x - 5 \\ x^2 - 5x + 3 \\ \hline \end{array}$$
Arrange like terms in columns.

2. Combine like terms ▶
$$\begin{array}{r} 4x^3 + 7x - 8 \\ 6x^3 + 2x^2 - 3x - 5 \\ +\ x^2 - 5x + 3 \\ \hline 10x^3 + 3x^2 - x - 10 \end{array}$$

Think: $-8 - 5 + 3 = -10$
$+7x - 3x - 5x = -x$
$+2x^2 + x^2 = +3x^2$
$4x^3 + 6x^3 = 10x^3$

Solution ▶ $(4x^3 + 7x - 8) + (6x^3 + 2x^2 - 3x - 5) + (x^2 - 5x + 3) = 10x^3 + 3x^2 - x - 10.$

In most cases it is helpful to write the polynomials in standard form first.

Example ▶ $(7xy^2 - 4x^2y^3 + 15xy^4 - 2x^3y) + (3xy^2 - 5x^3y - 6y^4x - 8y^3x^2) = ?$

1. Write standard form ▶ $7xy^2 - 4x^2y^3 + 15xy^4 - 2x^3y = -2x^3y - 4x^2y^3 + 15xy^4 + 7xy^2$ ⟵ the first polynomial in standard form

$3xy^2 - 5x^3y - 6y^4x - 8y^3x^2 = -5x^3y - 8x^2y^3 - 6xy^4 + 3xy^2$ ⟵ the second polynomial in standard form

2. Add as before ▶
$$\begin{array}{r} -2x^3y - 4x^2y^3 + 15xy^4 + 7xy^2 \\ -5x^3y - 8x^2y^3 - 6xy^4 + 3xy^2 \\ \hline -7x^3y - 12x^2y^3 + 9xy^4 + 10xy^2 \end{array}$$

Solution ▶ $(7xy^2 - 4x^2y^3 + 15xy^4 - 2x^3y) + (3xy^2 - 5x^3y - 6y^4x - 8y^3x^2)$
$= -7x^3y - 12x^2y^3 + 9xy^4 + 10xy^2$

Make Sure

Add polynomials.

See Example 1 ▶
1. $(2z^3 + 3z^2 + z) + (3z^2 - 4z - 3)$

2. $4x^3y + 2x^2y^2 - 3xy^3$
   $+\ x^3y - 3x^2y^2 + 2xy^3$

MAKE SURE ANSWERS: 1. $2z^3 + 6z^2 - 3z - 3$ 2. $5x^3y - x^2y^2 - xy^3$

The *additive inverse of a polynomial* is the sum of the additive inverses of each term of the polynomial.

Example ▶ The additive inverse of $4x^2y - 5xy^3 + 7x - 2$ is $-4x^2y + 5xy^3 + (-7x) + 2.$

Note ▶ To form the additive inverse of a polynomial, just write the opposite of each term.

172 Chapter 5 Polynomials

To subtract a polynomial, you add its additive inverse.

EXAMPLE 2: Subtract polynomials.

Problem ▶ $(3x^2 - 6x + 5) - (4x^2 + x - 8) = ?$

1. Write as a sum ▶ $(3x^2 - 6x + 5) - (4x^2 + x - 8) = (3x^2 - 6x + 5) + (-4x^2 - x + 8)$

(change to addition / write the additive inverse)

2. Add as before ▶ $= -x^2 - 7x + 13$

Solution ▶ $(3x^2 - 6x + 5) - (4x^2 + x - 8) = -x^2 - 7x + 13.$

You can also subtract polynomials using a vertical format.

Example ▶ $(3x^3 + 6x - 9) - (5x^3 + 4x^2 - 3x - 5) = ?$

1. Write vertical form ▶
$$\begin{array}{r} 3x^3 + 6x - 9 \\ -(5x^3 + 4x^2 - 3x - 5) \end{array}$$
⟵ minuend
⟵ subtrahend

2. Write as a sum ▶
$$\begin{array}{r} 3x^3 + 6x - 9 \\ +(-5x^3 - 4x^2 + 3x + 5) \end{array}$$
⟵ additive inverse of the subtrahend

3. Add as before ▶
$$\begin{array}{r} 3x^3 + 6x - 9 \\ -5x^3 - 4x^2 + 3x + 5 \\ \hline -2x^3 - 4x^2 + 9x - 4 \end{array}$$

Solution ▶ $(3x^3 + 6x - 9) - (5x^3 + 4x^2 - 3x - 5) = -2x^3 - 4x^2 + 9x - 4.$

Note ▶ To perform subtraction of polynomials mentally, change the sign of the terms in the subtrahend and then add.

Examples ▶

(a) $\begin{array}{r} 4x - 7 \\ -(x + 3) \quad \boxed{-x - 3} \\ \hline 3x - 10 \end{array}$ ⟵ difference

(b) $\begin{array}{r} 2x^2 + 5x - 5 \\ -(3x^2 - 2x - 4) \quad \boxed{-3x^2 + 2x + 4} \\ \hline -x^2 + 7x - 1 \end{array}$ ⟵ difference

(c) $\begin{array}{r} 4x^2 - 3 \\ -(-2x^2 + 5x + 4) \quad \boxed{2x^2 - 5x - 4} \\ \hline 6x^2 - 5x - 7 \end{array}$ ⟵ difference

Make Sure

Subtract polynomials.

See Example 2 ▶
1. $(6a^3 + 2a^2 - 3a - 4) - (4a^3 + 7a - 8)$
2. $\begin{array}{r} 5b^3 - 3b^2 + 6b - 7 \\ -(4b^3 + 4b^2 - b + 2) \end{array}$

MAKE SURE ANSWERS: 1. $2a^3 + 2a^2 - 10a + 4$ 2. $b^3 - 7b^2 + 7b - 9$

5.1 Practice

Name _____ Date _____ Class _____

Set 1: Add polynomials.

1. $(3a + 4) + (2a + 7)$
2. $(5b + 3) + (7b + 2)$

3. $(3c^2 + 2c + 5) + (5c^2 + 3c + 5)$
4. $(5d^2 + 4d + 7) + (2d^2 + 3d + 1)$

5. $(4e^3 - 2e^2 + 5) + (4e^2 - 2e - 9)$
6. $(7f^4 - 3f^2 + 1) + (3f^3 - 5f - 8)$

7. $\quad g^3 - g^2 - g + 5$
$\quad\underline{g^3 + g^2 + g - 3}$

8. $\quad h^4 + h^2 + h - 3$
$\quad\underline{h^4 - h^2 - h + 5}$

9. $\quad x^3 - 3x\ \ + 4x^2$
$\quad\underline{3x\ \ - 4x^2 + x^3 + 5}$

10. $\quad y\ \ - 3y^4 + 2y^2$
$\quad\underline{3y^2 + 4y^4 - y\ \ + 2}$

Set 2: Subtract polynomials.

11. $(3x^2 + 7) - (2x^2 + 5)$
12. $(6a + 3a^3) - (a + 2a^3)$
13. $(4y^2 + 7) - (2y^2 - 5)$

14. $(4r + 4) - (2r - 3)$
15. $(3k^2 - 3) - (5k^2 - 5)$
16. $(4h^3 - 5) - (7h^3 - 4)$

17. $\quad b^2 - 3b + 4$
$\quad\underline{-(2b^2 - 4b + 2)}$

18. $\quad 3c^3 - 4c - 5$
$\quad\underline{-(5c^3 + 6c - 7)}$

19. $\quad 2ab - 5a^2b - b$
$\quad\underline{-(5ab - 5a^2b + 3b)}$

20. $\quad 3xy^2 - 4x^2y + xy$
$\quad\underline{-(4xy^2 + 2x^2y - 3xy)}$

21. $\quad -5c^2d + 2c$
$\quad\underline{-(4c^2d - 3c)}$

22. $\quad 3ab + a$
$\quad\underline{-(-5ab + 3a)}$

174   Chapter 5   Polynomials

Mixed Practice: Perform the indicated operations.

23. $(2ab - 5a^2) + (b^2 - 5ab)$

24. $(3x^4 - 4xy) + (3xy - y^3)$

25. $(4abc - 7b^2c) - (5abc + b^2)$

26. $(7e - 4efg) - (3efg - 3e^2)$

27. $(c^2 + 3c + 5) + (4 - 5c - 3c^2)$

28. $(d^4 - 5d^2 + 2) + (8 - 3d^4 + 9d^2)$

29. $(p^8 - q^5) - (p^8 + 3p^3q - 5pq^3)$

30. $(r^3 - s^3) - (s^3 + 3rs - 4rs^3)$

31. $(x^2 - xy + yz) + (xy - zy - z^2)$

32. $(sr - st^2 + r^3) + (st^2 - r^3 - 3sr)$

Extra: Perform the indicated operations.

33. $(a - b - c - d) + (b - c + d - a) + (b - c)$

34. $(r + s - t + u) + (s + t - u - r) + (s - t)$

35. $(a - b + c - d) - (b - c + d - a) + (b - a)$

36. $(w - x - y + z) - (w - x + y - z) + (x - y)$

37. $(2x - 3y) + (4x + 3y) + (4x - 7y)$

38. $(a - b) + (3a - 5b) + (7a - 2b)$

39. $(3x^2 - y) + (2x^2 + 3y) - (x^2 + 5y)$

40. $(r - s^2) + (3r + 4s^2) - (5r - 3s^2)$

41. $(c^2 - d^2) - (c^2 + d^2) - (3d^2 - 2c^2)$

42. $(ab - bc) - (bc - ac) - (3ab + 5bc)$

Review: Work these problems on a separate sheet of paper. Attach your work to this page.

Multiply signed numbers. (See Lesson 1.3.)
43. $2(-3)(4)$
44. $3(-1)(-2)$
45. $-2(-4)(-5)$
46. $-3(-1)(-3)$
47. $-4(3)(-2)$
48. $-5(4)(-2)$
49. $-1(-2)(3)$
50. $-3(-1)(4)$

Remove the grouping symbols using the distributive properties. (See Lessons 1.5 and 1.6.)
51. $4(3z - 5)$
52. $5(2y - 7)$
53. $-3(5x + 4)$
54. $-5(7w + 3)$
55. $2v(3v - 2)$
56. $3u(4u - 5)$
57. $2t^2(t^2 - t - 1)$
58. $5s^3(s^2 + 4s - 2)$

Simplify each product of terms. (See Lesson 1.6.)
59. $z^2 \cdot z^5$
60. $y^3 \cdot y^4$
61. $x^3 \cdot x$
62. $w \cdot w^5$
63. $2v^3 \cdot v^4$
64. $4u^2 \cdot u^5$
65. $3t^4 \cdot 2t^3$
66. $4s^3 \cdot 5s^4$

Copyright © 1985 by Harcourt Brace Jovanovich, Inc. All rights reserved.

Problem Solving 6: Evaluate Polynomials

There are many important applications which can be solved by evaluating polynomials.

> **FACT:** For each dollar of principal (P) that is deposited in an account at interest rate r (compounded annually), the total amount (original principal plus earned interest) is:
> **(a)** $(r + 1)$ dollars at the end of a one-year time period (t),
> **(b)** $(r^2 + 2r + 1)$ dollars at the end of a two-year time period,
> **(c)** $(r^3 + 3r^2 + 3r + 1)$ dollars at the end of a three-year time period,
> **(d)** $(r^4 + 4r^3 + 6r^2 + 4r + 1)$ dollars at the end of a four-year time period,
>
> assuming that there are no further deposits or withdrawals during the given time period.

EXAMPLE: Evaluate the correct polynomial to help solve this problem.

1. Read and identify ▶ Principal of $500 is invested in an account that pays 10% interest compounded annually. Assuming that there are no further deposits or withdrawals, how much money will be in the account at the end of 2 years?

Remember, circle the facts and underline the question.

2. Understand ▶ The question asks you to find the total amount at the end of 2 years in an account that is compounded annually.

3. Decide ▶ To find the total amount, you **evaluate $r^2 + 2r + 1$ to find the total amount for each dollar at the end of 2 years, and then multiply that dollar amount by the number of dollars in the original principal.**

4. Evaluate ▶ $r^2 + 2r + 1 = (0.1)^2 + 2(0.1) + 1$ Think: 10% = 0.1
$= 0.01 + 0.2 + 1$
$= 1.21$ ← total amount for each dollar

5. Multiply ▶ (number of dollars in original principal)(1.21) = (500)(1.21) = 605

6. Interpret ▶ 605 means the total amount in savings at the end of 2 years is $605.

Note ▶ If $500 is replaced with x dollars, then the total amount at the end of 2 years is $1.21x$.

Practice: Evaluate the correct polynomial to solve each problem.

Find the total amount in each account at the end of the given time period. Assume that the principal is deposited at the beginning of each time period and that there are no further deposits or withdrawals. Round to the nearest cent when necessary.

	Principal	Interest Rate (compounded annually)	Time Period
1.	$100	10%	1 year
2.	$100	10%	2 years
3.	$100	10%	3 years
4.	$100	10%	4 years
5.	$1000	6%	3 years
6.	$1000	5%	4 years

1. _____
2. _____
3. _____
4. _____
5. _____
6. _____

Copyright © 1985 by Harcourt Brace Jovanovich, Inc. All rights reserved.

176 Chapter 5 Polynomials

> **FACT:** If one dollar of principal (P) is deposited in an account at the beginning of *each* year at interest rate r (compounded annually) and $x = r + 1$, the total amount in the account is:
>
> (a) x dollars at the end of a one-year time period (t),
> (b) $(x^2 + x)$ dollars at the end of a two-year time period,
> (c) $(x^3 + x^2 + x)$ dollars at the end of a three-year time period,
> (d) $(x^4 + x^3 + x^2 + x)$ dollars at the end of a four-year time period,
>
> assuming that there are no additional deposits and no withdrawals during the given time period.

Find the total amount in each account at the end of the given time period. Assume that the principal is deposited at the beginning of the year for each year in the time period and that there are no further deposits or withdrawals during the year.

	Principal	Interest Rate (compounded annually)	Time Period
7.	$100	10%	1 year
8.	$100	10%	2 years
9.	$100	10%	3 years
10.	$100	10%	4 years
11.	$1000	6%	3 years
12.	$1000	5%	4 years

> **FACT:** $0.0005h^3$ represents the weight in pounds of an average person in the United States whose height (h) is between 30 inches and 74 inches.

13. Complete Table 5.1 using the polynomial $0.0005h^3$. Round to the nearest whole pound when necessary.

TABLE 5.1 Height/Weight Table for an Average Person in the United States

Height	Weight	Height	Weight	Height	Weight
2ft 6in.		4ft 6in.		5ft 8in.	
3ft		5ft 2in.		5ft 10in.	
3ft 6in.		5ft 4in.		6ft	
4ft		5ft 6in.		6ft 2in.	

14. For an average person who grows from 3ft 4in. to 5ft during adolescence, what is the percent increase in a) height and b) weight to the nearest tenth of a percent?

15. For an average person who shrinks from 5ft 4in. to 5ft 3in. because of old age, what is the percent decrease in a) height and b) weight to the nearest tenth of a percent?

5.2 Multiply Polynomials

Recall ▶ A simplified polynomial consisting of exactly one term is called a monomial.

To multiply monomials, you multiply numerical coefficients and then multiply literal parts.

EXAMPLE 1: Multiply monomials.

Problem ▶ $(3a^3b)(-4a^2b)(5b^2) = ?$

1. Multiply coefficients ▶ $(3a^3b)(-4a^2b)(5b^2) = [3(-4)5](a^3a^2)(bbb^2)$ Group numbers and like bases together.

$\qquad\qquad\qquad\qquad\quad = -60(a^3a^2)(bbb^2)$ Think: $3(-4)5 = -12 \cdot 5 = -60$

2. Multiply literal parts ▶ $\qquad\qquad\quad = -60a^5b^4$ Think: $a^3a^2 = a^{3+2} = a^5$ and $bbb^2 = b^1b^1b^2 = b^4$

Solution ▶ $(3a^3b)(-4a^2b)(5b^2) = -60a^5b^4$

Other Examples ▶ (a) $(5xyz^2)(2x^3yz) = (5 \cdot 2)(xx^3)(yy)(z^2z) = 10x^4y^2z^3$

(b) $(3xw^3)(-7w^2)\left(\frac{1}{3}x\right) = \left[3(-7)\left(\frac{1}{3}\right)\right](xx)(w^3w^2) = -7x^2w^5$

(c) $(0.4mn)(1.5m^2n)(-3n^2) = [(0.4)(1.5)(-3)](mm^2)(nnn^2) = -1.8m^3n^4$

Make Sure

Multiply monomials.

See Example 1 ▶ 1. $3a^3(-4a^2)$ 2. $5b^2(3b)(-2b^3)$ 3. $-2a^2b(-3ab^3)(4a^2c)$

MAKE SURE ANSWERS: 1. $-12a^5$ 2. $-30b^6$ 3. $24a^5b^4c$

The distributive properties provide the basis for multiplying a polynomial by a monomial.

> **To multiply a polynomial by a monomial:**
> 1. Use the distributive properties to multiply each term of the polynomial by the monomial.
> 2. Simplify the sum of the products formed in Step 1.

EXAMPLE 2: Multiply a polynomial by a monomial.

Problem ▶ $3y(4y^2 - 5y + 7) = ?$

1. Distribute ▶ $3y(4y^2 - 5y + 7) = (3y)(4y^2) + (3y)(-5y) + (3y)7$

2. Simplify ▶ $\qquad\qquad\qquad\quad = 12y^3 + (-15y^2) + 21y$ Multiply the monomials.

$\qquad\qquad\qquad\qquad\quad = 12y^3 - 15y^2 + 21y$

Solution ▶ $3y(4y^2 - 5y + 7) = 12y^3 - 15y^2 + 21y$

Other Examples (a) $2ab(3a^2b - 4ab^2 + b^3) = (2ab)(3a^2b) + (2ab)(-4ab^2) + (2ab)(b^3)$
$= 6a^3b^2 - 8a^2b^3 + 2ab^4$

(b) $(4w^3 - 2w^2 - 5w + 7)(-3w) = (4w^3)(-3w) + (-2w^2)(-3w) + (-5w)(-3w) + (7)(-3w)$
$= -12w^4 + 6w^3 + 15w^2 - 21w$

You can also find the product of a polynomial and a monomial by using a vertical format.

Example ▶ Multiply $3x^2 + 5x - 7$ by $2x$ in vertical form.

1. Write vertical form ▶ $\begin{array}{r} 3x^2 + 5x - 7 \\ 2x \end{array}$ Write the monomial on the bottom as the multiplier.

2. Multiply ▶ $\begin{array}{r} 3x^2 + 5x - 7 \\ 2x \\ \hline 6x^3 + 10x^2 - 14x \end{array}$ Think: $2x(-7) = -14x$
$2x(+5x) = +10x^2$
$2x(3x^2) = 6x^3$

Solution ▶ $(2x)(3x^2 + 5x - 7) = 6x^3 + 10x^2 - 14x.$

Make Sure

Multiply a polynomial by a monomial.

See Example 2 ▶ 1. $-3a^2(4a^3 - 3a^2 - a + 1)$ 2. $(4b^3c - 3b^2c^2 + 2c^3 - 3)(-2b^2c)$

MAKE SURE ANSWERS: 1. $-12a^5 + 9a^4 + 3a^3 - 3a^2$ 2. $-8b^5c^2 + 6b^4c^3 + 6b^4c^3 - 4b^2c^4 + 6b^2c$

Repeated use of the distributive properties is the basis for the multiplication of polynomials.

> **To multiply a polynomial by a polynomial:**
> 1. Use the distributive properties to multiply each term of the first polynomial by each term of the second polynomial.
> 2. Combine like terms.

EXAMPLE 3: Multiply two polynomials.

Problem ▶ $(2x + 5)(3x + 4) = ?$

1. Distribute ▶ $(2x + 5)(3x + 4) = (2x + 5)3x + (2x + 5)4$ Distribute $(2x + 5)$ over $3x + 4$.

2. Distribute again ▶ $= (2x)(3x) + (5)(3x) + (2x)(4) + (5)(4)$ Distribute $3x$ over $2x + 5$.
Distribute 4 over $2x + 5$.

3. Combine like terms ▶ $= 6x^2 + 15x + 8x + 20$
$= 6x^2 + 23x + 20$

Solution ▶ $(2x + 5)(3x + 4) = 6x^2 + 23x + 20$

5.2 Multiply Polynomials

It is often convenient to multiply two polynomials using a vertical format.

Example ▶ Multiply $y + 4$ and $2y - 7$ in vertical form.

1. Write vertical form ▶

$$\begin{array}{r} 2y - 7 \\ y + 4 \end{array}$$

2. Distribute ▶

$$\begin{array}{r} 2y - 7 \\ y + 4 \\ \hline 8y - 28 \end{array} \longleftarrow 4(2y - 7)$$

3. Distribute again ▶

$$\begin{array}{r} 2y - 7 \\ y + 4 \\ \hline 8y - 28 \\ 2y^2 - 7y \end{array} \longleftarrow y(2y - 7)$$

Think: The partial product $2y^2 - 7y$ is indented so that only like terms are in the same column.

4. Combine like terms ▶

$$\begin{array}{r} 2y - 7 \\ y + 4 \\ \hline 8y - 28 \\ 2y^2 - 7y \\ \hline 2y^2 + y - 28 \end{array}$$

Solution ▶ $(y + 4)(2y - 7) = 2y^2 + y - 28$

The vertical format is especially useful for multiplying polynomials consisting of several terms.

Example ▶ $(a^2 - 3a + 6)(2a^3 + 4a^2 - 5a + 7) = ?$

Usually the polynomial consisting of the fewest terms is used as the multiplier.

1. Write vertical form ▶

$$\begin{array}{r} 2a^3 + 4a^2 - 5a + 7 \\ a^2 - 3a + 6 \end{array}$$

2. Distribute ▶

$$\begin{array}{r} 2a^3 + 4a^2 - 5a + 7 \\ a^2 - 3a + 6 \\ \hline 12a^3 + 24a^2 - 30a + 42 \end{array} \longleftarrow 6(2a^3 + 4a^2 - 5a + 7)$$

3. Distribute again ▶

$$\begin{array}{r} 2a^3 + 4a^2 - 5a + 7 \\ a^2 - 3a + 6 \\ \hline 12a^3 + 24a^2 - 30a + 42 \\ -6a^4 - 12a^3 + 15a^2 - 21a \end{array} \longleftarrow -3a(2a^3 + 4a^2 - 5a + 7)$$

4. Distribute again ▶

$$\begin{array}{r} 2a^3 + 4a^2 - 5a + 7 \\ a^2 - 3a + 6 \\ \hline 12a^3 + 24a^2 - 30a + 42 \\ -6a^4 - 12a^3 + 15a^2 - 21a \\ 2a^5 + 4a^4 - 5a^3 + 7a^2 \end{array} \longleftarrow a^2(2a^3 + 4a^2 - 5a + 7)$$

5. Combine like terms ▶

$$\begin{array}{r} 2a^3 + 4a^2 - 5a + 7 \\ a^2 - 3a + 6 \\ \hline 12a^3 + 24a^2 - 30a + 42 \\ -6a^4 - 12a^3 + 15a^2 - 21a \\ 2a^5 + 4a^4 - 5a^3 + 7a^2 \\ \hline 2a^5 - 2a^4 - 5a^3 + 46a^2 - 51a + 42 \end{array}$$

Solution ▶ $(a^2 - 3a + 6)(2a^3 + 4a^2 - 5a + 7) = 2a^5 - 2a^4 - 5a^3 + 46a^2 - 51a + 42$

Before you change to vertical form, make sure both polynomials are written in standard form. This will keep the like terms lined up in columns.

Example $(2 - w)(3w - w^2 + 5) = ?$

1. Write standard form
$$2 - w = -w + 2$$
$$3w - w^2 + 5 = -w^2 + 3w + 5$$

2. Write vertical form
$$\begin{array}{r} -w^2 + 3w + 5 \\ -w + 2 \end{array} \leftarrow \text{polynomial consisting of the fewest terms}$$

3. Multiply as before
$$\begin{array}{r} -w^2 + 3w + 5 \\ -w + 2 \\ \hline -2w^2 + 6w + 10 \quad \leftarrow 2(-w^2 + 3w + 5) \\ w^3 - 3w^2 - 5w \quad\quad\quad \leftarrow -w(-w^2 + 3w + 5) \\ \hline w^3 - 5w^2 + w + 10 \end{array}$$

Solution $(2 - w)(3w - w^2 + 5) = w^3 - 5w^2 + w + 10$

> CAUTION: When there is a missing term in one of the polynomials, you must be careful to keep only like terms in the same column.

Example $(5x^3 + 3xy^2 - 2y^3)(x - 4y) = ?$ Think: $5x^3 + 3xy^2 - 2y^3 = 5x^3 + 0x^2y + 3xy^2 - 2y^3$

$$\begin{array}{r} 5x^3 + 0x^2y + 3xy^2 - 2y^3 \\ x - 4y \\ \hline -20x^3y + 0 - 12xy^3 + 8y^4 \\ 5x^4 + 0 + 3x^2y^2 - 2xy^3 \\ \hline 5x^4 - 20x^3y + 3x^2y^2 - 14xy^3 + 8y^4 \end{array}$$

Although the vertical form is often used to compute the product of two polynomials, the horizontal form generally allows the work to be written more compactly.

Example Compute $(5x^3 + 3xy^2 - 2y^3)(x - 4y)$ using a horizontal format.

$$(5x^3 + 3xy^2 - 2y^3)(x - 4y) = (5x^3 + 3xy^2 - 2y^3)x - (5x^3 + 3xy^2 - 2y^3)4y$$
$$= 5x^4 + 3x^2y^2 - 2xy^3 - (20x^3y + 12xy^3 - 8y^4)$$
$$= 5x^4 + 3x^2y^2 - 2xy^3 - 20x^3y - 12xy^3 + 8y^4$$
$$= 5x^4 - 20x^3y + 3x^2y^2 - 14xy^3 + 8y^4$$

Make Sure

Multiply polynomials.

See Example 3 1. $(3a - 5)(4a + 3)$ 2. $(3a^2 + 4ab + b^2)(2a^3 - 3a^2b + b)$

MAKE SURE ANSWERS: 1. $12a^2 - 11a - 15$ 2. $6a^5 - a^4b - 10a^3b^2 - 3a^2b^3 + 3a^2b + 4ab^2 + b^3$

5.2 Practice

Set 1: Multiply monomials.

1. $(3a)(2a^2)$
2. $(3b^5)(4b^2)$
3. $(-3ab^3)(a^2b)(4a^3b^5)$
4. $(7c^4d^2)(-5c^3d^7)(2c^2d)$
5. $(-3x^3yz^2)(4xy^5z^4)(-xyz)$
6. $(rst^3)(-6r^2s^5t)(-3r^7s^2t^4)$

Set 2: Multiply polynomials by monomials.

7. $4x^2(3x^2 - 5x + 2)$
8. $3y^3(4y^3 - y^2 + 2y)$
9. $-3uv^3(4uv - 5u^5v^2 + 3u^3v^4)$
10. $-5s^3t^5(7st^2 - 3st + 4st^4)$
11. $(q^3 - 3q^2r + 2qr^4)(-3q^2r)$
12. $(3m^3 - 4m^2n + 5mn^4)(-5m^2n^5)$

Set 3: Multiply polynomials.

13. $(3x + 5)(4x + 3)$
14. $(5y - 3)(2y + 5)$
15. $(z^3 + 2z + 3)(2 - 2z - z^2)$
16. $(u^4 - 3u^2 - u)(u - 4u^2 + 5)$
17. $(v^2 + 2vw - w^2)(3v^3 - 4v^2w + 4vw)$
18. $(x + xy - y)(3x^3 - 4x^2 + 2xy)$

182 Chapter 5 Polynomials

Mixed Practice: Multiply polynomials.

19. $-3a^3b^2(5a^2b^5)(-3b^4c^2)$ 20. $2d^7e^4f(-5e^3f^4)(-3f^7d^2)$ 21. $(2g^2 - 3h)(5g^2 - 4h)$

22. $(m^2 - 3n)(m^2 - 5n)$ 23. $(p - q)(p^2 + pq + q^2)$ 24. $(r + s)(r^2 - rs + s^2)$

25. $(2x - 3y)(4x^2 + 6xy + 9y^2)$ 26. $(3a + b)(9a^2 - 3ab + b^2)$

27. $(a^2 + 2ab + b^2)(a^2 - 2ab + b^2)$ 28. $(2c^2 + 3cd + d^2)(c^2 + 4cd + 2d^2)$

Extra: Perform the indicated operations.

29. $(a - b)(a + b) + (a^2 + 2ab + b^2)$ 30. $(c + d)(c + d) + (c - d)(c - d)$

31. $(2e - 3f)(2e + 3f) - (4f^2 - 6e^2)$ 32. $(g - 3h)(3g + h) - (3g - h)(3g + h)$

Review: Work these problems on a separate sheet of paper. Attach your work to this page.

Simplify each expression by combining like terms. (See Lesson 1.5.)
33. $3z + 5z$ 34. $6y + 9y$ 35. $4x - 5x$ 36. $5w - 8w$
37. $-4v + 5v$ 38. $-7u + 9u$ 39. $-2t - 4t$ 40. $-5s - 7s$
41. $-3r - 5r + 3r$ 42. $-q - 5q + 4q$ 43. $-2p + 3p - 5p$ 44. $-7n + 2n - 4n$

Simplify each product of terms. (See Lesson 1.6.)
45. $3z(3z)$ 46. $5y(2y)$ 47. $4x(4x)$ 48. $5w(3w)$
49. $-3v(-3v)$ 50. $-4u(-4u)$ 51. $-t(3t)$ 52. $-s(5s)$
53. $2(3r)(-3r)$ 54. $2(-2q)(5q)$ 55. $2(-3p)(-5p)$ 56. $2(-5n)(-n)$

Copyright © 1985 by Harcourt Brace Jovanovich, Inc. All rights reserved.

5.3 Find Special Products

The problem of finding the product of two binomials occurs frequently in mathematics. Although you can find the product of two binomials by the methods shown in the previous section, you need to develop a technique that will produce the product quickly, without writing any intermediate steps.

If the terms of the binomials $(2x + 5)$ and $(3x - 4)$ are labeled as in Figure 5.1, then the product of the two binomials can be formed by the *FOIL method*.

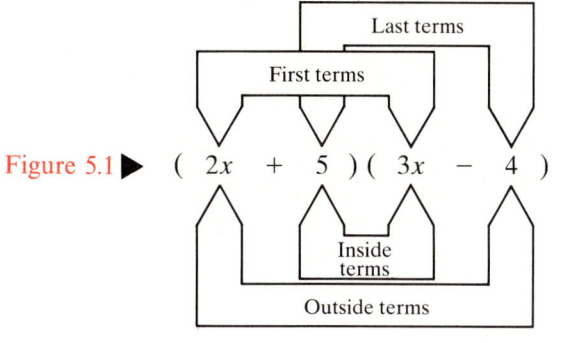

Figure 5.1 ▶ $(2x + 5)(3x - 4)$

The FOIL Method

To find the product of two binomials:

1. Find the product of the: First terms.
   Outside terms.
   Inside terms.
   Last terms.
2. Form the sum of the above products.
3. Combine like terms when possible.

EXAMPLE 1: Multiply two binomials using the FOIL method.

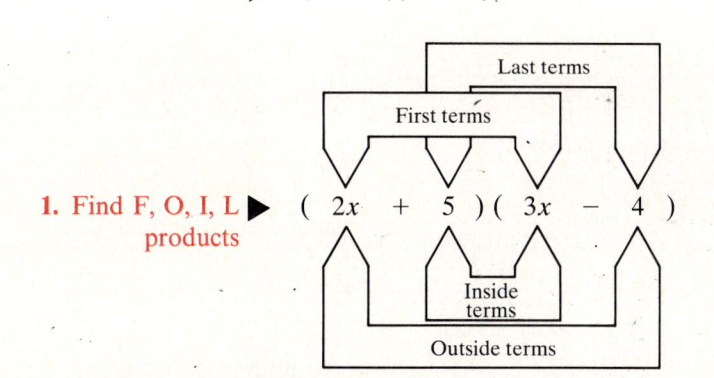

Problem ▶ $(2x + 5)(3x - 4) = ?$

1. Find F, O, I, L products ▶ $(2x + 5)(3x - 4)$

FOIL: First terms: $(2x)(3x) = 6x^2$
Outside terms: $(2x)(-4) = -8x$
Inside terms: $(5)(3x) = 15x$
Last terms: $(5)(-4) = -20$

2. Form sum ▶ $(2x + 5)(3x - 4) = 6x^2 + (-8x) + 15x + (-20)$

3. Combine like terms ▶ $\qquad = 6x^2 + 7x - 20$ Think: $(-8x) + 15x = 7x$

Solution ▶ $(2x + 5)(3x + 4) = 6x^2 + 7x - 20$

Note ▶ In Example 1, intermediate steps have been shown. However, with practice you can compute and combine the F, O, I, L products mentally.

184 Chapter 5 Polynomials

Example ▶ $(-6x + 5)(3x + 4) = -18x^2 - 24x + 15x + 20 = -18x^2 - 9x + 20$

Make Sure

Multiply two binomials using the FOIL method.

See Example 1 ▶ **1.** $(3m + 2n)(m + n)$ **2.** $(2p - q)(3p - 2q)$ **3.** $(r - 2s)(3r + s)$

MAKE SURE ANSWERS: **1.** $3m^2 + 5mn + 2n^2$ **2.** $6p^2 - 7pq + 2q^2$ **3.** $3r^2 - 5rs - 2s^2$

The problem of computing the square of a binomial occurs so often in mathematics that you should memorize its form and the form of its product.

Examples ▶ (a) $(r + s)^2 = ?$ (b) $(r - s)^2 = ?$

$(r + s)^2 = (r + s)(r + s)$ $(r - s)^2 = (r - s)(r - s)$

 F O I L F O I L
$= rr + rs + sr + ss$ $= rr + r(-s) + (-s)r + (-s)(-s)$
$= r^2 + 2rs + s^2$ $= r^2 - 2rs + s^2$

Both of the previous examples show that you can compute the square of a binomial by the following method.

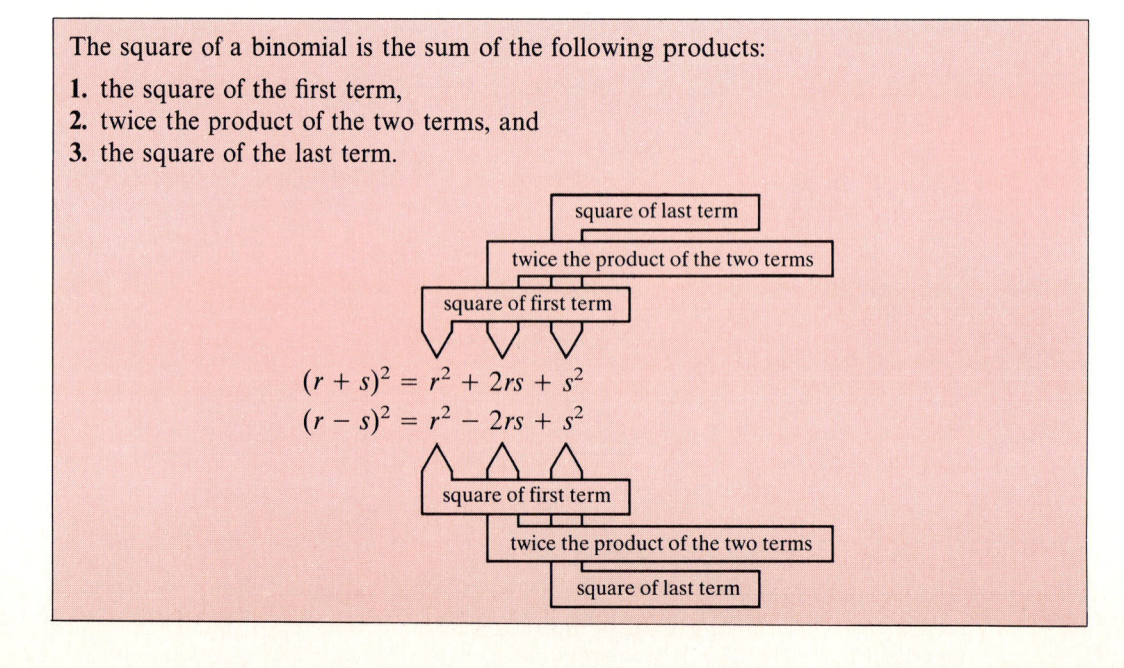

The square of a binomial is the sum of the following products:

1. the square of the first term,
2. twice the product of the two terms, and
3. the square of the last term.

$(r + s)^2 = r^2 + 2rs + s^2$
$(r - s)^2 = r^2 - 2rs + s^2$

EXAMPLE 2: Find the square of a binomial.

Problem ▶ $(3x - 4)^2 = ?$

1. Find products ▶ $(\underset{\text{first term}}{3x} - \underset{\text{second term}}{4})^2$

(1) The first term $(3x)$ squared is $9x^2$.
(2) Twice the product of the two terms $(3x)$ and (-4) is $-24x$.
(3) The second term (-4) squared is 16.

2. Form sum ▶ $(3x - 4)^2 = 9x^2 + (-24x) + 16$

Solution ▶ $(3x - 4)^2 = 9x^2 - 24x + 16$

Note ▶ In practice you should do all the computation mentally and only write the answer.

Other Examples ▶ (a) $(x - y)^2 = \underset{(x)^2}{x^2} - \underset{2 \cdot x \cdot (-y)}{2xy} + \underset{(-y)^2}{y^2}$

(b) $(4w + 5)^2 = \underset{(4w)^2}{16w^2} + \underset{2 \cdot 4w \cdot 5}{40w} + \underset{5^2}{25}$

(c) $(7c - 2d)^2 = \underset{(7c)^2}{49c^2} - \underset{2 \cdot 7c \cdot (-2d)}{28cd} + \underset{(-2d)^2}{4d^2}$

CAUTION: $(r + s)^2 \neq r^2 + s^2$

Example ▶ $(3 + 2)^2 = 5^2 = 25$ and $3^2 + 2^2 = 9 + 4 = 13$.

Make Sure

Find the square of a binomial.

See Example 2 ▶ 1. $(s + t)^2$ 2. $(3u - 4v)^2$ 3. $(2w^2 - 3x)^2$

MAKE SURE ANSWERS: 1. $s^2 + 2st + t^2$ 2. $9u^2 - 24uv + 16v^2$ 3. $4w^4 - 12w^2x + 9x^2$

When you multiply two binomials with form $r + s$ and $r - s$ together, the product is the binomial $r^2 - s^2$.

Example ▶
$$\overset{\text{F} \quad\quad \text{O} \quad\quad \text{I} \quad\quad \text{L}}{(r + s)(r - s) = r \cdot r + (-rs) + rs + s(-s)}$$
$$= r^2 + 0 + (-s^2) \quad\quad \text{Think: } -rs + rs = 0$$
$$= r^2 - s^2$$

If r and s are terms, then:

$(r + s)(r - s)$ [same, same, opposite signs] $= (r - s)(r + s) = r^2 - s^2$ ← difference of two squares [square of last term, square of first term, difference]

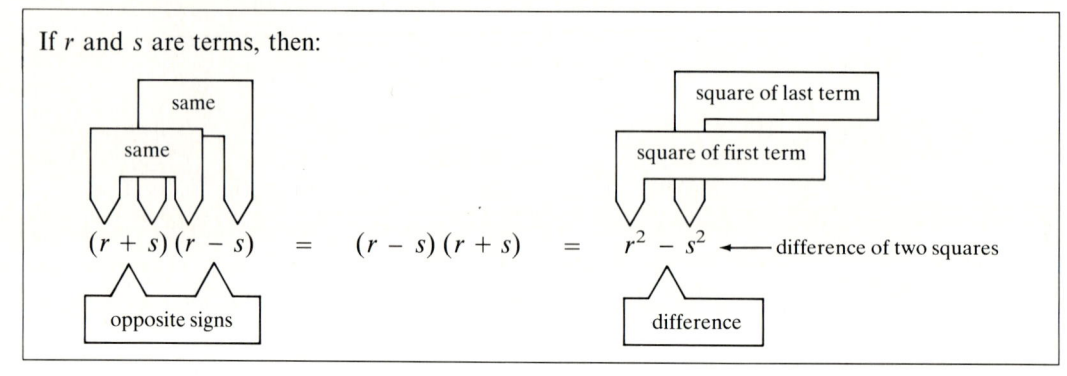

EXAMPLE 3: Find the product of the sum and the difference of the same two terms.

Problem ▶ $(4x + 7)(4x - 7) = ?$

1. Compute squares ▶ $(\boxed{4x + 7})(4x - 7)$ The square of the first term ($4x$) is $16x^2$.
↑   ↑   The square of the second term (7) is 49.
first   second
term   term

2. Form difference ▶ $16x^2 - 49$ ← the square of the first term minus the square of the second term

Solution ▶ $(4x + 7)(4x - 7) = 16x^2 - 49$

Note ▶ In practice you should do all the computation mentally and only write the answer.

Examples ▶
(a) $(3y + 15)(3y - 15) = 9y^2 - 225$
(b) $(9w + 1)(9w - 1) = 81w^2 - 1$
(c) $(13ab - 11)(13ab + 11) = 169a^2b^2 - 121$
(d) $(5m + 6n)(5m - 6n) = 25m^2 - 36n^2$

Make Sure

Multiply the sum and the difference of the same two terms.

See Example 3 ▶
1. $(y + z)(y - z)$
2. $(3w + 2x)(3w - 2x)$
3. $(u^2 + 3v)(u^2 - 3v)$

MAKE SURE ANSWERS: 1. $y^2 - z^2$ 2. $9w^2 - 4x^2$ 3. $u^4 - 9v^2$

Summary ▶ If r and s are terms, then:
$(r + s)^2 = r^2 + 2rs + s^2$, $(r - s)^2 = r^2 - 2rs + s^2$, and $(r + s)(r - s) = r^2 - s^2$.

5.3 Practice

Set 1: Multiply binomials using the FOIL method.

1. $(a + 3)(a + 5)$
2. $(b + 4)(b + 7)$
3. $(2c - 3)(5c + 7)$

4. $(4d + 5)(6d - 5)$
5. $(5e - 3f)(4e - 7f)$
6. $(2g - 3h)(5g - 4h)$

Set 2: Find the square of each binomial.

7. $(a + 3)^2$
8. $(b - 5)^2$
9. $(4c + 3d)^2$

10. $(3c - 5d)^2$
11. $(2e^2 - 3f)^2$
12. $(5g^2 - 7h)^2$

Set 3: Multiply the sum and the difference of the same two terms.

13. $(z + 3)(z - 3)$
14. $(y + 4)(y - 4)$
15. $(3 + 2x)(3 - 2x)$

16. $(5 + 6w)(5 - 6w)$
17. $(3u + 2v)(3u - 2v)$
18. $(4t + 3s)(4t - 3s)$

Mixed Practice: Perform the indicated operations.

19. $(2x + 5)(3x + 2)$
20. $(7y + 3)(4y + 5)$
21. $(8 + 5z)^2$

22. $(3 - 7u)^2$
23. $(4v - 3w)(4v + 3w)$
24. $(5s - t)(5s + t)$

25. $(9q - 4r)(9q + 4r)$
26. $(3n - 8m)(3n + 8m)$
27. $(h^2 - k^2)(h^2 + k^2)$

28. $(2g^2 - 3f^2)(2g^2 + 3f^2)$
29. $(3c^2 - 2d^2)^2$
30. $(a^2 - 4b^2)^2$

Extra: Perform the indicated operations.

31. $(e + 2f)^2 - (e - 3f)^2$
32. $(2g - 4h)^2 - (2g - 3h)(g - 4h)$

33. $(4m - 3n)(3m + 5n) + (2m - 3n)^2$
34. $(3p - 5q)(3p + 5q) + (5p - 3q)^2$

Review: Work these problems on a separate sheet of paper. Attach your work to this page.
Divide signed numbers. (See Lesson 1.3.)

35. $\dfrac{-5}{5}$
36. $\dfrac{-6}{6}$
37. $\dfrac{-8}{4}$
38. $\dfrac{-24}{12}$

39. $\dfrac{-36}{-12}$
40. $\dfrac{-48}{-6}$
41. $\dfrac{45}{-9}$
42. $\dfrac{42}{-3}$

Simplify each product of terms. (See Lesson 1.6.)

43. $z^3 \cdot \dfrac{1}{z^2}$
44. $y^5 \cdot \dfrac{1}{y^3}$
45. $x^6 \cdot \dfrac{1}{x^3}$
46. $w^3 \cdot \dfrac{1}{w}$

47. $v^7 \cdot \dfrac{1}{v}$
48. $-u^3 \cdot \dfrac{1}{-u^2}$
49. $-t^4 \cdot \dfrac{1}{-t^2}$
50. $s^5 \cdot \dfrac{1}{-s^3}$

Subtract polynomials. (See Lesson 5.1.)
51. $(3x - 4) - (5x - 3)$
52. $(5y + 6) - (3y - 3)$
53. $(3z - 4) - (3 - 5z)$
54. $(3u^2 - 5) - (5u + 2)$
55. $(3v^2 - 4v) - (2v^2 - 5v + 3)$
56. $(5w^3 - 3w + 4) - (4w^3 - 4w)$

5.4 Divide Polynomials

> To divide monomials, you divide the numerical coefficients and then divide the literal parts.

EXAMPLE 1: Divide two monomials.

Problem $\dfrac{28x^3y^4z}{-7xy^5z} = ?$

1. Divide coefficients $\dfrac{28x^3y^4z}{-7xy^5z} = \boxed{\left(\dfrac{28}{-7}\right)\left(\dfrac{x^3}{x}\right)\left(\dfrac{y^4}{y^5}\right)\left(\dfrac{z}{z}\right)}$

$= -4\left(\dfrac{x^3}{x}\right)\left(\dfrac{y^4}{y^5}\right)\left(\dfrac{z}{z}\right)$

2. Divide literals $= -4(x^2)(y^{-1})(z^0)$ Think: $\dfrac{x^3}{x} = x^{3-1} = x^2$ $\dfrac{y^4}{y^5} = y^{4-5} = y^{-1}$

$\dfrac{z^1}{z^1} = z^{1-1} = z^0$

3. Simplify $= \dfrac{-4x^2}{y}$ Write using only positive exponents.

Solution $\dfrac{28x^3y^4z}{-7xy^5z} = \dfrac{-4x^2}{y}$

Other Examples (a) $\dfrac{-12p^5}{3p} = -4p^4$ (b) $\dfrac{7y^3z}{14y^5z} = \dfrac{1}{2y^2}$ (c) $\dfrac{57wxy}{19wx^5} = \dfrac{3y}{x^4}$

Make Sure

Divide two monomials.

See Example 1 1. $\dfrac{k^7}{k^2}$ _____   2. $\dfrac{5m^5n^2}{15m^3n^6}$ _____   3. $\dfrac{-6p^4q^2}{12p^3}$ _____

4. $\dfrac{12r^3s^4t}{4rs^3t^2}$ _____   5. $\dfrac{64u^3v^2w}{-8u^2v^3w}$ _____   6. $\dfrac{-56x^4y^7z^3}{-7x^2y^4z^5}$ _____

MAKE SURE ANSWERS: 1. k^5 2. $\dfrac{m^2}{3n^4}$ 3. $-\dfrac{pq^2}{2}$ 4. $\dfrac{3r^2s}{t}$ 5. $-\dfrac{8u}{v}$ 6. $\dfrac{8x^2y^3}{z^2}$

190 Chapter 5 Polynomials

The distributive properties provide the basis for dividing a polynomial by a monomial.

> **To divide a polynomial by a monomial:**
> 1. Use the distributive properties to divide each term of the polynomial by the monomial.
> 2. Simplify the sum of the quotients formed in Step 1.

EXAMPLE 2: Divide a polynomial by a monomial.

Problem ▶ $\dfrac{4x^3y + 6x^2y^2 - 12xy^3}{-2xy} = ?$

1. Distribute ▶ $\dfrac{4x^3y + 6x^2y^2 - 12xy^3}{-2xy} = \dfrac{4x^3y}{-2xy} + \dfrac{6x^2y^2}{-2xy} + \dfrac{-12xy^3}{-2xy}$ Think: $-2xy$ is a common denominator.

2. Simplify ▶ $\qquad\qquad\qquad\qquad\qquad = -2x^2 + (-3)xy + 6y^2$

Solution ▶ $\dfrac{4x^3y + 6x^2y^2 - 12xy^3}{-2xy} = -2x^2 - 3xy + 6y^2$

Other Examples ▶ (a) $\dfrac{15a^4b - 20a^3b^2 + 10a^2b^3 + 5ab^4}{-5a^2b} = -3a^2 + 4ab - 2b^2 - \dfrac{b^3}{a}$

(b) $\dfrac{4mn - 3m^2n^3}{4m^2n^4} = \dfrac{1}{mn^3} - \dfrac{3}{4n}$

(c) $(14j^3 - 21j^2 + 7j) \div (7j) = 2j^2 - 3j + 1$

Make Sure

Divide polynomials by a monomial.

See Example 2 ▶ 1. $\dfrac{8x^5y^3 - 12x^4y^5 - 28x^3y^4 + 24x^3y^2}{4x^3y^2}$ 2. $(9u^4v^2 - 27u^3v^3 + 3u^2v^4 - 3uv^2) \div (-3uv^2)$

_____   _____

MAKE SURE ANSWERS: 1. $2x^2y - 3xy^3 - 7y^2 + 6$ **2.** $-3u^3 + 9u^2v - uv^2 + 1$

The method used to divide a polynomial by a polynomial is similar to the long division algorithm (method) used for natural numbers.

> **To divide a polynomial by a polynomial**
> 1. Write in division box form. Arrange both the divisor and the dividend in standard form. Insert 0 for any missing terms.
> 2. Divide the first term of the divisor into the first term of the dividend to obtain the first term of the quotient.
> 3. Multiply the first term of the quotient by each term in the divisor. Write this product under the dividend, aligning like terms.
> 4. Subtract like terms and bring down one or more terms as needed.
> 5. Compare. If the result in Step 4 is 0 or of a lower degree than the divisor, then the division is complete. Otherwise use the result in Step 4 as a new dividend and repeat Steps 2–5.

5.4 Divide Polynomials

EXAMPLE 3: Divide polynomials.

Problem ▶ $\dfrac{x^2 - 8 + 3x}{x - 2} = ?$

1. **Write division box form** ▶ $x - 2 \,\overline{\smash{\big)}\, x^2 + 3x - 8}$ Think: Arrange both the divisor and the dividend in standard form.

2. **Divide** ▶
$$\begin{array}{r} x \\ x - 2 \,\overline{\smash{\big)}\, x^2 + 3x - 8} \end{array}$$
$x^2 \div x = x$ ⟵ 1st term of the quotient
⟵ 1st term of the divisor
⟵ 1st term of the dividend

3. **Multiply** ▶
$$\begin{array}{r} x \\ x - 2 \,\overline{\smash{\big)}\, x^2 + 3x - 8} \\ x^2 - 2x \end{array}$$
$\begin{array}{r} x - 2 \\ \underline{\times x} \\ x^2 - 2x \end{array}$ ⟵ divisor
⟵ 1st term of the divisor

4. **Subtract and bring down** ▶
$$\begin{array}{r} x \\ x - 2 \,\overline{\smash{\big)}\, x^2 + 3x - 8} \\ \underline{x^2 - 2x } \downarrow \\ 5x - 8 \end{array}$$

5. **Compare** ▶
$$\begin{array}{r} x \\ x - 2 \,\overline{\smash{\big)}\, x^2 + 3x - 8} \\ \underline{x^2 - 2x } \\ 5x - 8 \end{array}$$
The degree of the remainder $5x - 8$ is 1.
The degree of the divisor $x - 2$ is 1.
Think: $1 \not< 1$ means the division process can continue.

6. **Repeat Steps 2–5** ▶
$$\begin{array}{r} x + 5 \\ x - 2 \,\overline{\smash{\big)}\, x^2 + 3x - 8} \\ \underline{x^2 - 2x } \\ 5x - 8 \end{array}$$
Divide: $5x \div x = 5$

$$\begin{array}{r} x + 5 \\ x - 2 \,\overline{\smash{\big)}\, x^2 + 3x - 8} \\ \underline{x^2 - 2x } \\ 5x - 8 \\ 5x - 10 \end{array}$$
Multiply: $\begin{array}{r} x - 2 \\ \underline{\times 5} \\ 5x - 10 \end{array}$ ⟵ divisor
⟵ 2nd term of the quotient

$$\begin{array}{r} x + 5 \\ x - 2 \,\overline{\smash{\big)}\, x^2 + 3x - 8} \\ \underline{x^2 - 2x } \\ 5x - 8 \\ \underline{5x - 10} \\ 2 \end{array}$$
Subtract.

$$\begin{array}{r} x + 5 \\ x - 2 \,\overline{\smash{\big)}\, x^2 + 3x - 8} \\ \underline{x^2 - 2x } \\ 5x - 8 \\ \underline{5x - 10} \\ 2 \end{array}$$
Compare: The degree of the remainder 2 is 0.
The degree of the divisor $x - 2$ is 1.
$0 < 1$ means the division is complete.

Solution ▶ $\dfrac{x^2 - 8 + 3x}{x - 2} = \overbrace{x + 5}^{\text{quotient}} + \dfrac{2}{x - 2}$ ⟵ remainder
⟵ divisor

Other Examples ▸ (a) $\dfrac{4x^2 + 2x - 11}{2x - 3} = ?$

$$\begin{array}{r}
2x + 4 \quad \leftarrow \text{quotient} \\
2x - 3 \overline{\smash{\big)}\, 4x^2 + 2x - 11} \\
\underline{4x^2 - 6x} \\
8x - 11 \\
\underline{8x - 12} \\
+1 \quad \leftarrow \text{remainder}
\end{array}$$

$$\dfrac{4x^2 + 2x - 11}{2x - 3} = 2x + 4 + \dfrac{1}{2x - 3}$$

(b) $\dfrac{2x^3 + 5x^2 - 7x + 5}{x^2 + 4x - 5} = ?$

$$\begin{array}{r}
2x - 3 \quad \leftarrow \text{quotient} \\
x^2 + 4x - 5 \overline{\smash{\big)}\, 2x^3 + 5x^2 - 7x + 5} \\
\underline{2x^3 + 8x^2 - 10x} \\
-3x^2 + 3x + 5 \\
\underline{-3x^2 - 12x + 15} \\
15x - 10 \quad \leftarrow \text{remainder}
\end{array}$$

$$\dfrac{2x^3 + 5x^2 - 7x + 5}{x^2 + 4x - 5} = 2x - 3 + \dfrac{15x - 10}{x^2 + 4x - 5}$$

(c) $(a^2 + 2ab - 4b^2) \div (a - b) = ?$

$$\begin{array}{r}
a + 3b \quad \leftarrow \text{quotient} \\
a - b \overline{\smash{\big)}\, a^2 + 2ab - 4b^2} \\
\underline{a^2 - ab} \\
3ab - 4b^2 \\
\underline{3ab - 3b^2} \\
-b^2 \quad \leftarrow \text{remainder}
\end{array}$$

$$(a^2 + 2ab - 4b^2) \div (a - b) = a + 3b - \dfrac{b^2}{a - b}$$

To check a division problem, you add the remainder to the product of the divisor and the quotient and see if this result is equal to the dividend.

Example ▸ Check the division: $\dfrac{x^2 - 8 + 3x}{x - 2} = x + 5 + \dfrac{2}{x - 2}$.

$$\begin{array}{r}
x - 2 \quad \leftarrow \text{divisor} \\
x + 5 \quad \leftarrow \text{quotient} \\
\hline
5x - 10 \\
x^2 - 2x \\
\hline
x^2 + 3x - 10 \\
+ 2 \quad \leftarrow \text{remainder} \\
\hline
x^2 + 3x - 8 \quad \leftarrow \text{original dividend}
\end{array}$$

Multiply the divisor by the quotient.

Add the remainder to the product.

Solution ▸ The division: $\dfrac{x^2 - 8 + 3x}{x - 2} = x + 5 + \dfrac{2}{x - 2}$ is correct.

If a divisor or a dividend has a missing term, then it may be helpful to insert a 0 for that term, so that like terms will be aligned in the proper column.

5.4 Divide Polynomials

Example ▶ $\dfrac{x^4 - 1}{x - 1} = ?$

$$\begin{array}{r} x^3 + x^2 + x + 1 \\ x-1\overline{\smash{)}x^4 + 0 + 0 + 0 - 1} \\ \underline{x^4 - x^3} \\ x^3 + 0 \\ \underline{x^3 - x^2} \\ x^2 + 0 \\ \underline{x^2 - x} \\ x - 1 \\ \underline{x - 1} \\ 0 \end{array}$$

← quotient

Think: $x^4 - 1 = x^4 + 0x^3 + 0x^2 + 0x - 1$

← remainder

Solution ▶ $\dfrac{x^4 - 1}{x - 1} = x^3 + x^2 + x + 1$

Note ▶ Since the remainder is 0, you can check the division by multiplying the divisor $x - 1$ and the quotient $x^3 + x^2 + x + 1$ and then compare the product with the dividend $x^4 - 1$.

Make Sure

Divide polynomials.

See Example 3 ▶ **1.** $(2x^3 - 6x^2 + x - 3) \div (x - 3)$ **2.** $(6y^3 + y^2 + 3y - 1) \div (2y + 1)$

MAKE SURE ANSWERS: 1. $2x^2 + 1$ 2. $3y^2 - y + 2 - \dfrac{3}{2y - 1}$

In any division problem, a zero remainder means that both the divisor and the quotient are factors of the dividend.

Example ▶ Determine whether $2m + n$ is a factor of $6m^3 + 11m^2n - 6mn^2 - 5n^3$.

1. Divide ▶
$$\begin{array}{r} 3m^2 + 4mn - 5n^2 \\ 2m+n\overline{\smash{)}6m^3 + 11m^2n - 6mn^2 - 5n^3} \\ \underline{6m^3 + 3m^2n} \\ 8m^2n - 6mn^2 \\ \underline{8m^2n + 4mn^2} \\ -10mn^2 - 5n^3 \\ \underline{-10mn^2 - 5n^3} \\ 0 \end{array}$$

2. Interpret ▶ A remainder of 0 means the divisor is a factor of the dividend.

Solution ▶ Yes: $2m + n$ is a factor of $6m^3 + 11m^2n - 6mn^2 - 5n^3$.

Note ▶ The quotient $3m^2 + 4mn - 5n^2$ is also a factor of $6m^3 + 11m^2n - 6mn^2 - 5n^3$.

Expressions that involve a combination of operations can often be simplified. It is important to perform the indicated operations in the proper order.

194   Chapter 5   Polynomials

EXAMPLE 4: Simplify an expression that involves polynomials and a combination of operations.

Problem $\dfrac{(2x + 3)^2 - (3x^2 + 2x)}{x + 1} = ?$

1. Simplify in grouping symbols $\dfrac{(2x + 3)^2 - (3x^2 + 2x)}{x + 1} = \dfrac{(4x^2 + 12x + 9) - (3x^2 + 2x)}{x + 1}$ The fraction bar is a grouping symbol.

2. Combine like terms $= \dfrac{4x^2 + 12x + 9 - 3x^2 - 2x}{x + 1}$

$= \dfrac{x^2 + 10x + 9}{x + 1}$ or
$$\begin{array}{r} x + 9 \\ x+1\overline{\smash{)}x^2 + 10x + 9} \\ \underline{x^2 + x} \\ 9x + 9 \\ \underline{9x + 9} \\ 0 \end{array}$$

Solution $\dfrac{(2x + 3)^2 - (3x^2 + 2x)}{x + 1} = x + 9$

Another Example $\dfrac{(y - 1)^3}{(2y + 3) - (y + 5)} = ?$

$\dfrac{(y - 1)^3}{(2y + 3) - (y + 5)} = \dfrac{(y - 1)(y - 1)(y - 1)}{(2y + 3) - (y + 5)}$

$= \dfrac{(y^2 - 2y + 1)(y - 1)}{y - 2}$

$= \dfrac{y^3 - 2y^2 + y - y^2 + 2y - 1}{y - 2}$

$= \dfrac{y^3 - 3y^2 + 3y - 1}{y - 2}$ or
$$\begin{array}{r} y^2 - y + 1 + \dfrac{1}{y - 2} \\ y-2\overline{\smash{)}y^3 - 3y^2 + 3y - 1} \\ \underline{y^3 - 2y^2} \\ -y^2 + 3y \\ \underline{-y^2 + 2y} \\ y - 1 \\ \underline{y - 2} \\ 1 \end{array}$$

Make Sure

Perform the indicated operations.

See Example 4   **1.** $\dfrac{(2x - 1)^2 - (2x - 3)(x + 1)}{x - 1}$ **2.** $\dfrac{(x - y)^2 + (x + y)^2}{x^2 - y^2}$

MAKE SURE ANSWERS: 1. $2x - 1 + \dfrac{3}{x - 1}$ **2.** $\dfrac{x^2 + y^2}{4y^2}$

5.4 Practice

Set 1: Divide two monomials.

1. $\dfrac{a^7}{a^4}$
2. $\dfrac{b^6}{b^3}$
3. $\dfrac{-c^5}{c^2}$

4. $\dfrac{-d^4}{d^2}$
5. $\dfrac{-w^6 z^5}{-w^4 z^2}$
6. $\dfrac{-x^4 y^3}{-x^2 y^2}$

7. $\dfrac{21 a^4 b^5 c^3}{7 a^2 b^5 c^2}$
8. $\dfrac{24 d^5 e^4 f^9}{6 d^3 e^4 f^6}$
9. $\dfrac{7 p^4 q r^2}{-49 p^7 q^3 r^4}$

10. $\dfrac{8 t^4 u^3 v}{-64 t^6 u^5 v^7}$
11. $\dfrac{-1.6 x^2 y^5}{-3.2 x^5 y^9 z^4}$
12. $\dfrac{-0.9 r^3 s^3}{-1.8 r^6 s^3 t}$

Set 2: Divide polynomials by monomials.

13. $\dfrac{4a^2 - 6}{2}$
14. $\dfrac{5 - 10 b^2}{5}$
15. $\dfrac{3c^2 + 6c}{-3c}$

16. $\dfrac{6d + 8d^2}{-2d}$
17. $\dfrac{14 a^3 - 7 a^2}{7 a^2}$
18. $\dfrac{16 b^4 - 8 b^3}{4 b^2}$

19. $\dfrac{12 a^2 b - 6 a b^2 + 9 a^3 b^4}{3ab}$
20. $\dfrac{14 x^3 y^2 - 21 x^3 y^4 + 28 x^4 y}{7xy}$

21. $\dfrac{10 w^5 z^4 - 30 w^2 z^5 + 25 w^3 z}{5 w^2 z^3}$
22. $\dfrac{6 a^4 b^3 - 12 a^5 b^6 + 18 a b^5}{6 a^3 b^3}$

23. $\dfrac{4 x^3 y^2 - 6 x^2 y^3 + 8 y^3}{-2 x^2 y}$
24. $\dfrac{15 c^4 d^5 - 9 c^3 d^4 - 12 c^6}{-3 c^3 d^4}$

Set 3: Divide polynomials.

25. $\dfrac{z^2 - 4z + 4}{z - 2}$

26. $\dfrac{y^2 - 5y + 6}{y - 3}$

27. $\dfrac{2x^2 - 5x - 3}{x - 3}$

28. $\dfrac{3w^2 + 5w - 2}{w + 2}$

29. $\dfrac{u^2 - v^2}{u - v}$

30. $\dfrac{s^2 - t^2}{s + t}$

31. $\dfrac{6r^5 + 18r^4 - 13r^3 - 22r^2 + 30r - 4}{2r^2 + 6r - 1}$

32. $\dfrac{2q^5 - q^4 - 6q^2 - 2}{q^2 - q - 1}$

Set 4: Perform the indicated operations.

33. $(2z - 3)^2 + (3z - 2)^2$

34. $(4y - 3)^2 + (5y - 3)^2$

35. $(2v - 5)^2 - (v^2 - 9) \div (v - 3)$

36. $(3u - 2)^2 - (u^2 - 4) \div (u + 2)$

37. $(3t - 5)^2 + (t - 3)(t + 2)$

38. $(2r - 3)(3r - 2) \div (r + 1)$

Review: Work these problems on a separate sheet of paper. Attach your work to this page.

Add signed numbers. (See Lesson 1.3, Example 1.)
39. $5 + (-3)$
40. $3 + (-6)$
41. $-5 + 6$
42. $-7 + 4$
43. $-7 + (-3)$
44. $-2 + (-7)$
45. $-5 + (-6)$
46. $-3 + (-1)$

Multiply signed numbers. (See Lesson 1.3, Example 3.)
47. $3(-2)$
48. $2(-4)$
49. $4(-1)$
50. $5(-1)$
51. $-1(3)$
52. $-1(5)$
53. $-2(-3)$
54. $-5(-2)$

Write polynomials in standard form. (See Lesson 5.1.)
55. $3 - 2z$
56. $5 - 3y$
57. $3x - 2 + x^2$
58. $5w + 3 + 2w^2$
59. $4v^2 - 3 + 5v - v^3$
60. $7u^2 - 3u + u^3 - 1$
61. $3 - 7t^3 - t$
62. $5 + 3s^3 - 2s$
63. $1 - r^3$

5.5 Use Synthetic Division

The division algorithm for dividing a polynomial in x (in standard form) by a binomial of the form $x - c$ can be condensed by a process called *synthetic division*.

The synthetic division process can best be understood by first examining a specific example of the division algorithm and then condensing the procedure in a systematic manner.

Example ▶

$$\begin{array}{r} 3x^2 - 2x + 4 \\ x - 3 \overline{\smash{)}3x^3 - 11x^2 + 10x + 1} \\ \underline{3x^3 - 9x^2} \\ -2x^2 + 10x \\ \underline{-2x^2 + 6x} \\ 4x + 1 \\ \underline{4x - 12} \\ 13 \end{array}$$

Division model.

1. Omit duplications in each column ▶

$$\begin{array}{r} 3x^2 - 2x + 4 \\ x - 3 \overline{\smash{)}3x^3 - 11x^2 + 10x + 1} \\ \underline{3x^3 - 9x^2} \\ -2x^2 + \cancel{10x} \\ \underline{-2x^2 + 6x} \\ 4x + \cancel{1} \\ \underline{4\cancel{x} - 12} \\ 13 \end{array}$$

No information will be lost if these terms are omitted because they are each duplications of the term directly above them.

2. Omit the variables ▶

$$\begin{array}{r} 3 \quad -2 \quad4 \\ -3\,\overline{\smash{|}\,3 \quad -11 \quad10 \quad 1} \\ \underline{-9} \\ -2 \\ 6 \\ \underline{4} \\ -12 \\ \underline{13} \end{array}$$

No information will be lost by omitting the variables because the polynomials are written in standard form. The position of a number indicates the power of the variable associated with it.

3. Omit the top row ▶

$$\begin{array}{r} \cancel{3} \quad \cancel{-2} \quad\cancel{4} \\ -3\,\overline{\smash{|}\,\boxed{3} \quad -11 \quad10 \quad 1} \\ -9 \\ \boxed{-2} \\ 6 \\ \boxed{4} \\ -12 \\ 13 \end{array}$$

No information will be lost if these numbers are omitted because they are duplicates of the shaded numbers.

4. Condense the notation ▶

$$\begin{array}{r|rrrr} -3 & 3 & -11 & 10 & 1 \\ & & -9 & 6 & -12 \\ \hline & 3 & -2 & 4 & 13 \end{array}$$

Condensing the vertical spacing allows the work to be written in a more compact form.

——— this duplicate of the above leading coefficient is written here as a matter of convenience

5. Convert from subtraction to addition ▶

$$\begin{array}{r|rrrr} +3 & 3 & -11 & 10 & 1 \\ & & +9 & -6 & +12 \\ \hline & 3 & -2 & 4 & 13 \end{array}$$

Changing the sign of the number in the divisor makes it possible to perform addition instead of subtraction on the numbers in the columns.

198   Chapter 5   Polynomials

6. Identify parts ▶

$$\begin{array}{r|rrrr} 3 & 3 & -11 & 10 & 1 \\ & & 9 & -6 & 12 \\ \hline & 3 & -2 & 4 & 13 \end{array}$$ ← coefficients of the terms in the dividend

↑ coefficients of the terms in the quotient   ↖ the remainder

Think: The quotient is $3x^2 - 2x + 4$.
The remainder is 13.

Solution ▶ $\dfrac{3x^3 - 11x^2 + 10x + 1}{x - 3} = 3x^2 - 2x + 4 + \dfrac{13}{x - 3}$

The necessary steps for using synthetic division are summarized as follows.

> **Synthetic Division**
> 1. Arrange the polynomials in standard form, inserting 0 for any missing terms.
> 2. Write c for the divisor $x - c$ followed by the coefficients of the dividend. Duplicate the leading coefficient of the dividend as the first number in the 3rd (bottom) row.
> 3. Compute the product of c and the number just written on the bottom row. Write this product in the next column to the right in the middle row.
> 4. Add the numbers in the column from Step 3 and write the sum in the bottom row. Alternate between Steps 3 and 4 until all columns are filled.
> 5. Identify the quotient and remainder from the bottom row.

EXAMPLE 1: Divide using synthetic division.

Problem ▶ $\dfrac{x^3 + 9 - 3x^2}{x - 2} = ?$

1. Arrange in standard form ▶
$x^3 + 9 - 3x^2 = x^3 - 3x^2 + 9$ ← standard form
$= x^3 - 3x^2 + 0 + 9$

Think: Insert 0 for the missing x term ($0 = 0x$).
Think: Write 2 for the divisor $x - 2$ followed by the coefficients of the dividend.

2. Write in synthetic division form ▶
$\begin{array}{r|rrrr} 2 & 1 & -3 & 0 & 9 \\ & & & & \\ \hline & 1 & & & \end{array}$ ← duplicate the leading coefficient of the dividend

3. Multiply ▶
$\begin{array}{r|rrrr} 2 & 1 & -3 & 0 & 9 \\ & & 2 & & \\ \hline & 1 & & & \end{array}$

4. Add ▶
$\begin{array}{r|rrrr} 2 & 1 & -3 & 0 & 9 \\ & & 2 & & \\ \hline & 1 & -1 & & \end{array}$

5. Alternate between Steps 3 and 4 ▶
$\begin{array}{r|rrrr} 2 & 1 & -3 & 0 & 9 \\ & & 2 & -2 & -4 \\ \hline & 1 & -1 & -2 & 5 \end{array}$

6. Identify the quotient and remainder ▶
$\begin{array}{r|rrrr} 2 & 1 & -3 & 0 & 9 \\ & & 2 & -2 & -4 \\ \hline & 1 & -1 & -2 & 5 \end{array}$

↑ the coefficients of the terms of the quotient   ↖ the remainder

Think: The quotient is $x^2 - x - 2$.
The remainder is 5.

Solution ▶ $\dfrac{x^3 + 9 - 3x^2}{x - 2} = x^2 - x - 2 + \dfrac{5}{x - 2}$

5.5 Use Synthetic Division

> CAUTION: Synthetic division can only be used when the divisor is a polynomial of the form $x - c$.

Other Examples ▶ (a) $\dfrac{2x^4 - 3x^3 + x^2 - 5x + 7}{x - 1} = ?$

$$\begin{array}{r|rrrrr} 1 & 2 & -3 & 1 & -5 & 7 \\ & & 2 & -1 & 0 & -5 \\ \hline & 2 & -1 & 0 & -5 & 2 \end{array}$$

Think: The quotient is $2x^3 - x^2 - 5$.
The remainder is 2.

$$\dfrac{2x^4 - 3x^3 + x^2 - 5x + 7}{x - 1} = 2x^3 - x^2 - 5 + \dfrac{2}{x - 1}$$

(b) $\dfrac{5x^3 + 6x^2 - 17x + 3}{x + 3} = ?$

$$\begin{array}{r|rrrr} -3 & 5 & 6 & -17 & 3 \\ & & -15 & 27 & -30 \\ \hline & 5 & -9 & 10 & -27 \end{array}$$

Think: The divisor $x + 3 = x - (-3)$.
The quotient is $5x^2 - 9x + 10$.
The remainder is -27.

$$\dfrac{5x^3 + 6x^2 - 17x + 3}{x + 3} = 5x^2 - 9x + 10 - \dfrac{27}{x + 3}$$

(c) $(y^3 + 1) \div (y + 1) = ?$

$$\begin{array}{r|rrrr} -1 & 1 & 0 & 0 & 1 \\ & & -1 & 1 & -1 \\ \hline & 1 & -1 & 1 & 0 \end{array}$$

Think: The quotient is $y^2 - y + 1$.
The remainder is 0.

$(y^3 + 1) \div (y + 1) = y^2 - y + 1$

(d) $\dfrac{16x^4 - x^2 - \frac{3}{4}}{x - \frac{1}{2}} = ?$

$$\begin{array}{r|rrrrr} \frac{1}{2} & 16 & 0 & -1 & 0 & -\frac{3}{4} \\ & & 8 & 4 & \frac{3}{2} & \frac{3}{4} \\ \hline & 16 & 8 & 3 & \frac{3}{2} & 0 \end{array}$$

Think: The quotient is $16x^3 + 8x^2 + 3x + \frac{3}{2}$.
The remainder is 0.

$$\dfrac{16x^4 - x^2 - \frac{3}{4}}{x - \frac{1}{2}} = 16x^3 + 8x^2 + 3x + \frac{3}{2}$$

Make Sure

Divide polynomials using synthetic division.

See Example 1 ▶ 1. $(y^3 - 2y^2 + 2y - 1) \div (y - 1)$ 2. $(4x^4 + 4x^2 - 18x + 18x^3 - 6) \div (x + 4)$

MAKE SURE ANSWERS: 1. $y^2 - y + 1$ 2. $4x^3 + 2x^2 - 4x - 2 + \dfrac{2}{x - 4}$

Recall ▶ In any division problem, a remainder of zero indicates that the divisor is a factor of the dividend.

Synthetic division is often used to determine if a polynomial is a factor of a given polynomial.

EXAMPLE 2: Find factors of a polynomial using synthetic division.

Problem ▶ Is $x - 3$ a factor of $2x^3 - 10x^2 + 17x - 15$?

1. Divide ▶
$$\begin{array}{r|rrrr} 3 & 2 & -10 & 17 & -15 \\ & & 6 & -12 & 15 \\ \hline & 2 & -4 & 5 & 0 \end{array}$$ ← the remainder

2. Interpret ▶ A remainder of 0 means the divisor is a factor of the dividend.

Solution ▶ Yes: $x - 3$ is a factor of $2x^3 - 10x^2 + 17x - 15$.

Note ▶ The quotient $2x^2 - 4x + 5$ is also a factor of $2x^3 - 10x^2 + 17x - 15$.

Another Example ▶ Is $x + 1$ a factor of $x^4 + 1$?

1. Divide ▶
$$\begin{array}{r|rrrrr} -1 & 1 & 0 & 0 & 0 & 1 \\ & & -1 & 1 & -1 & 1 \\ \hline & 1 & -1 & 1 & -1 & 2 \end{array}$$ ← the remainder

2. Interpret ▶ A nonzero remainder means the divisor is not a factor of the dividend.

Solution ▶ No: $x + 1$ is not a factor of $x^4 + 1$.

Make Sure

Determine if the binomial is a factor of the second polynomial.

See Example 2 ▶
1. $y + 2,\ y^5 - y^4 - 2y^3 + 3y^2 + 1$

2. $x + 1,\ 4x^4 - 18x^3 + 4x^2 + 18x - 8$

3. $u - 1,\ 3u^4 - 4u + 3u^2 - 4$

4. $v - 2,\ 2v^3 + 9v - 6 - 7v^2$

MAKE SURE ANSWERS: 1. no **2.** yes **3.** no **4.** yes

5.5 Practice

Set 1: Divide polynomials using synthetic division.

1. $(z^2 + 6z + 9) \div (z + 3)$

2. $(y^2 + 8y + 16) \div (y + 4)$

3. $(x^2 - 9) \div (x - 3)$

4. $(w^2 - 16) \div (w - 4)$

5. $(v^2 - 4v + 4) \div (v - 2)$

6. $(u^2 - 5u + 6) \div (u - 3)$

7. $(t^3 + 2t^2 - 4t - 8) \div (t + 2)$

8. $(s^3 - 2s^2 + 9s - 18) \div (s - 2)$

9. $(r^4 + 5r^2 - 36) \div (r + 2)$

10. $(q^4 - 3q^2 - 54) \div (q - 3)$

11. $(9p^2 - 64) \div (p - 4)$

12. $(4n^2 - 25) \div (n - 5)$

13. $(m^3 - 27) \div (m - 3)$

14. $(n^3 + 64) \div (n + 4)$

15. $(6p^2 - p - 1) \div (p - \tfrac{1}{2})$

16. $(12q^2 - 5q - 2) \div (q + \tfrac{1}{4})$

17. $(1 - r^4) \div (1 + r)$

18. $(16 - s^4) \div (2 + s)$

202   Chapter 5   Polynomials

Set 2: Determine if the binomial is a factor of the second polynomial using synthetic division.

19. $z + 3, z^2 + 5z + 6$
20. $y + 4, y^2 + 2y - 8$
21. $x - 5, x^2 - 2x - 15$

22. $w - 6, w^2 - 2w - 24$
23. $v + 3, 2v^2 + 11v + 4$
24. $u + 5, 3u^2 + 17u + 16$

25. $t + 7, t^2 - 49$
26. $s - 4, s^2 - 16$
27. $r + 3, r^2 + 9$

28. $q - 5, q^2 + 25$
29. $p + 8, p^3 + 8p^2 - 4p - 32$

30. $m + 4, m^3 + 4m^2 - 8m - 32$
31. $p - 4, 5p^4 - 4p^3 + 2p^2 - 5p + 12$

32. $q - 3, 8q^4 - 62q^2 - 42q + 40$
33. $r - 3, r^3 + 27$

34. $s + 4, s^3 - 64$
35. $t - 2, t^4 - 16$
36. $u + 3, u^4 - 81$

Review: Work these problems on a separate sheet of paper. Attach your work to this page.

Factor each composite number as a product of primes.
37. 6
38. 12
39. 16
40. 24
41. 36
42. 45
43. 54
44. 64
45. 70
46. 72
47. 96
48. 320

Divide two monomials. (See Lesson 5.4.)

49. $\dfrac{z^8}{z^3}$
50. $\dfrac{y^7}{y^2}$
51. $\dfrac{x^6}{x}$
52. $\dfrac{w^4}{w}$

53. $\dfrac{6v^3}{3v^2}$
54. $\dfrac{8u^5}{2u^2}$
55. $\dfrac{-12t^6}{4t^2}$
56. $\dfrac{-18s^7}{3s^3}$

57. $\dfrac{28m^5n^7}{-4m^2n^3}$
58. $\dfrac{42p^3q^6}{-7pq^3}$
59. $\dfrac{-48rs^3}{-12rs}$
60. $\dfrac{-56t^5u^2v}{-14t^2u^2v}$

Name _____ Date _____ Class _____

Chapter 5 Review

			What to Review if You Have Trouble	
Objectives		Lesson	Example	Page
Add polynomials ▶	1. $(-3x^2 + 5x - 2) + (5x^2 + 7x - 8)$ _____	5.1	1	170
Subtract polynomials ▶	2. $(2x^2 - 7x + 11) - (3x^2 + x - 9)$ _____	5.1	2	172
Multiply monomials ▶	3. $(2a^2b^3)(-3ab^2)(-b)$ _____	5.2	1	177
Multiply polynomials ▶	4. $3x(2x^2 - 7x + 8)$ _____	5.2	2	177
	5. $(3x + 4)(2x - 5)$ _____	5.2	3	178
Multiply two binomials by the FOIL method ▶	6. $(3x - 5)(2x + 7)$ _____	5.3	1	183
Find special products ▶	7. $(4y - 3z)^2$ _____	5.3	2	185
	8. $(5x + 9)(5x - 9)$ _____	5.3	3	186

Copyright © 1985 by Harcourt Brace Jovanovich, Inc. All rights reserved.

204 Chapter 5 Polynomials

Divide polynomials ▶ 9. $\dfrac{-51x^4y^3z}{17x^5yz}$ _____ 5.4 1 189

10. $\dfrac{8x^2y^3 - 12x^2y^2 + 14xy^3}{-2xy^2}$ _____ 5.4 2 190

11. $\dfrac{x^2 + 10x - 9}{x - 3}$ _____ 5.4 3 191

Simplify expressions involving polynomials ▶ 12. $\dfrac{(7x^2 - 13x + 4) - (x - 3)^2}{2x + 1}$ _____ 5.4 4 194

Divide using synthetic division ▶ 13. $\dfrac{x^3 - 8 - 2x^2}{x + 3}$ _____ 5.5 1 198

Find factors using synthetic division ▶ 14. Is $x - 2$ a factor of $x^3 + 2x^2 - 13x - 10$? _____ 5.5 2 200

Evaluate polynomials ▶ 15. Principal of $100 is invested in an account that pays 10% interest compounded annually. Assuming that there are no further deposits or withdrawals, how much money will be in the account at the end of 3 years? _____ PS 6 — 175

CHAPTER 5 REVIEW ANSWERS: 1. $2x^2 + 12x - 10$ 2. $-x^2 - 8x + 20$ 3. $6a^3b^6$ 4. $6x^3 - 21x^2 + 24x$ 5. $6x^2 - 7x - 20$ 6. $6x^2 + 11x - 35$ 7. $16y^2 - 24yz + 9z^2$ 8. $25x^2 - 81$ 9. $-\dfrac{3y^2}{x}$ 10. $-4xy + 6x - 7y$ 11. $x + 13 + \dfrac{30}{x-3}$ 12. $3x - 5$ 13. $x^2 - 5x + 15 - \dfrac{53}{x+3}$ 14. $x - 2$ is not a factor of $x^3 + 2x^2 - 13x - 10$ 15. $133.10

POLYNOMIALS AND FACTORING

6 Factoring

6.1 Factor Using the GCF

6.2 Factor $x^2 + bx + c$

6.3 Factor $ax^2 + bx + c$

6.4 Factor Special Polynomials

6.5 Factor Using the General Strategy

6.6 Factor to Solve Equations

PS 7: Factor to Solve Problems

Introduction to Factoring

To *factor a polynomial* means to write the polynomial as the product of other polynomials. A polynomial whose numerical coefficients and constant term are integers is called an *integral polynomial*. This chapter will discuss factoring integral polynomials.

An integer is a factor of another integer if it divides the other integer *evenly* (with a zero remainder).

Examples ▶ 1, −1, 2, −2, 5, −5, 10, −10 are all factors of 10.   3 is not a factor of 10 because 3 does not divide 10 evenly.

To *factor a monomial* means to write it as a product.

Examples ▶ (a) $10 = 2 \cdot 5$ (b) $10 = (-2)(-5)$ (c) $10 = (-1)(-10)$

A *prime number* is a natural number greater than 1 that has no natural number factors other than itself and 1.

Examples ▶ 2, 3, 5, 7, 11, 13, 17, 19, 23, 29 ⟵ the ten smallest prime numbers

A *composite number* is a natural number greater than 1 that is not a prime number.

Examples ▶ 4, 6, 8, 9, 10, 12, 14, 15, 16, 18 ⟵ the ten smallest composite numbers

Each composite number has one and only one *prime factorization* (disregarding the order of the factors). This important fact is known as the *Fundamental Theorem of Arithmetic*.

Examples ▶ Prime factorizations: (a) $10 = 2 \cdot 5$ (b) $30 = 2 \cdot 3 \cdot 5$ (c) $12 = 2^2 \cdot 3$

Because each negative integer can be written as the product of (−1) and its opposite, you can write a *prime factorization of a negative integer* as the product of (−1) and the prime factors of its opposite.

Examples ▶ (a) $-6 = -1 \cdot 2 \cdot 3$ (b) $-42 = -1 \cdot 2 \cdot 3 \cdot 7$ (c) $-9 = -1 \cdot 3^2$

6.1 Factor Using the GCF

The *greatest common factor* (GCF) of two or more exponential expressions with the same base is the exponential expression that has the smallest exponent.

Examples ▶ (a) The GCF of 3^4, 3^2, and 3^7 is 3^2. (b) The GCF of a^3, a^5, a, and a^2 is a.

The GCF of two or more monomials is the product of the GCF of each common base.

EXAMPLE 1: Find the GCF of two or more monomials.

Problem ▶ Find the GCF of $-18x^2y^3z$, $24x^3y$, and $30x^2y^2z^4$.

1. Factor ▶ $-18x^2y^3z = -1 \cdot 2 \cdot 3^2 x^2 y^3 z$ Think: Use the prime factorization of each
$24x^3y = 2^3 \cdot 3 x^3 y$ of the numerical coefficients.
$30x^2y^2z^4 = 2 \cdot 3 \cdot 5 x^2 y^2 z^4$

2. Identify ▶ $2 \cdot 3 x^2 y$ Write each common base with its smallest exponent.

Solution ▶ The GCF of $-18x^2y^3z$, $24x^3y$, and $30x^2y^2z^4$ is $6x^2y$.

6.1 Factor Using the GCF

Other Examples ▸ (a) The GCF of $7x$ and 14 is 7. (b) The GCF of $-3x$ and 3 is 3.
(c) The GCF of $6m$ and $5n$ is 1. (d) The GCF of $12a^2b$, $18ab^2$, and $40a^2b^2$ is $2ab$.

Make Sure

Find the GCF of two or more monomials.

See Example 1 ▸ 1. $12ab^2, 9a^2b$ 2. $8c^2d^3, -12c^3d, -16c^2d^4$ 3. $36e^3f^4g^2, 16e^4f$

MAKE SURE ANSWERS: 1. $3ab$ 2. $4c^2d$ 3. $4e^3f$

To find the GCF of a polynomial, you find the GCF of its terms.

Example ▸ The GCF of $12x^2y + 18xy^2 + 40x^2y^2$ is $2xy$.

You can use the distributive properties to factor out the GCF from a polynomial.

EXAMPLE 2: Factor out the GCF from a polynomial.

Problem ▸ Factor out the GCF from $24x^3y^2 - 15x^2y^2 + 27xy^3$.

1. Find GCF ▸
$$\left.\begin{array}{l} 24x^3y^2 = 2^3 \cdot 3x^3y^2 \\ -15x^2y^2 = -1 \cdot 3 \cdot 5x^2y^2 \\ 27xy^3 = 3^3xy^3 \end{array}\right\} \longrightarrow \text{The GCF is } 3xy^2.$$

2. Factor each term ▸ $24x^3y^2 - 15x^2y^2 + 27xy^3 = 3xy^2(8x^2) + 3xy^2(-5x) + 3xy^2(9y)$ Factor each term using the GCF.

3. Factor out GCF ▸ $= 3xy^2(8x^2 - 5x + 9y)$ Use a distributive property.

4. Check ▸ Does $8x^2 - 5x + 9y$ have any common factors other than 1? No.

Does $3xy^2(8x^2 - 5x + 9y)$ equal the original polynomial? Yes:

$$3xy^2(8x^2 - 5x + 9y) = 3xy^2(8x^2) - 3xy^2(5x) + 3xy^2(9y)$$
$$= 24x^3y^2 - 15x^2y^2 + 27xy^3 \longleftarrow \text{ original polynomial}$$

Solution ▸ $24x^3y^2 - 15x^2y^2 + 27xy^3$ factored is $3xy^2(8x^2 - 5x + 9y)$.

Note ▸ In Example 2, you could factor out $-3xy^2$ from $24x^3y^2 - 15x^2y^2 + 27xy^3$ as follows:

$$24x^3y^2 - 15x^2y^2 + 27xy^3 = -3xy^2(-8x^2) - 3xy^2(5x) - 3xy^2(-9y)$$
$$= -3xy^2(-8x^2 + 5x - 9y)$$

Both $3xy^2(8x^2 - 5x + 9y)$ and $-3xy^2(-8x^2 + 5x - 9y)$ are considered correct factored forms for $24x^3y^2 - 15x^2y^2 + 27xy^3$. However, $3xy^2(8x^2 - 5x + 9y)$ is the *simplest factored form* because it has fewer negative signs than $-3xy^2(-8x^2 + 5x - 9y)$.

Many times you will be able to factor out the GCF from a polynomial without writing any intermediate steps.

Example ▶ $3m^3n + 12m^2n^2 - 6mn^3 = 3mn \cdot m^2 + 3mn \cdot 4mn + 3mn(-2n^2)$
$= 3mn(m^2 + 4mn - 2n^2)$

Sometimes a common binomial factor can be factored out of an algebraic expression.

Examples ▶ (a) $(p + q)x + (p + q)y = (p + q)(x + y)$
(b) $(3x + 4y)2z + (3x + 4y) = (3x + 4y)2z + (3x + 4y) \cdot 1$
$= (3x + 4y)(2z + 1)$

Make Sure

Factor out the GCF from each polynomial.

See Example 2 ▶ **1.** $18a^4b^3 - 12a^3b^5 - 9ab^4$ **2.** $12x^3yz^2 + 15xy^4z - 36x^2y^3 - 24x^2y^3z^4$

MAKE SURE ANSWERS: **1.** $3ab^3(6a^3 - 4a^2b^2 - 3b)$ **2.** $3xy(4xz^2 + 5y^3z - 12xy^2 - 8xy^2z^4)$

Some polynomials consisting of 4 terms can be factored by grouping the terms in pairs.

EXAMPLE 3: Factor a polynomial consisting of 4 terms by grouping the terms in pairs.

Problem ▶ Factor $x^3 + 4x^2 + 2x + 8$.

1. Regroup ▶ $x^3 + 4x^2 + 2x + 8 = (x^3 + 4x^2) + (2x + 8)$ Regroup as the sum of two binomials.

2. Factor out ▶ $= x^2(x + 4) + 2(x + 4)$ Factor out the GCF in each binomial.

3. Factor out again ▶ $= (x + 4)(x^2 + 2)$ Factor out the common binomial factor.

Solution ▶ $x^3 + 4x^2 + 2x + 8 = (x + 4)(x^2 + 2)$ Check as before.

Another Example ▶ $5pr + 3qr - 10ps - 6qs = (5pr + 3qr) + (-10ps - 6qs)$
$= r(5p + 3q) + (-2s)(5p + 3q)$ The GCF of $-10ps$ and $-6qs$ is $-2s$.
$= (5p + 3q)(r - 2s)$

Make Sure

Factor polynomials consisting of four terms by grouping the terms in pairs.

See Example 3 ▶ **1.** $12z^2 - 45z + 4z - 15$ **2.** $16y^3 - 6y^2 + 72y - 27$

MAKE SURE ANSWERS: **1.** $(3z + 1)(4z - 15)$ **2.** $(2y^2 + 9)(8y - 3)$

6.1 Practice

Set 1: Find the GCF of two or more monomials.

1. $2x^2, 4x^3, 6x$
2. $3y^3, 6y^4, 9y^2$
3. $z^4, 3z^2, 5z^3$

4. $7w^4, 5w^2, w^3$
5. $4u^3v^5, 3u^2v^4, 7v^3$
6. $6st^3, 4s^3t^2, 7t^4$

7. $4q^3r^5, 8q^2r^3, 6qr^2$
8. $6m^5n, 12m^2n^3, 15mn^4$

9. $-5h^3k^7, -4h^2k^5, -3hk^5$
10. $-6f^3g^2, -4f^5g^7, -3fg^3$

11. $-24d^7e^8, -8d^4e^3, -12d^2e^4$
12. $-36a^5b, -24a^2b^7, -15a^7b^3$

Set 2: Factor out the GCF from each polynomial.

13. $2a^5 + 4a^3 + 6a^4$
14. $3b^3 - 6b + 9b^2$

15. $12c^4 + 15c^3 - 9c^2$
16. $32d^5 - 24d^3 - 56d^4$

17. $6e^3f^2 - 8e^4f^5 + 12e^2f$
18. $8g^3h^5 - 16g^4h^3 + 20g^5h$

19. $5m^4n^7 - 15m^3n^5 + 12m^2n^3$
20. $9p^5q^7 - 12p^4q^3 + 13p^3q^5$

21. $4rs^4t^3 - 9r^3s^5 + 12r^4s^2t$
22. $8u^3v^4w^2 - 15u^5v^7 + 10u^4v^2w$

23. $24x^3y^5 - 36x^4z^3 + 40y^3z^2$
24. $15a^3b^7 - 21a^5c^3 + 30b^4c^2$

Chapter 6 Factoring

Set 3: Factor polynomials consisting of four terms by grouping in pairs.

25. $12x^2 + 15x + 4x + 5$
26. $12x^2 + 6x + 10x + 5$
27. $36x^2 + 9x + 44x + 11$

28. $36d^2 + 6d + 66d + 11$
29. $5e^3 + 7e^2 + 5e + 7$
30. $5f^3 + 35f^2 - f - 7$

31. $8g^2 - 14g + 4g - 7$
32. $8h^2 - 28h + 2h - 7$
33. $3k^3 + 3k^2 + 2k + 2$

34. $3m^3 + 6m^2 + m + 2$
35. $12n^2 + 15n - 8n - 10$
36. $12p^2 + 24p - 5p - 10$

37. $9q^3 - 21q^2 + 6q - 14$
38. $9r^3 - 18r^2 + 7r - 14$
39. $10s^2 - 15s - 8s + 12$

40. $5t^2 - 10t - 3t + 6$
41. $8u^2 + 24u - 5u - 15$
42. $8v^2 + 12v - 10v - 15$

43. $8w^3 - 20w^2 + 6w - 15$
44. $12x^3 + 9x^2 - 20x - 15$
45. $12y^2 - 10y + 18y - 15$

46. $12z^2 + 18z - 14z - 21$
47. $5 + 20x - 3x - 12x^2$
48. $12 - 9x + 32x - 24x^2$

Review: Work these problems on a separate sheet of paper. Attach your work to this page.

Multiply two binomials. (See Lesson 5.2.)
49. $(z + 1)(z + 1)$
50. $(y + 3)(y + 5)$
51. $(x - 2)(x + 4)$
52. $(w - 3)(w + 2)$
53. $(v + 5)(v - 3)$
54. $(u + 6)(u - 2)$
55. $(t - 3)(t + 5)$
56. $(s - 5)(s + 7)$
57. $(r - 2)(r - 4)$
58. $(q - 3)(q - 4)$
59. $(p + 5)(p + 3)$
60. $(n + 4)(n + 2)$
61. $(m - 2)(m - 7)$
62. $(k - 6)(k - 8)$
63. $(h - 4)(h + 8)$
64. $(g + 3)(g - 4)$
65. $(f + 1)(f - 2)$
66. $(e - 3)(e - 1)$
67. $(d - 2)(d - 3)$
68. $(z - 4)(z - 5)$
69. $(y - 5)(y - 1)$
70. $(x + 2)(x + 5)$
71. $(w - 7)(w + 1)$
72. $(v - 2)(v + 7)$

6.2 Factor $x^2 + bx + c$

Trinomials of the form $x^2 + bx + c$ often result when you multiply two binomials.

Example ▶ $(x - 2)(x + 5) = x^2 + \underbrace{5x - 2x} + \underbrace{(-2)(5)}$

$\qquad\qquad\qquad = x^2 + 3x - 10 \longleftarrow$ in $x^2 + bx + c$ form, with $b = 3$ and $c = -10$

Note ▶ b is the sum of -2 and 5. c is the product of -2 and 5.

The above results are not coincidental. Consider the following general case.

$$\begin{array}{c} \quad\text{F}\quad\text{O}\quad\text{I}\quad\text{L} \\ (x + d)(x + e) = x^2 + ex + dx + de \end{array}$$

$\qquad\qquad\qquad = x^2 + (e + d)x + de \longleftarrow$ in $x^2 + bx + c$ form, with $b = e + d$ and $c = de$

To factor a trinomial of the form $x^2 + bx + c$ means to express the trinomial as a product of two binomials of the form $(x + d)$ and $(x + e)$, where $de = c$ and $e + d = b$.

Example ▶ Factor $x^2 - 8x + 15$.

1. Identify b and c ▶ $x^2 - 8x + 15 \longleftarrow$ in $x^2 + bx + c$ form, with $b = -8$ and $c = 15$.

2. Factor c ▶

Factor pairs of 15	Sums of factors
1, 15	16
−1, −15	−16
3, 5	8
−3, −5	−8
↑ ↑	↑
d e	c

Stop: -8 is the correct coefficient of x. Let $d = -3$ and $e = -5$.

3. Write factored form ▶ $x^2 - 8x + 15 = (x + d)(x + e)$

$\qquad\qquad\qquad\qquad = (x + (-3))(x + (-5))$

Solution ▶ $\qquad\qquad\qquad = (x - 3)(x - 5)$ Use the FOIL method to check.

You can reduce the number of factor pairs under consideration if you use relationships between the signs of the terms to establish a sign pattern.

> If the constant term of the trinomial is positive, the constant terms of the binomials have the same sign:
>   The signs are both positive if the coefficient of x is positive.
>   The signs are both negative if the coefficient of x is negative.
> If the constant term of the trinomial is negative, the constant terms of the binomials have opposite signs.

To visualize these relationships, examine the following cases, where B and C are natural numbers.

	Trinomial	Sign Pattern	
Case 1:	$x^2 + Bx + C$	$= (x + ?)(x + ?)$	See Example 1.
Case 2:	$x^2 - Bx + C$	$= (x - ?)(x - ?)$	See Example 2.
Case 3:	$x^2 + Bx - C$	$= (x + ?)(x - ?)$	See Example 3.
Case 4:	$x^2 - Bx - C$	$= (x + ?)(x - ?)$	See Example 4.

Chapter 6 Factoring

EXAMPLE 1: Factor a trinomial of the form $x^2 + Bx + C$.

Problem ▶ Factor $x^2 + 12x + 20$.

1. Factor C ▶

          Factor pairs of 20
          1,   20
          2,   10
          4,   5

2. Write sign pattern ▶ $x^2 \;+\; 12x \;+\; 20 = (x + \blacksquare)(x + \blacksquare)$ Case 1: $x^2 + Bx + C$, with $B = 12$ and $C = 20$.

3. Use trial factors ▶

Trial Factors	Check Middle Term
$(x + 1)(x + 20)$ ⟶	$20x + 1x = 21x$
$(x + 2)(x + 10)$ ⟶	$10x + 2x = 12x$ Stop! $12x$ is the correct middle term.

      ⎿factors⏌
        of 20

Solution ▶ $x^2 + 12x + 20 = (x + 2)(x + 10)$ Check as before.

Make Sure

Factor trinomials of the form $x^2 + Bx + C$. Check each solution.

See Example 1 ▶   **1.** $x^2 + 4x + 3$ **2.** $x^2 + 5x + 6$ **3.** $y^2 + 13y + 36$

_____   _____   _____

MAKE SURE ANSWERS: 1. $(x + 1)(x + 3)$ **2.** $(x + 2)(x + 3)$ **3.** $(y + 4)(y + 9)$

EXAMPLE 2: Factor a trinomial of the form $x^2 - Bx + C$.

Problem ▶ Factor $x^2 - 10x + 24$.

1. Factor C ▶

          Factor pairs of 24
          1,   24
          2,   12
          3,   8
          4,   6

2. Write sign pattern ▶ $x^2 \;-\; 10x \;+\; 24 = (x - \blacksquare)(x - \blacksquare)$ Case 2: $x^2 - Bx + C$, with $A = 1$, $B = 10$ and $C = 24$.

3. Use trial factors ▶

Trial Factors	Check Middle Term
$(x - 1)(x - 24)$	$-24x + (-1)x = -25x$
$(x - 2)(x - 12)$	$-12x + (-2)x = -14x$
$(x - 3)(x - 8)$	$-8x + (-3)x = -11x$
$(x - 4)(x - 6)$	$-6x + (-4)x = -10x$ Stop! $-10x$ is the correct middle term.

Solution ▶ $x^2 - 10x + 24 = (x - 4)(x - 6)$ Check as before.

Make Sure

Factor trinomials of the form $x^2 - Bx + C$. Check each solution.

See Example 2 ▶
1. $x^2 - 4x + 3$
2. $x^2 - 7x + 12$
3. $z^2 - 12z + 36$

MAKE SURE ANSWERS: 1. $(x-1)(x-3)$ 2. $(x-3)(x-4)$ 3. $(z-6)(z-6)$

EXAMPLE 3: Factor a trinomial of the form $x^2 + Bx - C$.

Problem ▶ Factor $x^2 + 7x - 18$.

1. Factor C ▶

Factor pairs of 18

1, 18
2,   9
3,   6
         |
      larger
      factor

2. Write sign pattern ▶ $x^2 \;+\; 7x \;-\; 18 = (x + \blacksquare)(x - \blacksquare)$

Case 3: $x^2 + Bx - C$, with $B = 7$ and $C = 18$.
Think: The larger factor goes in the $+\blacksquare$ because the middle term is positive ($+7x$).

3. Use trial factors ▶

Trial Factors	Check Middle Term
$(x + 18)(x - 1)$	$18x + (-1x) = 17x$
$(x + 9)(x - 2)$	$9x + (-2x) = 7x$ Stop! $7x$ is the correct middle term.

Solution ▶ $x^2 + 7x - 18 = (x + 9)(x - 2)$ Check as before.

Note ▶ If the trial factors produce a middle term that is the opposite of the correct middle term, then switch the signs in each of the trial factors.

Make Sure

Factor trinomials of the form $x^2 + Bx - C$. Check each solution.

See Example 3 ▶
1. $x^2 + 2x - 3$
2. $x^2 + 2x - 8$
3. $x^2 + 2x - 48$

MAKE SURE ANSWERS: 1. $(x+3)(x-1)$ 2. $(x+4)(x-2)$ 3. $(x+8)(x-6)$

When you factor a trinomial of the form $x^2 - Bx - C$, you can reduce the number of trial factors to be considered by noting that the larger factor of C must be negative because the middle term is negative.

EXAMPLE 4: Factor a trinomial of the form $x^2 - Bx - C$.

Problem ▶ Factor $x^2 - 13x - 30$.

1. Factor C ▶

Factor pairs of 30

1,	30
2,	15
3,	10
5,	6

↓ larger factor

2. Write sign pattern ▶ $x^2 \;\blacksquare\; 13x \;\blacksquare\; 30 = (x + \blacksquare)(x - \blacksquare)$

Case 4: $x^2 - Bx - C$, with $B = 13$ and $C = 30$.
Think: The larger factor goes in $-\blacksquare$ because the middle term is negative ($-13x$).

3. Use trial factors ▶

Trial Factors	Check Middle Term	
$(x + 1)(x - 30)$	$-30x + 1x = -29x$	
$(x + 2)(x - 15)$	$-15x + 2x = -13x$	Stop! $-13x$ is the correct middle term.

Solution ▶ $x^2 - 13x - 30 = (x + 2)(x - 15)$ Check as before.

Some trinomials cannot be factored using only integers. For example, $x^2 + 7x + 2$ cannot be factored using integers. The only positive integral factors of 2 are 1 and 2, and their sum does not equal 7. This trinomial is *irreducible over the integers*.

Make Sure

Factor trinomials of the form $x^2 - Bx - C$. Check each solution.

See Example 4 ▶ 1. $x^2 - 3x - 4$ _____   2. $x^2 - 7x - 8$ _____   3. $y^2 - 13y - 48$ _____

MAKE SURE ANSWERS: 1. $(x + 1)(x - 4)$ 2. $(x + 1)(x - 8)$ 3. $(y + 3)(y - 16)$

Some 4th-degree polynomials can be factored by the methods presented in this lesson.

Example ▶ Factor $x^4 + 7x^2 + 10$.

1. Factor C ▶

Factor pairs of 10

1,	10
2,	5

2. Write sign pattern ▶ $(x^2)^2 \;+\; 7x^2 \;+\; 10 = (x^2 + \blacksquare)(x^2 + \blacksquare)$

3. Use trial factors ▶

Trial Factors	Check Middle Term	
$(x^2 + 1)(x^2 + 10)$	$10x^2 + 1x^2 = 11x^2$	
$(x^2 + 2)(x^2 + 5)$	$5x^2 + 2x^2 = 7x^2$	Stop! $7x^2$ is the correct middle term.

Solution ▶ $x^4 + 7x^2 + 10 = (x^2 + 2)(x^2 + 5)$ Check as before.

Name _____ Date _____ Class _____

6.2 Practice: *Check each solution.*

Set 1: Factor trinomials of the form $x^2 + Bx + C$.

1. $x^2 + 3x + 2$
2. $y^2 + 4y + 3$
3. $z^2 + 6z + 8$

4. $w^2 + 7w + 12$
5. $u^2 + 12u + 36$
6. $v^2 + 14v + 45$

Set 2: Factor trinomials of the form $x^2 - Bx + C$.

7. $x^2 - 4x + 4$
8. $x^2 - 6x + 9$
9. $x^2 - 8x + 15$

10. $d^2 - 9d + 18$
11. $e^2 - 13e + 36$
12. $f^2 - 18f + 45$

Set 3: Factor trinomials of the form $x^2 + Bx - C$.

13. $p^2 + p - 2$
14. $q^2 + 4q - 5$
15. $r^2 + r - 6$

16. $s^2 + 2s - 8$
17. $t^2 + 6t - 27$
18. $m^2 + m - 56$

Set 4: Factor trinomials of the form $x^2 - Bx - C$.

19. $z^2 - 4z - 5$
20. $y^2 - 8y - 9$
21. $x^2 - 6x - 16$

22. $w^2 - w - 20$
23. $v^2 - 8v - 48$
24. $u^2 - 6u - 72$

Mixed Practice: Factor each trinomial.

25. $x^2 + 17x + 72$
26. $x^2 + 18x + 72$
27. $72 - 27x + x^2$

28. $72 - 22d + d^2$
29. $e^2 + 3ef - 70f^2$
30. $g^2 + 8gh - 48h^2$

31. $m^2 - 8mn - 84n^2$
32. $p^2 - 8pq - 48q^2$
33. $r^4 + 17r^2 + 72$

34. $s^4 + 15s^2 + 56$
35. $t^4 + 2t^2 - 8$
36. $u^4 - 4u^2 - 12$

Extra: Factor each trinomial.

37. $(z + 2)^2 + 4(z + 2) + 4$
38. $(y - 3)^2 + 3(y - 3) + 2$
39. $(x + 4)^2 + (x + 4) - 6$

40. $(w - 1)^2 + 5(w - 1) - 6$
41. $(v - 5)^2 - (v - 5) - 6$
42. $(u + 1)^2 - 6(u + 1) + 8$

Review: Work these problems on a separate sheet of paper. Attach your work to this page.

Multiply two binomials. (See Lesson 5.4.)
43. $(2z + 3)(3z + 4)$
44. $(3y + 5)(4y + 3)$
45. $(5x - 2)(3x - 4)$
46. $(3w - 5)(4w - 3)$
47. $(2v - 3)(4v + 3)$
48. $(4u - 5)(3u + 2)$
49. $(4t + 3)(3t - 2)$
50. $(3s + 4)(2s - 3)$
51. $(5r - 3)(3r + 1)$
52. $(3q - 5)(2q + 1)$
53. $(4p + 3)(2p - 1)$
54. $(3n + 4)(2n - 1)$
55. $(2m - 3)(2m - 3)$
56. $(3k - 4)(3k - 4)$
57. $(2h + 5)(2h + 5)$
58. $(3g - 1)(2g + 1)$
59. $(5f - 2)(3f + 4)$
60. $(2z + 3)(5z - 3)$
61. $(3y + 1)(2y - 5)$
62. $(7x + 1)(3x + 5)$
63. $(6w + 5)(2w + 3)$

6.3 Factor $ax^2 + bx + c$

To factor trinomials of the form $ax^2 + bx + c$, you need to form trial binomial factors whose first terms are factors of a and whose last terms are factors of c. Then check the trial binomial factors to see if the correct middle term is produced.

You can use the following sign patterns to reduce the number of trial binomial factor pairs to be considered. In each of the following cases, A, B, and C represent natural numbers.

	Trinomial	Sign Pattern	
Case 1:	$Ax^2 + Bx + C =$	$(?x + ?)(?x + ?)$	See Example 1.
Case 2:	$Ax^2 - Bx + C =$	$(?x - ?)(?x - ?)$	See Example 2.
Case 3:	$Ax^2 + Bx - C =$	$(?x + ?)(?x - ?)$	See Example 3.
Case 4:	$Ax^2 - Bx - C =$	$(?x + ?)(?x - ?)$	See Example 4.

EXAMPLE 1: Factor a trinomial of the form $Ax^2 + Bx + C$.

Problem ▶ Factor $2x^2 + 13x + 15$.

1. Factor A and C ▶

Factors of 2: 1, 2

Factors of 15: 1, 15 ; 3, 5

2. Choose sign pattern ▶ $2x^2 + 13x + 15 = (\blacksquare x + \blacksquare)(\blacksquare x + \blacksquare)$ Case 1: $Ax^2 + Bx + C$, with $A = 2$, $B = 13$, and $C = 15$.

3. Use trial factors ▶

Trial Factors	Check Middle Term	
$(1x + 1)(2x + 15)$	$15x + 2x = 17x$	
$(1x + 15)(2x + 1)$	$1x + 30x = 31x$	
$(1x + 3)(2x + 5)$	$5x + 6x = 11x$	
$(1x + 5)(2x + 3)$	$3x + 10x = 13x$	Stop! $13x$ is the correct middle term.

4. Check ▶ $(x + 5)(2x + 3) = 2x^2 + 3x + 10x + 15$
$= 2x^2 + 13x + 15$

Solution ▶ $2x^2 + 13x + 15 = (x + 5)(2x + 3)$

Make Sure

Factor trinomials of the form $Ax^2 + Bx + C$. Check each solution.

See Example 1 ▶
1. $2x^2 + 5x + 3$
2. $4x^2 + 11x + 6$
3. $6y^2 + 25y + 24$

MAKE SURE ANSWERS: 1. $(2x + 3)(x + 1)$ 2. $(4x + 3)(x + 2)$ 3. $(3y + 8)(2y + 3)$

EXAMPLE 2: Factor a trinomial of the form $Ax^2 - Bx + C$.

Problem ▶ Factor $4x^2 - 8x + 3$.

1. Factor A and C ▶

Factors of 4	Factors of 3
1, 4	1, 3
2, 2	

2. Choose sign pattern ▶ $4x^2 - 8x + 3 = (\blacksquare x - \blacksquare)(\blacksquare x - \blacksquare)$ Case 2: $Ax^2 - Bx + C$, with $A = 4$, $B = 8$, and $C = 3$.

3. Use trial factors ▶

Trial Factors	Check Middle Term	
$(1x - 1)(4x - 3)$	$-3x - 4x = -7x$	
$(1x - 3)(4x - 1)$	$-1x - 12x = -13x$	
$(2x - 1)(2x - 3)$	$-6x - 2x = -8x$	Stop! $-8x$ is the correct middle term.

Solution ▶ $4x^2 - 8x + 3 = (2x - 1)(2x - 3)$ Check as before.

Make Sure

Factor trinomials of the form $Ax^2 - Bx + C$. Check each solution.

See Example 2 ▶
1. $2x^2 - 7x + 3$
2. $12x^2 - 11x + 2$
3. $12 - 19x + 4x^2$

MAKE SURE ANSWERS: **1.** $(2x - 1)(x - 3)$ **2.** $(3x - 2)(4x - 1)$ **3.** $(3 - 4x)(4 - x)$

EXAMPLE 3: Factor a trinomial of the form $Ax^2 + Bx - C$.

Problem ▶ Factor $21y^2 + 25y - 4$.

1. Factor A and C ▶

Factors of 21	Factors of 4
1, 21	1, 4
3, 7	2, 2

2. Choose sign pattern ▶ $21y^2 + 25y - 4 = (\blacksquare y + \blacksquare)(\blacksquare y - \blacksquare)$ Case 3: $Ax^2 + Bx - C$, with $A = 21$, $B = 25$, $C = 4$, and $x = y$.

3. Use trial factors ▶

Trial Factors	Check Middle Term	
$(1y + 1)(21y - 4)$	$-4y + 21y = 17y$	
$(1y + 4)(21y - 1)$	$-1y + 84y = 83y$	
$(1y + 2)(21y - 2)$	$-2y + 42y = 40y$	
$(3y + 1)(7y - 4)$	$-12y + 7y = -5y$	
$(3y + 4)(7y - 1)$	$-3y + 28y = 25y$	Stop! $25y$ is the correct middle term.

Solution ▶ $21y^2 + 25y - 4 = (3y + 4)(7y - 1)$ Check as before.

Make Sure

Factor trinomials of the form $Ax^2 + Bx - C$. Check each solution.

See Example 3 ▶ **1.** $2x^2 + x - 6$ **2.** $2x^2 + 5x - 12$ **3.** $6y^2 + 7y - 24$

_____ _____ _____

MAKE SURE ANSWERS: 1. $(x + 2)(2x - 3)$ **2.** $(x + 4)(2x - 3)$ **3.** $(3y + 8)(2y - 3)$

SHORTCUT 6.1: If a trinomial does not have a factor that is common to all of its terms, then neither of its binomial factors will have a common factor.

EXAMPLE 4: Factor a trinomial of the form $Ax^2 - Bx - C$.

Problem ▶ Factor $6z^2 - 11z - 10$.   Observe: The GCF of each of the terms is 1.

1. Factor A and C ▶ **Factors of 6**   **Factors of 10**
   1, 6   1, 10
   2, 3   2, 5

2. Choose sign pattern ▶ $6z^2 - 11z - 10 = (\blacksquare z - \blacksquare)(\blacksquare z - \blacksquare)$ Case 4: $Ax^2 - Bx - C$, with $A = 6$. $B = -11$, $C = 10$, and $x = z$.

3. Use trial factors ▶ **Trial Factors**   **Check Middle Term**
   $(1z + 1)(6z - 10)$ No! $6z$ and 10 have a common factor of 2.
   $(1z + 10)(6z - 1)$ $-1z + 60z = 59z$
   $(1z + 2)(6z - 5)$ $-5z + 12z = 7z$
   $(1z + 5)(6z - 2)$ No! $6z$ and 2 have a common factor of 2.
   $(2z + 1)(3z - 10)$ $-20z + 3z = -17z$
   $(2z + 10)(3z - 1)$ No! $2z$ and 10 have a common factor of 2.
   $(2z + 2)(3z - 5)$ No! $2z$ and 2 have a common factor of 2.
   $(2z + 5)(3z - 2)$ $-4z + 15z = +11z$ ⟵ the opposite of the correct middle term
   $(2z - 5)(3z + 2)$ $4z - 15z = -11z$ ⟵ the correct middle term

Solution ▶ $6z^2 - 11z - 10 = (2z - 5)(3z + 2)$ Check as before.

Recall ▶ If the trial factors produce a middle term that is the opposite of the correct middle term, then switch the signs in each of the trial binomial factors.

Make Sure

Factor trinomials of the form $Ax^2 - Bx - C$. Check each solution.

See Example 4 ▶ **1.** $2x^2 - 5x - 3$ **2.** $6x^2 - x - 12$ **3.** $12z^2 - 23z - 24$

_____ _____ _____

MAKE SURE ANSWERS: 1. $(2x + 1)(x - 3)$ **2.** $(3x + 4)(2x - 3)$ **3.** $(4z + 3)(3z - 8)$

Sometimes the methods introduced in this lesson can be used to factor trinomials in two variables.

Example ▶ Factor $12x^2 + 16xy - 35y^2$.

1. Factor first and last coefficient ▶

Factors of 12	Factors of 35
1, 12	1, 35
2, 6	5, 7
3, 4	

2. Choose sign pattern ▶ $12x^2 \;\boxed{+}\; 16xy \;\boxed{-}\; 35y^2 = (\boxed{}x + \boxed{}y)(\boxed{}x - \boxed{}y)$ Case 3: $Ax^2 + Bx - C$, with $A = 12$, $B = 16y$, and $C = 35y^2$

3. Use trial factors ▶

Trial Factors	Check Middle Term	
$(1x + 1y)(12x - 35y)$	$-35xy + 12xy = -23xy$	
$(1x + 35y)(12x - 1y)$	$-1xy + 420xy = 419xy$	
$(1x + 5y)(12x - 7y)$	$-7xy + 60xy = 53xy$	
$(1x + 7y)(12x - 5y)$	$-5xy + 84xy = 79xy$	
$(2x + 1y)(6x - 35y)$	$-70xy + 6xy = -64xy$	
$(2x + 35y)(6x - 1y)$	$-2xy + 210xy = 208xy$	
$(2x + 5y)(6x - 7y)$	$-14xy + 30xy = 16xy$	Stop! $16xy$ is the correct middle term.

4. Check ▶ $(2x + 5y)(6x - 7y) = 12x^2 - 14xy + 30xy - 35y^2$
$= 12x^2 + 16xy - 35y^2$

Solution ▶ $12x^2 + 16xy - 35y^2 = (2x + 5y)(6x - 7y)$

Another Example ▶ Factor $25v^2 - 5vw - 2w^2$.

1. Factor first and last coefficient ▶

Factors of 25	Factors of 2
1, 25	1, 2
5, 5	

2. Choose sign pattern ▶ $25v^2 \;\boxed{-}\; 5vw \;\boxed{-}\; 2w^2 = (\boxed{}v + \boxed{}w)(\boxed{}v - \boxed{}w)$

3. Use trial factors ▶

Trial Factors	Check Middle Term	
$(1v + 1w)(25v - 2w)$	$-2vw + 25vw = 23vw$	
$(1v + 2w)(25v - 1w)$	$-1vw + 50vw = 49vw$	
$(5v + 1w)(5v - 2w)$	$-10vw + 5vw = -5vw$	Stop.

Solution ▶ $25v^2 - 5vw - 2w^2 = (5v + w)(5v - 2w)$ Check as before.

If $ax^2 + bx + c$ can be factored as the product of two binomials using integers, then you can factor it by grouping when you find the integers m and n so that $mn = ac$ and $m + n = b$. This result leads to the *ac method of factoring* $ax^2 + bx + c$.

To factor $ax^2 + bx + c$ using the *ac* method:

1. Find integers m and n so that $mn = ac$ and $m + n = b$.
2. Substitute $mx + nx$ for bx to get $ax^2 + mx + nx + c$.
3. Factor $ax^2 + mx + nx + c$ by grouping. (See Lesson 6.1.)
4. Check to see if the product of the factors from Step 3 is the original polynomial.

6.3 Factor $ax^2 + bx + c$

EXAMPLE 5: Factor $ax^2 + bx + c$ using the *ac* method.

Problem ▶ Factor $4x^2 + 21x - 18$.

1. Find *m* and *n* ▶ In $4x^2 + 21x - 18$, $a = 4$, $b = 21$, and $c = -18$.

$mn = ac = 4(-18) = -72$

$m + n = b = 21$

$mn = -72$	$m + n = 21$
$-1(72)$	$-1 + 72 = -71$
$-2(36)$	$-2 + 36 = 34$
$-3(24)$	$-3 + 24 = 21$ Stop: $m = -3$ and $n = 24$ will work.

2. Substitute ▶ $4x^2 + 21x - 18 = 4x^2 - 3x + 24x - 18$ Think: $21x = -3x + 24x$

3. Factor ▶ $ = (4x^2 - 3x) + (24x - 18)$ Factor by grouping in pairs.

$ = x(4x - 3) + 6(4x - 3)$

$ = (4x - 3)(x + 6)$

4. Check ▶ $(4x - 3)(x + 6) = 4x^2 + 24x - 3x - 18$ Use FOIL method.

$ = 4x^2 + 21x - 18$ ⟵ original polynomial

Solution ▶ $4x^2 + 21x - 18 = (4x - 3)(x + 6)$

Note 1 ▶ The only trial-and-error procedure occurs in finding *m* and *n* in Step 1.

Note 2 ▶ If the values of *m* and *n* are reversed, you will still get the same result. Try it.

Other Examples ▶ (a) Factor $-3y^2 - 7y + 40$ using the *ac* method.

In $-3y^2 - 7y + 40$, $a = -3$, $b = -7$, and $c = 40$.

$mn = ac = -3(40) = -120$	$m + n = b = -7$
$1(-120)$	$1 + (-120) = -119$
$2(-60)$	$2 + (-60) = -58$
$3(-20)$	$3 + (-20) = -17$
$4(-30)$	$4 + (-30) = -26$
$5(-24)$	$5 + (-24) = -19$
$6(-20)$	$6 + (-20) = -14$
$8(-15)$	$8 + (-15) = -7$ Stop: $8y$ and $-15y$ will work.

$-3y^2 - 7y + 40 = -3y^2 + 8y - 15y + 40$

$ = (-3y^2 + 8y) + (-15y + 40)$ Factor by grouping in pairs.

$ = y(-3y + 8) + 5(-3y + 8)$

$ = (-3y + 8)(y + 5)$ Check as before.

(b) Factor $4x^2 + 2x - 5$ using the *ac* method.

In $4x^2 + 2x - 5$, $a = 4$, $b = 2$, and $c = -5$.

$mn = ac = 4(-5) = -20$	$m + n = b = 2$
$-1(20)$	$-1 + 20 = 19$
$-2(10)$	$-2 + 10 = 8$
$-4(5)$	$-4 + 5 = 1$

Think: There are no integers m and n such that $mn = -20$ and $m + n = 2$.

$4x^2 + 2x - 5$ is irreducible over the integers.

The *ac* method can sometimes be used to factor trinomials of the form $ax^2 + bxy + y^2$.

Example ▶ Factor $6x^2 + xy - 12y^2$ using the *ac* method.

In $6x^2 + xy - 12y^2$, $a = 6$, $b = 1$, and $c = -12$.

$mn = ac = 6(-12) = -72$	$m + n = b = 1$
$-1(72)$	$-1 + 72 = 71$
$-2(36)$	$-2 + 36 = 34$
$-3(24)$	$-3 + 24 = 21$
$-4(18)$	$-4 + 18 = 14$
$-6(12)$	$-6 + 12 = 6$
$-8(9)$	$-8 + 9 = 1$

Stop: $-8xy$ and $9xy$ will work.

$$\begin{aligned} 6x^2 + xy - 12y^2 &= 6x^2 - 8xy + 9xy - 12y^2 \\ &= (6x^2 - 8xy) + (9xy - 12y^2) && \text{Factor by grouping in pairs.} \\ &= 2x(3x - 4y) + 3y(3x - 4y) \\ &= (3x - 4y)(2x + 3y) && \text{Check as before.} \end{aligned}$$

Make Sure

Factor each trinomial using the *ac* method. Check each solution.

See Example 5 ▶
1. $12z^2 - 41z - 15$
2. $24y^2 + 38y + 10$
3. $16x^2 + 66xy - 27y^2$

MAKE SURE ANSWERS: 1. $(3z + 1)(4z - 15)$ 2. $(8y + 10)(3y + 1)$ 3. $(2x + 9y)(8x - 3y)$

Name _____ Date _____ Class _____

6.3 Practice: *Check each solution.*

Set 1: Factor trinomials of the form $Ax^2 + Bx + C$.

1. $3x^2 + 5x + 2$
2. $3x^2 + 7x + 2$
3. $2y^2 + 9y + 9$

4. $2y^2 + 9y + 10$
5. $12z^2 + 19z + 5$
6. $12z^2 + 16z + 5$

Set 2: Factor trinomials of the form $Ax^2 - Bx + C$.

7. $2p^2 - 11p + 5$
8. $2q^2 - 7q + 5$
9. $7r^2 - 16r + 9$

10. $7s^2 - 24s + 9$
11. $36t^2 - 53t + 11$
12. $36u^2 - 72u + 11$

Set 3: Factor trinomials of the form $Ax^2 + Bx - C$.

13. $5v^2 + 2v - 7$
14. $5w^2 + 34w - 7$
15. $3x^2 + 2x - 8$

16. $3y^2 + 10y - 8$
17. $6z^2 + z - 5$
18. $6x^2 + 29x - 5$

Chapter 6 Factoring

Set 4: Factor trinomials of the form $Ax^2 - Bx - C$.

19. $2x^2 - 13x - 7$
20. $2x^2 - 5x - 7$
21. $5y^2 - 13y - 6$

22. $5y^2 - 7y - 6$
23. $8z^2 - 10z - 7$
24. $8z^2 - 26z - 7$

Set 5: Factor each trinomial using the *ac* method.

25. $3z^2 + 5z + 2$
26. $3y^2 + 7y + 2$
27. $12x^2 + 7x - 10$

28. $12w^2 + 19w - 10$
29. $9v^2 - 15v - 14$
30. $9u^2 - 11u - 14$

31. $4t^2 + 5t - 6$
32. $4s^2 + 15s - 4$
33. $10r^2 - 23r + 12$

34. $5q^2 - 13q + 6$
35. $8p^2 + 19p - 15$
36. $8x^2 + 2x - 15$

Name _____ Date _____ Class _____

Mixed Practice: Factor each trinomial.

37. _____

37. $8x^2 - 14x - 15$ 38. $12k^2 - 11k - 15$ 39. $12h^2 + 8h - 15$

38. _____

39. _____

40. _____

40. $12g^2 + 4g - 21$ 41. $12f^2 - 7f + 1$ 42. $12e^2 - 8e + 1$

41. _____

42. _____

43. _____

43. $10 + 51d + 27d^2$ 44. $10 + 33x + 27x^2$ 45. $5 + 17x - 12x^2$

44. _____

45. _____

46. $10 + 31x - 14x^2$ 47. $12 + 23z - 24z^2$ 48. $20 + 9y - 18y^2$

46. _____

47. _____

48. _____

49. $12x^2 - 36y^2 - 11xy$ 50. $16w^2 - 27v^2 - 24vw$ 51. $27u^2 - 8v^2 + 6uv$

49. _____

50. _____

51. _____

52. $18s^2 - 12t^2 + 19st$ 53. $27q^2 + 8p^2 - 30pq$ 54. $18x^2 + 12y^2 - 35xy$

52. _____

53. _____

54. _____

Extra: Factor each trinomial.

55. $2z^4 - 13z^2 - 7$
56. $2y^4 - 5y^2 - 7$
57. $5x^4 + 13x^2 - 6$

58. $5w^4 + 7w^2 - 6$
59. $12v^4 + 7v^2 - 10$
60. $12u^4 + 19u^2 - 10$

61. $9s^4 - 15s^2t^2 - 14t^4$
62. $9q^4 - 11q^2r^2 - 14r^4$
63. $12x^4 + 8x^2y^2 - 15y^4$

64. $12h^4 + 4h^2k^2 - 21k^4$
65. $12e^4f^4 + 19e^2f^2 + 4$
66. $7x^4y^4 + 16x^2y^2 + 9$

67. $2x^6 - 13x^3y^3 - 7y^6$
68. $12x^6 + 7x^3y^3 - 10y^6$
69. $2(x-4)^2 - 7(x-4) + 5$

70. $7(w+2)^2 - 16(w+2) + 9$
71. $4(v-1)^4 + 15(v-1)^2 - 4$
72. $8(u+1)^4 - 26(u+1)^2 - 7$

Review: Work these problems on a separate sheet of paper. Attach your work to this page.

Multiply polynomials. (See Lesson 5.3.)
73. $(z + 3)(z + 3)$
74. $(y + 4)(y + 4)$
75. $(3x + 4)(3x + 4)$
76. $(2w + 3)(2w + 3)$
77. $(4v - 3)(4v - 3)$
78. $(5u - 2)(5u - 2)$
79. $(3t - 2)(3t + 2)$
80. $(5s - 3)(5s + 3)$
81. $(4r + 1)(4r - 1)$

Factor out the GCF from each polynomial. (See Lesson 6.1.)
82. $4(q - 3) - 2q(q - 3)$
83. $4p(p + 3) - 5(p + 3)$
84. $3m(2m - 3) - 4(2m - 3)$
85. $4(a - 3) + a(a - 3)$
86. $5(b - 4) - b(b - 4)$
87. $c(2c + 3) - 3(2c + 3)$
88. $3d(2d + 4) - 2(2d + 4)$
89. $5e(3 - 2e) - 2(3 - 2e)$
90. $4f(1 - 3f) - 3(1 - 3f)$

6.4 Factor Special Polynomials

The square of a monomial is called a *perfect square*.

	Monomial	Square of Monomial		Perfect Square
Examples ▶ (a)	$2x$	$(2x)^2$	$=$	$4x^2$
(b)	$5xy^2z^3$	$(5xy^2z^3)^2$	$=$	$25x^2y^4z^6$

Note ▶ Perfect squares are always nonnegative and have variables raised only to even powers.

The binomial $x^2 - y^2$ is the *difference of two squares*. To factor the difference of two squares, you use the special product $(x + y)(x - y) = x^2 - y^2$ in reverse.

Difference of Two Squares		Sum and Difference of Two Terms
$x^2 - y^2$	$=$	$(x + y)(x - y)$

EXAMPLE 1: Factor the difference of two squares.

Problem ▶ Factor $a^2 - 25$.

1. Identify ▶ $a^2 - 25 \boxed{= a^2 - 5^2}$ ⟵ difference of two squares

2. Factor ▶ $= (a + 5)(a - 5)$

Solution ▶ $a^2 - 25 = (a + 5)(a - 5)$ Use the FOIL method to check.

Other Examples ▶ (a) $4m^2 - 9n^2 = (2m + 3n)(2m - 3n)$ (b) $144x^2 - y^4 = (12x + y^2)(12x - y^2)$

Note ▶ The sum of two squares, $x^2 + y^2$, cannot be factored using integer coefficients unless x and y have a common factor.

Examples ▶ (a) $9a^2 + 16b^2$ is not factorable using integers.
(b) $4p^2 + 36q^2 = 4(p^2 + 9q^2)$, which does not factor any further using integers.

Make Sure

Factor the difference of two squares. Check each solution.

See Example 1 ▶ **1.** $a^2 - 49$ **2.** $9b^2 - 25$ **3.** $36c^2 - 49d^2$

_____   _____   _____

MAKE SURE ANSWERS: **1.** $(a + 7)(a - 7)$ **2.** $(3b + 5)(3b - 5)$ **3.** $(6c + 7d)(6c - 7d)$

A trinomial is called a *perfect square trinomial* (PST) if it is the square of a binomial. To recognize a perfect square trinomial, you make use of the following special products.

Square of a Binomial		PST
$(r + s)^2$	$=$	$r^2 + 2rs + s^2$
$(r - s)^2$	$=$	$r^2 - 2rs + s^2$

228 Chapter 6 Factoring

Every PST can be factored by the trial and error procedure, but it is generally less time-consuming to identify the PST form $r^2 \pm 2rs + s^2$ and then write it in its factored form as $(r \pm s)^2$.

EXAMPLE 2: Factor perfect square trinomials.

Problems ▶ Factor $9x^2 + 30x + 25$ and $16x^2 - 56xy + 49y^2$.

1. Identify ▶ $9x^2 + 30x + 25 = (3x)^2 + 2(3x)(5) + 5^2$ $16x^2 - 56xy + 49y^2 = (4x)^2 - 2(4x)(7y) + (7y)^2$

2. Factor ▶ $ = (3x + 5)^2$ $ = (4x - 7y)^2$

Solutions ▶ $9x^2 + 30x + 25 = (3x + 5)^2$ $16x^2 - 56xy + 49y^2 = (4x - 7y)^2$

Note ▶ You can check the factorization by mentally squaring your answer.

Make Sure

Factor each PST. Check each solution.

See Example 2 ▶ **1.** $x^2 + 4x + 4$ **2.** $4y^2 - 12y + 9$ **3.** $81u^2 - 144uv + 64v^2$

MAKE SURE ANSWERS: **1.** $(x + 2)^2$ **2.** $(2y - 3)^2$ **3.** $(9u - 8v)^2$

Recall ▶ a^3 is called the cube of a.

a is called the *cube root* of a^3.

The following products are special because the first one consists of the sum of two cubes and the second one consists of the difference of two cubes.

$$\begin{array}{r} x^2 - xy + y^2 \\ x + y \\ \hline x^2y - xy^2 + y^3 \\ x^3 - x^2y + xy^2 \\ \hline x^3 + y^3 \end{array}$$ ← sum of two cubes

$$\begin{array}{r} x^2 + xy + y^2 \\ x - y \\ \hline -x^2y - xy^2 - y^3 \\ x^3 + x^2y + xy^2 \\ \hline x^3 - y^3 \end{array}$$ ← difference of two cubes

The above special products show you how to factor the sum or difference of two cubes.

$$x^3 + y^3 = (x + y)(x^2 - xy + y^2)$$
$$x^3 - y^3 = (x - y)(x^2 + xy + y^2)$$

Factoring the Sum or Difference of Two Cubes

1. Identify the polynomial as a sum or difference of two cubes.
2. Write the binomial factor that is the sum or difference of two cube roots.
3. Use the binomial factor to write the trinomial factor as follows:
   (a) The first term of the trinomial is the square of the first term of the binomial.
   (b) The second term of the trinomial is the opposite of the product of the two terms of the binomial.
   (c) The third term of the trinomial is the square of the second term of the binomial.

6.4 Factor Special Polynomials

EXAMPLE 3: Factor the sum or difference of two cubes.

Problems ▶ Factor $27a^3 + 8b^3$ and $y^3 - 64z^3$.

1. Identify ▶ $27a^3 + 8b^3 = (3a)^3 + (2b)^3$ | $y^3 - 64z^3 = y^3 - (4z)^3$

2. Write binomial factor ▶ $= (3a + 2b)(\ ?\ -\ ?\ +\ ?\ )$ | $= (y - 4z)(\ ?\ +\ ?\ +\ ?\ )$

3. Write trinomial factor ▶ $= (3a + 2b)(\underbrace{9a^2} - \underbrace{6ab} + \underbrace{4b^2})$ | $= (y - 4z)(\underbrace{y^2} + \underbrace{4yz} + \underbrace{16z^2})$

square of the first term

opposite of the product of the two terms

square of the last term

Solutions ▶ $27a^3 + 8b^3 = (3a + 2b)(9a^2 - 6ab + 4b^2)$ | $y^3 - 64z^3 = (y - 4z)(y^2 + 4yz + 16z^2)$

Make Sure

Factor the sum or difference of two cubes. Check each solution.

See Example 3 ▶ **1.** $a^3 + 8$ _____ **2.** $27b^3 - 64$ _____ **3.** $c^3 - 125d^3$ _____

MAKE SURE ANSWERS: **1.** $(a + 2)(a^2 - 2a + 4)$ **2.** $(3b - 4)(9b^2 + 12b + 16)$ **3.** $(c - 5d)(c^2 + 5cd + 25d^2)$

Some polynomials consisting of four or more terms can be factored by grouping the polynomial into groups of special polynomials.

EXAMPLE 4: Factor each polynomial with four or more terms by grouping.

Problem ▶ Factor $x^2 - 2xy + y^2 - 49$.

1. Regroup ▶ $x^2 - 2xy + y^2 - 49 = (x^2 - 2xy + y^2) - 49$ Think: The trinomial is a PST.

2. Factor trinomial ▶ $= (x - y)^2 - 49$

3. Factor difference of squares ▶ $= (x - y)^2 - 7^2$

$= [(x - y) + 7][(x - y) - 7]$

$= [x - y + 7][x - y - 7]$

Solution ▶ $x^2 - 2xy + y^2 - 49 = [x - y + 7][x - y - 7]$ Check as before.

Examples ▶ (a) Factor $aw^2 + 7aw + 12a + w^2 - 9$.

$$aw^2 + 7aw + 12a + w^2 - 9 = (aw^2 + 7aw + 12a) + (w^2 - 9)$$
$$= a(w^2 + 7w + 12) + (w + 3)(w - 3)$$
$$= a(w + 3)(w + 4) + (w + 3)(w - 3)$$
$$= (w + 3)[a(w + 4) + (w - 3)]$$
$$= (w + 3)[aw + 4a + w - 3]$$
$$= (w + 3)(aw + w + 4a - 3)$$

(b) Factor $a^3 + b^3 + a^2 - ab + b^2$.

$$a^3 + b^3 + a^2 - ab + b^2 = (a^3 + b^3) + (a^2 - ab + b^2)$$
$$= (a + b)(a^2 - ab + b^2) + (a^2 - ab + b^2)(1)$$
$$= (a^2 - ab + b^2)[(a + b) + (1)]$$
$$= (a^2 - ab + b^2)(a + b + 1)$$

It is sometimes necessary to rearrange the terms of a polynomial to factor by grouping.

Example ▶ Factor $a^2 + a - b^2 + b$.

1. Regroup ▶ $a^2 + a - b^2 + b = (a^2 + a) + (-b^2 + b)$

                    $= a(a + 1) + b(-b + 1)$ Think: There is no common binomial factor.

2. Interpret ▶ The above grouping does not lead to a factorization. Try a different grouping.

3. Rearrange ▶ $a^2 + a - b^2 + b = a^2 - b^2 + a + b$

4. Regroup ▶ $= (a^2 - b^2) + (a + b)$

5. Factor ▶ $= (a + b)(a - b) + (a + b)$

                             $= (a + b)(a - b) + (a + b)(1)$

Solution ▶ $= (a + b)(a - b + 1)$

It is not always possible to determine which grouping will lead to a factorization without some experimenting. As a guide, however, it is best to group terms that have a common factor or terms that form one of the special polynomials from this lesson.

Make Sure

Factor polynomials consisting of four or more terms by grouping. Check each solution.

See Example 4 ▶   **1.** $a^3 - 5a^2 + a - 5$ **2.** $b^3 + 5b^2 - 9b - 45$ **3.** $12c^3 + 18d^3 - 27cd^2 - 8c^2d$

MAKE SURE ANSWERS: **1.** $(a - 5)(a^2 + 1)$ **2.** $(b + 5)(b + 3)(b - 3)$ **3.** $(3c - 2d)(2c - 3d)(2c + 3d)$

Name _____ Date _____ Class _____

6.4 Practice: *Check each solution.*

Set 1: Factor each difference of two squares.

1. $r^2 - 36$
2. $q^2 - 25$
3. $4p^2 - 49$

4. $25n^2 - 64$
5. $81k^4 - 64m^4$
6. $49g^4 - 36h^4$

Set 2: Factor each PST.

7. $z^2 + 6z + 9$
8. $y^2 + 10y + 25$
9. $4x^2 - 20x + 25$

10. $16w^2 - 24w + 9$
11. $4u^2 - 44uv + 121v^2$
12. $25s^2 - 120st + 144t^2$

Set 3: Factor the sum or difference of two cubes.

13. $f^3 + 27$
14. $e^3 + 64$
15. $d^3 - 1$

16. $c^3 - 125$
17. $8a^6 - 27b^6$
18. $125b^6 + 64a^6$

Set 4: Factor polynomials consisting of four or more terms by grouping.

19. $z^3 + z^2 + 2z + 2$
20. $y^3 + 3y^2 + 2y + 6$

21. $w^2 + 2wx + x^2 - v^2$
22. $9r^2 - 6rs + s^2 - 4t^2$

23. $2p^4 - 16pq^3 - 3p^3 + 24q^3$
24. $3m^4 + 27n^3 - 3mn^3 - 27m^3$

Mixed Practice: Factor each polynomial.

25. $a^2 - 2ab + b^2 - c^2$
26. $d^2 + 4de + 4e^2 - 9f^2$
27. $16g^4 - 72g^2 + 81$

28. $81h^4 + 72h^2 + 16$
29. $16k^4 - 81$
30. $81m^4 - 16$

31. $6n^4 + 17n^2 + 7$
32. $10p^4 - 7p^2 + 1$
33. $27q^3 - 64r^3$

34. $8s^3 - 125t^3$
35. $64u^6 - 1$
36. $v^6 - 64$

37. $w^2 - x^2 - 4x - 4$
38. $y^2 - 16 - 8z - z^2$

39. $a^3 + b - a - b^3$
40. $c^3 - d + c - d^3$

41. $x^2 + 2xy + y^2 + ax^2 - ay^2$
42. $x^2 - 2xy + y^2 + ax^2 - ay^2$

Review: Work these problems on a separate sheet of paper. Attach your work to this page.

Factor out the GCF from the polynomial. (See Lesson 6.1.)
43. $5z^3 - 10z^2 + 25z$
44. $4y^3 - 16y^2 + 24y$
45. $3w^2x + 9wx^2 - 12wx^3$
46. $6u^3v^2 - 12u^2v^3 - 15uv^4$
47. $24s^4t^2 - 36s^3t^3 - 18s^2t^4$
48. $56q^5r^3 + 35q^3r^4 + 42q^2r^5$

Factor each trinomial. (See Lesson 6.3.)
49. $p^2 - 4p + 4$
50. $n^2 - 6n + 9$
51. $m^2 + 7m + 12$
52. $6k^2 + k - 12$
53. $6h^2 - h - 12$
54. $12g^2 - 7g - 12$
55. $12f^2 + 7f - 12$
56. $12e^2 - 25e + 12$
57. $12d^2 + 25d + 12$

6.5 Factor Using the General Strategy

To *factor polynomials completely*, you need a general strategy.

> **General Factoring Strategy**
> To completely factor a polynomial over the integers:
> 1. Factor out the greatest common factor (GCF).
> 2. If the polynomial is a binomial, then try to factor it using:
>   (a) the difference of two squares.
>   (b) the sum of two cubes.
>   (c) the difference of two cubes.
> 3. If the polynomial is a trinomial, then try to factor it:
>   (a) as a perfect square trinomial.
>   (b) using trial and error or the ac method.
> 4. If the polynomial has four or more terms, then try to factor it by grouping.
> 5. After each factorization, examine each new factor to see if it can be factored.

EXAMPLE 1: Factor a binomial using the general strategy.

Problem ▶ Factor $96y^4 - 6$.

1. Factor out the GCF ▶ $96y^4 - 6 = 6 \cdot 16y^4 - 6 \cdot 1$ Think: $96 = 2^5 \cdot 3$, $6 = 2 \cdot 3$. The GCF is $2 \cdot 3 = 6$.
$= 6(16y^4 - 1)$

2. Factor the binomial ▶ $= 6[(4y^2)^2 - 1^2]$ Think: $16y^4 - 1 = (4y^2)^2 - 1^2$ ⟵ difference of two squares
$= 6(4y^2 + 1)(4y^2 - 1)$

3. Factor again, if possible ▶ $= 6(4y^2 + 1)[(2y)^2 - 1^2]$ Think: $4y^2 - 1 = (2y)^2 - 1^2$ ⟵ difference of two squares
$= 6(4y^2 + 1)(2y + 1)(2y - 1)$

Solution ▶ $96y^4 - 6 = 6(4y^2 + 1)(2y + 1)(2y - 1)$ Check as before.

Recall ▶ $4y^2 + 1$ is the sum of two squares. It is irreducible over the integers.

Another Example ▶ Factor $16a^4b - 2ab^4$ completely.

1. Factor out the GCF ▶ $16a^4b - 2ab^4 = 2ab(8a^3 - b^3)$ Think: $2ab$ is the GCF.

2. Factor the binomial ▶ $2ab[(2a)^3 - b^3)]$ Think: $8a^3 - b^3 = (2a)^3 - b^3$ ⟵ difference of two cubes

$2ab(2a - b) \cdot (? + ? + ?)$

$2ab(2a - b)(4a^2 + 2ab + b^2)$

— the square of the last term $(-b)$
— the opposite of the product of the two terms $2a$ and $(-b)$
— the square of the first term $2a$

3. Factor again, if possible ▶ $2ab(2a - b)(4a^2 + 2ab + b^2)$ ⟵ each of these factors is irreducible over the integers

Solution ▶ $16a^4b - 2ab^4 = 2ab(2a - b)(4a^2 + 2ab + b^2)$

Some binomials can be factored in more than one way.

Example ▶ Factor $a^6 - b^6$.

$a^6 - b^6 = (a^3)^2 - (b^3)^2$ ⟵ difference of two squares

$\qquad = (a^3 + b^3)(a^3 - b^3)$ Think: The first factor is the sum of two cubes. The second factor is the difference of two cubes.

$\qquad = (a + b)(a^2 - ab + b^2)(a - b)(a^2 + ab + b^2)$

$a^6 - b^6 = (a + b)(a^2 - ab + b^2)(a - b)(a^2 + ab + b^2)$

If you try to factor $a^6 - b^6$ as the difference of two cubes, you will find:

$a^6 - b^6 = (a^2)^3 - (b^2)^3$ ⟵ difference of two cubes

$\qquad = (a^2 - b^2)(a^4 + a^2b^2 + b^4)$

$\qquad = (a + b)(a - b)(a^4 + a^2b^2 + b^4).$

At this point it appears that the two different methods of factoring have produced different answers; however, the following approach can be used to factor $(a^4 + a^2b^2 + b^4)$ and bring the answers into agreement.

$a^4 + a^2b^2 + b^4 = a^4 + 2a^2b^2 + b^4 - a^2b^2$ Think: The trinomial would be a PST if the middle terms were $2a^2b^2$. Add and subtract a^2b^2.

$\qquad = (a^4 + 2a^2b^2 + b^4) - a^2b^2$ Regroup.

$\qquad = (a^2 + b^2)^2 - a^2b^2$ Factor the PST.

$\qquad = [(a^2 + b^2) + ab][(a^2 + b^2) - ab]$ Factor the difference of two squares.

$\qquad = (a^2 + ab + b^2)(a^2 - ab + b^2)$

Although both methods have produced the same factorization, the difference-of-two-squares method was much more direct and easier to apply.

Make Sure

Factor each binomial using the general strategy. Check each solution.

See Example 1 ▶ **1.** $8a^3 - 216$ **2.** $32b^5 - 2bc^4$ **3.** $6d^4e - 48de^4$

_____ _____ _____

MAKE SURE ANSWERS: **1.** $8(a - 3)(a^2 + 3a + 9)$ **2.** $2b(4b^2 + c^2)(2b + c)(2b - c)$
3. $6de(d - 2e)(d^2 + 2de + 4e^2)$

6.5 Factor Using the General Strategy

EXAMPLE 2: Factor a trinomial using the general strategy.

Problem ▶ Factor $12a^2b - 60ab^2 + 75b^3$.

1. Factor out the GCF ▶ $12a^2b - 60ab^2 + 75b^3 = 3b(4a^2 - 20ab + 25b^2)$ $3b$ is the GCF.

2. Factor the trinomial ▶ $= 3b((2a)^2 - 2(2a)(5b) + (5b)^2)$ $4a^2 - 20ab + 25b^2$ is a PST.

3. Factor again, if possible ▶ $= 3b(2a - 5b)^2$ ⟵ each factor is irreducible over the integers

Solution ▶ $12a^2b - 60ab^2 + 75b^3 = 3b(2a - 5b)^2$ Check as before.

Another Example ▶ Completely factor $15x^3y + 20x^2y - 20xy$.

1. Factor out the GCF ▶ $15x^3y + 20x^2y - 20xy = 5xy(3x^2 + 4x - 4)$ The GCF is $5xy$.

    $= 5xy(3x^2 + 4x - 4)$ $3x^2 + 4x - 4$ is not a PST. Try trial and error.

2. Factor the trinomial ▶

Factors of 3	Factors of 4
1, 3	1, 4
	2, 2

$3x^2 + 4x - 4 = (?x + ?)(?x - ?)$

Trial Factors	Check Middle Term	
$(1x + 1)(3x - 4)$	$-4x + 3x = -1x$	
$(1x + 4)(3x - 1)$	$-1x + 12x = 11x$	
$(1x + 2)(3x - 2)$	$-2x + 6x = 4x$	Stop! $4x$ is the correct middle term.

3. Factor again, if possible ▶ $5xy$, $(1x + 2)$ and $(3x - 2)$ are all irreducible over the integers.

Solution ▶ $15x^3y + 20x^2y - 20xy = 5xy(x + 2)(3x - 2)$

Make Sure

Factor a trinomial using the general strategy. Check each solution.

See Example 2 ▶   **1.** $6x^3 - 13x^2y + 6xy^2$ **2.** $48x^2y - 72xy^2 + 27y^3$ **3.** $30u^3v + 5u^2v^2 - 75uv^3$

MAKE SURE ANSWERS: 1. $x(3x - 2y)(2x - 3y)$ **2.** $3y(4x - 3y)^2$ **3.** $5uv(2u - 3v)(3u + 5v)$

EXAMPLE 3: Factor a polynomial consisting of four or more terms using the general strategy.

Problem ▶ Factor $15a^2cx + 10abcx + 60a^2x + 40abx$.

1. Factor out the GCF ▶ $5a^2cx + 10abcx + 60a^2x + 40abx = 5ax[3ac + 2bc + 12a + 8b]$

2. Factor by grouping ▶
$\qquad = 5ax[(3ac + 2bc) + (12a + 8b)]$ ⟵ group in pairs
$\qquad = 5ax[c(3a + 2b) + 4(3a + 2b)]$ ⟵ remove common monomial factors

3. Factor again, if possible ▶ $\qquad = 5ax(3a + 2b)(c + 4)$ ⟵ each of these factors is irreducible over the integers

Solution ▶ $5a^2cx + 10abcx + 60a^2x + 40abx = 5ax(3a + 2b)(c + 4)$

Another Example ▶ Factor $a^3 - b^3 + a^3b + a^2b^2 + ab^3$.

$a^3 - b^3 + a^3b + a^2b^2 + ab^3 = (a^3 - b^3) + (a^3b + a^2b^2 + ab^3)$

$\qquad = (a^3 - b^3) + ab(a^2 + ab + b^2)$

$\qquad = (a - b)(a^2 + ab + b^2) + ab(a^2 + ab + b^2)$

$\qquad = (a^2 + ab + b^2)(a - b + ab)$

Make Sure

Factor polynomials consisting of four or more terms using the general strategy. Check each solution.

See Example 3 ▶ **1.** $a^5b^2 - 2a^4b^3 + 2a^3b^4 - a^2b^5$ **2.** $8c^3de^2 - 32cd^3e^2 - 20c^3def + 80cd^3ef$

MAKE SURE ANSWERS: **1.** $a^2b^2(a - b)(a^2 - ab + b^2)$ **2.** $4cde(2e - 5f)(c - 2d)(c + 2d)$

6.5 Practice: Check each solution.

Set 1: Factor each binomial using the general strategy.

1. $16a^2 - 100b^2$
2. $72c^2 - 288d^2$
3. $48e^2f - 27f^3$
4. $324m^2n - 196n^3$
5. $9p^4 - 144q^4$
6. $64r^4 - 324s^4$

Set 2: Factor each trinomial using the general strategy.

7. $y^2z - 6yz - 27z$
8. $wx^2 + 14wx + 48w$
9. $v^4 - 7v^2 + 12$
10. $u^4 - 9u^2 + 20$
11. $t^4 + 2t^2u^2 - 3u^4$
12. $r^4 - 10r^2s^2 + 24s^4$

Set 3: Factor polynomials consisting of four or more terms using the general strategy.

13. $8ad + 2ac - 5bc - 20bd$
14. $3ef - 9eg + 8fh - 24gh$
15. $3k^4 + 12k^3 - 3k^2 - 12k$
16. $4m^4 - 36m^3 - 4m^2 + 36m$
17. $3p^4q + 3p^3q^2 - 9p^2q^3 - 9pq^4$
18. $16rs^4 + 16r^2s^3 - 36r^3s^2 - 36r^4s$

Mixed Practice: Factor each polynomial using the general strategy.

19. $a^3 - ab^2 + a^2b - b^3$
20. $c^3 + d^3 - cd^2 - c^2d$
21. $5e^5f + 20e^4f^3 + 15e^2f^5$
22. $6g^5h^2 + 60g^4h^4 + 96g^2h^5$
23. $48k^5m - 243km^5$
24. $324n^6p^2 - 1024n^2p^6$
25. $e^4 - 5e^2 - 5ef + ef^3$
26. $g^5 + 3g^3 - 3g^2h - g^2h^3$
27. $m^2 + 2mn + n^2 - 4$
28. $p^2 + 4pq + 4q^2 - 9$
29. $9r^3 + 12r^2s + 4rs^2 - 16rt^2$
30. $16u^3 - 24u^2v + 9uv^2 - 4uw^2$
31. $3x^2 + 4x + 1 - y^2$
32. $4y^2 - 3y - 1 - z^2$
33. $a^3b^4 + 6a^2b^2 - 16a$
34. $c^5d^2 + c^3d - 12c$
35. $e^6 - 64f^6$
36. $64g^6 - 729h^6$

Review: Work these problems on a separate sheet of paper. Attach your work to this page.

Solve each equation using the zero-product property. (See Lesson 2.2.)
37. $(z - 3)(z + 2) = 0$
38. $(y - 4)(y + 3) = 0$
39. $(x - 5)(x - 3) = 0$
40. $(2w - 3)(3w + 2) = 0$
41. $(4v - 3)(3v - 4) = 0$
42. $(5u + 3)(4u + 3) = 0$
43. $(t - 3)(2t + 1)(3t - 4) = 0$
44. $(2s - 1)(3s + 2)(s + 2) = 0$
45. $(1 - r)(3 - 2r)(4r + 3) = 0$

Factor out the GCF from each polynomial. (See Lesson 6.1.)
46. $3q^2 - 6q$
47. $4p^2 - 6p$
48. $6n^3 - 12n + 3n^2$
49. $12m^3 - 8m^2 + 16m$
50. $6h^3k - 9hk^3 + 12h^2k^3$
51. $10e^3fg^2 + 15e^2f^3 + 20e^4f^3g$

6.6 Factor to Solve Equations

> A polynomial equation is in *standard form* if:
> 1. one member is zero,
> 2. the polynomial is in descending powers, and
> 3. the first term has a positive coefficient.

Examples ▶
(a) $6x^2 + 4x = 0$ is in standard form.
(b) $9y^2 - 16 = 0$ is in standard form.
(c) $0 = 6z^2 + 11z - 35$ is in standard form.
(d) $w^3 + 5w^2 - 9w - 45 = 0$ is in standard form.

Note ▶ Not all equations are in standard form.

Examples ▶
(a) $6x^2 = -4x$ is not in standard form because neither member is zero.
(b) $0 = 11z + 6z^2 - 35$ is not in standard form because $11z + 6z^2 - 35$ is not in descending powers.
(c) $-9y^2 + 16 = 0$ is not in standard form because -9 is not a positive coefficient.

Note ▶ $ax^2 + bx + c = 0$ is in standard form if a is positive.

To solve an equation containing a common factor in each term, you can sometimes use the *Zero-Product Property for Polynomials*.

> **The Zero-Product Property for Polynomials**
> If p and q are any polynomials, then $p \cdot q = 0$ means $p = 0$ or $q = 0$.

EXAMPLE 1: Solve an equation containing a common factor in each term by factoring.

Problem ▶ Solve $6x^2 = -4x$.

1. Write standard form ▶
$6x^2 = -4x$
$6x^2 + 4x = 0$ ⟵ standard form

2. Factor ▶ $2x(3x + 2) = 0$ Factor out $2x$.

3. Use the zero-product property ▶ $2x = 0$ or $3x + 2 = 0$ Think: $p \cdot q = 0$ means $p = 0$ or $q = 0$.

4. Solve ▶
$\frac{1}{2} \cdot 2x = \frac{1}{2} \cdot 0$ or $3x + 2 + (-2) = 0 + (-2)$
$x = 0$ or $3x = -2$
$x = 0$ or $\frac{1}{3} \cdot 3x = \frac{1}{3} \cdot (-2)$
$x = 0$ or $x = -\frac{2}{3}$

5. Check ▶

$6x^2$	$= -4x$		$6x^2$	$= -4x$	⟵ original equation
$6(0)^2$	$-4(0)$		$6(-\frac{2}{3})^2$	$-4(-\frac{2}{3})$	
$6(0)$	0		$6(\frac{4}{9})$	$-4(-\frac{2}{3})$	
0	0		$\frac{8}{3}$	$\frac{8}{3}$	

Solutions ▶ $x = 0$ or $-\frac{2}{3}$

Chapter 6 Factoring

Make Sure

Solve equations containing a common factor in each term by factoring. Check each solution.

See Example 1 ▶ **1.** $a^2 + 5a = 0$ _____   **2.** $12b^2 = 18b$ _____   **3.** $\frac{2}{3}c = \frac{3}{4}c^2$ _____

___

MAKE SURE ANSWERS: **1.** $0, -5$ **2.** $0, \frac{3}{2}$ **3.** $0, \frac{8}{9}$

To solve equations with form $r^2 - s^2 = 0$, use the Zero-Product Property for Polynomials.

EXAMPLE 2: Solve an equation with form $r^2 - s^2 = 0$ by factoring.

Problem ▶ Solve $-9y^2 + 16 = 0$.

1. Write standard form ▶
$-9y^2 + 16 = 0$
$-1 \cdot (-9y^2 + 16) = -1 \cdot 0$ Multiply by -1 to make the first coefficient positive.
$9y^2 - 16 = 0$ ⟵ standard form

2. Factor ▶ $(3y + 4)(3y - 4) = 0$

3. Use the zero-product property ▶ $3y + 4 = 0$ or $3y - 4 = 0$ Think: $p \cdot q = 0$ means $p = 0$ or $q = 0$.

4. Solve ▶
$3y = -4$ or $3y = 4$
$y = -\frac{4}{3}$ or $y = \frac{4}{3}$

5. Check ▶

$-9y^2 + 16 = 0$		$-9y^2 + 16 = 0$		⟵ original equation
$-9(-\frac{4}{3})^2 + 16$	0	$-9(\frac{4}{3})^2 + 16$	0	
$-16 + 16$	0	$-16 + 16$	0	
0	0	0	0	

Solutions ▶ $y = -\frac{4}{3}$ or $\frac{4}{3}$

Make Sure

Solve equations with form $r^2 - s^2 = 0$ by factoring. Check each solution.

See Example 2 ▶ **1.** $a^2 - 25 = 0$ **2.** $16b^2 - 49 = 0$ **3.** $36c^2 = 81$

___

MAKE SURE ANSWERS: **1.** $-5, 5$ **2.** $-\frac{7}{4}, \frac{7}{4}$ **3.** $-\frac{3}{2}, \frac{3}{2}$

Sometimes you can factor to solve an equation of the form $ax^2 + bx + c = 0$.

EXAMPLE 3: Solve an equation of the form $ax^2 + bx + c = 0$ by factoring.

Problem ▶ Solve $-1.1z = 0.6z^2 - 3.5$.

1. Write standard form ▶
$$-1.1z = 0.6z^2 - 3.5$$
$$-11z = 6z^2 - 35 \qquad \text{Multiply by 10 to clear decimals.}$$
$$0 = 6z^2 + 11z - 35$$

2. Factor ▶
$$0 = (2z + 7)(3z - 5)$$

3. Use the zero-product property ▶
$$2z + 7 = 0 \quad \text{or} \quad 3z - 5 = 0$$

4. Solve ▶
$$2z = -7 \quad \text{or} \quad 3z = 5$$
$$z = -\tfrac{7}{2} \quad \text{or} \quad z = \tfrac{5}{3}$$

5. Check ▶

$-1.1z = 0.6z^2 - 3.5$		$-1.1z = 0.6z^2 - 3.5$		← original equation
$-1.1(-\tfrac{7}{2})$	$0.6(-\tfrac{7}{2})^2 - 3.5$	$-1.1(\tfrac{5}{3})$	$0.6(\tfrac{5}{3})^2 - 3.5$	
$-\tfrac{11}{10}(-\tfrac{7}{2})$	$\tfrac{6}{10}(\tfrac{49}{4}) - \tfrac{35}{10}$	$-\tfrac{11}{10}(\tfrac{5}{3})$	$\tfrac{6}{10}(\tfrac{25}{9}) - \tfrac{35}{10}$	
$\tfrac{77}{20}$	$\tfrac{77}{20}$	$-\tfrac{11}{6}$	$-\tfrac{11}{6}$	

Solutions ▶ $z = -\tfrac{7}{2}$ or $\tfrac{5}{3}$

Make Sure

Solve equations of the form $ax^2 + bx + c = 0$ by factoring. Check each solution.

See Example 3 ▶   **1.** $x^2 + 5x + 4 = 0$ **2.** $6y^2 - y - 12 = 0$ **3.** $7z = 45 - 12z^2$

MAKE SURE ANSWERS: **1.** $-4, -1$ **2.** $-\tfrac{4}{3}, \tfrac{3}{2}$ **3.** $-\tfrac{9}{4}, \tfrac{5}{3}$

The zero-product property can be extended to include the product of several polynomials.

> **The Extended Zero-Product Property for Polynomials**
>
> If $p_1, p_2, p_3, \cdots, p_n$ are all polynomials, then:
>
> $p_1 \cdot p_2 \cdot p_3 \cdot \cdots \cdot p_n = 0$ means $p_1 = 0$ or $p_2 = 0$ or $p_3 = 0$ or $\cdots p_n = 0$.

The Extended Zero-Product Property for Polynomials states that if the product of a finite number of polynomials equals zero, then at least one of those polynomials must equal zero.

242 Chapter 6 Factoring

You can use the Extended Zero-Product Property for Polynomials to solve equations that can be written as the product of first degree polynomials set equal to zero.

EXAMPLE 4: Solve an equation that can be written in the form $ax^3 + bx^2 + cx + d = 0$ by factoring.

Problem ▶ Solve $w^3 + 5w^2 - 9w = 45$.

1. Write standard form ▶
$$w^3 + 5w^2 - 9w = 45$$
$$w^3 + 5w^2 - 9w - 45 = 0$$

2. Factor ▶ $(w^3 + 5w^2) + (-9w - 45) = 0$ Think: Factor a four-term polynomial by grouping.

$$w^2(w + 5) - 9(w + 5) = 0$$
$$(w + 5)(w^2 - 9) = 0$$
$$(w + 5)(w + 3)(w - 3) = 0$$

3. Use the extended zero-product property ▶ $w + 5 = 0$ or $w + 3 = 0$ or $w - 3 = 0$

4. Solve ▶ $w = -5$ or $w = -3$ or $w = 3$

Solutions ▶ $w = -5$ or -3 or 3 Check as before.

Make Sure

Solve equations that can be written in the form $ax^3 + bx^2 + cx + d = 0$ by factoring. Check each solution.

See Example 4 ▶ 1. $x^3 - 7x^2 - 4x + 28 = 0$ _____   2. $y^3 + 8 = 4y + 2y^2$ _____

MAKE SURE ANSWERS: 1. $-2, 2, 7$ 2. $-2, 2$

Summary ▶ To solve equations by factoring:
1. Write standard form for the given equation.
2. Factor the equation in standard form.
3. Use the Zero-Product Property for Polynomials on the factored equation from Step 2.
4. Solve each equation from Step 3.
5. Check each solution from Step 4 in the original equation.

6.6 Practice: *Check each solution.*

Set 1: Solve equations containing a common factor in each term by factoring.

1. $3a^2 + 6a = 0$
2. $5b^2 - 20b = 0$
3. $8c^2 = 4c$

4. $5d = 15d^2$
5. $0.34e^2 = 1.7e$
6. $1.5f^2 = 6f$

7. $\frac{1}{9}g^2 = \frac{1}{4}g$
8. $\frac{1}{2}h^2 = -\frac{2}{3}h$
9. $\frac{5}{9}m^2 = 20m$

Set 2: Solve equations with form $r^2 - s^2$ by factoring.

10. $a^2 - 4 = 0$
11. $b^2 - 9 = 0$
12. $c^2 = 81$

13. $49 = d^2$
14. $4e^2 = 0.49$
15. $0.09 = 0.16f^2$

16. $0.15g^2 = 3.75$
17. $0.4h^2 = 6.4$
18. $m^2 = \frac{49}{16}$

Set 3: Solve equations of the form $ax^2 + bx + c = 0$ by factoring.

19. $a^2 + 10a + 16 = 0$

20. $b^2 - 10b + 24 = 0$

21. $c^2 = 36 - 5c$

22. $d^2 = 3d + 28$

23. $6f^2 = 20f - 16$

24. $7g = 36 - 4g^2$

25. $4h = 6h^2 - 16$

26. $2m^2 - 2.5m = 3$

Set 4: Solve equations of the form $ax^3 + bx^2 + cx + c = 0$ by factoring.

27. $a^3 + 3a^2 - 4a - 12 = 0$

28. $b^3 + 6b^2 - 9b - 54 = 0$

29. $c^3 - 7c^2 - 16c + 112 = 0$

30. $d^3 - 4d^2 - 4d + 16 = 0$

31. $3e^3 + 15e^2 - 12e - 60 = 0$

32. $4f^3 - 20f^2 - 100f + 500 = 0$

33. $4g^3 - 12g^2 = 36g - 108$

34. $4h^3 = 12h^2 + h - 3$

6.6 Practice 245

Name _____ Date _____ Class _____

Mixed Practice: Solve each equation by factoring.

35. $27a = \dfrac{3}{4}a^2$

36. $6b = \dfrac{2}{3}b^2$

37. $\dfrac{2}{3}c = \dfrac{5}{6}c^2$

38. $\dfrac{4}{5}d = \dfrac{3}{4}d^2$

39. $0.13e^2 = 0.0208$

40. $4f^2 = 1.96$

41. $g^2 + 9g + 18 = 0$

42. $h^2 + 10h + 21 = 0$

43. $16k^2 + 24k - 27 = 0$

44. $27 + 6m = 16m^2$

45. $\dfrac{1}{9}n^2 = \dfrac{1}{12}n + \dfrac{1}{8}$

46. $\dfrac{1}{6}p^2 = \dfrac{17}{72}p - \dfrac{1}{12}$

47. $\dfrac{1}{9}q + \dfrac{1}{3} = \dfrac{1}{4}q^3 + \dfrac{3}{4}q^2$

48. $\dfrac{1}{36}r^3 - \dfrac{1}{4}r = \dfrac{1}{6}r^2 - \dfrac{3}{2}$

49. $12.25 - 4s^2 = -0.49s + 0.16s^3$

50. $t^2 - 0.36 = 0.25t^3 - 0.09t$

51. $0.16u^2 - 0.49 = 0$

52. $0.25 - 0.64v^2 = 0$

Copyright © 1985 by Harcourt Brace Jovanovich, Inc. All rights reserved.

Extra: Solve each equation by factoring.

53. $(a - 3)^2 = 4$ **54.** $(b + 4)^2 = 9$ **55.** $(c - 3)^2 = 9$

56. $(d + 5)^2 = 25$ **57.** $e^3 = 9e$ **58.** $f^3 = 16f$

59. $(g - 3)^3 = g - 3$ **60.** $(h + 4)^3 = h + 4$ **61.** $6k^3 = 36k - 15k^2$

62. $12m^3 - 32m = -8m^2$ **63.** $6n^4 + 30n^3 = 24n^2 + 120n$ **64.** $12p^4 - 36p^3 = 3p^2 - 9p$

Review: Work these problems on a separate sheet of paper. Attach your work to this page.

Solve each equation. (See Lesson 2.1.)
65. $z - 2 = 0$ **66.** $y + 4 = 0$ **67.** $2x - 6 = 0$
68. $3w + 4 = 0$ **69.** $4v - 2 = 0$ **70.** $5u + 7 = 0$

Factor each trinomial. (See Lesson 6.3.)
71. $z^2 - 7z - 12$ **72.** $y^2 - 2y - 15$ **73.** $6x^2 - x - 12$
74. $6w^2 + 17w + 12$ **75.** $24v^2 + 2v - 15$ **76.** $24u^2 - 38u + 15$

Factor each binomial. (See Lesson 6.4.)
77. $t^2 - 25$ **78.** $4s^2 - 49$ **79.** $16r^2 - 1$
80. $9q^2 - 64$ **81.** $36p^2 - 25$ **82.** $4n^2 - 9$

Problem Solving 7: Factor to Solve Problems

To solve a word problem by factoring, you will first need to translate to an equation.

EXAMPLE: Factor to solve this word problem.

1. Read and identify ▶ The polynomial $\frac{1}{20}r^2 + r$ represents the stopping distance in feet for an average car moving at a rate of r miles per hour (mph) on an average road. By how many miles per hour was a car traveling under or over the speed limit if the skid marks are 120 feet in a 30 mph zone? (Assume both the car and road to be average.)

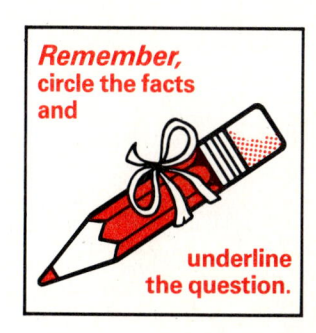

Remember, circle the facts and underline the question.

2. Understand ▶ The question asks you to find the rate given that both $\frac{1}{20}r^2 + r$ and 120 feet represent the stopping distance.

3. Decide ▶ To find the rate r given that $\frac{1}{20}r^2 + r$ and 120 feet are equal, you **translate to an equation** and **then solve the equation.**

4. Translate ▶ The stopping distance in feet is 120 feet.

$$\frac{1}{20}r^2 + r = 120$$

5. Solve as before ▶ The LCD of $\frac{1}{20}$, 1, and 120 is 20.   Clear fractions.

$$20(\tfrac{1}{20}r^2 + r) = 20(120)$$

$$20(\tfrac{1}{20}r^2) + 20(r) = 20(120)$$

$$r^2 + 20r = 2400 \longleftarrow \text{fractions cleared}$$

$$r^2 + 20r - 2400 = 0 \qquad \text{Write standard form.}$$

$$(r + 60)(r - 40) = 0 \qquad \text{Factor: } +60(-40) = -2400 \text{ and } +60 + (-40) = +20$$

$$r + 60 = 0 \text{ or } r - 40 = 0 \qquad \text{Use the zero-product property.}$$

$$r = -60 \text{ or } \qquad r = 40 \longleftarrow \text{proposed solutions}$$

6. Interpret ▶ $r = -60$ means the rate of the car was -60 mph. **Wrong:** Car speeds are never negative.
$r = 40$ means the rate of the car was 40 mph.

7. Check ▶ $\frac{1}{20}r^2 + r = \frac{1}{20}(40)^2 + (40)$ Substitute the proposed solution into the original polynomial to see if you get the given stopping distance.

$$= \tfrac{1}{20}(1600) + 40$$

$$= 80 + 40$$

$$= 120 \longleftarrow 40 \text{ is correct}$$

Solution ▶ 40 mph in a 30 mph speed zone is 10 mph over the speed limit.

Note ▶ Of the two solutions for r (-60 and 40), only one solution (40) made sense as an answer to the original question. (See Step 6.)

248   Chapter 6   Factoring

1. _____
2. _____
3. _____
4. _____
5. _____
6. _____
7. _____
8. _____
9. _____
10. _____
11. _____
12. _____
13. _____
14. _____
15. _____
16. _____
17a. _____
  b. _____
18a. _____
  b. _____

Practice: Factor to solve each word problem.

How many miles per hour was a car traveling under or over the speed limit if the skid marks are:

1. 75 feet in a 35 mph zone? (Use the polynomial in the Example problem.)
2. 240 feet in a 55 mph zone?

FACT: $\frac{1}{2}n^2 + \frac{1}{2}n$ represents the sum of all positive integers 1 to n.

How many of the first positive integers would it take to get a sum of:

3. 6   4. 15   5. 36   6. 55

7. 210   8. 325   9. 1275   10. 5050

Extra: Solve find-the-number problems.

11. The square of a number plus twice the number is 35. Find the number.
    Hint: Let n = the number
         then n^2 = the number squared
         and $2n$ = twice the number

12. The square of a number minus three times the number is 54. Find the number.

13. Six times the square of a number plus twelve times the number is 48. Find the number.

14. Eight times the square of a number minus twelve times the number is 36. Find the number.

Extra: Solve geometry problems.

15. The height (h) of a triangle is 4 in. longer than the base (b). The area of the triangle is 48 in.² Find the height and base. (Hint: area of a triangle = $\frac{1}{2}bh$.)

16. The height (h) of a triangle is 8 ft shorter than the base (b). The area of the triangle is 120 ft². Find the height and base.

17. In Figure 6.1, the volume of the larger box is three times the volume of the smaller box. Find the dimensions of a) the smaller box b) the larger box.

Figure 6.1

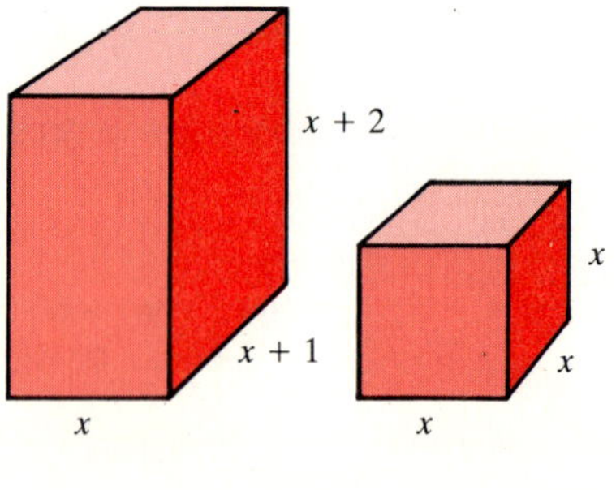

18. For Figure 6.2, show that a) the area of the shaded cross-section is $\pi(a+b)(a-b)$ and b) the volume of the shaded portion of the cylinder is $\pi l(a+b)(a-b)$.

Figure 6.2

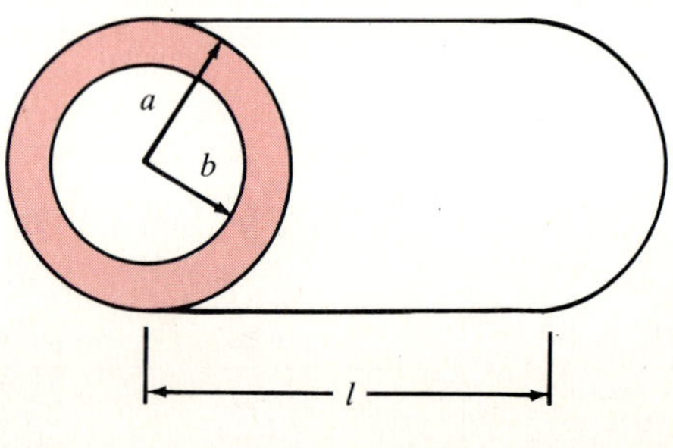

Copyright © 1985 by Harcourt Brace Jovanovich, Inc. All rights reserved.

Chapter 6 Review

		What to Review if You Have Trouble		
Objectives		Lesson	Example	Page
Find the GCF of monomials ▶	1. Find the GCF of $-36x^3y^2z^5$, $42x^4y^2z^4$, and $54x^3y^2z$.	6.1	1	206
Factor out the GCF from a polynomial ▶	2. Factor out the GCF from $30x^2y^3 - 27x^3y^2 + 18x^2y^2$.	6.1	2	207
Factor by grouping ▶	3. Factor $x^3 - 5x^2 + 3x - 15$.	6.1	3	208
Factor by trial and error ▶	4. Factor $x^2 + 14x + 24$.	6.2	1	212
	5. Factor $x^2 - 11x + 24$.	6.2	2	212
	6. Factor $x^2 + 9x - 36$.	6.2	3	213
	7. Factor $x^2 - 6x - 27$.	6.2	4	214
	8. Factor $6x^2 + 19x + 10$.	6.3	1	217
	9. Factor $15x^2 - 41x + 12$.	6.3	2	218
	10. Factor $12x^2 + 5x - 3$.	6.3	3	218
	11. Factor $27x^2 - 12x - 20$.	6.3	4	219
Factor using the *ac* method ▶	12. Factor $-4w^2 + 43w - 30$.	6.3	5	221

250 Chapter 6 Factoring

Factor a difference of two squares
13. Factor $16y^2 - 1$.   6.4   1   227

Factor perfect square trinomials
14. Factor **a.** $9x^2 - 30x + 25$ and **b.** $4x^2 + 36xy + 81$.

a. _____ 6.4   2   228

b. _____

Factor the sum or difference of two cubes
15. Factor **a.** $27a^3 + 1$ and **b.** $8m^3 - 125$.   6.4   3   229

a. _____

b. _____

Factor by grouping
16. Factor $2y^3 + 3y^2 + 6y + 9$.   6.4   4   229

Factor using the general strategy
17. Factor $32w^4 - 162$.   6.5   1   233

18. Factor $6x^2y^3 + 3xy^3 - 30y^3$.   6.5   2   235

19. Factor $4cx^3 - 16cx - 2cx^2 + 8c$.   6.5   3   236

Solve equations by factoring
20. Solve $4w^2 = -9w$.   6.6   1   239

21. Solve $-25y^2 + 49 = 0$.   6.6   2   240

22. Solve $-1.4x = 0.8x^2 - 1.5$.   6.6   3   241

23. Solve $z^3 + 2z^2 - 25z - 50 = 0$.   6.6   4   242

Factor to solve problems
24. By how much was a car traveling over a 55 mph speed limit if the skid marks are 400 feet long?   PS 7   —   247

CHAPTER 6 REVIEW ANSWERS: 1. $6x^3y^2z$ 2. $3x^2y^2z(10y^2 - 9x + 6)$ 3. $(x - 5)(x^2 + 3)$ 4. $(x + 2)(x + 12)$ 5. $(x - 3)(x - 8)$ 6. $(x + 3)(x - 9)$ 7. $(x + 3)(x - 4)$ 8. $(3x + 2)(2x + 5)$ 9. $(3x - 1)(5x - 12)$ 10. $(4x + 3)(3x - 1)$ 11. $(3x + 2)(9x - 10)$ 12. $(-4w + 3)(w - 10)$ 13. $(4y + 1)(4y - 1)$ 14a. $(3x - 5)^2$ b. $(2x + 9)^2$ 15a. $(3a + 1)(9a^2 - 3a + 1)$ b. $(2m - 5)(4m^2 + 10m + 25)$ 16. $(2y + 3)(y^2 + 3)$ 17. $2(4w^2 + 9)(2w + 3)(2w - 3)$ 18. $3y^3(2x + 5)(x - 2)$ 19. $2c(2x - 1)(x + 2)(x - 2)$ 20. $0, -\frac{9}{4}$ 21. $-\frac{7}{5}, \frac{7}{5}$ 22. $-\frac{5}{2}, \frac{3}{4}$ 23. $-5, -2, 5$ 24. 25 mph

NONLINEAR RELATIONS

7 Rational Expressions

7.1 Simplify Rational Expressions

7.2 Multiply and Divide Rational Expressions

7.3 Add and Subtract Rational Expressions

7.4 Simplify Complex Fractions

7.5 Solve Equations Containing Rational Expressions

PS 8: Solve Problems Using Rational Expressions

PS 9: Solve Formulas Containing Rational Expressions

7.1 Simplify Rational Expressions

A fraction that has a polynomial for both its numerator and its denominator is called a *rational expression*.

Examples ▶ (a) $\dfrac{3}{4}$ (b) $\dfrac{x}{x+1}$ (c) $\dfrac{y^2 + 4y + 3}{y^2 + 7y + 12}$ (d) $\dfrac{x^2 + 2xy + y^2}{x^2 - y^2}$ ⟵ rational expressions

(e) $\dfrac{7}{\sqrt{v} + 1}$ is not a rational expression because its denominator is not a polynomial.

(f) $\dfrac{u^{2/3}}{u^2 - 5}$ is not a rational expression because its numerator is not a polynomial.

Every rational expression represents a quotient of two polynomials. Since division by 0 is not allowed, the denominator of any rational expression cannot equal 0.

Every numerical replacement of the variable(s) that causes the denominator of a rational expression to equal zero is called an *excluded value* of the rational expression.

> **To find excluded value(s) of a rational expression:**
> 1. Set the denominator equal to 0.
> 2. Solve the equation formed in Step 1.

EXAMPLE 1: Find the excluded value(s) of a rational expression.

Problem ▶ Find the excluded value(s) of $\dfrac{5}{2x^2 - 7x}$.

1. Set denominator equal to 0 ▶ $2x^2 - 7x = 0$

2. Solve ▶ $x(2x - 7) = 0$ Factor out the GCF.

$x = 0$ or $2x - 7 = 0$ Use the zero-product property.

$x = 0$ or $2x = 7$

$x = 0$ or $x = \tfrac{7}{2}$

Solution ▶ The excluded values of $\dfrac{5}{2x^2 - 7x}$ are 0 and $\dfrac{7}{2}$.

Other Examples ▶ (a) $\dfrac{3}{w}$ has 0 as an excluded value because if $w = 0$, the denominator equals 0.

(b) $\dfrac{5}{2v + 9}$ has $-\dfrac{9}{2}$ as an excluded value because $2v + 9 = 0$ when $v = -\dfrac{9}{2}$.

(c) $\dfrac{8x}{5}$ has no excluded values because the denominator 5 is never equal to 0.

(d) $\dfrac{4u^2 - 11}{(u + 5)(u - 2)(3u - 7)}$ has $-5, 2,$ and $\dfrac{7}{3}$ as excluded values.

(e) $\dfrac{2}{x - y}$ has an infinite number of excluded values because the denominator will equal 0 whenever $x = y$.

7.1 Simplify Rational Expressions   253

Make Sure

Find the excluded value(s) of each rational expression.

See Example 1 ▶ 1. $\dfrac{4a}{a^2 - 16}$ 2. $\dfrac{2b - 3}{3b^3 + b^2 - 2b}$

MAKE SURE ANSWERS: 1. $-4, 4$ 2. $-1, 0, \tfrac{2}{3}$

Agreement ▶ When you write a rational expression, you must understand that the value(s) of the variable(s) that make the denominator equal to 0 are excluded. Sometimes the necessary restrictions are listed to the right of the rational expression as a reminder.

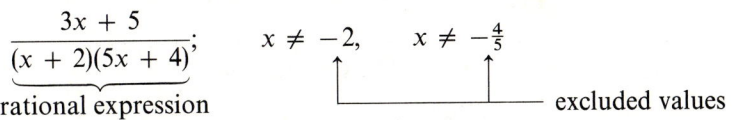

Example ▶ $\underbrace{\dfrac{3x + 5}{(x + 2)(5x + 4)}}_{\text{rational expression}};\quad x \neq -2,\quad x \neq -\tfrac{4}{5}\quad \text{excluded values}$

The excluded values are said to make the rational expression *undefined*.

There are three signs associated with every fraction. They are the sign of the numerator, the sign of the denominator, and the sign preceding the fraction.

Example ▶ sign preceding the fraction ⟶ $+\dfrac{-3}{+5}$ — sign of the numerator / sign of the denominator

The rules for multiplication and division of signed numbers state that if any two of the three signs of a fraction are changed, the value of the fraction is unchanged.

Examples ▶
(a) $+\dfrac{-3}{+5} = -\dfrac{+3}{+5} = +\dfrac{+3}{-3} = -\dfrac{-3}{-5}$ (b) $+\dfrac{-p}{+q} = -\dfrac{+p}{+q} = +\dfrac{+p}{-q} = -\dfrac{-p}{-q}$

(c) $+\dfrac{-3}{-5} = -\dfrac{-3}{+5} = -\dfrac{+3}{-5} = +\dfrac{+3}{+5}$ (d) $+\dfrac{-p}{-q} = -\dfrac{-p}{+q} = -\dfrac{+p}{-q} = +\dfrac{+p}{+q}$

It is general practice not to write the plus signs associated with a fraction.

In your work with rational expressions, it is very important to notice that $x - y$ and $y - x$ are opposites of each other. To verify this, you can use the distributive properties.

Verification ▶ $-(x - y) = -1(x - y)$

$= (-1)x - (-1)y$ Use a distributive property.

$= -x + y$

$= y + (-x)$ Use the commutative property for addition.

$= y - x$ Use the definition of subtraction.

To form the opposite of the difference of two terms, just change the order of the terms.

254   Chapter 7   Rational Expressions

Examples ▶ (a) $-(2-7) = 7-2$ (b) $-(x-y) = y-x$ (c) $-(6w-1) = 1-6w$

To determine whether two rational expressions are equal, you can use the sign rules and the concept of opposites.

EXAMPLE 2: Determine if the two given rational expressions are equal.

Problem ▶ $\dfrac{2x-3}{5x-1} \stackrel{?}{=} -\dfrac{3-2x}{5x-1}$

1. Identify changes ▶ $+\dfrac{\boxed{2x-3}}{5x-1} \stackrel{?}{=} -\dfrac{\boxed{3-2x}}{5x-1}$ (opposites; different signs)

Think: The numerators $2x-3$ and $3-2x$ are opposites of each other. The signs preceding the rational expression are different. The denominators are the same.

2. Interpret ▶ The rational expressions are the same except for two sign changes. This means the rational expressions are equal.

Solution ▶ $\dfrac{2x-3}{5x-1} = -\dfrac{3-2x}{5x-1}$

Another Example ▶ $-\dfrac{x^2+7x+12}{x^2-3x+15} \stackrel{?}{=} -\dfrac{x^2+7x+12}{-x^2+3x-15}$

1. Identify ▶ $-\dfrac{x^2+7x+12}{\boxed{x^2-3x+15}} \stackrel{?}{=} -\dfrac{x^2+7x+12}{\boxed{-x^2+3x-15}}$ (opposites)

Think: The numerators are the same.
The signs preceding the rational expressions are the same.
The denominators are opposites of each other.

2. Interpret ▶ The rational expressions are the same except for one sign change. This means that the rational expressions are not equal, but are instead opposites of each other.

Solution ▶ $-\dfrac{x^2+7x+12}{x^2-3x+15} \neq -\dfrac{x^2+7x+12}{-x^2+3x-15}$

Make Sure

Determine if the two given rational expressions are equal; write yes or no.

See Example 2 ▶ 1. $\dfrac{3a-5}{a^2-12} \stackrel{?}{=} -\dfrac{5-3a}{a^2-12}$ ____   2. $\dfrac{b^2-16}{b^2-3bc-4c^2} \stackrel{?}{=} -\dfrac{16-b^2}{4c^2+3bc-b^2}$ ____

MAKE SURE ANSWERS: 1. yes 2. no

7.1 Simplify Rational Expressions 255

> **The Fundamental Rule of Rational Expressions**
>
> If p, q, and r are polynomials, then: $\dfrac{pr}{qr} = \dfrac{p}{q}$, provided $q \neq 0$ and $r \neq 0$.

The Fundamental Rule is useful for reducing rational expressions. It states that if a nonzero polynomial is a common factor in both the numerator and the denominator of a rational expression, then that rational expression is equal to the reduced rational expression formed by eliminating the common polynomial factor.

A rational expression is said to be *in lowest terms* if the numerator and the denominator have no common factors other than 1 or -1.

> **To reduce a rational expression to lowest terms:**
> 1. Factor the numerator and the denominator completely.
> 2. Use the Fundamental Rule of Rational Expressions to eliminate all nonzero factors common to both the numerator and the denominator.

EXAMPLE 3: Reduce a rational expression to lowest terms.

Problem ▶ Reduce $\dfrac{2x^2 - 18}{2x^2 + 4x - 30}$ to lowest terms.

1. Factor ▶ $\dfrac{2x^2 - 18}{2x^2 + 4x - 30} = \dfrac{2(x^2 - 9)}{2(x^2 + 2x - 15)}$ Factor out the GCF from both the numerator and the denominator.

$$= \dfrac{2(x + 3)(x - 3)}{2(x - 3)(x + 5)}$$

2. Use the Fundamental Rule ▶ $= \dfrac{\cancel{2}(x + 3)\cancel{(x - 3)}}{\cancel{2}\cancel{(x - 3)}(x + 5)}$ The slash marks indicate the common factors.

$$= \dfrac{x + 3}{x + 5}$$

Solution ▶ $\dfrac{2x^2 - 18}{2x^2 + 4x - 30} = \dfrac{x + 3}{x + 5}$ in lowest terms.

Other Examples ▶

(a) $\dfrac{5u^2 - u}{5u^2 + u} = \dfrac{u(5u - 1)}{u(5u + 1)}$

$= \dfrac{\cancel{u}(5u - 1)}{\cancel{u}(5u + 1)}$

$= \dfrac{5u - 1}{5u + 1}$

(b) $\dfrac{3x + 3}{21x} = \dfrac{3(x + 1)}{3 \cdot 7 \cdot x}$

$= \dfrac{\cancel{3}(x + 1)}{\cancel{3} \cdot 7 \cdot x}$

$= \dfrac{x + 1}{7x}$

(c) $\dfrac{a^2 + 2a - 15}{a^2 + 4a - 21} = \dfrac{(a - 3)(a + 5)}{(a - 3)(a + 7)}$

$= \dfrac{\cancel{(a - 3)}(a + 5)}{\cancel{(a - 3)}(a + 7)}$

$= \dfrac{a + 5}{a + 7}$

(d) $\dfrac{25m^3 - 25mn^2}{5m^3 - 10m^2n + 5mn^2} = \dfrac{25m(m^2 - n^2)}{5m(m^2 - 2mn + n^2)}$

$= \dfrac{25m(m + n)(m - n)}{5m(m - n)(m - n)}$

$= \dfrac{5 \cdot \cancel{5} \cdot \cancel{m}(m + n)\cancel{(m - n)}}{\cancel{5} \cdot \cancel{m}\cancel{(m - n)}(m - n)}$

$= \dfrac{5(m + n)}{m - n}$ or $\dfrac{5m + 5n}{m - n}$

Because $x - y$ and $y - x$ are opposites of each other, their quotient is -1. The following shortcut uses this result to simplify rational expressions involving binomials of the form $\dfrac{x-y}{y-x}$.

> **SHORTCUT 7.1:** If p and q are polynomials: $\dfrac{p(x-y)}{q(y-x)} = -\dfrac{p}{q}$ provided $q \neq 0$ and $x \neq y$.

Example ▶ Reduce $\dfrac{2x^2 - 2y^2}{5y - 5x}$ to lowest terms.

1. Factor ▶ $\dfrac{2x^2 - 2y^2}{5y - 5x} = \dfrac{2(x^2 - y^2)}{5(y - x)}$ Factor out the GCF from both the numerator and the denominator.

$= \dfrac{2(x + y)(x - y)}{5(y - x)}$ Think: $x^2 - y^2$ is the difference of two squares.

2. Simplify ▶ $= \dfrac{2(x + y)(x - y)}{5(y - x)}$ Use Shortcut 7.1.

$= -\dfrac{2(x + y)}{5}$

Solution ▶ $\dfrac{2x^2 - 2y^2}{5y - 5x} = -\dfrac{2(x + y)}{5}$

Note ▶ The answer in the above example $-\dfrac{2(x + y)}{5}$ can be written in several equivalent forms:

$$-\dfrac{2(x + y)}{5} = \dfrac{-2(x + y)}{5} = \dfrac{-2x - 2y}{5} = -\dfrac{2x + 2y}{5}.$$

Make Sure

Reduce each rational expression to lowest terms.

See Example 3 ▶
1. $\dfrac{a^2 + 5a + 4}{a^2 - 16}$ _____

2. $\dfrac{b^3 + 2b^2c + 2bc^2 + c^3}{b^3 - c^3}$ _____

MAKE SURE ANSWERS: 1. $\dfrac{a+1}{a-4}$ 2. $\dfrac{b+c}{b-c}$

Name _____ Date _____ Class _____

7.1 Practice

Set 1: Find the excluded value(s) of each rational expression.

1. $\dfrac{5}{a}$
2. $\dfrac{-3}{b}$
3. $\dfrac{4}{c-2}$

4. $\dfrac{-1}{d-5}$
5. $\dfrac{-2e}{e(e+3)}$
6. $\dfrac{4f^2}{f(f+5)}$

7. $\dfrac{2g-3}{3g-g^2}$
8. $\dfrac{6-3h}{4h-h^2}$
9. $\dfrac{2m-3}{m^2-4}$

10. $\dfrac{4-5n^2}{9-n^2}$
11. $\dfrac{5+7p}{36-4p^2}$
12. $\dfrac{4-3q}{5-5q^2}$

13. $\dfrac{4r^2-1}{3-r-2r^2}$
14. $\dfrac{3s-s^2}{12-5s-2s^2}$
15. $\dfrac{t^2-t-1}{4t^2+4t+1}$

16. $\dfrac{u-5u^2}{6u^2-13u+6}$
17. $\dfrac{v+2}{v^3-v^2-6v}$
18. $\dfrac{w-3w^2}{w-5w^2+6w^3}$

Set 2: Determine if the two given rational expressions are equal; write yes or no.

19. $\dfrac{-a}{5} \stackrel{?}{=} \dfrac{a}{-5}$
20. $\dfrac{b}{-3} \stackrel{?}{=} \dfrac{-b}{3}$
21. $\dfrac{-c}{-3} \stackrel{?}{=} -\dfrac{c}{3}$

22. $-\dfrac{d}{5} \stackrel{?}{=} \dfrac{-d}{-5}$
23. $\dfrac{-e+5}{4e-3} \stackrel{?}{=} \dfrac{5-e}{3-4e}$
24. $\dfrac{f+3}{4f-5} \stackrel{?}{=} \dfrac{3+f}{5-4f}$

25. $\dfrac{2g-3}{g^2-4} \stackrel{?}{=} -\dfrac{2g-3}{4-g^2}$
26. $\dfrac{-4h+3}{h^2-9} \stackrel{?}{=} -\dfrac{3-4h}{9-h^2}$

27. $\dfrac{18-7r-r^2}{r^2-9r+18} \stackrel{?}{=} \dfrac{r^2-7r-18}{9r-r^2-18}$
28. $\dfrac{15-2s-s^2}{s^2-2s-15} \stackrel{?}{=} \dfrac{s^2+2s-15}{15+2s-s^2}$

29. $\dfrac{w^2-x^2}{w^3-x^3} \stackrel{?}{=} -\dfrac{w^2-x^2}{x^3-w^3}$
30. $\dfrac{y^2+z^2}{y^3-z^3} \stackrel{?}{=} -\dfrac{y^2+z^2}{z^3-y^3}$

Chapter 7 Rational Expressions

Set 3: Reduce each rational expression to lowest terms.

31. $\dfrac{4}{2z - 6}$
32. $\dfrac{-3}{6y + 9}$
33. $\dfrac{6x}{3x^2 - 12x}$

34. $\dfrac{8w}{4w^2 + 16w}$
35. $\dfrac{v + 1}{v^2 - v - 2}$
36. $\dfrac{u + 3}{u^2 + 2u - 3}$

37. $\dfrac{6 - 3t}{2t^2 - 5t + 2}$
38. $\dfrac{20 - 5s}{3s^2 - 17s + 20}$
39. $\dfrac{r - p}{p^3 - r^3}$

40. $\dfrac{p - r}{r^3 - p^3}$
41. $\dfrac{4m^2 + 12m + 9}{9 - 4m^2}$
42. $\dfrac{9n^2 - 25}{25 - 30n + 9n^2}$

43. $\dfrac{3h^2 + 7h - 6}{24h^3 - 34h^2 + 12h}$
44. $\dfrac{3g^2 + 6g - 24}{3g^3 - 18g^2 + 24g}$
45. $\dfrac{e^2 - 4f^2}{8f^3 - e^3}$

46. $\dfrac{9c^2 - d^2}{d^3 + 27c^3}$
47. $\dfrac{9b - 3b^2 - 2b^3}{3b + 7b^2 - 6b^3}$
48. $\dfrac{a^3 - 9a}{3a + 5a^2 - 2a^3}$

Review: Work these problems on a separate sheet of paper. Attach your work to this page.

Simplify each expression using the Order of Operations Rule. (See Lesson 1.4.)

49. $(a + b) \cdot (a - b) \div (a - b)$
50. $(a^2 - b^2) \div (a - b) \cdot (a + b)$
51. $(a^2 - b^2) \div (a - b) \div (a + b)$
52. $(a^2 - b^2) \div (a^2 - b^2) \div (a + b)$
53. $(a - b) \cdot [(a^2 - b^2) \div (a + b)]$
54. $(a - b) \div [(a^2 - b^2) \div (a + b)]$

Factor each trinomial. (See Lesson 6.3.)

55. $z^2 + 4z + 4$
56. $3y^2 + 5y + 2$
57. $12x^2 + 19x + 5$
58. $2w^2 - 11w + 5$
59. $6v^2 + 29v - 5$
60. $5u^2 - 7u - 6$
61. $8t^2 - 10t - 7$
62. $12s^2 - 11s - 15$

Factor each binomial. (See Lesson 6.4.)

63. $4a^2 - 9$
64. $9b^2 - 25$
65. $c^2 - 49d^2$
66. $16e^2 - 81f^2$
67. $g^3 - 8$
68. $h^3 + 64$
69. $27m^3 - 8n^3$
70. $64p^3 - 27n^3$

7.2 Multiply and Divide Rational Expressions

Because the variables in a rational expression represent real numbers, the operations of addition, subtraction, multiplication, and division with rational expressions are defined in a manner consistent with those of arithmetic fractions.

> **Multiplication of Rational Expressions**
>
> If p, q, r, and s are polynomials, then: $\dfrac{p}{q} \cdot \dfrac{r}{s} = \dfrac{pr}{qs}$ provided $q \neq 0$ and $s \neq 0$.

The above definition states that the product of two rational expressions is the rational expression whose numerator is the product of the numerators of the two rational expressions, and whose denominator is the product of the denominators of the two rational expressions. The product should be reduced if its numerator and denominator contain common factors.

> **To multiply rational expressions:**
> 1. Use the definition of multiplication of rational expressions.
> 2. Factor the numerator and the denominator of the product.
> 3. Reduce, if possible, using the Fundamental Rule of Rational Expressions.

EXAMPLE 1: Find the product of two rational expressions.

Problem ▶ $\dfrac{2y + 6}{3y - 12} \cdot \dfrac{y^2 - 3y - 4}{2y^2 - 4y - 30} = ?$

1. Use definition ▶ $\dfrac{2y + 6}{3y - 12} \cdot \dfrac{y^2 - 3y - 4}{2y^2 - 4y - 30} = \dfrac{(2y + 6)(y^2 - 3y - 4)}{(3y - 12)(2y^2 - 4y - 30)}$

2. Factor ▶ $= \dfrac{2(y + 3)(y + 1)(y - 4)}{3(y - 4)2(y - 5)(y + 3)}$

3. Reduce ▶ $= \dfrac{\cancel{2}\cancel{(y + 3)}(y + 1)\cancel{(y - 4)}}{3\cancel{(y - 4)}\cancel{2}(y - 5)\cancel{(y + 3)}}$

$= \dfrac{y + 1}{3(y - 5)}$

Solution ▶ $\dfrac{2y + 6}{3y - 12} \cdot \dfrac{y^2 - 3y - 4}{2y^2 - 4y - 30} = \dfrac{y + 1}{3(y - 5)}$ or $\dfrac{y + 1}{3y - 15}$; $y \neq -3, 4,$ or 5.

Make Sure

Find the product of two rational expressions.

See Example 1 ▶ 1. $\dfrac{a^2 + 4a + 4}{a^2 - 4} \cdot \dfrac{a^2 - 4a + 4}{a^3 + 8}$ 2. $\dfrac{b^3 - 5b^2 - 6b}{b^2 + b - 6} \cdot \dfrac{b^2 - b - 12}{b^2 - 10b + 24}$

MAKE SURE ANSWERS: 1. $\dfrac{a^2 - 2a + 4}{a - 2}$ 2. $\dfrac{b^2 + b}{b - 2}$

Division of Rational Expressions

If p, q, r, and s are polynomials, then: $\dfrac{p}{q} \div \dfrac{r}{s} = \dfrac{p}{q} \cdot \dfrac{s}{r}$ provided $q \neq 0$, $r \neq 0$, and $s \neq 0$.

The above definition states that the quotient of two rational expressions is the product of the dividend and the reciprocal of the divisor.

To divide rational expressions:
1. Use the definition of division of rational expressions.
2. Multiply using the multiplication procedure for rational expressions.

EXAMPLE 2: Find the quotient of two rational expressions.

Problem ▶ $\dfrac{2x}{x+2y} \div \dfrac{6xy}{3x+6y} = ?$

1. Use definition ▶ $\dfrac{2x}{x+2y} \div \dfrac{6xy}{3x+6y} = \dfrac{2x}{x+2y} \cdot \dfrac{3x+6y}{6xy}$

2. Multiply as before ▶

$= \dfrac{2x(3x+6y)}{(x+2y)6xy}$ Use the definition of multiplication.

$= \dfrac{2x(3)(x+2y)}{(x+2y)2 \cdot 3xy}$ Factor.

$= \dfrac{\cancel{2x}(\cancel{3})(\cancel{x+2y})}{(\cancel{x+2y})\cancel{2} \cdot \cancel{3} \cdot \cancel{x}y}$ Reduce.

$= \dfrac{1}{y}$

Solution ▶ $\dfrac{2x}{x+2y} \div \dfrac{6xy}{3x+6y} = \dfrac{1}{y}$; $x \neq 0$ and $y \neq 0$ or $-\dfrac{x}{2}$.

Note ▶ The restriction $y \neq -\dfrac{x}{2}$ is found by solving $x + 2y = 0$ for y: $x + 2y = 0$

$2y = -x$

$y = -\dfrac{x}{2}$

Make Sure

Find the quotient of two rational expressions.

See Example 2 ▶ 1. $\dfrac{a^2 - 4a}{a^2 + a - 6} \div \dfrac{a^3 - 2a^2 - 8a}{a^2 - 4}$ 2. $\dfrac{\dfrac{b^2 - 10bc + 24c^2}{b^2 - 2bc - 15c^2}}{\dfrac{b^2 - 2bc - 8c^2}{b^2 + 5bc + 6c^2}}$

MAKE SURE ANSWERS: 1. $\dfrac{1}{a+3}$ 2. $\dfrac{b-6c}{b-5c}$

7.2 Multiply and Divide Rational Expressions

If a problem involves more than one operation, it is important to perform the operations in the proper order.

EXAMPLE 3: Perform the indicated operations on the given rational expressions.

Problem ▶ $\dfrac{y^2 - 3y}{y - 5} \div \dfrac{y^2 + 4y}{y^2 - 25} \cdot \dfrac{y + 4}{3 - y} = ?$

1. Use the Order of Operations Rule ▶

$\dfrac{y^2 - 3y}{y - 5} \div \dfrac{y^2 + 4y}{y^2 - 25} \cdot \dfrac{y + 4}{3 - y} = \left[\dfrac{y^2 - 3y}{y - 5} \div \dfrac{y^2 + 4y}{y^2 - 25}\right] \cdot \dfrac{y + 4}{3 - y}$ The operations of × and ÷ are performed from left to right.

$= \left[\dfrac{y^2 - 3y}{y - 5} \cdot \dfrac{y^2 - 25}{y^2 + 4y}\right] \cdot \dfrac{y + 4}{3 - y}$

$= \left[\dfrac{y(y - 3)(y + 5)(y - 5)}{(y - 5)y(y + 4)}\right] \cdot \dfrac{y + 4}{3 - y}$

2. Reduce ▶

$= \dfrac{(y - 3)(y + 5)(y + 4)}{(y + 4)(3 - y)}$

$= -\dfrac{(y - 3)(y + 5)(y + 4)}{(y + 4)(3 - y)}$ Think: $\dfrac{y - 3}{3 - y} = -1$

Solution ▶ $\dfrac{y^2 - 3y}{y - 5} \div \dfrac{y^2 + 4y}{y^2 - 25} \cdot \dfrac{y + 4}{3 - y} = -(y + 5)$ or $-y - 5$

Another Example ▶ $\dfrac{3x^2y + 3xy^2}{4a^2} \div \left(\dfrac{8y}{5a} \div \dfrac{4y}{x + y}\right) = \dfrac{3x^2y + 3xy^2}{4a^2} \div \left(\dfrac{8y}{5a} \cdot \dfrac{x + y}{4y}\right)$ Clear grouping symbols first.

$= \dfrac{3x^2y + 3xy^2}{4a^2} \div \left(\dfrac{2 \cdot 4y(x + y)}{5 \cdot 4ay}\right)$

$= \dfrac{3x^2y + 3xy^2}{4a^2} \div \dfrac{2(x + y)}{5a}$

$= \dfrac{3xy(x + y)}{4a^2} \cdot \dfrac{5a}{2(x + y)}$

$= \dfrac{3xy(x + y)5a}{8aa(x + y)}$

$= \dfrac{15xy}{8a}$

Make Sure

Perform the indicated operations.

See Example 3 ▶ 1. $\dfrac{a + 4}{a - 2} \div \dfrac{a^2 + 2a + 4}{a^2 + a - 6} \cdot \dfrac{a^3 - 8}{a^2 + 7a + 12}$ _____

2. $\dfrac{a^3 + b^3}{a^2 - b^2} \cdot \dfrac{a^2 - 2ab + b^2}{a^3 - b^3} \div \dfrac{a^2 - ab + b^2}{a^2 + ab + b^2}$ _____

MAKE SURE ANSWERS: 1. $a - 2$ 2. 1

262   Chapter 7   Rational Expressions

Since $\dfrac{p}{q} = \dfrac{pr}{qr} = \dfrac{p}{q} \cdot \dfrac{r}{r}$ $(q \neq 0, r \neq 0)$, you may raise any rational expression to higher terms by multiplying its numerator and its denominator by the same nonzero polynomial r.

Examples ▶ (a) $\dfrac{3}{5x} = \dfrac{3}{5x} \cdot \dfrac{2}{2}$

$= \dfrac{6}{10x}$ ⟵ higher terms

(b) $\dfrac{2x - 3}{x + 5} = \dfrac{2x - 3}{x + 5} \cdot \dfrac{3x - 1}{3x - 1}$

$= \dfrac{6x^2 - 11x + 3}{3x^2 + 14x - 5}$ ⟵ higher terms

To rename a rational expression as a rational expression in higher terms with a given denominator, you first compare their denominators in factored form.

EXAMPLE 4: Rename a rational expression in given higher terms.

Problem ▶ $\dfrac{x - 2}{3x - 1} = \dfrac{?}{6x^2 + 13x - 5}$

1. Factor ▶ $\dfrac{x - 2}{3x - 1} = \dfrac{?}{(3x - 1)(2x + 5)}$

2. Compare ▶ $\dfrac{x - 2}{\boxed{3x - 1}} = \dfrac{?}{\boxed{(3x - 1)}(2x + 5)}$

   └ same ┘

   Interpret: The new factor is $(2x + 5)$.

3. Multiply ▶ $\dfrac{x - 2}{3x - 1} = \dfrac{(x - 2)}{(3x - 1)} \cdot \dfrac{(2x + 5)}{(2x + 5)}$

$= \dfrac{2x^2 + x - 10}{6x^2 + 13x - 5}$ ⟵ higher terms

Solution ▶ $\dfrac{x - 2}{3x - 1} = \dfrac{2x^2 + x - 10}{6x^2 + 13x - 5}$

Make Sure

Rename each rational expression in the given higher terms.

See Example 4 ▶ 1. $\dfrac{a + 1}{a - 1} = \dfrac{?}{a^2 - 2a + 1}$

2. $\dfrac{2b - 3}{b^2 - 4b} = \dfrac{?}{b^3 - b^2 - 12b}$

MAKE SURE ANSWERS: 1. $\dfrac{a^2 - 1}{a^2 - 2a + 1}$ 2. $\dfrac{2b^2 + 3b - 9}{b^3 - b^2 - 12b}$

7.2 Practice

Set 1: Multiply rational expressions.

1. $\dfrac{4z - 3}{2 - 3z} \cdot \dfrac{2 - 3z}{4 - 3z}$ 2. $\dfrac{5y - 4}{6 - y} \cdot \dfrac{6 - y}{5 - 4y}$

3. $\dfrac{5x - 3}{3x + 2} \cdot \dfrac{2 - 3x}{3 - 5x}$ 4. $\dfrac{3 + 2w}{5w - 7} \cdot \dfrac{7 - 5w}{2w + 3}$

5. $\dfrac{v^2 - 2v - 3}{v^2 - 3v - 4} \cdot \dfrac{v^2 - 6v + 8}{v^2 - v - 6}$ 6. $\dfrac{u^2 - 2u - 15}{u^2 + 2u - 15} \cdot \dfrac{u^2 + 3u - 18}{u^2 + 9u + 18}$

7. $\dfrac{t^2 - 2t + 4}{t^2 - 5t + 6} \cdot \dfrac{t^2 - 9}{t^3 - 8}$ 8. $\dfrac{3r^2 - 27s^2}{2r^2 - 5rs - 3s^2} \cdot \dfrac{4r^2 - s^2}{6r^2 - 21rs + 9s^2}$

Set 2: Divide rational expressions.

9. $\dfrac{5a - 3}{a^2 - 4} \div \dfrac{3 - 5a}{a - 2}$ 10. $\dfrac{b^2 - 9}{4 - 3b} \div \dfrac{b + 3}{3b - 4}$

11. $\dfrac{e^2 - f^2}{f^2 - e^2} \div \dfrac{2e^2 - ef + 3f^2}{5ef + 3f^2 - 2e^2}$ 12. $\dfrac{4g^2 - 9h^2}{6g^2 - 5gh - 6h^2} \div \dfrac{4g^2 + 12gh + 9h^2}{2g^2 + gh - 3h^2}$

13. $\dfrac{m^2 + 8mn + 16n^2}{m^2 - 16n^2} \div \dfrac{16n^2 - m^2}{m^2 - 8mn + 16n^2}$ 14. $\dfrac{25p^2 - 10pq + q^2}{q^2 - 25p^2} \div \dfrac{25p^2 - q^2}{25p^2 + 10pq + q^2}$

264 Chapter 7 Rational Expressions

Set 3: Perform the indicated operations and simplify.

15. $\dfrac{15c^2 - 20c}{6c^2 + c - 12} \cdot \dfrac{8c + 12}{c^2 - 2c - 15} \div \dfrac{5c^2 + 25c}{c^2 - 25}$

16. $\dfrac{6d^2 + 15d}{3d^2 + 14d + 8} \cdot \dfrac{d^2 - 16}{12d^2 + 42d} \div \dfrac{2d^2 - 3d - 20}{28 - 6d - 4d^2}$

17. $\dfrac{e^3 - 27}{e^2 - 16} \div \dfrac{e^2 + 4e - 21}{4e^2 - 13e - 12} \cdot \dfrac{e^2 + 11e + 28}{4e^2 - 9e - 9}$

18. $\dfrac{f^2 - 25}{f^3 + 64} \div \dfrac{f^2 + 10f + 25}{f^2 - 4f + 16} \cdot \dfrac{f^2 + 9f + 20}{10f - f^2 - 25}$

19. $\dfrac{m^2 - 25}{m^2 - 16} \cdot \left[\dfrac{m^2 + 2m - 35}{m^2 - 9} \div \dfrac{m^2 - 25}{m^2 - m - 12} \right]$

20. $\dfrac{n^2 - 16}{n^2 - 25} \div \left[\dfrac{n^2 - 9}{n^2 + 6n + 9} \cdot \dfrac{n^2 + 7n + 12}{n^2 - 10n + 25} \right]$

Set 4: Rename each rational expression in the given higher terms.

21. $\dfrac{a}{a + 3} = \dfrac{?}{a^2 + 5a + 6}$

22. $\dfrac{-b}{3 - b} = \dfrac{?}{9 - 6b + b^2}$

23. $\dfrac{c + 5}{c - 5} = \dfrac{?}{c^2 - 25}$

24. $\dfrac{d - 3}{d + 3} = \dfrac{?}{d^2 - 9}$

25. $\dfrac{2e + 3}{3e + 2} = \dfrac{?}{9e^2 + 12e + 4}$

26. $\dfrac{3f - 4}{4f - 3} = \dfrac{?}{16f^2 - 24f + 9}$

27. $\dfrac{4g^2 + 9}{2g - 3} = \dfrac{?}{12g - 9 - 4g^2}$

28. $\dfrac{9h^2 - 4}{3h - 4} = \dfrac{?}{24h - 9h^2 - 16}$

Review: Work these problems on a separate sheet of paper. Attach your work to this page.

Simplify. (See Lesson 1.5.)
29. $z^2 + z + z - 1$
30. $y^2 - y + 2y - 1$
31. $x^2 - x - 1 + x$
32. $(1 - w) + (w^2 - w)$

Subtract polynomials. (See Lesson 5.1, Example 2.)
33. $(z^2 + z) - (z - 1)$
34. $(5y + 7) - (3y + 5)$
35. $(3x - 4) - (3 - 2x)$

Reduce each rational expression to lowest terms. (See Lesson 7.1.)

36. $\dfrac{a^2 - b^2}{a - b}$

37. $\dfrac{a^3 + b^3}{a + b}$

38. $\dfrac{c^2 + 4c + 4}{c^2 - 4}$

39. $\dfrac{d^2 - d - 6}{d^2 - 9}$

Copyright © 1985 by Harcourt Brace Jovanovich, Inc. All rights reserved.

7.3 Add and Subtract Rational Expressions

Addition (or subtraction) of rational expressions is accomplished by generalizing the definitions used to add (or subtract) rational numbers.

> **Addition or Subtraction of Rational Expressions Containing Common Denominators**
> If p, q, and r are polynomials, then:
> $$\frac{p}{q} + \frac{r}{q} = \frac{p+r}{q} \quad \text{and} \quad \frac{p}{q} - \frac{r}{q} = \frac{p-r}{q} \quad \text{provided } q \neq 0.$$

The above definition states that to add (or subtract) rational expressions containing common denominators, you add (or subtract) their numerators and place the result over the common denominator. This result may then reduce as shown in the following example.

Example ▸ $\dfrac{3y^2 - 5y + 7}{y^2 + 2y - 15} - \dfrac{2y^2 + 2y - 5}{y^2 + 2y - 15} = ?$

1. Use definition ▸ $\dfrac{3y^2 - 5y + 7}{y^2 + 2y - 15} - \dfrac{2y^2 + 2y - 5}{y^2 + 2y - 15} = \dfrac{(3y^2 - 5y + 7) - (2y^2 + 2y - 5)}{y^2 + 2y - 15}$ ⟵ common denominator

2. Simplify ▸ $= \dfrac{3y^2 - 5y + 7 - 2y^2 - 2y + 5}{y^2 + 2y - 15}$

$= \dfrac{y^2 - 7y + 12}{y^2 + 2y - 15}$ Combine like terms.

3. Reduce ▸ $= \dfrac{\cancel{(y-3)}(y-4)}{\cancel{(y-3)}(y+5)}$

$= \dfrac{y-4}{y+5}$

Solution ▸ $\dfrac{3y^2 - 5y + 7}{y^2 + 2y - 15} - \dfrac{2y^2 + 2y - 5}{y^2 + 2y - 15} = \dfrac{y-4}{y+5};\quad y \neq -5 \text{ or } 3$

Note ▸ It is important to note the use of parentheses in the above subtraction. A common mistake is to write:

$\dfrac{3y^2 - 5y + 7}{y^2 + 2y - 15} - \dfrac{2y^2 + 2y - 5}{y^2 + 2y - 15}$ as $\dfrac{3y^2 - 5y + 7 - 2y^2 \boxed{+} 2y \boxed{-} 5}{y^2 + 2y - 15}$. ⟵ errors

The entire numerator, $2y^2 + 2y - 5$, is to be subtracted. Using parentheses around the numerators to be subtracted prevents this mistake.

To add or subtract rational expressions containing different denominators, you must first express each rational expression as an equivalent rational expression, making sure that they all have a common denominator. As with rational numbers, you could use any common denominator, but it will generally be easier to use the *least common denominator (LCD) of the rational expressions*.

> **To find the LCD of a set of rational expressions:**
> 1. Factor each denominator completely. Express repeated factors in exponential form.
> 2. Identify each factor to the largest power that occurs in any single factorization.
> 3. The LCD is the product of the powers formed in Step 2.

266 Chapter 7 Rational Expressions

Example ▶ Find the LCD of $\dfrac{3w}{4w^2 - 36}$ and $\dfrac{-5w + 2}{2w^2 + 12w + 18}$.

1. Factor ▶
$$4w^2 - 36 = 4(w^2 - 9) \qquad\qquad 2w^2 + 12w + 18 = 2(w^2 + 6w + 9)$$
$$ = 4(w + 3)(w - 3) \qquad\qquad = 2(w + 3)^2$$
$$ = 2^2(w + 3)(w - 3)$$

2. Identify ▶

the largest power of 2

$4w^2 - 36 = \boxed{2^2}(w + 3)\boxed{(w - 3)} \qquad 2w^2 + 12w + 18 = \boxed{2}\boxed{(w + 3)^2}$

the largest power of $w - 3$ ⤴ the largest power of $w + 3$ ⤴

3. Form the product ▶ $2^2(w + 3)^2(w - 3)$ ⟵ the indicated product of the powers identified in Step 2

Solution ▶ The LCD of $\dfrac{3w}{4w^2 - 36}$ and $\dfrac{-5w + 2}{2w^2 + 12w + 18}$ is $2^2(w + 3)^2(w - 3)$.

Note ▶ The powers 2^2 and $(w + 3)^2$ could be expressed as 4 and $w^2 + 6w + 9$, but it is generally better to leave them in their exponential form.

> **To add (subtract) rational expressions containing unlike denominators:**
> 1. Determine the LCD.
> 2. Rename each rational expression as an equivalent rational expression whose denominator is the LCD from Step 1.
> 3. Add (subtract) the rational expressions developed in Step 2, using the procedure for adding (subtracting) rational expressions with a common denominator.
> 4. Reduce, if possible.

EXAMPLE 1: Add rational expressions.

Problem ▶ $\dfrac{3x}{2x + 2} + \dfrac{4}{x - 5} = ?$

1. Determine the LCD ▶
the only 2   the only $(x + 1)$

$2x + 2 = \boxed{2}\boxed{(x + 1)} \qquad\qquad\Longrightarrow$ The LCD is $2(x + 1)(x - 5)$.

$x - 5 = \boxed{(x - 5)}$

⟵ the only $(x - 5)$

2. Rename ▶ $\dfrac{3x}{2x + 2} + \dfrac{4}{x - 5} = \dfrac{3x}{2(x + 1)} \cdot \dfrac{x - 5}{x - 5} + \dfrac{4}{x - 5} \cdot \dfrac{2(x + 1)}{2(x + 1)}$

$$= \dfrac{3x^2 - 15x}{2(x - 5)(x + 1)} + \dfrac{8x + 8}{2(x - 5)(x + 1)} \quad\longleftarrow \text{ like denominators}$$

3. Add as before ▶ $= \dfrac{(3x^2 - 15x) + (8x + 8)}{2(x - 5)(x + 1)}$

4. Reduce, if possible ▶ $= \dfrac{3x^2 - 7x + 8}{2(x - 5)(x + 1)} \qquad$ Think: $3x^2 - 7x + 8$ does not factor using integers.

Solution ▶ $\dfrac{3x}{2x + 2} + \dfrac{4}{x - 5} = \dfrac{3x^2 - 7x + 8}{2(x - 5)(x + 1)}$

7.3 Add and Subtract Rational Expressions

Another Example ▶ $\dfrac{x}{(x+3)^2(2x-1)^2} + \dfrac{x+2}{(x+3)(2x-1)^3} = ?$

the largest power of $(x+3)$

$(x+3)^2(2x-1)^2$

$(x+3)(2x-1)^3$ ⟹ The LCD is $(x+3)^2(2x-1)^3$.

the largest power of $(2x-1)$

$\dfrac{x}{(x+3)^2(2x-1)^2} + \dfrac{x+2}{(x+3)(2x-1)^3} = \dfrac{x}{(x+3)^2(2x-1)^2} \cdot \dfrac{(2x-1)}{(2x-1)} + \dfrac{(x+2)}{(x+3)(2x-1)^3} \cdot \dfrac{(x+3)}{(x+3)}$

$= \dfrac{2x^2 - x}{(x+3)^2(2x-1)^3} + \dfrac{x^2 + 5x + 6}{(x+3)^2(2x-1)^3}$

$= \dfrac{3x^2 + 4x + 6}{(x+3)^2(2x-1)^3}$

Make Sure

Add rational expressions.

See Example 1 ▶   **1.** $\dfrac{a}{a+1} + \dfrac{a^2 - a}{a^3 + 1}$ **2.** $\dfrac{2b - c}{2b^2 + 4bc} + \dfrac{4c - b}{b^2 - 4c^2}$

MAKE SURE ANSWERS: 1. $\dfrac{a^3}{a^3 + 1}$ 2. $\dfrac{3bc + 2c^2}{2b^3 - 8bc^2}$

EXAMPLE 2: Subtract rational expressions.

Problem ▶ $\dfrac{3x - 5}{x^2 + x - 6} - \dfrac{2x - 1}{x^2 + 2x - 8} = ?$

the only $(x+3)$

1. Determine the LCD ▶ $x^2 + x - 6 = (x - 2)(x + 3)$

$x^2 + 2x - 8 = (x - 2)(x + 4)$ ⟹ The LCD is $(x-2)(x+3)(x+4)$.

the only $(x+4)$

the most $(x-2)$s

2. Rename ▶ $\dfrac{3x - 5}{x^2 + x - 6} - \dfrac{2x - 1}{x^2 + 2x - 8} = \dfrac{3x - 5}{(x-2)(x+3)} \cdot \dfrac{x+4}{x+4} - \dfrac{2x - 1}{(x-2)(x+4)} \cdot \dfrac{x+3}{x+3}$

$= \dfrac{3x^2 + 7x - 20}{(x+4)(x-2)(x+3)} - \dfrac{2x^2 + 5x - 3}{(x+4)(x-2)(x+3)}$

3. Subtract ▶ $= \dfrac{(3x^2 + 7x - 20) - (2x^2 + 5x - 3)}{(x+4)(x-2)(x+3)}$

4. Reduce, if possible ▶ $= \dfrac{x^2 + 2x - 17}{(x+4)(x-2)(x+3)}$ Think: There are no common factors.

Solution ▶ $\dfrac{3x - 5}{x^2 + x - 6} - \dfrac{2x - 1}{x^2 + 2x - 8} = \dfrac{x^2 + 2x - 17}{(x+4)(x-2)(x+3)}$

268   Chapter 7   Rational Expressions

Make Sure

Subtract rational expressions.

See Example 2 ▶ 1. $\dfrac{-(5a+1)}{a^2-1} - \dfrac{a-1}{a+1}$ _____   2. $\dfrac{2b-c}{2b^2-4bc} - \dfrac{b-2c}{b^2-4c^2}$ _____

MAKE SURE ANSWERS: 1. $\dfrac{-(a+2)}{a-1}$ 2. $\dfrac{7bc-2c^2}{2b^3-8bc^2}$

If more than one operation is involved, it is important to perform the operations in the proper order.

EXAMPLE 3: Perform the indicated operations on the given rational expressions.

Problem ▶ $\dfrac{z+5}{z+3} + \dfrac{6}{z^2-z-12} \cdot \dfrac{z-4}{z} = ?$

1. Use the Order of Operations Rule ▶ $\dfrac{z+5}{z+3} + \dfrac{6}{z^2-z-12} \cdot \dfrac{z-4}{z} = \dfrac{z+5}{z+3} + \left[\dfrac{6}{(z+3)(z-4)} \cdot \dfrac{z-4}{z}\right]$ Perform the multiplication before the addition.

$= \dfrac{z+5}{z+3} + \dfrac{6(z-4)}{(z+3)(z-4)z}$

$= \dfrac{z+5}{z+3} + \dfrac{6}{(z+3)z}$

$= \dfrac{z+5}{z+3} \cdot \dfrac{z}{z} + \dfrac{6}{(z+3)z}$ Think: The LCD is $z(z+3)$.

$= \dfrac{z^2+5z}{(z+3)z} + \dfrac{6}{(z+3)z}$

2. Reduce, if possible ▶ $= \dfrac{z^2+5z+6}{(z+3)z}$

$= \dfrac{(z+3)(z+2)}{(z+3)z}$

Solution ▶ $\dfrac{z+5}{z+3} + \dfrac{6}{z^2-z-12} \cdot \dfrac{z-4}{z} = \dfrac{z+2}{z}$

Make Sure

Perform the indicated operations.

See Example 3 ▶ 1. $\dfrac{5}{a-b} + \dfrac{4}{a+b} - \dfrac{8a}{a^2-b^2}$ 2. $\dfrac{a-2}{a^2-a-6} - \dfrac{a-2}{a^2+a-2} + \dfrac{1}{a^2-4a+3}$

MAKE SURE ANSWERS: 1. $\dfrac{1}{a-b}$ 2. $\dfrac{3a-2}{a^3-2a^2-5a+6}$

7.3 Practice

Set 1: Add rational expressions.

1. $\dfrac{a^2 + 5a}{2a + 3} + \dfrac{a^2 - 6a}{2a + 3}$
2. $\dfrac{b^2 - 4b}{7 - 3b} + \dfrac{5b - 2b^2}{7 - 3b}$
3. $\dfrac{4c}{2c - 3} + \dfrac{3c}{3 - 2c}$

4. $\dfrac{d}{3d - 5} + \dfrac{5d}{5 - 3d}$
5. $\dfrac{5e}{7e^2} + \dfrac{4}{3e}$
6. $\dfrac{4f}{3f^2} + \dfrac{3}{4f}$

7. $\dfrac{g^2}{g^2 - 9} + \dfrac{g}{3 - g}$
8. $\dfrac{h - 3}{3 - h} + \dfrac{h^2}{h^2 - 9}$
9. $\dfrac{3k - 4}{4k^2 - 9} + \dfrac{3 - 2k}{6k^2 + 5k - 6}$

Set 2: Subtract rational expressions.

10. $\dfrac{5a + 7}{2a^2} - \dfrac{3a + 5}{2a^2}$
11. $\dfrac{2b - 3}{9b^2} - \dfrac{5b + 3}{9b^2}$
12. $\dfrac{4}{3c} - \dfrac{3}{5c}$

13. $\dfrac{5}{4d} - \dfrac{7}{6d}$
14. $\dfrac{2e - 3}{4e - 5} - \dfrac{3e + 3}{5 - 4e}$
15. $\dfrac{4f - 7}{3f - 7} - \dfrac{5 - 4f}{7 - 3f}$

16. $\dfrac{g^2}{g^2 - 4} - \dfrac{g^3 - 8}{g + 2}$
17. $\dfrac{h - 2}{h^2 - 9} - \dfrac{3h + 9 - 2h^2}{h - 3}$
18. $\dfrac{k^2 + k - 6}{k^2 - k - 6} - \dfrac{6k - 12}{k^2 - 4}$

270 Chapter 7 Rational Expressions

Mixed Practice: Perform the indicated operations.

19. $\dfrac{z^2 + z}{z^2 - 1} + \dfrac{z + 1}{1 - z^2}$

20. $\dfrac{y^2 - y}{y^2 - 1} + \dfrac{1 - y}{1 - y^2}$

21. $\dfrac{x^2 - x}{x^2 - 2x + 1} - \dfrac{1 - x}{x^2 - 2x + 1}$

22. $\dfrac{1 - w}{w^2 + 2w + 1} - \dfrac{w^2 - w}{w^2 + 2w + 1}$

23. $\dfrac{1}{v^2 - 9} - \dfrac{v}{v^2 - v - 6}$

24. $\dfrac{2u}{u^2 + 5u + 6} - \dfrac{3u}{u^2 - u - 6}$

25. $\dfrac{2}{t^2 - 4t + 3} + \dfrac{3t}{t^2 + t - 2}$

26. $\dfrac{s - 5}{3 + 2s - s^2} + \dfrac{5 + 3s}{3 + 4s + s^2}$

Set 3: Perform the indicated operations.

27. $2a + 4 \div \dfrac{6a + 12}{5}$

28. $b + 2 \div \dfrac{2b + 4}{3}$

29. $\dfrac{2c - 3}{2c + 3} - \dfrac{2c + 3}{3 - 2c} + \dfrac{24c}{4c^2 - 9}$

30. $\dfrac{4d - 1}{4 - 3d} + \dfrac{3d + 2}{2d + 3} - \dfrac{18 - 7d - d^2}{6d^2 + d - 12}$

31. $\left(1 + \dfrac{1}{e}\right)^2 \div \left[\left(1 + \dfrac{1}{e}\right)\left(1 - \dfrac{1}{e}\right)\right]$

32. $\left(f + 1 + \dfrac{1}{f - 1}\right) \div \left[\left(f + \dfrac{1}{f}\right) \div \left(f - \dfrac{1}{f}\right)\right]$

Review: Work these problems on a separate sheet of paper. Attach your work to this page.

Multiply polynomials. (See Lesson 5.3.)

33. $yz\left(\dfrac{1}{y} + \dfrac{1}{z}\right)$

34. $wx\left(\dfrac{1}{w} - \dfrac{1}{x}\right)$

35. $uv\left(\dfrac{u}{v} - \dfrac{v}{u}\right)$

36. $s^2t^2\left(\dfrac{1}{s^2} - \dfrac{1}{t^2}\right)$

37. $p^2q^2\left(\dfrac{1}{p^2} + \dfrac{1}{q^2}\right)$

38. $m^2n^2\left(\dfrac{m}{n^2} - \dfrac{n}{m^2}\right)$

39. $hk\left(\dfrac{h}{k} - \dfrac{1}{h}\right)$

40. $e^2f^2\left(\dfrac{e^2}{f^2} - \dfrac{1}{e^2}\right)$

41. $c^2d^2\left(\dfrac{c}{d^2} - \dfrac{1}{c^2}\right)$

7.4 Simplify Complex Fractions

> A *complex fraction* is a fraction that has a fraction in its numerator, its denominator, or both.

Examples ▶ (a) $\dfrac{\frac{a}{b}}{\frac{c}{d}}$ (b) $\dfrac{\frac{4}{x} - x}{2x - 1}$ (c) $\dfrac{1 + \frac{a+b}{a-b}}{\frac{2a-b}{a-2b}}$ ⟵ complex fractions

If a complex fraction is written using bar notation, the longest fraction bar is called the *main fraction bar*.

The main fraction bar separates the complex fraction into two parts. The part above the main fraction bar is called the *numerator of the complex fraction*, and the part below the main fraction bar is called the *denominator of the complex fraction*.

Example ▶
$\dfrac{a}{b}$ ⟵ numerator of the complex fraction

— ⟵ main fraction bar

$\dfrac{c}{d}$ ⟵ denominator of the complex fraction

The following method is often used to simplify complex fractions that do not contain another complex fraction in either their numerator or denominator.

Simplify Complex Fractions

Method 1: **The Fundamental Rule Method**

1. Find the LCD of the fractions in the numerator and in the denominator of the complex fraction.

2. Multiply both the numerator and the denominator of the complex fraction by the LCD from Step 1.

3. Simplify, if possible.

EXAMPLE 1: Simplify a complex fraction using the Fundamental Rule Method.

Problem ▶ Simplify $\dfrac{\frac{a}{b} - \frac{b}{a}}{\frac{1}{a} + \frac{1}{b}}$.

1. Find LCD of the fractions ▶ The LCD of $\dfrac{a}{b}$, $-\dfrac{b}{a}$, $\dfrac{1}{a}$, and $\dfrac{1}{b}$ is ab.

2. Multiply ▶ $\dfrac{\frac{a}{b} - \frac{b}{a}}{\frac{1}{a} + \frac{1}{b}} = \dfrac{\frac{a}{b} - \frac{b}{a}}{\frac{1}{a} + \frac{1}{b}} \cdot \dfrac{ab}{ab}$ Use the Fundamental Rule.

Chapter 7 Rational Expressions

3. Simplify $\quad = \dfrac{\dfrac{a}{b}\cdot ab - \dfrac{b}{a}\cdot ab}{\dfrac{1}{a}\cdot ab + \dfrac{1}{b}\cdot ab}\quad$ Distribute the LCD in the numerator and the denominator.

$$= \dfrac{\dfrac{a}{\cancel{b}}\cdot a\cancel{b} - \dfrac{b}{\cancel{a}}\cdot \cancel{a}b}{\dfrac{1}{\cancel{a}}\cdot \cancel{a}b + \dfrac{1}{\cancel{b}}\cdot a\cancel{b}}$$

$$= \dfrac{a^2 - b^2}{b + a}$$

$$= \dfrac{(a+b)(a-b)}{a+b}$$

$$= \dfrac{\cancel{(a+b)}(a-b)}{\cancel{(a+b)}}\qquad \text{Reduce when possible.}$$

$$= a - b$$

Solution $\quad \dfrac{\dfrac{a}{b} - \dfrac{b}{a}}{\dfrac{1}{a} + \dfrac{1}{b}} = a - b;\quad a \ne 0,\ b \ne 0,\ \text{and}\ a \ne -b$

Other Examples (a) $\dfrac{1}{\dfrac{1}{r}+\dfrac{1}{s}} = \dfrac{1}{\dfrac{1}{r}+\dfrac{1}{s}}\cdot\dfrac{rs}{rs}$

$$= \dfrac{1\cdot rs}{\dfrac{1}{r}\cdot rs + \dfrac{1}{s}\cdot rs}$$

$$= \dfrac{rs}{s+r}$$

(b) $\dfrac{\dfrac{x}{y}+3}{\dfrac{1}{y}} = \dfrac{\dfrac{x}{y}+3}{\dfrac{1}{y}}\cdot\dfrac{y}{y}$

$$= \dfrac{\dfrac{x}{\cancel{y}}\cdot\cancel{y} + 3y}{\dfrac{1}{\cancel{y}}\cdot\cancel{y}}$$

$$= \dfrac{x+3y}{1}$$

$$= x + 3y$$

Make Sure

Simplify each complex fraction using the Fundamental Rule Method.

See Example 1 1. $\dfrac{\dfrac{a}{b}-\dfrac{b}{a}}{\dfrac{a}{b}+\dfrac{b}{a}}$ _____ 2. $\dfrac{\dfrac{c}{d^2}+\dfrac{d}{c^2}}{\dfrac{d^2}{c}+\dfrac{c^2}{d}}$ _____

MAKE SURE ANSWERS: 1. $\dfrac{a^2-b^2}{a^2+b^2}$ 2. $\dfrac{1}{cd}$

7.4 Simplify Complex Fractions

> **Simplify Complex Fractions**
>
> Method 2: **The Division Method**
>
> 1. Simplify both the numerator and the denominator of the complex fraction to single fractions.
> 2. Divide the single fraction in the numerator by the single fraction in the denominator.
> 3. Reduce, if possible.

EXAMPLE 2: Simplify a complex fraction by the Division Method.

Problem ▶ Simplify $\dfrac{1 + \dfrac{1}{y}}{1 - \dfrac{1}{y^2}}$.

1. Simplify numerator and denominator ▶

$$\dfrac{1 + \dfrac{1}{y}}{1 - \dfrac{1}{y^2}} = \dfrac{\dfrac{1}{1} + \dfrac{1}{y}}{\dfrac{1}{1} - \dfrac{1}{y^2}}$$

$$= \dfrac{\dfrac{1}{1} \cdot \dfrac{y}{y} + \dfrac{1}{y}}{\dfrac{1}{1} \cdot \dfrac{y^2}{y^2} - \dfrac{1}{y^2}} \qquad \text{The LCD of } \dfrac{1}{1} \text{ and } \dfrac{1}{y} \text{ is } y.$$

$$\qquad\qquad\qquad\qquad\qquad \text{The LCD of } \dfrac{1}{1} \text{ and } -\dfrac{1}{y^2} \text{ is } y^2.$$

$$= \dfrac{\boxed{\dfrac{y + 1}{y}}}{\boxed{\dfrac{y^2 - 1}{y^2}}} \begin{array}{l} \leftarrow \text{ a single fraction in the numerator} \\[1em] \leftarrow \text{ a single fraction in the denominator} \end{array}$$

2. Divide ▶

$$= \dfrac{y + 1}{y} \div \dfrac{y^2 - 1}{y^2} \qquad \text{Think: The main fraction bar indicates a division of the numerator by the denominator of the complex fraction.}$$

$$= \dfrac{y + 1}{y} \cdot \dfrac{y^2}{y^2 - 1} \qquad \text{Multiply by the reciprocal of the divisor.}$$

$$= \dfrac{(y + 1) \cdot y^2}{y(y^2 - 1)}$$

$$= \dfrac{(y + 1)yy}{y(y + 1)(y - 1)} \qquad \text{Factor.}$$

$$= \dfrac{\cancel{(y + 1)}\cancel{y}y}{\cancel{y}\cancel{(y + 1)}(y - 1)}$$

$$= \dfrac{y}{y - 1}$$

Solution ▶ $\dfrac{1 + \dfrac{1}{y}}{1 - \dfrac{1}{y^2}} = \dfrac{y}{y - 1}; \qquad y \neq 0, -1, \text{ or } 1$

274 Chapter 7 Rational Expressions

Other Examples ▶ (a) $\dfrac{\frac{1}{2}}{\frac{3}{5}} = \dfrac{1}{2} \div \dfrac{3}{5}$ (b) $\dfrac{\frac{x}{x+y}}{\frac{x^2}{x^2-y^2}} = \dfrac{x}{x+y} \div \dfrac{x^2}{x^2-y^2}$

$\qquad\qquad\qquad = \dfrac{1}{2} \cdot \dfrac{5}{3}$

$\qquad\qquad\qquad = \dfrac{5}{6}$

$\qquad\qquad\qquad\qquad\qquad\qquad = \dfrac{x}{x+y} \cdot \dfrac{x^2-y^2}{x^2}$

$\qquad\qquad\qquad\qquad\qquad\qquad = \dfrac{x(x^2-y^2)}{(x+y)x^2}$

$\qquad\qquad\qquad\qquad\qquad\qquad = \dfrac{x(x+y)(x-y)}{(x+y)xx}$

$\qquad\qquad\qquad\qquad\qquad\qquad = \dfrac{x\cancel{(x+y)}(x-y)}{\cancel{(x+y)}xx}$

$\qquad\qquad\qquad\qquad\qquad\qquad = \dfrac{x-y}{x}$

(c) $\dfrac{\dfrac{1}{m} + \dfrac{2}{n}}{2m+n} = \dfrac{\dfrac{1}{m}\cdot\dfrac{n}{n} + \dfrac{2}{n}\cdot\dfrac{m}{m}}{2m+n}$ (d) $\dfrac{\dfrac{a}{b} - \dfrac{b}{a}}{\dfrac{1}{a} + \dfrac{1}{b}} = \dfrac{\dfrac{a}{b}\cdot\dfrac{a}{a} - \dfrac{b}{a}\cdot\dfrac{b}{b}}{\dfrac{1}{a}\cdot\dfrac{b}{b} + \dfrac{1}{b}\cdot\dfrac{a}{a}}$

$\qquad\quad = \dfrac{\dfrac{n+2m}{mn}}{2m+n}$ $\qquad\qquad\qquad\qquad = \dfrac{\dfrac{a^2-b^2}{ab}}{\dfrac{b+a}{ab}}$

$\qquad\quad = \dfrac{n+2m}{mn} \div (2m+n)$ $\qquad\qquad = \dfrac{a^2-b^2}{ab} \div \dfrac{b+a}{ab}$

$\qquad\quad = \dfrac{n+2m}{mn} \cdot \dfrac{1}{2m+n}$ $\qquad\qquad = \dfrac{(a+b)(a-b)}{ab} \cdot \dfrac{ab}{b+a}$

$\qquad\quad = \dfrac{\cancel{2m+n}}{mn} \cdot \dfrac{1}{\cancel{2m+n}}$ $\qquad\qquad = \dfrac{\cancel{(a+b)}(a-b)}{\cancel{ab}} \cdot \dfrac{\cancel{ab}}{\cancel{b+a}}$

$\qquad\quad = \dfrac{1}{mn}$ $\qquad\qquad\qquad\qquad\quad = a - b$

Note ▶ You should compare the division method used in Example (d) with the Fundamental Rule method used in Example 1.

Make Sure

Simplify each complex fraction using the Division Method.

See Example 2 ▶ 1. $\dfrac{\dfrac{b}{a} - \dfrac{a}{b}}{\dfrac{1}{a} + \dfrac{1}{b}}$ 2. $\dfrac{\dfrac{c}{d^2} - \dfrac{d}{c^2}}{\dfrac{d^2}{c} - \dfrac{c^2}{d}}$

MAKE SURE ANSWERS: 1. $b - a$ 2. $-\dfrac{1}{cd}$

7.4 Simplify Complex Fractions 275

The Division Method is often used to simplify a complex fraction containing another complex fraction.

EXAMPLE 3: Simplify a complex fraction containing another complex fraction.

Problem ▶ Simplify $\dfrac{1 - \dfrac{1}{y}}{1 + \dfrac{1}{1 - \dfrac{1}{y}}}$.

1. Simplify numerator and denominator ▶
$$\dfrac{1 - \dfrac{1}{y}}{1 + \dfrac{1}{1 - \dfrac{1}{y}}} = \dfrac{\dfrac{y}{y} - \dfrac{1}{y}}{1 + \dfrac{1}{\dfrac{y}{y} - \dfrac{1}{y}}}$$

$$= \dfrac{\dfrac{y-1}{y}}{1 + \dfrac{1}{\dfrac{y-1}{y}}}$$

$$= \dfrac{\dfrac{y-1}{y}}{1 + \dfrac{y}{y-1}} \qquad \text{Think: } \dfrac{1}{\dfrac{y-1}{y}} \text{ is the reciprocal of } \dfrac{y-1}{y}.$$

$$= \dfrac{\dfrac{y-1}{y}}{\dfrac{y-1}{y-1} + \dfrac{y}{y-1}}$$

$$= \dfrac{\dfrac{y-1}{y}}{\dfrac{y-1+y}{y-1}}$$

$$= \dfrac{\dfrac{y-1}{y}}{\dfrac{2y-1}{y-1}}$$

2. Divide ▶
$$= \dfrac{y-1}{y} \div \dfrac{2y-1}{y-1}$$

$$= \dfrac{y-1}{y} \cdot \dfrac{y-1}{2y-1}$$

$$= \dfrac{(y-1)^2}{y(2y-1)}$$

Solution ▶ $\dfrac{1 - \dfrac{1}{y}}{1 + \dfrac{1}{1 - \dfrac{1}{y}}} = \dfrac{(y-1)^2}{y(2y-1)}$

Chapter 7 Rational Expressions

Other Examples ▶ (a) $\dfrac{x}{2+\dfrac{1}{1+\frac{1}{3}}} = \dfrac{x}{2+\dfrac{1}{\frac{4}{3}}}$

$= \dfrac{x}{2+\frac{3}{4}}$

$= \dfrac{x}{2\frac{3}{4}}$

$= \dfrac{x}{\frac{11}{4}}$

$= x \div \dfrac{11}{4}$

$= x \cdot \dfrac{4}{11}$

$= \dfrac{4x}{11}$

(b) $1 - \dfrac{t}{1 - \dfrac{1}{1 - \dfrac{1}{1-t}}} = 1 - \dfrac{t}{1 - \dfrac{1}{\dfrac{1-t}{1-t} - \dfrac{1}{1-t}}}$

$= 1 - \dfrac{t}{1 - \dfrac{1}{\dfrac{1-t-1}{1-t}}}$

$= 1 - \dfrac{t}{1 - \dfrac{1}{\dfrac{-t}{1-t}}}$

$= 1 - \dfrac{t}{1 - \dfrac{1-t}{-t}}$

$= 1 - \dfrac{t}{1 + \dfrac{1-t}{t}}$

$= 1 - \dfrac{t}{\dfrac{t}{t} + \dfrac{1-t}{t}}$

$= 1 - \dfrac{t}{\dfrac{t+1-t}{t}}$

$= 1 - \dfrac{t}{\dfrac{1}{t}}$

$= 1 - t^2$

Make Sure

Simplify complex fractions containing other complex fractions.

See Example 3 ▶ **1.** $\dfrac{1 + \dfrac{1}{a}}{1 + \dfrac{1}{1 + \dfrac{1}{a}}}$ **2.** $\dfrac{\dfrac{b}{1 - \dfrac{1}{b}} - b}{\dfrac{b}{1 - \dfrac{1}{1 - \dfrac{1}{b}}} - 1}$

MAKE SURE ANSWERS: **1.** $\dfrac{a^2 + 2a + 1}{2a^2 + a}$ **2.** $\dfrac{b^3 - 2b^2 + 2b - 1}{-b}$

7.4 Practice

Set 1: Simplify each complex fraction using the Fundamental Rule Method.

1. $\dfrac{1 + \dfrac{1}{a}}{1 - \dfrac{1}{a}}$

2. $\dfrac{\dfrac{1}{b} - 1}{\dfrac{1}{b} + 1}$

3. $\dfrac{\dfrac{1}{c} + \dfrac{1}{c^2}}{\dfrac{1}{c} - \dfrac{1}{c^2}}$

4. $\dfrac{\dfrac{1}{f^2} - \dfrac{1}{f}}{\dfrac{1}{f} - \dfrac{1}{f^2}}$

5. $\dfrac{\dfrac{q}{p} - \dfrac{q}{p+1}}{\dfrac{q}{p} - \dfrac{q}{p-1}}$

6. $\dfrac{\dfrac{r}{s+1} + \dfrac{r}{s}}{\dfrac{r}{s-1} - \dfrac{r}{s}}$

Set 2: Simplify each complex fraction using the Division Method.

7. $\dfrac{\dfrac{1}{x} - x}{x + \dfrac{1}{x}}$

8. $\dfrac{w - \dfrac{1}{w}}{\dfrac{1}{w} + w}$

9. $\dfrac{\dfrac{1}{v^2} + \dfrac{1}{v}}{\dfrac{1}{v} - \dfrac{1}{v^2}}$

10. $\dfrac{\dfrac{1}{u} - \dfrac{1}{u^2}}{\dfrac{1}{u^3} - \dfrac{1}{u}}$

11. $\dfrac{\dfrac{4}{2f - 3} + f - 5}{5 - f - \dfrac{4}{2f - 3}}$

12. $\dfrac{\dfrac{12 - 11g}{4g - 3} - 3g + 2}{2g - 3 + \dfrac{4g^2 + 3}{4g + 3}}$

Mixed Practice: Simplify each complex fraction.

13. $\dfrac{\dfrac{6z+3}{z^2-1}}{\dfrac{2z+1}{z^2-2z+1}}$

14. $\dfrac{\dfrac{y^2+2y+1}{y^2+4y+3}}{\dfrac{y^2-1}{y^2+2y-3}}$

15. $\dfrac{\dfrac{2x^2-18}{x^2-2x-15}}{\dfrac{x^2-6x+9}{x^2+x-12}}$

16. $\dfrac{\dfrac{w+2}{w^2+7w+10}}{\dfrac{w^2+2w-3}{w^2+8w+15}}$

17. $\dfrac{\dfrac{m^2+6mn-7n^2}{2m^2+mn-n^2}}{\dfrac{m^2-n^2}{4m^2-n^2}}$

18. $\dfrac{\dfrac{g^2-9h^2}{g^2-h^2}}{\dfrac{g^2-3gh-18h^2}{g^2-5gh-6h^2}}$

Set 3: Simplify complex fractions containing other complex fractions.

19. $\dfrac{c - \dfrac{1}{c^2}}{c - \dfrac{1}{c - \dfrac{1}{1-\dfrac{1}{c}}}}$

20. $\dfrac{\dfrac{1}{d^2} - d}{d + \dfrac{1}{\dfrac{1}{1-\dfrac{1}{d}} - d}}$

21. $\dfrac{g+1+\dfrac{1-g}{g-1}}{g-1+\dfrac{1}{1+\dfrac{1}{g-1}}}$

22. $\dfrac{\dfrac{2-h}{h-2}+1+h}{h-1+\dfrac{1}{1+\dfrac{2}{h-2}}}$

23. $\dfrac{k+1-\dfrac{1}{1-\dfrac{1}{k}}}{\dfrac{1}{1-\dfrac{1}{1-\dfrac{1}{k}}} - \dfrac{1}{1-\dfrac{1}{k}}}$

24. $\dfrac{\dfrac{1}{1+\dfrac{1}{m}}+m-1}{\dfrac{1}{1-\dfrac{1}{1+\dfrac{1}{m}}} - \dfrac{1}{1+\dfrac{1}{m}}}$

Review: Work these problems on a separate sheet of paper. Attach your work to this page.

Solve fractional equations. (See Lesson 2.3.)

25. $\dfrac{z}{3} + \dfrac{z}{2} = \dfrac{1}{6}$

26. $\dfrac{y}{4} - \dfrac{y}{3} = \dfrac{1}{6}$

27. $\dfrac{w-4}{6} = \dfrac{3-w}{5}$

28. $2 = \dfrac{5}{6} + \dfrac{3u}{5}$

29. $\dfrac{t-3}{5} = \dfrac{t+4}{6}$

30. $\dfrac{2s-3}{4} = \dfrac{s+1}{6}$

Find the excluded value(s) of each rational expression. (See Lesson 7.1.)

31. $\dfrac{3}{z}$

32. $\dfrac{b}{b-2}$

33. $\dfrac{2e+3}{e^2-5e-6}$

34. $\dfrac{g^2-3g}{g^3-g^2-6g}$

7.5 Solve Equations Containing Rational Expressions

Recall ▶ In Lesson 2.3, you solved fractional equations containing constants in the denominators. You cleared the equation of fractions by multiplying both members of the equation by the LCD of all the denominators.

Example ▶ Solve $\dfrac{7y}{3} - 5 = \dfrac{3y}{2}$.

1. Find the LCD ▶ The LCD of $\dfrac{7y}{3}, -\dfrac{5}{1},$ and $\dfrac{3y}{2}$ is 6.

2. Clear fractions ▶ $6\left[\dfrac{7y}{3} - 5\right] = 6 \cdot \dfrac{3y}{2}$ Multiply both members by the LCD.

$6 \cdot \dfrac{7y}{3} - 6(5) = 6 \cdot \dfrac{3y}{2}$ Distributive rule.

$14y - 30 = 9y$ ⟵ an equation clear of fractions

3. Solve as before ▶ $5y = 30$

$y = 6$

4. Check ▶ $\dfrac{7y}{3} - 5 = \dfrac{3y}{2}$ ⟵ original equation

$\dfrac{7(6)}{3} - 5 \;\Big|\; \dfrac{3(6)}{2}$

$14 - 5 \;\Big|\; 9$

$9 \;\Big|\; 9$ ⟵ 6 checks

Solution ▶ $y = 6$

Recall ▶ Every numerical replacement of the variable(s) that causes the denominator of a rational expression to equal zero is called an excluded value of the rational expression.

Equations that contain rational expressions may have excluded values.

Example ▶ Find the excluded value(s) of $\dfrac{3x}{x^2 + 3x} + \dfrac{2}{x - 1} = \dfrac{5x}{x^2 + 5x + 6}$.

     first denominator   second denominator   third denominator

1. Set each denominator equal to 0. ▶ $x^2 + 3x = 0$ | $x - 1 = 0$ | $x^2 + 5x + 6 = 0$

2. Solve each equation ▶ $x(x + 3) = 0$ | $x = 1$ | $(x + 2)(x + 3) = 0$

$x = 0$ or $x + 3 = 0$ | $x + 2 = 0$ or $x + 3 = 0$

$x = 0$ or $x = -3$ | $x = -2$ or $x = -3$

Solution ▶ The excluded values of $\dfrac{3x}{x^2 + 3x} + \dfrac{2}{x - 1} = \dfrac{5x}{x^2 + 5x + 6}$ are $-3, -2, 0,$ and 1.

Chapter 7 Rational Expressions

If an equation contains rational expressions, you may solve the equation using the method of clearing fractions, provided that you restrict the variable(s) from any excluded value(s).

> **To solve an equation containing rational expressions:**
> 1. Determine the excluded value(s).
> 2. Multiply both members of the equation by the LCD of all the denominators.
> 3. Solve the resulting equation. The solution(s) of this equation are called proposed solution(s).
> 4. Compare the proposed solution(s) with the excluded value(s). Reject any proposed solution that is also an excluded value, and check any proposed solution that is *not* an excluded value in the original equation.

EXAMPLE 1: Solve an equation containing rational expressions that has one proposed solution.

Solve ▶ $\dfrac{4}{n} + \dfrac{2}{n-1} = \dfrac{5n-2}{n(n-1)}$

1. Determine excluded values(s) ▶

$n = 0$	$n - 1 = 0$	$n(n-1) = 0$	Set each denominator equal to 0.
	$n = 1$	$n = 0$ or $n - 1 = 0$	Solve each equation.
		$n = 0$ or $\quad n = 1$	

The excluded values are 0 and 1.

2. Clear fractions ▶

$n(n-1)\left[\dfrac{4}{n} + \dfrac{2}{n-1}\right] = n(n-1)\dfrac{5n-2}{n(n-1)}$ Multiply both members by the LCD.

$n(n-1)\dfrac{4}{n} + n(n-1)\dfrac{2}{n-1} = n(n-1)\dfrac{(5n-2)}{n(n-1)}$ The Distributive Rule.

$\cancel{n}(n-1)\dfrac{4}{\cancel{n}} + n\cancel{(n-1)}\dfrac{2}{\cancel{(n-1)}} = \cancel{n(n-1)}\dfrac{(5n-2)}{\cancel{n(n-1)}}$ Simplify.

$4n - 4 + 2n = 5n - 2$ ⟵ an equation clear of fractions

3. Solve as before ▶ $6n - 4 = 5n - 2$

$n = 2$ ⟵ proposed solution

4. Check ▶ The proposed solution 2 is not equal to either of the excluded values 0, 1.

$\dfrac{4}{n} + \dfrac{2}{n-1} = \dfrac{5n-2}{n(n-1)}$ ⟵ original equation

$\dfrac{4}{2} + \dfrac{2}{2-1}$	$\dfrac{5(2)-2}{2((2)-1)}$
$2 + 2$	$\dfrac{8}{2}$
4	4 ⟵ 2 checks

Solution ▶ $n = 2$

> **CAUTION:** If a proposed solution is also an excluded value, then it is not a solution.

7.5 Solve Equations Containing Rational Expressions

Example ▶ Solve $\dfrac{5x}{x-3} - 2 = \dfrac{2x+9}{x-3}$.

1. Determine excluded value(s) ▶ $x - 3 = 0$ means $x = 3$ is the only excluded value.

2. Clear fractions ▶ $(x-3)\left[\dfrac{5x}{x-3} - 2\right] = (x-3)\dfrac{(2x+9)}{(x-3)}$ Provided $x \neq 3$.

$$\cancel{(x-3)}\dfrac{5x}{\cancel{(x-3)}} - (x-3)2 = \cancel{(x-3)}\dfrac{(2x+9)}{\cancel{(x-3)}}$$

$$5x - 2x + 6 = 2x + 9$$

3. Solve as before ▶ $3x + 6 = 2x + 9$

$x = 3$ ⟵ proposed solution

4. Check ▶ $x = 3$ is not a solution of the original equation because 3 is an excluded value.

Solution ▶ There is no value of x such that: $\dfrac{5x}{x-3} - 2 = \dfrac{2x+9}{x-3}$. The solution set is the empty set.

Make Sure

Solve equations containing rational expressions that have one proposed solution. Check each solution for excluded values.

See Example 1 ▶ 1. $\dfrac{4}{a-3} - \dfrac{3}{a-4} = 0$ 2. $\dfrac{b}{2b+1} + \dfrac{b}{4b^2+8b+3} = \dfrac{b+3}{2b+3}$

MAKE SURE ANSWERS: 1. 7   2. −1

An equation containing rational expressions may have more than one proposed solution.

EXAMPLE 2: Solve an equation containing rational expressions that has more than one proposed solution.

Solve ▶ $\dfrac{5}{x-4} + 1 = \dfrac{20}{x(x-4)}$

1. Determine excluded value(s) ▶ $x - 4 = 0$ | $x(x-4) = 0$

$x = 4$ | $x = 0$ or $x - 4 = 0$

$x = 0$ or $x = 4$ Think: 0 and 4 are the excluded values.

2. Clear fractions ▶ $x(x-4)\left[\dfrac{5}{x-4} + 1\right] = x(x-4)\dfrac{20}{x(x-4)}$ (provided $x \neq 0$, $x \neq 4$)

$\cancel{x(x-4)}\dfrac{5}{\cancel{(x-4)}} + x(x-4)1 = \cancel{x(x-4)}\dfrac{20}{\cancel{x(x-4)}}$

282   Chapter 7   Rational Expressions

3. Solve as before ▶

$$5x + x^2 - 4x = 20$$
$$x^2 + x - 20 = 0 \longleftarrow \text{standard form}$$
$$(x - 4)(x + 5) = 0 \qquad \text{Factor.}$$
$$x - 4 = 0 \text{ or } x + 5 = 0 \qquad \text{Zero-product property.}$$
$$x = 4 \text{ or } \quad x = -5$$

proposed solutions

4. Check ▶ $x = 4$ is not a solution of the original equation because 4 is an excluded value.
$x = -5$ is not an excluded value, and it checks in the original equation.

Solution ▶ $x = 5$

Another Example ▶   Solve $\dfrac{x + 1}{x + 2} = \dfrac{x^2 - 2x - 3}{x^2 - x - 6}$.

1. Determine excluded value(s) ▶

$$\begin{array}{l|l} x + 2 = 0 & x^2 - x - 6 = 0 \\ x = -2 & (x + 2)(x - 3) = 0 \\ & x + 2 = 0 \text{ or } x - 3 = 0 \\ & x = -2 \text{ or } \quad x = 3 \end{array}$$

Think: -2 and 3 are the excluded values.

2. Clear fractions ▶

$$(x + 2)(x - 3) \cdot \dfrac{x + 1}{x + 2} = (x + 2)(x - 3) \cdot \dfrac{x^2 - 2x - 3}{(x + 2)(x - 3)}$$

$$\dfrac{\cancel{(x + 2)}(x - 3)x + 1}{\cancel{x + 2}} = \dfrac{\cancel{(x + 2)}\cancel{(x - 3)}(x^2 - 2x - 3)}{\cancel{(x + 2)}\cancel{(x - 3)}}$$

$$(x - 3)(x + 1) = x^2 - 2x - 3$$
$$x^2 - 2x - 3 = x^2 - 2x - 3 \qquad \text{Think: This is an identity.}$$

Solution ▶   The solution set is the set of all real numbers except for the excluded values -2 and 3.

Make Sure

Solve equations containing rational expressions that have more than one proposed solution. Check each solution for excluded values.

See Example 2 ▶   **1.** $\dfrac{2}{3a} - \dfrac{1}{2a} = \dfrac{3}{4}$ _____   **2.** $\dfrac{7}{2b - 3} + \dfrac{1}{2b - 2} = \dfrac{3}{b - 1}$ _____

MAKE SURE ANSWERS: 1. $\dfrac{2}{9}$ 2. 1, $-\dfrac{1}{4}$

7.5 Practice: *Check for excluded values.*

Set 1: Solve equations containing rational expressions that have one proposed solution.

1. $\dfrac{4}{a-3} = \dfrac{3}{a+2}$
2. $\dfrac{2}{b+4} = \dfrac{1}{b-3}$

3. $\dfrac{c}{c-2} = \dfrac{c}{c+4}$
4. $\dfrac{d+1}{d-3} = \dfrac{d-2}{d+4}$

5. $\dfrac{1}{e^2} = \dfrac{3}{e} - \dfrac{4}{e}$
6. $\dfrac{4}{f} - \dfrac{5}{f} = \dfrac{6}{f^2}$

7. $\dfrac{3}{2g-3} = \dfrac{4}{3g-2}$
8. $\dfrac{5}{3h-4} = \dfrac{2}{4h-3}$

Set 2: Solve equations containing rational expressions and having more than one proposed solution.

9. $\dfrac{z^2+6}{z^2-3z} = \dfrac{5}{z-3} - \dfrac{2}{z}$
10. $\dfrac{y^2-8}{y^2+4y} = \dfrac{-2y}{y+4} - \dfrac{2}{y}$

11. $\dfrac{x^2-x-10}{x^2-4} = \dfrac{3}{x+2} - \dfrac{2}{x-2}$
12. $\dfrac{w^2-2w+15}{w^2-9} = \dfrac{3}{w-3} - \dfrac{2}{w+3}$

13. $\dfrac{2}{v} + \dfrac{3}{2v} = \dfrac{21}{6v}$
14. $\dfrac{6}{4u} - \dfrac{2}{3u} = \dfrac{5}{6u}$

15. $\dfrac{4}{t^2} - \dfrac{1}{t} = 3$
16. $4 = \dfrac{4}{s} - \dfrac{1}{s^2}$

Mixed Practice: Solve equations containing rational expressions.

17. $\dfrac{5}{a-3} + \dfrac{3}{a+4} = \dfrac{3a}{a^2+a-12}$

18. $\dfrac{2b}{2b-5} - \dfrac{3b}{4b+3} = \dfrac{13}{8b^2-14b-15}$

19. $\dfrac{3c}{6c-1} - \dfrac{2c}{4c-3} = \dfrac{4}{24c^2-22c+3}$

20. $\dfrac{3d-1}{2d+3} - \dfrac{d^2+11}{6d^2+5d-6} = \dfrac{4d-2}{3d-2}$

21. $\dfrac{5e}{7-e} - \dfrac{5e}{e+7} = 0$

22. $\dfrac{-3}{f-7} - \dfrac{3}{7-f} = 0$

23. $g = \dfrac{24}{g} + 5$

24. $h = 7 - \dfrac{12}{h}$

25. $\dfrac{k}{k+3} + \dfrac{3k}{2k-3} = \dfrac{-1}{2k^2+3k-9}$

26. $\dfrac{3m}{2m+3} - \dfrac{m}{m-3} = \dfrac{28}{2m^2-3m-9}$

27. $\dfrac{3}{n+3} - \dfrac{3}{n-3} = 0$

28. $\dfrac{-3}{p+4} - \dfrac{3}{4-p} = 0$

29. $\dfrac{-16}{3q^2-13q+12} = \dfrac{2q}{3-q} + \dfrac{2-5q}{4-3q}$

30. $\dfrac{8}{8-10r-3r^2} = \dfrac{r+5}{4+r} + \dfrac{4r}{2-3r}$

Review: Work these problems on a separate sheet of paper. Attach your work to this page.

Simplify each product of terms. (See Lesson 1.6, Example 2.)

31. $z^4 z^3$
32. $y^8 y^3$
33. $t^3 t^{-2}$
34. $s^{-4} s^{-3}$
35. $p^{-4} p$
36. $m m^{-3}$
37. $k^3 \cdot \dfrac{1}{k^2}$
38. $\dfrac{1}{g^3} \cdot g^4$

Simplify each term raised to a power. (See Lesson 1.6, Example 3.)

39. $(a^3)^2$
40. $(b^4)^3$
41. $(c^{-3})^2$
42. $(d^{-2})^4$
43. $(e^3)^{-1}$
44. $(f^4)^{-2}$
45. $(g^{-2})^{-1}$
46. $(h^{-1})^{-3}$

Problem Solving 8: Solve Problems Using Rational Expressions

You can solve many types of *work problems* by using rational expressions.

EXAMPLE 1: Solve this work problem using rational expressions.

1. **Read and identify** ▶ A large pipe can fill a tank in 6 hours. A smaller pipe can fill the same tank in 15 hours. A third pipe can empty the same tank in 10 hours. How long will it take to fill the tank if all three pipes are left open?

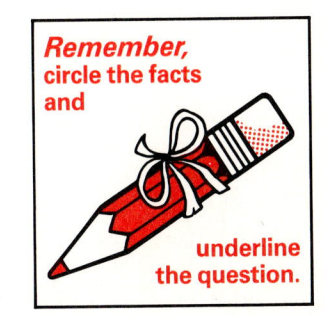

   Remember, circle the facts and underline the question.

2. **Understand** ▶ The unknown is **the number of hours it takes to fill the empty pool with all 3 pipes open.**

3. **Decide** ▶ Let x = the number of hours to fill the empty pool with all 3 pipes open.

4. **Make a table** ▶

	unit rate (per hour)	time (in hours)	work (unit rate × time)
1st inlet pipe	$\frac{1}{6}$	x	$\frac{1}{6}x$ or $\frac{x}{6}$
2nd inlet pipe	$\frac{1}{15}$	x	$\frac{1}{15}x$ or $\frac{x}{15}$
outlet pipe	$\frac{1}{10}$	x	$\frac{1}{10}x$ or $\frac{x}{10}$

5. **Translate** ▶ The water from the inlet pipes minus the water from the outlet pipe equals the total work.

$$\frac{x}{6} + \frac{x}{15} - \frac{x}{10} = 1$$

6. **Solve as before** ▶
$$30 \cdot \frac{x}{6} + 30 \cdot \frac{x}{15} - 30 \cdot \frac{x}{10} = 30(1)$$

   Think: The LCD of $\frac{x}{6}, \frac{x}{15},$ and $\frac{x}{10}$ is 30.

$$5x + 2x - 3x = 30$$
$$4x = 30$$
$$x = \frac{30}{4} \text{ or } 7\frac{1}{2}$$

7. **Interpret** ▶ $x = 7\frac{1}{2}$ means the pool can be filled with all 3 pipes open in $7\frac{1}{2}$ hours.

Practice: Solve each work problem using rational expressions.

1. An inlet pipe can fill a pool in 8 hours. An outlet pipe can empty the same pool in 12 hours. How long will it take to fill the empty pool using the inlet pipe, if the outlet pipe is mistakenly left open?

2. One machine can do a certain job in 5 hours. Two other machines can do the same job in 4 hours and 10 hours, respectively. How long will it take to do the same job if all three machines work together?

286 Chapter 7 Rational Expressions

3. _____

4. _____

   _____

   _____

3. The hot-water faucet takes 8 minutes longer than the cold-water faucet to fill a certain bathtub. If both faucets are turned on, the tub is filled in 15 minutes. If both faucets are turned on, how long would it take the hot-water faucet to fill the tub if the cold-water faucet is turned off after 5 minutes?

4. Barbara can do half as much work as Sue, and Sue can do $1\frac{1}{2}$ times as much work as Deirdre. Working together, they can build a garage in 10 hours. Find the time needed for each to build the garage separately.

You can solve many types of *uniform motion problems* by using rational expressions.

EXAMPLE 2: Solve this uniform motion problem using rational expressions.

1. Read and identify ▶ The *airspeed* (the rate in still air) of an airplane is 180 mph. The airplane can travel 800 miles with the wind in the same time that it can travel 640 miles against the same wind. What is the velocity of the wind?

2. Understand ▶ The unknowns are $\begin{cases} \text{the velocity of the wind} \\ \text{the rate of the airplane with the wind} \\ \text{the rate of the airplane against the wind} \end{cases}$.

3. Decide ▶ Let x = the velocity of the wind
then $180 + x$ = the rate of the airplane with the wind
and $180 - x$ = the rate of the airplane against the wind

4. Make a table ▶

	distance (in miles)	rate (in mph)	time $\left(\dfrac{distance}{rate}\right)$
with the wind	800	$180 + x$	$\dfrac{800}{180 + x}$
against the wind	640	$180 - x$	$\dfrac{640}{180 - x}$

5. Translate ▶ The time flying with the wind equals the time flying against the wind.

$$\frac{800}{180 + x} = \frac{640}{180 - x}$$

6. Solve as before ▶ $(180 + x)(180 - x) \cdot \dfrac{800}{180 + x} = (180 + x)(180 - x) \cdot \dfrac{640}{180 - x}$ The LCD is $(180 + x)(180 - x)$.

$$144{,}000 - 800x = 115{,}200 + 640x$$
$$28{,}800 = 1440x$$
$$20 = x$$

7. Interpret ▶ $x = 20$ means the wind velocity is 20 mph.

Practice: Solve each uniform motion problem using rational expressions.

5. _____

6. _____

5. The river current is 2 km/h. It takes the same amount of time for a man to paddle upstream 24 km and back again as it does for him to paddle 50 km in still water. What is his rate in still water?

6. One car travels 620 miles in the same time that another car travels 500 miles. If the first car's rate is 10 mph faster than the second car's rate, what is the speed of the second car?

Copyright © 1985 by Harcourt Brace Jovanovich, Inc. All rights reserved.

7. A man can paddle at a constant rate of 5 mph. He paddles 4 miles upstream and returns. The upstream trip takes 4 times as long as the downstream trip. What is the rate of the current?

8. The difference between the maximum speed of two cars is 12 mph. The faster car can travel 217 miles in the same time that the slower car can travel 175 miles. What is the maximum speed of each car?

Mixed Practice: Solve each problem using rational expressions.

9. The hot-water faucet can fill a bathtub in 15 minutes. The cold-water faucet can fill the same bathtub in 10 minutes. How long will it take to fill the bathtub with both faucets on?

10. One pipe fills a tank in 10 minutes. Two other pipes take 15 minutes and 30 minutes, respectively, to fill the same tank. How long will it take all three pipes to fill the tank together? (Hint: Your table should have three rows.)

11. Chris can type a paper in 5 hours. Chris and Jan together can type the same paper in 2 hours. How long will it take Jan to type the paper alone?

12. Three machines can do a certain job in 1 hour working together. Two of the machines take 5 hours and 4 hours, respectively, to do the same job. How long will the third machine take working alone?

13. An airplane travels 600 miles with a tailwind of 25 mph. On the return trip against the same wind, it takes the airplane the same amount of time to travel only 450 miles. What is the plane's airspeed?

14. A man can row a boat at the rate of 4 mph in still water. Rowing upstream, it takes him as long to cover 4 miles as it does to cover 12 miles rowing downstream. What is the rate of the current?

15. A train took 4 hours longer to travel 750 miles than it took to travel 500 miles. The train traveled at a constant speed on both trips. Find the time for each trip.

16. Eleanor travelled 1800 miles by airplane and 90 miles by bus. The constant rate of the airplane was 8 times that of the bus. The complete trip took 7 hours. How long did Eleanor travel by bus?

17. An inlet pipe can fill a tank in 40 minutes. An outlet pipe can empty the tank in 1 hour. How long will it take to fill the empty tank if both pipes are open?

18. The sink faucets can fill the sink in 15 minutes. The sink drain can empty it in 10 minutes. How long will it take to empty one-half of a sink full of water with the faucets on and the sink drain open?

19. Kristina can build a wall in 6 days. Stephanie can build the same wall in 8 days. Stephanie works on the wall for 3 days alone before Kristina joins her. Together they complete the wall. How long does it take to build the wall?

20. Ryan can paint a car in 5 hours. Sean can paint the same car in 8 hours. They paint the car together for 2 hours and then Sean quits to paint another car. How long does it take Ryan to finish the job?

21. The air distance from Washington, D.C., to Los Angeles is about 2300 miles. If a cargo plane has a cruising speed of 300 mph and a headwind of 40 mph, how far from Washington, D.C., on the way to Los Angeles, is *the point of no return* (the point when the time it takes to fly to Los Angeles is the same as the time it takes to return to Washington, D.C.)?

22. A car will have a 12% gain in fuel with an 18 mph tailwind and a 10% loss in fuel with an 18 mph headwind. A car travels at a constant rate of 50 mph with an 18 mph tailwind for part of a trip and an 18 mph headwind for the other part of the trip. The total trip takes 6 hours with an 8% fuel gain over the entire trip. How far was each part of the trip?

288   Chapter 7   Rational Expressions

Solve each number problem using rational expressions.

23. The difference between the reciprocals of two consecutive even integers is $\frac{1}{12}$. What are the integers?

24. The sum of the reciprocal and twice the square of the reciprocal of a number is 3. Find the number.

25. The reciprocal of 2 less than a number is twice the reciprocal of the number. What is the number?

26. What number added to the numerator and subtracted from the denominator of $\frac{8}{9}$ makes the resulting fraction equal $\frac{3}{4}$?

27. The numerator of a fraction is 5 less than the denominator. When the numerator is decreased by 3 and the denominator is increased by 1, the resulting fraction equals $\frac{1}{2}$. What is the original fraction?

28. The denominator of a fraction is 1 greater than the numerator. When 6 is added to both numerator and denominator, the resulting fraction equals $\frac{8}{9}$. Find the original fraction.

29. One-fourth of a number is equal to $\frac{1}{2}$ the sum of the number and 12. What is the number?

30. A certain fraction equals $\frac{3}{4}$. If 3 is added to both the numerator and denominator, the resulting fraction equals $\frac{4}{5}$. Find the original fraction.

31. Three tired and hungry men had a bag of apples. When they were asleep one of them awoke, ate $\frac{1}{3}$ of the apples, and then went back to sleep. Later, a second man awoke, ate $\frac{1}{3}$ of the remaining apples, and then went back to sleep. Finally, the third man awoke and ate $\frac{1}{3}$ of the remaining apples, leaving 8 apples in the bag. How many apples were in the bag originally?

32. Diophantus (about 275 A.D.) was reported to be the first man to study equations with two or more unknowns. Not much more is known about his life; however, the age at which he died is known because of the following inscription on his tomb: "Diophantus passed $\frac{1}{6}$ of his life in childhood, $\frac{1}{12}$ in youth, and $\frac{1}{7}$ more as a bachelor; 5 years after his marriage a son was born who died four years before his father, at half his father's age." How old was Diophantus when he died?

According to Guinness

FARTHEST AND NEAREST TO CITY CENTER
THE AIRPORT FARTHEST FROM THE CITY CENTER IT ALLEGEDLY SERVES IS VIRACOPOS, BRAZIL, WHICH IS 60 MILES FROM SÃO PAULO. THE GIBRALTAR AIRPORT IS 800 YARDS FROM THE CENTER.

Assume the man going to the São Paulo city center takes a taxi and the man going to Gibraltar walks. If it takes the taxi 84 minutes longer than it takes the walker to reach the respective city centers, and the constant speed of the taxi is 8.8 times that of the walker, then:

33. How long does it take the walker?

34. What is the constant speed of the taxi?

Copyright © 1985 by Harcourt Brace Jovanovich, Inc. All rights reserved.

Problem Solving 9: Solve Formulas Containing Rational Expressions

To solve a formula containing rational expressions for a given letter, you first use the LCD of all the rational expressions to clear fractions.

EXAMPLE: Solve a formula containing rational expressions for a given letter.

Problem ▶ Solve $P = \dfrac{A}{1 + rt}$ (finance formula) for r.

Remember, clear fractions and isolate the given letter.

1. Find the LCD ▶ The LCD of $\dfrac{P}{1}$ and $\dfrac{A}{1 + rt}$ is $1 + rt$.

2. Clear fractions ▶
$$(1 + rt)P = (1 + rt) \cdot \dfrac{A}{1 + rt}$$
$$(1 + rt)P = A \longleftarrow \text{fractions cleared}$$

3. Isolate given letter ▶
$$1 \cdot P + rt \cdot P = A \qquad \text{Clear parentheses.}$$
$$P + rtP = A \longleftarrow \text{parentheses cleared}$$
$$rtP = A - P \qquad \text{To get the term containing } r \text{ by itself, add the opposite of } P \text{ to both members.}$$

given letter $\longrightarrow r = \dfrac{A - P}{tP}$ To get r by itself, multiply both members by the reciprocal of tP.

Solution ▶ $r = \dfrac{A - P}{tP}$ or $\dfrac{A - P}{Pt}$

Another Example ▶ Solve $\dfrac{1}{R} = \dfrac{1}{R_1} + \dfrac{1}{R_2}$ (electricity formula) for R_1.

The LCD of $\dfrac{1}{R}, \dfrac{1}{R_1},$ and $\dfrac{1}{R_2}$ is RR_1R_2.

$$RR_1R_2 \cdot \dfrac{1}{R} = RR_1R_2\left(\dfrac{1}{R_1} + \dfrac{1}{R_2}\right) \qquad \text{Clear fractions.}$$

$$RR_1R_2 \cdot \dfrac{1}{R} = RR_1R_2 \cdot \dfrac{1}{R_1} + RR_1R_2 \cdot \dfrac{1}{R_2}$$

$$R_1R_2 = RR_2 + RR_1 \longleftarrow \text{fractions cleared}$$

$$R_1R_2 - RR_1 = RR_2 \qquad \text{Collect all the terms with the given letter } R_1 \text{ in one member.}$$

$$R_1(R_2 - R) = RR_2 \qquad \text{Factor out } R_1.$$

$$R_1(R_2 - R) \cdot \dfrac{1}{R_2 - R} = RR_2 \cdot \dfrac{1}{R_2 - R} \qquad \text{To get } R_1 \text{ by itself, multiply both members by the reciprocal of } R_2 - R.$$

$$R_1(R_2 - R) \cdot \dfrac{1}{R_2 - R} = \dfrac{RR_2}{R_2 - R} \qquad \text{Eliminate the like factors of } R_2 - R.$$

given letter $\longrightarrow R_1 = \dfrac{R_2R}{R_2 - R}$

Chapter 7 Rational Expressions

Practice: Solve each rational formula for the indicated letter. State all restrictions.

Solve $I = \dfrac{E}{R_1 + R_2}$ (electricity formula) for: 1. E 2. R_2

Solve $s = \dfrac{a}{1 - r}$ (series formula) for: 3. a 4. r

Solve $\dfrac{1}{f} = \dfrac{1}{a} + \dfrac{1}{b}$ (optics formula) for: 5. f 6. a

Solve $f = \dfrac{1}{1 - m + mt}$ (finance formula) for: 7. t 8. m

Solve $I = \dfrac{E}{R + \dfrac{r}{n}}$ (electricity formula) for: 9. R 10. n

Solve $v = \dfrac{v_1 + v_2}{1 + \dfrac{v_1 v_2}{c^2}}$ (Einstein's velocity formula) for: 11. c^2 12. v_1

Solve $T = \dfrac{24I}{B(n + 1)}$ (interest formula) for: 13. B 14. n

Solve $R = \dfrac{S + F + P}{S + P}$ (space shuttle formula) for: 15. F 16. P

According to Guinness

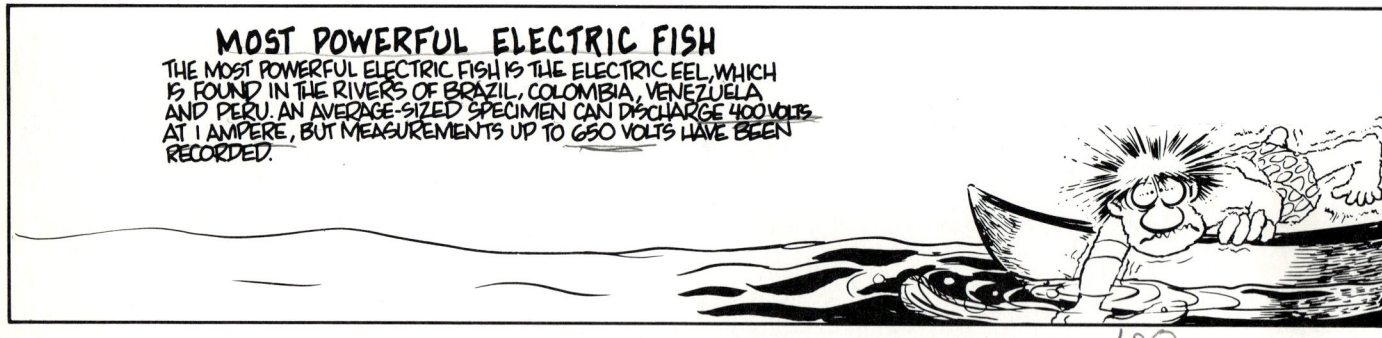

MOST POWERFUL ELECTRIC FISH
THE MOST POWERFUL ELECTRIC FISH IS THE ELECTRIC EEL, WHICH IS FOUND IN THE RIVERS OF BRAZIL, COLOMBIA, VENEZUELA AND PERU. AN AVERAGE-SIZED SPECIMEN CAN DISCHARGE 400 VOLTS AT 1 AMPERE, BUT MEASUREMENTS UP TO 650 VOLTS HAVE BEEN RECORDED.

Ohm's Law is: $I = \dfrac{E}{R}$ where $\begin{cases} I \text{ is } current \text{ measured in } amperes\ (A) \\ E \text{ is } voltage \text{ measured in } volts\ (V) \\ R \text{ is } resistance \text{ measured in } ohm\ (\Omega) \end{cases}$

The *Power Law* is: $I^2 = \dfrac{P}{R}$ where P is *power* measured in *watts* (W).

17. Solve Ohm's Law for R.

18. Solve the Power Law for P.

19. What is the resistance of an average-sized electric eel found in many rivers of South America? (See Problem 17.)

20. How much power was generated when the greatest measure of volts for an electric eel was recorded? (See Problem 18.)

Name _____ Date _____ Class _____

Chapter 7 Review

What to Review if You Have Trouble

Objectives		Lesson	Example	Page
Find excluded value(s) of rational expressions	1. Find the excluded value(s) of $\dfrac{3}{4x^2 - 5x}$.	7.1	1	252
Determine if two rational expressions are equal	2. $\dfrac{3y - 7}{2x - 1} \stackrel{?}{=} \dfrac{7 - 3y}{1 - 2x}$.	7.1	2	254
Reduce rational expressions to lowest terms	3. Reduce $\dfrac{3y^2 - 12}{y^2 - 3y - 10}$ to lowest terms.	7.1	3	255
Multiply rational expressions	4. $\dfrac{6x - 9}{2x + 1} \cdot \dfrac{2x^2 - 9x - 5}{2x^2 - 13x + 15}$	7.2	1	259
Divide rational expressions	5. $\dfrac{x + 2y}{3x} \div \dfrac{x^2 + xy - 2y^2}{5x}$	7.2	2	260
Multiply and divide with rational expressions	6. $\dfrac{x^3 - 4x}{x^2 - 3x} \div \dfrac{x}{x - 2} \cdot \dfrac{2 + x}{2 - x}$	7.2	3	261
Rename rational expressions in higher terms	7. $\dfrac{3x - 1}{2x + 5} = \dfrac{?}{6x^2 + 11x - 10}$.	7.2	4	262
Add rational expressions	8. $\dfrac{3x}{2x + 5} + \dfrac{7x - 1}{3x - 2}$	7.3	1	266
Subtract rational expressions	9. $\dfrac{4z - 11}{3z - 5} - \dfrac{3z - 7}{z + 8}$	7.3	2	267

292 Chapter 7 Rational Expressions

Perform a combination of operations with rational expressions	10. $\dfrac{w+4}{w+2} - \dfrac{3}{w^2 - w - 6} \div \dfrac{6}{w-3}$	7.3	3	268
Simplify complex fractions using the fundamental rule	11. $\dfrac{\dfrac{1}{x} - \dfrac{1}{y}}{\dfrac{x}{y} + \dfrac{y}{x}}$	7.4	1	271
Simplify complex fractions using the division method	12. $\dfrac{3 - \dfrac{2}{y}}{4 + \dfrac{1}{y^2}}$	7.4	2	273
Simplify complex fractions containing other complex fractions	13. $\dfrac{2 - \dfrac{1}{y}}{3 + \dfrac{2}{5 - \dfrac{3}{y}}}$	7.4	3	275
Solve rational equations	14. $\dfrac{3}{n} + \dfrac{1}{n-2} = \dfrac{3n-1}{n^2 - 2n}$	7.5	1	280
	15. $\dfrac{3}{x-3} + 2 = \dfrac{5}{x(x-3)}$	7.5	2	281
Solve problems using rational expressions	16. The maximum rate of a motorboat is 4 times as fast as the rate of the river current. It takes 4 hours to go 12 miles upstream and then back again. Find the maximum rate of the motorboat.	PS 8	2	286
Solve formulas containing rational expressions	17. Solve $I = \dfrac{nE}{R + nr}$ (electricity formula) for n.	PS 9	—	289

CHAPTER 7 REVIEW ANSWERS: 1. $0, \tfrac{2}{5}$ 2. Yes, they are equal 3. $\dfrac{3(y-2)}{y-5}$ 4. 3 5. $\dfrac{5}{3(x-y)}$ 6. $-\dfrac{x}{(x-2)(x+2)^2}$ 7. $\dfrac{9x^2 - 9x + 2}{6x^2 + 11x - 10}$ 8. $\dfrac{23x^2 + 27x - 5}{(2x+5)(3x-2)}$ 9. $\dfrac{-5z^2 + 57z - 123}{(3z-5)(z+8)}$ 10. $\dfrac{2w+7}{2(w+2)}$ 11. $\dfrac{y-x}{x^2+y^2}$ 12. $\dfrac{y(3y-2)}{4y^2+1}$ 13. $\dfrac{(2y-1)(5y-3)}{y(17y-9)}$ 14. 5 15. $-1, \tfrac{2}{5}$ 16. 15 mph 17. $\dfrac{IR}{E-Ir}$; $R+nr \neq 0$, $E-In \neq 0$

NONLINEAR RELATIONS

8 Exponents and Radicals

8.1 Compute with Rational Exponents

8.2 Simplify Radicals

8.3 Compute with Radicals

8.4 Solve Radical Equations

PS 10: Solve Formulas Containing Radicals

PS 11: Evaluate Formulas Containing Radicals

8.5 Add and Subtract Complex Numbers

8.6 Multiply and Divide Complex Numbers

Introduction to Rational Exponents

If n is a natural number, then b is an *nth root* of a provided $b^n = a$.

Examples ▶ (a) 3 is a 2nd root (square root) of 9, because $3^2 = 9$.
(b) -3 is a square root of 9, because $(-3)^2 = 9$.
(c) 2 is a 3rd root (cube root) of 8, because $2^3 = 8$.
(d) -4 is a cube root of -64, because $(-4)^3 = -64$.
(e) 3 is a 5th root of 243, because $3^5 = 243$.

Every positive real number has a positive square root and a negative square root. The positive square root is called the *principal square root*.

Example ▶ Both 4 and -4 are square roots of 16 because: $4^2 = 16$ and: $(-4)^2 = 16$. However, the principal square root of 16 is 4.

Negative real numbers do not have a real-number square root because there is no real number b such that b^2 is a negative real number.

Every real number a has a single real-number cube root.

Examples ▶ (a) 2 is the cube root of 8. (b) -2 is the cube root of -8. (c) 0 is the cube root of 0.

Roots are often denoted by using *radical notation*.

The notation $\sqrt[n]{a}$ is called a *radical expression*.

Index ⟶ $\sqrt[n]{a}$ ⟵ Radicand
         └ Radical

The number a is called the *radicand*.
The number n is called the *index* or *order of the radical*. It can be any natural number.

If n is even, and: $a > 0$, then $\sqrt[n]{a}$ is a positive real number called the *principal nth root* of a.
$a = 0$, then $\sqrt[n]{a} = 0$.
$a < 0$, then $\sqrt[n]{a}$ does not represent a real number.

If n is odd, and: $a > 0$, then $\sqrt[n]{a}$ is a positive real number.
$a = 0$, then $\sqrt[n]{a} = 0$.
$a < 0$, then $\sqrt[n]{a}$ is a negative real number.

Note ▶ If $n = 2$ then $\sqrt[2]{a}$ is written as simply $\sqrt{a}$.

Examples ▶ (a) $\sqrt{9} = 3$ because $3^2 = 9$ and 3 is positive.
(b) $\sqrt{9} \neq -3$ because -3 is not the principal root.
(c) $\sqrt[3]{125} = 5$ because $5^3 = 125$.
(d) $\sqrt[3]{-64} = -4$ because $(-4)^3 = -64$.
(e) $\sqrt{0} = 0$ because $0^2 = 0$.
(f) $\sqrt{-4}$ does not represent a real number, since there is no real number b such that $b^2 = -4$.

The symbol $\sqrt{}$ is used to represent the nonnegative square root. To represent the negative square root of a real number, you write a negative sign in front of the radical.

Examples ▶ (a) $\sqrt{49} = 7$ (b) $-\sqrt{49} = -7$

8.1 Compute with Rational Exponents

You might be tempted to think $\sqrt{x^2} = x$, but the following examples show that this is not always the case.

Examples ▶ (a) If $x = 3$, then $\sqrt{x^2} = \sqrt{3^2} = \sqrt{9} = 3$ or x.
(b) If $x = -3$, then $\sqrt{x^2} = \sqrt{(-3)^2} = \sqrt{9} = 3$ or $-x$.
(c) If $x = 0$, then $\sqrt{x^2} = \sqrt{0^2} = \sqrt{0} = 0$ or x.

A close analysis of the previous examples shows that in every case,

if $x \geq 0$, then $\sqrt{x^2} = x$, and if $x < 0$, then $\sqrt{x^2} = -x$.

> For any real number x, $\sqrt{x^2} = |x|$.

Examples ▶ (a) $\sqrt{6^2} = |6|$ (b) $\sqrt{(-6)^2} = |-6|$ (c) $\sqrt{(5m)^2} = |5m|$ (d) $\sqrt{(-3n)^2} = |-3n|$
$\phantom{(a)\ \sqrt{6^2}} = 6$ $\phantom{(b)\ \sqrt{(-6)^2}} = 6$ $\phantom{(c)\ \sqrt{(5m)^2}} = |5|\,|m|$ $\phantom{(d)\ \sqrt{(-3n)^2}} = |-3|\,|n|$
$\phantom{(c)\ \sqrt{(5m)^2} = 555} = 5|m|$ $\phantom{(d)\ \sqrt{(-3n)^2} = 555} = 3|n|$

(e) $\sqrt{(-mn^2)^2} = |-mn^2|$ (f) $\sqrt{n^2 - 6n + 9} = \sqrt{(n-3)^2}$
$\phantom{(e)\ \sqrt{(-mn^2)^2}} = |-m|\,|n^2|$ $\phantom{(f)\ \sqrt{n^2 - 6n + 9}} = |n - 3|$
$\phantom{(e)\ \sqrt{(-mn^2)^2}} = |m|n^2$ Think: Since $n^2 \geq 0$, $|n^2| = n^2$.

To compute roots that involve an odd index, you will not need to use absolute value notation.

Examples ▶ (a) $\sqrt[3]{x^3} = x$ (b) $\sqrt[5]{y^5} = y$

> For any real number x: If n is odd $\sqrt[n]{x^n} = x$. If n is even $\sqrt[n]{x^n} = |x|$.

It may seem surprising that this introduction to rational exponents begins with a discussion on radicals, but the following concepts will demonstrate the relationship between rational exponents and radicals.

The definition of exponents, as stated in Chapter 1, gave meaning to the notation a^n, provided a is a real number and n is an integer. The rules for exponents were also stated in terms of integral exponents. At this point, $2^{1/2}$ was meaningless because the exponent is not an integer. The rules for integral exponents, however, can be extended to include rational exponents. Consider the following:

Examples ▶ If m and n are rational numbers, and $a^m a^n = a^{m+n}$ then:
(a) $2^{1/2} \cdot 2^{1/2}$ must equal 2^1. This implies that $2^{1/2}$ is a square root of 2.
(b) $8^{1/3} \cdot 8^{1/3} \cdot 8^{1/3}$ must equal 8^1. This implies that $8^{1/3}$ is a cube root of 8.

These examples suggest that expressions of the form $a^{1/n}$ can be defined in terms of roots.

> **DEFINITION 8.1:** If n is a natural number, then: $a^{1/n} = \sqrt[n]{a}$. If n is an even number, a is restricted to nonnegative real numbers.

Examples ▶ Evaluate $16^{1/2}$, $-16^{1/2}$, $(-8)^{1/3}$, and $81^{1/4}$.

1. Use Definition 8.1 ▶ $16^{1/2} = \sqrt{16}$ | $-16^{1/2} = -\sqrt{16}$ | $(-8)^{1/3} = \sqrt[3]{-8}$ | $81^{1/4} = \sqrt[4]{81}$

2. Simplify ▶ $\phantom{16^{1/2}} = 4$ | $\phantom{-16^{1/2}} = -4$ | $\phantom{(-8)^{1/3}} = -2$ | $\phantom{81^{1/4}} = 3$

Solutions ▶ $16^{1/2} = 4$ | $-16^{1/2} = -4$ | $(-8)^{1/3} = -2$ | $81^{1/4} = 3$

Chapter 8 Exponents and Radicals

> CAUTION: $-a^{1/n}$ does not mean the same as $(-a)^{1/n}$.

Examples ▶ (a) $-9^{1/2} = -(9^{1/2}) = -3$. (b) $(-9)^{1/2}$ does not represent a real number.

8.1 Compute with Rational Exponents

Definition 8.1 only defines $a^{1/n}$ for exponents of the form $\dfrac{1}{n}$, where n is a natural number. However, if the rules of exponents are to hold, then:

$$a^{m/n} = a^{m \cdot 1/n} \quad \text{or} \quad a^{m/n} = a^{(1/n)m}$$
$$= (a^m)^{1/n} \qquad\qquad\quad = (a^{1/n})^m$$
$$= \sqrt[n]{a^m} \qquad\qquad\qquad = (\sqrt[n]{a})^m$$

> **DEFINITION 8.2:** If m is an integer and n is a natural number, then:
> $$a^{m/n} = \sqrt[n]{a^m} = (\sqrt[n]{a})^m$$
> If n is an even number, a is restricted to nonnegative real numbers.

Note ▶ Definition 8.2 does not apply to $(-4)^{3/2}$ because $(\sqrt{-4})^3$ does not represent a real number.

As shown above, expressions that contain rational exponents do not always represent real numbers when the expression has a negative number as its base. It is for this reason that the following restriction is stated.

Agreement ▶ > All variables in this chapter are assumed to be positive numbers unless otherwise stated.

EXAMPLE 1: Write rational exponential notation as radical notation.

Problems ▶ Write $5^{2/6}$, $y^{0.5}$, $(uv)^{6/8}$ and $uv^{6/8}$ in radical notation.   Think: y, u, and v are positive numbers.

1. Simplify ▶ $5^{2/6} = 5^{1/3}$ | $y^{0.5} = y^{1/2}$ | $(uv)^{6/8} = (uv)^{3/4}$ | $uv^{6/8} = uv^{3/4}$
2. Use Definition 8.2 ▶ $= \sqrt[3]{5^1}$ | $= \sqrt{y^1}$ | $= \sqrt[4]{(uv)^3}$ | $= u\sqrt[4]{v^3}$
   Solutions ▶ $5^{2/6} = \sqrt[3]{5}$ | $y^{0.5} = \sqrt{y}$ | $(uv)^{6/8} = \sqrt[4]{(uv)^3}$ | $uv^{6/8} = u\sqrt[4]{v^3}$

Note ▶ All of the radicals represent real numbers because the variables are assumed to be positive.

Make Sure

Write each rational exponential notation as radical notation.

See Example 1 ▶ 1. $7^{9/12}$ _____   2. $y^{0.75}$ _____   3. $(ab)^{8/12}$ _____   4. $cd^{21/24}$ _____

MAKE SURE ANSWERS: 1. $\sqrt[4]{7^3}$ 2. $\sqrt[4]{y^3}$ 3. $\sqrt[3]{(ab)^2}$ 4. $c\sqrt[8]{d^7}$

For computational purposes, it is often more convenient to use $a^{m/n} = (\sqrt[n]{a})^m$.

EXAMPLE 2: Evaluate expressions containing rational exponents.

Problems ▸ Evaluate $27^{2/3}$, $9^{3/2}$, $(-8)^{2/3}$, and $-8^{2/3}$.

1. Use Definition 8.2 ▸
$27^{2/3} = (\sqrt[3]{27})^2$ | $9^{3/2} = (\sqrt{9})^3$ | $(-8)^{2/3} = (\sqrt[3]{-8})^2$ | $-8^{2/3} = -(\sqrt[3]{8})^2$

2. Simplify ▸
$= 3^2$ | $= 3^3$ | $= (-2)^2$ | $= -(2)^2$
$= 9$ | $= 27$ | $= 4$ | $= -4$

Solutions ▸ $27^{2/3} = 9$ | $9^{3/2} = 27$ | $(-8)^{2/3} = 4$ | $-8^{2/3} = -4$

Make Sure

Evaluate expressions containing rational exponents.

See Example 2 ▸ 1. $64^{5/6}$ _____   2. $-81^{3/4}$ _____   3. $(-27)^{2/3}$ _____   4. $8^{-2/3}$ _____

MAKE SURE ANSWERS: 1. 32 2. −27 3. 9 4. $\frac{1}{4}$

Recall ▸ An exponential expression is simplified if each different base occurs only once, no power is raised to a power, and each exponent is positive.

Since the rules for integral exponents from Chapter 1 have been extended to include rational exponents, you can now simplify expressions containing rational exponents.

EXAMPLE 3: Simplify expressions containing rational exponents.

Problems ▸ Simplify $v^{1/6} \cdot v^{1/3}$, $\dfrac{x^{1/4}}{x^{2/3}}$, and $(y^{-1/2}z^{1/3})^{-3/4}$. Assume v, x, y, and z to be positive numbers.

1. Use rules for exponents ▸
$v^{1/6} \cdot v^{1/3} = v^{1/6 + 1/3}$ | $\dfrac{x^{1/4}}{x^{2/3}} = x^{1/4 - 2/3}$ | $(y^{-1/2}z^{1/3})^{-3/4} = (y^{-1/2})^{-3/4} \cdot (z^{1/3})^{-3/4}$

2. Simplify ▸
$= v^{1/6 + 2/6}$ | $= x^{3/12 - 8/12}$ | $= y^{(-1/2)(-3/4)} \cdot z^{(1/3)(-3/4)}$
$= v^{3/6}$ | $= x^{-(5/12)}$ | $= y^{3/8} z^{-1/4}$
$= v^{1/2}$ | $= \dfrac{1}{x^{5/12}}$ | $= \dfrac{y^{3/8}}{z^{1/4}}$

Solutions ▸ $v^{1/6} \cdot v^{1/3} = v^{1/2}$ | $\dfrac{x^{1/4}}{x^{2/3}} = \dfrac{1}{x^{5/12}}$ | $(y^{-1/2}z^{1/3})^{-3/4} = \dfrac{y^{3/8}}{z^{1/4}}$

Another Example ▸
$\dfrac{(3x^{1/2}y^{-1/3})^2}{(27x^{-2/3}y^{3/2})^{-1/3}} = \dfrac{3^2 xy^{-2/3}}{(3^3)^{-1/3}x^{2/9}y^{-1/2}}$ Think: Write 27 as 3^3. Use the Extended Power Rule.

$= 3^3 x^{7/9} y^{-1/6}$

$= \dfrac{27x^{7/9}}{y^{1/6}}$

Make Sure

Simplify expressions containing rational exponents.

See Example 3 ▶ **1.** $a^{1/6} \cdot a^{3/4}$ _____ **2.** $\dfrac{b^{5/6}}{b^{2/3}}$ _____

3. $(c^{3/4}d^{-5/8})^{1/2}$ _____ **4.** $e^{1/6}(e^{2/3} - 3e^{1/2} + 5e^{-1/3})$ _____

MAKE SURE ANSWERS: **1.** $a^{11/12}$ **2.** $b^{1/6}$ **3.** $\dfrac{d^{5/16}}{c^{3/8}}$ **4.** $e^{5/6} - 3e^{2/3} + \dfrac{5}{e^{1/6}}$

You can also use Definition 8.2 to write radical expressions in rational exponential notation.

EXAMPLE 4: Write radical expressions in rational exponential notation.

Problems ▶ Write $\sqrt{x^2}$, $\sqrt[6]{u^4 v}$ and $\dfrac{w^2}{\sqrt{w}}$ in rational exponential notation. Assume x, u, v, and z to be positive numbers.

1. Use Definition 8.2 ▶ $\sqrt{x^2} = x^{2/2}$ | $\sqrt[6]{u^4 v} = (u^4 v)^{1/6}$ | $\dfrac{w^2}{\sqrt{w}} = \dfrac{w^2}{w^{1/2}}$

2. Simplify ▶ $= x^1$ | $= (u^4)^{1/6}(v^1)^{1/6}$ | $= w^{2-1/2}$

$= u^{4/6} \cdot v^{1/6}$ | $= w^{3/2}$

Solutions ▶ $\sqrt{x^2} = x$ | $\sqrt[6]{u^4 v} = u^{2/3} v^{1/6}$ | $\dfrac{w^2}{\sqrt{w}} = w^{3/2}$

Note ▶ Absolute value symbols have not been used because the variables are positive.

Make Sure

Write each radical expression in rational exponential notation.

See Example 4 ▶ **1.** $\sqrt[6]{a^3}$ **2.** $\sqrt[8]{b^2 c^4}$ **3.** $\dfrac{\sqrt[4]{d^3}}{\sqrt[3]{d^2}}$ **4.** $(\sqrt[5]{e^3 f^4})^2$

MAKE SURE ANSWERS: **1.** $a^{1/2}$ **2.** $b^{1/4} c^{1/2}$ **3.** $d^{1/12}$ **4.** $e^{6/5} f^{8/5}$

8.1 Practice

Set 1: Write expressions containing rational exponents in radical notation.

1. $2^{2/3}$
2. $5^{3/4}$
3. $4^{2/3}$
4. $3^{3/5}$
5. $a^{1/6}$
6. $b^{2/5}$
7. $c^{4/5}$
8. $d^{3/7}$
9. $(ef)^{2/5}$
10. $(gh)^{2/3}$
11. $mn^{3/4}$
12. $pq^{4/5}$

Set 2: Evaluate expressions containing rational exponents.

13. $16^{1/4}$
14. $27^{1/3}$
15. $125^{2/3}$
16. $32^{3/5}$
17. $8^{-2/3}$
18. $81^{-3/4}$
19. $(-125)^{4/3}$
20. $(-32)^{4/5}$
21. $-216^{2/3}$
22. $-81^{3/4}$
23. $(-128)^{-4/7}$
24. $(-128)^{-3/7}$

Set 3: Simplify expressions containing rational expressions.

25. $a^{2/3}a^{5/6}$

26. $b^{2/3}b^{6/7}$

27. $(c^{4/5})^{1/2}$

28. $(d^{2/3})^{3/5}$

29. $(e^{4/9}f^{4/3})^{-3/2}$

30. $(g^{3/2}h^{9/4})^{-2/9}$

31. $\dfrac{12m^{5/2}n^{3/2}}{-4m^{7/2}n^{5/2}}$

32. $\dfrac{-18p^{4/3}q^{3/4}}{6p^{2/3}q^{1/4}}$

33. $r^{3/4}(r^{1/2} + 2r^{4/9} - 3r^3)$

34. $s^{5/6}(s^{3/5} - 3s^{3/10} + 5s^{2/5})$

35. $\dfrac{(u^{3/4}v^{5/6})^{8/15}(u^{2/3}v^{3/4})^{5/6}}{(u^{2/5}v^{2/3})^{-3/4}}$

36. $\dfrac{(x^{3/4}y^{3/8}z^{3/2})^{2/3}(x^{5/6}y^{3/4}z^{3/2})^{1/3}}{(x^{2/3}y^{3/2}z)^{-1/2}}$

Set 4: Write each radical expression in rational exponential notation.

37. $\sqrt{36a^2}$

38. $\sqrt{49b^4}$

39. $\sqrt[3]{27c^6}$

40. $\sqrt[3]{64d^9}$

41. $\sqrt[3]{e^2} \cdot \sqrt[3]{e}$

42. $\sqrt[4]{f} \cdot \sqrt[4]{f^3}$

43. $\dfrac{\sqrt[3]{g^2}}{\sqrt[4]{g^3}}$

44. $\dfrac{\sqrt[6]{h^4}}{\sqrt[5]{h^3}}$

45. $(\sqrt[3]{m^7n^5})^2$

46. $(\sqrt[4]{p^6q^2})^3$

47. $\left(\dfrac{\sqrt[4]{r^3s^5}}{\sqrt[3]{r^4s^2}}\right)^2$

48. $\left(\dfrac{\sqrt[5]{u^2v^3}}{\sqrt[4]{u^2v^3}}\right)^3$

Review: Work these problems on a separate sheet of paper. Attach your work to this page.

Factor each composite as a product of primes.
49. 24
50. 56
51. 96
52. 144
53. 360
54. 648
55. 1024
56. 91

Write each product of repeated factors in exponential notation. (See Lesson 1.4.)
57. $2 \cdot 2 \cdot 2 \cdot 2$
58. $3 \cdot 3 \cdot 3 \cdot 3 \cdot 3$
59. $2 \cdot 2 \cdot 2 \cdot 3 \cdot 3$
60. $3 \cdot 3 \cdot 5 \cdot 5 \cdot 5 \cdot 5$
61. $2 \cdot 3 \cdot 3 \cdot 3 \cdot 5 \cdot 5$
62. $2 \cdot 3 \cdot 3 \cdot 5 \cdot 5 \cdot 5 \cdot 7 \cdot 7$
63. $5 \cdot 7 \cdot 7 \cdot 7 \cdot 11 \cdot 11$
64. $2 \cdot 2 \cdot 5 \cdot 5 \cdot 5 \cdot 5 \cdot 7$
65. $2 \cdot 3 \cdot 3 \cdot 3 \cdot 5 \cdot 5 \cdot 7 \cdot 7$

8.2 Simplify Radicals

The rules for rational exponents can be used to write radicals in different but equivalent forms.

Examples ▶ (a) $\sqrt{2 \cdot 3} = (2 \cdot 3)^{1/2} = 2^{1/2} \cdot 3^{1/2} = \sqrt{2} \cdot \sqrt{3}$ (b) $\sqrt[3]{\dfrac{4}{5}} = \left(\dfrac{4}{5}\right)^{1/3} = \dfrac{4^{1/3}}{5^{1/3}} = \dfrac{\sqrt[3]{4}}{\sqrt[3]{5}}$

└── equivalent radical forms ──┘ └── equivalent radical forms ──┘

These examples suggest that each of the rules of rational exponents can be stated in terms of radicals. In radical form, they are called the *rules of radicals*. The following rules of radicals are particularly useful for simplifying radicals.

> **Rules of Radicals**
>
> If a and b are positive real numbers, and n is a natural number, then:
>
> RULE 8.1 $\sqrt[n]{ab} = \sqrt[n]{a}\,\sqrt[n]{b}$ ⟵ the Product Rule for Radicals
>
> RULE 8.2 $\sqrt[n]{\dfrac{a}{b}} = \dfrac{\sqrt[n]{a}}{\sqrt[n]{b}}$ ⟵ the Quotient Rule for Radicals
>
> RULE 8.3 $\sqrt[n]{a^n} = a$

In $\sqrt[n]{a^p}$, the radicand a^p is called a *perfect nth power* if p is exactly divisible by n.

Examples ▶ (a) The radicand x^2 of $\sqrt{x^2}$ is a perfect 2nd power because 2 is exactly divisible by 2.
(b) The radicand 2^{15} of $\sqrt[3]{2^{15}}$ is a perfect 3rd power because 15 is exactly divisible by 3.

If the radicand is a perfect nth power, or a product of perfect nth powers, then you can use the rules of exponents and radicals to write the radical in an equivalent form that does not have a radical. It is generally helpful to first factor the radicand into prime factored form.

Examples ▶ (a) $\sqrt[4]{81a^4} = \sqrt[4]{3^4 a^4}$ assume $a > 0$ (b) $\sqrt[3]{64x^3} = \sqrt[3]{2^6 x^3}$
$\phantom{(a)\ \sqrt[4]{81a^4}} = \sqrt[4]{3^4}\sqrt[4]{a^4}$ Rule 8.1 $\phantom{(b)\ \sqrt[3]{64x^3}} = \sqrt[3]{(2^2)^3 x^3}$
$\phantom{(a)\ \sqrt[4]{81a^4}} = 3a$ Rule 8.3 $\phantom{(b)\ \sqrt[3]{64x^3}} = \sqrt[3]{(2^2)^3}\,\sqrt[3]{x^3}$ Rule 8.1
$\phantom{(b)\ \sqrt[3]{64x^3}} = 2^2 x$ or $4x$ Rule 8.3

If a radicand has powers larger than the index of the radical, you can simplify by factoring. Write the radicand as a product of perfect powers of the index and of powers less than the index.

EXAMPLE 1: Simplify radicals by factoring.

Problems ▶ Simplify $\sqrt[3]{16}$ and $\sqrt{12x^3 y^8}$. Assume that the variables are positive real numbers.

1. Factor ▶ $\sqrt[3]{16} = \sqrt[3]{2^4}$ $\sqrt{12x^3 y^8} = \sqrt{2^2 \cdot 3 x^3 y^8}$

2. Form powers of the index and simplify ▶ $= \sqrt[3]{2^3 \cdot 2} = \sqrt[3]{2^3}\sqrt[3]{2}$ $= \sqrt{2^2 \cdot 3 \cdot x^2 x y^8} = \sqrt{2^2}\sqrt{3}\sqrt{x^2}\sqrt{x}\sqrt{y^8}$
$\phantom{= \sqrt[3]{2^3 \cdot 2}} = 2\sqrt[3]{2}$ $\phantom{= \sqrt{2^2 \cdot 3 \cdot x^2 x y^8}} = 2xy^4 \sqrt{3}\sqrt{x}$

Solutions ▶ $\sqrt[3]{16} = 2\sqrt[3]{2}$ $\sqrt{12x^3 y^8} = 2xy^4 \sqrt{3x}$

> **CAUTION:** Even though the variables in this chapter are assumed to be positive, you will need to use absolute value symbols when you take the square root of some variable expressions.

Example ▶ $x - 3$ is negative if $x < 3$ and $x - 3$ is nonnegative if $x \geq 3$, so $\sqrt{(x-3)^2} = |x - 3|$.

302 Chapter 8 Exponents and Radicals

Make Sure

Simplify each radical by factoring. Assume all variables are positive.

See Example 1 ▶ 1. $\sqrt{64a^2b^3c^7}$ _____ 2. $\sqrt[3]{432d^3e^5f^{13}}$ _____ 3. $\sqrt[5]{1215g^5h^9k^{28}}$ _____

MAKE SURE ANSWERS: 1. $8abc^3\sqrt{bc}$ 2. $6def^4\sqrt[3]{2e^2f}$ 3. $3ghk^5\sqrt[5]{5h^4k^3}$

You can write some radicals in an equivalent form containing a radical with a smaller index number. The process of writing a radical in an equivalent form that contains a radical with a smaller index number is called *reducing the order of the radical*.

You can use the rules of radicals to reduce the order of a radical; however, the process is generally easier to understand when you use the rules of exponents.

EXAMPLE 2: Reduce the order of a radical.

Problem ▶ Reduce the order of $\sqrt[6]{81}$.

1. Factor ▶ $\sqrt[6]{81} = \sqrt[6]{3^4}$

2. Write exponential form ▶ $= 3^{4/6}$ Definition 8.2

3. Simplify ▶ $= 3^{2/3}$

4. Write radical form ▶ $= \sqrt[3]{3^2}$ Definition 8.2

Solution ▶ $\sqrt[6]{81} = \sqrt[3]{9}$

Other Examples ▶ (a) $\sqrt[4]{144a^8} = \sqrt[4]{2^4 3^2 a^8}$

$= (2^4 3^2 a^8)^{1/4}$ Definition 8.1

$= 2^{4/4} 3^{2/4} a^{8/4}$ The Extended Power Rule.

$= 2^1 \cdot 3^{1/2} a^2$

$= 2a^2\sqrt{3}$

(b) $\sqrt[8]{x^{16}y^6z^2} = (x^{16}y^6z^2)^{1/8}$

$= x^{16/8} y^{6/8} z^{2/8}$

$= x^2 y^{3/4} z^{1/4}$

$= x^2 \sqrt[4]{y^3 z}$

Make Sure

Reduce the order of each radical. Assume all variables are positive.

See Example 2 ▶ 1. $\sqrt[6]{2304}$ _____ 2. $\sqrt[4]{5184a^4b^2c^{14}}$ _____ 3. $\sqrt[6]{512e^6f^9g^{15}}$ _____

MAKE SURE ANSWERS: 1. $2\sqrt[3]{6}$ 2. $6ac^3\sqrt{2bc}$ 3. $2efg^2\sqrt{2fg}$

To *rationalize the denominator* of an expression, you multiply both the numerator and the denominator by the same factor so that you can use Rule 8.3 ($\sqrt[n]{a^n} = a$) to clear the radical from the denominator.

EXAMPLE 3: Rationalize the denominator of a radical expression.

Problem ▶ Rationalize the denominator of $\dfrac{2}{\sqrt[3]{a}}$.

1. Multiply ▶ $\dfrac{2}{\sqrt[3]{a}} = \dfrac{2}{\sqrt[3]{a^1}} \cdot \dfrac{\sqrt[3]{a^2}}{\sqrt[3]{a^2}}$ Think: a^1 must be multiplied by a^2 to make it a perfect 3rd power.

2. Simplify ▶ $= \dfrac{2\sqrt[3]{a^2}}{\sqrt[3]{a^3}}$

$= \dfrac{2\sqrt[3]{a^2}}{a}$ The denominator is rationalized because radicals are cleared.

Solution ▶ $\dfrac{2}{\sqrt[3]{a}} = \dfrac{2\sqrt[3]{a^2}}{a}$

Other Examples ▶ (a) $\dfrac{2}{\sqrt{2}} = \dfrac{2}{\sqrt{2}} \cdot \dfrac{\sqrt{2}}{\sqrt{2}}$

$= \dfrac{2\sqrt{2}}{2}$

$= \sqrt{2}$

(b) $\sqrt{\dfrac{4}{5}} = \dfrac{\sqrt{4}}{\sqrt{5}}$

$= \dfrac{2}{\sqrt{5}} \cdot \dfrac{\sqrt{5}}{\sqrt{5}}$

$= \dfrac{2\sqrt{5}}{5}$

(c) $\dfrac{1}{\sqrt[3]{25w}} = \dfrac{1}{\sqrt[3]{25w}} \cdot \dfrac{\sqrt[3]{5w^2}}{\sqrt[3]{5w^2}}$

$= \dfrac{\sqrt[3]{5w^2}}{\sqrt[3]{125w^3}}$

$= \dfrac{\sqrt[3]{5w^2}}{5w}$

Make Sure

Rationalize the denominator of each radical expression. Assume all variables are positive.

See Example 3 ▶ 1. $\dfrac{5}{\sqrt{2a}}$ 2. $\dfrac{4}{\sqrt[3]{2b^2}}$ 3. $\dfrac{2}{\sqrt[5]{4cd^3e^4}}$

MAKE SURE ANSWERS: 1. $\dfrac{5\sqrt{2a}}{2a}$ 2. $\dfrac{2\sqrt[3]{4b}}{b}$ 3. $\dfrac{\sqrt[5]{8c^4d^2e}}{cde}$

Simplest Radical Form

A radical expression is in *simplest radical form* if it involves only one radical and:
1. the radicand contains only powers less than the index number,
2. the order of the radical is reduced as far as possible,
3. no radical appears in a denominator, and
4. no fraction appears in the radicand.

Chapter 8 Exponents and Radicals

EXAMPLE 4: Write a radical expression in simplest radical form.

Problem ▶ Simplify $\sqrt{\dfrac{35a^7}{90}}$. Assume $a > 0$.

1. Factor ▶ $\sqrt{\dfrac{35a^7}{90}} = \sqrt{\dfrac{5 \cdot 7a^7}{2 \cdot 3^2 \cdot 5}}$

2. Reduce ▶ $= \sqrt{\dfrac{7a^7}{2 \cdot 3^2}}$

3. Remove perfect powers of the index ▶ $= \dfrac{\sqrt{7a^7}}{\sqrt{2 \cdot 3^2}}$

$= \dfrac{a^3\sqrt{7a}}{3\sqrt{2}}$

4. Rationalize denominator ▶ $= \dfrac{a^3\sqrt{7a}}{3\sqrt{2}} \cdot \dfrac{\sqrt{2}}{\sqrt{2}}$

$= \dfrac{a^3\sqrt{14a}}{3\sqrt{2^2}}$

5. Simplify ▶ $= \dfrac{a^3\sqrt{14a}}{3 \cdot 2}$

Solution ▶ $\sqrt{\dfrac{35a^7}{90}} = \dfrac{a^3\sqrt{14a}}{6}$

Other Examples ▶ (a) $\dfrac{3\sqrt{xv^2}}{\sqrt{3y}} = \dfrac{3v\sqrt{x}}{\sqrt{3y}}$ (b) $\dfrac{\sqrt{2rs}}{\sqrt{3r}} = \dfrac{\sqrt{2rs}}{\sqrt{3r}} \cdot \dfrac{\sqrt{3r}}{\sqrt{3r}}$

$= \dfrac{3v\sqrt{x}}{\sqrt{3y}} \cdot \dfrac{\sqrt{3y}}{\sqrt{3y}}$ $= \dfrac{\sqrt{6r^2s}}{3r}$

$= \dfrac{3v\sqrt{3xy}}{3y}$ $= \dfrac{r\sqrt{6s}}{3r}$

$= \dfrac{v\sqrt{3xy}}{y}$ $= \dfrac{\sqrt{6s}}{3}$

Make Sure

Write each radical expression in simplest radical form. Assume all variables are positive.

See Example 4 ▶ **1.** $\sqrt{\dfrac{13a}{32}}$ **2.** $\sqrt[3]{\dfrac{45}{81b^2}}$ **3.** $\sqrt[4]{\dfrac{32\,c^7d^3}{81\,c^3d^6}}$

MAKE SURE ANSWERS: **1.** $\dfrac{\sqrt{26a}}{8}$ **2.** $\dfrac{\sqrt[3]{15b}}{3b}$ **3.** $\dfrac{2c\sqrt[4]{2d}}{3d}$

8.2 Practice: Assume all radicands are positive.

Set 1: Simplify each radical by factoring.

1. $\sqrt{32a^5}$
2. $\sqrt[3]{54b^8}$
3. $\sqrt[3]{72c^{10}d^6}$
4. $\sqrt[4]{64e^{11}f^8}$
5. $\sqrt[3]{-27g^6h^7}$
6. $\sqrt[5]{-32m^{10}n^4}$
7. $\sqrt[3]{-8p^{10}q^5}$
8. $\sqrt[5]{-128r^3s^6}$
9. $\sqrt[3]{(t-3)^4}$
10. $\sqrt[4]{(u+8)^5}$
11. $\sqrt{v^2 - 4v + 4}$
12. $\sqrt{9 - 6w + w^2}$

Set 2: Reduce the order of each radical.

13. $\sqrt[6]{8}$
14. $\sqrt[8]{81}$
15. $\sqrt[4]{a^2}$
16. $\sqrt[6]{b^3}$
17. $\sqrt[6]{64c^{12}d^{18}}$
18. $\sqrt[5]{243e^{10}f^{20}}$

Set 3: Rationalize the denominator of each radical expression.

19. $\sqrt{\dfrac{7}{3}}$
20. $\sqrt{\dfrac{5}{7}}$
21. $\dfrac{2}{\sqrt{2a}}$
22. $\dfrac{3}{\sqrt{3b}}$
23. $\dfrac{4}{\sqrt[3]{4}}$
24. $\dfrac{9}{\sqrt[3]{9}}$
25. $\sqrt[3]{\dfrac{27}{4c}}$
26. $\sqrt[3]{\dfrac{16}{9d}}$
27. $\dfrac{e}{\sqrt[3]{2e}}$
28. $\dfrac{f}{\sqrt[4]{4f}}$
29. $\dfrac{\sqrt[3]{81g^8}}{\sqrt[3]{3g^4}}$
30. $\dfrac{\sqrt[4]{64h^6}}{\sqrt[4]{4h}}$
31. $\sqrt{\dfrac{25y^3}{64z^2}}$
32. $\sqrt{\dfrac{49w^5}{81x^2}}$
33. $\sqrt[3]{\dfrac{z^4}{27v^3}}$

306   Chapter 8   Exponents and Radicals

Set 4: Write each radical expression in simplest radical form.

34. $\sqrt{\dfrac{175a^2}{180a^3}}$ 35. $\sqrt{\dfrac{171b^4}{114b^5}}$ 36. $\sqrt[3]{\dfrac{64c^2}{81c^4}}$

37. $\sqrt[3]{\dfrac{54d^3}{324d^5}}$ 38. $\sqrt{\dfrac{8e^5f^3}{27e^3f^4}}$ 39. $\sqrt{\dfrac{25g^2h}{160gh^4}}$

40. $\sqrt{\dfrac{36m^7n^4}{20mn^7}}$ 41. $\sqrt{\dfrac{49p^9q^2}{28p^3q^5}}$ 42. $\sqrt[3]{\dfrac{8r^8s^3}{135r^5s^5}}$

43. $\sqrt[3]{\dfrac{27u^9v^2}{56u^6v^4}}$ 44. $\sqrt[4]{\dfrac{16w^7x}{405w^4x^2}}$ 45. $\sqrt[4]{\dfrac{32y^3z}{567y^2z^3}}$

Extra: Simplify each expression. Assume all variables and radicands are positive.

46. $\sqrt[3]{(x-2)^4}$ 47. $\sqrt[4]{(x+1)^5}$ 48. $\sqrt[6]{(x+3)^8}$

49. $\sqrt[8]{(2x-1)^{17}}$ 50. $\sqrt{x^2 - 4x + 4}$ 51. $\sqrt{x^2 + 6x + 9}$

52. $\sqrt{4x^2 + 12x + 9}$ 53. $\sqrt{x^4 - 2x^2 + 1}$ 54. $\sqrt{x^4 + 4x^2 + 4}$

Review: Work these problems on a separate sheet of paper. Attach your work to this page.

Multiply polynomials. (See Lesson 5.3.)
55. $z(z - 3)$ 56. $y(y + 3)$ 57. $x(x - 5)$
58. $(w - 5)(w + 3)$ 59. $(u - v)(u + v)$ 60. $(s - t)(s - t)$
61. $(3p - 4q)(2p + 3q)$ 62. $(2m - n)(5m + 3n)$ 63. $(3g + 4k)(2h + 3k)$
64. $(e + 2f)(e + 3f)$ 65. $(3c - d)(4c - d)$ 66. $(4a - 3b)(2a - 5b)$

8.3 Compute with Radicals

You have used addition and subtraction to combine like terms of polynomials. A similar process will now be developed to add and subtract *like radicals*.

> Radicals with the same index number and the same radicand are called like radicals.

Examples ▶ (a) $\sqrt{3}$, $-2\sqrt{3}$, and $\frac{1}{2}\sqrt{3}$ are like radicals.   Index = 2, radicand = 3

(b) $\sqrt[3]{x^2y}$, $5\sqrt[3]{x^2y}$, and $-7\sqrt[3]{x^2y}$ are like radicals.   Index = 3, radicand = x^2y

(c) $\sqrt{3}$ and $\sqrt{5}$ are unlike radicals because they have different radicands.

(d) $\sqrt{3}$ and $\sqrt[3]{3}$ are unlike radicals because they have different indexes.

You can add or subtract like radicals by using a distributive property.

Examples ▶ $2\sqrt{3} + 5\sqrt{3} = ?$, $5\sqrt[3]{2x} - 2\sqrt[3]{2x} = ?$, $5y\sqrt[4]{y} + 3y\sqrt[4]{y} - y\sqrt[4]{y} = ?$

1. Factor out the common radical ▶ $2\sqrt{3} + 5\sqrt{3} = (2+5)\sqrt{3}$ | $5\sqrt[3]{2x} - 2\sqrt[3]{2x} = (5-2)\sqrt[3]{2x}$ | $5y\sqrt[4]{y} + 3y\sqrt[4]{y} - y\sqrt[4]{y} = (5y + 3y - 1y)\sqrt[4]{y}$

2. Simplify ▶ $= 7\sqrt{3}$ | $= 3\sqrt[3]{2x}$ | $= 7y\sqrt[4]{y}$

Solutions ▶ $2\sqrt{3} + 5\sqrt{3} = 7\sqrt{3}$ | $5\sqrt[3]{2x} - 2\sqrt[3]{2x} = 3\sqrt[3]{2x}$ | $5y\sqrt[4]{y} + 3y\sqrt[4]{y} - y\sqrt[4]{y} = 7y\sqrt[4]{y}$

Note ▶ The sum $3\sqrt{2} + \sqrt[3]{4}$ cannot be simplified. The terms are unlike radicals.

Just as addition and subtraction of like terms are referred to as combining like terms, addition and subtraction of like radicals are referred to as *combining like radicals*. Some radicals can be combined if you simplify the radicals first.

EXAMPLE 1: Simplify and combine like radicals.

Problem ▶ $2x\sqrt{8x} + 3\sqrt{18x^3} = ?$

1. Simplify radicals ▶ $2x\sqrt{8x} + 3\sqrt{18x^3} = 2x\sqrt{2^3 x} + 3\sqrt{2 \cdot 3^2 x^3}$
$= 2x \cdot 2\sqrt{2x} + 3 \cdot 3x\sqrt{2x}$

2. Combine like radicals ▶ $= 4x\sqrt{2x} + 9x\sqrt{2x}$
$= (4x + 9x)\sqrt{2x}$

Solution ▶ $2x\sqrt{8x} + 3\sqrt{18x^3} = 13x\sqrt{2x}$

Other Examples ▶ (a) $\sqrt{54} - 2\sqrt{24} + \sqrt{150} = \sqrt{2 \cdot 3^3} - 2\sqrt{2^3 \cdot 3} + \sqrt{2 \cdot 3 \cdot 5^2}$
$= 3\sqrt{2 \cdot 3} - 2 \cdot 2\sqrt{2 \cdot 3} + 5\sqrt{2 \cdot 3}$
$= 3\sqrt{6} - 4\sqrt{6} + 5\sqrt{6}$
$= (3 - 4 + 5)\sqrt{6}$
$= 4\sqrt{6}$

(b) $2\sqrt[3]{a^4} + 4a\sqrt[3]{27a} - \frac{1}{a}\sqrt[3]{a^7} = 2\sqrt[3]{a^4} + 4a\sqrt[3]{3^3 a} - \frac{1}{a}\sqrt[3]{a^7}$

$\phantom{(b) 2\sqrt[3]{a^4} + 4a\sqrt[3]{27a} - \frac{1}{a}\sqrt[3]{a^7}} = 2a\sqrt[3]{a} + 4 \cdot a \cdot 3\sqrt[3]{a} - \frac{1}{a} \cdot a^2 \sqrt[3]{a}$

$\phantom{(b) 2\sqrt[3]{a^4} + 4a\sqrt[3]{27a} - \frac{1}{a}\sqrt[3]{a^7}} = 2a\sqrt[3]{a} + 12a\sqrt[3]{a} - a\sqrt[3]{a}$

$\phantom{(b) 2\sqrt[3]{a^4} + 4a\sqrt[3]{27a} - \frac{1}{a}\sqrt[3]{a^7}} = (2a + 12a - a)\sqrt[3]{a}$

$\phantom{(b) 2\sqrt[3]{a^4} + 4a\sqrt[3]{27a} - \frac{1}{a}\sqrt[3]{a^7}} = 13a\sqrt[3]{a}$

Make Sure

Simplify and combine like radicals. Assume radicands are nonnegative.

See Example 1 ▶ 1. $\sqrt{72} - \sqrt{162} + \sqrt[4]{64}$ _____ 2. $\sqrt[3]{81a^7 b^5} - ab\sqrt[3]{192a^4 b^2}$ _____

MAKE SURE ANSWERS: 1. $-\sqrt{2}$ 2. $-a^2 b \sqrt[3]{3ab^2}$

To multiply radicals containing the same index, you use the Product Rule for Radicals: $\sqrt[n]{a}\sqrt[n]{b} = \sqrt[n]{ab}$.

Examples ▶ (a) $\sqrt{3}\sqrt{5} = \sqrt{15}$ (b) $\sqrt[3]{x}\sqrt[3]{xy} = \sqrt[3]{x^2 y}$ (c) $\sqrt{2}\sqrt{20} = \sqrt{40}$

$\phantom{Examples (a) \sqrt{3}\sqrt{5} = \sqrt{15} (b) \sqrt[3]{x}\sqrt[3]{xy} = \sqrt[3]{x^2 y} (c) \sqrt{2}\sqrt{20}} = \sqrt{2^3 \cdot 5}$ Simplify when possible.

$\phantom{Examples (a) \sqrt{3}\sqrt{5} = \sqrt{15} (b) \sqrt[3]{x}\sqrt[3]{xy} = \sqrt[3]{x^2 y} (c) \sqrt{2}\sqrt{20}} = 2\sqrt{2 \cdot 5}$

$\phantom{Examples (a) \sqrt{3}\sqrt{5} = \sqrt{15} (b) \sqrt[3]{x}\sqrt[3]{xy} = \sqrt[3]{x^2 y} (c) \sqrt{2}\sqrt{20}} = 2\sqrt{10}$

> CAUTION: A common error is to omit the index when writing the product of two radicals.

Example ▶ $\sqrt[3]{a}\sqrt[3]{b} = \sqrt[3]{ab}$ — do not forget to write this index number

> CAUTION: To use the Product Rule for Radicals, the index numbers must be the same for each of the radicals.

Examples ▶ (a) $\sqrt[3]{a}\sqrt{b} \ne \sqrt[3]{ab}$ (b) $\sqrt[3]{a}\sqrt{b} \ne \sqrt{ab}$

Sometimes you can use a distributive property to multiply radical expressions.

EXAMPLE 2: Multiply radical expressions.

Problem ▶ $\sqrt{2}(3\sqrt{10} - 4\sqrt{20}) = ?$

1. Use a distributive property ▶ $\sqrt{2}(3\sqrt{10} - 4\sqrt{20}) = \sqrt{2} \cdot 3\sqrt{10} - \sqrt{2} \cdot 4\sqrt{20}$

2. Use Product Rule for Radicals ▶ $\phantom{\sqrt{2}(3\sqrt{10} - 4\sqrt{20})} = 3\sqrt{20} - 4\sqrt{40}$

3. Simplify ▶ $\phantom{\sqrt{2}(3\sqrt{10} - 4\sqrt{20})} = 3\sqrt{2^2 \cdot 5} - 4\sqrt{2^3 \cdot 5}$

$\phantom{\sqrt{2}(3\sqrt{10} - 4\sqrt{20})} = 3 \cdot 2\sqrt{5} - 4 \cdot 2\sqrt{2 \cdot 5}$

Solution ▶ $\sqrt{2}(3\sqrt{10} - 4\sqrt{20}) = 6\sqrt{5} - 8\sqrt{10}$

If the radical expressions are both binomials, you can multiply using the FOIL method.

Example ▶ $(3\sqrt{2} - \sqrt{5})(\sqrt{2} + \sqrt{5}) = ?$

$$\phantom{(3\sqrt{2} - \sqrt{5})(\sqrt{2} + \sqrt{5}) =}\ \ \text{F} \quad\ \ \text{O} \quad\ \ \text{I} \quad\ \ \text{L}$$

1. Use FOIL method ▶ $(3\sqrt{2} - \sqrt{5})(\sqrt{2} + \sqrt{5}) = 3\sqrt{2}\sqrt{2} + 3\sqrt{2}\sqrt{5} - \sqrt{5}\sqrt{2} - \sqrt{5}\sqrt{5}$

2. Simplify ▶
$$= 3 \cdot 2 + 3\sqrt{10} - \sqrt{10} - 5$$
$$= 6 + 2\sqrt{10} - 5$$

Solution ▶ $(3\sqrt{2} - \sqrt{5})(\sqrt{2} + \sqrt{5}) = 1 + 2\sqrt{10}$

Other Examples ▶ (a) $(\sqrt{11} + \sqrt{3})(\sqrt{11} - \sqrt{3}) = (\sqrt{11})^2 - (\sqrt{3})^2$
$$= 11 - 3$$
$$= 8$$

(b) $(\sqrt{3a} - \sqrt{5})^2 = (\sqrt{3a})^2 - 2\sqrt{3a}\sqrt{5} + (\sqrt{5})^2$
$$= 3a - \sqrt{15a} + 5$$

Note ▶ The above example is worthy of special attention because the product of the radical expressions does not contain a radical. This will be the case if the radical expressions are *rationalizing factors* of each other.

> The radical expressions $(\sqrt{a} + \sqrt{b})$ and $(\sqrt{a} - \sqrt{b})$ are called rationalizing factors of each other.

Examples ▶ (a) The rationalizing factor of $\sqrt{5} + \sqrt{3}$ is $\sqrt{5} - \sqrt{3}$.
(b) The rationalizing factor of $3\sqrt{y} - 2\sqrt{y}$ is $3\sqrt{y} + 2\sqrt{y}$.
(c) The rationalizing factor of $\sqrt{2a} + 5$ is $\sqrt{2a} - 5$.   Think: $5 = \sqrt{5^2}$

> $(\sqrt{a} + \sqrt{b})(\sqrt{a} - \sqrt{b}) = (\sqrt{a})^2 - (\sqrt{b})^2 = a - b$ ⟵ a rational expression

Make Sure

Multiply radical expressions. Assume radicands are nonnegative.

See Example 2 ▶ 1. $\sqrt{5a}(2\sqrt{5a} - \sqrt{2a})$ _____   2. $(\sqrt{3b} - \sqrt{2c})(\sqrt{2b} + \sqrt{3c})$ _____

MAKE SURE ANSWERS: 1. $10a - a\sqrt{10}$ 2. $b\sqrt{6} - c\sqrt{6} + \sqrt{bc}$

310 Chapter 8 Exponents and Radicals

To divide by a radical factor, you rationalize the denominator.

Examples ▶ (a) $\dfrac{5}{\sqrt[3]{2}} = \dfrac{5}{\sqrt[3]{2}} \cdot \dfrac{\sqrt[3]{2^2}}{\sqrt[3]{2^2}}$ (b) $\dfrac{3}{\sqrt{3}} = \dfrac{3}{\sqrt{3}} \cdot \dfrac{\sqrt{3}}{\sqrt{3}}$ (c) $\dfrac{x\sqrt{3x}}{\sqrt{5x}} = \dfrac{x\sqrt{3x}}{\sqrt{5x}} \cdot \dfrac{\sqrt{5x}}{\sqrt{5x}}$

$\qquad\qquad\quad = \dfrac{5\sqrt[3]{4}}{2} \qquad\qquad\quad = \dfrac{3\sqrt{3}}{3} \qquad\qquad\quad = \dfrac{x\sqrt{15x^2}}{5x}$

$\qquad\qquad\qquad\qquad\qquad\qquad\quad = \sqrt{3} \qquad\qquad\qquad\quad = \dfrac{x^2\sqrt{15}}{5x}$

$\qquad\qquad\qquad\qquad\qquad\qquad\qquad\qquad\qquad\qquad\quad = \dfrac{x\sqrt{15}}{5}$

To divide by a binomial expression containing square roots, you rationalize the denominator.

EXAMPLE 3: Rationalize the denominator of a fraction that has a binomial radical expression as its denominator.

Problem ▶ Rationalize the denominator of $\dfrac{2}{\sqrt{3} + \sqrt{2}}$.

1. Multiply ▶ $\dfrac{2}{\sqrt{3} + \sqrt{2}} = \dfrac{2}{\sqrt{3} + \sqrt{2}} \cdot \dfrac{\sqrt{3} - \sqrt{2}}{\sqrt{3} - \sqrt{2}}$ Multiply numerator and denominator by the rationalizing factor of the denominator.

2. Simplify ▶ $\qquad\qquad = \dfrac{2(\sqrt{3} - \sqrt{2})}{3 - 2}$

Solution ▶ $\dfrac{2}{\sqrt{3} + \sqrt{2}} = 2\sqrt{3} - 2\sqrt{2}$

Another Example ▶ $\dfrac{\sqrt{7} + \sqrt{3x}}{\sqrt{7} - \sqrt{3x}} = \dfrac{\sqrt{7} + \sqrt{3x}}{\sqrt{7} - \sqrt{3x}} \cdot \dfrac{\sqrt{7} + \sqrt{3x}}{\sqrt{7} + \sqrt{3x}}$

$\qquad\qquad\quad = \dfrac{7 + 2\sqrt{21x} + 3x}{7 - 3x}$

Make Sure

Rationalize the denominator of each rational expression. Assume radicands are positive.

See Example 3 ▶ 1. $\dfrac{3}{3 - \sqrt{3}}$ 　　　 2. $\dfrac{\sqrt{3} - \sqrt{2}}{\sqrt{3} + \sqrt{2}}$ 　　　 3. $\dfrac{2\sqrt{a} - 3\sqrt{b}}{3\sqrt{a} - 2\sqrt{b}}$

_____ 　　 _____ 　　 _____

MAKE SURE ANSWERS: 1. $\dfrac{3 + \sqrt{3}}{2}$ 2. $5 - 2\sqrt{6}$ 3. $\dfrac{6a - 6b - 5\sqrt{ab}}{9a - 4b}$

8.3 Practice: Assume all radicands are positive.

Set 1: Simplify and combine like radical expressions.

1. $5\sqrt{3} + 7\sqrt{3}$
2. $4\sqrt[3]{5} + 3\sqrt[3]{5}$
3. $3a\sqrt{80a^5} + 2\sqrt{45a^7}$

4. $3\sqrt{48b^3} + 2b\sqrt{27b}$
5. $4\sqrt[3]{81c^5} - 3c\sqrt[6]{576c^4}$
6. $4d\sqrt[4]{243d^3} - 6\sqrt[8]{9d^{14}}$

7. $\sqrt[3]{54e^2} + \sqrt[3]{128e^2}$
8. $\sqrt[4]{32f^3} + \sqrt[4]{162f^3}$
9. $\sqrt{16g^3h} + \sqrt[4]{16g^6h^2}$

Set 2: Multiply radical expressions.

10. $\sqrt{3}\sqrt{15}$
11. $\sqrt{6}\sqrt{12}$
12. $\sqrt{3}(2 - \sqrt{3})$

13. $\sqrt{7}(5 - \sqrt{7})$
14. $\sqrt{3a}(\sqrt{3a} + \sqrt{27a^3})$
15. $\sqrt{5b}(\sqrt{5b} + \sqrt{45b^3})$

16. $(4\sqrt{3c} - 3\sqrt{2d})(\sqrt{3c} + \sqrt{2d})$
17. $(3\sqrt{2e} - \sqrt{3f})(3\sqrt{2e} + 4\sqrt{3f})$

18. $(2\sqrt{3f} - 4\sqrt{2g})^2$
19. $(5\sqrt{2h} + 2\sqrt{3k})^2$

20. $(\sqrt{3m} - \sqrt{2n})(\sqrt{3m} + \sqrt{2n})$
21. $(\sqrt{4p} - 2\sqrt{3q})(\sqrt{4p} + 2\sqrt{3q})$

Chapter 8 Exponents and Radicals

Set 3: Rationalize the denominator of each rational expression.

22. $\dfrac{6}{2 - \sqrt{2}}$

23. $\dfrac{4}{3 + \sqrt{3}}$

24. $\dfrac{3}{\sqrt{2} - \sqrt{3}}$

25. $\dfrac{5}{\sqrt{2} + \sqrt{3}}$

26. $\dfrac{4}{\sqrt{a} - \sqrt{3}}$

27. $\dfrac{5}{\sqrt{b} + \sqrt{3}}$

28. $\dfrac{\sqrt{c}}{\sqrt{c} - \sqrt{d}}$

29. $\dfrac{\sqrt{f}}{\sqrt{e} + \sqrt{f}}$

30. $\dfrac{3\sqrt{2} + 4\sqrt{3}}{5\sqrt{3} - 6\sqrt{2}}$

31. $\dfrac{6\sqrt{5} - 5\sqrt{3}}{2\sqrt{3} + 4\sqrt{5}}$

32. $\dfrac{\sqrt{g} - \sqrt{h}}{\sqrt{g} + \sqrt{h}}$

33. $\dfrac{\sqrt{m} + 2\sqrt{n}}{2\sqrt{m} - \sqrt{n}}$

34. $\dfrac{2\sqrt{p} - 3\sqrt{q}}{3\sqrt{p} + 2\sqrt{q}}$

35. $\dfrac{4\sqrt{r} + 3\sqrt{s}}{3\sqrt{r} - 2\sqrt{s}}$

36. $\dfrac{3\sqrt{2t} - 4\sqrt{3u}}{4\sqrt{3t} + 5\sqrt{2u}}$

37. $\dfrac{5\sqrt{3v} + 2\sqrt{5w}}{2\sqrt{5v} - 5\sqrt{3w}}$

38. $\dfrac{\sqrt{x - 3} - \sqrt{x}}{\sqrt{x - 3} + \sqrt{x}}$

39. $\dfrac{\sqrt{x - y} + \sqrt{x + y}}{\sqrt{x - y} - \sqrt{x + y}}$

Review: Work these problems on a separate sheet of paper. Attach your work to this page.

Find the square of each binomial. (See Lesson 5.3.)
40. $(z + 2)^2$
41. $(y - 3)^2$
42. $(\sqrt{x} - 2)^2$
43. $(\sqrt{w - 2} + 3)^2$
44. $(\sqrt{v + 3} - 4)^2$
45. $(\sqrt{3u - 2} - 3)^2$

Solve equations of the form $ax^2 + bx + c = 0$ by factoring. (See Lesson 6.6.)
46. $t^2 - 3t - 4 = 0$
47. $s^2 - s - 2 = 0$
48. $3r^2 - 5r - 2 = 0$
49. $2q^2 - q - 3 = 0$
50. $4p^2 + 4p + 1 = 0$
51. $4n^2 - 4n + 1 = 0$
52. $m^2 - 9m + 20 = 0$
53. $k^2 + 7k + 12 = 0$
54. $12h^2 - 7h + 1 = 0$

8.4 Solve Radical Equations

An equation that has one or more radicals that contains a variable in the radicand is called a *radical equation*.

Examples ▶ (a) $4 + \sqrt{3x} = 10$ (b) $\sqrt[3]{2y + 55} = 5$ (c) $\sqrt{z^2 + 30} = \sqrt{z}$ ⟵ radical equations

To solve a radical equation, you first *clear the radical(s)*. You can do this by using the *Power Principle for Equations*.

> **The Power Principle for Equations**
> If x is a solution of the equation: $a = b$, then x is a solution of the equation: $a^n = b^n$ for any natural number n.

If an equation has only one radical term, you can clear the radical by isolating it as one member of the equation, and then apply the Power Principle for Equations using the index of the radical as n.

EXAMPLE 1: Solve an equation containing a single radical term.

Solve ▶ $\sqrt[3]{3x - 10} + 5 = 7$ Agreement: The radicands in this lesson are assumed to be nonnegative.

1. Isolate radical ▶ $\sqrt[3]{3x - 10} + 5 = 7$

$\sqrt[3]{3x - 10} = 2$ The radical is isolated as the left member.

2. Clear radical ▶ $(\sqrt[3]{3x - 10})^3 = 2^3$ Power Principle: The radical is a cube root, so $n = 3$.

$3x - 10 = 8$ ⟵ radicals cleared

3. Solve as before ▶ $3x = 18$

$x = 6$ ⟵ proposed solution

4. Check ▶ $\sqrt[3]{3x - 10} + 5 = 7$ ⟵ original equation

$\sqrt[3]{3(6) - 10} + 5$	7
$\sqrt[3]{8} + 5$	7
$2 + 5$	7
7	7	⟵ 6 checks

Solution ▶ $x = 6$

> **CAUTION:** The converse of the Power Principle may not be true. That is, if x is a solution of the equation: $a^n = b^n$, x may not be a solution of the equation: $a = b$.

Example ▶ -3 is the only solution of: $x = -3$, but: $x^2 = (-3)^2$ has both 3 and -3 as solutions. The squaring of both members of $x = -3$ has produced the *extraneous solution* 3.

A solution of the equation: $a^n = b^n$, which is not a solution of equation: $a = b$, is called an *extraneous solution*.

> Proposed solutions obtained by using the Power Principle must be checked to determine whether they are actual solutions of the original equation or whether they are extraneous solutions.

314　Chapter 8　Exponents and Radicals

Agreement ▶ In this section, the word *solve* means to find proposed solutions and check to eliminate all extraneous solutions.

Example ▶ Solve $x = 3 + \sqrt{x - 1}$. 　　　Assume $x - 1$ is nonnegative.

1. Isolate radical ▶ 　　　$x = 3 + \sqrt{x - 1}$

　　　　　　　　　　$x - 3 = \sqrt{x - 1}$ 　　　The radical is isolated as the right member.

2. Clear radical ▶ 　　$(x - 3)^2 = (\sqrt{x - 1})^2$ 　　　Power Principle: The radical is a square root, so $n = 2$.

　　　　　　　　　$x^2 - 6x + 9 = x - 1$ ⟵ radical cleared

3. Solve as before ▶ 　$x^2 - 7x + 10 = 0$ 　　　Write standard form.

　　　　　　　　　$(x - 2)(x - 5) = 0$ 　　　Factor.

　　　　　　　$x - 2 = 0$ or $x - 5 = 0$ 　　　Use the zero-product property.

　　　　　　　　$x = 2$ or 　　$x = 5$ ⟵ proposed solutions

4. Check ▶

$x = 3 + \sqrt{x - 1}$	⟵ original equation ⟶	$x = 3 + \sqrt{x - 1}$
2　|　$3 + \sqrt{2 - 1}$		5　|　$3 + \sqrt{5 - 1}$
2　|　$3 + \sqrt{1}$		5　|　$3 + \sqrt{4}$
2　|　4		5　|　5

5. Interpret ▶ $2 \neq 4$ means 2 is not a solution. 　　　$5 = 5$ means 5 is a solution.

Solution ▶ $x = 5$

Some radical equations have no solution.

Example ▶ Solve $\sqrt{2x + 5} = -7$.

1. Clear radical ▶ 　$(\sqrt{2x + 5})^2 = (-7)^2$ 　　　The radical is already isolated.

　　　　　　　　　　$2x + 5 = 49$ ⟵ radical cleared

2. Solve as before ▶ 　　　$2x = 44$

　　　　　　　　　　　$x = 22$ ⟵ proposed solution

3. Check ▶

$\sqrt{2x + 5} = -7$	⟵ original equation
$\sqrt{2(22) + 5}$	-7
$\sqrt{49}$	-7
7	-7 ⟵ 22 does not check

Solution ▶ The equation $2x + 5 = -7$ has no solution.

Note ▶ You could have determined that the above equation has no solution by observing that the principle square root of a number is never negative.

Make Sure

Solve equations containing a single radical term. Check for extraneous solutions.

See Example 1 ▶
1. $\sqrt{2a} = 6$ _____
2. $\sqrt{3b + 1} = 5$ _____
3. $\sqrt[3]{1 - c^2} = 2$ _____

MAKE SURE ANSWERS: 1. 18 2. 8 3. no solution

If an equation has more than one term containing a radical, you may need to use the Power Principle more than once.

> **To solve a radical equation containing more than one radical:**
> 1. Isolate one of the radical terms.
> 2. Use the Power Principle for Equations.
> 3. If the equation is not yet clear of radicals, repeat Steps 1 and 2.
> 4. Solve the resulting equation for the unknown.
> 5. Check proposed solutions in the original equation.

EXAMPLE 2: Solve a radical equation containing more than one radical expression.

Solve ▶ $\sqrt{3y + 7} - \sqrt{y - 5} = 4$

1. Isolate one of the radicals ▶
$\sqrt{3y + 7} - \sqrt{y - 5} = 4$
$\sqrt{3y + 7} = \sqrt{y - 5} + 4$ ← $\sqrt{3y + 7}$ is isolated

2. Clear radicals ▶
$(\sqrt{3y + 7})^2 = (\sqrt{y - 5} + 4)^2$ — Use the Power Principle.
$3y + 7 = y - 5 + 8\sqrt{y - 5} + 16$ — do not forget this term

3. Repeat Steps 1 and 2 ▶
$2y - 4 = 8\sqrt{y - 5}$ Isolate the remaining radical term.
$y - 2 = 4\sqrt{y - 5}$ Divide all terms by 2.
$(y - 2)^2 = (4\sqrt{y - 5})^2$ Use the Power Principle.
$y^2 - 4y + 4 = 16(y - 5)$ ← radicals cleared

4. Solve as before ▶
$y^2 - 4y + 4 = 16y - 80$
$y^2 - 20y + 84 = 0$ Write standard form.
$(y - 6)(y - 14) = 0$ Factor.
$y - 6 = 0$ or $14 = 0$ The zero-product property.
$y = 6$ or $y = 14$ ← proposed solutions

5. Check ▶

$\sqrt{3y + 7} - \sqrt{y - 5} = 4$ ← original equation → $\sqrt{3y + 7} - \sqrt{y - 5} = 4$

$\sqrt{3(6) + 7} - \sqrt{6 - 5}$	4		$\sqrt{3(14) + 7} - \sqrt{14 - 5}$	4
$\sqrt{25} - \sqrt{1}$	4		$\sqrt{49} - \sqrt{9}$	4
4	4		4	4

Solution ▶ $y = 6$ or 14

CAUTION: When you square a binomial do not forget the middle term.

Example ▶ A common error is to write: $(\sqrt{y+5}+4)^2$ as $(\sqrt{y+5})^2+4^2$ instead of as the correct result: $(\sqrt{y+5})^2+8\sqrt{y+5}+4^2$.

You can solve some equations containing two radicals with only one use of the Power Principle.

Examples ▶ (a) $\quad \sqrt{x+4} = \sqrt{-x+10}$

$\quad\quad (\sqrt{x+4})^2 = (\sqrt{-x+10})^2 \quad$ Use the Power Principle.

$\quad\quad x+4 = -x+10$

$\quad\quad 2x = 6$

$\quad\quad x = 3 \longleftarrow$ proposed solution

$$\begin{array}{c|c} \sqrt{x+4} & = \sqrt{-x+10} \\ \sqrt{3+4} & \sqrt{-3+10} \\ \sqrt{7} & \sqrt{7} \end{array}$$ Check as before.

The solution is 3.

(b) $\quad \sqrt{x}\sqrt{x+5} = 6$

$\quad\quad (\sqrt{x}\sqrt{x+5})^2 = 6^2 \quad$ Use the Power Principle.

$\quad\quad x(x+5) = 36$

$\quad\quad x^2 + 5x = 36$

$\quad\quad x^2 + 5x - 36 = 0$

$\quad\quad (x+9)(x-4) = 0$

$\quad\quad x+9 = 0 \text{ or } x-4 = 0$

$\quad\quad x = -9 \text{ or } \quad x = 4 \longleftarrow$ proposed solutions

$$\begin{array}{c|c} \sqrt{x}\sqrt{x+5} = 6 & \sqrt{x}\sqrt{x+5} = 6 \\ \sqrt{-9}\sqrt{-9+5} \mid 6 & \sqrt{4}\sqrt{4+5} \mid 6 \\ & 2 \cdot 3 \mid 6 \\ & 6 \mid 6 \longleftarrow 4 \text{ checks} \end{array}$$

Stop! These radicals do not represent real numbers.

The only solution is 4.

Make Sure

Solve equations containing more than one radical term. Check for extraneous solutions.

See Example 2 ▶ 1. $\sqrt{3a+7} + \sqrt{a+3} = 2$ _____ 2. $\sqrt{b} + \sqrt{b-1} = 3$ _____

MAKE SURE ANSWERS: 1. -2 2. $\frac{25}{9}$

Name _____ Date _____ Class _____

8.4 Practice: *Check for extraneous solutions.*

Set 1: Solve equations containing a single radical term.

1. $\sqrt{z-1} = 4$
2. $\sqrt{y+3} = 1$
3. $\sqrt{x^2-7} = 3$

4. $\sqrt{w^2-9} = 4$
5. $\sqrt[3]{v^2+15} = 4$
6. $\sqrt[3]{u^2-1} = 2$

Set 2: Solve equations containing more than one radical term.

7. $\sqrt{a+2} = \sqrt{2a}$
8. $\sqrt{3-2b} = \sqrt{b}$
9. $\sqrt{2c-3} + \sqrt{3-2c} = 0$

10. $\sqrt{d-1} + \sqrt{2-2d} = 0$
11. $\sqrt{e+2} - \sqrt{2e-2} = 0$
12. $\sqrt{4f-1} - \sqrt{2f+3} = 0$

13. $\sqrt{g-3} + \sqrt{g+2} = 5$
14. $\sqrt{h-5} + \sqrt{h+3} = 4$
15. $\sqrt{2k-1} - \sqrt{2k+3} = 0$

1. _____
2. _____
3. _____
4. _____
5. _____
6. _____
7. _____
8. _____
9. _____
10. _____
11. _____
12. _____
13. _____
14. _____
15. _____

Mixed Practice: Solve each equation.

16. $\sqrt{z+2} = z$
17. $\sqrt{4y-4} = y$
18. $\sqrt{x^2+4} = x+1$

19. $\sqrt{w^2-8} = w+4$
20. $\sqrt{v+5} - \sqrt{v-3} = 2$
21. $\sqrt{u+14} - \sqrt{u+2} = 2$

22. $\sqrt{5t-6} - \sqrt{t-2} = 2$
23. $\sqrt{10s-5} - \sqrt{2s-5} = 4$
24. $\sqrt{8r-7} - 3 = \sqrt{2r-4}$

25. $\sqrt{7q-12} - 3 = \sqrt{q-3}$
26. $\sqrt{6p+28} - 2 = \sqrt{2p+8}$

27. $\sqrt{7n+30} - 3 = \sqrt{n+3}$
28. $\sqrt{3m+13} + 3 = \sqrt{9m+52}$

29. $\sqrt{k+5} + 4 = \sqrt{9k+61}$
30. $\sqrt{h-3} = 2\sqrt{h} - \sqrt{h+5}$

Review: Work these problems on a separate sheet of paper. Attach your work to this page.

Solve each second-order system. (See Lesson 4.2.)

31. $\begin{cases} x + y - 4 = 0 \\ 2x - 3y - 3 = 0 \end{cases}$
32. $\begin{cases} 4x - 3y - 7 = 0 \\ 2x - 3y + 1 = 0 \end{cases}$
33. $\begin{cases} 2x - 3y + 7 = 0 \\ 5x + y + 9 = 0 \end{cases}$

Add or subtract polynomials. (See Lesson 5.2.)
34. $(3 + x) + (5 + 3x)$
35. $(2 - 3y) + (3 + 7y)$
36. $(3 - 5z) + (5z - 5)$
37. $(4 - 2u) - (3 + 5u)$
38. $(3 - 5v) - (4 - 3v)$
39. $(5 - 2w) - (5w - 7)$

Problem Solving 10: Solve Formulas Containing Radicals

To solve a formula containing a radical for a given letter, you first use the Power Principle.

EXAMPLE: Solve a formula containing a radical for a given letter.

Problem ▶ Solve $f = \dfrac{1}{2\pi\sqrt{\dfrac{LC_1C_2}{C_1+C_2}}}$ (frequency formula) for C_2.

Remember, clear radicals and isolate the given letter.

1. Clear radical ▶
$$f^2 = \left(\dfrac{1}{2\pi\sqrt{\dfrac{LC_1C_2}{C_1+C_2}}}\right)^2$$

$$f^2 = \dfrac{1^2}{2^2\pi^2\left(\sqrt{\dfrac{LC_1C_2}{C_1+C_2}}\right)^2}$$

$$f^2 = \dfrac{1}{4\pi^2 \cdot \dfrac{LC_1C_2}{C_1+C_2}} \quad \longleftarrow \text{radical cleared}$$

2. Clear fractions ▶
$$f^2 = \dfrac{1}{\dfrac{4\pi^2 LC_1C_2}{C_1+C_2}} \quad \longleftarrow \text{simplify denominator}$$

$$f^2 = \dfrac{C_1+C_2}{4\pi^2 LC_1C_2} \qquad \text{Think: } \dfrac{1}{\dfrac{a}{b}} = 1 \div \dfrac{a}{b} = 1 \cdot \dfrac{b}{a} = \dfrac{b}{a}$$

$$4\pi^2 LC_1C_2 \cdot f^2 = 4\pi^2 LC_1C_2 \cdot \dfrac{C_1+C_2}{4\pi^2 LC_1C_2} \qquad \text{Think: The LCD is } 4\pi^2 LC_1C_2.$$

$$4\pi^2 LC_1C_2 f^2 = C_1 + C_2 \longleftarrow \text{fractions cleared}$$

3. Isolate given letter ▶ $\quad 4\pi^2 LC_1C_2 f^2 - C_2 = C_1 \qquad$ Collect all terms containing the given letter C_2 in one member.

$\qquad C_2(4\pi^2 LC_1 f^2 - 1) = C_1 \qquad$ Factor out the given letter C_2.

Solution ▶ given letter ⟶ $C_2 = \dfrac{C_1}{4\pi^2 LC_1 f^2 - 1}$

Practice: Solve for each given letter.

1. _____

2. _____ $\qquad$ Solve $E = \sqrt{PR}$ (electricity formula) for: $\qquad$ 1. P $\quad$ 2. R

3. _____ $\qquad$ Solve $I = \sqrt{\dfrac{P}{R}}$ (electricity formula) for: $\qquad$ 3. P $\quad$ 4. R

4. _____

Chapter 8 Exponents and Radicals

Solve $X = \sqrt{Z^2 - R^2}$ (electricity formula) for: 5. Z^2 6. R^2

Solve $l = \sqrt{d^2 + 5.3s^2}$ (wire sag formula) for: 7. d^2 8. s^2

Solve $T = 2\pi\sqrt{\dfrac{L}{g}}$ (pendulum formula) for: 9. L 10. g

Solve $f = \dfrac{1}{2\pi\sqrt{LC}}$ (electronics formula) for: 11. L 12. C

Solve $T = 2\pi\sqrt{\dfrac{m}{k}}$ (spring period formula) for: 13. m 14. k

Solve $v = r\sqrt{\dfrac{g}{r + h}}$ (orbital speed formula) for: 15. g 16. h

Solve $M = \dfrac{m}{\sqrt{1 - \dfrac{v^2}{c^2}}}$ (relativity formula) for: 17. m 18. v^2

Solve $X = \dfrac{\theta}{\sqrt{1 + (2\pi ft)^2}}$ (heat transfer formula) for: 19. θ 20. t^2

Solve $I = \dfrac{E}{\sqrt{R^2 + (\omega L)^2}}$ (electricity formula) for: 21. E 22. L^2

Solve $f = \dfrac{1}{2\pi\sqrt{L\dfrac{1}{C_1} + \dfrac{1}{C_2}}}$ (frequency formula) for: 23. L 24. C_1

Solve $A = \sqrt{s(s-a)(s-b)(s-c)}$ (Hero's formula) for: 25. a 26. b

Copyright © 1985 by Harcourt Brace Jovanovich, Inc. All rights reserved.

Problem Solving 11: Evaluate Formulas Containing Radicals

You can use the following formula to find the *period for a pendulum* (the time it takes a pendulum to complete one swing from left to right and then back again).

$$T = 2\pi \sqrt{\frac{L}{g}} \text{ where } \begin{cases} T \text{ is the period in seconds.} \\ L \text{ is the length of the pendulum in feet.} \\ g \text{ is the gravitational pull in feet per second per second.} \end{cases}$$

To find the period for a pendulum on earth, you can use $g = 32$ because the gravitational pull near the earth's surface is about 32 feet per second per second.

EXAMPLE: Solve this problem by evaluating the pendulum formula.

1. Read and identify ▶ In 1851, a physicist named J.B.L. Foucault used a (200-foot pendulum) to measure the earth's gravitational pull more accurately. What was the period of the Foucault pendulum, using $\pi \approx$ (3.14)?

Remember, circle the facts and underline the question.

2. Understand ▶ The question asks you to find the time it takes for a pendulum on earth to complete one swing from left to right and back again.

3. Decide ▶ To find the period of a pendulum on earth, you **use $g = 32$.**

length = 200 ft
period = ?

4. Evaluate ▶ $T = 2\pi \sqrt{\frac{L}{g}}$

$\approx 2(3.14) \sqrt{\frac{200}{32}}$

$= 6.28 \sqrt{\frac{25}{4}}$

$= 6.28 \cdot \frac{5}{2}$

$= 15.7$

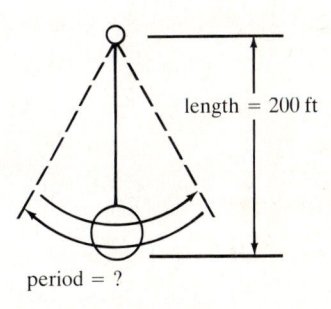

5. Intercept ▶ 15.7 means the period of the Foucault pendulum was about 15.7 seconds.

Practice: Round to the nearest tenth when necessary. Use $\pi \approx 3.14$.

1. Find the period of a pendulum on earth if its length is 8 feet.

2. Find the length of a pendulum on earth if its period is 8 seconds.

3. What should the length of a pendulum be in order for it to "beat seconds" on earth? (Hint: To "beat seconds" means to take one second to move from left to right and another full second to move back again.)

4. The gravitational pull on the moon is about $\frac{1}{6}$ that of earth. a) What length must a pendulum be in order to "beat seconds" on the moon? b) A pendulum with a normal 2-second period on earth would have what period on the moon?

5. By how much must the length of the pendulum in Problem 3 be increased to double its period on earth?

6. The United Nations building in New York City has a Foucault-type pendulum that is 75 feet long. What is its period?

1. _____
2. _____
3. _____
4a. _____
b. _____
5. _____
6. _____

Copyright © 1985 by Harcourt Brace Jovanovich, Inc. All rights reserved.

322   Chapter 8   Exponents and Radicals

The distance that a person can see to the horizon is given by the formula:

$$d = \sqrt{\frac{3h}{2}} \text{ where } \begin{cases} h \text{ is the height in feet above the sea or level ground.} \\ d \text{ is the distance in miles that a person can see to the horizon.} \end{cases}$$

7. How far can a person see to the horizon out of a jet airliner that is flying at 40,000 feet?

8. How high must a person be to see 15 miles to the horizon?

9. How far can a person see to the horizon while standing at the ocean shore if the eye level is 6 feet from the ground?

10. How high is the eye level of a person who can see 2 miles to the horizon while standing at the ocean shore?

An object that circles a planet without using power is called a *satellite* of the planet and is said to be in *orbit* around the planet. The speed needed to keep a satellite in orbit is called the *orbital speed* (S_o). The speed needed for an object to break away from a planet's gravitational pull and fly into space is called *escape speed* (S_e). The orbital speed and escape speed can be calculated by evaluating the following two formulas:

$$S_o = r\sqrt{\frac{g}{r+h}} \text{ where } \begin{cases} S_o \text{ is the orbital speed in miles per second (mps).} \\ g \text{ is the gravitational pull of the planet near the surface in miles per second per second (mps}^2\text{).} \\ r \text{ is the radius of the planet in miles (mi).} \\ h \text{ is the height of the object above the planet's surface in miles (mi).} \end{cases}$$

$$S_e = r\sqrt{\frac{2g}{r+h}} \text{ where } S_e \text{ is the escape speed in miles per second (mps).}$$

11. An orbit of 2.23×10^4 mi above the earth's surface is needed for a satellite to stay in *geostationary orbit* (hover over the exact same stationary point on the earth's surface). The earth's radius is 3.96×10^3 mi and the gravitational pull near the surface is 6.09×10^{-3} mps^2. What is the orbital speed necessary for a *geostationary orbit* in miles per hour? (Hint: 1 mps = 3.600×10^3 mph.)

12. On a future mission, the crew of the *U.S. Columbia* space vehicle will place a space station in orbit at 6.00×10^2 miles above the earth's surface. In miles per hour: a) What will be the orbital speed for the space station? b) What is the escape speed needed to fly into space from the space station? c) What is the percent decrease in escape speed needed to fly into space from the space station as compared with the escape speed from the earth's surface?

According to Guinness

TALLEST LIGHTHOUSE
THE WORLD'S TALLEST LIGHTHOUSE IS THE 348-FOOT-TALL TOWER IN YOKOHAMA, JAPAN, WITH A POWER OF 600,000 CANDLES.

PERIGEE AND APOGEE
THE CLOSEST APPROACH OF THE MOON TO THE EARTH IS 216,420 MILES. THE MOST EXTREME DISTANCE BETWEEN THE TWO IS 247,667 MILES.

13. To the nearest tenth mile, how far away can a person on a ship be and still see the light in the world's tallest lighthouse?

14. Using the average of the perigee and apogee measures as the height (h) of the moon above the earth, what is the moon's orbital speed around the earth?

8.5 Add and Subtract Complex Numbers

The system of real numbers has developed through a series of extensions. For example:

The equation

$$x + 7 = 5$$

has no solution among the natural numbers. It does have -2 as a solution if you extend the numbers under consideration to include integers.

The equation

$$2x = 3$$

has no solution among the integers. It does have $\frac{3}{2}$ as a solution if you extend the numbers under consideration to include rational numbers.

The equation

$$x^2 = 2$$

has no solution among the rational numbers. It does have $\sqrt{2}$ and $-\sqrt{2}$ as solutions if you extend the numbers under consideration to include irrational numbers.

The equation

$$x^2 = -1$$

has no solution among the real numbers. Recall that $x^2 \geq 0$ for every real number x. If $x^2 = -1$ is to have a solution, then the numbers under consideration must be extended beyond the set of real numbers. This extension will require a new kind of number whose square is a negative number.

> **DEFINITION 8.3:** The symbol i, called the *imaginary unit*, is a number whose square is -1.
> $$i^2 = -1$$

The following definition allows you to express the square root of any negative real number as the product of the imaginary unit i and a positive real number.

> **DEFINITION 8.4:** For any positive real number a: $\sqrt{-a} = i\sqrt{a}$

EXAMPLE 1: Write a radical expression in terms of i.

Problems ▶ Write $\sqrt{-4}$, $\sqrt{-27}$, and $-\sqrt{-12}$ in terms of i.

1. Use Definition 8.4 ▶ $\sqrt{-4} = i\sqrt{4}$ | $\sqrt{-27} = i\sqrt{27}$ | $-\sqrt{-12x^3} = -i\sqrt{12x^3}$

2. Simplify ▶ $= i2$ | $= i3\sqrt{3}$ | $= -i2x\sqrt{3x}$

Solutions ▶ $\sqrt{-4} = 2i$ | $\sqrt{-27} = 3i\sqrt{3}$ | $-\sqrt{-12x^3} = -2xi\sqrt{3x}$

324 Chapter 8 Exponents and Radicals

Make Sure

Write each radical expression in terms of i.

See Example 1 ▶   1. $\sqrt{-25}$ 2. $\sqrt{-45}$ 3. $\sqrt{-72a^3}$ 4. $\sqrt{-81bc^5}$

MAKE SURE ANSWERS: 1. $5i$ 2. $3i\sqrt{5}$ 3. $-6ai\sqrt{2a}$ 4. $9c^2i\sqrt{bc}$

Any number of the form bi, where b is a nonzero real number is called an *imaginary number*.

Examples ▶   (a) $2i$ (b) $5.3i$ (c) $-6i$ (d) $\sqrt{3}i$ ⟵ imaginary numbers

> If a and b are real numbers, the expression $a + bi$ is called a *complex number*. The number a is called the *real part* and the number b is called the *imaginary part* of the complex number $a + bi$.
>
> $a + bi$ ⟵ **complex number**
>   ↑ **imaginary part**
> ↑ **real part**

Examples ▶   (a) $4 + 3i$ (b) $-5 + i$ (c) $-\dfrac{3}{4} + 2i$ (d) 3 (e) $2i$ ⟵ complex numbers

Note 1 ▶   Every real number is a complex number (use $b = 0$): $a = a + 0 = a + 0i$.

Note 2 ▶   Every imaginary number is a complex number (use $a = 0$): $bi = 0 + bi$.

Examples ▶   Identify a (the real part) and b (the imaginary part) of $-5i + 2$, $4i$, and 3.

1. Write $a + bi$ form ▶ $-5i + 2 = 2 + (-5)i$ | $4i = 0 + 4i$ | $3 = 3 + 0i$

2. Label the real part with the letter a ▶ a | a | a Think: The real part is the left term of a complex number in $a + bi$ form.

3. Label the imaginery part with the letter b ▶ b | b | b Think: The imaginary part is the coefficient of i.

Solutions ▶   In $-5i + 2$, $a = 2$ and $b = -5$. | In $4i$, $a = 0$ and $b = 4$. | In 3, $a = 3$ and $b = 0$.

Since every complex number can be written as a binomial, it is natural to define arithmetic operations with complex numbers in a way that is consistent with the operations defined for binomials.

Addition of Complex Numbers	$(a + bi) + (c + di) = (a + c) + (b + d)i$
Subtraction of Complex Numbers	$(a + bi) - (c + di) = (a - c) + (b - d)i$

You can apply the above definitions by using the following procedures.

To add (subtract) two complex numbers:

1. Add (subtract) their real parts. Add (subtract) their imaginary parts.
2. Write the sum of the above results in $a \pm bi$ form.

EXAMPLE 2: Add (subtract) complex numbers.

Problem ▶ $(3 + 4i) + (-7 + 2i) = ?$

1. Add (subtract) real ▶ $(3 + 4i) + (-7 + 2i) = [3 + (-7)] + [4 + 2]i$
and imaginary parts

2. Simplify ▶ $= -4 + 6i$ ⟵ $a + bi$ form

Solution ▶ $(3 + 4i) + (-7 + 2i) = -4 + 6i$

Other Examples ▶ (a) $(-2 + 3i) - (7 - 5i) = [(-2) - (7)] + [3 - (-5)]i$
$= -9 + 8i$

(b) $(5 + 4i) - 3 = [5 - 3] + [4 - 0]i$
$= 2 + 4i$

(c) $(0.75 - 0.5i) + (0.25 + 3.75i) = [0.75 + 0.25] + [(-0.5) + (3.75)]i$
$= 1 + 3.25i$

(d) $4 + (2 - 3i) = [4 + 2] + [0 + (-3)]i$
$= 6 + (-3)i$
$= 6 - 3i$

Make Sure

Add or subtract complex numbers.

See Example 2 ▶ **1.** $(7 - 3i) + (5i - 4)$ _____ **2.** $(5 - 2i) - (3 - 5i)$ _____

MAKE SURE ANSWERS: 1. $3 + 2i$ 2. $2 + 3i$

Two complex numbers are *equal* if and only if both their real parts and their imaginary parts are equal.

EXAMPLE 3: Determine real values of a and b so that given complex numbers are equal.

Problem ▶ Determine the real values of a and b so that $5 + b^3 i = (a - 1) + 8i$.

1. Identify real and imaginary parts ▶

   real parts: 5 and $(a-1)$

   imaginary parts: $b^3 i$ and $8i$

2. Equate respective parts ▶ $5 = a - 1$ **and** $b^3 = 8$

3. Solve both equations ▶ $5 + 1 = a$ **and** $b = \sqrt[3]{8}$

   $6 = a$ **and** $b = 2$ Check as before.

Solution ▶ $5 + b^3 i = (a - 1) + 8i$ provided $a = 6$ and $b = 2$.

Make Sure

Determine the real values of a and b so that the given complex numbers are equal.

See Example 3 ▶ 1. $5 - a + (b^2 + 1)i = 7 + 5i$ _____   2. $5 - a^2 + b^2 i = 3 + 5i$ _____

MAKE SURE ANSWERS: 1. $a = -2, b = 2$ or -2 2. $a = \sqrt{2}$ or $-\sqrt{2}, b = \sqrt{5}$ or $-\sqrt{5}$

According to Guinness

HIGHEST WATERFALL

THE HIGHEST WATERFALL IN THE WORLD IS THE SALTO ANGEL (ANGEL FALLS), IN VENEZUELA, WITH A TOTAL DROP OF 3212 FEET.

3. _____   3. The time t (in seconds) that it takes an object to fall a distance d (in feet) is given by: $t = \sqrt{d}/4$. If it takes 12.865 seconds for water to drop from one level of Salto Angel to the next, how long is this portion of the waterfall, to the nearest whole foot?

8.5 Practice

Set 1: Write each radical expression in terms of *i*.

1. $\sqrt{-2}$
2. $\sqrt{-7}$
3. $\sqrt{-8} + \sqrt{-8}$

4. $\sqrt{-28} + \sqrt{-28}$
5. $\sqrt{-16}$
6. $\sqrt{-256}$

7. $\sqrt{-9} - \sqrt{-16}$
8. $\sqrt{-64} - \sqrt{-36}$
9. $\sqrt{-a^3 b^6}$

10. $\sqrt{-c^{10} d^{15}}$
11. $\sqrt{-8e^3} + \sqrt{-32e^3}$
12. $\sqrt{-125f^6} - \sqrt{-405f^6}$

Set 2: Perform the indicated operations.

13. $(3 + 5i) + (4 + 2i)$
14. $(7 + 3i) + (2 + 5i)$
15. $(3 - 7i) + (5 + 4i)$

16. $(6 - 5i) + (2 + 3i)$
17. $(5 - 2i) - (4 + 3i)$
18. $(6 - 4i) - (3 + 5i)$

19. $(2 + 4i) - (3 - 5i)$
20. $(6 - 2i) - (5 - 4i)$
21. $(6 - 3i) - (5i - 2)$

22. $(2 - 3i) - (6i - 4)$
23. $(4 - 5i) + (3i - 2)$
24. $(3 - 4i) + (6i - 5)$

25. $(6 - 5i) + (2i + 3)$
26. $(6 - 4i) + (5i + 3)$
27. $(3i - 4) - (3 - 5i)$

28. $(7 - 3i) + (2 - 4i) + (7i - 2)$
29. $(3 - 5i) - (2i - 3) + (4 - 3i)$

30. $(2i - 3) + (5 - 4i) - (3 - 2i)$
31. $(2i - 1) + (4 - 3i) - (i - 1)$

32. $(3 + 2i) - (2i - 5) - (3i + 4)$
33. $(4 + 3i) - (3i - 7) - (5i + 6)$

Set 3: Find the real values of a and b so that the given complex numbers are equal.

34. $5 + bi = a + 3i$

35. $a - 5i = 2 - bi$

36. $a + 3i = -5 - bi$

37. $-4 - bi = a + 5i$

38. $a + 3 + 5i = 7 + (b + 2)i$

39. $a + 4 + 2i = 5 + (b - 2)i$

40. $a - 2 + 7i = 5 - (b - 2)i$

41. $a - 5 - 5i = 7 + (b + 2)i$

42. $4(a - 3) + 3bi = 12 - (b - 4)i$

43. $5(a + 3) + 4bi = 20 - (3 - b)i$

44. $3a - 4bi = 16i$

45. $5a - bi = 4i$

46. $7a - 5bi = 14$

47. $6a - 4bi = 3$

48. $3 - 2i - (4 - 5i) = a + bi$

49. $4 + 6i - (3 - 4i) = a + bi$

Extra: Find the real values of x and y that satisfy the following equations.

50. $x + y - 4 + (2x - 3y - 3)i = 0$

51. $x - y + 2 + (2x - y)i = 0$

Review: Work these problems on a separate sheet of paper. Attach your work to this page.

Determine if the given ordered pair is a solution of the given equation. (See Lesson 3.1.)
52. $(2, 3)$, $2x + 3y = 13$ **53.** $(2, 3)$, $3x + 2y = 12$ **54.** $(-2, 3)$, $3x - 2y = 12$

Multiply radical expressions. (See Lesson 8.3, Example 2.)
55. $(2 + 3\sqrt{5})(5 - 2\sqrt{5})$ **56.** $(3\sqrt{2} - 4\sqrt{5})(3\sqrt{2} + 4\sqrt{5})$ **57.** $(2\sqrt{3} - 5\sqrt{2})(2\sqrt{3} + 5\sqrt{2})$

Divide by a binomial containing square roots. (See Lesson 8.3, Example 3.)

58. $\dfrac{4}{2 - \sqrt{3}}$

59. $\dfrac{2}{3 - \sqrt{5}}$

60. $\dfrac{3}{3 - \sqrt{2}}$

61. $\dfrac{4}{5 + \sqrt{3}}$

62. $\dfrac{3 + 4\sqrt{5}}{3 + 2\sqrt{5}}$

63. $\dfrac{2 - \sqrt{5}}{2 + 3\sqrt{5}}$

8.6 Multiply and Divide Complex Numbers

Multiplication of two complex numbers can be accomplished by the definition
$$(a + bi)(c + di) = (ac - bd) + (ad + bc)i;$$
however, the method shown in Example 1 below is generally used because it uses the FOIL method and does not require additional memorization.

EXAMPLE 1: Multiply complex numbers.

Problem ▶ $(-2 + 5i)(3 - 4i) = ?$

1. **Use FOIL method** ▶ $(-2 + 5i)(3 - 4i) = (-6) + 8i + 15i - 20i^2$

2. **Rename i^2 as -1** ▶ $\qquad = (-6) + 8i + 15i - 20(-1)$

3. **Write $a + bi$ form** ▶ $\qquad = [-6 + 20] + [8 + 15]i$

Solution ▶ $(-2 + 5i)(3 - 4i) = 14 + 23i$

If the complex numbers are not in $a \pm bi$ form, then they should be written in $a \pm bi$ form before computing the product.

Example ▶ $(6i^2 + i)(7 - \sqrt{-4}) = (-6 + i)(7 - 2i)$
$\qquad = (-42) + 12i + 7i - 2i^2$
$\qquad = (-42) + 12i + 7i - 2(-1)$
$\qquad = -40 + 19i$

To multiply a complex number by a real number, you can use a distributive property.

Examples ▶ (a) $4(-3 + 7i) = 4(-3) + 4(7i)$ (b) $i^6 = (i^4)(i^2)$ (c) $\sqrt{2} \cdot \sqrt{-3} = \sqrt{2} \cdot i\sqrt{3}$
$\qquad\qquad\qquad = -12 + 28i \qquad\qquad = (1)(-1) \qquad\qquad = i\sqrt{6}$
$\qquad\qquad\qquad\qquad\qquad\qquad\qquad = -1 \text{ or } -1 + 0i \quad = i\sqrt{6} \text{ or } 0 + i\sqrt{6}$

> CAUTION: The Product Rule for Radicals $\sqrt{a}\sqrt{b} = \sqrt{ab}$ does not hold when both radicands are negative numbers.

Example ▶ $\sqrt{-2} \cdot \sqrt{-3} = i\sqrt{2} \cdot i\sqrt{3} = -\sqrt{6}$ is correct. If you multiply the radicals using the Product Rule for Radicals, however, you get $\sqrt{-2} \cdot \sqrt{-3} = \sqrt{(-2)(-3)} = \sqrt{6}$ which is not correct!

> To compute $\sqrt{a}\sqrt{b}$ when both a and b are negative numbers, express each radical in terms of i before multiplying.

Make Sure

Multiply complex numbers.

See Example 1 ▶ 1. $(4 - 3i)(5 - 2i)$ _____ 2. $(\sqrt{3} + i\sqrt{2})(\sqrt{3} - i\sqrt{2})$ _____

MAKE SURE ANSWERS: 1. $14 - 23i$ 2. 5 or $5 + 0i$

330 Chapter 8 Exponents and Radicals

> The *conjugate* of $a + bi$ (denoted by $\overline{a + bi}$) is $a - bi$.
> The *conjugate* of $a - bi$ (denoted by $\overline{a - bi}$) is $a + bi$.

To find the conjugate of a complex number in $a \pm bi$ form, you change the sign of the imaginary part.

Example ▶

$$\overline{-5 - \sqrt{-9}} = \overline{-5 - 3i} \quad \Big| \quad \overline{-4i^2 + i} = \overline{4 + 1i} \quad \Big| \quad \overline{7} = \overline{7 + 0i} \quad \Big| \quad \overline{-8i} = \overline{0 - 8i}$$
$$= -5 + 3i \qquad\qquad = 4 - 1i \qquad\qquad = 7 - 0i \qquad\qquad = 0 + 8i$$

Note ▶ The conjugate of any real number a is a. The conjugate of any imaginary number bi is $-bi$.

> The product of a complex number and its conjugate is always a real number.

Examples ▶
(a) $(3 + 4i)(3 - 4i) = 3^2 - (4i)^2$
$\qquad\qquad\qquad = 9 - 16i^2$
$\qquad\qquad\qquad = 9 - (-16)$
$\qquad\qquad\qquad = 25 \longleftarrow$ a real number

(b) $(1 - 5i)(1 + 5i) = 1^2 - (5i)^2$
$\qquad\qquad\qquad = 1 - 25i^2$
$\qquad\qquad\qquad = 1 - (-25)$
$\qquad\qquad\qquad = 26 \longleftarrow$ a real number

To find the quotient of two complex numbers, you can use conjugates.

EXAMPLE 2: Divide complex numbers.

Problem ▶ $\dfrac{3 + 2i}{4 - i} = ?$

1. Multiply ▶ $\dfrac{3 + 2i}{4 - i} = \dfrac{3 + 2i}{4 - i} \cdot \dfrac{4 + i}{4 + i}$

Multiply both numerator and denominator by the conjugate of the denominator. This will make the denominator a real number.

2. Write $a \pm bi$ form ▶ $= \dfrac{(3 + 2i)(4 + i)}{(4 - i)(4 + i)}$

$= \dfrac{12 + 3i + 8i + 2i^2}{16 - i^2}$

$= \dfrac{12 + 11i - 2}{16 + 1}$

$= \dfrac{10 + 11i}{17}$

Solution ▶ $\dfrac{3 + 2i}{4 - i} = \dfrac{10}{17} + \dfrac{11}{17}i$

Other Examples ▶
(a) $\dfrac{1 - i}{1 + i} = \dfrac{1 - i}{1 + i} \cdot \dfrac{1 - i}{1 - i}$

$= \dfrac{1 - 2i + i^2}{1 + 1}$

$= \dfrac{-2i}{2}$

$= -i$ or $0 - 1i$

(b) $\dfrac{1}{-i} = \dfrac{1}{-i} \cdot \dfrac{i}{i}$

$= \dfrac{i}{-i^2}$

$= \dfrac{i}{1}$

$= i$ or $0 + 1i$

Make Sure

Divide complex numbers. Write your answers in $a \pm bi$ form.

See Example 2 1. $\dfrac{3 + 2i}{2 - 3i}$ _____   2. $\dfrac{5 - 4i}{6 - 5i}$ _____

MAKE SURE ANSWERS: 1. $0 + 1i$; 2. $\dfrac{50}{61} + \dfrac{1}{61}i$

If an expression involves more than one operation, it is important to perform the operations in the proper order.

EXAMPLE 3: Simplify an expression that involves complex numbers and a combination of operations. Write your answer in $a \pm bi$ form.

Problem ▶ $(3 + 4i) - \dfrac{(2 - 3i)^2}{1 + 5i} = \ ?$

1. Use the Order of Operations Rule ▶
$$(3 + 4i) - \dfrac{(2 - 3i)^2}{1 + 5i} = (3 + 4i) - \dfrac{4 - 12i + 9i^2}{1 + 5i}$$ Compute $(2 - 3i)^2$ first.

$$= (3 + 4i) - \dfrac{4 - 12i - 9}{1 + 5i}$$

$$= (3 + 4i) - \dfrac{-5 - 12i}{1 + 5i}$$

$$= (3 + 4i) - \dfrac{-5 - 12i}{1 + 5i} \cdot \dfrac{1 - 5i}{1 - 5i}$$ To divide by $1 + 5i$, use its conjugate.

$$= (3 + 4i) - \dfrac{-5 + 13i + 60i^2}{1 + 25}$$

$$= (3 + 4i) - \dfrac{-5 + 13i - 60}{26}$$

$$= (3 + 4i) - \left(\dfrac{-65}{26} + \dfrac{13}{26}i\right)$$

2. Simplify ▶
$$= \left(3 - \dfrac{-65}{26}\right) + \left(4 - \dfrac{13}{26}\right)i$$ Write answer in $a + bi$ form.

$$= \left(3 + \dfrac{5}{2}\right) + \left(4 - \dfrac{1}{2}\right)i$$

Solution ▶ $(3 + 4i) - \dfrac{(2 - 3i)^2}{1 + 5i} = \dfrac{11}{2} + \dfrac{7}{2}i$

332 Chapter 8 Exponents and Radicals

Make Sure

Perform the indicated operations.

See Example 3 ▶ 1. $\dfrac{(2 - 3i)^2}{(3 + 2i)^2}$ _____ 2. $\dfrac{(1 - i)^3 - (2 - 3i)(3 - 2i)}{(2 + 3i)(2 - 3i)}$ _____

MAKE SURE ANSWERS: 1. $-1 + 0i$ **2.** $-\dfrac{7}{11} + \dfrac{13}{11}i$

Some equations have complex numbers as solutions.

EXAMPLE 4: Determine if a given complex number is a solution of an equation.

Problem ▶ Determine if $2 + 3i$ is a solution of $x^2 - 4x + 13 = 0$.

1. Substitute ▶

$x^2 - 4x + 13 = 0$	← original equation
$(2 + 3i)^2 - 4(2 + 3i) + 13$ $\mid$ 0	Substitute $2 + 3i$ for x.

2. Compute ▶

$4 + 12i + 9i^2 - 8 - 12i + 13 \mid 0$

$4 + 12i - 9 - 8 - 12i + 13 \mid 0$

$(4 - 9 - 8 + 13) + (12i - 12i) \mid 0$

3. Compare ▶ $0 \mid 0$ Think: $0 = 0$ means $2 + 3i$ is a solution.

Solution ▶ $2 + 3i$ is a solution of $x^2 - 4x + 13 = 0$.

Another Example ▶ Determine if i is a solution of $2x^2 + 4x - 5 = 0$.

$2x^2 + 4x - 5 = 0$

$2(i)^2 + 4(i) - 5 \mid 0$

$-2 + 4i - 5 \mid 0$

$-7 + 4i \mid 0$ ← i is not a solution of $2x^2 + 4x - 5 = 0$

Make Sure

Determine if the given complex number is a solution of the given equation; write yes or no.

See Example 4 ▶ 1. $1 + i$, $x^2 + 1 = 0$ _____ 2. i, $x^3 + 3x^2 + 1x + 3 = 0$ _____

MAKE SURE ANSWERS: 1. no **2.** yes

8.6 Practice

Set 1: Multiply complex numbers.

1. $3(4 + 5i)$
2. $-5(2 + 3i)$
3. $-2i(5 - 3i)$

4. $3i(4 - 3i)$
5. $(2 + 3i)(4 + 5i)$
6. $(3 + 2i)(4 + 3i)$

7. $(3 + 4i)(2 - 5i)$
8. $(2 - i)(3 + 2i)$
9. $(3 + 2i)(3 - 2i)$

10. $(5 + 3i)(5 - 3i)$
11. $(2 + 5i)^2$
12. $(3 - 2i)^2$

Set 2: Divide complex numbers.

13. $\dfrac{4 - 2i}{2i}$
14. $\dfrac{6 - 9i}{3i}$
15. $\dfrac{4i}{2 - 3i}$

16. $\dfrac{3i}{1 - 2i}$
17. $\dfrac{1 - i}{1 + i}$
18. $\dfrac{1 + i}{1 - i}$

19. $\dfrac{2 - i}{3 + 2i}$
20. $\dfrac{3 + i}{2 + 3i}$
21. $\dfrac{3 - 4i}{1 - i}$

22. $\dfrac{5 - 3i}{2 - i}$
23. $\dfrac{2 - 5i}{2 + 3i}$
24. $\dfrac{4 - 2i}{3 + 2i}$

25. $\dfrac{3 + 2i}{3 - 2i}$
26. $\dfrac{1 + 2i}{1 - 2i}$
27. $\dfrac{2 + 5i}{5 + 2i}$

334 Chapter 8 Exponents and Radicals

Set 3: Perform the indicated operations.

28. $\dfrac{(1-i)(1+2i)}{2-i}$

29. $\dfrac{(1+i)(1-2i)}{2+i}$

30. $\dfrac{(2+3i)(3-2i)}{3+2i}$

31. $\dfrac{(3+2i)(3-2i)}{2-3i}$

32. $\dfrac{(1+i)^2}{(1-i)^2}$

33. $\dfrac{(2-i)^2}{(2+i)^2}$

34. $\dfrac{(3-i)^2-(2-i)(1+i)}{(2+i)^2}$

35. $\dfrac{(2+i)^2-(1-2i)(3-i)}{(i-3i)^2}$

36. $\dfrac{(i+1)^3}{(1-i)^2}$

Mixed Practice: Perform the indicated operations.

37. $(1-3i)^2(1+3i)^{-1}$

38. $(2+5i)^2(2-5i)^{-1}$

39. $\dfrac{3+2i}{2i(1-i)}$

40. $\dfrac{1-3i}{3i(2+i)}$

41. $(3-4i)^{-2}$

42. $(3+2i)^{-2}$

43. $\dfrac{3-i\sqrt{2}}{3+i\sqrt{2}}$

44. $\dfrac{3+2i\sqrt{3}}{3-2i\sqrt{3}}$

45. $i^6 + i^9 + i^{11} - i^{25}$

Set 4: Determine if the given complex number is a solution of the given equation; write yes or no.

46. i, $a^2 + 1 = 0$

47. $2i$, $b^2 + 4 = 0$

48. $1 - 3i$, $c^2 - 2c + 10 = 0$

49. $1 + 3i$, $d^2 - 2d + 10 = 0$

50. $3 - 2i$, $e^2 - 6e + 13 = 0$

51. $3 + 2i$, $f^2 - 6f + 13 = 0$

Review: Work these problems on a separate sheet of paper. Attach your work to this page.

Factor out the GCF from each polynomial. (See Lesson 6.1.)

52. $3z^2 + 6z$
53. $6y^2 + 12y$
54. $4x^2 - 2x$
55. $6w^2 - 3w$
56. $4v^2 + 6v$
57. $6u^2 + 9u$
58. $8t^2 - 12t$
59. $12s^2 - 18s$

Solve equations containing a common factor in each term by factoring. (See Lesson 6.6.)

60. $3a^2 + 6a = 0$
61. $6b^2 + 6b = 0$
62. $4c^2 - 2c = 0$
63. $6d^2 - 3d = 0$
64. $4e^2 + 6e = 0$
65. $6f^2 + 9f = 0$
66. $8g^2 - 12g = 0$
67. $12h^2 - 18h = 0$

Name _____ Date _____ Class _____

Chapter 8 Review

What to Review if You Have Trouble

Objectives		Lesson	Example	Page
Write rational exponent notation in radical notation	1. Write **a.** $3^{4/6}$ **b.** $x^{0.25}$ **c.** $(xy)^{2/6}$ and **d.** $xy^{2/6}$ in radical notation. **a.** _____ **b.** _____ **c.** _____ **d.** _____	8.1	1	296
Evaluate expressions containing rational exponents	2. Evaluate **a.** $8^{2/3}$ **b.** $32^{2/5}$ **c.** $(-64)^{2/3}$ and **d.** $-64^{2/3}$. **a.** _____ **b.** _____ **c.** _____ **d.** _____	8.1	2	297
Simplify expressions containing rational exponents	3. Simplify **a.** $w^{1/3} \cdot w^{2/5}$ **b.** $\dfrac{x^{1/2}}{x^{1/3}}$ and **c.** $(y^{2/5}z^{-1/4})^{3/2}$. **a.** _____ **b.** _____ **c.** _____	8.1	3	297
Write radical expressions in rational exponential notation	4. Write **a.** $\sqrt[6]{x^4}$ **b.** $\sqrt[4]{a^2b^3}$ and **c.** $\dfrac{w^5}{\sqrt{w}}$ in rational exponential notation. **a.** _____ **b.** _____ **c.** _____	8.1	4	298
Simplify radicals by factoring	5. Simplify **a.** $\sqrt{24a^{10}b^7}$ and **b.** $\sqrt[3]{54x^{12}y^{32}}$. **a.** _____ **b.** _____	8.2	1	301
Reduce the order of radicals	6. Reduce the order of $\sqrt[6]{16}$. _____	8.2	2	302
Rationalize the denominator of radical expressions	7. Rationalize the denominator of $\dfrac{5a}{\sqrt[3]{a^2b}}$. _____	8.2	3	303
Write radical expressions in simplest radical form	8. Write $\sqrt{\dfrac{42r^8s^5}{12t^3}}$ in simplest radical form. _____	8.2	4	304
Simplify and combine like radicals	9. $4x\sqrt{27x} + 5\sqrt{18x^3}$ _____	8.3	1	307
Multiply radical expressions	10. $(3\sqrt{x} - 2\sqrt{y})(\sqrt{x} + 5\sqrt{y})$ _____	8.3	2	308

Copyright © 1985 by Harcourt Brace Jovanovich, Inc. All rights reserved.

336 Chapter 8 Exponents and Radicals

Rationalize the denominator of fractions containing binomial denominators ▶	11. Rationalize the denominator of $\dfrac{\sqrt{x}+\sqrt{3}}{\sqrt{x}-\sqrt{3}}$.	_____	8.3 3	310
Solve radical equations ▶	12. Solve $\sqrt{4x+13}+2=7$.	_____	8.4 1	313
	13. Solve $\sqrt{5y+9}-\sqrt{y-7}=6$.	_____	8.4 2	315
Write radical expressions in terms of i ▶	14. Write a. $\sqrt{-9}$ b. $-\sqrt{288}$ and c. $\sqrt{-50}$ in terms of i. a. _____ b. _____ c. _____		8.5 1	323
Add (subtract) complex numbers ▶	15. $(7+2i)-(-5-3i)=?$ Write your answer in $a+bi$ form.	_____	8.5 2	325
Determine the values of unknowns that make complex numbers equal ▶	16. Determine the real values of a and b so that $2+b^3i=(a+3)+27i$.	_____	8.5 3	326
Multiply complex numbers ▶	17. $(-5-7i)(2+3i)=?$ Write your answer in $a+bi$ form.	_____	8.6 1	329
Divide complex numbers ▶	18. $(5-3i) \div (2+7i)=?$ Write your answer in $a+bi$ form.	_____	8.6 2	330
Simplify expressions containing complex numbers ▶	19. $(2-3i)+\dfrac{(3-4i)^2}{2-i}=?$ Write your answer in $a+bi$ form.	_____	8.6 3	331
Determine if a given complex number is a solution of an equation ▶	20. Determine if $1+i$ is a solution of $x^2-2x+2=0$, write yes or no.	_____	8.6 4	332
Solve radical formulas ▶	21. Solve $v=r\sqrt{\dfrac{2g}{r+h}}$ for h.	_____	PS 10	319
Evaluate radical formulas ▶	22. At an altitude of 60 miles from the earth, the earth's atmosphere ends. This is also the very lowest altitude a satellite can stay in orbit around the earth. What is the orbital speed, in mph, of a satellite that is just skimming over the earth's atmosphere? (Assume 60 is accurate to 3 significant digits.)	_____	PS 11	321

CHAPTER 8 REVIEW ANSWERS: 1a. $\sqrt[3]{32}$ b. $\sqrt[4]{xy}$ c. $\sqrt[3]{x}$ d. $x\sqrt[3]{y}$ 2a. 4 b. 4 c. 16 d. 16 3a. $w^{11/15}$
b. $x^{1/6}$ c. $\dfrac{y^{3/5}}{z^{3/8}}$ 4a. $x^{2/3}$ b. $a^{1/2}b^{3/4}$ c. $w^{9/2}$ 5a. $2a^5b^3\sqrt{6b}$ b. $3x^4y^{10}\sqrt{2y^3}$ 6. $\sqrt[4]{4}$ 7. $\dfrac{5y\sqrt[3]{ab^2}}{b}$ 8. $\dfrac{r^4s^2\sqrt{14st}}{2t^2}$
9. $27x\sqrt{3x}$ 10. $3x+13\sqrt{xy}-10y$ 11. $\dfrac{x+2\sqrt{3x}+3}{x-3}$ 12. 3 13. 8, 11 14a. $3i$ b. $-12i\sqrt{2}$ c. $5i\sqrt{2}$
15. $12+5i$ 16. $a=-1, b=3$ 17. $11+(-29)i$ 18. $-\tfrac{5}{3}+(-\tfrac{33}{3})i$ 19. $4+(-14)i$ 20. Yes
21. $h=\dfrac{2r^2g-rv^2}{v^2}$ 22. 17,500 mph

NONLINEAR RELATIONS

9 Quadratic Equations

9.1 Solve Incomplete Quadratic Equations

9.2 Solve by Completing the Square

9.3 Solve by the Quadratic Formula

9.4 Rename and Solve Equations

PS 12: Solve Problems Using Quadratic Equations

9.5 Solve Inequalities

Introduction to Quadratic Equations

Polynomial equations whose highest degree term is two are called *quadratic equations*.

Examples ▶ (a) $x^2 - 9 = 0$ (b) $3y^2 + 4y + 5 = 0$ (c) $5z^2 = z$ ⟵ quadratic equations in one variable

(d) $x^2 + y^2 = 16$ (e) $xy = 8$ (f) $y = 2x^2 + 3x - 7$ ⟵ quadratic equations in two variables

The work in the next chapter will involve quadratic equations in two variables. The remainder of this chapter will only involve quadratic equations in one variable.

> An equation with form $ax^2 + bx + c = 0$ is called a *quadratic equation* in x.
>
> $ax^2 + bx + c = 0$ ⟵ quadratic equation
> - constant term
> - linear term
> - quadratic term

Every quadratic equation in this textbook will have real coefficients. The coefficient of the quadratic term of a quadratic equation in one variable cannot be zero because every quadratic equation in one variable must have a squared term. The term "quadratic" was derived from the Latin word for square, as in x^2.

In Chapter 6, you solved some quadratic equations by factoring. This chapter will introduce methods for solving all quadratic equations in one variable, whether they factor or not.

9.1 Solve Incomplete Quadratic Equations

The numerical coefficients of a quadratic equation play a major role in the determination of its solution(s). To identify the numerical coefficients of a quadratic equation, you first write it in standard form.

Recall ▶ A polynomial equation is in standard form if:
1. one member is zero,
2. the other member is a polynomial written in descending powers, and
3. the first term has a positive coefficient.

Note ▶ If a quadratic equation has all rational coefficients, then it is convenient to also require that its standard form contain only numerical coefficients that are integers.

EXAMPLE 1: Write a quadratic equation in standard form, and write each value for a, b, and c.

Problem ▶ Write $\dfrac{x^2}{3} = 3x - \dfrac{1}{4}$ in standard form and then write each value for a, b, and c.

1. Write standard form ▶ $\dfrac{1}{3} x^2 = 3 x - \dfrac{1}{4}$ Clear fractions when possible.

The LCD of $\frac{1}{3}$, 3, and $-\frac{1}{4}$ is 12.

$$12\left(\frac{x^2}{3}\right) = 12\left(3x - \frac{1}{4}\right)$$

$$\frac{12 \cdot x^2}{3} = 12 \cdot 3x - \frac{12}{4} \qquad \text{Use a distributive property.}$$

$$4x^2 = 36x - 3 \qquad \text{Think: The numerical coefficients are all integers.}$$

$$4x^2 - 36x + 3 = 0 \longleftarrow \text{descending powers set equal to zero}$$

$$a = 4, b = -36, c = 3 \qquad \begin{array}{l}\text{Think: } a \text{ is the numerical coefficient of the quadratic term.} \\ b \text{ is the numerical coefficient of the linear term.} \\ c \text{ is the constant term.}\end{array}$$

2. Write each value for a, b, and c

Solution ▶ A standard form of $\frac{x^2}{3} = 3x - \frac{1}{4}$ is $4x^2 - 36x + 3 = 0$ with $a = 4$, $b = -36$, and $c = 3$.

Other Examples ▶
(a) $2x^2 - 4x = 15$ is a quadratic equation in x. It can be written in standard form as $2x^2 - 4x - 15 = 0$ with $a = 2$, $b = -4$, and $c = -15$.

(b) $3y^2 = 10$ is a quadratic equation in y. It can be written in standard form as $3y^2 + 0y - 10 = 0$ with $a = 3$, $b = 0$, and $c = -10$.

(c) $\frac{w^2}{2} = 3w$ is a quadratic equation in w. It can be written in standard form as $w^2 - 6w + 0 = 0$ with $a = 1$, $b = -6$, and $c = 0$.

(d) $3z^2 + \sqrt{5}z = -1$ is a quadratic equation in z. It can be written in standard form as $3z^2 + \sqrt{5}z + 1 = 0$ with $a = 3$, $b = \sqrt{5}$, and $c = 1$.

Make Sure

Write each equation in standard form, and write each value for a, b, and c.

See Example 1 ▶ 1. $3z = 5 - 2z^2$ _____ 2. $\frac{y^2}{2} = 3 + \frac{y}{3}$ _____

_____ _____

MAKE SURE ANSWERS: 1. $2z^2 + 3z - 5 = 0$, $a = 2$, $b = 3$, $c = -5$ 2. $3y^2 - 2y - 18 = 0$, $a = 3$, $b = -2$, $c = -18$

A quadratic equation in which $b = 0$ or $c = 0$ is called an *incomplete quadratic equation*.

Examples ▶
(a) $3y^2 - 10 = 0$ is an incomplete quadratic equation with $b = 0$.
(b) $w^2 - 6w = 0$ is an incomplete quadratic equation with $c = 0$.
(c) $5x^2 = 0$ is an incomplete quadratic equation with both $b = 0$ and $c = 0$.

Note ▶ Any incomplete quadratic equation with both $b = 0$ and $c = 0$ simplifies to $ax^2 = 0$ form, which has 0 as its only solution.

340 Chapter 9 Quadratic Equations

Every equation of the form $ax^2 + bx = 0$ can be solved by factoring.

EXAMPLE 2: Solve an incomplete quadratic equation of the form $ax^2 + bx = 0$ by factoring.

Solve ▶ $\quad 8y^2 = 6y$

1. Write standard form ▶ $\quad 8y^2 - 6y = 0$

2. Factor ▶ $\quad 2y(4y - 3) = 0 \qquad$ Think: The GCF is $2y$.

3. Use zero-product property ▶ $\quad 2y = 0$ or $4y - 3 = 0$

4. Solve each equation ▶ $\quad y = 0$ or $\quad 4y = 3$

$\qquad\qquad\qquad\qquad y = 0$ or $\quad y = \frac{3}{4}\; \longleftarrow\;$ proposed solutions

Solution ▶ $\quad y = 0$ or $\frac{3}{4}$

Other Examples ▶

(a) $\qquad 2w^2 = -7w \qquad\qquad$ (b) $\qquad \sqrt{3}x^2 = 10x$

$\qquad 2w^2 + 7w = 0 \qquad\qquad\qquad\qquad \sqrt{3}x^2 - 10x = 0$

$\qquad w(2w + 7) = 0 \qquad\qquad\qquad\qquad x(\sqrt{3}x - 10) = 0$

$\qquad w = 0$ or $2w + 7 = 0 \qquad\qquad x = 0$ or $\sqrt{3}x - 10 = 0$

$\qquad w = 0$ or $\qquad 2w = -7 \qquad\qquad x = 0$ or $\qquad x = \dfrac{10}{\sqrt{3}}$

$\qquad w = 0$ or $\qquad w = -\frac{7}{2} \qquad\qquad x = 0$ or $\qquad x = \dfrac{10\sqrt{3}}{3}$

Make Sure

Solve incomplete quadratic equations of the form $ax^2 + bx = 0$ by factoring.

See Example 2 ▶ **1.** $6z - z^2 = 0$ _____ **2.** $\dfrac{2}{3}y = \dfrac{3}{4}y^2$ _____ **3.** $0.49x = 0.7x^2$ _____

MAKE SURE ANSWERS: 1. 0, 6   2. 0, $\frac{8}{9}$ 3. 0, 0.7

CAUTION: Dividing both members of an equation by an expression containing a variable may not produce an equivalent equation.

Example ▶ $8y^2 = 6y \longleftarrow$ original equation with roots of 0 and $\frac{3}{4}$ $\qquad$ See Example 2.

$\qquad\qquad 4y = 3 \qquad$ Divide both members by $2y$.

$\qquad\qquad y = \frac{3}{4} \longleftarrow$ new equation with only $\frac{3}{4}$ as a root

Division by $2y$ has produced an equation whose only root is $\frac{3}{4}$. The root 0 has been lost by the division process involving the variable y.

CAUTION: Do not solve equations by dividing by expressions containing a variable because you may lose solutions.

9.1 Solve Incomplete Quadratic Equations

You can solve the incomplete quadratic equation $x^2 = 4$ by factoring as in the following:

$$x^2 = 4$$
$$x^2 - 4 = 0$$
$$(x + 2)(x - 2) = 0$$
$$x + 2 = 0 \text{ or } x - 2 = 0$$
$$x = -2 \text{ or } x = 2$$

However, it is easier to solve $x^2 = 4$ using the *Square Root Rule*.

> **The Square Root Rule**
> If $p^2 = q$, then: $p = \sqrt{q}$ or $-\sqrt{q}$.

Note ▶ $p = \sqrt{q}$ or $p = -\sqrt{q}$ can be written as $p = \pm\sqrt{q}$, where the symbol $\pm$ is read "plus or minus."

Examples ▶

(a) If $x^2 = 4$ then: $x = \pm\sqrt{4}$
$x = \pm 2$

(b) If $x^2 = 7$ then: $x = \pm\sqrt{7}$

(c) If $x^2 = -16$ then: $x = \pm\sqrt{-16}$
$x = \pm 4i$

(d) If $x^2 = 8$ then: $x = \pm\sqrt{8}$
$x = \pm 2\sqrt{2}$

(e) If $x^2 = 0$ then: $x = \pm\sqrt{0}$
$x = 0$

(f) If $x^2 = -75$ then: $x = \pm\sqrt{-75}$
$x = \pm 5i\sqrt{3}$

To solve any incomplete quadratic equation of the form $ax^2 + c = 0$, you use the Square Root Rule.

EXAMPLE 3: Solve an incomplete quadratic equation of the form $ax^2 + c = 0$ using the Square Root Rule.

Solve ▶ $3x^2 - 36 = 0$

1. Solve for x^2 ▶ $3x^2 - 36 = 0$
$3x^2 = 36$
$x^2 = 12$

2. Use Square Root Rule ▶ $x = \pm\sqrt{12}$ Be sure to insert the $\pm$ sign in front of the radical.
$x = \pm 2\sqrt{3}$ Think: $\pm\sqrt{12} = \pm\sqrt{2^2 \cdot 3} = \pm 2\sqrt{3}$
$x = 2\sqrt{3}$ or $x = -2\sqrt{3}$ ⟵ proposed solutions

3. Check ▶

$3x^2 - 36 = 0$		$3x^2 - 36 = 0$	⟵ original equation
$3(-2\sqrt{3})^2 - 36$	0	$3(2\sqrt{3})^2 - 36$	0
$3[(-2)^2(\sqrt{3})^2] - 36$	0	$3[2^2(\sqrt{3})^2] - 36$	0
$3[4 \cdot 3] - 36$	0	$3[4 \cdot 3] - 36$	0
$36 - 36$	0	$36 - 36$	0
0	0	0	0

Solution ▶ $x = \pm 2\sqrt{3}$

Another Example ▶ Solve $27x^2 = -4$.

$$x^2 = \frac{-4}{27}$$

$$x = \pm\sqrt{\frac{-4}{27}} \qquad \text{Use the Square Root Rule.}$$

$$x = \pm\frac{\sqrt{-4}}{\sqrt{27}}$$

$$x = \pm\frac{2i}{3\sqrt{3}}$$

$$x = \pm\frac{2i}{3\sqrt{3}} \cdot \frac{\sqrt{3}}{\sqrt{3}} \qquad \text{Rationalize the denominator.}$$

$$x = \pm\frac{2i\sqrt{3}}{9}$$

$$x = \frac{2i\sqrt{3}}{9} \text{ or } x = -\frac{2i\sqrt{3}}{9} \longleftarrow \text{proposed solutions}$$

$27x^2 = -4$		$27x^2 = -4$	$\longleftarrow$ original equation
$27\left(\frac{2i\sqrt{3}}{9}\right)^2$	-4	$27\left(-\frac{2i\sqrt{3}}{9}\right)^2$	-4
$27\left(\frac{2^2 i^2 \sqrt{3}^2}{9^2}\right)$	-4	$27\left(\frac{(-2)^2 i^2 \sqrt{3}^2}{9^2}\right)$	-4
$27\left(\frac{4(-1)3}{81}\right)$	-4	$27\left(\frac{4(-1)3}{81}\right)$	-4
-4	-4	-4	-4

$$x = \pm\frac{2i\sqrt{3}}{9} \longleftarrow \text{solutions check}$$

Make Sure

Solve incomplete quadratic equations of the form $ax^2 + c = 0$ using the Square Root Rule.

See Example 3 ▶ 1. $5z^2 - 45 = 0$ _____ 2. $3y^2 - 24 = 0$ _____ 3. $8x^2 + 27 = 0$ _____

MAKE SURE ANSWERS: 1. $-3, 3$ 2. $-2\sqrt{2}, 2\sqrt{2}$ 3. $-\frac{3i\sqrt{6}}{4}, \frac{3i\sqrt{6}}{4}$

9.1 Practice

Set 1: Write each equation in standard form and write each value for *a*, *b*, and *c*.

1. $4z = 2z^2 - 5$
2. $2 = 3y - 4y^2$
3. $x^2 = \frac{4}{3}x - 2$
4. $\frac{3}{4}w^2 = w - 2$
5. $v = \frac{5}{7}v^2 - \frac{3}{7}$
6. $\frac{7}{5}u = \frac{3}{5} - u^2$
7. $3t + \frac{5}{t} = 4$
8. $\frac{2}{s} = 3 - 4s$
9. $5r = 4 + \frac{7}{r}$
10. $3 = \frac{5}{q} - 2q$
11. $4 = \frac{5}{p} - \frac{6}{p^2}$
12. $\frac{3}{n} = \frac{2}{n^2} - 5$

Set 2: Solve incomplete quadratic equations of the form $ax^2 + bx = 0$ by factoring.

13. $3z^2 - 6z = 0$
14. $6y^2 - 12y = 0$
15. $8x + 4x^2 = 0$
16. $12w + 3w^2 = 0$
17. $16v^2 = 12v$
18. $18u^2 = 12u$

Set 3: Solve incomplete quadratic equations of the form $ax^2 + c = 0$ using the Square Root Rule.

19. $z^2 - 9 = 0$ **20.** $y^2 - 25 = 0$ **21.** $x^2 + 8 = 0$

22. $w^2 + 12 = 0$ **23.** $9v^2 + 16 = 0$ **24.** $25u^2 + 49 = 0$

Mixed Practice: Solve incomplete quadratic equations.

25. $27 + 18t^2 = 0$ **26.** $0 = 28 + 84s^2$ **27.** $21r - 35r^2 = 0$

28. $63q - 84q^2 = 0$ **29.** $0 = 75 - 32p^2$ **30.** $0 = 27 - 8n^2$

31. $0 = 36m + 42m^2$ **32.** $24k + 40k^2 = 0$ **33.** $\frac{3}{4}h = \frac{1}{2}h^2$

34. $\frac{5}{6}g = \frac{2}{3}g^2$ **35.** $12f^2 + 45 = 0$ **36.** $27e^2 + 50 = 0$

Review: Work these problems on a separate sheet of paper. Attach your work to this page.

Multiply fractions.
37. $\frac{1}{2} \cdot \frac{3}{4}$ **38.** $\frac{1}{2} \cdot \frac{4}{5}$ **39.** $\frac{1}{2} \cdot \frac{5}{6}$ **40.** $\frac{1}{2} \cdot \frac{6}{7}$
41. $\frac{1}{2} \cdot \frac{2}{3}$ **42.** $\frac{1}{2} \cdot \frac{1}{2}$ **43.** $\frac{1}{2} \cdot 3$ **44.** $\frac{1}{2} \cdot \frac{b}{a}$

Factor each PST. (See Lesson 6.4.)
45. $z^2 - 4z + 4$ **46.** $y^2 + 6y + 9$ **47.** $x^2 - 8x + 16$ **48.** $w^2 + 9w + \frac{81}{4}$

49. $v^2 + 5v + \frac{25}{4}$ **50.** $t^2 - \frac{3}{2}t + \left(\frac{3}{4}\right)^2$ **51.** $s^2 - \frac{5}{4}s + \left(\frac{5}{8}\right)^2$ **52.** $r^2 + \frac{4}{3}r + \left(\frac{2}{3}\right)^2$

9.2 Solve by Completing the Square

The Square Root Rule (if $p^2 = q$, then: $p = \pm\sqrt{q}$) can also be used when p is a binomial.

Example ▶ If $(x + d)^2 = e$, then: $x + d = \pm\sqrt{e}$.

EXAMPLE 1: Solve an equation of the form $(x + d)^2 = e$ using the Square Root Rule.

Solve ▶ $(x + 5)^2 = 16$

1. Use Square Root Rule ▶ $x + 5 = \pm\sqrt{16}$ Think: $(x + d)^2 = e$ means: $x + d = \pm\sqrt{e}$

2. Solve for x ▶ $x + 5 = \pm 4$

$x = -5 \pm 4$

$x = -5 + 4$ or $x = -5 - 4$

$x = -1$ or $x = -9$ ⟵ proposed solutions

3. Check ▶

$(x + 5)^2$	$= 16$		$(x + 5)^2$	$= 16$	⟵ original equation
$[(-1) + 5]^2$	16		$[(-9) + 5]^2$	16	
4^2	16		$(-4)^2$	16	
16	16		16	16	

Solution ▶ $x = -1$ or -9

Other Examples ▶ (a) Solve $(w + 3)^2 = 5$.   (b) Solve $(x - 2)^2 = -9$.

$w + 3 = \pm\sqrt{5}$ $x - 2 = \pm\sqrt{-9}$ Use the Square Root Rule.

$w = -3 \pm \sqrt{5}$ $x - 2 = \pm 3i$

$x = 2 \pm 3i$ Check as before.

Make Sure

Solve equations of the form $(x + d)^2 = e$ using the Square Root Rule.

See Example 1 ▶ **1.** $(z - 3)^2 = 4$ **2.** $(y + 4)^2 = 8$ **3.** $(x - \sqrt{3})^2 + 27 = 0$

_____   _____   _____

MAKE SURE ANSWERS: 1. 5, 1   2. $-4 + 2\sqrt{2}, -4 - 2\sqrt{2}$ 3. $\sqrt{3} + 3i\sqrt{3}, \sqrt{3} - 3i\sqrt{3}$

Chapter 9 Quadratic Equations

You can solve every equation of the form $(x + d)^2 = e$ by using the Square Root Rule. If a quadratic equation is not in $(x + d)^2 = e$ form, you can transform it into this form by *completing the square*.

Completing the square involves transforming polynomials of the form $x^2 \pm bx$ into perfect square trinomials (PST). Examine the following binomial squares and their PST products.

Recall ▶ (a) $(x + n)^2 = \underbrace{x^2 + 2nx + n^2}_{\text{a PST}}$ (b) $(x - n)^2 = \underbrace{x^2 - 2nx + n^2}_{\text{a PST}}$

In both cases, it is important to note that the last term of the PST is the square of one-half of the coefficient of the linear term.

Examples ▶ (a) $(x + 6)^2 = x^2 + \underset{[\frac{1}{2} \text{ of } 12]^2}{12x} + 36$ (b) $\left(x - \frac{2}{5}\right)^2 = x^2 - \frac{4}{5}x + \underset{[\frac{1}{2} \text{ of } -\frac{4}{5}]^2}{\frac{4}{25}}$

Completing the Square of a Binomial

To change a binomial of the form $x^2 + bx$ to a perfect square trinomial, you add $c = \left(\frac{1}{2}b\right)^2$ to it. This is called completing the square.

Example ▶ Complete the square of $x^2 + 7x$ and then factor.

1. Identify b ▶ In $x^2 + 7x$, $b = 7$.

2. Compute c ▶ $c = \left(\frac{1}{2}b\right)^2$

$= \left(\frac{1}{2} \cdot 7\right)^2$

3. Add c ▶ $x^2 + 7x + \left(\frac{7}{2}\right)^2$

4. Factor ▶ $\left(x + \frac{7}{2}\right)^2$

Solution ▶ Completing the square of $x^2 + 7x$ produces $x^2 + 7x + \left(\frac{7}{2}\right)^2$, which factors into $\left(x + \frac{7}{2}\right)^2$.

CAUTION: The preceding steps for completing the square apply only to polynomials of the form $x^2 + bx$. In particular, the coefficient of the quadratic term must be 1.

Another Example ▶ $x^2 - \frac{1}{3}x + \left(\frac{1}{2}b\right)^2 = x^2 - \frac{1}{3}x + \left[\frac{1}{2}\left(-\frac{1}{3}\right)^2\right]$

$= x^2 - \frac{1}{3}x + \frac{1}{36}$

$= (x)^2 - 2 \cdot \frac{1}{6}x + \left(\frac{1}{6}\right)^2$

$= \left(x - \frac{1}{6}\right)^2$

> The completing-the-square method can be used to transform any quadratic equation with form
>
> $$ax^2 + bx + c = 0$$
>
> into an equation with form
>
> $$(x + d)^2 = e,$$
>
> which can then be solved using the Square Root Rule.

EXAMPLE 2: Solve a quadratic equation by completing the square.

Solve ▸ $\quad 3x^2 - 5x - 2 = 0$

1. Write $ax^2 + bx = c$ form ▸ $\quad 3x^2 - 5x = 2$

2. Divide by a ▸ $\quad x^2 - \dfrac{5}{3}x = \dfrac{2}{3}$

   Think: Completing the square can be applied only if the coefficient of x^2 is 1.

3. Complete the square ▸ $\quad x^2 - \dfrac{5}{3}x + \left(-\dfrac{5}{6}\right)^2 = \dfrac{2}{3} + \left(-\dfrac{5}{6}\right)^2$

   Think: Add the square of one-half of $-\dfrac{5}{3}$, $\left(-\dfrac{5}{6}\right)^2$, to both members.

4. Factor left member ▸ $\quad (x)^2 - 2 \cdot \dfrac{5}{6}x + \left(\dfrac{5}{6}\right)^2 = \dfrac{2}{3} + \dfrac{25}{36}$

   $$\left(x - \dfrac{5}{6}\right)^2 = \dfrac{49}{36}$$

5. Use Square Root Rule ▸ $\quad x - \dfrac{5}{6} = \pm\sqrt{\dfrac{49}{36}}$

   See Example 1.

   $$x - \dfrac{5}{6} = \pm\dfrac{7}{6}$$

6. Solve for x ▸ $\quad x = \dfrac{5}{6} \pm \dfrac{7}{6}$

   $$x = \dfrac{5}{6} + \dfrac{7}{6} \text{ or } x = \dfrac{5}{6} - \dfrac{7}{6}$$

   $$x = \dfrac{12}{6} \text{ or } x = -\dfrac{2}{6}$$

   $$x = 2 \text{ or } x = -\dfrac{1}{3}$$

Solution ▸ $\quad x = 2 \text{ or } -\dfrac{1}{3} \qquad$ Check as before.

If you can factor a quadratic equation using integers, then solving by factoring will generally be easier than solving by completing the square. The quadratic equations in the next two examples are not factorable using integers, so completing the square is the only way to solve them at this time.

Other Examples ▶ (a) Solve $x^2 + 8x + 4 = 0$.

$$x^2 + 8x = -4$$

$$x^2 + 8x + 16 = -4 + 16$$

$$(x + 4)^2 = 12$$

$$x + 4 = \pm\sqrt{12}$$

$$x = -4 \pm \sqrt{12}$$

$$x = -4 \pm 2\sqrt{3}$$

(b) Solve $2x^2 + 2x + 5 = 0$.

$$x^2 + x + \frac{5}{2} = 0$$

$$x^2 + x = -\frac{5}{2}$$

$$x^2 + x + \frac{1}{4} = -\frac{5}{2} + \frac{1}{4}$$

$$\left(x + \frac{1}{2}\right)^2 = -\frac{9}{4}$$

$$x + \frac{1}{2} = \pm\sqrt{-\frac{9}{4}}$$

$$x + \frac{1}{2} = \pm\frac{3}{2}i$$

$$x = -\frac{1}{2} \pm \frac{3}{2}i$$

Though the method of completing the square is somewhat laborious, it is an important technique. It will be used in the next lesson to develop an important formula and also in Chapter 10 to graph conic sections.

Make Sure

Solve each quadratic equation by completing the square.

See Example 2 ▶ 1. $z^2 - 2z - 8 = 0$ 2. $4y^2 + 3y - 2 = 0$ 3. $3x^2 - 2x + 1 = 0$

_____   _____   _____

MAKE SURE ANSWERS: 1. 4, −2   2. $-\frac{3}{8} - \frac{\sqrt{41}}{8}, -\frac{3}{8} + \frac{\sqrt{41}}{8}$ 3. $\frac{1}{3} - \frac{\sqrt{2}}{3}i, \frac{1}{3} + \frac{\sqrt{2}}{3}i$

9.2 Practice

Set 1: Solve equations of the form $(x + d)^2 = e$ using the Square Root Rule.

1. $(z + 1)^2 = 1$
2. $(y - 1)^2 = 1$
3. $(x + 3)^2 = 4$

4. $(w - 3)^2 = 4$
5. $(v + 6)^2 = 9$
6. $(u + 5)^2 = 9$

7. $(t - 7)^2 = 25$
8. $(s - 2)^2 = 36$
9. $(r + 4)^2 = 4$

10. $(q + 3)^2 = 9$
11. $(2p - 3)^2 = 2$
12. $(3k - 4)^2 = 3$

Set 2: Solve each quadratic equation by completing the square.

13. $z^2 + 2z - 3 = 0$
14. $y^2 + 4y - 5 = 0$
15. $x^2 - 6x - 7 = 0$

16. $w^2 - 8w - 9 = 0$
17. $v^2 + 6v + 4 = 0$
18. $u^2 + 8u + 8 = 0$

Mixed Practice: Solve each quadratic equation.

19. $z^2 + 4z + 5 = 0$
20. $y^2 + 6y + 13 = 0$
21. $x^2 + 5x - 3 = 0$

22. $w^2 + 7w - 4 = 0$
23. $2v^2 + 3v + 1 = 0$
24. $2u^2 + 5u - 3 = 0$

25. $(3 - t)^2 = 4$
26. $(5 - s)^2 = 3$
27. $3r^2 = 5r - 1$

28. $4q^2 = 4q + 1$
29. $5p^2 = 3p - 1$
30. $7n^2 = 4n - 2$

Review: Work these problems on a separate sheet of paper. Attach your work to this page.

Multiply polynomials. (See Lesson 5.3.)
31. $(z - 3)(z + 2)$
32. $(y - 2i)(y + 2i)$
33. $(x - 3i)(x + 3i)$
34. $(w + \frac{2}{3})(w - \frac{3}{4})$
35. $(v - 1 - i)(v - 1 + i)$
36. $(u + 3 - 2i)(u + 3 + 2i)$

Simplify each radical. (See Lesson 8.2.)
37. $\sqrt{48}$
38. $\sqrt{72}$
39. $\sqrt{-24}$
40. $\sqrt{-50}$
41. $\sqrt{16 - 24}$
42. $\sqrt{36 - 48}$
43. $\sqrt{25 - 45}$
44. $\sqrt{1 - 8}$

Write each quadratic equation in standard form and each value a, b, and c. (See Lesson 9.1.)
45. $2z^2 = 4z - 1$
46. $4y = y^2 + 3$
47. $3x^2 = 5x$
48. $6 - 3w^2 = 0$
49. $2 = 3v - v^2$
50. $3 - 2u = u^2$

9.3 Solve by the Quadratic Formula

To solve any quadratic equation you can use the method of completing the square. However, you will generally find it easier to solve quadratic equations by using the *quadratic formula*. To develop the quadratic formula, you complete the square on the quadratic equation in standard form $ax^2 + bx + c = 0$.

Example ▶ Solve $ax^2 + bx + c = 0$ ($a \neq 0$) by completing the square.

1. Write $ax^2 + bx = -c$ form ▶
$$ax^2 + bx + c = 0$$
$$ax^2 + bx = -c$$

2. Divide by a ▶
$$x^2 + \frac{b}{a}x = -\frac{c}{a}$$

3. Complete the square ▶
$$x^2 + \frac{b}{a}x + \left(\frac{b}{2a}\right)^2 = -\frac{c}{a} + \left(\frac{b}{2a}\right)^2$$

4. Factor left member ▶
$$\left(x + \frac{b}{2a}\right)^2 = -\frac{c}{a} + \left(\frac{b}{2a}\right)^2$$

5. Simplify right member ▶
$$\left(x + \frac{b}{2a}\right)^2 = -\frac{c}{a} + \frac{b^2}{4a^2}$$
$$\left(x + \frac{b}{2a}\right)^2 = \frac{-4ac}{4a^2} + \frac{b^2}{4a^2}$$
$$\left(x + \frac{b}{2a}\right)^2 = \frac{b^2 - 4ac}{4a^2}$$

6. Use Square Root Rule ▶
$$x + \frac{b}{2a} = \pm\sqrt{\frac{b^2 - 4ac}{4a^2}}$$

7. Solve for x ▶
$$x = -\frac{b}{2a} \pm \sqrt{\frac{b^2 - 4ac}{4a^2}}$$
$$x = -\frac{b}{2a} \pm \frac{\sqrt{b^2 - 4ac}}{2a}$$

Solution ▶
$$x = \frac{-b \pm \sqrt{b^2 - 4ac}}{2a}$$

The Quadratic Formula

If $ax^2 + bx + c = 0$ ($a \neq 0$), then: $x = \dfrac{-b \pm \sqrt{b^2 - 4ac}}{2a}$.

EXAMPLE 1: Solve a quadratic equation using the quadratic formula.

Solve ▶ $x^2 - 3 = 2x$

1. Write standard form ▶ $x^2 - 2x - 3 = 0$

2. Identify a, b, and c ▶ In $x^2 - 2x - 3 = 0$, $a = 1$, $b = -2$, and $c = -3$.

3. Use quadratic formula ▶ $$x = \frac{-b \pm \sqrt{b^2 - 4ac}}{2a}$$ ⟵ quadratic formula

$$x = \frac{-(-2) \pm \sqrt{(-2)^2 - 4(1)(-3)}}{2(1)}$$ Substitute: $a = 1$
$b = -2$
$c = -3$

$$x = \frac{2 \pm \sqrt{4 + 12}}{2}$$ Simplify.

$$x = \frac{2 \pm 4}{2}$$

$$x = \frac{2 + 4}{2} \text{ or } x = \frac{2 - 4}{2}$$

$$x = 3 \quad \text{ or } x = -1$$

Solution ▶ $x = 3$ or -1 Check as before.

Note ▶ The quadratic equation in Example 1 could have been solved more readily by factoring.

To solve a quadratic equation, you first determine whether you can solve it by factoring as in Lesson 6.6. If the equation cannot be factored using integers, then you use the quadratic formula.

Example ▶ Solve $2y^2 + 4y = 1$.

1. Write standard form ▶ $2y^2 + 4y - 1 = 0$

2. Try to factor ▶ $2y^2 + 4y - 1 = 0$ does not factor using integers.

3. Identify a, b, and c ▶ In $2y^2 + 4y - 1 = 0$: $a = 2$, $b = 4$, and $c = -1$.

4. Use quadratic formula ▶ $$y = \frac{-b \pm \sqrt{b^2 - 4ac}}{2a}$$

$$y = \frac{-(4) \pm \sqrt{(4)^2 - 4(2)(-1)}}{2(2)}$$ Substitute: $a = 2$
$b = 4$
$c = -1$

$$y = \frac{-4 \pm \sqrt{16 + 8}}{4}$$ Simplify.

$$y = \frac{-4 \pm 2\sqrt{6}}{4}$$

$$y = \frac{-4}{4} \pm \frac{2\sqrt{6}}{4}$$

Solution ▶ $y = -1 \pm \dfrac{\sqrt{6}}{2}$ Check as before.

Note ▶ $y = -1 \pm \dfrac{\sqrt{6}}{2}$ is often written as $\dfrac{-2 \pm \sqrt{6}}{2}$.

Another Example ▶ Solve $z^2 = 2z - 6$.

1. Write standard form ▶ $z^2 - 2z + 6 = 0$

2. Try to factor ▶ $z^2 - 2z + 6 = 0$ does not factor using integers.

3. Identify a, b, and c ▶ In $z^2 - 2z + 6 = 0$, $a = 1$, $b = -2$, and $c = 6$.

4. Use quadratic formula ▶ $z = \dfrac{-b \pm \sqrt{b^2 - 4ac}}{2a}$

$$z = \dfrac{-(-2) \pm \sqrt{(-2)^2 - 4(1)(6)}}{2(1)}$$

Substitute: $a = 1$
$b = -2$
$c = 6$

$z = \dfrac{2 \pm \sqrt{4 - 24}}{2}$ Simplify.

$z = \dfrac{2 \pm \sqrt{-20}}{2}$

$z = \dfrac{2 \pm 2i\sqrt{5}}{2}$

$z = \dfrac{2}{2} \pm \dfrac{2i\sqrt{5}}{2}$

$z = 1 \pm i\sqrt{5}$ ⟵ proposed solutions

5. Check ▶

$z^2 = 2z - 6$		$z^2 = 2z - 6$ ⟵ original equation	
$(1 + i\sqrt{5})^2$	$2(1 + i\sqrt{5}) - 6$	$(1 - i\sqrt{5})^2$	$2(1 - i\sqrt{5}) - 6$
$1 + 2i\sqrt{5} - 5$	$2 + 2i\sqrt{5} - 6$	$1 - 2i\sqrt{5} - 5$	$2 - 2i\sqrt{5} - 6$
$-4 + 2i\sqrt{5}$	$-4 + 2i\sqrt{5}$	$-4 - 2i\sqrt{5}$	$-4 - 2i\sqrt{5}$

Solution ▶ $z = 1 \pm i\sqrt{5}$

To solve a quadratic equation with real (rational and/or irrational) coefficients, you can use the quadratic formula.

Example ▶ Solve $w^2 - 8\sqrt{3}w + 48 = 0$.

$w^2 + (-8\sqrt{3})w + 48 = 0$ ⟵ standard form

In $w^2 + (-8\sqrt{3})w + 48 = 0$, $a = 1$, $b = -8\sqrt{3}$, and $c = 48$.

$w = \dfrac{-b \pm \sqrt{b^2 - 4ac}}{2a}$ ⟵ quadratic formula

$w = \dfrac{-(-8\sqrt{3}) \pm \sqrt{(-8\sqrt{3})^2 - 4(1)(48)}}{2(1)}$

$w = \dfrac{8\sqrt{3} \pm \sqrt{192 - 192}}{2}$

$w = \dfrac{8\sqrt{3} \pm 0}{2}$

$w = 4\sqrt{3}$ Check as before.

Note ▶ You should memorize the quadratic formula because it is a useful formula and you will need it throughout the remaining chapters.

Make Sure

Solve each quadratic equation using the quadratic formula.

See Example 1 ▶ 1. $12z^2 + z - 6 = 0$ 2. $2y^2 - 4y - 7 = 0$ 3. $5x^2 - 4x + 2 = 0$

MAKE SURE ANSWERS: 1. $-\frac{3}{4}, \frac{2}{3}$ 2. $1 - \frac{\sqrt{3}}{2}, 1 + \frac{\sqrt{3}}{2}$ 3. $\frac{2}{5} - \frac{\sqrt{6}}{5}i, \frac{2}{5} + \frac{\sqrt{6}}{5}i$

The quadratic formula indicates that the solutions of the quadratic equation $ax^2 + bx + c = 0$, where a, b, and c are real numbers, are given by:

$$x = \frac{-b + \sqrt{b^2 - 4ac}}{2a} \text{ or } x = \frac{-b - \sqrt{b^2 - 4ac}}{2a}.$$

The expression $b^2 - 4ac$ under the radical is called the *discriminant* of the quadratic equation $ax^2 + bx + c = 0$.

> $b^2 - 4ac$ is called the discriminant because it is used to determine the type of solutions (roots) that the quadratic equation $ax^2 + bx + c = 0$ has.

Examples ▶ (a) If $b^2 - 4ac = 0$, then: $x = \dfrac{-b + \sqrt{0}}{2a}$ or $x = \dfrac{-b - \sqrt{0}}{2a}$

$$x = \frac{-b + 0}{2a} \text{ or } x = \frac{-b - 0}{2a}$$

$$x = \frac{-b}{2a} \text{ or } x = \frac{-b}{2a}$$

Observation: If $b^2 - 4ac = 0$, then both solutions equal the same real number $-\dfrac{b}{2a}$.

(b) If $b^2 - 4ac > 0$, then $\sqrt{b^2 - 4ac}$ is a real number, and both solutions are distinct real numbers.

(c) If $b^2 - 4ac < 0$, then $\sqrt{b^2 - 4ac}$ is an imaginary number, and both solutions are nonreal solutions (conjugates).

Summary

If the discriminant $b^2 - 4ac$ is	The equation $ax^2 + bx + c = 0$ (where a, b, and c are all real numbers) will have
zero	one real solution.
positive	two real solutions.
negative	two nonreal number solutions (conjugates).

EXAMPLE 2: Classify the solutions of a quadratic equation using the discriminant.

Problem ▶ Does $2x^2 + 7x = -3$ have $\begin{cases} \text{one real solution} \\ \text{two real solutions} \\ \text{two nonreal solutions} \end{cases}$?

1. Write standard form ▶ $2x^2 + 7x + 3 = 0$

2. Identify a, b, and c ▶ In $2x^2 + 7x + 3 = 0$; $a = 2$, $b = 7$, and $c = 3$.

3. Compute ▶ $b^2 - 4ac = 7^2 - 4(2)(3)$
$= 49 - 24$
$= 25$ ⟵ value of the discriminant

4. Interpret ▶ A quadratic equation with a positive discriminant has two real solutions.

Solution ▶ $2x^2 + 7x = -3$ has two real solutions.

Other Examples ▶ Classify the solutions of the following quadratic equations.

(a) $5x^2 + 4 = -6x$

In $5x^2 + 6x + 4 = 0$
$a = 5$, $b = 6$, and $c = 4$.
$b^2 - 4ac = 6^2 - 4(5)(4)$
$= 36 - 80$
$= -44$

Interpret: A negative discriminant means there are two nonreal solutions.

(b) $9y^2 + 6y = 1$

In $9y^2 + 6y - 1 = 0$
$a = 9$, $b = 6$, and $c = -1$.
$b^2 - 4ac = 6^2 - 4(9)(-1)$
$= 36 - 36$
$= 0$

Interpret: A discriminant of 0 means there is one real solution.

Make Sure

Classify the solutions of each quadratic equation using the discriminant.

See Example 2 ▶ **1.** $9z^2 + 12z + 4 = 0$ **2.** $3y^2 - 6y + 2 = 0$ **3.** $5x^2 + 4x + 2 = 0$

MAKE SURE ANSWERS: 1. one real solution. 2. two real solutions. 3. two nonreal solutions.

To find a quadratic equation given the specific roots, you reverse the method of solving a quadratic equation by factoring.

EXAMPLE 3: Write a quadratic equation in standard form for the given pair of roots.

Problem ▶ Write a quadratic equation in standard form that has 2 and $-\frac{1}{3}$ as roots.

1. Equate unknown with each root ▶ $x = 2$ or $x = -\frac{1}{3}$

2. Get zero as one member ▶ $x - 2 = 0$ or $x + \frac{1}{3} = 0$

3. Form a product ▶ $(x - 2)\left(x + \frac{1}{3}\right) = 0$
   Think: If $(x - 2)$ or $(x + \frac{1}{3})$ equal zero, then their product is also equal to zero.

4. Multiply ▶ $x^2 - \frac{5}{3}x - \frac{2}{3} = 0$

5. Write standard form ▶ $3x^2 - 5x - 2 = 0$ Multiply each term by the LCD 3.

Solution ▶ A quadratic equation in standard form that has 2 and $-\frac{1}{3}$ as roots is $3x^2 - 5x - 2 = 0$.

> **SHORTCUT 9.1:** If r_1 and r_2 are the given roots of a quadratic equation, you can find the quadratic equation by beginning with $(x - r_1)(x - r_2) = 0$.

Example ▶ Find a quadratic equation (in standard form) that has $\frac{-3}{2} + \frac{\sqrt{3}}{2}$ and $\frac{-3}{2} - \frac{\sqrt{3}}{2}$ as roots.

$$\left[x - \left(\frac{-3}{2} + \frac{\sqrt{3}}{2}\right)\right]\left[x - \left(\frac{-3}{2} - \frac{\sqrt{3}}{2}\right)\right] = 0$$

$$x^2 - x\left(\frac{-3}{2} - \frac{\sqrt{3}}{2}\right) - \left(\frac{-3}{2} + \frac{\sqrt{3}}{2}\right)x + \left(\frac{-3}{2} + \frac{\sqrt{3}}{2}\right)\left(\frac{-3}{2} - \frac{\sqrt{3}}{2}\right) = 0$$

$$x^2 + \frac{3}{2}x + \frac{\sqrt{3}}{2}x + \frac{3}{2}x - \frac{\sqrt{3}}{2}x + \frac{9}{4} + \frac{3\sqrt{3}}{4} - \frac{3\sqrt{3}}{4} - \frac{3}{4} = 0$$

$$x^2 + 3x + \frac{3}{2} = 0$$

$$2x^2 + 6x + 3 = 0$$

Make Sure

Write a quadratic equation in standard form for each given pair of roots.

See Example 3 ▶ 1. $\frac{2}{3}, -\frac{3}{4}$ 2. $2 - \sqrt{3}, 2 + \sqrt{3}$ 3. $\frac{3 + 2i\sqrt{3}}{4}, \frac{3 - 2i\sqrt{3}}{4}$

MAKE SURE ANSWERS: 1. $12z^2 + z - 6 = 0$ 2. $y^2 - 4y + 1 = 0$ 3. $16x^2 - 24x + 21 = 0$

9.3 Practice

Set 1: Solve each quadratic equation using the quadratic formula.

1. $z^2 + 4z + 4 = 0$
2. $y^2 + 6y + 9 = 0$
3. $4x^2 - 4x - 3 = 0$

4. $6w^2 - w - 2 = 0$
5. $12v^2 = 17v - 6$
6. $12u^2 = u + 6$

7. $t^2 = t + 1$
8. $s^2 = 1 - s$
9. $1 - 2r = r^2$

10. $2q = q^2 - 1$
11. $3p = 2p^2 + 2$
12. $4n = 3n^2 + 2$

13. $k^2 + 2\sqrt{2}k + 2 = 0$
14. $h^2 - 6\sqrt{3}h + 27 = 0$

358   Chapter 9   Quadratic Equations

Set 2: Classify the solutions of each quadratic equation using the discriminant.

15. $4z^2 + 4z + 1 = 0$ **16.** $9y^2 - 6y + 1 = 0$ **17.** $3x^2 + 5x + 2 = 0$

18. $4w^2 - 6w + 3 = 0$ **19.** $3v^2 - 6v - 4 = 0$ **20.** $2u^2 - 6u - 3 = 0$

21. $2t^2 + 5t + 1 = 0$ **22.** $5s^2 - 6s + 1 = 0$ **23.** $9r^2 = 6r - 4$

24. $4q^2 = 5q - 3$ **25.** $4p = 3p^2 + 2 = 0$ **26.** $7n = 5n^2 + 2$

Set 3: Write a quadratic equation in standard form for each pair of given roots.

27. $2, -3$ **28.** $-3, 4$ **29.** $\frac{2}{3}, -\frac{1}{4}$

30. $\frac{3}{4}, -\frac{1}{5}$ **31.** $-3i, 3i$ **32.** $1 - 2i, 1 + 2i$

33. $-\sqrt{3}, \sqrt{3}$ **34.** $-\sqrt{5}, \sqrt{5}$ **35.** $-5, -5$

36. $7, 7$ **37.** $2 + 3i, 2 - 3i$ **38.** $3 - 2i, 3 + 2i$

Review: Work these problems on a separate sheet of paper. Attach your work to this page.

Solve equations containing rational expressions. (See Lesson 7.5.)

39. $\dfrac{3}{z+3} = \dfrac{4}{z-5}$ **40.** $\dfrac{2}{3y} - \dfrac{5}{y^2-4} = 0$ **41.** $\dfrac{8}{8-10x-3x^2} = \dfrac{x+5}{4+x} + \dfrac{2x}{2-3x}$

Solve equations containing radical terms. (See Lesson 8.4.)

42. $\sqrt{x+14} = 2 + \sqrt{x+2}$ **43.** $\sqrt{y+2} = \sqrt{2y-2}$ **44.** $\sqrt{16z-15} = 3\sqrt{z} + \sqrt{z-3}$

9.4 Rename and Solve Equations

Many equations that are not quadratic equations can be renamed as quadratic equations by simplification or substitution. Once you have renamed them as quadratic equations, you can solve by factoring or by using the quadratic formula.

EXAMPLE 1: Solve a rational equation that can be renamed as a quadratic equation.

Solve ▶ $\dfrac{x-2}{x} = \dfrac{15}{x^2}$

1. Rename as a quadratic equation ▶ $x^2\left(\dfrac{x-2}{x}\right) = x^2\left(\dfrac{15}{x^2}\right)$ Clear fractions.

$x^2 - 2x = 15$ ⟵ quadratic equation

2. Write standard form ▶ $x^2 - 2x - 15 = 0$

3. Solve ▶ $(x+3)(x-5) = 0$

$x + 3 = 0$ or $x - 5 = 0$

$x = -3$ or $x = 5$

Solution ▶ $x = -3$ or 5 Check as before.

Recall ▶ An excluded value cannot be a solution.

Example ▶ Solve $1 = \dfrac{28}{(w+4)^2} + \dfrac{7w}{(w+4)^2}$.   Think: $w = -4$ is an excluded value. It causes division by zero.

$(w+4)^2 \cdot 1 = (w+4)^2\left[\dfrac{28}{(w+4)^2} + \dfrac{7w}{(w+4)^2}\right]$ Clear fractions.

$(w+4)^2 = 28 + 7w$

$w^2 + 8w + 16 = 28 + 7w$

$w^2 + w - 12 = 0$

$(w+4)(w-3) = 0$ Factor.

$w + 4 = 0$ or $w - 3 = 0$

$w = -4$ or $w = 3$ ⟵ proposed solutions

-4 cannot be a solution because it is an excluded value.

$1 = \dfrac{28}{(w+4)^2} + \dfrac{7w}{(w+4)^2}$ Check $w = 3$ in the original equation.

$1 \mid \dfrac{28}{(3+4)^2} + \dfrac{7 \cdot (3)}{(3+4)^2}$ Substitute 3 for w.

$1 \mid \dfrac{28}{49} + \dfrac{21}{49}$

$1 \mid 1$ ⟵ $w = 3$ checks

$w = 3$ ⟵ solution

Another Example ▶ Solve $\frac{y}{4} - 1 = \frac{1}{4y}$. Think: $y = 0$ is an excluded value; it causes division by 0.

$$4y\left[\frac{y}{4} - 1\right] = 4y\left(\frac{1}{4y}\right)$$

$$y^2 - 4y = 1$$

$y^2 - 4y - 1 = 0$ means $a = 1, b = -4, c = -1$.

$$y = \frac{-b \pm \sqrt{b^2 - 4ac}}{2a}$$

Think: $y^2 - 4y - 1$ does not factor; use the quadratic formula.

$$y = \frac{-(-4) \pm \sqrt{(-4)^2 - 4(1)(-1)}}{2(1)}$$

$$y = \frac{4 \pm \sqrt{16 + 4}}{2}$$

$$y = 2 \pm \sqrt{5}$$ Check as before.

Make Sure

Solve each rational equation by renaming it as a quadratic equation.

See Example 1 ▶ **1.** $\frac{3z - 4}{z} = \frac{2}{z^2}$ _____ **2.** $1 = \frac{4}{3 - y} - \frac{6}{(3 - y)^2}$ _____

MAKE SURE ANSWERS: 1. $\frac{2 - \sqrt{10}}{3}, \frac{2 + \sqrt{10}}{3}$; 2. $1 - i\sqrt{2}, 1 + i\sqrt{2}$

You can rename many radical equations as quadratic equations.

EXAMPLE 2: Solve a radical equation that can be renamed as a quadratic equation.

Solve ▶ $1 + \sqrt{y} = \sqrt{2y - 7}$

1. Rename ▶ $(1 + \sqrt{y})^2 = (\sqrt{2y - 7})^2$ Use the Power Principle.

$1 + 2\sqrt{y} + y = 2y - 7$

$2\sqrt{y} = y - 8$ Isolate the remaining radical.

$4y = y^2 - 16y + 64$ Use the Power Principle.

2. Write standard form ▶ $0 = y^2 - 20y + 64$

3. Solve ▶ $0 = (y - 4)(y - 16)$ Think: First try to solve by factoring.

$y - 4 = 0$ or $y - 16 = 0$

$y = 4$ or $y = 16$ ⟵ proposed solutions

4. Check ▶

$$\frac{\sqrt{2y-7} = 1 + \sqrt{y}}{\sqrt{2(4)-7} \mid 1 + \sqrt{4}}$$
$$\sqrt{1} \mid 1 + 2$$
$$1 \mid 3 \longleftarrow y = 4 \text{ does not check}$$

$$\frac{\sqrt{2y-7} = 1 + \sqrt{y}}{\sqrt{2(16)-7} \mid 1 + \sqrt{16}}$$
$$\sqrt{25} \mid 1 + 4$$
$$5 \mid 5 \longleftarrow y = 16 \text{ checks}$$

Solution ▶ $y = 16$

> CAUTION: Use of the Power Principle may introduce extraneous solutions. After using the Power Principle, you must check each proposed solution in the original equation.

Make Sure

Solve each radical equation by renaming it as a quadratic equation.

See Example 2 ▶ **1.** $\sqrt{2z} = \sqrt{z-1} + 1$ **2.** $\sqrt{2y-1} - \sqrt{y+3} = 2$

MAKE SURE ANSWERS: 1. 2, $16 - 4\sqrt{15}$ 2. $16 + 4\sqrt{15}$

> An equation in any variable is *quadratic in form* if it can be written as $au^2 + bu + c = 0$ where u is an expression involving that variable.

Equations that are quadratic in form can be renamed as quadratic equations in terms of another variable by making an appropriate substitution. Many trinomials that are quadratic in form can be renamed as quadratic equations by making a substitution as described in the following examples.

Example ▶ Rename $4x^4 = -5x^2 - 3$ as a quadratic equation by using substitution.

1. Write standard form ▶ $4x^4 + 5x^2 + 3 = 0$ ⟵ descending powers of x set equal to 0.

2. Determine u ▶ $u = x^2$ Think: Set u equal to the literal part of the middle term.

3. Compute u^2 ▶ $u^2 = (x^2)^2$
$$= x^4$$

4. Substitute ▶ $4x^4 + 5x^2 + 3 = 0$
$4u^2 + 5u + 3 = 0$ Substitute u for x^2 and u^2 for x^4.

Solution ▶ $4x^4 = -5x^2 - 3$ can be renamed as $4u^2 + 5u + 3 = 0$ by using the substitution $u = x^2$.

Chapter 9 Quadratic Equations

Another Example ▶ Rename $\left(\dfrac{z}{z+3}\right)^2 = 2\left(\dfrac{z}{z+3}\right) - 7$ as a quadratic equation by using substitution.

$\left(\dfrac{z}{z+3}\right)^2 - 2\left(\dfrac{z}{z+3}\right) + 7 = 0$ ⟵ descending powers of $\dfrac{z}{z+3}$ set equal to zero.

$u = \dfrac{z}{z+3}$ Set u equal to the literal part of the middle term.

$u^2 = \left(\dfrac{z}{z+3}\right)^2$

$\left(\dfrac{z}{z+3}\right)^2 - 2\left(\dfrac{z}{z+3}\right) + 7 = 0$

$u^2 - 2u + 7 = 0$ Substitute u for $\left(\dfrac{z}{z+3}\right)$ and u^2 for $\left(\dfrac{z}{z+3}\right)^2$.

$\left(\dfrac{z}{z+3}\right)^2 = 2\left(\dfrac{z}{z+3}\right) - 7$ can be renamed as $u^2 - 2u + 7 = 0$ by using substitution.

EXAMPLE 3: Solve equations that are quadratic in form.

Solve ▶ $w^4 - 7w^2 = -12$

1. Rename ▶ $w^4 - 7w^2 + 12 = 0$ Determine u: $u = w^2$.

$w^4 - 7w^2 + 12 = 0$

$u^2 - 7u + 12 = 0$ Substitute u for w^2 and u^2 for w^4.

2. Solve for u ▶ $(u-3)(u-4) = 0$

$u - 3 = 0$ or $u - 4 = 0$

$u = 3$ or $u = 4$

3. Substitute ▶ $w^2 = 3$ or $w^2 = 4$ Substitute: $u = w^2$.

4. Solve for w ▶ $w = \pm\sqrt{3}$ or $w = \pm 2$

Solution ▶ $w = \pm\sqrt{3}$ or ± 2 Check as before.

Make Sure

Solve each equation by renaming it in quadratic form.

See Example 3 ▶ 1. $4z^4 = 4z^2 + 3$ 2. $2 = 3\left(\dfrac{x+1}{x}\right)^2 + 2\left(\dfrac{x+1}{x}\right)^4$

MAKE SURE ANSWERS: 1. $\pm\dfrac{i\sqrt{2}}{2}, \pm\dfrac{\sqrt{6}}{2}$ 2. $-\dfrac{1}{3} \pm \dfrac{i\sqrt{2}}{3}, -2 \pm \sqrt{2}$

Name _____ Date _____ Class _____

9.4 Practice

Set 1: Solve each rational equation by renaming it as a quadratic equation.

1. $\dfrac{5}{z} = 1 + \dfrac{6}{z^2}$

2. $\dfrac{3}{y} = 2 + \dfrac{2}{y^2}$

3. $1 = \dfrac{6}{x+1} - \dfrac{4}{(x+1)^2}$

4. $1 = \dfrac{2}{w-3} + \dfrac{3}{(w-3)^2}$

5. $\dfrac{4}{(v-3)^2} = 1 + \dfrac{3v}{(v-3)^2}$

6. $2 = \dfrac{3}{(u+2)^2} + \dfrac{2u}{(u+2)^2}$

Set 2: Solve each radical equation by renaming it as a quadratic equation.

7. $\sqrt{3z+2} - \sqrt{z} = 1$

8. $\sqrt{3y} - \sqrt{y+1} = 1$

9. $\sqrt{2x+3} - \sqrt{x+1} = 1$

10. $\sqrt{3w-1} + \sqrt{w+2} = 1$

11. $\sqrt{3v-1} + \sqrt{v+1} = \sqrt{4v-1}$

12. $\sqrt{3u-1} - \sqrt{2u+1} = \sqrt{3u+1}$

Copyright © 1985 by Harcourt Brace Jovanovich, Inc. All rights reserved.

364 Chapter 9 Quadratic Equations

Set 3: Solve equations that are quadratic in form.

13. $6z^4 - 5z^2 = 6$
14. $y^4 - y^2 = 6$
15. $(5 - \sqrt{w})^2 + 4(5 - \sqrt{w}) = -4$

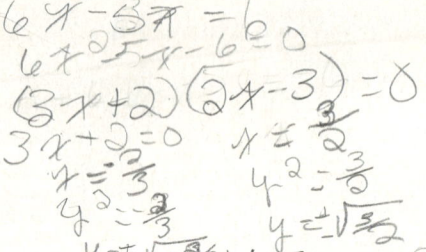

16. $3(\sqrt{x} + 3)^2 + (\sqrt{x} + 3) = 2$
17. $v^{2/3} - 2v^{1/3} - 8 = 0$
18. $u^{2/5} - 3u^{1/5} - 4 = 0$

Mixed Practice: Solve each equation.

19. $\dfrac{z^2 + 6}{z^2} = \dfrac{8}{z}$
20. $\dfrac{2y^2 + 2}{y^2} = \dfrac{3}{y}$
21. $\sqrt{2 - 3x} = \sqrt{5x + 4} + 2$

22. $\sqrt{3w - 2} = 2 + \sqrt{2 - 3w}$
23. $2v - 7\sqrt{v} - 4 = 0$
24. $3u = 3 - 8\sqrt{u}$

Review: Work these problems on a separate sheet of paper. Attach your work to this page.

Solve each equation using the zero-product property. (See Lesson 2.2.)
25. $(a - 3)(a + 3) = 0$
26. $(b + 5)b = 0$
27. $(2c - 3)(3c + 5) = 0$
28. $(3d - 4)(3d - 4) = 0$
29. $(e - \sqrt{3})(e + \sqrt{2}) = 0$
30. $f(f - \sqrt{5})(f + 3) = 0$

Find each excluded value of the given equation. (See Lesson 7.5.)
31. $\dfrac{z}{z - 3} = 4$
32. $\dfrac{y}{y^2 - 4} = 3$
33. $\dfrac{4}{u^2 - 9} = \dfrac{2}{u + 4} - \dfrac{1}{u^2}$
34. $\dfrac{3}{w^2 - 4} = \dfrac{5}{w^2 - w - 6}$
35. $\dfrac{3}{v^2 - 5v + 6} = \dfrac{5}{v^2 - 7v + 12}$
36. $\dfrac{5}{4x} = \dfrac{2}{x^2 - 9}$

Copyright © 1985 by Harcourt Brace Jovanovich, Inc. All rights reserved.

Problem Solving 12: Solve Problems Using Quadratic Equations

To solve certain problems that require geometry formulas, you must solve a quadratic equation.

EXAMPLE 1: Solve a geometry problem by solving a quadratic equation.

1. Read and identify ▶ A rectangular lawn is (120 feet) by (80 feet). How wide must a uniform strip be mowed around the lawn in order for the lawn to be (one-fourth) mowed, to the nearest inch?

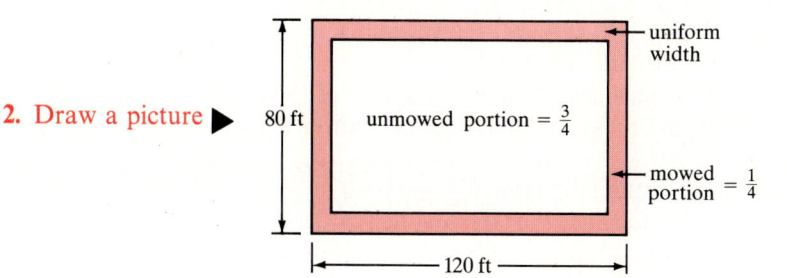

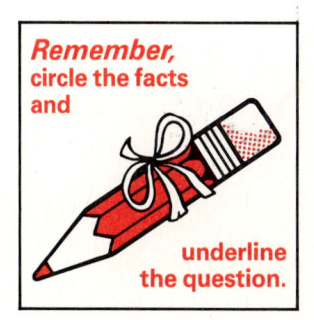

   Remember, circle the facts and underline the question.

2. Draw a picture ▶ 80 ft, unmowed portion = $\frac{3}{4}$, uniform width, mowed portion = $\frac{1}{4}$, 120 ft

3. Understand ▶ The unknowns are $\begin{cases} \text{the uniform width of the mowed strip} \\ \text{the length of the unmowed portion} \\ \text{the width of the unmowed portion} \end{cases}$

4. Decide ▶ Let $\quad x =$ the uniform width of the mowed strip
   then $\quad 120 - 2x =$ the length of the unmowed portion because $120 - x - x = 120 - 2x$
   and $\quad 80 - 2x =$ the width of the unmowed portion because $80 - x - x = 80 - 2x$.

5. Translate ▶ The area of the unmowed portion is three-fourths of the whole area.

   $$(120 - 2x)(80 - 2x) = \tfrac{3}{4} \cdot 80 \cdot 120$$

6. Solve as before ▶
   $(120 - 2x)(80 - 2x) = 7200$ Multiply.
   $9600 - 400x + 4x^2 = 7200$ Clear parentheses.
   $4x^2 - 400x + 2400 = 0$ Write standard form.
   $x^2 - 100x + 600 = 0$ Divide each term by the GCF 4.

   $$x = \frac{-b \pm \sqrt{b^2 - 4ac}}{2a}$$ Evaluate the quadratic formula.

   $$= \frac{-(-100) \pm \sqrt{(-100)^2 - 4(1)(600)}}{2(1)}$$

   $$= \frac{100 \pm \sqrt{7600}}{2}$$

   $$\approx 93.59 \text{ or } 6.41$$

7. Interpret ▶ $x = 93.59$ or 6.41 means the proposed solutions are 93.59 feet and 6.41 feet.

8. Round ▶ $93.59 = 93$ ft $+ 0.59(12)$ in. ≈ 93 ft 7 in. or 93′ 7″ ⟵
   $6.41 = 6$ ft $+ 0.41(12)$ in. ≈ 6 ft 5 in. or 6′ 5″ ⟵ nearest inch

9. Check $x \approx 93′7″$ ▶ Can the uniform width of the mowed strip be 93′ 7″? No: If the width of the whole lawn is only 80′, the uniform width of the mowed strip cannot be greater than 40′.

10. Check $x \approx 6′5″$ ▶ Can the uniform width of the mowed strip be 6′5″? Yes: Check and see.

366 Chapter 9 Quadratic Equations

To solve certain uniform motion problems, you must solve a quadratic equation.

EXAMPLE 2: Solve a uniform motion problem by solving a quadratic equation.

1. Read and identify ▶ An airplane completes a round trip between two cities 720 miles apart in 10 hours at a constant airspeed. The airplane flies the first half of the trip against a 30 mph headwind and then returns with a 40 mph tailwind. What is the airspeed of the airplane, to the nearest mile per hour?

2. Understand ▶ The unknowns are $\begin{cases} \text{the constant airspeed} \\ \text{the ground speed* for the first half of the trip} \\ \text{the ground speed for the second half of the trip} \end{cases}$.

3. Decide ▶ Let $r =$ the constant airspeed
then $r - 30 =$ the ground speed for the first half of the trip
and $r + 40 =$ the ground speed for the second half of the trip

4. Make a table ▶

	distance (d)	rate (r)	time $\left(t = \dfrac{d}{r}\right)$
first half	720	$r - 30$	$\dfrac{720}{r - 30}$
second half	720	$r + 40$	$\dfrac{720}{r + 40}$

5. Translate ▶ The time for the first half of the trip plus the time for the second half is 10 hours.

$$\dfrac{720}{r - 30} + \dfrac{720}{r + 40} = 10$$

6. Solve as before ▶ $(r - 30)(r + 40)\dfrac{720}{r - 30} + (r - 30)(r - 40)\dfrac{720}{r + 40} = (r - 30)(r + 40)10$

$(r + 40)720 + (r - 30)720 = (r^2 + 10r - 1200)10$

$720r + 28{,}800 + 720r - 21{,}600 = 10r^2 + 100r - 12{,}000$

$0 = 10r^2 - 1340r - 19{,}200$

$0 = 1r^2 - 134r - 1920$

$r = \dfrac{-b \pm \sqrt{b^2 - 4ac}}{2a}$

$= \dfrac{-(-134) \pm \sqrt{(-134)^2 - 4(1)(-1920)}}{2(1)}$

$\approx \dfrac{134 \pm 160}{2}$

$= 147 \text{ or } -13$

7. Interpret ▶ $r \approx -13$ means the airspeed was about -13 mph. Wrong: Rates are never negative.
$r \approx 147$ means the airspeed was about 147 mph.

*The ground speed is the true speed of the airplane with respect to the ground.

PS 12: Solve Problems Using Quadratic Equations

Practice

Set 1: Solve each geometry problem by solving a quadratic equation.

1. A square picture frame is uniformly 2 cm wide. The square picture inside the frame is two-thirds the area of the frame and picture together. Find the dimensions of the picture.

2. A rectangular lawn has an area of 500 ft^2 and requires 100 ft of fencing to enclose it. What are the dimensions of the lawn?

Set 2: Solve each uniform motion problem.

3. The speed of a freight train is 30 mph slower than the speed of a passenger train. It takes the same amount of time for the freight train to travel 150 miles as it does for the passenger train to travel 200 miles. Find the speed of each train.

4. A stream is flowing at 2 mph. A woman, paddling at a constant rate, takes just as long to paddle 12 miles downstream as she does to paddle 4 miles upstream. What is the woman's paddling rate in still water?

Mixed Practice: Solve each problem.

5. A box is made from a square piece of cardboard by cutting one-foot squares from each corner and then folding up the sides, as shown in Figure 9.1. The volume of the box is 16 ft^3. What are the dimensions of the box?

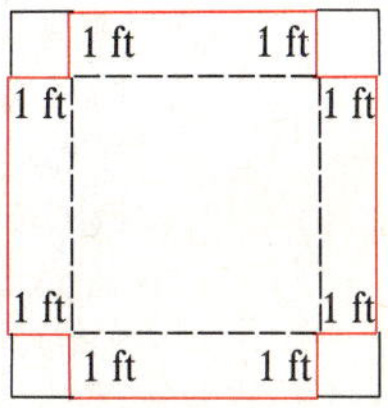

Figure 9.1

6. The length of a rectangle is 4 m longer than its width. When the length is decreased by 1 m and the width increased by 2 m, the area is doubled. Find the area of the original rectangle.

7. The area of one square is 360 cm^2 more than the area of another square. The side of the larger square is 10 cm longer than the side of the other square. Find the dimensions of each square.

8. If a car had traveled 4 mph faster than it actually did over a 224-mile trip, the time saved would have been 1 hour. How long did it actually take the car to make the trip?

9. Bert drove 192 miles and then back 192 miles in 14 hours of actual driving time. His rate going was 8 mph faster than his rate returning. How long did it take him to return?

10. A storm caused a train to travel 6 km/h slower than usual for 252 kilometers. If the trip took 1 hour longer than usual, what was the rate of the train?

11. An airplane has a 6-hour supply of fuel. How far can it fly at 180 mph, and then return at 140 mph, before running out of fuel?

12. A ship leaves port at noon, traveling at 30 mph. At 8 P.M. that night, an airplane flying at 180 mph flies over the port on its way to catch up to the ship. What time does the plane catch up to the ship?

13. A bus traveled at its normal speed to complete a 400-mile trip on schedule. On the return trip, weather caused the bus to reduce its normal speed by 10 mph, and the bus was 2 hours late getting back. Find the normal speed of the bus.

14. The length of a volleyball court is twice its width and has an area of 1800 ft^2. Find the dimensions of the volleyball court.

15. The length of a rectangle is twice its width. The area is 50 square units. Find the dimensions.

Chapter 9 Quadratic Equations

16. According to the ancient Greeks, the "Golden Rectangle" has the most pleasing shape to the human eye. A "Golden Rectangle" always satisfies the following proportion:

$$\frac{l}{w} = \frac{w}{l-w}$$ (See Figure 9.2.)

For a "Golden Rectangle":
a) Express l in terms of w.
b) Find l if $w = 1$.

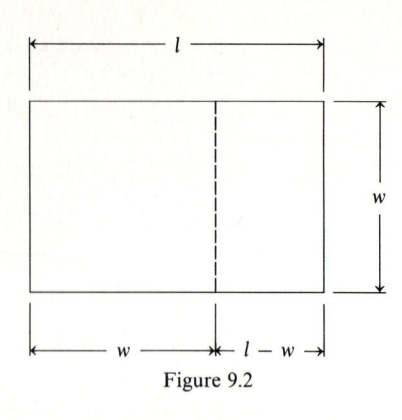

Figure 9.2

17. It takes a certain boat 1 hour more to go 30 miles upstream than it does to go 40 miles downstream. The speed of the boat in still water is 15 mph. Find the rate of the current.

18. A certain person drives 50 km to the airport and then flies 300 km aboard a plane. The plane travels 100 km/h faster than the car. The total travel time is 3 hours. Find the rate of the a) car b) plane.

19. A woman can row 18 km downstream and make the return trip upstream in a total time of 8 hours. If the rate of the current is 3 km/h, find a) the woman's rowing rate in still water, b) the rowing rate downstream.

20. The maximum speed of a certain fishing boat in still water is 10 mph. It takes the boat 4 hours longer to travel 96 miles up river against the current than it does to return down the river with the same current. What is the rate of the current?

Extra

21. The distance (d) traveled by a falling object in t seconds is given by: $d = 16t^2$. The speed of sound is 1080 feet per second. If it takes 3 seconds to hear the sound of a dropped rock hitting the bottom of a well, how deep is the well to the nearest whole foot?

According to Guinness

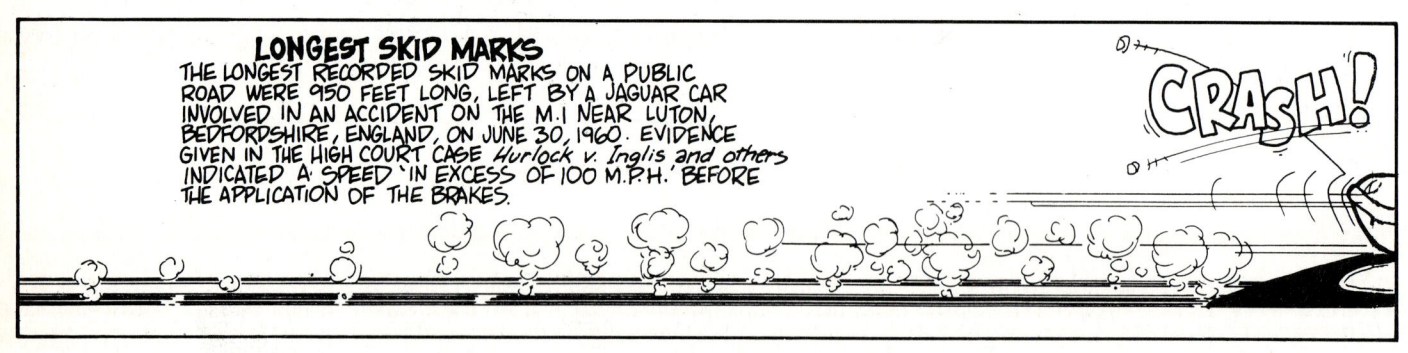

LONGEST SKID MARKS
THE LONGEST RECORDED SKID MARKS ON A PUBLIC ROAD WERE 950 FEET LONG, LEFT BY A JAGUAR CAR INVOLVED IN AN ACCIDENT ON THE M.1 NEAR LUTON, BEDFORDSHIRE, ENGLAND, ON JUNE 30, 1960. EVIDENCE GIVEN IN THE HIGH COURT CASE *Hurlock v. Inglis and others* INDICATED A SPEED 'IN EXCESS OF 100 M.P.H.' BEFORE THE APPLICATION OF THE BRAKES.

CRASH!

Recall ▶ The stopping distance (d) in feet for an average car on an average road traveling at r miles per hour (mph) is given by the formula: $d = \frac{1}{20}r^2 + r$.

22. Assuming the Jaguar is considered an average car and that it was traveling on an average road, by how much "in excess of 100 mph" was the Jaguar traveling before the application of the brakes, to the nearest whole mile per hour?

9.5 Solve Inequalities

The methods you have used to solve equations can be extended to solve many inequalities.

> A *critical number* of a rational inequality is a value of the variable that causes:
> 1. division by zero (called an excluded value), or
> 2. the left member to equal the right member (called an *equality value*).

Example ▶ Find the critical number(s) of $x - 4 > \dfrac{3x}{x + 3}$.

1. Find excluded value(s) ▶

$x + 3 = 0$ Think: What value(s) of x causes division by zero?

$x = -3$ ⟵ excluded value

2. Find equality value(s) ▶

$x - 4 = \dfrac{3x}{x + 3}$ Think: Replace the inequality symbol with an equality symbol and then solve.

$(x + 3)(x - 4) = (x + 3)\left(\dfrac{3x}{x + 3}\right)$ Clear fractions.

$x^2 - x - 12 = 3x$

$x^2 - 4x - 12 = 0$

$(x + 2)(x - 6) = 0$

$x + 2 = 0$ or $x - 6 = 0$

$x = -2 \quad\quad x = 6$ ⟵ equality values

Solution ▶ $x - 4 > \dfrac{3x}{x + 3}$ has critical numbers of -3, -2, and 6.

Rational inequalities have the property that all numbers in an interval formed by the critical numbers will make the inequality true, or all will make it false. You can use this property to solve any rational inequality for which you can determine all of its critical numbers.

> **The Critical Point Method of Solving Rational Inequalities**
> 1. Find all critical numbers and identify the intervals they form on the number line.
> 2. Substitute a test number from each interval formed by the critical numbers into the inequality to determine which intervals make the inequality true.
> 3. Graph the union of all intervals that satisfy the inequality.

EXAMPLE 1: Solve a rational inequality using the critical point method.

Solve ▶ $1 < \dfrac{x^2 - 5}{x^2 - x - 12}$

1. Find critical number(s) ▶

Find excluded value(s).

$x^2 - x - 12 = 0$

$(x + 3)(x - 4) = 0$

$x + 3 = 0$ or $x - 4 = 0$

$x = -3$ or $\quad x = 4$

⟵ excluded values ⟶

Find equality value(s).

$1 = \dfrac{x^2 - 5}{x^2 - x - 12}$

$(x^2 - x - 12)1 = (x^2 - x - 12)\left(\dfrac{x^2 - 5}{x^2 - x - 12}\right)$

$x^2 - x - 12 = x^2 - 5$

$x = -7$ ⟵ the equality value

2. Identify intervals ▶

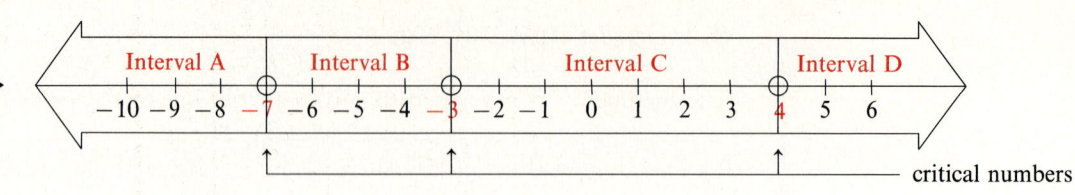

```
     Interval A      Interval B        Interval C          Interval D
<――|――|――|――⊕――|――|――|――⊕――|――|――|――|――|――⊕――|――|――>
  −10 −9 −8 −7 −6 −5 −4 −3 −2 −1  0  1  2  3  4  5  6
           ↑              ↑                    ↑
                                                    critical numbers
```

3. Test ▶ −8 is in interval A; −5 is in interval B; 0 is in interval C; and 6 is in interval D.

$$1 < \frac{x^2 - 5}{x^2 - x - 12} \quad\bigg|\quad 1 < \frac{x^2 - 5}{x^2 - x - 12} \quad\bigg|\quad 1 < \frac{x^2 - 5}{x^2 - x - 12} \quad\bigg|\quad 1 < \frac{x^2 - 5}{x^2 - x - 12}$$

$$1 < \frac{(-8)^2 - 5}{(-8)^2 - (-8) - 12} \quad\bigg|\quad 1 < \frac{(-5)^2 - 5}{(-5)^2 - (-5) - 12} \quad\bigg|\quad 1 < \frac{0^2 - 5}{0^2 - 0 - 12} \quad\bigg|\quad 1 < \frac{6^2 - 5}{6^2 - 6 - 12}$$

$$1 < \frac{64 - 5}{64 + 8 - 12} \quad\bigg|\quad 1 < \frac{25 - 5}{25 + 5 - 12} \quad\bigg|\quad 1 < \frac{0 - 5}{0 - 0 - 12} \quad\bigg|\quad 1 < \frac{36 - 5}{36 - 6 - 12}$$

$$1 < \frac{59}{60} \quad \text{false} \quad\bigg|\quad 1 < \frac{20}{18} \quad \text{true} \quad\bigg|\quad 1 < \frac{5}{12} \quad \text{false} \quad\bigg|\quad 1 < \frac{31}{18} \quad \text{true}$$

4. Interpret ▶ All numbers in Interval A are not solutions because the test value −8 is not a solution.

All numbers in Interval B are solutions because the test value −5 is a solution.

All numbers in Interval C are not solutions because the test value 0 is not a solution.

All numbers in Interval D are solutions because the test value 6 is a solution.

5. Graph solutions ▶

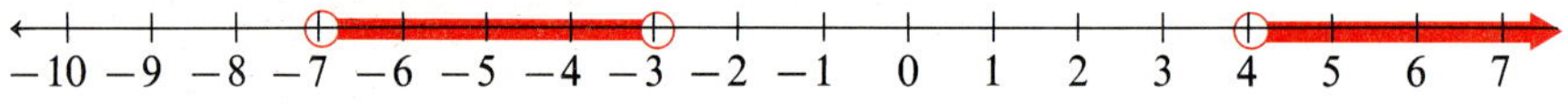

```
<――|――|――⊕▬▬▬▬▬⊕――|――|――|――|――|――⊕▬▬▬▬▬▶
 −10 −9 −8 −7 −6 −5 −4 −3 −2 −1  0  1  2  3  4  5  6  7
```

Solution ▶ The solution set for $1 < \frac{x^2 - 5}{x^2 - x - 12}$ is $\{x \mid -7 < x < -3 \text{ or } x > 4\}$.

Make Sure

Solve each rational inequality using the critical point method.

See Example 1 ▶ **1.** $16 \leq \frac{84}{y^2 - 1}$ **2.** $1 > \frac{2x + 2}{3x^2 - x - 4}$

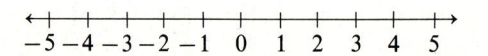

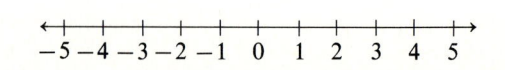

```
<―|―|―|―|―|―|―|―|―|―|―>        <―|―|―|―|―|―|―|―|―|―|―>
 −5 −4 −3 −2 −1 0 1 2 3 4 5     −5 −4 −3 −2 −1 0 1 2 3 4 5
```

MAKE SURE ANSWERS: 1. $\{y \mid -\frac{5}{2} \leq y \leq -1 \text{ or } 1 \leq y \leq \frac{5}{2}\}$ **2.** $\{x \mid x < -1 \text{ or } -\frac{2}{3} < x < \frac{4}{3} \text{ or } x > 2\}$

You can solve some quadratic inequalities by the critical point method.

EXAMPLE 2: Solve a quadratic inequality using the critical point method.

Solve ▶ $2x^2 + 5x + 3 \leq 0$

1. Find critical number(s) ▶ Find equality value(s).   Think: There are no excluded values.

$2z^2 + 5z + 3 = 0$

$(2z + 3)(z + 1) = 0$

$2z + 3 = 0$ or $z + 1 = 0$

$2z = -3$ or $z = -1$

$z = -\frac{3}{2}$ or $z = -1$

2. Identify intervals ▶

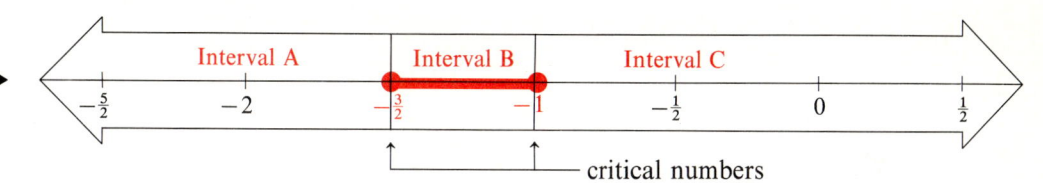

Interval A | Interval B | Interval C

critical numbers

3. Test ▶

-2 is in Interval A;

$2z^2 + 5z + 3 \leq 0$

$2(-2)^2 + 5(-2) + 3 \leq 0$

$8 - 10 + 3 \leq 0$

$1 \leq 0$ false

$-\frac{5}{4}$ is in Interval B;

$2z^2 + 5z + 3 \leq 0$

$2(-\frac{5}{4})^2 + 5(-\frac{5}{4}) + 3 \leq 0$

$\frac{25}{8} - \frac{25}{4} + 3 \leq 0$

$-\frac{1}{8} \leq 0$ true

0 is in Interval C.

$2z^2 + 5z + 3 \leq 0$

$2(0)^2 + 5(0) + 3 \leq 0$

$0 + 0 + 3 \leq 0$

$3 \leq 0$ false

4. Interpret ▶ All numbers in Interval A are not solutions because the test value -2 is not a solution.
All numbers in Interval B are solutions because the test value $-\frac{5}{4}$ is a solution.
All numbers in Interval C are not solutions because the test value 0 is not a solution.

5. Graph solutions ▶

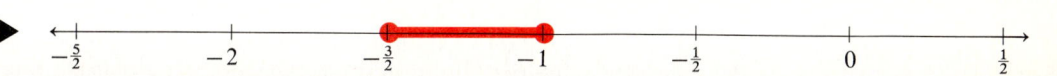

Solution ▶ The solution set for $2z^2 + 5z + 3 \leq 0$ is $\{z \mid -\frac{3}{2} \leq z \leq -1\}$.

Make Sure

Solve each quadratic inequality using the critical point method.

See Example 2 ▶ 1. $y^2 > 2y - 1$ 2. $x \leq \dfrac{8}{x - 2}$

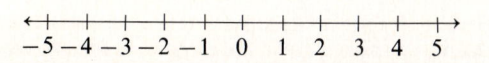

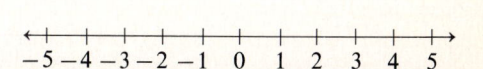

MAKE SURE ANSWERS: 1. $\{y \mid y < 1 \text{ or } y > 1\}$ 2. $\{x \mid x \leq -2 \text{ or } 2 < x \leq 4\}$

Chapter 9 Quadratic Equations

To solve some inequalities that involve a product of linear factors, you can also use the critical point method.

EXAMPLE 3: Solve an inequality involving a product of linear factors.

Solve ▶ $(3y - 2)(y + 4)(y - 5) > 0$

1. Find critical numbers ▶ Find equality value(s).   *Think: There are no excluded values.*

$(3y - 2)(y + 4)(y - 5) = 0$

$3y - 2 = 0$ or $y + 4 = 0$ or $y - 5 = 0$ *Use the zero-product property extended.*

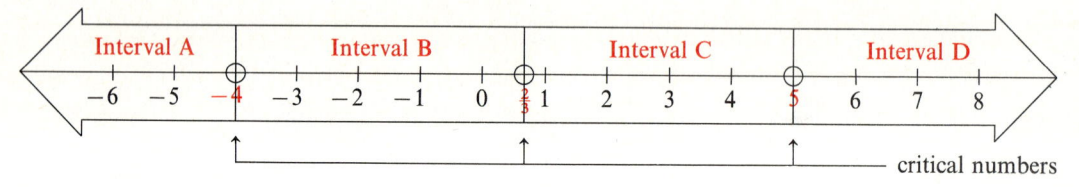

$y = \frac{2}{3}$ or $y = -4$ or $y = 5$

2. Identify intervals ▶

Interval A | Interval B | Interval C | Interval D, with critical numbers at -4, $\frac{2}{3}$, and 5.

3. Test ▶ -5 is in Interval A. 0 is in Interval B. 1 is in Interval C. 6 is in Interval D.

If $y = -5$, then $(3y - 2) < 0$, $(y + 4) < 0$, and $(y - 5) < 0$. This means the product $(3y - 2)(y + 4)(y - 5)$ is negative when $y = -5$.

If $y = 0$, then $(3y - 2) < 0$, $(y + 4) > 0$, and $(y - 5) < 0$. This means the product $(3y - 2)(y + 4)(y - 5)$ is positive when $y = 0$.

If $y = 1$, then $(3y - 2) > 0$, $(y + 4) > 0$, and $(y - 5) < 0$. This means the product $(3y - 2)(y + 4)(y - 5)$ is negative when $y = 1$.

If $y = 6$, then $(3y - 2) > 0$, $(y + 4) > 0$, and $(y - 5) > 0$. This means the product $(3y - 2)(y + 4)(y - 5)$ is positive when $y = 6$.

4. Interpret ▶ All numbers in Interval A are not solutions because the test value -5 is not a solution.
All numbers in Interval B are solutions because the test value 0 is a solution.
All numbers in Interval C are not solutions because the test value 1 is not a solution.
All numbers in Interval D are solutions because the test value 6 is a solution.

5. Graph solutions ▶

Solution ▶ The solution set for $(3y - 2)(y + 4)(y - 5) > 0$ is $\{x \mid -4 < x < \frac{2}{3} \text{ or } x > 5\}$.

Make Sure

Solve inequalities involving a product of linear factors.

See Example 3 ▶ **1.** $z(z + 2)(z - 3) < 0$ **2.** $(2y - 3)(3y + 4)(y - 3) > 0$

MAKE SURE ANSWERS: **1.** $\{z \mid z > -2 \text{ or } 0 < z < 3\}$ **2.** $\{y \mid -\frac{4}{3} < y < \frac{3}{2} \text{ or } y > 3\}$

9.5 Practice

Name _____ Date _____ Class _____

9.5 Practice: *Use number lines to solve each of the following problems.*

Set 1: Solve each rational inequality using the critical point method.

1. $5z > \dfrac{45}{z}$
2. $7y > \dfrac{28}{y}$
3. $1 \leq \dfrac{1}{x+1}$

4. $2 \leq \dfrac{2}{w-2}$
5. $\dfrac{1}{v+2} \geq \dfrac{2}{v^2-4}$
6. $\dfrac{3}{u^2-9} \geq \dfrac{1}{u-3}$

Set 2: Solve each quadratic inequality using the critical point method.

7. $z^2 - 5z + 6 \geq 0$
8. $y^2 - 5y - 6 \geq 0$
9. $x^2 + x < 1$

10. $w^2 + 1 < w$
11. $2v^2 \leq 3v + 2$
12. $3u^2 \leq 4u - 1$

Set 3: Solve inequalities involving a product of linear factors.

13. $z(z + 1)(z - 1) > 0$
14. $y(y + 2)(y - 3) \geq 0$

15. $(2w - 3)(3w + 4)(w - 2) \leq 0$
16. $(2v + 3)(v - 3)(v - 5) < 0$

Mixed Practice: Solve each inequality using the critical point method.

17. $\dfrac{z^2 - 3z}{2z^2 - 7z + 3} < 1$
18. $\dfrac{y^2}{2y^2 + 7y + 6} < 1$

19. $9x^2 - 12x \geq -4$
20. $w^2 + 6w \geq -9$

21. $(v - 2)(v + 2)(v - 4) > 0$
22. $(u + 5)(u - 5)(u - 1) > 0$

Review: Work these problems on a separate sheet of paper. Attach your work to this page.

Evaluate each equation. (See Lesson 3.1.)

23. $x = -\dfrac{b}{2a}$ for $a = 3, b = 2$
24. $x = \dfrac{-b}{2a}$ for $a = 3, b = -4$

25. $x = \dfrac{-b}{2a}$ for $a = 5, b = -2$
26. $y = b^2 - 4ac$ for $a = 4, b = 4$, and $c = -1$

Solve each quadratic equation by factoring. (See Lesson 6.6.)
27. $4x^2 - 4x + 1 = 0$
28. $9x^2 - 6x + 1 = 0$
29. $12x^2 - 17x + 6 = 0$
30. $12x^2 + x - 6 = 0$
31. $12x^2 - 7x + 1 = 0$
32. $12x^2 + x - 1 = 0$

Solve each quadratic equation using the quadratic formula. (See Lesson 9.3.)
33. $x^2 - x + 1 = 0$
34. $x^2 - x - 1 = 0$
35. $2x^2 - 3x - 2 = 0$
36. $3x^2 - 4x + 1 = 0$
37. $2x^2 - 3x + 2 = 0$
38. $2x^2 - 3x - 2 = 0$

Name _____ Date _____ Class _____

Chapter 9 Review

What to Review if You Have Trouble

Objectives		Lesson	Example	Page
Write standard form and identify coefficients	**1.** Write $\frac{3}{4}x^2 - \frac{2}{3}x = 5$ in standard form and write each value for a, b, and c. _____ $a = $ ____ $b = $ ____ $c = $ ____	9.1	1	338
Solve incomplete quadratic equations	**2.** $4y^2 = -6y$ _____	9.1	2	340
	3. $4x^2 - 36 = 0$ _____	9.1	3	341
Solve using the Square Root Rule	**4.** $(x - 3)^2 = 4$ _____	9.2	1	345
Solve by completing the square	**5.** $3x^2 - 14x - 5 = 0$ _____	9.2	2	347
Solve using the quadratic formula	**6.** $2x^2 + 5x = 12$ _____	9.3	1	351
Classify solutions using the discriminant. Circle either a), b), or c)	**7.** Does $3x^2 + x = -4$ have: a) two real solutions? b) one real solution? c) two nonreal solutions?	9.3	2	355

Copyright © 1985 by Harcourt Brace Jovanovich, Inc. All rights reserved.

376 Chapter 9 Quadratic Equations

Write a quadratic equation given its solutions ▶ 8. Write a quadratic equation in standard form that has solutions of -4 and $\frac{1}{3}$. ____ 9.3 3 356

Solve rational equations that simplify to quadratic equations ▶ 9. $\dfrac{x+3}{2x} = \dfrac{5}{x^2}$ ____ 9.4 1 359

Solve radical equations that simplify to quadratic equations ▶ 10. $\sqrt{3x-5} = 1 + \sqrt{x+2}$ ____ 9.4 2 360

Solve equations that are quadratic in form ▶ 11. $w^4 + 36 = 13w^2$ ____ 9.4 3 362

Solve rational inequalities ▶ 12. $1 \geq \dfrac{x^2 + 2x}{x^2 + 4x + 3}$ ____ 9.5 1 369

Solve quadratic inequalities ▶ 13. $12x^2 - 5x - 3 \geq 0$ ____ 9.5 2 371

Solve inequalities involving linear factors ▶ 14. $(3w + 5)(w - 1)(w + 4) < 0$ ____ 9.5 3 372

Solve problems using quadratic equations ▶ 15. A bus traveling 12 mph faster than a car completes a trip of 260 miles in $1\frac{1}{2}$ hours less time than the car does. What is the constant speed of the bus? ____ PS 12 2 366

CHAPTER 9 REVIEW ANSWERS: 1. $9x^2 - 8x - 60 = 0, a = 9, b = -8, c = -60$ **2.** $-\frac{2}{3}, 0$ **3.** $-3, 3$ **4.** $1, 5$ **5.** $-\frac{3}{5}$ **6.** $-4, \frac{4}{3}$ **7.** c) two nonreal solutions **8.** $3x^2 + 11x - 4 = 0$ **9.** $-5, 2$ **10.** 7 **11.** $-3, -2, 2, 3$ **12.** $\{x \mid -3 \leq x \leq -\frac{3}{2} \text{ or } x > -1\}$ **13.** $\{x \mid x \leq -\frac{1}{3} \text{ or } x \geq \frac{3}{4}\}$ **14.** $\{w \mid w < -4 \text{ or } -\frac{5}{3} < w < 1\}$ **15.** 52 mph

Copyright © 1985 by Harcourt Brace Jovanovich, Inc. All rights reserved.

NONLINEAR RELATIONS

10 Conics and Systems

10.1 Graph Parabolas

PS 13: Find Maximum and Minimum Values

10.2 Graph Circles

PS 14: Use the Pythagorean Theorem

10.3 Graph Ellipses

10.4 Graph Hyperbolas

10.5 Solve Quadratic Systems

PS 15: Solve Problems Using Quadratic Systems

378 Chapter 10 Conics and Systems

Introduction to Conic Sections

The geometric figures introduced in this chapter are called *conic sections*. A conic section is a curve formed by the intersection of a double-napped circular cone and a plane, as shown in Figure 10.1.

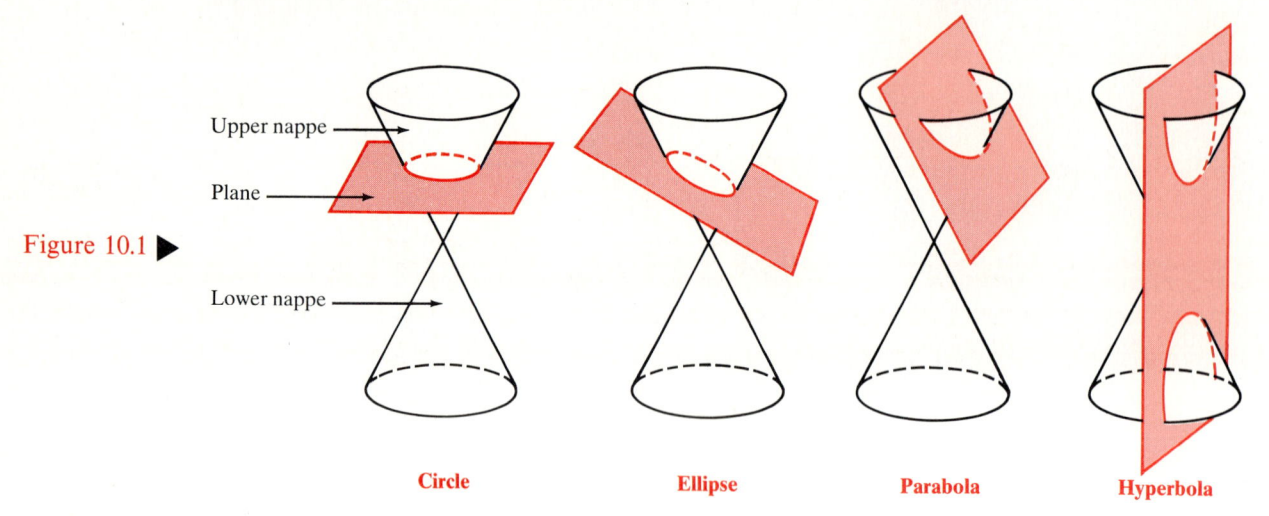

Figure 10.1 ▶

Upper nappe
Plane
Lower nappe

Circle **Ellipse** **Parabola** **Hyperbola**

Every conic section is the graph of a second-degree equation, and every second-degree equation in this text graphs as one of the conic sections in Figure 10.1.

10.1 Graph Parabolas

The graph of $y = x^2$ is a parabola. To graph $y = x^2$, you make a table of ordered pairs that satisfy the equation and then draw a smooth curve through the graphs of the ordered pair solutions. See Figure 10.2.

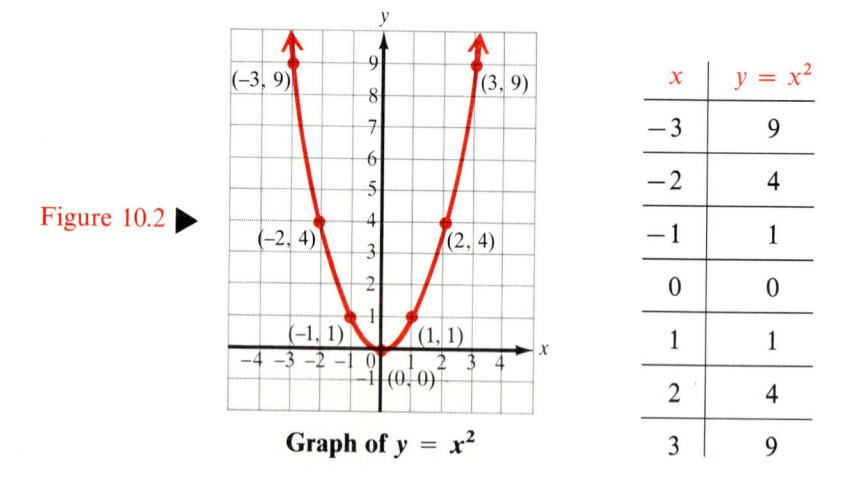

Figure 10.2 ▶

Graph of $y = x^2$

x	$y = x^2$
-3	9
-2	4
-1	1
0	0
1	1
2	4
3	9

Every parabola is symmetrical about a line called its *axis of symmetry*.

Example ▶ In Figure 10.2, the axis of symmetry is the y-axis.

The intersection of a parabola and its axis of symmetry is a point called the *vertex*.

Example ▶ In Figure 10.2, the vertex is the point located at the origin.

If a parabola has a vertical axis of symmetry, then the vertex will be the lowest point on the parabola if the parabola opens upward, or it will be the highest point on the parabola if the parabola opens downward. See Figure 10.3.

Figure 10.3 ▶

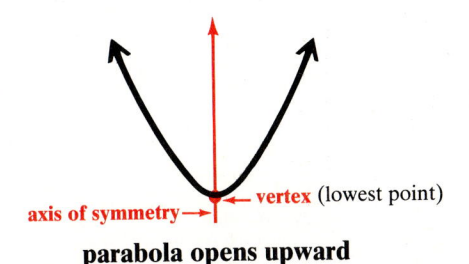

parabola opens upward — vertex (lowest point), axis of symmetry

parabola opens downward — vertex (highest point), axis of symmetry

The graph of every quadratic equation of the form $y = ax^2 + bx + c$ is a parabola with a vertical axis of symmetry. The parabola opens upward if $a > 0$ and downward if $a < 0$.

Examples ▶ (a) The graph of $y = x^2 + 4x - 6$ opens upward because $a = 1\ (1 > 0)$.
(b) The graph of $y = -3x^2 + 7x + 2$ opens downward because $a = -3\ (-3 < 0)$.

Recall ▶ Every point on the y-axis has an x-coordinate of 0. Any point where a graph intersects the y-axis is called a *y-intercept*.

To find the y-intercept(s) for the graph of an equation, you substitute 0 for x and then solve the resulting equation for y.

Examples ▶ Find the y-intercept(s) for $y = 4x^2 - 7x + 3$, and $y = ax^2 + bx + c$.

1. Substitute 0 for x ▶
$y = 4x^2 - 7x + 3$ 　　　　　$y = ax^2 + bx + c$
$y = 4(0)^2 - 7(0) + 3$ 　　　$y = a(0)^2 + b(0) + c$

2. Solve for y ▶
$y = 0 - 0 + 3$ 　　　　　　　$y = 0 + 0 + c$
$y = 3$ 　　　　　　　　　　　$y = c$

Solution ▶ The y-intercept for $y = 4x^2 - 7x + 3$ is $(0, 3)$. 　The y-intercept for $y = ax^2 + bx + c$ is $(0, c)$.

Recall ▶ Every point on the x-axis has a y value of 0. Any point where a graph intersects the x-axis is called an *x-intercept*.

The graph of an equation of the form $y = ax^2 + bx + c$ is a parabola, which has either zero, one, or two x-intercepts, as shown in Figure 10.4.

Possible x-intercepts of $y = ax^2 + bx + c.\ (a > 0)$

Figure 10.4 ▶

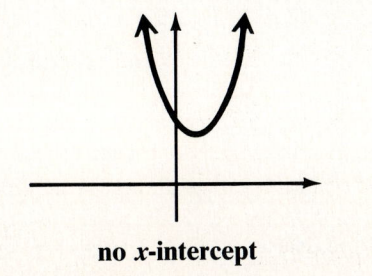

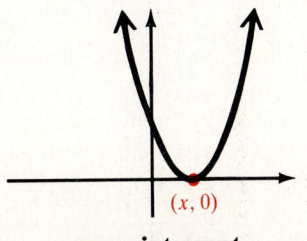

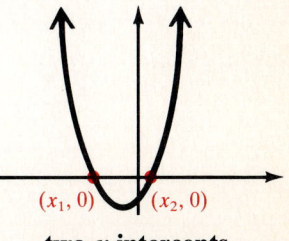

no x-intercept　　　　one x-intercept $(x, 0)$　　　　two x-intercepts $(x_1, 0)\ (x_2, 0)$

To determine the x-intercept(s) for the graph of $y = ax^2 + bx + c$, you substitute 0 for y and then find the real roots of the resulting equation ($0 = ax^2 + bx + c$).

EXAMPLE 1: Find each x-intercept of a parabola of the form $y = ax^2 + bx + c$.

Problems ▶ Find the x-intercept(s) for $y = 2x^2 + 5x - 3$ and $y = x^2 + 8x + 16$.

1. Substitute 0 for y ▶

$$y = 2x^2 + 5x - 3$$
$$0 = 2x^2 + 5x - 3$$

$$y = x^2 + 8x + 16$$
$$0 = x^2 + 8x + 16$$

2. Solve for x ▶

$$0 = (2x - 1)(x + 3)$$
$0 = 2x - 1$ or $0 = x + 3$
$1 = 2x$ or $-3 = x$
$\frac{1}{2} = x$ or $-3 = x$

$$0 = (x + 4)^2$$
$0 = x + 4$ or $0 = x + 4$
$-4 = x$ or $-4 = x$

3. Interpret ▶ The two real roots $\frac{1}{2}$ and -3 mean that $(\frac{1}{2}, 0)$ and $(-3, 0)$ are both x-intercepts. | The single real root -4 means that $(-4, 0)$ is the only x-intercept.

Solutions ▶ The x-intercepts for $y = 2x^2 + 5x - 3$ are $(-3, 0)$ and $(\frac{1}{2}, 0)$. | The x-intercept for $y = x^2 + 8x + 16$ is $(-4, 0)$.

Another Example ▶ Find the x-intercept(s) for the graph of $y = x^2 - 2x + 10$.

1. Substitute 0 for y ▶

$$y = x^2 - 2x + 10$$
$$0 = x^2 - 2x + 10$$

2. Solve for x ▶

$$x = \frac{-(-2) \pm \sqrt{(-2)^2 - 4(1)(10)}}{2(1)}$$
$$= \frac{2 \pm \sqrt{-36}}{2}$$
$$= 1 \pm 3i$$

Because $x^2 - 2x + 10$ does not factor over the set of integers, you use the quadratic formula.

3. Interpret ▶ There are no x-intercepts, because $0 = x^2 - 2x + 10$ has no real roots.

Solution ▶ There are no x-intercepts for $y = x^2 - 2x + 10$.

Make Sure

Find the x-intercept(s) of the graph of each equation.

See Example 1 ▶ 1. $y = 4x^2 - 12x + 9$ 2. $y = 2x^2 + x - 6$ 3. $y = x^2 - x + 1$

MAKE SURE ANSWERS: 1. $(\frac{3}{2}, 0)$ 2. $(-2, 0), (\frac{3}{2}, 0)$ 3. no x-intercept

To develop a formula for finding the vertex of a parabola, you complete the square on the general equation $y = ax^2 + bx + c$.

Example ▶ Find the vertex of the parabola for $y = ax^2 + bx + c$, where $a > 0$.

1. Complete the square ▶
$$y = ax^2 + bx + c$$
$$y = (ax^2 + bx) + c$$
$$y = a\left(x^2 + \frac{b}{a}x\right) + c$$
$$y = a\left(x^2 + \frac{b}{a}x + \frac{b^2}{4a^2}\right) + c - \frac{b^2}{4a}$$
Think: The $\frac{b^2}{4a^2}$ inside the parentheses is being multiplied by a, so you need to subtract $\frac{b^2}{4a}$ to maintain the equality.

$$y = a\left(x + \frac{b}{2a}\right)^2 + c - \frac{b^2}{4a}$$
Factor the PST.

$$y = a\left(x + \frac{b}{2a}\right)^2 + \frac{4ac - b^2}{4a}$$

2. Identify ▶ The graph of $y = ax^2 + bx + c$ (with $a > 0$) is a parabola that opens upward.

3. Understand ▶ Of all the points on a parabola that open upward, the vertex is the lowest point and has the smallest y-coordinate.

4. Decide ▶ To make the y in $y = a\left(x + \frac{b}{2a}\right)^2 + \frac{4ac - b^2}{4a}$ as small as possible, you must make $\left(x + \frac{b}{2a}\right)^2$ as small as possible. The smallest value of $\left(x + \frac{b}{2a}\right)^2$ is 0.

5. Solve for x ▶ $\left(x + \frac{b}{2a}\right)^2 = 0$ means $x = -\frac{b}{2a}$

6. Substitute ▶ $x = -\frac{b}{2a}$ means $y = a\left(-\frac{b}{2a} + \frac{b}{2a}\right)^2 + \frac{4ac - b^2}{4a}$

$$y = a(0)^2 + \frac{4ac - b^2}{4a}$$

$$y = \frac{4ac - b^2}{4a}$$

7. Interpret ▶ $x = -\frac{b}{2a}$ and $y = \frac{4ac - b^2}{4a}$ means the lowest point on $y = ax^2 + bx + c$ ($a > 0$) is $\left(-\frac{b}{2a}, \frac{4ac - b^2}{4a}\right)$.

Solution ▶ The vertex of the parabola for $y = ax^2 + bx + c$ ($a > 0$) is $\left(-\frac{b}{2a}, \frac{4ac - b^2}{4a}\right)$.

A similar analysis will show that $\left(-\frac{b}{2a}, \frac{4ac - b^2}{4a}\right)$ is also the vertex point of the parabola for $y = ax^2 + bx + c$ with $a < 0$.

The graph of $y = ax^2 + bx + c$ has its vertex at:

$$(x, y) = \left(-\frac{b}{2a}, \frac{4ac - b^2}{4a}\right).$$

Because the graph of $y = ax^2 + bx + c$ has a vertical axis of symmetry that passes through its vertex, the equation of its axis of symmetry is $x = -\dfrac{b}{2a}$.

EXAMPLE 2: Find the vertex and axis of symmetry of a parabola.

Problem ▶ Find the vertex and axis of symmetry of the graph of $y = -3x^2 + 2x - 7$.

1. Identify ▶ In $y = x^2 + 4x - 12$, $a = 1$, $b = 4$, and $c = -12$.

2. Evaluate ▶

$$x = -\dfrac{b}{2a} \qquad y = \dfrac{4ac - b^2}{4a}$$

$$= -\dfrac{4}{2(1)} \qquad = \dfrac{4(1)(-12) - 4^2}{4(1)}$$

$$= -2 \qquad = \dfrac{-48 - 16}{4}$$

$$\qquad\qquad = -16$$

3. Interpret ▶ $x = -2$ and $y = -16$ means $(x, y) = (-2, -16)$.

Solution ▶ The vertex of $y = x^2 + 4x - 12$ is $(-2, -16)$. The axis of symmetry is $x = -2$.

Although the y-coordinate of the vertex can be found by evaluating $\dfrac{4ac - b^2}{4a}$, it is often easier to compute the y-value of the vertex point by substituting the x-value of the vertex into the equation for the parabola ($y = ax^2 + bx + c$).

Example ▶ Find the vertex of $y = -3x^2 + 2x - 7$.

In $y = -3x^2 + 2x - 7$, $a = -3$, $b = 2$, and $c = -7$.

$$x = -\dfrac{b}{2a} \qquad y = -3x^2 + 2x - 7 \longleftarrow \text{the original equation}$$

$$= -\dfrac{2}{2(-3)} \qquad = -3\left(\dfrac{1}{3}\right)^2 + 2\left(\dfrac{1}{3}\right) - 7 \qquad \text{Substitute } \tfrac{1}{3} \text{ for } x.$$

$$= \dfrac{1}{3} \qquad = -\dfrac{1}{3} + \dfrac{2}{3} - 7$$

$$\qquad\qquad = -\dfrac{20}{3}$$

The vertex of $y = -3x^2 + 2x - 7$ is $\left(\tfrac{1}{3}, -\tfrac{20}{3}\right)$. The axis of symmetry is $x = \tfrac{1}{3}$.

Make Sure

Find the vertex and axis of symmetry of each parabola.

See Example 2 ▶ 1. $y = 4x^2 - 12x + 9$ 2. $y = -x^2 + x + 1$ 3. $y = 2x^2 + 6x - 7$

MAKE SURE ANSWERS: 1. $\left(\tfrac{3}{2}, 0\right)$, $x = \tfrac{3}{2}$ 2. $\left(\tfrac{1}{2}, \tfrac{5}{4}\right)$, $x = \tfrac{1}{2}$ 3. $\left(-\tfrac{3}{2}, -\tfrac{23}{2}\right)$, $x = -\tfrac{3}{2}$

10.1 Graph Parabolas

> **Graphing Parabolas**
> To graph a parabola that has an equation of the form $y = ax^2 + bx + c$:
> 1. Identify whether the parabola opens upward or downward.
> 2. Find the vertex using $(x, y) = \left(-\dfrac{b}{2a}, \dfrac{4ac - b^2}{4a}\right)$.
> 3. Find the y-intercept $(0, c)$.
> 4. Find the x-intercept(s) by solving $0 = ax^2 + bx + c$.
> 5. Find additional solutions of $y = ax^2 + bx + c$.
> 6. Draw a smooth curve through the graphs of all of the ordered pairs found in Steps 2–5.

EXAMPLE 3: Graph a parabola.

Graph ▶ $y = -x^2 - 4x + 5$

1. Identify ▶ In $y = -x^2 - 4x + 5$, $a = -1$, $b = -4$, and $c = 5$.   Because $a < 0$, the parabola opens downward.

2. Find vertex ▶
$$x = -\dfrac{b}{2a} \qquad y = \dfrac{4ac - b^2}{4a}$$
$$= -\dfrac{-4}{2(-1)} \qquad = \dfrac{4(-1)(5) - (-4)^2}{4(-1)}$$
$$= -2 \qquad\qquad = 9 \qquad \text{The vertex is } (-2, 9).$$

3. Find y-intercept ▶ $(0, c) = (0, 5)$ ⟵ the y-intercept

4. Find x-intercept(s) ▶
$$0 = -x^2 - 4x + 5$$
$$0 = (-x + 1)(x + 5)$$
$$0 = -x + 1 \text{ or } 0 = x + 5$$
$$x = 1 \qquad \text{or } x = -5 \qquad \text{The } x\text{-intercepts are } (1, 0), \text{ and } (-5, 0).$$

5. Find additional solutions ▶

x	-6	-4	-3	-1	2
$y = -x^2 - 4x + 5$	-7	5	8	8	-7

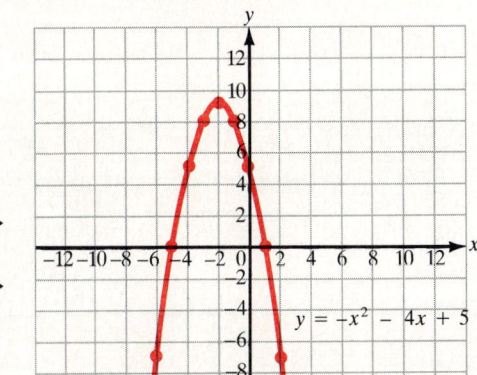

6. Sketch graph ▶

Solution ▶

Draw a smooth curve through the graphs of all of the ordered pairs found in Steps 2–5.

Note ▶ To find additional solutions, you can use the symmetry of the parabola.

Example ▶ The point $(0, 5)$ is two units to the right of the axis of symmetry and is at a height of 5. Therefore, a point two units to the left of the axis of symmetry must also be at a height of 5. That is, the point $(-4, 5)$ is also a solution of $y = -x^2 - 4x + 5$.

384   Chapter 10   Conics and Systems

Make Sure

Graph each parabola.

See Example 3

1. $y = x^2 - 3$

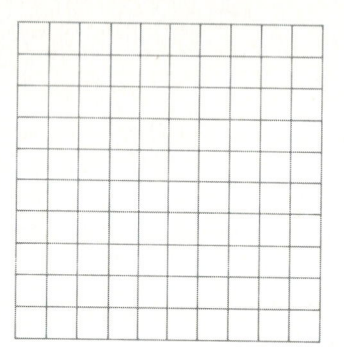

2. $y = -x^2 + 2x$

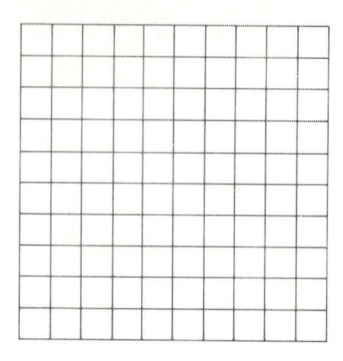

3. $y = x^2 - 4x + 4$

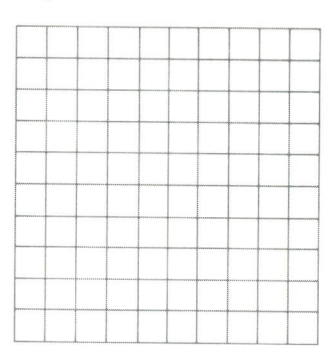

4. $y = -x^2 - 3x + 4$

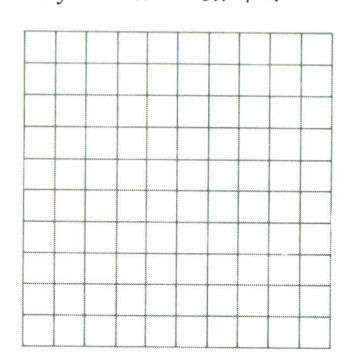

MAKE SURE ANSWERS: See Appendix Selected Answers.

According to Guinness

MOTORCYCLE STUNT
THE LONGEST DISTANCE FOR MOTORCYCLE LONG-JUMPING IS 212 FEET OVER 16 BUSES, BY ALAIN JEAN PRIEUR AT MONTLHERY, NEAR PARIS, FRANCE.

Assuming the motorcycle and rider followed a parabolic path during the motorcycle stunt and that the highest point reached during the jump was 20 feet, find an equation of the parabola if the y-axis goes through the highest point and:

5. _____

5. the x-axis is parallel to the ground and also goes through the highest point.

6. the x-axis is at ground level.

6. _____

Name _____ Date _____ Class _____

10.1 Practice

Set 1: Find the x-intercept(s) of the graph of each equation.

1. $y = 3x^2 - 6x$
2. $y = 2x^2 - 6x$
3. $y = 4x^2 - 9$

4. $y = 9x^2 - 5$
5. $y = 2x^2 - 3x - 2$
6. $y = 6x^2 + x - 1$

Set 2: Find the vertex and the axis of symmetry of each parabola.

7. $y = x^2 + 3$
8. $y = 5x^2 + 4$
9. $y = 2x^2 - 8x + 5$

10. $y = 3x^2 + 6x + 2$
11. $x^2 + y - 2x + 1 = 0$
12. $x^2 + 4x - y - 3 = 0$

Set 3: Graph each parabola.

13. $y = 2x^2 + 8$

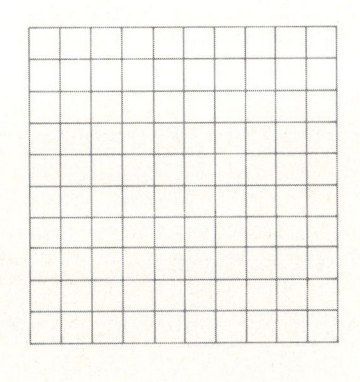

14. $y = -3x^2 + 12$

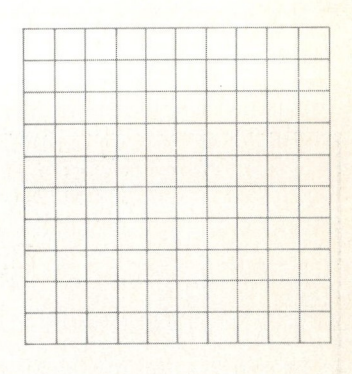

15. $y = x^2 - 4x - 5$

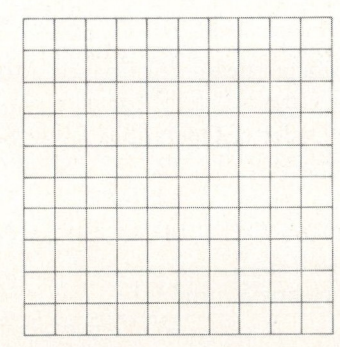

16. $y = x^2 + 6x - 7$

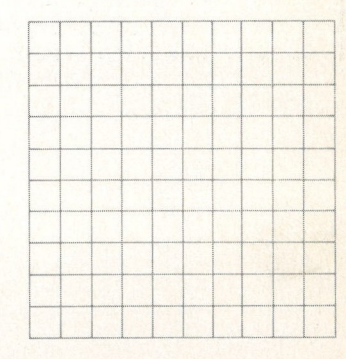

Graph each parabola on graph paper. Attach your work to this page.

17. $y = 2x^2$
18. $y = 4x^2$
19. $y = -3x^2$

20. $y = -5x^2$
21. $y = 3x^2 - 12x$
22. $y = 2x^2 - 8x$

23. $y = -x^2 + 4$
24. $y = -3x^2 + 6$
25. $y = x^2 + 4x + 6$

26. $y = x^2 + 6x + 5$
27. $y = 3x^2 + 2x - 8$
28. $y = 2x^2 - 3x - 5$

29. $y = 2x^2 + 2x + 1$
30. $y = 2x^2 + 3x - 3$
31. $y = 3x^2 - x - 4$

32. $y = 6x^2 - 5x - 6$
33. $y = 4 + x - 3x^2$
34. $y = 6 + 5x - 6x^2$

Extra: Graph each parabola on graph paper. Attach your work to this page.

35. $x^2 - 8x + 2y + 4 = 0$
36. $x^2 - 12x - 2y - 30 = 0$
37. $x^2 - 4x + 2y + 2 = 0$

38. $x^2 - 6x + 3y - 3 = 0$
39. $2x^2 + 4x - y + 3 = 0$
40. $2x^2 + 2x - y + 1 = 0$

Review: Work these problems on a separate sheet of paper. Attach your work to this page.

Evaluate each expression using the Order of Operations Rule. (See Lesson 1.4.)
41. $(3 - 4)^2 + (5 - 2)^2$
42. $(5 - 2)^2 + (1 - 6)^2$
43. $(0 - 2)^2 + (-2 - 3)^2$
44. $(-2 - (-3))^2 + (0 - 3)^2$
45. $(3 - (-4))^2 + (2 - 4)^2$
46. $(-3 - (-3))^2 + (2 - (-4))^2$

Solve for c in each equation. (See Lessons 3.1 and 9.1.)
47. $c^2 = a^2 + b^2$ for $a = 3, b = 4$.
48. $c^2 = a^2 + b^2$ for $a = 5, b = 12$.
49. $c^2 = a^2 + b^2$ for $a = 2, b = 2$.
50. $c^2 = a^2 + b^2$ for $a = 3, b = 5$.

Problem Solving 13: Find Maximum and Minimum Values

Recall ▶ The vertex of $y = ax^2 + bx + c$ ($a \neq 0$) is at $\left(-\dfrac{b}{2a}, \dfrac{4ac - b^2}{4a}\right)$.

> If $a > 0$, then the vertex is the lowest point on the graph of $y = ax^2 + bx + c$ and is called the *minimum value*. If $a < 0$, then the vertex is the highest point on the graph of $y = ax^2 + bx + c$ and is called the *maximum value*.

EXAMPLE: Solve this problem by finding the maximum or minimum value.

1. Read and identify ▶ A farmer has (2 miles) of fencing. What is the (largest) rectangular field that he can enclose with the fencing?

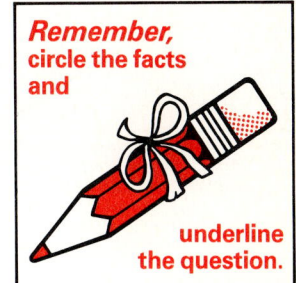

Remember, circle the facts and underline the question.

2. Understand ▶ The unknowns are $\begin{cases} \text{the length of the field with maximum area} \\ \text{the width of the field with maximum area} \end{cases}$.

3. Decide ▶ Let l = the length of the field with maximum area
then w = the width of the field with maximum area

4. Draw a picture ▶

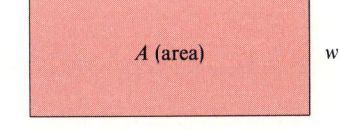

A (area), w, l

5. Use geometry formulas to get a quadratic equation ▶

$P = 2l + 2w$ ⟵ perimeter formula for a rectangle

$2 = 2l + 2w$ Substitute: $P = 2$ (miles)

$1 = l + w$ Divide each term by 2.

$l = 1 - w$ Solve for l or w.

$A = lw$ ⟵ area formula for a rectangle

$A = (1 - w)w$ Substitute: $l = 1 - w$

$A = w - w^2$ Clear parentheses.

$A = -w^2 + w$ Think: $a = -1 < 0$ means the vertex of $A = -w^2 + w$ is a maximum value.

6. Find the vertex ▶

vertex $= \left(-\dfrac{b}{2a}, \dfrac{4ac - b^2}{4a}\right)$

vertex $= \left(-\dfrac{1}{2(-1)}, \dfrac{4(-1)(0) - (1)^2}{4(-1)}\right)$ Think: $A = -w^2 + w$ means $a = -1$, $b = 1$, and $c = 0$.

vertex $= (\tfrac{1}{2}, \tfrac{1}{4})$

7. Interpret ▶ vertex $= (\tfrac{1}{2}, \tfrac{1}{4})$ means the maximum value occurs when $w = \tfrac{1}{2}$ and $A = \tfrac{1}{4}$, or a width of $\tfrac{1}{2}$ mile will maximize the area at $\tfrac{1}{4}$ square mile.
$l = 1 - w = 1 - \tfrac{1}{2} = \tfrac{1}{2}$ means the length of the maximum area is also $\tfrac{1}{2}$ mile.

8. Check ▶ Is the perimeter of the maximum rectangular area 2 miles? Yes:
$P = 2l + 2w = 2(\tfrac{1}{2}) + 2(\tfrac{1}{2}) = 1 + 1 = 2$
Does the length times the width equal the maximum area? Yes: $A = lw = \tfrac{1}{2} \cdot \tfrac{1}{2} = \tfrac{1}{4}$

Note ▶ The largest rectangular field that can be enclosed with 2 miles of fencing is a square measuring $\tfrac{1}{2}$ mile on each side with an area of $\tfrac{1}{4}$ square mile.

Copyright © 1985 by Harcourt Brace Jovanovich, Inc. All rights reserved.

Chapter 10 Conics and Systems

Practice: Solve each problem by finding the maximum or minimum value.

1. What is the area of the largest rectangular garden that can be enclosed with 64 feet of fencing?

2. What is the answer to Problem 1 if only 3 sides are fenced and an existing wall is used as the fourth side?

3. The height (*h*) at time (*t*) of an object thrown vertically upward is given by:

$$h = -16t^2 + v_0 t + h_0$$

where v_0 is the initial velocity and h_0 is the initial height. Find the maximum height of a ball that is thrown upward with an initial velocity of 32 feet per second from 5 feet above the ground.

4. a) Find the maximum height of a ball that is thrown vertically upward with an initial velocity of 48 feet per second from 6 feet above the ground. b) How long does it take the ball to reach its maximum height? c) How long does it take the ball to reach the ground?

5. A farmer has 2 miles of fencing. If two adjacent rectangular fields are to be completely fenced so that the areas are separated, what is the maximum combined area of the two fields?

6. A piece of wire 1 yard long is to be cut once and each piece is to be bent into the shape of a square. How should the wire be cut so that the sum of their areas will be minimized?

7. The perimeter of a rectangle is to be 16 m. Find its area if the diagonal is to be as short as possible.

8. Find two numbers whose sum is 1 and whose product is as large as possible.

9. Find two numbers whose difference is 1 and whose product is as small as possible.

10. A farmer has 2 miles of fencing. What is the largest rectangular field that can be enclosed if a straight river is used as one side instead of fencing?

Extra

11. A farmer has 2 miles of fencing. He wants to fence the two legs of a right triangular area so that a straight river serves as the hypotenuse of the triangle. Find the minimum river frontage that can be used. See Lesson 10.2.

12. What is the largest area that can be fenced in Problem 11 if the length of the river frontage is not a factor in the problem?

According to Guinness

THE LARGEST ANCIENT CARPET WAS THE GOLD-ENRICHED SILK CARPET OF HASHIM (DATED 743 A.D.) OF THE ABBASID CALIPHATE IN BAGHDAD IRAQ. IT IS REPUTED TO HAVE MEASURED 960 FEET IN PERIMETER.

13. What is the maximum floor space that such a carpet could cover?

14. What is the floor space covered by such a carpet if the diagonal is minimized?

10.2 Graph Circles

A triangle that contains one 90° angle (square corner) is called a *right triangle*. The side opposite the 90° angle is called the *hypotenuse*. The other two sides are called *legs*. See Figure 10.5.

Figure 10.5 ▶

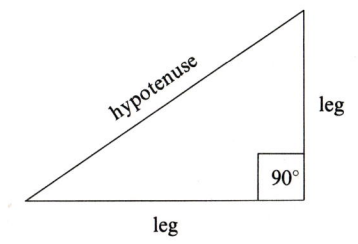

One of the most famous of all the theorems from geometry is the *Pythagorean Theorem*. Named after the Greek mathematician Pythagoras (c.580–c.500 B.C.), it states that the square of the hypotenuse of a right triangle is equal to the sum of the squares of the two legs. See Figure 10.6.

The Pythagorean Theorem

In any right triangle $\quad a^2 + b^2 = c^2$
where a and b are the lengths of the legs, and
c is the length of the hypotenuse.

Figure 10.6 ▶

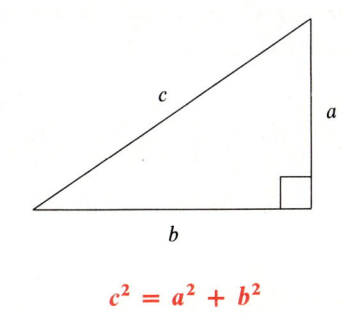

$$c^2 = a^2 + b^2$$

If you are given the lengths of two sides of a right triangle, you can use the Pythagorean Theorem to find the length of the third side.

Example ▶ Find the length of the hypotenuse c, given that the legs have lengths of $a = 3$ and $b = 4$.

1. Use Pythagorean Theorem ▶ $c^2 = a^2 + b^2$

2. Evaluate ▶ $c^2 = 3^2 + 4^2$
$c^2 = 9 + 16$
$c^2 = 25$

3. Solve for c ▶ $c = \sqrt{25}$ Think: The length of the hypotenuse must be nonnegative.
$c = 5$

Solution ▶ A right triangle with legs $a = 3$ and $b = 4$ has an hypotenuse of $c = 5$.

You can apply the Pythagorean Theorem to find the distance between points in a Cartesian coordinate system. Consider a line segment with end points $A(x_1, y_1)$ and $B(x_2, y_2)$. Locate a third point C with coordinates (x_2, y_1) as shown in the following Figure 10.7.

390   Chapter 10   Conics and Systems

Figure 10.7 ▶

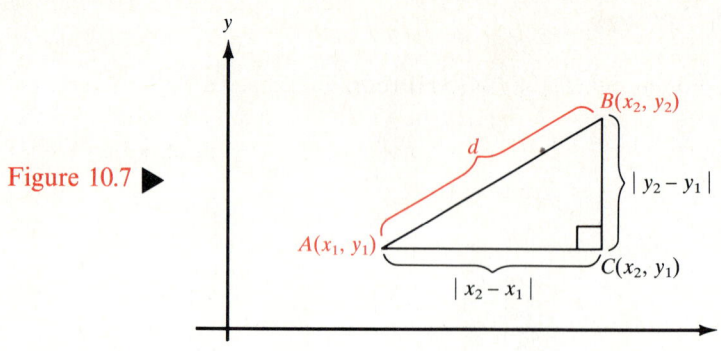

[Figure: Right triangle ABC with A(x_1, y_1), B(x_2, y_2), C(x_2, y_1); hypotenuse d from A to B; vertical side |y_2 − y_1|; horizontal side |x_2 − x_1|.]

The triangle ABC is a right triangle. Let d be the length of the hypotenuse $\overline{AB}$. The lengths of the sides can be represented by:

$BC = |y_2 - y_1|$ Recall: $|y_2 - y_1|$ is the distance between y_2 and y_1.

$AC = |x_2 - x_1|$ $|x_2 - x_1|$ is the distance between x_2 and x_1.

Applying the Pythagorean Theorem to triangle ABC produces:

$$d^2 = |x_2 - x_1|^2 + |y_2 - y_1|^2$$
$$= (x_2 - x_1)^2 + (y_2 - y_1)^2 \quad \text{Recall: } |x^2| = x^2$$

or $d = \sqrt{(x_2 - x_1)^2 + (y_2 - y_1)^2}$ Take the positive square root of both members. Think: Since d is nonnegative $\sqrt{d^2} = d$.

Distance-Between-Two-Points Formula

The distance d between two points with coordinates (x_1, y_1) and (x_2, y_2) is given by:
$$d = \sqrt{(x_2 - x_1)^2 + (y_2 - y_1)^2}.$$

EXAMPLE 1: Find the distance between two given points.

Problem ▶ Find the distance between the points $P_1(-1, 3)$ and $P_2(6, -2)$.

1. Use distance formula ▶ $d = \sqrt{(x_2 - x_1)^2 + (y_2 - y_1)^2}$

2. Evaluate ▶ $= \sqrt{(6 - (-1))^2 + (-2 - 3)^2}$ Substitute: $P_1(-1, 3)$ means $x_1 = -1$ and $y_1 = 3$.
$P_2(6, -2)$ means $x_2 = 6$ and $y_2 = -2$.
$= \sqrt{7^2 + (-5)^2}$
$= \sqrt{49 + 25}$
$= \sqrt{74}$ $\sqrt{74}$ is approximately 8.602 (See Appendix Table 1)

Solution ▶ $\sqrt{74}$ or ≈ 8.602 is the distance between the points $P_1(-1, 3)$ and $P_2(6, -2)$.

Note ▶ Because $(x_2 - x_1)^2 = (x_1 - x_2)^2$ and $(y_2 - y_1)^2 = (y_1 - y_2)^2$ it makes no difference which point is labeled (x_1, y_1) and which point is labeled (x_2, y_2).

Other Examples ▶ (a) The distance between $P_1(-3, 2)$ and $P_2(5, -2)$ is $d = \sqrt{(5 - (-3))^2 + (-2 - 2)^2}$
$= \sqrt{64 + 16}$
$= \sqrt{80}$
$= 4\sqrt{5} \approx 8.944$

(b) The distance between $P_1(0, 0)$ and $P_2(5, 4)$ is $d = \sqrt{(5 - 0)^2 + (4 - 0)^2}$
$= \sqrt{25 + 16}$
$= \sqrt{41} \approx 6.403$

10.2 Graph Circles 391

Make Sure

Find the distance between each given pair of points.

See Example 1 ▶ **1.** (2, 3) and (6, 6)   **2.** (6, −2) and (−6, 3)   **3.** (−1, 2) and (5, −1)

MAKE SURE ANSWERS: 1. 5   2. 13   3. $3\sqrt{5}$

A *circle* is a set of points in a plane that are all the same distance from a fixed point in the plane. The fixed point is called the *center* of the circle. The distance from the center to any point on the circle is called a *radius r* of the circle.

Examples ▶ (a)   (b)   (c)

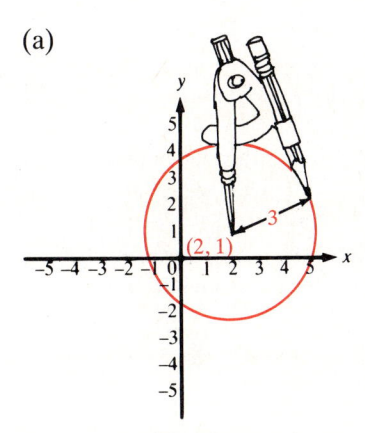

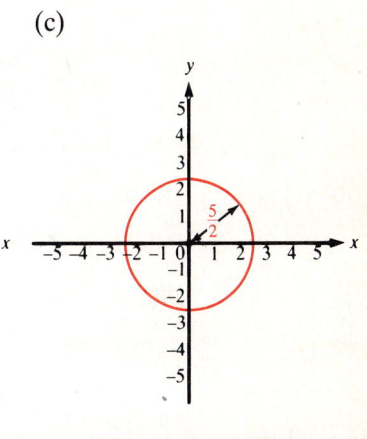

A circle with center (2, 1) and radius 3.   A circle with center (−3, 3) and radius 2.   A circle with center (0, 0) and radius $\frac{5}{2}$.

To find an equation for a circle with radius *r* and center (*h*, *k*), you use the definition of a circle and the distance formula.

Example ▶ Find an equation of a circle with radius *r* and center (*h*, *k*).

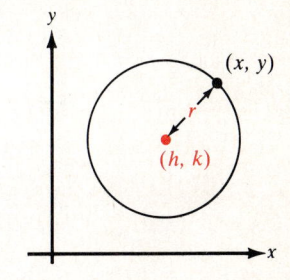

1. Use distance formula ▶ $d = \sqrt{(x_2 - x_1)^2 + (y_2 - y_1)^2}$

2. Substitute ▶ $r = \sqrt{(x - h)^2 + (y - k)^2}$

Substitute: $d = r$, $(x_1, y_1) = (h, k)$, and $(x_2, y_2) = (x, y)$. If (x, y) is any point on the circle, then it is a distance of *r* units from the center (h, k).

3. Square both members ▶ $r^2 = (x - h)^2 + (y - k)^2$

Solution ▶ The equation of the circle with center (*h*, *k*) and radius *r* is $(x - h)^2 + (y - k)^2 = r^2$.

Chapter 10 Conics and Systems

Summary ▶ The *standard equation of a circle* with center (h, k) and radius r is:
$$(x - h)^2 + (y - k)^2 = r^2.$$

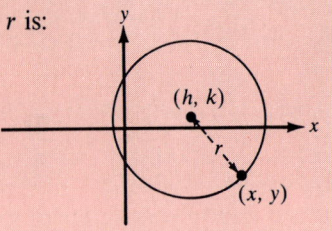

To write an equation of a circle for a given center and a given radius, you use the standard equation.

EXAMPLE 2: Write an equation of a circle given its center and radius.

Problem ▶ Write an equation of a circle with center $(2, -5)$ and radius 4.

1. Use standard equation ▶ $(x - h)^2 + (y - k)^2 = r^2$

2. Substitute ▶ $(x - 2)^2 + [y - (-5)]^2 = 4^2$ Think: Center $(2, -5)$ means $h = 2$ and $k = -5$. Radius 4 means $r = 4$.

Solution ▶ An equation of a circle with center $(2, -5)$ and radius 4 is $(x - 2)^2 + [y - (-5)]^2 = 4^2$.

Note ▶ The equation of the circle in the previous example can be written in several forms.

Examples ▶ $(x - 2)^2 + [y - (-5)]^2 = 4^2$ can be written as $(x - 2)^2 + (y + 5)^2 = 16$
or $x^2 + y^2 - 4x + 10y + 13 = 0$.

Other Examples ▶ (a) An equation of a circle with center $(-3, 7)$ and radius 5 is: $(x + 3)^2 + (y - 7)^2 = 5^2$.
(b) An equation of a circle with center $(0, 0)$ and radius 9 is: $x^2 + y^2 = 9^2$.

Make Sure

Write the standard equation of each circle given its center and radius.

See Example 2 ▶ 1. $(-3, 2), r = 2$ _____   2. $(1, 0), r = 4$ _____

3. $(0, -2), r = 3$ _____   4. $(4, -3), r = 5$ _____

MAKE SURE ANSWERS: 1. $[x - (-3)]^2 + (y - 2)^2 = 2^2$ 2. $(x - 1)^2 + (y - 0)^2 = 4^2$
3. $(x - 0)^2 + [y - (-2)]^2 = 3^2$ 4. $(x - 4)^2 + [y - (-3)]^2 = 5^2$

10.2 Graph Circles

To graph a circle you first use its equation to identify its center (h, k) and its radius r.

EXAMPLE 3: Graph a circle given its equation.

Graph ▶ $(x - 2)^2 + (y + 1)^2 = 9$

1. Write in standard form ▶ $(x - 2)^2 + (y + 1)^2 = 9$

$(x - 2)^2 + [y - (-1)]^2 = 3^2$ ⟵ standard form

2. Identify h, k, and r ▶ $(x - h)^2 + (y - k)^2 = r^2$

Think: $h = 2$ and $k = -1$ means the center is $(2, -1)$.
$r = 3$ means the radius is 3.

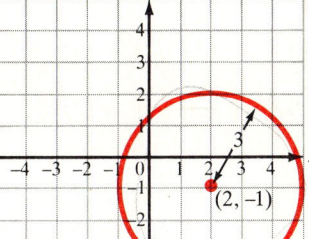

3. Graph ▶

Solution ▶

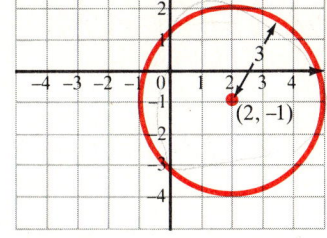

Draw a circle with center $(2, -1)$ and radius 3.

$(x - 2)^2 + (y + 1)^2 = 9$

Make Sure

Graph each circle given its equation.

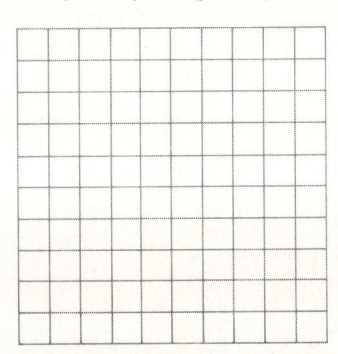

See Example 3 ▶
1. $(x + 1)^2 + (y - 2)^2 = 9$
2. $(x - 3)^2 + (y + 2)^2 = 4$

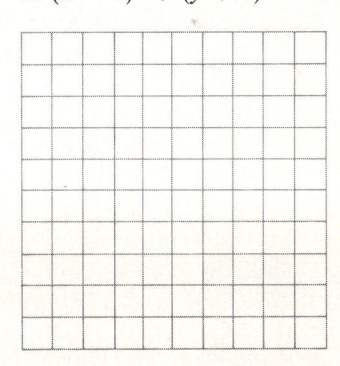

MAKE SURE ANSWERS: See Appendix Selected Answers.

To find the center and radius of a circle from its equation, you first write the equation in standard form. The process of writing the equation in standard form is often accomplished by completing the square.

394 Chapter 10 Conics and Systems

EXAMPLE 4: Find the center and radius of a circle from its equation.

Problem ▶ Find the center and radius of the graph of $x^2 + y^2 + 8x - 2y + 8 = 0$.

1. Regroup ▶

$$x^2 + y^2 + 8x - 2y + 8 = 0$$

$$(x^2 + 8x) + (y^2 - 2y) = -8$$

Think: Group x terms together, group y terms together, and then isolate the constant term.

2. Complete the square ▶

$$(x^2 + 8x + 16) + (y^2 - 2y + 1) = -8 + 16 + 1$$

Think: $[\tfrac{1}{2}(8)]^2 = 16$. $[\tfrac{1}{2}(-2)]^2 = 1$. To maintain the equality you must add the same quantities to both members.

$$(x^2 + 8x + 16) + (y^2 - 2y + 1) = 9$$

3. Factor each PST ▶

$$(x + 4)^2 + (y - 1)^2 = 9$$

4. Write standard form ▶

$$[x - (-4)]^2 + (y - 1)^2 = 3^2$$

5. Identify h, k, and r ▶ $h = -4 \quad k = 1 \quad r = 3$

Solution ▶ The graph of $x^2 + y^2 + 8x - 2y + 8 = 0$ is a circle with center $(-4, 1)$ and radius 3.

Another Example ▶ Find the center and radius of the graph of $x^2 + y^2 + 3x - 10 = 0$.

1. Regroup ▶

$$x^2 + y^2 + 3x - 10 = 0$$

$$(x^2 + 3x) + y^2 = 10$$

Think: Group x terms together, y terms together, and isolate the constant term.

2. Complete the square ▶

$$\left(x^2 + 3x + \frac{9}{4}\right) + y^2 = 10 + \frac{9}{4}$$

Think: $[\tfrac{1}{2}(3)]^2 = \tfrac{9}{4}$. To maintain the equality, add $\tfrac{9}{4}$ to both members.

3. Factor the PST ▶

$$\left(x + \frac{3}{2}\right)^2 + y^2 = \frac{49}{4}$$

4. Write standard form ▶

$$\left[x - \left(-\frac{3}{2}\right)\right]^2 + (y - 0)^2 = \left(\frac{7}{2}\right)^2$$

Think: $y^2 = (y - 0)^2$

5. Identify h, k, and r ▶ $h = -\dfrac{3}{2} \quad k = 0 \quad r = \dfrac{7}{2}$

Solution ▶ The graph of $x^2 + y^2 + 3x - 10 = 0$ is a circle with center $(-\tfrac{3}{2}, 0)$ and radius $\tfrac{7}{2}$.

Make Sure

Find the center and radius of each circle.

See Example 4 ▶ **1.** $x^2 + y^2 - 6x + 8y = 0$ **2.** $x^2 + y^2 - 12x + 16y + 19 = 0$

MAKE SURE ANSWERS: **1.** $(3, -4)$, $r = 5$ **2.** $(6, -8)$, $r = 9$

Name _____ Date _____ Class _____

10.2 Practice

Set 1: Find the distance between each given pair of points.

1. (1, 4) and (5, 7)   2. (−3, 2) and (1, 5)   3. (4, −2) and (−8, 3)

4. (−2, 5) and (3, −2)   5. (−3, 3) and (4, 3)   6. (0, −5) and (0, 2)

Set 2: Write the standard equation of each circle given its center and radius.

7. (0, 0), $r = 3$ 8. (0, 0), $r = 5$ 9. (2, 3), $r = 2$

10. (4, 3), $r = 4$ 11. (−5, 3), $r = 1$ 12. (−3, −5), $r = 3$

Set 3: Graph each equation of a circle.

13. $x^2 + y^2 = 36$

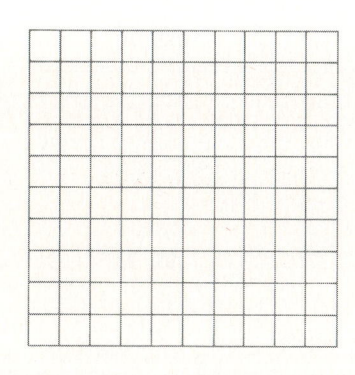

14. $x^2 + y^2 = 25$

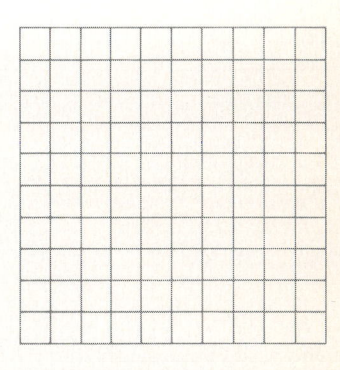

15. $(x − 3)^2 + y^2 = 4$

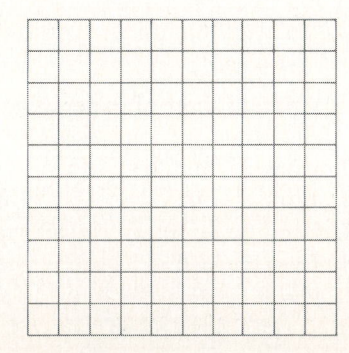

16. $(x − 2)^2 + y^2 = 9$

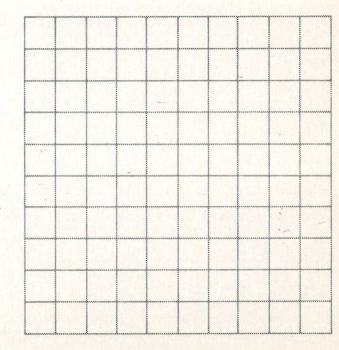

29. _____

30. _____

31. _____

32. _____

33. _____

34. _____

35. _____

36. _____

37. _____

38. _____

39. _____

40. _____

Graph each equation on graph paper. Attach your work to this page.

17. $x^2 + (y - 1)^2 = 1$
18. $x^2 + (y - 4)^2 = 1$
19. $(x - 2)^2 + (y - 4)^2 = 4$

20. $(x - 1)^2 + (y - 3)^2 = 4$
21. $(x - 2)^2 + (y + 1)^2 = 9$
22. $(x + 1)^2 + (y + 2)^2 = 9$

23. $(x - 1)^2 + (y + 2)^2 = 4$
24. $(x - 3)^2 + (y + 1)^2 = 4$
25. $(x + 2)^2 + (y - 1)^2 = 9$

26. $(x + 3)^2 + (y - 2)^2 = 4$
27. $(x - 1)^2 + y^2 = 16$
28. $(x + 3)^2 + y^2 = 1$

Set 4: Find the center and radius of each circle.

29. $x^2 + y^2 - 4 = 0$
30. $x^2 + y^2 - 9 = 0$
31. $x^2 + y^2 - 6x = 0$

32. $x^2 + y^2 + 8x = 0$
33. $x^2 + y^2 - 8x - 6y = 0$
34. $x^2 + y^2 + 6x - 8y = 0$

35. $x^2 + y^2 + 2x + 4y = 20$
36. $x^2 + y^2 + 4x + 6y = 12$

37. $x^2 + y^2 + 4x - 8y = 5$
38. $x^2 + y^2 + 10x - 8y = -16$

39. $9x^2 + 9y^2 = 12x + 6y + 40$
40. $9x^2 + 9y^2 = 12x - 6y + 31$

Review: Work these problems on a separate sheet of paper. Attach your work to this page.

Graph each linear equation. (See Lessons 3.2 and 3.3.)
41. $x + 4y = -4$
42. $4x - y = 4$
43. $2x + 3y = 6$
44. $3x - 2y = -6$
45. $5x + 2y = 10$
46. $2x - 5y = -10$
47. $3x + 5y = 0$
48. $4x - 5y = 0$
49. $x = 5$
50. $x = -3$
51. $y = 3$
52. $y = -2$

Problem Solving 14: Use the Pythagorean Theorem

Recall ▶ Pythagorean Theorem

In any right triangle $a^2 + b^2 = c^2$ where a and b are the lengths of the legs, and c is the length of the hypotenuse.

To solve a problem using the Pythagorean Theorem, you must also use the Square Root Rule.

EXAMPLE: Solve this problem using the Pythagorean Theorem.

1. Read and identify ▶ A baseball diamond is (90 feet square). How far must a baseball be thrown from home plate to second base, to the nearest tenth of a foot?

Remember, circle the facts and underline the question.

2. Understand ▶ The question asks you to find the length of the hypotenuse of a right triangle given the lengths of the legs as 90 feet each.

3. Decide ▶ To find the length of one side of a right triangle given the lengths of the other two sides, you **use the Pythagorean Theorem.**

4. Use Pythagorean Theorem ▶
$c^2 = a^2 + b^2$
$c^2 = 90^2 + 90^2$
$c^2 = 8100 + 8100$
$c^2 = 16{,}200$

5. Use Square Root Rule ▶ $c = \sqrt{16{,}200}$ Think: Measures are never negative.
≈ 127.27922 Use a calculator.
≈ 127.3 Round to the nearest tenth.

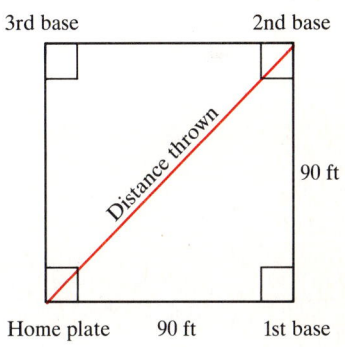

6. Interpret ▶ $c = +127.3$ means the baseball must be thrown **127.3 feet**.

7. Check ▶
$c^2 = a^2 + b^2$
$(127.3)^2 = 90^2 + 90^2$
$16{,}205.29 \approx 16{,}200$ ⟵ 127.3 checks

Substitute the rounded proposed solution and known measures into the Pythagorean Theorem to see if you get two numbers that are approximately equal.

Practice: Solve each problem using the Pythagorean Theorem. Round your answer to the nearest tenth when necessary. Assume the "ground" to be flat.

1. _____
2. _____
3. _____
4. _____
5. _____
6. _____

1. A 17-foot ladder is leaning against a building. The base of the building is 8 feet from the base of the ladder. How high on the building does the ladder reach?

2. A guy wire reaches from the top of a vertical 50-foot flag pole to a ground anchor. The base of the flag pole is 16 feet from the anchor. How long is the guy wire?

3. Two vertical poles are 38 m and 46 m high, respectively. The bases of the poles are 15 m apart. How far is it from the top of one pole to the top of the other?

4. A rectangular field is 84 meters long and 63 meters wide. What is the distance from one corner to the corner diagonally opposite?

5. One leg of a right triangle is 1 m less than twice the other leg. If the hypotenuse is 17 m, find the dimensions of the triangle.

6. The hypotenuse of a right triangle is 13 cm. One leg of the right triangle is 7 cm longer than the other leg. Find the area of the right triangle.

Copyright © 1985 by Harcourt Brace Jovanovich, Inc. All rights reserved.

398 Chapter 10 Conics and Systems

7a. _____

b. _____

8a. _____

b. _____

9. _____

10. _____

11. _____

12. _____

13. _____

14. _____

15. _____

16. _____

17. _____

18. _____

7. Let the length of each side of a square be denoted by *s*. Let the length of the diagonal of the square be denoted by *d*. a) Express *d* in terms of *s*. b) Express *s* in terms of *d*.

8. Let the length of each side of an equilateral triangle (a triangle with 3 equal sides) be denoted by *s*. Let the altitude (the shortest distance from a side to the vertex opposite) be denoted by *h*. a) Express the altitude *h* in terms of *s*. b) Express the area *A* of the triangle in terms of *s*.

9. The length of an edge of a cube is denoted by *e*. Find a formula for length of a diagonal (*d*) of a cube in terms of *e*. (Hint: Use the Pythagorean Theorem twice.)

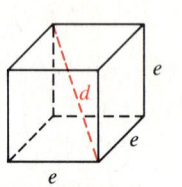

10. A house gable has a rise of 8 feet and a run of 15 feet. How long is each rafter?

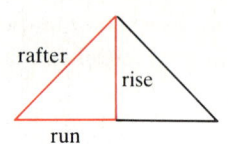

11. What is the rise of a gable if the rafter is 20 feet long and the run is 16 feet?

12. The pitch of a roof is $\frac{3}{4}$. The rafter is 24 feet long. Find the area of the gable. (Hint: pitch = rise ÷ run.)

13. A square is inscribed in a circle of radius 8 cm. Find the length of each side of the square. (A square is inscribed in a circle if all four of the square's vertices are on the circle.)

14. A circle is circumscribed around a square. The area of the square is 25 cm². What is the radius of the circle? (A circle is circumscribed around a square if the square is inscribed in the circle. See Problem 13.)

15. A 1-foot-square bar is to be cut from round stock. Stock with diameters of 1 foot, $1\frac{1}{4}$ feet, and $1\frac{1}{2}$ feet are available. Which size stock should be used?

16. Which is larger, the area of an equilateral triangle with a side of 4 m or the area of a square inscribed in a circle of radius 4 m?

17. A certain building is 2 miles west and 5 miles north of another building. Find the shortest distance between the two buildings.

18. A pilot wants to fly 300 km due west. He takes the wrong course and flies in a straight line to end up 50 km due south of his planned destination. How far did the pilot fly?

According to Guinness

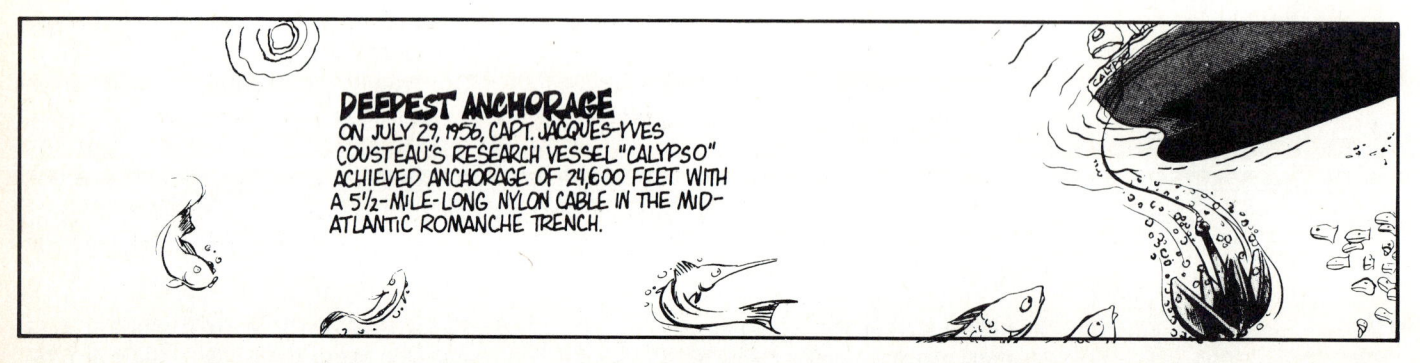

DEEPEST ANCHORAGE
ON JULY 29, 1956, CAPT. JACQUES-YVES COUSTEAU'S RESEARCH VESSEL "CALYPSO" ACHIEVED ANCHORAGE OF 24,600 FEET WITH A 5½-MILE-LONG NYLON CABLE IN THE MID-ATLANTIC ROMANCHE TRENCH.

19. _____

19. Assuming the 5½-mile-long nylon cable formed a straight line and the ocean bottom was flat around the anchorage, how far from the anchorage was the point on the ocean bottom directly beneath the boat, to the nearest hundred feet?

Copyright © 1985 by Harcourt Brace Jovanovich, Inc. All rights reserved.

10.3 Graph Ellipses

> An *ellipse* is the set of all points in a plane such that the sum of the distances from two fixed points in the plane is constant.

To draw an ellipse, you use a pencil, a string, two tacks, and the technique shown in Figure 10.8.

Figure 10.8 ▶

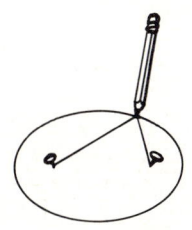

It can be shown that the graph of $4x^2 + y^2 = 16$ is an ellipse. You can graph $4x^2 + y^2 = 16$ by solving the equation for y and then computing some solutions.

Example ▶ Graph $4x^2 + y^2 = 16$.

1. Solve for y ▶
$$y^2 = 16 - 4x^2$$
$$y = \pm\sqrt{16 - 4x^2}$$
$$y = \pm 2\sqrt{4 - x^2}$$

2. Find solutions ▶

x	$y = \pm 2\sqrt{4 - x^2}$
-2	0
-1	$\pm 2\sqrt{3} \approx \pm 3.464$ (See Appendix Table 1.)
0	± 4
1	$\pm 2\sqrt{3} \approx \pm 3.464$
2	0

Think: To keep the radicand nonnegative, use only values of x such that $-1 \leq x \leq 1$.

3. Plot solutions ▶

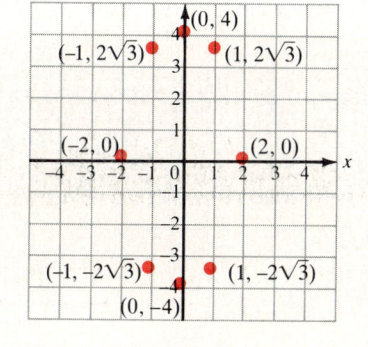

4. Sketch graph ▶

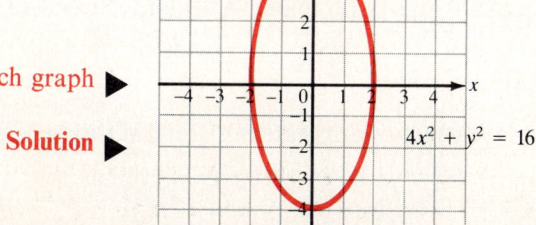

Solution ▶

The following definitions will permit you to identify and graph ellipses by using an easier procedure. A *chord* of an ellipse is a line segment with its end points on the ellipse. The longest chord of an ellipse is called the *major axis*. The shortest chord of an ellipse is called the *minor axis*. The end points of the major and minor axes are called the *vertices* of the ellipse. The point of intersection of the major and minor axes is called the *center* of the ellipse.

Figure 10.9 shows an ellipse with major axis $\overline{AC}$, minor axis $\overline{BD}$, and vertices A, B, C, and D. The center of the ellipse is point 0.

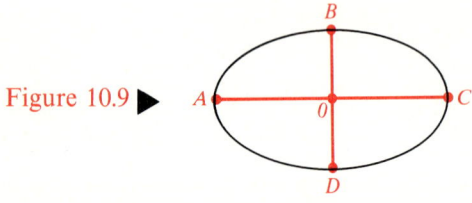

Figure 10.9

Note ▶ If an ellipse has its major and minor axes on the coordinate axes, then the vertices are also the x- and y-intercepts.

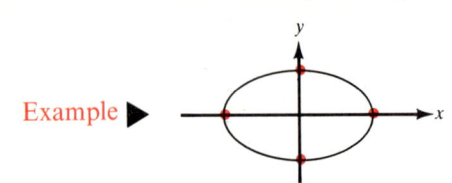

Example ▶

The *standard equation of an ellipse* with center (0, 0) and vertices $(-a, 0)$, $(a, 0)$, $(0, -b)$, and $(0, b)$ is: $\dfrac{x^2}{a^2} + \dfrac{y^2}{b^2} = 1$.

Case 1: $a > b$

Case 2: $a < b$

EXAMPLE 1: Write the equation in standard form of an ellipse with center (0, 0) and vertices on the coordinate axes.

Problem ▶ Write the equation in standard form of an ellipse with center (0, 0) and vertices $(-3, 0)$, $(3, 0)$, $(0, -2)$, and $(0, 2)$.

1. Identify a and b ▶ $a = 3$ Think: a equals the absolute value of the x components of the x-intercepts.
$b = 2$ b equals the absolute value of the y components of the y-intercepts.

2. Use standard equation ▶ $\dfrac{x^2}{a^2} + \dfrac{y^2}{b^2} = 1$ ⟵ standard equation

3. Substitute ▶ $\dfrac{x^2}{3^2} + \dfrac{y^2}{2^2} = 1$ Substitute 3 for a and 2 for b.

Solution ▶ The equation in standard form of an ellipse with center (0, 0) and vertices $(-3, 0)$, $(3, 0)$, $(0, -2)$, and (0, 2) is $\dfrac{x^2}{3^2} + \dfrac{y^2}{2^2} = 1$.

Another Example ▶ The equation in standard form of an ellipse with center (0, 0) and vertices $(-7, 0)$, $(7, 0)$, $(0, -5)$, and (0, 5) is $\dfrac{x^2}{7^2} + \dfrac{y^2}{5^2} = 1$.

10.3 Graph Ellipses 401

Make Sure

Write the standard equation of an ellipse with center (0, 0) and the given vertices.

See Example 1 ▶
1. $(-4, 0), (4, 0) (0, 2) (0, -2)$
2. $(-\frac{2}{3}, 0), (\frac{2}{3}, 0) (0, \frac{4}{5}), (0, -\frac{4}{5})$

MAKE SURE ANSWERS: 1. $\dfrac{x^2}{4^2} + \dfrac{y^2}{2^2} = 1$ 2. $\dfrac{x^2}{(\frac{2}{3})^2} + \dfrac{y^2}{(\frac{4}{5})^2} = 1$

To graph an ellipse with center (0, 0) and vertices on the coordinate axes, you first write its equation in standard form.

EXAMPLE 2: Graph ellipses with center (0, 0) and vertices on the coordinate axes.

Graph ▶ $\dfrac{x^2}{25} + \dfrac{y^2}{4} = 1$ and $\dfrac{x^2}{9} + \dfrac{y^2}{16} = 1$

1. Write standard form ▶

Case 1: $a > b$ $\dfrac{x^2}{25} + \dfrac{y^2}{4} = 1$

$\dfrac{x^2}{5^2} + \dfrac{y^2}{2^2} = 1$

Case 2: $a < b$ $\dfrac{x^2}{9} + \dfrac{y^2}{16} = 1$

$\dfrac{x^2}{3^2} + \dfrac{y^2}{4^2} = 1$

2. Identify a and b ▶ $a = 5$ and $b = 2$. $a = 3$ and $b = 4$.

3. Plot intercepts ▶

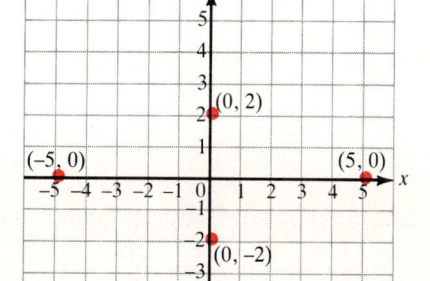

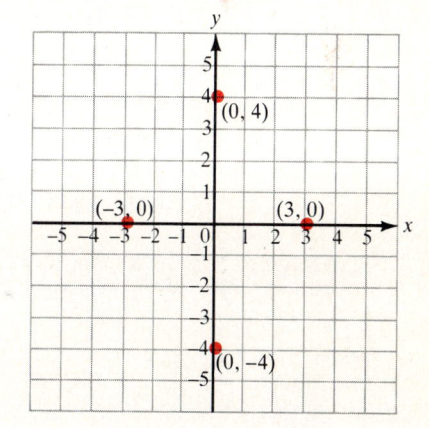

4. Sketch smooth curve through intercepts ▶

Solutions ▶

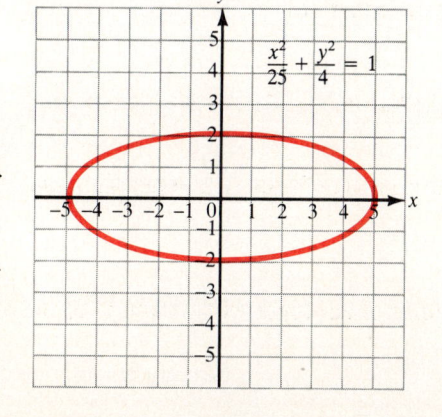

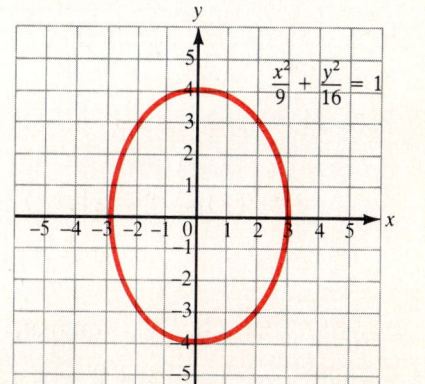

Chapter 10 Conics and Systems

To graph an ellipse with an equation of the form $Ax^2 + By^2 = C$, you first write the equation in standard form.

Example ▶ Graph $64x^2 + 9y^2 = 144$.

1. Rename ▶ $\dfrac{64x^2}{144} + \dfrac{9y^2}{144} = \dfrac{144}{144}$ Think: The standard equation has 1 as its constant term.

$\dfrac{4x^2}{9} + \dfrac{y^2}{16} = 1$

$\dfrac{x^2}{\frac{9}{4}} + \dfrac{y^2}{16} = 1$ Think: The standard equation has *1* as the numerators of the variable terms, so divide the numerator and the denominator of the x^2 term by 4.

$\dfrac{x^2}{(\frac{3}{2})^2} + \dfrac{y^2}{4^2} = 1$ ⟵ standard form

2. Identify a and b ▶ $a = \dfrac{3}{2}$ and $b = 4$.

3. Sketch graph ▶   Think: The vertices are $(\pm a, 0) = (\pm\frac{3}{2}, 0)$ and $(0, \pm b) = (0, \pm 4)$.

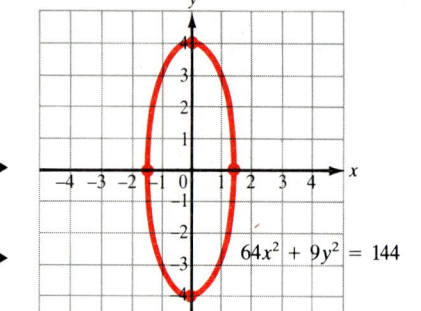

Solution ▶

Make Sure

Graph each ellipse.

See Example 2 ▶   **1.** $\dfrac{x^2}{4} + \dfrac{y^2}{25} = 1$ **2.** $\dfrac{x^2}{16} + \dfrac{y^2}{9} = 1$

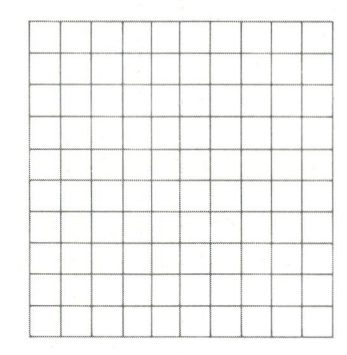

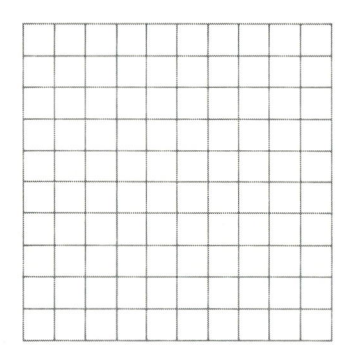

MAKE SURE ANSWERS: See Appendix Selected Answers.

10.3 Practice

Set 1: Write the standard equation of ellipses with center (0, 0) and the given vertices.

1. $(1, 0), (-1, 0), (0, 2), (0, -2)$
2. $(2, 0), (-2, 0), (0, 3), (0, -3)$
3. $(3, 0), (-3, 0), (0, 8), (0, -8)$
4. $(4, 0), (-4, 0), (0, 3), (0, -3)$
5. $(\sqrt{2}, 0), (-\sqrt{2}, 0), (0, 2), (0, -2)$
6. $(\sqrt{5}, 0), (-\sqrt{5}, 0), (0, 5), (0, -5)$

Set 2: Graph each ellipse.

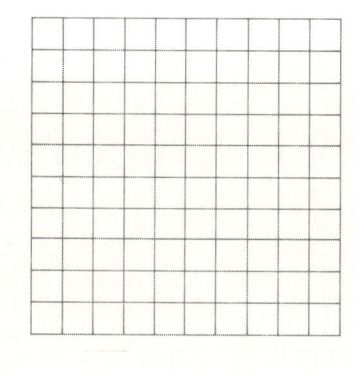

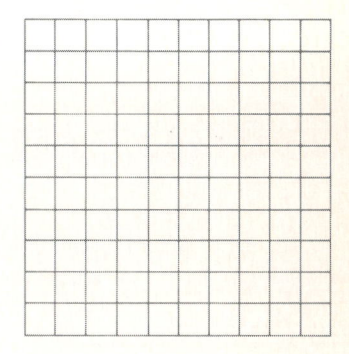

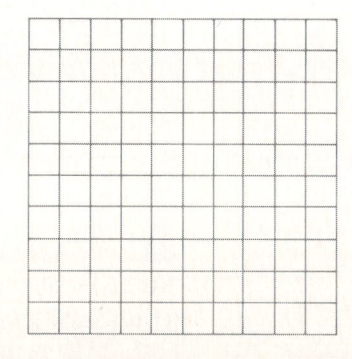

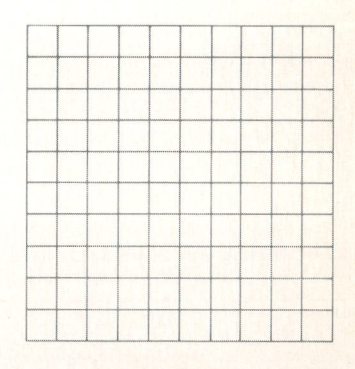

7. $\dfrac{x^2}{1} + \dfrac{y^2}{16} = 1$
8. $\dfrac{x^2}{4} + \dfrac{y^2}{16} = 1$
9. $\dfrac{x^2}{16} + \dfrac{y^2}{4} = 1$
10. $\dfrac{x^2}{25} + \dfrac{y^2}{1} = 1$

404 Chapter 10 Conics and Systems

11. $4x^2 + 9y^2 = 36$

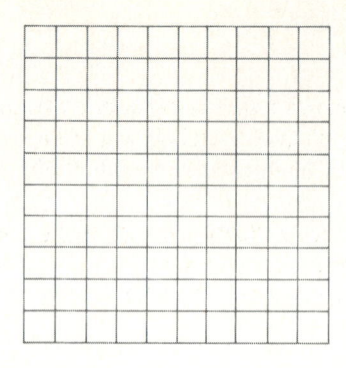

12. $9x^2 + 16y^2 = 144$

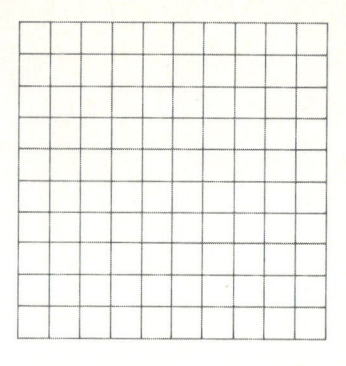

13. $16x^2 + 25y^2 = 400$

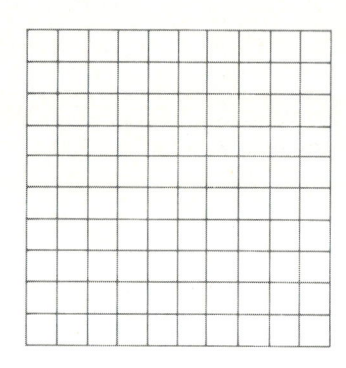

14. $9x^2 + 4y^2 = 36$

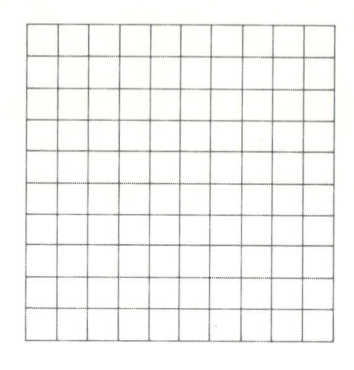

Mixed Practice: Graph each ellipse on graph paper. Attach your work to this page.

15. $\dfrac{x^2}{16} + \dfrac{y^2}{1} = 1$

16. $\dfrac{x^2}{9} + \dfrac{y^2}{25} = 1$

17. $\dfrac{x^2}{16} + \dfrac{y^2}{25} = 1$

18. $\dfrac{x^2}{1} + \dfrac{y^2}{9} = 1$

19. $\dfrac{x^2}{4} + \dfrac{y^2}{9} = 1$

20. $\dfrac{x^2}{4} + \dfrac{y^2}{1} = 1$

21. $4x^2 + y^2 = 4$

22. $16x^2 + y^2 = 16$

23. $16x^2 + 9y^2 = 144$

24. $4x^2 + 16y^2 = 16$

25. $x^2 + 25y^2 = 25$

26. $9x^2 + 25y^2 = 225$

Review: Work these problems on a separate sheet of paper. Attach your work to this page.

Solve each equation for y. (See Lesson 3.1.)
27. $xy = 8$ for $x = 8, 4, 2, 1$.
28. $xy = 6$ for $x = 6, 3, 2, 1$.
29. $xy = 4$ for $x = 4, 2, 1, \tfrac{1}{2}$.
30. $xy = 2$ for $x = 2, 1, \tfrac{1}{2}, \tfrac{1}{4}$.
31. $xy = 1$ for $x = 1, \tfrac{1}{2}, \tfrac{1}{4}, -1$.
32. $xy = -2$ for $x = -2, -1, -\tfrac{1}{2}, \tfrac{1}{4}, \tfrac{1}{2}, 1$.

Copyright © 1985 by Harcourt Brace Jovanovich, Inc. All rights reserved.

10.4 Graph Hyperbolas

The hyperbola is the only conic section that consists of two nonintersecting parts. Each part is called a *branch*. Although a branch of a hyperbola has the appearance of a parabola, it is not a parabola.

The shortest line segment that connects the two branches of a hyperbola is called its *transverse axis*. The midpoint of the transverse axis is called the *center* of the hyperbola. The end points of the transverse axis are called the *vertices* of the hyperbola.

Examples ▶ (a), (b), (c)

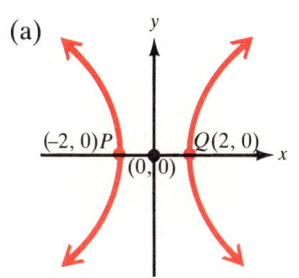

(a) A hyperbola with center (0, 0) and vertices (−2, 0) and (2, 0). The line segment PQ is the transverse axis.

(b) A hyperbola with center (0, 0) and vertices (0, −3) and (0, 3). The line segment RS is the transverse axis.

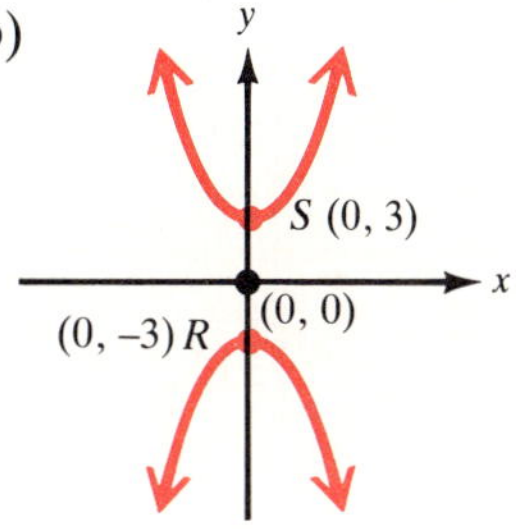

(c) A hyperbola with center (0, 0) and vertices (−3, −3) and (3, 3). The line segment TU is the transverse axis.

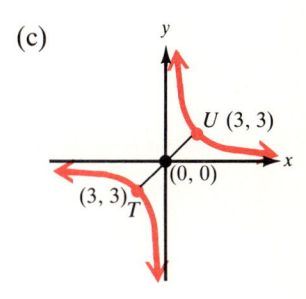

Associated with every hyperbola is a pair of intersecting lines called *asymptotes*. See Figure 10.10.

Figure 10.10 ▶

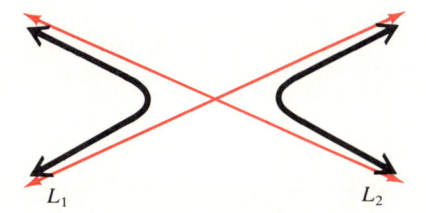

Lines L_1 and L_2 are asymptotes for the hyperbola.

The asymptotes of a hyperbola intersect at its center. A hyperbola does not intersect either of its asymptotes, but a point moving along a branch of the hyperbola gets closer and closer to an asymptote as the point gets further and further from the center point. Understanding this relationship between a hyperbola and its asymptotes is helpful when sketching the graph of a hyperbola.

The *standard equation of a hyperbola* with center (0, 0) and vertices:

(−a, 0) and (a, 0) is $\dfrac{x^2}{a^2} - \dfrac{y^2}{b^2} = 1$

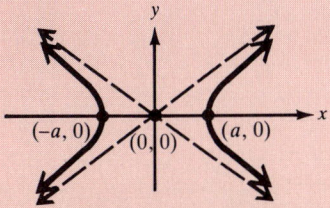

(0, −b) and (0, b) is $-\dfrac{x^2}{a^2} + \dfrac{y^2}{b^2} = 1$

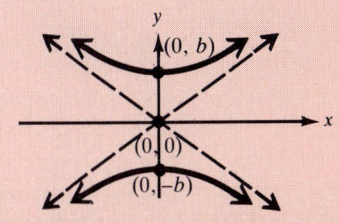

In both cases, the asymptotes of the hyperbolas have the equations $y = -\dfrac{b}{a}x$ and $y = \dfrac{b}{a}x$.

Chapter 10 Conics and Systems

EXAMPLE 1: Graph a hyperbola with center (0, 0) and vertices on the x-axis.

Problem ▶ Graph $\dfrac{x^2}{9} - \dfrac{y^2}{4} = 1$.

1. Write standard form ▶ $\dfrac{x^2}{3^2} - \dfrac{y^2}{2^2} = 1$

2. Identify a and b ▶ $a = 3$ and $b = 2$.

3. Graph asymptotes and vertices ▶

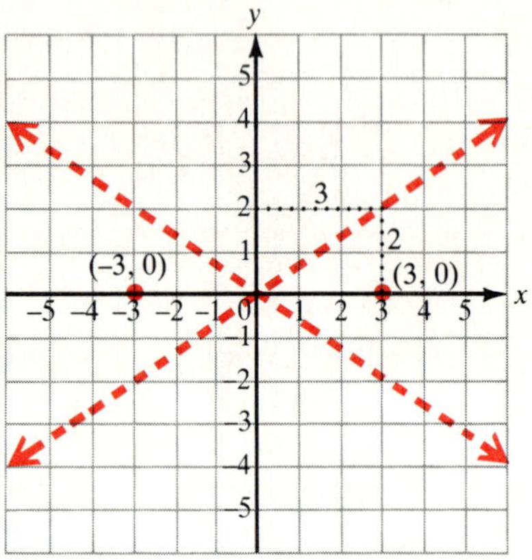

Think: The asymptotes are lines that intersect at the center (0, 0). One line has slope $-\dfrac{b}{a} = -\dfrac{2}{3}$ and the other has slope $\dfrac{b}{a} = \dfrac{2}{3}$.

Think: The standard equation $\dfrac{x^2}{a^2} - \dfrac{y^2}{b^2} = 1$ has vertices $(-a, 0) = (-3, 0)$ and $(a, 0) = (3, 0)$.

4. Sketch graph ▶

Solution ▶

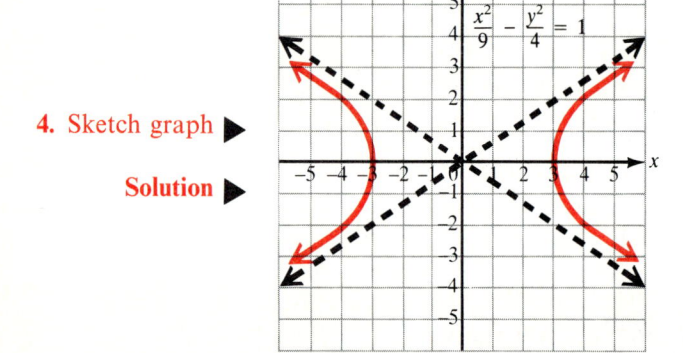

Think: Sketch each branch as a smooth curve passing through its vertex and approaching the asymptotes.

Note ▶ The asymptotes are shown as dotted lines to indicate that they are not part of the graph of the hyperbola, but that they have been used as guide lines in sketching the graph.

Make Sure

Graph each hyperbola.

See Example 1 ▶ **1.** $\dfrac{x^2}{16} - \dfrac{y^2}{25} = 1$ **2.** $\dfrac{x^2}{16} - \dfrac{y^2}{4} = 1$

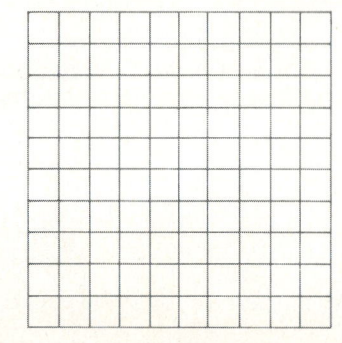

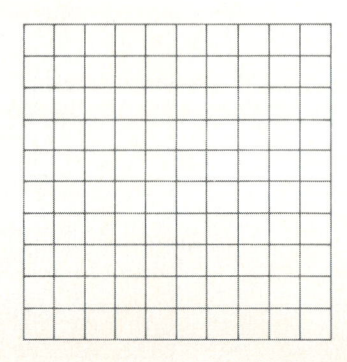

MAKE SURE ANSWERS: See Appendix Selected Answers.

10.4 Graph Hyperbolas   407

EXAMPLE 2: Graph a hyperbola with center (0, 0) and vertices on the y-axis.

Problem ▶ Graph $-\dfrac{x^2}{25} + \dfrac{y^2}{16} = 1$.

1. Write standard form ▶ $-\dfrac{x^2}{5^2} + \dfrac{y^2}{4^2} = 1$

2. Identify a and b ▶ $a = 5$ and $b = 4$.

3. Graph asymptotes and vertices ▶

Think: The asymptotes are lines that intersect at the center (0, 0). One has slope $\dfrac{b}{a} = \dfrac{4}{5}$. The other has slope $-\dfrac{b}{a} = -\dfrac{4}{5}$.

Think: The standard equation $-\dfrac{x^2}{a^2} + \dfrac{y^2}{b^2} = 1$ has vertices $(0, -b) = (0, -4)$ and $(0, b) = (0, 4)$.

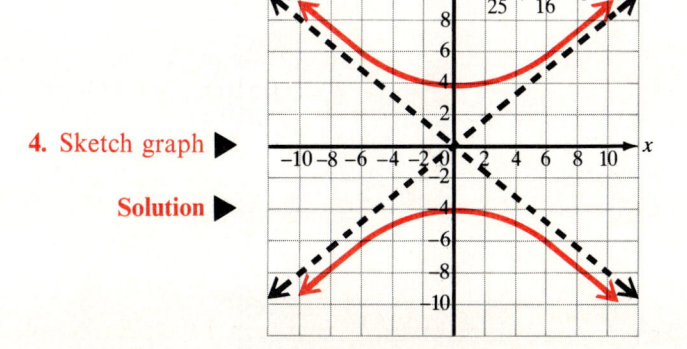

4. Sketch graph

Solution ▶

Make Sure

Graph each hyperbola.

See Example 2 ▶ **1.** $-\dfrac{x^2}{9} + \dfrac{y^2}{25} = 1$ **2.** $-\dfrac{x^2}{16} + \dfrac{y^2}{1} = 1$

MAKE SURE ANSWERS: See Appendix Selected Answers.

408 Chapter 10 Conics and Systems

The equation $xy = k$, where k is a nonzero constant, graphs to be a hyperbola with center $(0, 0)$ and has the coordinate axes as its asymptotes.

If $k > 0$, the branches of the hyperbola are in Quadrants I and III.

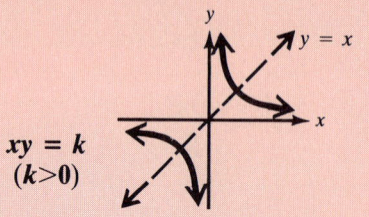

$xy = k$
$(k>0)$

The vertices are on the line $y = x$.

If $k < 0$, the branches of the hyperbola are in Quadrants II and IV.

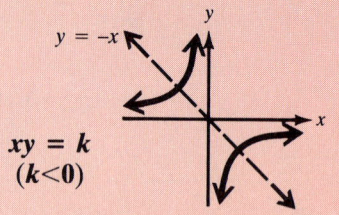

$xy = k$
$(k<0)$

The vertices are on the line $y = -x$.

To graph an equation of the form $xy = k$, you first find several solutions.

EXAMPLE 3: Graph hyperbolas given their equations in $xy = k$ form.

Problem ▶ Graph $xy = 8$ and $xy = -4$.

1. Find solutions ▶ $xy = 8$ means the product of x and y is 8.

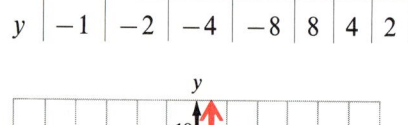

x	-8	-4	-2	-1	1	2	4	8
y	-1	-2	-4	-8	8	4	2	1

$xy = -4$ means the product of x and y is -4.

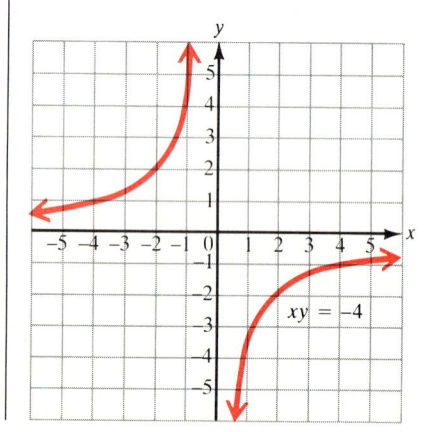

x	-4	-2	-1	1	2	4
y	1	2	4	-4	-2	-1

2. Sketch the graph ▶

Solutions ▶

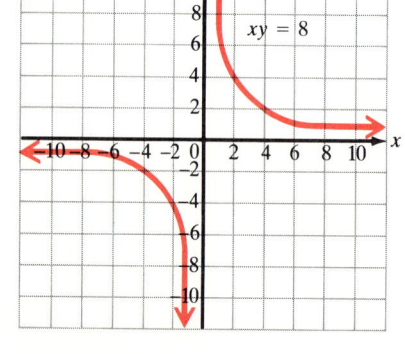

Make Sure

Graph each hyperbola.

See Example 3 ▶ **1.** $xy = 6$

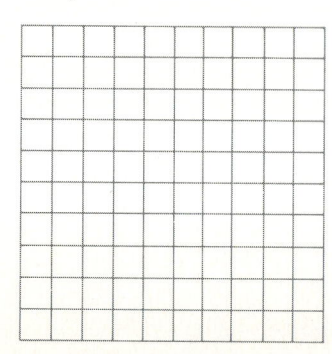

2. $xy = -6$

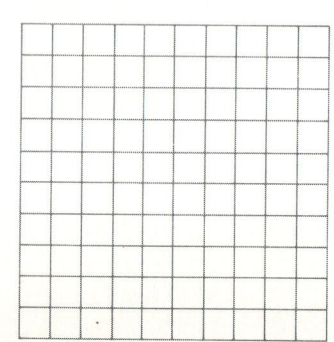

MAKE SURE ANSWERS: See Appendix Selected Answers.

10.4 Practice

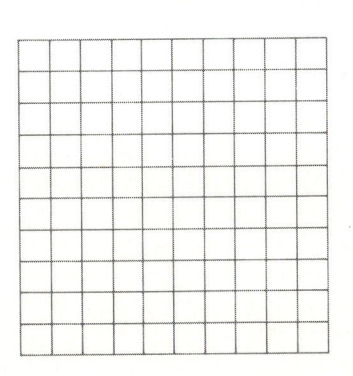

Set 1: Graph each hyperbola.

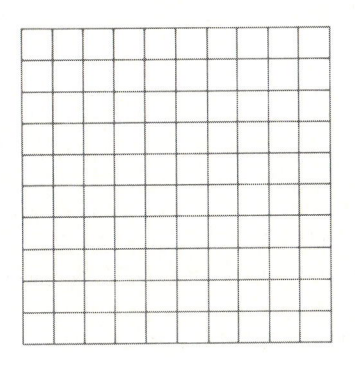

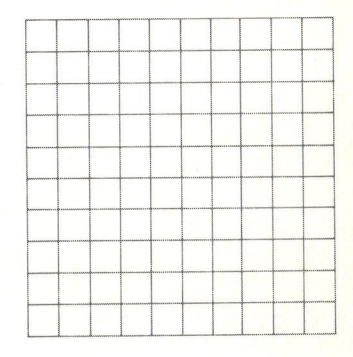

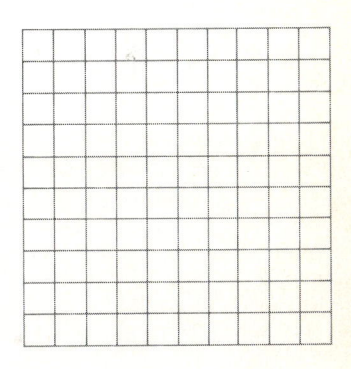

1. $\dfrac{x^2}{4} - \dfrac{y^2}{9} = 1$

2. $\dfrac{x^2}{1} - \dfrac{y^2}{4} = 1$

3. $\dfrac{x^2}{9} - \dfrac{y^2}{4} = 1$

4. $\dfrac{x^2}{4} - \dfrac{y^2}{1} = 1$

Set 2: Graph each hyperbola.

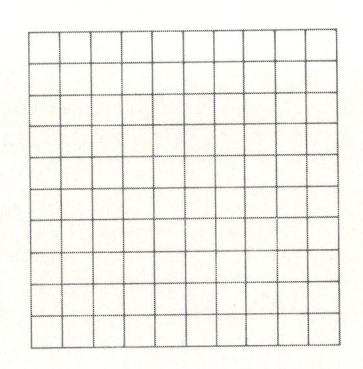

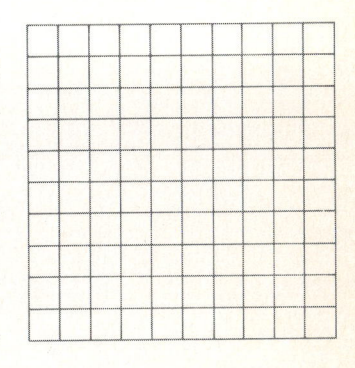

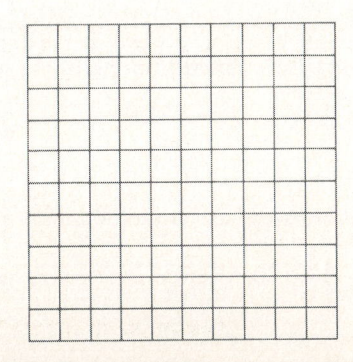

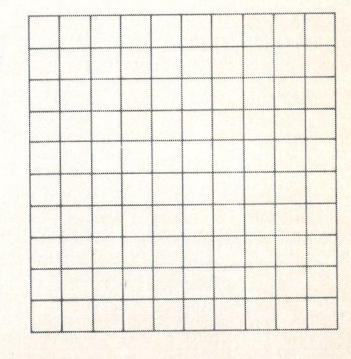

5. $-\dfrac{x^2}{4} + \dfrac{y^2}{16} = 1$

6. $-\dfrac{x^2}{1} + \dfrac{y^2}{25} = 1$

7. $-\dfrac{x^2}{9} + \dfrac{y^2}{4} = 1$

8. $-\dfrac{x^2}{25} + \dfrac{y^2}{4} = 1$

Set 3: Graph each hyperbola.

9. $xy = 4$

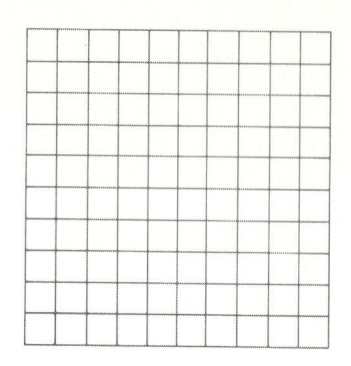

10. $xy = 12$

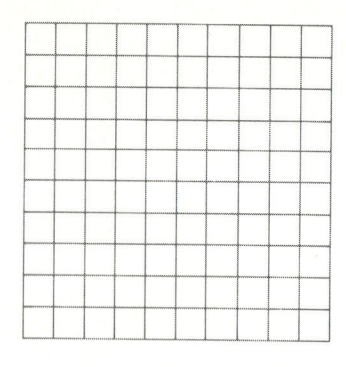

11. $xy = -12$

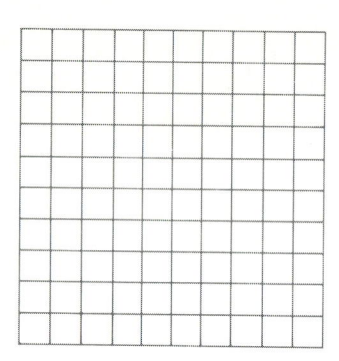

12. $xy = -8$

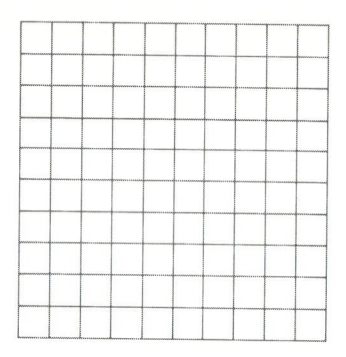

Mixed Practice: Graph each hyperbola on graph paper. Attach your work to this page.

13. $\dfrac{x^2}{25} = 1 + \dfrac{y^2}{9}$

14. $\dfrac{x^2}{16} = 1 + \dfrac{y^2}{9}$

15. $\dfrac{x^2}{4} = 1 + \dfrac{y^2}{4}$

16. $x^2 = 1 + y^2$

17. $\dfrac{y^2}{9} = 1 + \dfrac{x^2}{4}$

18. $\dfrac{y^2}{9} = 1 + \dfrac{x^2}{16}$

19. $\dfrac{y^2}{9} = 1 + \dfrac{x^2}{9}$

20. $y^2 = 1 + x^2$

21. $xy - 3 = 0$

22. $xy + 3 = 0$

23. $xy - \dfrac{1}{4} = 0$

24. $xy + \dfrac{1}{2} = 0$

Review: Work these problems on a separate sheet of paper. Attach your work to this page.

Graph each linear inequality. (See Lesson 3.5.)

25. $3x - 4y \geq 12$

26. $2x \leq 3y$

27. $1 < y \leq 4$

Solve each second-order system using the substitution method. (See Lesson 4.2.)

28. $\begin{cases} x + y = 5 \\ x - y = 3 \end{cases}$

29. $\begin{cases} x + y = 5 \\ 2x + 3y = 12 \end{cases}$

30. $\begin{cases} 3x + 4y = 20 \\ 2x - y = 6 \end{cases}$

Solve each second-order system using the addition method. (See Lesson 4.3.)

31. $\begin{cases} 2x + y = 7 \\ 2x - y = 1 \end{cases}$

32. $\begin{cases} 2x + y = 7 \\ 3x - 2y = 0 \end{cases}$

33. $\begin{cases} 2x - 3y = 12 \\ 3x + 2y = 5 \end{cases}$

Introduction to Quadratic Systems

A system of equations in which the highest-degree equation is of second degree, is called a *quadratic system of equations*.

Examples ▶ (a) $\begin{cases} x^2 + y^2 = 16 \\ x^2 - y^2 = 1 \end{cases}$ (b) $\begin{cases} xy = 9 \\ 2x + 3y = 5 \end{cases}$ ⟵ quadratic systems of equations

Recall ▶ Any ordered pair that is a common solution of all equations in a system is called a solution of the system, or, solution.

Example ▶ (0, 1) is a solution of $\begin{cases} y = x^2 + 4x + 1 \\ 4x - y = -1 \end{cases}$ because it is a solution of both equations.

If both coordinates of a solution of a system are real numbers, then the solution is called a *real solution*. If one or both of the coordinates of a solution of a system are complex numbers, then the solution is called a *complex solution*.

Example ▶ (0, 6) is a real solution and both $(i\sqrt{13}, -7)$ and $(-i\sqrt{13}, -7)$ are complex solutions of $\begin{cases} y = x^2 + 6 \\ x^2 + y^2 = 36 \end{cases}$.

If the graphs of the two equations in a system are drawn on the same set of axes, then the real solutions are the ordered pairs represented by the points where the graphs intersect. Although it is generally not possible to locate exact solutions by graphing, you can approximate the real solutions of a system by estimating the coordinates of any point(s) where the graphs of the equations intersect.

Example ▶ Approximate the real solutions of $\begin{cases} x^2 + y^2 = 25 \\ y = 2x + 2 \end{cases}$ by graphing.

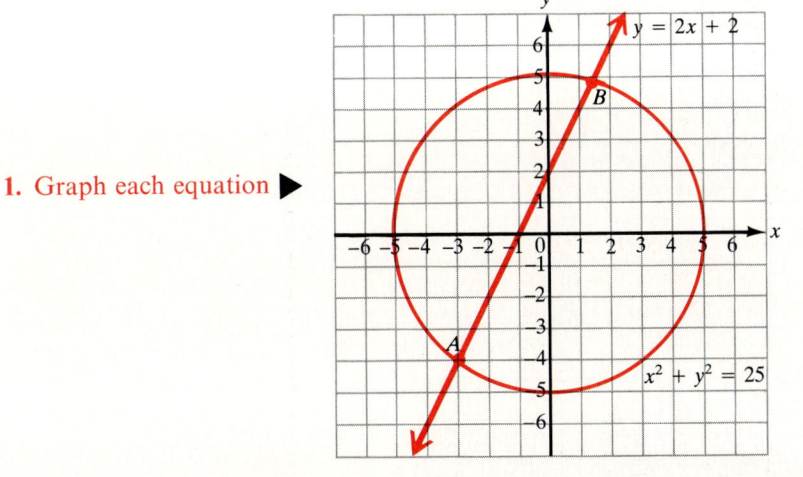

1. Graph each equation ▶

Think: $x^2 + y^2$ graphs to be a circle with center (0, 0) and radius 5.
$y = 2x + 2$ graphs to be a line with y-intercept (0, 2) and slope 2.

2. Estimate coordinates ▶ The nearest point with integral coordinates to: (a) point A is $(-3, -4)$. (b) point B is $(1, 5)$.

Solution ▶ $(-3, -4)$ and $(1, 5)$ are approximate solutions of $\begin{cases} x^2 + y^2 = 25 \\ y = 2x + 2 \end{cases}$.

Note ▶ Substitution of -3 for x and -4 for y into both of the equations of the above system shows that $(-3, -4)$ is an actual solution of the system. The substitution of 1 for x and 5 for y shows that $(1, 5)$ is not a solution of the system.

10.5 Solve Quadratic Systems

Recall ▶ To solve a system means to find all of its solutions.

Many quadratic systems cannot be solved by graphing because the complex solutions cannot be determined by graphing and often graphing only gives approximations of the real solutions.

Some quadratic systems can be solved by the following substitution method. The substitution method is particularly useful in solving systems with one linear equation and one quadratic equation.

> **To solve a quadratic system consisting of one linear equation and one quadratic equation:**
> 1. Solve the linear equation for one of the variables.
> 2. Substitute the expression from Step 1 into the quadratic equation.
> 3. Solve the resulting equation for the other variable.
> 4. Evaluate the expression from Step 1 for the solution(s) from Step 3.
> 5. Interpret the solution(s) from Steps 3 and 4 as ordered pair(s).
> 6. Check the proposed solution(s) from Step 5 in both original system equations.

EXAMPLE 1: Solve a quadratic system using the substitution method.

Problem ▶ Solve $\begin{cases} 9x^2 + 4y^2 = 36 \\ x + 2y = -2 \end{cases}$ using the substitution method.

1. Solve for a variable in the linear equation ▶

$x + 2y = -2$

$x = -2y - 2$

2. Substitute ▶

$9x^2 + 4y^2 = 36$ ⟵ the quadratic equation

$9(-2y - 2)^2 + 4y^2 = 36$

3. Solve for the other variable ▶

$9(4y^2 + 8y + 4) + 4y^2 = 36$ Think: The variable x is eliminated. Solve for y.

$36y^2 + 72y + 36 + 4y^2 = 36$

$40y^2 + 72y = 0$

$8y(5y + 9) = 0$

$8y = 0$ or $5y + 9 = 0$ Use the Zero-Product Property.

$y = 0$ or $5y = -9$

$y = 0$ or $y = -\dfrac{9}{5}$

4. Evaluate ▶

$x = -2y - 2$	$x = -2y - 2$ ⟵ equation from Step 1
$x = -2(0) - 2$	$x = -2(-\frac{9}{5}) - 2$ Substitute 0 for y and $-\frac{9}{5}$ for y.
$x = 0 - 2$	$x = \frac{18}{5} - 2$ Solve for x.
$x = -2$	$x = \frac{8}{5}$

5. Interpret ▶ $y = 0$ and $x = -2$ means $(-2, 0)$ is a proposed solution.
$y = -\frac{9}{5}$ and $x = \frac{8}{5}$ means $(\frac{8}{5}, -\frac{9}{5})$ is a proposed solution.

Solution ▶ The solutions of $\begin{cases} 9x^2 + 4y^2 = 36 \\ x + 2y = -2 \end{cases}$ are $(-2, 0)$ and $(\frac{8}{5}, -\frac{9}{5})$.   Check in both original system equations.

Note ▶ The following is a graph of the system in Example 1.

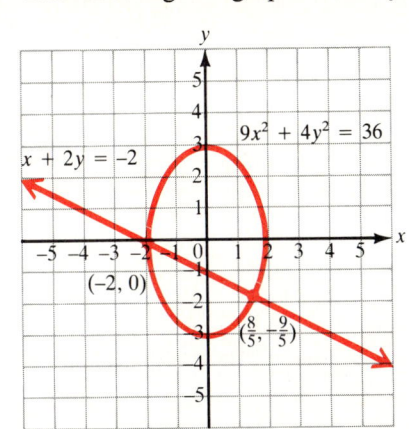

Another Example ▶ Solve $\begin{cases} x^2 + y^2 = 6 \\ x + y = 4 \end{cases}$ using the substitution method.

1. Solve for a variable in the linear equation ▶
$x + y = 4$
$y = 4 - x$

2. Substitute ▶
$x^2 + y^2 = 6$ ⟵ the quadratic equation
$x^2 + (4 - x)^2 = 6$

3. Solve for the other variable ▶
$x^2 + 16 - 8x + x^2 = 6$
$2x^2 - 8x + 10 = 0$
$x^2 - 4x + 5 = 0$ Divide all terms by 2.

$$x = \frac{-(-4) \pm \sqrt{(-4)^2 - 4 \cdot 1 \cdot 5}}{2 \cdot 1}$$ Think: Use the quadratic formula to solve for x.

$$= \frac{4 \pm \sqrt{16 - 20}}{2}$$

$$= \frac{4 \pm \sqrt{-4}}{2}$$

$$= \frac{4 \pm 2i}{2}$$

$$= 2 \pm i$$

4. Evaluate ▶
$y = 4 - x$	$y = 4 - x$ ⟵ equation from Step 1
$= 4 - (2 - i)$	$= 4 - (2 + i)$
$= 4 - 2 + i$	$= 4 - 2 - i$
$= 2 + i$	$= 2 - i$

5. Interpret ▶ $x = 2 - i$ and $y = 2 + i$ means $(2 - i, 2 + i)$ is a proposed solution.
$x = 2 + i$ and $y = 2 - i$ means $(2 + i, 2 - i)$ is a proposed solution.

Solution ▶ The solutions of $\begin{cases} x^2 + y^2 = 6 \\ x + y = 4 \end{cases}$ are $(2 - i, 2 + i)$ and $(2 + i, 2 - i)$.   Check as before.

Note ▶ A graph of the system in the previous example also shows that the system does not have any real solutions.

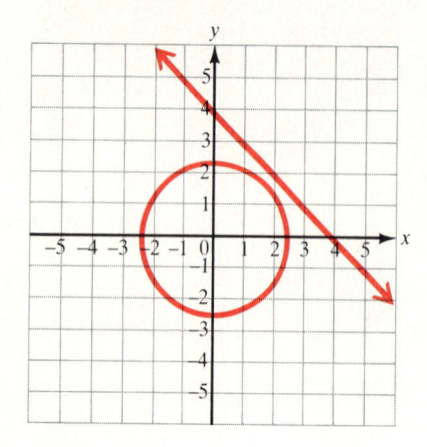

Make Sure

Solve each quadratic system using the substitution method.

See Example 1 ▶ 1. $\begin{cases} x^2 + y^2 = 25 \\ x - y = 1 \end{cases}$ 2. $\begin{cases} 9x^2 + 4y^2 = 40 \\ 2x - y = 3 \end{cases}$ 3. $\begin{cases} x^2 - y = -2 \\ x + y = 5 \end{cases}$

MAKE SURE ANSWERS: 1. $(4, 3), (-3, -4)$ 2. $(2, 1), (-\frac{2}{3}, -\frac{25}{9})$ 3. $\left(\frac{-1 + \sqrt{13}}{2}, \frac{11 - \sqrt{13}}{2} \right), \left(\frac{-1 - \sqrt{13}}{2}, \frac{11 + \sqrt{13}}{2} \right)$

The addition method is often used to solve a quadratic system of two equations when each equation can be written in $Ax^2 + By^2 = C$ form.

> **To solve a quadratic system consisting of two quadratic equations:**
> 1. Write each equation in $Ax^2 + By^2 = C$ form.
> 2. Multiply each equation by the appropriate constants so that the coefficients of one of the variables are opposites.
> 3. Add equations from Step 2 to eliminate a variable.
> 4. Solve the equation from Step 3 for the remaining variable.
> 5. Evaluate either of the original system equations for the solution(s) from Step 4.
> 6. Interpret the solution(s) from Steps 4 and 5 as an ordered pair(s).
> 7. Check the proposed solution(s) from Step 6 in both original system equations.

EXAMPLE 2: Solve a quadratic system using the addition method.

Problem ▶ Solve $\begin{cases} 3x^2 + 4y^2 = 16 \\ 2x^2 = 3y^2 + 5 \end{cases}$ using the addition method.

1. Write $Ax^2 + By^2 = C$ ▶ $3x^2 + 4y^2 = 16$
   form $2x^2 - 3y^2 = 5$

10.5 Solve Quadratic Systems

2. Multiply ▶ $3[3x^2 + 4y^2] = 3 \cdot 16 \longrightarrow 9x^2 + 12y^2 = 48$

$4[2x^2 - 3y^2] = 4 \cdot 5 \longrightarrow 8x^2 - 12y^2 = 20$

3. Add ▶ $\overline{17x^2 + 0 = 68}$

4. Solve as before ▶ $x^2 = 4$

$x = \pm 2$

5. Evaluate ▶

$3x^2 + 4y^2 = 16$	$3x^2 + 4y^2 = 16$ ⟵ 1st system equation
$3(-2)^2 + 4y^2 = 16$	$3(2)^2 + 4y^2 = 16$ Substitute -2 for x and 2 for x.
$12 + 4y^2 = 16$	$12 + 4y^2 = 16$
$4y^2 = 4$	$4y^2 = 4$
$y^2 = 1$	$y^2 = 1$
$y = \pm 1$	$y = \pm 1$

6. Interpret ▶ $x = -2$ and $y = \pm 1$ means both $(-2, -1)$ and $(-2, 1)$ are proposed solutions.
$x = 2$ and $y = \pm 1$ means both $(2, -1)$ and $(2, 1)$ are proposed solutions.

Solution ▶   The solutions of $\begin{cases} 3x^2 + 4y^2 = 16 \\ 2x^2 = 3y^2 + 5 \end{cases}$ are $(-2, -1), (-2, 1), (2, -1),$ and $(2, 1)$.   Check as before.

Recall ▶   A system that reduces to a false statement has no solutions.

Example ▶   Solve $\begin{cases} x^2 + y^2 = 4 \\ x^2 + y^2 = 1 \end{cases}$. $\begin{array}{r} x^2 + y^2 = 4 \\ -x^2 - y^2 = -1 \\ \hline 0 + 0 = 3 \end{array}$ Think: Multiply the 2nd equation by -1 and add to the 1st equation.

$0 = 3$ ⟵ a false statement

The system $\begin{cases} x^2 + y^2 = 4 \\ x^2 + y^2 = 1 \end{cases}$ has no solutions.

Make Sure

Solve each quadratic system using the addition method.

See Example 2 ▶   **1.** $\begin{cases} x^2 + y^2 = 6y \\ x^2 + y^2 = 9 \end{cases}$ **2.** $\begin{cases} 2x^2 - 3y^2 = 6 \\ x^2 + y^2 = 13 \end{cases}$ **3.** $\begin{cases} 16x^2 + 9y^2 = 144 \\ 4x^2 + y^2 = 4 \end{cases}$

MAKE SURE ANSWERS: 1. $\left(-\frac{3\sqrt{3}}{2}, \frac{3}{2}\right), \left(\frac{3\sqrt{3}}{2}, \frac{3}{2}\right)$ **2.** $(-3, -2), (-3, 2), (3, -2), (3, 2)$ **3.** no solutions

416   Chapter 10   Conics and Systems

A system of inequalities in which the highest-degree inequality is of second degree, is called a *quadratic system of inequalities*.

Examples ▶ (a) $\begin{cases} x^2 + y^2 < 10 \\ y > x^2 \end{cases}$ (b) $\begin{cases} 4x^2 + 5y^2 \leq 20 \\ x^2 - y^2 > 1 \\ y < 2 \end{cases}$ ⟵ quadratic systems of inequalities

The graph of each quadratic inequality to be considered in this lesson consists of one or two regions. To graph a quadratic inequality, you use a method similar to the one you used to graph a linear inequality.

If the boundary is included in the graph of a region, the region is called **a** *closed region*, and the boundary is drawn as a *solid curve*. If the boundary is not included in the graph of a region, then the region is called an *open region*, and the boundary is drawn as a *broken curve*.

Example ▶ Graph $xy > 6$.

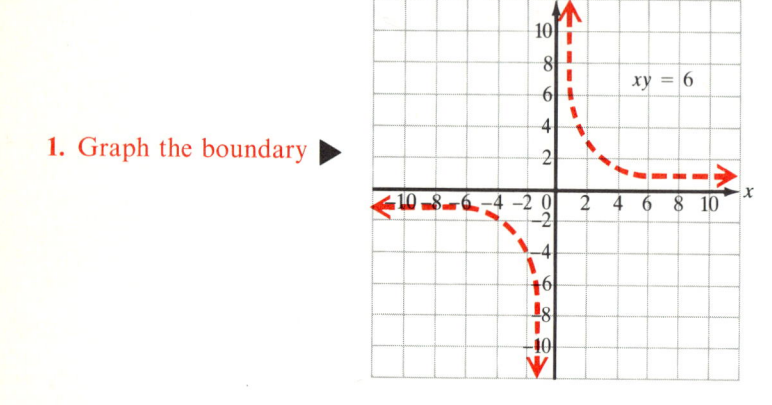

1. Graph the boundary ▶

Think: Graph the associated equation $xy = 6$ as:
(a) a broken curve for $<$ or $>$.
(b) a solid curve for $\leq$ or $\geq$.

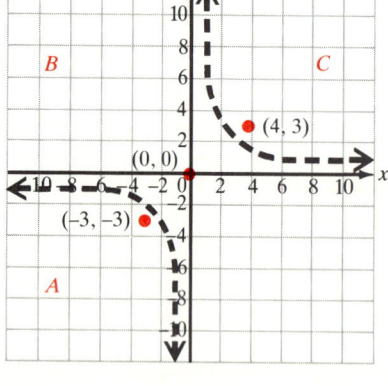

2. Locate check points in each region ▶

Think: $(-3, -3)$ is in region A.
$(0, 0)$ is in region B.
$(4, 3)$ is in region C.

3. Check ▶

For $(-3, -3)$: $xy > 6$
$(-3)(-3) > 6$
$9 > 6$ true

For $(0, 0)$: $xy > 6$
$(0)(0) > 6$
$0 > 6$ false

For $(4, 3)$: $xy > 6$
$(4)(3) > 6$
$12 > 6$ true

4. Interpret ▶ Region A is part of the graph.

Region B is not part of the graph.

Region C is part of the graph.

Think: (a) A true statement means every point in the region that includes the check point is a member of the graph.
(b) A false statement means every point in the region that includes the check point is not a member of the graph.

5. Shade the correct region(s)

Solution

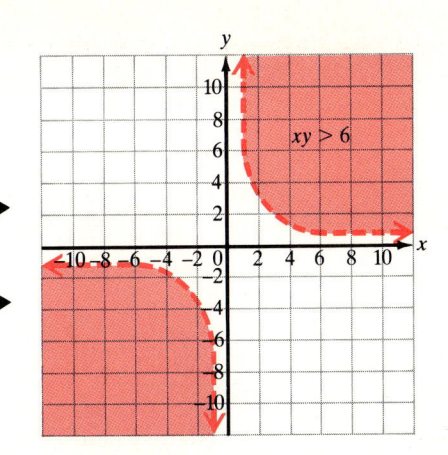

Recall ▶ Any ordered pair that is a common solution of all inequalities in a system of inequalities is called a solution of the system, or, solution.

If the graphs of the inequalities in a system are drawn on the same set of axes, then the solution(s) of the system is the ordered pair(s) represented by the point(s) where the graphs intersect.

EXAMPLE 3: Graph a quadratic system of inequalities.

Problem ▶ Graph $\begin{Bmatrix} x^2 + y^2 \leq 4 \\ -x + y > -2 \end{Bmatrix}$.

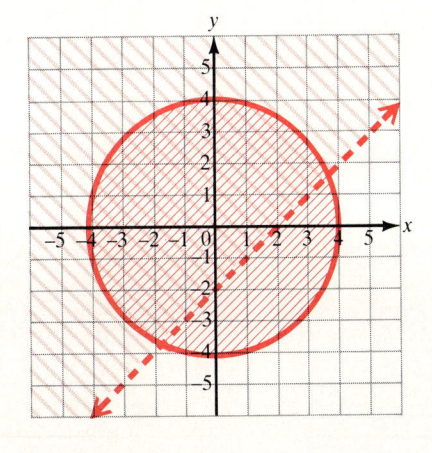

1. Graph each inequality

2. Shade intersection

Solution

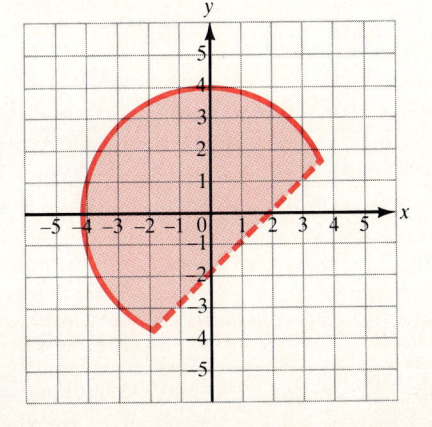

Think: The intersection consists of only those points that lie on or inside the circle and above the line.

The graph of $\begin{Bmatrix} x^2 + y^2 \leq 4 \\ -x + y > -2 \end{Bmatrix}$.

Chapter 10 Conics and Systems

Another Example ▶ Graph $\begin{cases} 16x^2 + 9y^2 \leq 144 \\ y > -x^2 \\ x \leq 1 \end{cases}$.

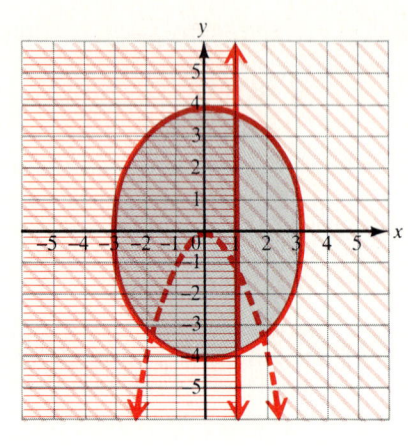

Graph each inequality. Shade intersection.

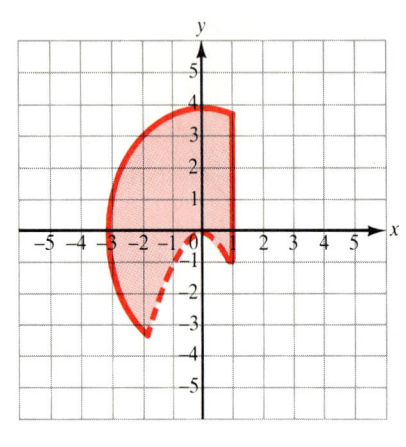

The graph of $\begin{cases} 16x^2 + 9y^2 \leq 144 \\ y > -x^2 \\ x \leq 1 \end{cases}$.

Make Sure

Graph each quadratic system of inequalities.

See Example 3 ▶ 1. $\begin{cases} xy \geq 3 \\ x + y \leq 4 \end{cases}$

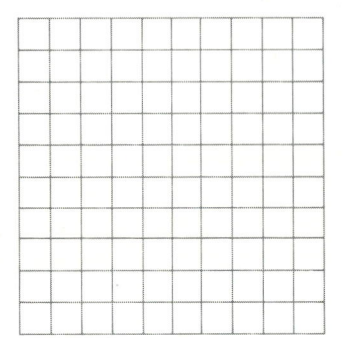

2. $\begin{cases} x^2 - 4y^2 \geq 4 \\ x^2 - y < 9 \end{cases}$

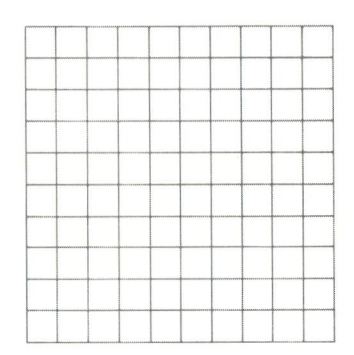

MAKE SURE ANSWERS: See Appendix Selected Answers.

10.5 Practice

Set 1: Solve each quadratic system using the substitution method.

1. $\begin{cases} x^2 + y^2 = 25 \\ 3x + 4y = 0 \end{cases}$
2. $\begin{cases} x^2 + y^2 = 25 \\ 4x - 3y = 0 \end{cases}$
3. $\begin{cases} x^2 - 6x - y + 9 = 0 \\ 3x + y - 9 = 0 \end{cases}$

4. $\begin{cases} x^2 - y + 9 = 0 \\ 3x + y - 9 = 0 \end{cases}$
5. $\begin{cases} x^2 + y^2 + 8y = 0 \\ 5x + y = 16 \end{cases}$
6. $\begin{cases} x^2 + y^2 - 6x = 0 \\ 4x - y = 9 \end{cases}$

7. $\begin{cases} 4x^2 + 9y^2 = 36 \\ x + y - 6 = 0 \end{cases}$
8. $\begin{cases} 16y^2 - 9x^2 = 144 \\ 3x - 4y = 0 \end{cases}$
9. $\begin{cases} xy + 4y = 12 \\ xy = 4 \end{cases}$

Chapter 10 Conics and Systems

Set 2: Solve each quadratic system using the addition method.

10. $\begin{cases} 9x^2 - y^2 = 9 \\ 9x^2 + y^2 = 9 \end{cases}$

11. $\begin{cases} x^2 - 4y^2 = 4 \\ x^2 + 4y^2 = 4 \end{cases}$

12. $\begin{cases} x^2 + y^2 = 6 \\ x^2 - y = 0 \end{cases}$

13. $\begin{cases} x^2 + y^2 = 8 \\ x^2 + y = 6 \end{cases}$

14. $\begin{cases} 9x^2 - 144 = -16y^2 \\ x^2 + y^2 = 16 \end{cases}$

15. $\begin{cases} 9x^2 + 4y^2 = 36 \\ x^2 - 9 = -y^2 \end{cases}$

16. $\begin{cases} 4x^2 + 9y^2 = 36 \\ x^2 - y = 2 \end{cases}$

17. $\begin{cases} 9x^2 + 4y^2 = 36 \\ x^2 + y = 3 \end{cases}$

18. $\begin{cases} 9x^2 - 16y^2 = 144 \\ x^2 + y^2 = 4 \end{cases}$

10.5 Practice **421**

Name _____ Date _____ Class _____

19. _____

Mixed Practice: Solve each quadratic system.

19. $\begin{cases} x^2 + y^2 = 6y \\ 3x - 4y = 0 \end{cases}$ 20. $\begin{cases} x^2 + y^2 = 4x \\ 3x + 4y = 0 \end{cases}$ 21. $\begin{cases} xy + 4y = 4 \\ xy - 4y = -4 \end{cases}$

20. _____

21. _____

22. $\begin{cases} xy - 4y = 4 \\ xy + 4y = -4 \end{cases}$ 23. $\begin{cases} x^2 + 6x - y = 0 \\ 9x + y = 0 \end{cases}$ 24. $\begin{cases} x^2 - 6x - y = -9 \\ x - y = 0 \end{cases}$

22. _____

23. _____

24. _____

Set 3: Graph each quadratic system of inequalities.

25. $\begin{cases} x^2 - y \geq 0 \\ x^2 - y \leq 9 \end{cases}$ 26. $\begin{cases} x^2 - 6x - y + 9 \leq 0 \\ x^2 - 6x + y \leq 0 \end{cases}$

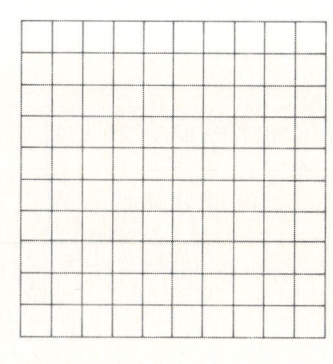

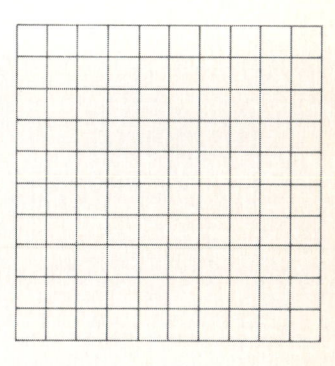

27. $\begin{cases} x^2 + y^2 \leq 16 \\ x^2 - y^2 > 9 \end{cases}$ 28. $\begin{cases} x^2 + y^2 + 6x \leq 16 \\ x^2 + y^2 + 6x > 7 \end{cases}$

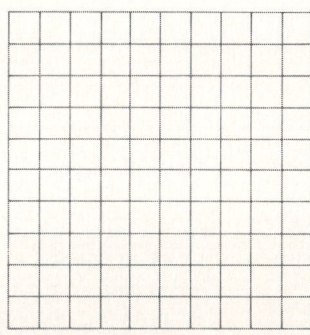

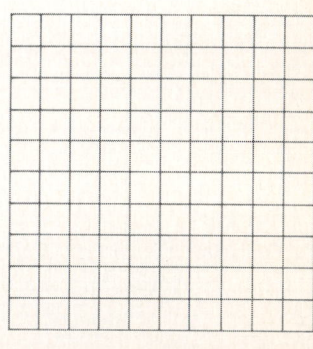

Copyright © 1985 by Harcourt Brace Jovanovich, Inc. All rights reserved.

29. $\begin{cases} 4x^2 + 9y^2 \geq 36 \\ 9x^2 + 4y^2 < 36 \end{cases}$

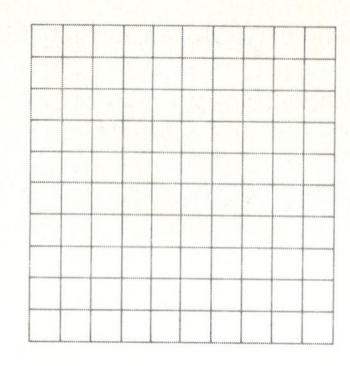

30. $\begin{cases} 9x^2 + 16y^2 < 144 \\ 16x^2 + 9y^2 \leq 144 \end{cases}$

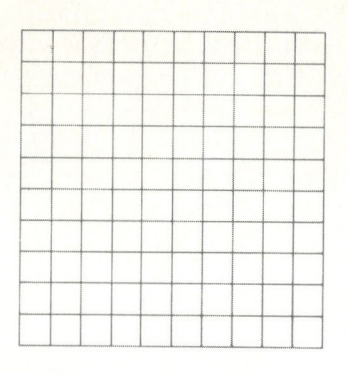

31. $\begin{cases} 9x^2 - 4y^2 > 36 \\ xy \geq 12 \end{cases}$

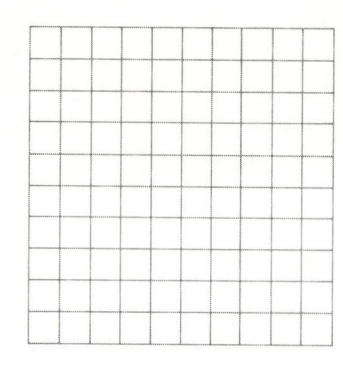

32. $\begin{cases} 9x^2 - 16y^2 \leq 144 \\ xy \leq 12 \end{cases}$

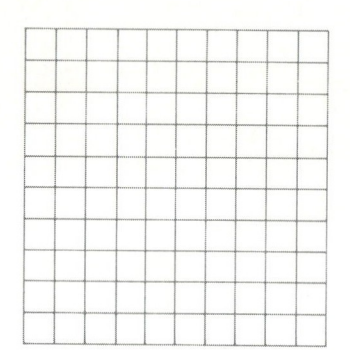

33. $\begin{cases} x^2 + y^2 < 16 \\ x - y \geq -2 \\ x + y \leq 2 \end{cases}$

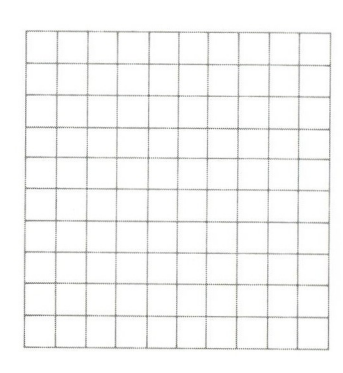

34. $\begin{cases} x^2 + y^2 \leq 9 \\ x - y \leq -2 \\ x + y < 2 \end{cases}$

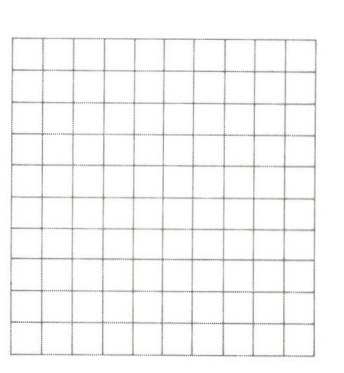

35. $\begin{cases} x^2 + y^2 \leq 9 \\ x^2 + y^2 - 6y < 0 \\ x^2 + y^2 - 6x < 0 \end{cases}$

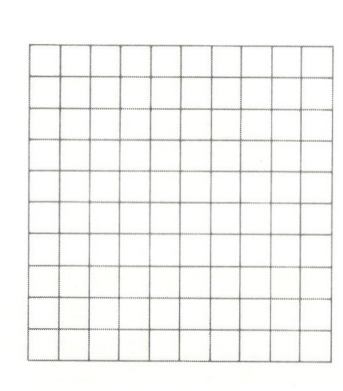

36. $\begin{cases} x^2 + y^2 \geq 4 \\ x^2 + y^2 + 2y \leq 0 \\ x^2 + y^2 + 2x \leq 0 \end{cases}$

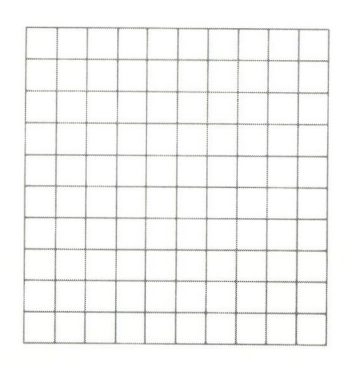

Review: Work these problems on a separate sheet of paper. Attach your work to this page.

Solve each equation for y. (See Lesson 3.1.)

37. $2x + y = 7$ for $x = -7, 0, 3$

38. $2x - 3y = 6$ for $x = -3, 0, 3$

39. $2y = 3x + 4$ for $x = -2, 1, 3$

40. $4x + 3 = 5y$ for $x = -2, 1, 3$

41. $3x = 2 - 4y$ for $x = -4, -3, 1$

42. $2x = 3 - 4y$ for $x = -4, 1, 2$

Problem Solving 15: Solve Problems Using Quadratic Systems

To solve a number problem that requires a quadratic system, you first choose a different variable for each different unknown number and then translate to get the system equations.

EXAMPLE 1: Solve this number problem using a quadratic system.

1. Read and identify ▶ The difference between (two numbers) is 1. The product of the same two numbers is also 1. What are the numbers?

   Remember, circle the unknowns and underline the question.

2. Understand ▶ The unknowns are $\begin{cases} \text{the larger number} \\ \text{the smaller number} \end{cases}$.

3. Decide ▶ Let x = the larger number
   then y = the smaller number

4. Translate to get system equations ▶ The difference between the two numbers is 1.
   $$x - y = 1$$

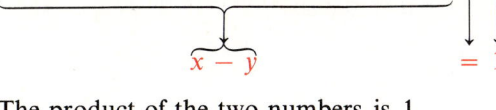

   The product of the two numbers is 1.
   $$xy = 1$$

   System Equations: $\begin{matrix} x - y = 1 \\ xy = 1 \end{matrix}$ ⟵ quadratic system

5. Solve as before ▶ Solve $x - y = 1$ for x or y.

   $x = y + 1$ ⟵ solved equation
   $xy = 1$ ⟵ other equation

   $(y + 1)y = 1$ Substitute to eliminate the variable x.
   $y^2 + y = 1$ Clear parentheses.
   $y^2 + y - 1 = 0$ Write standard form.

   $y = \dfrac{-b \pm \sqrt{b^2 - 4ac}}{2a}$ Use the quadratic formula.

   $= \dfrac{-(1) \pm \sqrt{(1)^2 - 4(1)(-1)}}{2(1)}$

   $= \dfrac{-1 \pm \sqrt{5}}{2}$

   $x = y + 1$ ⟵ solved equation | $x = y + 1$ ⟵ solved equation
   $= \dfrac{-1 + \sqrt{5}}{2} + \dfrac{2}{2}$ Substitute. | $= \dfrac{-1 - \sqrt{5}}{2} + \dfrac{2}{2}$ Substitute.
   $= \dfrac{1 + \sqrt{5}}{2}$ ⟵ proposed solution for x | $= \dfrac{1 - \sqrt{5}}{2}$ ⟵ proposed solution for x

6. Interpret ▶ The proposed solutions are: $x = \dfrac{1 + \sqrt{5}}{2}$ and $y = \dfrac{-1 + \sqrt{5}}{2}$

   or: $x = \dfrac{1 - \sqrt{5}}{2}$ and $y = \dfrac{-1 - \sqrt{5}}{2}$.

7. Check ▶ Does $x - y = 1$ and $xy = 1$ for the proposed solutions? Yes: Substitute and see.

424   Chapter 10   Conics and Systems

To solve a geometry problem that requires a quadratic system, you first choose a different variable for each unknown dimension and then use geometry formulas to get system equations.

EXAMPLE 2: Solve this geometry problem using a quadratic system.

1. Read and identify ▶ The area of a rectangle is (588 m²). The diagonal of the rectangle is (35 m). What are the dimensions of the rectangle?

2. Understand ▶ The unknowns are $\begin{cases} \text{the length of the rectangle} \\ \text{the width of the rectangle} \end{cases}$.

3. Decide ▶ Let l = the length of the rectangle
then w = the width of the rectangle

4. Use geometry formulas to get system equations ▶

$A = lw$ ⟵ area of a rectangle
$588 = lw$

$a^2 + b^2 = c^2$ ⟵ Pythagorean Theorem
$l^2 + w^2 = 35^2$
$l^2 + w^2 = 1225$ Simplify when possible.

System Equations: $l^2 + w^2 = 1225$ ⟵⎤
$lw = 588$ ⟵⎦ quadratic system

5. Solve as before ▶

$l = \dfrac{588}{w}$ ⟵ solved equation   Solve $lw = 588$ for l or w.

$l^2 + w^2 = 1225$ ⟵ other equation

$\left(\dfrac{588}{w}\right)^2 + w^2 = 1225$ Substitute to eliminate a variable.

$w^2 \cdot \dfrac{345{,}744}{w^2} + w^2(w^2) = w^2(1225)$ Clear parentheses and fractions.

$345{,}744 + w^4 = 1225w^2$

$w^4 - 1225w^2 + 345{,}744 = 0$ Write standard form.

$u^2 - 1225u + 345{,}744 = 0$ Substitute $u = w^2$ and $u^2 = w^4$.

$u = \dfrac{-(-1225) \pm \sqrt{(-1225)^2 - 4(1)(345{,}744)}}{2(1)}$ Use the quadratic formula.

$u = \dfrac{1225 \pm 343}{2}$

$u = 784$ or 441

$w^2 = 784$ or 441 Substitute $w^2 = u$.

$w = \pm 28$ or ± 21 ⟵ proposed solutions for w (See the following Note 1.)

$l = \dfrac{588}{w}$ ⟵ solved equation $l = \dfrac{588}{w}$ ⟵ solved equation

$= \dfrac{588}{28}$ Substitute $w = 28$. $= \dfrac{588}{21}$ Substitute $w = 21$.

$= 21$ ⟵ proposed solution for l $= 28$ ⟵ proposed solution for l

6. Interpret ▶ $l = 28$ and $w = 21$ mean the length is 28 m and the width is 21 m.

Note 1 ▶ The proposed solutions $w = -21$ and $w = -28$ must be rejected because measures, such as length and width, cannot be negative.

Note 2 ▶ The solutions $l = 21$ and $w = 28$ are equivalent to the solutions $l = 28$ and $w = 21$. When equivalent solutions are found, only one of the equivalent solutions need be checked.

Practice: Solve problems using quadratic systems.

1. The sum of two numbers is 5. The product of the same two numbers is also 5. What are the numbers?

2. The sum of two numbers is 1. The product of the same two numbers is also 1. Find the numbers.

3. The difference between two numbers is 2. The product of the same two numbers is 4. What are the numbers?

4. The sum of the squares of two positive numbers is 25. The difference of the squares of the same two numbers is 7. Find the numbers.

5. The area of a rectangle is 216 ft². Its perimeter is 60 ft. What are the dimensions?

6. The area of a rectangle is 27 m². Its diagonal is 7.5 m. Find the perimeter.

7. The hypotenuse of a right triangle is 2 cm longer than one leg and 6 cm longer than the other leg. Find the perimeter.

8. The area of a right triangle is 210 m². The perimeter of the triangle is 70 m. What is the length of the hypotenuse?

9. Four times the sum of the digits of a two-digit number is 52. The sum of the squares of the same digits is 89. Find the two possible numbers.

10. The difference between the digits of a two-digit number is 3. The product of the same digits is 28. Find the two possible numbers.

11. The area of a right triangle is 96 yd². Its hypotenuse is 20 yd. Find the perimeter.

12. The area of a right triangle is 20 m². The sum of the legs is 13 m. Find the length of the hypotenuse.

13. The perimeter of a rectangle is 28 ft. Its diagonals measure 10 ft each. Find the area.

14. A rectangular plot of land requires 168 m of fencing to enclose it. Its area is 1728 m². Find the diagonal distance.

15. The sum of the area of two squares is 832 ft². The difference between their areas is 320 ft². Find the perimeter of each square.

16. The sum of two numbers is 15. The difference between the squares of the same two numbers is also 15. What are the numbers?

17. The area of a right triangle is 7 yd². One leg is 5 yd longer than the other. Find the perimeter.

18. The area of a rectangular plot is 180 ft². It is divided into two rectangular plots that require 66 ft of fencing to separate and surround them. What are the two possible perimeters for the plot?

19. The area of a rectangular pasture is 15 mi². One side of the pasture borders a river. If it takes 11 miles of fencing for the other 3 sides, what is the perimeter of the pasture?

20. The value of a two-digit number is 1 greater than the sum of the squares of its digits. If the digits are reversed, the new number is 18 more than the original number. Find the original number.

21. Find two pairs of numbers that have a difference of 3 and a product of 9.

22. Find two numbers whose product is -20 and whose sum is 5.

23. A rectangular piece of cardboard has an area of 200 cm². An open box with a volume of 168 cm³ is formed when a 3 cm square is cut from each corner and the flaps are folded up. Find the perimeter of the original piece of cardboard.

24. When a certain rectangle's width is decreased by 1 and the length is increased by 2, the new rectangle has the same area as the original one. When the original rectangle's length is decreased by 4 and the width is increased by 4, the new rectangle also has the same area as the original one. Find the area of the original rectangle.

25. The area of a rectangle is $\sqrt{2}$ ft². Its diagonal is $\sqrt{3}$ ft. Find the perimeter.

26. The area of a rectangle is $\sqrt{6}$ m². Its perimeter is $2(\sqrt{2} + \sqrt{3})$ m. Find the length of a diagonal.

Extra: Solve each problem using a quadratic system.

27. A car travels at a constant rate for 75 km. Traffic reduces the constant rate by 10 km/h for the next 20 km. The total time required to travel the 95 km is 2 hours. Find the original constant rate.

28. The annual income from an investment at simple interest is $90 per year. If the simple interest rate is increased by $\frac{1}{2}$%, then the investment would earn $97.50 per year. Find the amount of the investment and the original interest rate.

29. A bus travels 200 miles at a constant speed. A car traveling 10 mph faster completed the same distance in one hour less time. a) How many hours did the bus take? b) What was the speed of the car?

30. A plane completed a 1500-mile trip with a 75 mph tailwind. The trip would have taken one hour longer in still air. a) What is the plane's airspeed? b) How long did the 1500-mile trip take?

31. The surface area (SA) of a sphere is given by: $SA = 4\pi r^2$. The difference between the radii of two spheres is 5 cm. The difference between the surface areas is 690.8 cm². Find the radius of each sphere using 3.14 for π.

32. Given the area (A) and perimeter (P) of a rectangle, show on a separate piece of paper that the length (l) and width (w) are:
a) $l = \frac{1}{4}(P + \sqrt{P^2 - 16A})$ and
b) $w = \frac{1}{4}(P - \sqrt{P^2 - 16A})$.

According to Guinness

LARGEST SWIMMING POOL
THE LARGEST SWIMMING POOL IN THE WORLD IS THE SALT-WATER ORTHLIEB POOL IN CASABLANCA, MOROCCO. THE PERIMETER OF THE POOL IS 1214 yd (1110 m) AND ITS AREA IS 43,050 yd² (36,000 m²).

What are the dimensions of the pool in:

33. feet?

34. meters?

Name _____ Date _____ Class _____

Chapter 10 Review

		What to Review if You Have Trouble		
Objectives		Lesson	Example	Page
Find each *x*-intercept of ▶ $y = ax^2 + bx + c$	1. Find each *x*-intercept of the graph of $y = 2x^2 - 5x - 12$. _____	10.1	1	380
Find the vertex and axis ▶ of symmetry of $y = ax^2 + bx + c$	2. Find the **a.** vertex and **b.** axis of symmetry of the graph of $y = -3x^2 + 6x - 7$. a. _____ b. _____	10.1	2	382
Graph parabolas ▶	3. Graph $y = x^2 - 6x + 8$ on Grid A.	10.1	3	383
Find the distance ▶ between two points	4. Find the distance between the points $P_1(5, -7)$ and $P_2(8, 2)$.	10.2	1	390
Write the equation of a ▶ circle in standard form, given its center and radius	5. Write an equation of a circle with center $(-3, 7)$ and radius 2. _____	10.2	2	392
Graph circles ▶	6. Graph $(x + 2)^2 + (y - 3)^2 = 9$ on Grid B. Label its center and indicate its radius.	10.2	3	393
Find the center and ▶ radius of a circle	7. Find the center and radius of the graph of $x^2 + y^2 + 4x - 6y - 12 = 0$. _____	10.2	4	394
Write the equation of an ▶ ellipse given its vertices	8. Write the equation of an ellipse with vertices $(-16, 0)$, $(16, 0)$, $(0, -1)$, and $(0, 1)$. _____	10.3	1	400
Graph ellipses ▶	9a. Graph $\dfrac{x^2}{81} + \dfrac{y^2}{36} = 1$ on Grid C.	10.3	2	401
	b. Graph $\dfrac{x^2}{25} + \dfrac{y^2}{49} = 1$ on Grid D.	10.3	2	401

Grid A   Grid B   Grid C   Grid D

Copyright © 1985 by Harcourt Brace Jovanovich, Inc. All rights reserved.

428 Chapter 10 Conics and Systems

Graph hyperbolas	10. Graph $\dfrac{x^2}{5^2} - \dfrac{y^2}{3^2} = 1$ on Grid E.	10.4	1	406
	11. Graph $-\dfrac{x^2}{3^2} + \dfrac{y^2}{2^2} = 1$ on Grid F.	10.4	2	407
	12. Graph $xy = 4$ on Grid G.	10.4	3	408
Solve a quadratic system using the substitution method	13. Solve $\begin{cases} x^2 + y^2 = 10 \\ 3x + y = 6 \end{cases}$ using the substitution method.	10.5	1	412
Solve a quadratic system using the addition method	14. Solve $\begin{cases} x^2 + y^2 = 5 \\ 2x^2 - 3y^2 = 5 \end{cases}$ using the addition method.	10.5	2	414
Graph a quadratic system of inequalities	15. Graph $\begin{cases} x^2 + y^2 < 20 \\ y \leq x^2 \end{cases}$ on Grid H.	10.5	3	417
Find maximum and minimum values	16. What is the maximum rectangular area can be enclosed with 20 feet of fencing?	PS 13	—	387
Use the Pythagorean Theorem	17. What is the exact length of the diagonal for the maximum rectangular area in Problem 16, in simplest form?	PS 14	—	397
Solve problems using quadratic systems	18. The area of a rectangle is 34 ft². The perimeter of the rectangle is 25 ft. Find the length of the diagonal to the nearest tenth foot.	PS 15	2	424

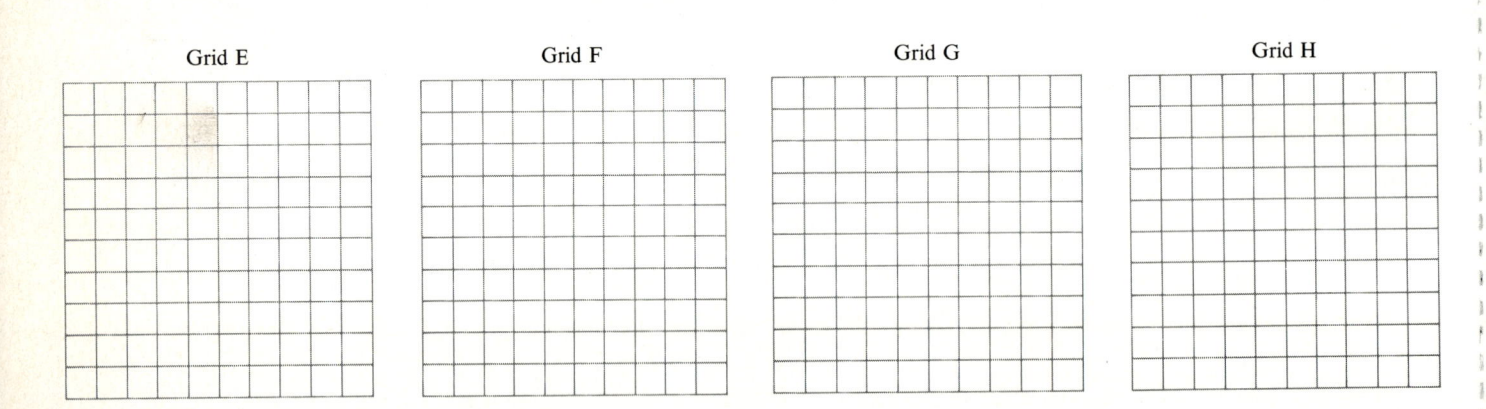

Grid E   Grid F   Grid G   Grid H

CHAPTER 10 REVIEW ANSWERS: 1. $(-\tfrac{2}{3}, 0)$, $(4, 0)$ **2a.** $(1, -4)$ **b.** $x = 1$ **3.** See Appendix Selected Answers **4.** $3\sqrt{10}$ **5.** $(x - (-3)) + (y - 7)^2 = 2^2$ **6.** See Appendix Selected Answers **7.** $(-2, 3)$, $r = 5$ **8.** $\dfrac{x^2}{16^2} + y^2 = 1$ **9.–12.** See Appendix Selected Answers **13.** $(1, 3)$, $(\tfrac{13}{5}, -\tfrac{9}{5})$ **14.** $(-2, -1)$, $(-2, 1)$, $(2, -1)$, $(2, 1)$ **15.** See Appendix Selected Answers **16.** 25 ft² **17.** $5\sqrt{2}$ ft **18.** 9.4 ft

Copyright © 1985 by Harcourt Brace Jovanovich, Inc. All rights reserved.

FUNCTIONS

11 Functions

11.1 Find the Domain and Range

11.2 Identify Functions

11.3 Evaluate Functions

11.4 Graph Functions

11.5 Find the Inverse of a Function

11.6 Solve Variation Problems

PS 16: Solve Applied Variation Problems

Introduction to Relations

The concept of a correspondence between the members of two sets plays a powerful role in the understanding and use of algebra. You are already familiar with correspondences.

Examples ▶ (a) To each person —there corresponds→ a birth date.

(b) To each earthquake —there corresponds→ a magnitude.

(c) To each day on the stock exchange —there corresponds→ a Dow Jones Industrial Average.

(d) To each number —there corresponds→ an absolute value.

(e) To each car owner —there corresponds→ at least one license plate number.

The correspondence concept in mathematics is called a *relation*.

A relation is a correspondence that pairs every member from a first set D, called the *domain*, with at least one member in a second set R, called the *range*.

Arrow diagrams can be used to illustrate relations.

Examples ▶ (a) $\quad D \quad\quad R \quad$ This correspondence is a relation with
$\quad\quad 0 \longrightarrow 0 \quad$ domain $D = \{0, 1, 2, 3\}$ and
$\quad\quad\quad\quad\quad\quad\quad$ range $R = \{0, 1, 8, 27\}$.
$\quad\quad 1 \longrightarrow 1$

$\quad\quad 2 \longrightarrow 8$

$\quad\quad 3 \longrightarrow 27$

(b) $\quad D \quad\quad R \quad$ This correspondence is a relation with
$\quad\quad 0 \longrightarrow 0 \quad$ domain $D = \{0, 1, 4\}$ and
$\quad\quad\quad\quad\quad\quad\quad$ range $R = \{-2, -1, 0, 1, 2\}$.
$\quad\quad 1 \rightrightarrows \begin{matrix}1\\-1\end{matrix}$

$\quad\quad 4 \rightrightarrows \begin{matrix}2\\-2\end{matrix}$

The arrow diagrams illustrate the concept that a relation is a correspondence between the members of two sets. Ordered pair notation is another way to specify, or illustrate, relations.

Example ▶ **Arrow diagram**   **Set of ordered pairs**
$\quad\quad D \quad\quad R \quad\quad \{(1, 1), (2, 3), (2, 6), (3, 8)\} \quad$ Both notations specify a relation with domain $D = \{1, 2, 3\}$ and range $R = \{1, 3, 6, 8\}$. The arrows and the parentheses indicate how the members are paired.
$\quad\quad 1 \longrightarrow 1$

$\quad\quad 2 \rightrightarrows \begin{matrix}3\\6\end{matrix}$

$\quad\quad 3 \longrightarrow 8$

An equation, with its set of ordered pairs in its solution set, is a very concise way to specify a relation. Another way to specify a relation is by graphing a set of ordered pairs.

11.1 Find the Domain and Range   431

The following chart illustrates the most common methods used to specify relations.

Summary ▶

Method of specifying a relation	Example	A specific correspondence
A set of ordered pairs	{(1, 2), (2, 3), (3, 5), (4, 7)}	$x = 3$ corresponds to $y = 5$
A table	x | 1 | 2 | 3 | 4 y | 2 | 3 | 5 | 7	$x = 4$ corresponds to $y = 7$
A graph	(graph of bell-shaped curve centered at origin, peak at y=1)	$x = 0$ corresponds to $y = 1$
An equation	$y = \pm\sqrt{x^3 + 1}$	$x = 2$ corresponds to $y = \pm 3$

Sometimes it is helpful to specify a relation using more than one method as in the following free-fall experiment.

Example ▶ Use the previously explained methods to specify the relation between the time t (in seconds) an object falls and the distance d (in feet) that it has fallen.

Arrow Method

$t = 0 \longrightarrow d = 0$
$t = \frac{1}{2} \longrightarrow d = 4$

$t = 1 \longrightarrow d = 16$

$t = \frac{3}{2} \longrightarrow d = 36$

Ordered Pair Method
{(0, 0), ($\frac{1}{2}$, 4), (1, 16), ($\frac{3}{2}$, 36)}

Table Method

time t (in seconds)	distance d (in feet)
0	0
$\frac{1}{2}$	4
1	16
$\frac{3}{2}$	36

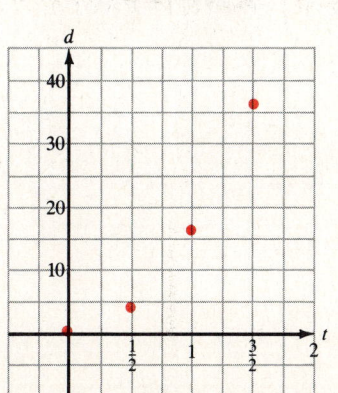

Graphing Method

(graph with points plotted at (0,0), (½,4), (1,16), (3/2,36))

Equation Method
$d = 16t^2$, where $t = 0, \frac{1}{2}, 1,$ or $\frac{3}{2}$

Note 1 ▶ In the graphing method, it is common practice to associate the domain values with the horizontal axis and the range values with the vertical axis.

Note 2 ▶ A different scale is used on each axis of the previous graph.

A variable that is used to represent a domain value is called the *independent variable*.
A variable that is used to represent a range value is called the *dependent variable*.

Example ▶ In the free-fall experiment, t was the independent variable and d the dependent variable.

11.1 Find the Domain and Range

If a relation is specified by a set of ordered pairs, then the set of all first components form the domain D, and the set of all second components form the range R.

EXAMPLE 1: Find the domain and range of a relation given by a set of ordered pairs.

Problem ▶ Find the domain and range of $\{(1, 1), (2, 3), (3, 6), (4, 6)\}$.

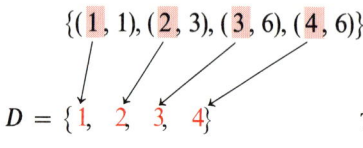

1. Find domain ▶ $\{(\boxed{1}, 1), (\boxed{2}, 3), (\boxed{3}, 6), (\boxed{4}, 6)\}$

$D = \{1, 2, 3, 4\}$ Think: The domain is the set of all first components.

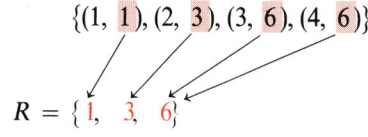

2. Find range ▶ $\{(1, \boxed{1}), (2, \boxed{3}), (3, \boxed{6}), (4, \boxed{6})\}$

$R = \{1, 3, 6\}$ Think: The range is the set of all second components. The common number 6 is only listed once.

Solution ▶ The domain of $\{(1, 1), (2, 3), (3, 6), (4, 6)\}$ is: $D = \{1, 2, 3, 4\}$.
The range of $\{(1, 1), (2, 3), (3, 6), (4, 6)\}$ is: $R = \{1, 3, 6\}$.

Make Sure

Find the domain and range of each relation given by the sets of ordered pairs.

See Example 1 ▶   **1.** $\{(3, 0), (0, -2), (-2, -4), (-4, 5)\}$ **2.** $\{(3, 5), (-2, 5), (-4, 7), (1, 7)\}$

_____ , _____   _____ , _____

MAKE SURE ANSWERS: 1. $D = \{-4, -2, 0, 3\}$, $R = \{-4, -2, 0, 5\}$ **2.** $D = \{-4, -2, 1, 3\}$, $R = \{5, 7\}$

11.1 Find the Domain and Range   433

To find the domain of a relation from its graph, you project each point of the graph vertically onto the *x*-axis. To find the range, you project each point horizontally onto the *y*-axis.

EXAMPLE 2: Find the domain and range of a relation from its graph.

Problem ▶ Find the domain and range of the relation given by the graph on Grid A.

Grid A

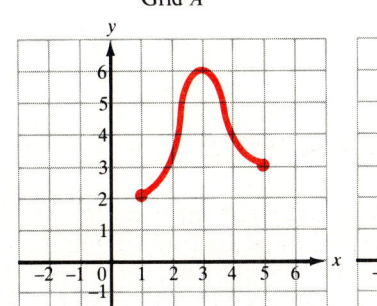

1. Project graph vertically onto *x*-axis ▶

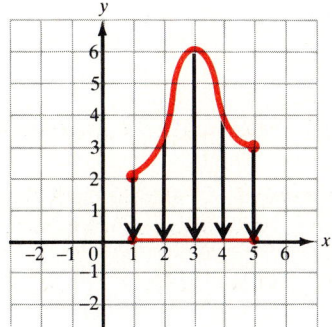

2. Project graph horizontally onto *y*-axis ▶

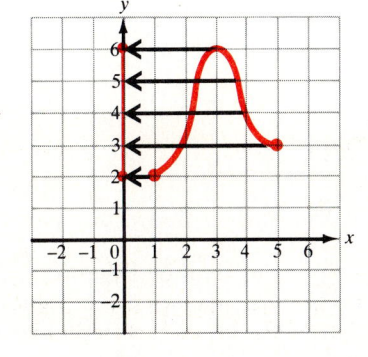

Solution ▶ The domain of the relation given by the graph on Grid A is $D = \{x \mid 1 \leq x \leq 5\}$.
The range of the relation given by the graph on Grid A is $R = \{y \mid 2 \leq y \leq 6\}$.

Make Sure

Find the domain and range of each relation given by the graphs.

See Example 2 ▶ 1.   2.

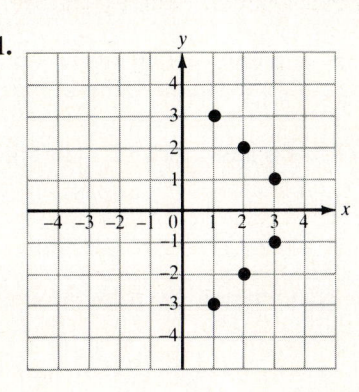

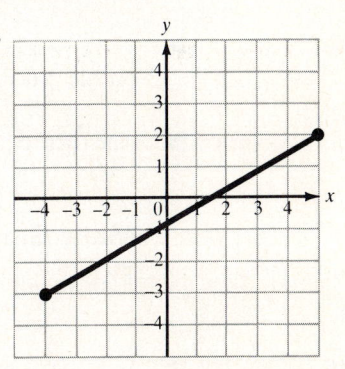

MAKE SURE ANSWERS: 1. $D = \{1, 2, 3\}$, $R = \{-3, -2, -1, 1, 2, 3\}$ 2. $D = \{x \mid -4 \leq x \leq 5\}$, $R = \{y \mid -3 \leq y \leq 2\}$

You can use a single equation to specify different relations by changing its domain. Figure 11.1 shows the graph of two different relations specified by $y = -x + 2$.

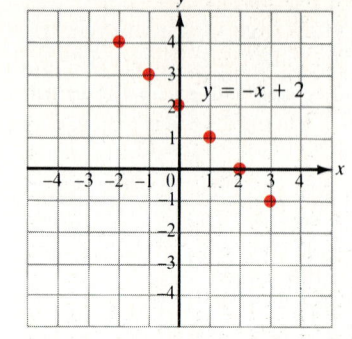

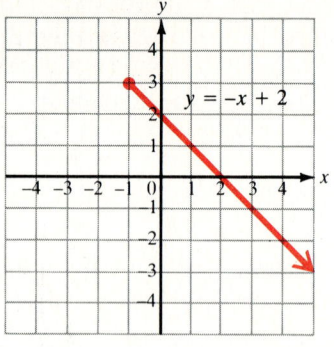

Figure 11.1 ▶

Domain = $\{-2, -1, 0, 1, 2, 3\}$ Domain = $\{x \mid -1 \leq x\}$

Agreement ▶ If a relation is specified by an equation and the domain is not stated, then the domain shall include all the real numbers that can be substituted for the independent variable and shall produce at least one real value of the dependent variable. The range shall consist of the real numbers that the dependent variable assumes in the above substitution process.

EXAMPLE 3: Find the domain and range of a relation given by an equation.

Problem ▶ Find the domain and range of: $y = \sqrt{x - 4} + 2$.

1. Find domain ▶ $y = \sqrt{x - 4} + 2$ is a real number only if $x - 4 \geq 0$.

$x - 4 \geq 0$ means $x \geq 4$ ⟵ values to be included in the domain

2. Find range ▶ Because $\sqrt{x - 4}$ assumes all values greater than or equal to 0, $y = \sqrt{x - 4} + 2$ assumes all values greater than or equal to 2.

Solution ▶ The relation given by $y = \sqrt{x - 4} + 2$ has domain $D = \{x \mid x \geq 4\}$ and range $R = \{y \mid y \geq 2\}$.

Sometimes you can find the range by considering the graph of the equation.

Example ▶ Find the domain and range of: $y = -x^2 + x - 3$.

1. Find domain ▶ $y = -x^2 + x - 3$ is a real number for all real values of x.

2. Find range ▶ $y = -x^2 + x - 3$ graphs to be a parabola that opens downward. Since the highest point on the parabola is the vertex $(\frac{1}{2}, -\frac{11}{4})$, y assumes all real numbers less than or equal to $-\frac{11}{4}$.

Solution ▶ The relation given by $y = -x^2 + x - 3$ has domain $D = \{x \mid x \in \mathcal{R}\}$ and range $R = \{y \mid y \leq -\frac{11}{4}\}$.

Make Sure

Find the domain and range of each relation given by the equations.

See Example 3 ▶ 1. $y = \dfrac{1}{x}$ 2. $y = x^2 - 2$ 3. $y = \sqrt{25 - x^2}$

_____ , _____   _____ , _____   _____ , _____

MAKE SURE ANSWERS: 1. $D = \{x \mid x \neq 0\}$, $R = \{y \mid y \neq 0\}$ **2.** $D = \{x \mid x \in \mathcal{R}\}$, $R = \{y \mid y \geq -2\}$ **3.** $D = \{x \mid -5 \leq x \leq 5\}$, $R = \{y \mid 0 \leq y \leq 5\}$

11.1 Practice

Set 1: Find the domain and range of each relation given by the sets of ordered pairs.

1. {(2, 3), (4, 5), (1, 2), (5, 2)}
2. {(1, 2), (2, 3), (3, 4), (4, 5)}
3. {(−2, 2), (−3, 3), (−4, 4), (−1, 1)}
4. {(0, 2), (−1, 3), (−2, 4), (1, 1)}
5. {(1, 2), (1, 3), (1, 4), (1, 5)}
6. {(0, 0), (0, 1), (0, 2), (0, 3)}
7. {(1, 1), (2, 1), (3, 1), (4, 1)}
8. {(0, 0), (1, 0), (1, 0), (3, 0)}
9. {(1, 1), (1, 2), (2, 3), (2, 4)}
10. {(1, 3), (1, 4), (0, 1), (0, 2)}
11. {(1, 1), (1, 2), (2, 1), (2, 2)}
12. {(0, 3), (3, 0), (0, 0), (3, 3)}

Set 2: Find the domain and range of each relation given by the graphs.

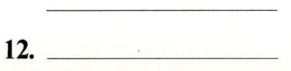

13.

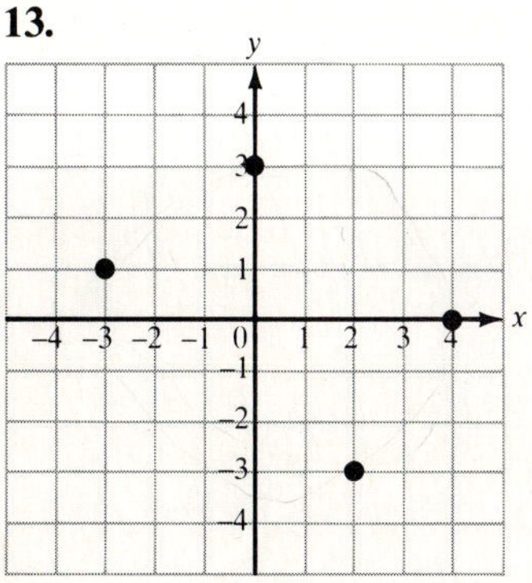

14.

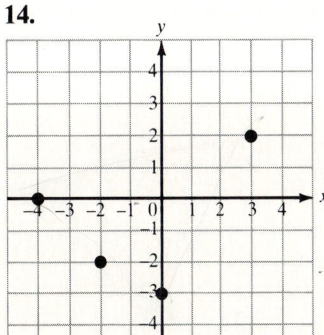

15.

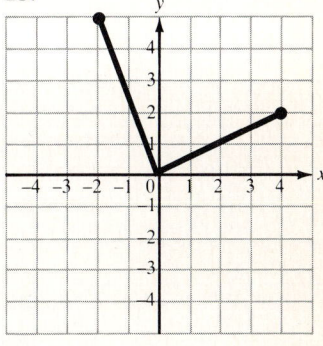

16.

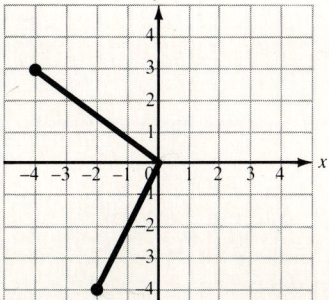

17.

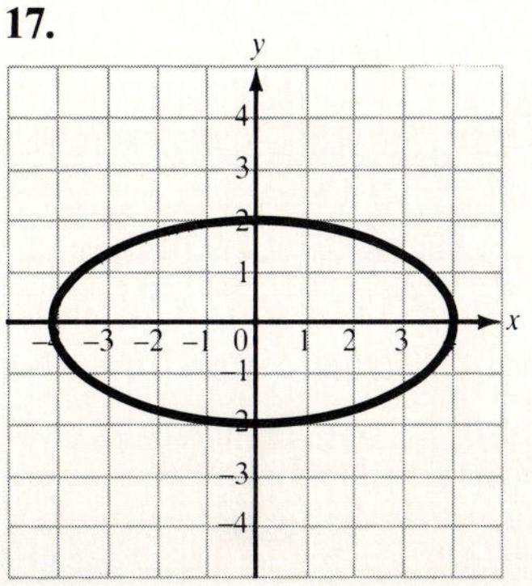

18.

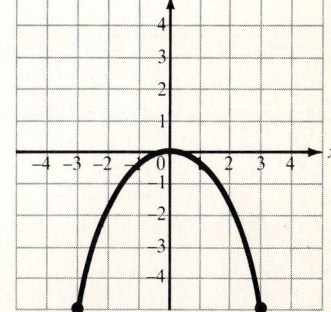

Set 3: Find the domain and range of each relation given by the equations.

19. $y = 2x + 1$ **20.** $y = 2x + 3$ **21.** $y = x^2$

22. $y = -x^2$ **23.** $y = x^2 + 4$ **24.** $y = x^2 - 3$

25. $y = 3$ **26.** $x = -2$ **27.** $y = |x|$

28. $y = -|x|$ **29.** $x = |y|$ **30.** $x = -|y|$

31. $y = \sqrt{x}$ **32.** $x = \sqrt{y}$ **33.** $y = \sqrt{x - 3}$

34. $y = \sqrt{x^2 - 4}$ **35.** $y = \sqrt{4 - x^2}$ **36.** $y = \sqrt{3 - x^2}$

Extra: Find the domain and range when the domain has excluded values.

37. $y = \dfrac{1}{x - 1}$ **38.** $y = \dfrac{1}{1 - x}$ **39.** $y = \dfrac{x^2}{x^2 - 4}$

40. $y = \dfrac{x^2}{4 - x^2}$ **41.** $y = \dfrac{x}{x + 1}$ **42.** $y = \dfrac{x + 1}{x}$

Review: Work each problem on a separate sheet of paper. Attach your work to this page.

Solve each equation for *y*. (See Lesson 2.6.)
43. $3x + 2y = 6$ **44.** $2x - 3y = 5$ **45.** $x^2 - 6x + y = 0$
46. $x^2 + 6x + y = 4$ **47.** $xy - 3y = 1$ **48.** $xy - 3x = 1$
49. $9y - 5x + 160 = 0$ **50.** $9x - 5y + 150 = 0$ **51.** $x^2 + y^2 = 25$
52. $x^2 - y^2 = 9$ **53.** $4x^2 + 9y^2 = 36$ **54.** $4x^2 - 25y^2 = 100$

11.2 Identify Functions

The concept of a *function* is one of the most important concepts in mathematics.

> A function is a relation that pairs each member in the domain with exactly one member in the range.

When a relation is given by a small number of ordered pairs, you can usually determine by inspection if the relation is a function.

EXAMPLE 1: Determine if a relation given by a set of ordered pairs is a function.

Problem ▸ Does $\{(-1, 5), (2, 5), (3, 7)\}$ represent a function?

1. Examine domain and corresponding range members ▸ $\{(-1, 5), (2, 5), (3, 7)\}$

The domain member -1 is paired with exactly one range member 5.

$\{(-1, 5), (2, 5), (3, 7)\}$

The domain member 2 is paired with exactly one range member 5.

$\{(-1, 5), (2, 5), (3, 7)\}$

The domain member 3 is paired with exactly one range member 7.

2. Interpret ▸ Each domain member is paired with exactly one range member, which means $\{(-1, 5), (2, 5), (3, 7)\}$ represents a function.

Solution ▸ $\{(-1, 5), (2, 5), (3, 7)\}$ represents a function.

> CAUTION: All functions are relations, but some relations are not functions.

Example ▸ Does $\{(2, 3), (3, -5), (3, \frac{7}{2}), (4, 11)\}$ represent a function?

1. Examine domain and corresponding range members ▸ $\{(2, 3), (3, -5), (3, \frac{7}{2}), (4, 11)\}$

The domain member 2 is paired with exactly one range member 3.

$\{(2, 3), (3, -5), (3, \frac{7}{2}), (4, 11)\}$ Stop!

2. Interpret ▸ The domain member 3 is paired with two different range members -5 and $\frac{7}{2}$, which means $\{(2, 3), (3, -5), (3, \frac{7}{2}), (4, 11)\}$ does not represent a function.

Solution ▸ $\{(2, 3), (3, -5), (3, \frac{7}{2}), (4, 11)\}$ does not represent a function.

438   Chapter 11   Functions

Make Sure

Determine if each relation given by a set of ordered pairs is a function.

See Example 1 ▶ **1.** $\{(1, -1), (2, -2), (3, -3)\}$ **2.** $\{(-1, 1), (-1, -1), (2, -2), (2, 2)\}$

MAKE SURE ANSWERS: 1. function   2. not a function

Recall ▶ A function is a relation that pairs each member in the domain with exactly one member in the range.

Graphically, this means that any vertical line will intersect the graph of a function at no more than one point. To determine if a relation given by a graph is a function, you can use the following *vertical line test*.

> **The Vertical Line Test**
>
> A given graph is the graph of a function if each vertical line in the coordinate system intersects the graph in no more than one point (no points or one point).

EXAMPLE 2: Determine if a relation given by a graph is a function.

Problem ▶ Which of the following graphs are graphs of functions?

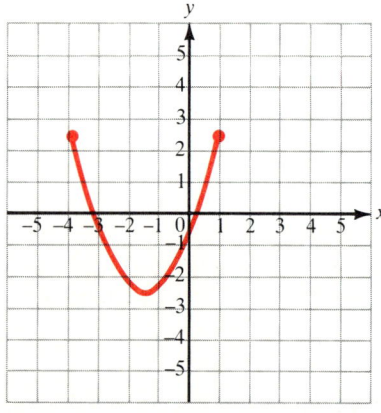

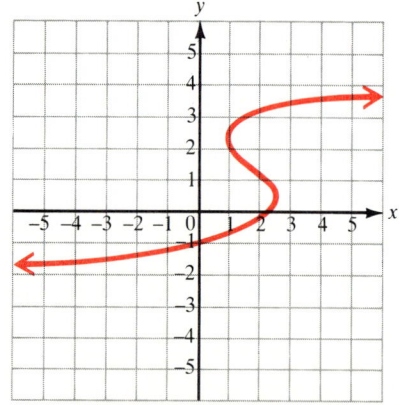

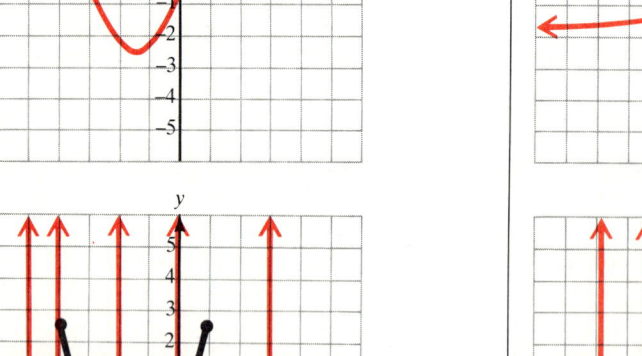

1. Draw vertical lines ▶

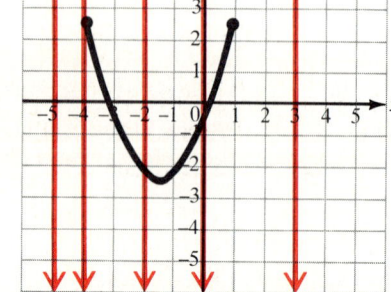

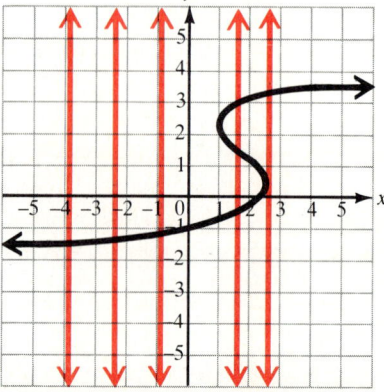

2. Interpret ▶ All vertical lines intersect the graph in no more than one point.

Some vertical lines intersect the graph at more than one point.

Solution ▶ This is the graph of a function.

This is not the graph of a function.

Make Sure

Determine if each relation given by a graph is a function.

See Example 2 ▶ 1.

2.

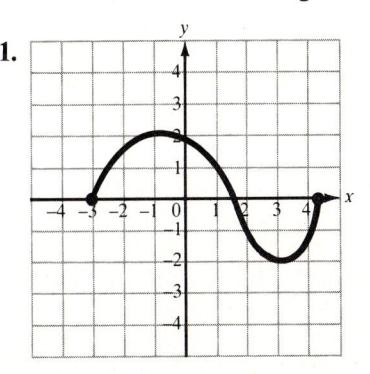

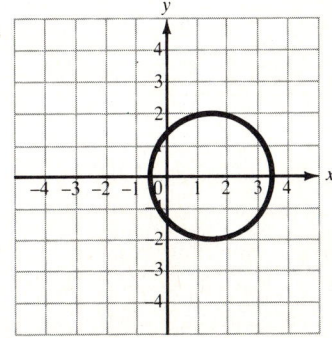

MAKE SURE ANSWERS: 1. function 2. not a function

To determine if a relation given by an equation is a function, you examine the number of values of the dependent variable for each value of the independent variable.

EXAMPLE 3: Determine if a relation given by an equation is a function.

Problem ▶ Which equations $x^2 + y^2 = 4$, $3x^2 + y = 5$, $x^3 + y^3 = 0$ represent functions?

1. Solve for dependent variable ▶

$x^2 + y^2 = 9$	$3x^2 + y = 5$	$x^3 + y^3 = 0$
$y^2 = 9 - x^2$	$y = -3x^2 + 5$	$y^3 = -x^3$
$y = \pm\sqrt{9 - x^2}$		$y = \sqrt[3]{-x^3}$
		$y = -x$

2. Interpret ▶

All domain members greater than -3 and less than 3 are paired with two range members ($\pm\sqrt{9 - x^2}$).	Every domain member x is paired with a single range member ($-3x^2 + 5$).	Every domain member is paired with a single range member ($-x$).

Solution ▶

$x^2 + y^2 = 9$ does not represent a function.	$3x^2 + y = 5$ does represent a function.	$x^3 + y^3 = 0$ does represent a function.

Note ▶ To show that $x^2 + y^2 = 9$ does not represent a function, you can also use the vertical line test.

Example ▶

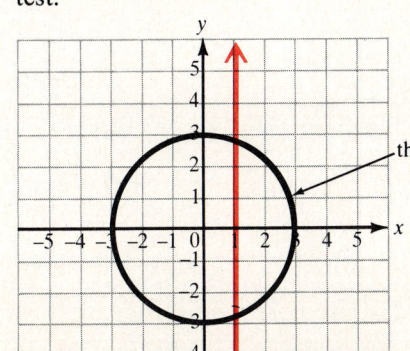

Think: Because some vertical lines intersect the graph at two points, $x^2 + y^2 = 9$ does not represent a function.

the graph of $x^2 + y^2 = 9$

440   Chapter 11   Functions

Similarly, because $3x^2 + y = 5$ graphs to be a parabola that opens downward, you also know by the vertical line test that it represents a function.

Example ▶

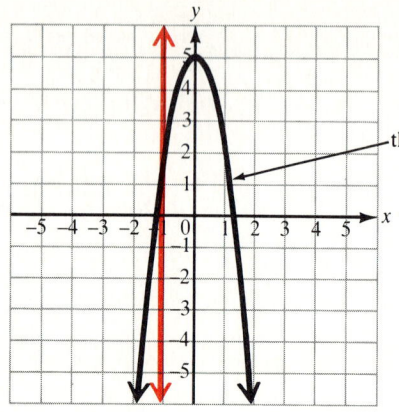

Think: Every vertical line will intersect the graph of the parabola in exactly one point, which means $3x^2 + y = 5$ represents a function.

the graph of $3x^2 + y = 5$

Make Sure

Determine if each relation given by an equation is a function.

See Example 3 ▶   **1.** $3x - y = 2$ _____   **2.** $3x^2 + y = 12$ _____   **3.** $3x^2 - y^2 = 0$ _____

MAKE SURE ANSWERS: 1. function   2. function   3. not a function

According to Guinness

WORLD'S LARGEST PIZZA PIE
THE LARGEST PIZZA EVER BAKED WAS ONE MEASURING 80 FEET 1 INCH IN DIAMETER AND 18,664 LBS. IN WEIGHT AT THE OMA PIZZA RESTAURANT, GLENS FALLS, NEW YORK, OWNED BY LORENZO AMATO, ON OCTOBER 8, 1978.

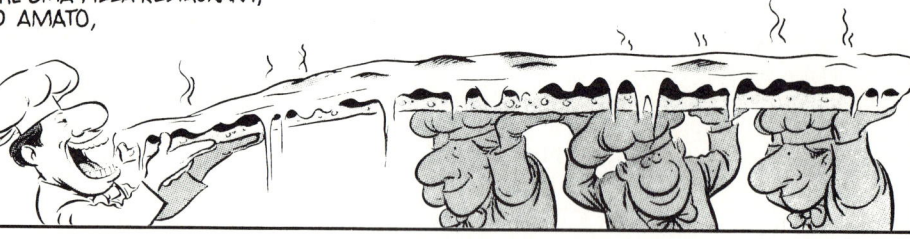

The formulas for the circumference and area of a circle are $C = \pi d$ and $A = \pi r^2$, respectively, where $\pi \approx 3.14$ and r is the radius of the circle.

4. _____

4. Does $C = \pi d$ represent a function? Why?   **5.** Does $A = \pi r^2$ represent a function? Why?

5. _____

Assuming the world's largest pizza is round:

6. _____

6. Find the circumference to the nearest whole foot.   **7.** Find the area to the nearest whole square foot.

7. _____

8. _____

8. Find the weight per square foot, using your answer to Problem 7, to the nearest whole pound.   **9.** Find the area of a $\frac{1}{2}$-pound serving of pizza using your answer to Problem 8.

9. _____

11.2 Practice

Set 1: Determine if each relation given by a set of ordered pairs is a function.

1. $\{(2, 1), (3, 2), (4, 3)\}$
2. $\{(-1, 0), (2, 1), (1, 3)\}$
3. $\{(1, 2), (2, 3), (1, 3)\}$
4. $\{(1, 1), (2, 2), (1, 2)\}$
5. $\{(3, 1), (2, 2), (1, 1)\}$
6. $\{(-1, 1), (0, 0), (1, 1)\}$
7. $\{(1, 0), (2, 0), (3, 0)\}$
8. $\{(-2, 2), (0, 2), (2, 2)\}$
9. $\{(-\frac{2}{3}, 1), (0, 1), (\frac{2}{3}, 1)\}$
10. $\{(\frac{1}{3}, \frac{1}{4}), (\frac{1}{4}, \frac{1}{3}), (\frac{1}{4}, \frac{1}{4})\}$
11. $\{(\frac{1}{2}, \frac{3}{4}), (\frac{3}{4}, \frac{4}{5}), (\frac{4}{5}, \frac{1}{2})\}$
12. $\{(-\frac{3}{4}, -\frac{1}{2}), (0, -\frac{1}{2}), (\frac{3}{5}, -\frac{1}{2})\}$

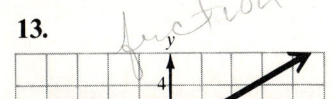

Set 2: Determine if relations given by the graph are functions.

13.

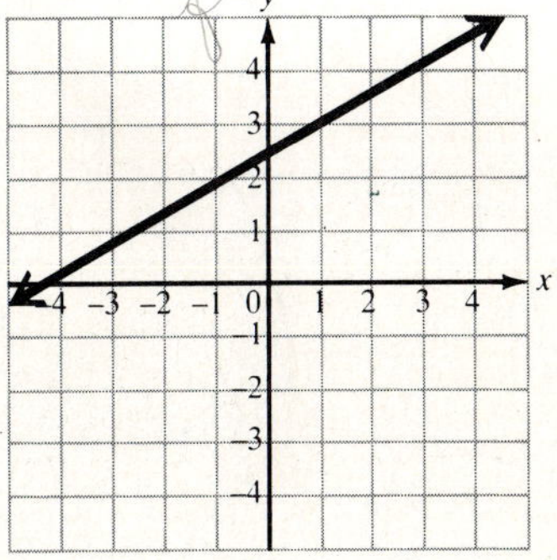

14.

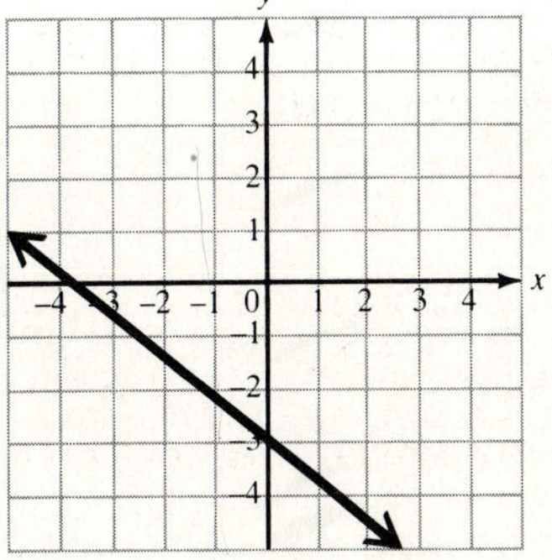

15.

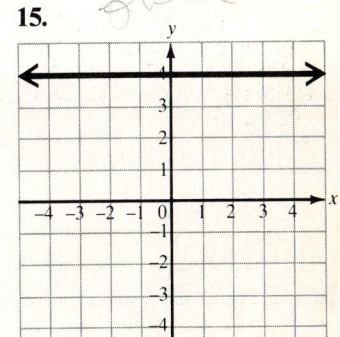

16.

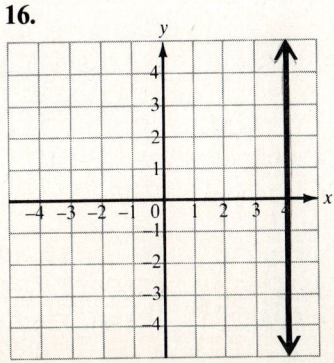

17.

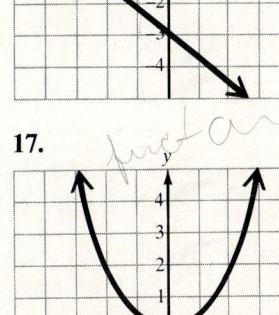

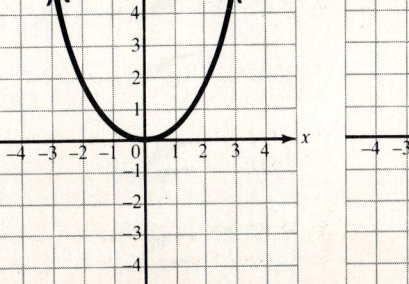

18.

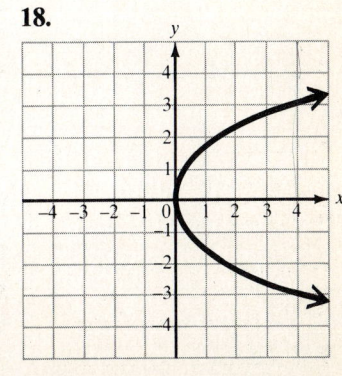

442 Chapter 11 Functions

19. _____

20. _____

21. _____

22. _____

23. _____

24. _____

25. _____

26. _____

27. _____

28. _____

29. _____

30. _____

31. _____

32. _____

33. _____

34. _____

35. _____

36. _____

Set 3: Determine if each relation given by an equation is a function.

19. $y = 3x + 4$ 20. $y = 2x - 3$ 21. $y = x^2 - 3$

22. $y = x^2 + 4$ 23. $x = y^2 - 3$ 24. $x = y^2 + 4$

25. $x = 3$ 26. $y = -3$ 27. $y = |x|$

28. $x = |y|$ 29. $x^2 + y^2 = 9$ 30. $x^2 + y^2 = 4$

31. $4x^2 - 9y^2 = 36$ 32. $4x^2 + 9y^2 = 36$ 33. $y = \sqrt{x}$

34. $y = -\sqrt{x}$ 35. $y = \sqrt{4 - x}$ 36. $y = \sqrt{4 - x^2}$

Review: Work these problems on a separate sheet of paper. Attach your work to this page.

Evaluate each equation. (See Lesson 3.1.)
37. $y = 3x + 2$ for $x = -2, 1, 3, 5$.
38. $y = 2x - 5$ for $x = 3, 1, 0, -2, -4$.
39. $y = 2x^2 - 3$ for $x = -2, -1, 0, 3$.
40. $y = 3x^3 - 4$ for $x = 4, 2, 0, -1, -2$.
41. $y = 2x^2 - 5x + 1$ for $x = -2, 0, 3, 7$.
42. $y = 3x^2 - 6x - 2$ for $x = 4, 2, 0, -1, -5$.

Copyright © 1985 by Harcourt Brace Jovanovich, Inc. All rights reserved.

11.3 Evaluate Functions

It is sometimes helpful to think of a function as a machine. The function machine accepts members from the domain of a function, and produces the corresponding range member. See Figure 11.2.

Figure 11.2 ▶

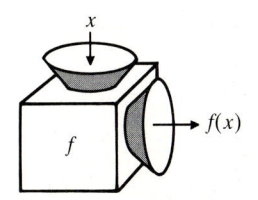

The function machine concept can be represented in a more compact form by an equation using *functional notation*.

> **Functional Notation**
> If f is a function and x is a member of the domain of f, then $f(x)$ is the corresponding range member.
> $f(x)$ is read as "f of x" or "the value of f at x."

In Figure 11.2, the function machine has been named f; x is the input, and $f(x)$ is the output.

> CAUTION: The symbols f and $f(x)$ do not have the same meaning.

The symbol f is used to represent a function. It is neither in the domain nor the range. However, $f(x)$ is a member of the range.

It is often convenient to replace the range variable of an equation with the symbol $f(x)$.

Examples ▶ Using functional notation:
(a) $y = 2x + 5$ is written as $f(x) = 2x + 5$. (b) $y = \dfrac{1}{x+3}$ is written as $f(x) = \dfrac{1}{x+3}$.

> CAUTION: The notation $f(3)$ does not mean f times 3.

The notation $f(3)$ represents the range value associated with the domain value of 3 in the function f. To evaluate $f(3)$ in an equation, you first substitute 3 for the independent variable throughout the equation.

EXAMPLE 1: Evaluate a function for given domain values.

Problems ▶ Find $f(3)$ and $f(h + 2)$ given $f(x) = x^2 - 3x + 1$.

1. Substitute ▶
$f(x) = x^2 - 3x + 1$ $f(x) = x^2 - 3x + 1$
$f(3) = 3^2 - 3(3) + 1$ $f(h + 2) = (h + 2)^2 - 3(h + 2) + 1$

2. Evaluate ▶
$\quad\quad = 9 - 9 + 1$ $\quad\quad = h^2 + 4h + 4 - 3h - 6 + 1$

Solutions ▶ $f(3) = 1$ $f(h + 2) = h^2 + h - 1$

Note ▶ Any letter can be used to represent a function; however, f, g, h, F, G, and H are the letters most frequently used.

444 Chapter 11 Functions

Another Example ▶ Given $g(x) = 3x^2 - 7x$, $h(t) = 100 + 80t - 16t^2$, and $A(r) = \frac{22}{7} \cdot r^2$, find $g(5)$, $h(2)$, and $A(7)$.

$g(x) = 3x^2 - 7x$	$h(t) = 100 + 80t - 16t^2$	$A(r) = \frac{22}{7} \cdot r^2$
$g(5) = 3(5)^2 - 7(5)$	$h(2) = 100 + 80(2) - 16(2)^2$	$A(7) = \frac{22}{7} \cdot (7)^2$
$= 3 \cdot 25 - 35$	$= 100 + 160 - 16 \cdot 4$	$= \frac{22}{7} \cdot 49$
$= 40$	$= 196$	$= 154$

Make Sure

Evaluate each function for given domain values.

See Example 1 ▶
1. If $f(x) = 3x - 2$, ____, ____, ____
   find $f(-3)$, $f(0)$, and $f(2)$.

2. If $G(x) = 3x^2 - 2x + 3$, ____, ____,
   find $G(-4)$, $G(-1)$, and $G(x + 2)$. ____

MAKE SURE ANSWERS: **1.** $-11, -2, 4$ **2.** $59, 8, 3x^2 + 10x + 11$

On a graph, $f(2)$ represents the height (depth) of the graph at $x = 2$. Sometimes you can use the graph of a function to estimate functional values.

EXAMPLE 2: Estimate functional values using a graph.

Problem ▶ Estimate $g(-1)$, $g(1)$, and $g(4)$ given the graph $g(x)$ on Grid B.

Grid B

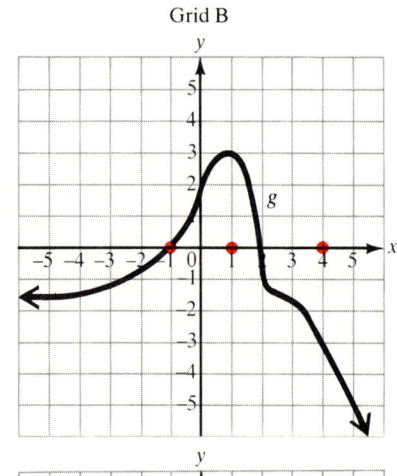

1. Locate domain values ▶

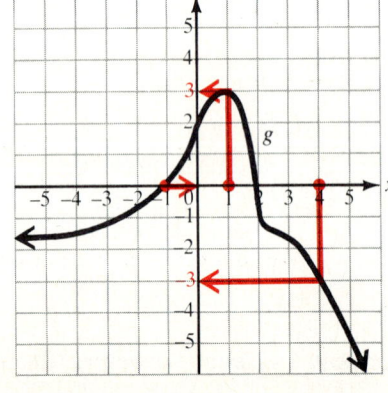

2. Draw vertical line segments and estimate range values ▶

Draw a vertical line segment from each given domain value to the graph of g.

Read the height (depth) of the intersection of the line segment and the graph of g on the y-axis.

Solution ▶ $g(-1) \approx 0$, $g(1) \approx 3$, and $g(4) \approx -3$.

Make Sure

Estimate functional values using a graph.

See Example 2

1. $f(-2)$, $f(0)$, and $f(1)$ ____ , ____ , ____

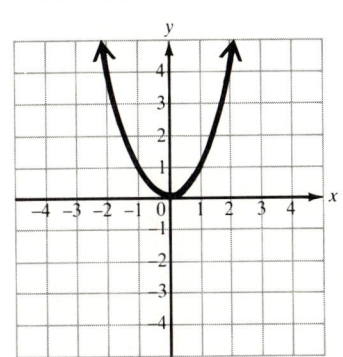

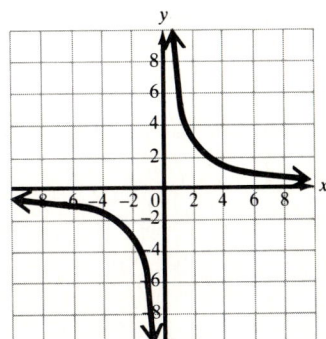

2. $g(1)$, $g(3)$, and $g(-2)$ ____ , ____ , ____

MAKE SURE ANSWERS: 1. 4, 0, 1   2. 5, 2, −3

Sometimes you will need to evaluate a function of a function. Using the function machine concept, you can visualize the process as shown in Figure 11.3.

Figure 11.3

To evaluate $f(g(c))$ for some constant c, it is generally easiest to evaluate $g(c)$ first and then substitute this result for x in the f function.

EXAMPLE 3: Evaluate a function of a function.

Problem ▶ Find $f(g(5))$ given $f(x) = 2x + 3$ and $g(x) = x^2 - 4$.

1. Evaluate $g(c)$ ▶ $f(g(5)) = f((5)^2 - 4)$ Think: Substitute 5 for x in the formula for $g(x)$.

                 $= f(21)$

2. Substitute ▶ $= 2(21) + 3$ Think: Substitute 21 for x in the formula for $f(x)$.

                 $= 45$

Solution ▶ $f(g(5)) = 45$

You can evaluate $g(f(c))$ by first evaluating $f(c)$.

Example ▶ Find $g(f(5))$ given $f(x) = 2x + 3$ and $g(x) = x^2 - 4$.

1. Evaluate $f(c)$ ▶ $g(f(5)) = g(2(5) + 3)$ Think: Substitute 5 for x in the formula for $f(x)$.

$ = g(13)$

2. Substitute ▶ $ = (13)^2 - 4$ Think: Substitute 13 for x in the formula for $g(x)$.

$ = 165$

Solution ▶ $g(f(5)) = 165$

> **CAUTION:** In general $f(g(x)) \neq g(f(x))$.

Example ▶ In the previous two examples, $f(g(5)) = 45$ and $g(f(5)) = 165$.

If f and g are given by formulas, you can use the previous techniques to find a formula for $f(g(x))$.

Example ▶ Find a formula for $f(g(x))$ given $f(x) = 2x + 3$ and $g(x) = x^2 - 4$.

1. Substitute for $g(x)$ ▶ $f(g(x)) = f(x^2 - 4)$ Think: Substitute $x^2 - 4$ for $g(x)$.

2. Substitute in $f(x)$ ▶ $ = 2(x^2 - 4) + 3$ Think: Substitute $x^2 - 4$ for x in the formula for $f(x)$.

$ = 2x^2 - 8 + 3$

Solution ▶ $f(g(x)) = 2x^2 - 5$

> **CAUTION:** The notation $f(g(x))$ is only defined when $g(x)$ is in the domain of f.

Examples ▶ (a) If $f(x) = \dfrac{1}{x}$ and $g(x) = x - 2$, then $f(g(2))$ is undefined because $g(2) = 2 - 2 = 0$ and 0 is not in the domain of f. (Division by 0 is undefined.)

(b) If $F(x) = \sqrt{x}$ and $G(x) = x - 5$ then $F(G(1))$ is undefined because $G(1) = 1 - 5 = -4$ and -4 is not in the domain of F. Recall the agreement from Lesson 11.1, which stated that if the domain of a function is not given, then it consists of all real numbers that produce a real number when they are substituted for the independent variable. ($\sqrt{-4}$ is not a real number.)

Make Sure

Evaluate each function of a function.

See Example 3 ▶
1. If $f(x) = x^2$, and $g(x) = 2x - 1$, find $f(g(1))$, and $f(g(-2))$.   _____, _____

2. If $f(x) = x^2 + 1$, and $g(x) = 1 - 2x$, find $f(g(2))$, and $f(g(x + 1))$.   _____, _____

MAKE SURE ANSWERS: 1. 1, 25   2. 10, $4x^2 + 4x + 2$

Name _____ Date _____ Class _____

11.3 Practice

Set 1: Evaluate each function for the given domain values.

1. If $f(x) = 3x^2$, find $f(-2)$, $f(0)$, and $f(3)$.

2. If $f(x) = 2x^3$, find $f(-3)$, $f(0)$, and $f(2)$.

3. If $f(x) = 1 - 2x^2$, find $f(-3)$, $f(0)$, and $f(2)$.

4. If $f(x) = 1 + 2x^2$, find $f(-1)$, $f(1)$, and $f(3)$.

5. If $g(x) = x^2 - 3x + 1$, find $g(-2)$, $g(0)$, and $g(3)$.

6. If $h(x) = 1 + 3x + x^2$, find $h(-1)$, $h(1)$, and $h(3)$.

7. If $g(x) = x^3 - 2x^2 - 1$, find $g(0)$, $g(2)$, and $g(y)$.

8. If $p(x) = 2x^2 - x + 3$, find $p(1)$, $p(a)$, and $p(x + 1)$.

9. If $V(x) = x^2 + 3$, find $V(2)$, $V(x + 1)$, and $V(x - 1)$.

Set 2: Estimate functional values using a graph.

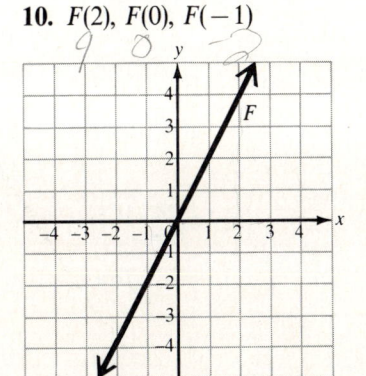

10. $F(2)$, $F(0)$, $F(-1)$

11. $G(3)$, $G(0)$, $G(-2)$

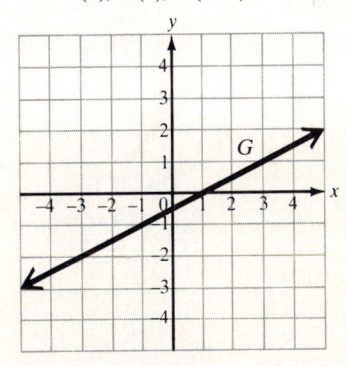

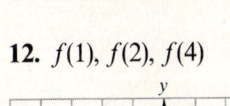

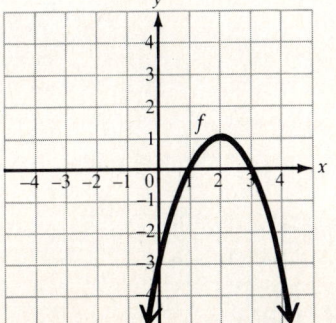

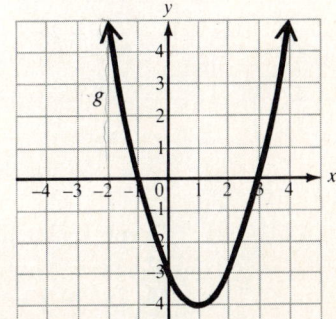

12. $f(1)$, $f(2)$, $f(4)$

13. $g(-2)$, $g(0)$, $g(3)$

Set 3: Evaluate each function of a function.

14. If $f(x) = x + 1$,
    $g(x) = x^2$, find
    $f(g(1)), g(f(1))$.

15. If $f(x) = 1 - x$,
    $g(x) = x + 4$, find
    $f(g(2)), g(f(2))$.

16. If $f(x) = x^2 + 1$,
    $g(x) = 2x + 3$, find
    $f(g(0)), g(f(0))$.

17. If $f(x) = 2x^2 + 1$,
    $g(x) = 2x + 1$, find
    $f(g(3)), g(f(3))$.

18. If $f(x) = 2x^2 + 3$,
    $g(x) = 2 - x$, find
    $f(g(-1)), g(f(-1))$.

19. If $f(x) = 3x^2 - 2$,
    $g(x) = 1 - 2x$, find
    $f(g(-2)), g(f(-2))$.

20. If $f(x) = 2 + x^2$,
    $g(x) = 2 - x^2$, find
    $f(g(-2)), g(f(-2))$.

21. If $f(x) = 1 - x^2$,
    $g(x) = x^2 - 1$, find
    $f(g(2)), g(f(2))$.

22. If $f(x) = 4x^2 - 1$,
    $g(x) = x^2$, find
    $f(g(-2)), f(g(2))$.

23. If $f(x) = x^2$,
    $g(x) = 4 - x^2$, find
    $f(g(-1)), f(g(1))$.

24. If $f(x) = x^2$,
    $g(x) = \sqrt{x}$, find
    $f(g(9)), f(g(25))$.

25. If $f(x) = \sqrt{x}$,
    $g(x) = x^2$, find
    $f(g(4)), g(f(1))$.

Extra: Evaluate each of the following.

26. If $f(x) = x^2$,
    $g(x) = x$, find
    $f(g(a + 1)), g(f(a + 1))$.

27. If $f(x) = x + 2$,
    $g(x) = x^2$, find
    $f(g(a + 1)), g(f(a + 1))$.

28. If $f(x) = x^2$,
    $g(x) = x + 1$, find
    $f(g(h - 1)), g(f(h - 1))$.

29. If $f(x) = x^2$,
    $g(x) = x - 1$, find
    $f(g(x - 1)), g(f(x - 1))$.

30. If $f(x) = 2x - 1$,
    $g(x) = x + 3$, find
    $f(g(x^2)), g(f(x^2))$.

31. If $f(x) = 2x + 1$,
    $g(x) = 2x - 1$, find
    $f(g(x^2)), g(f(x^2))$.

Extra: Work these problems on a separate sheet of paper. Attach your work to this page.

Graph each equation.
32. $2x + 3y = 6$
33. $2x + y = 8$
34. $2x^2 - y = 6$
35. $2x^2 + y = 6$
36. $y = |x|$
37. $-|x| = y$
38. $y = |x - 2|$
39. $y = |x + 2|$
40. $y = |x - 5|$

11.4 Graph Functions

Some functions occur so often in mathematics that you will find it helpful to study them in more detail.

> If m and b are real numbers, then the function $f(x) = mx + b$ is called a *linear function*.

Recall ▶ An equation of the form $y = mx + b$ graphs to be a straight line with slope m and y-intercept b. See Lesson 3.2.

> The function $f(x) = x$ is called the *identity function*.

The identity function is the special linear function obtained by letting $m = 1$ and $b = 0$ in the equation $f(x) = mx + b$. The identity function pairs each real number with itself. The graph of the identity function is shown in Figure 11.4.

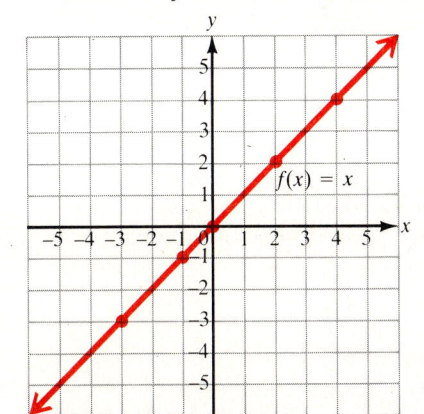

Figure 11.4 ▶

x	$y = f(x)$
-3	-3
-1	-1
0	0
2	2
4	4

> If b is a real number, then the function $f(x) = b$ is called a *constant function*.

Example ▶ For the constant function $f(x) = 3$, (a) $f(-2) = 3$, (b) $f(0) = 3$, (c) $f(3.1) = 3$.

The following graphs are both constant functions.

Examples ▶

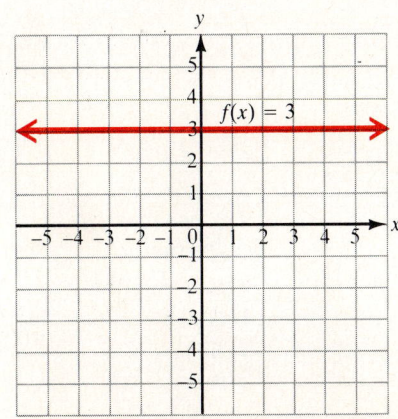

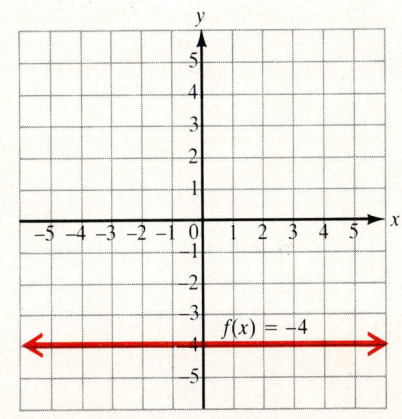

> If a, b, and c are real numbers ($a \neq 0$), then the function $f(x) = ax^2 + bx + c$ is called a *quadratic function*.

Recall ▶ An equation of the form $y = ax^2 + bx + c$ graphs to be a parabola that opens upward if $a > 0$ and downward if $a < 0$. See Lesson 10.1.

450   Chapter 11   Functions

> The function $f(x) = |x|$ is called the *absolute value function*.

Recall ▶ The definition of $|x|$ is given by $|x| = \begin{cases} x & \text{if } x \geq 0 \\ -x & \text{if } x < 0 \end{cases}$.

The graph of the absolute value function is given in Figure 11.5.

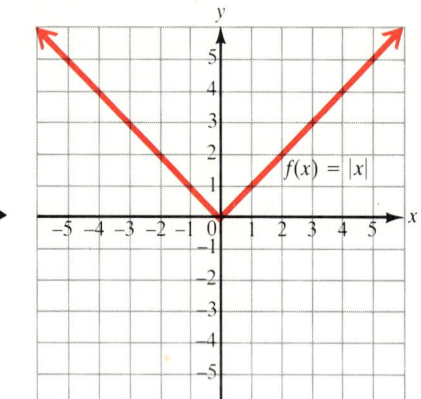

Figure 11.5 ▶

> The *greatest integer function* is denoted by the symbol $[x]$ and is defined by the rule: $[x]$ is equal to the greatest integer that is less than or equal to x.

Examples ▶
(a) $[2.63] = 2$ Think: The greatest integer less than or equal to 2.63 is 2.
(b) $[4] = 4$ The greatest integer less than or equal to 4 is 4.
(c) $[0.23] = 0$ The greatest integer less than or equal to 0.23 is 0.
(d) $[-3.4] = -4$ The greatest integer less than or equal to -3.4 is -4.

Note ▶ For any number x on the real number line, the $[x]$ will be:
(a) x if x is an integer, or
(b) the nearest integer to the left of x if x is not an integer.

The greatest integer function is often called a *step function* because its graph resembles a set of steps. See Figure 11.6.

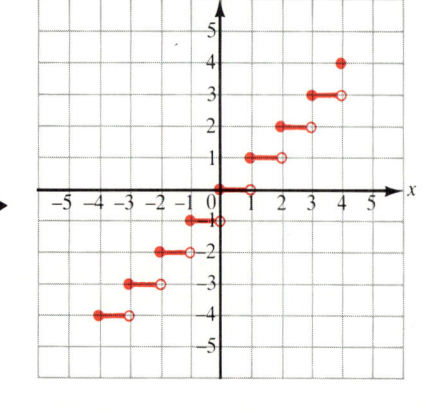

Figure 11.6 ▶

x	$y = [x]$
$-4 \leq x < -3$	-4
$-3 \leq x < -2$	-3
$-2 \leq x < -1$	-2
$-1 \leq x < 0$	-1
$0 \leq x < 1$	0
$1 \leq x < 2$	1
$2 \leq x < 3$	2
$3 \leq x < 4$	3
4	4

The graph of $y = [x]$ for $-4 \leq x \leq 4$

Recall ▶ A solid dot indicates that the point is included in the graph.
An open dot indicates that the point is not included in the graph.

11.4 Graph Functions

You can graph many functions by plotting several points.

EXAMPLE 1: Graph a function by plotting points.

Problem ▶ Graph $f(x) = |x + 2|$ for $-4 \leq x \leq 4$.

1. Make a table and plot points ▶

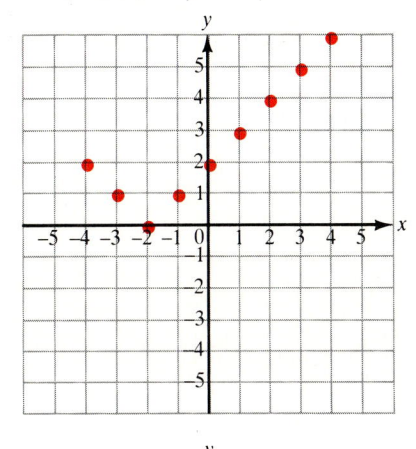

x	$f(x) = \|x + 2\|$
-4	2
-3	1
-2	0
-1	1
0	2
1	3
2	4
3	5
4	6

2. Sketch graph ▶

Solution ▶ $f(x) = |x + 2|$ for $-4 \leq x \leq 4$

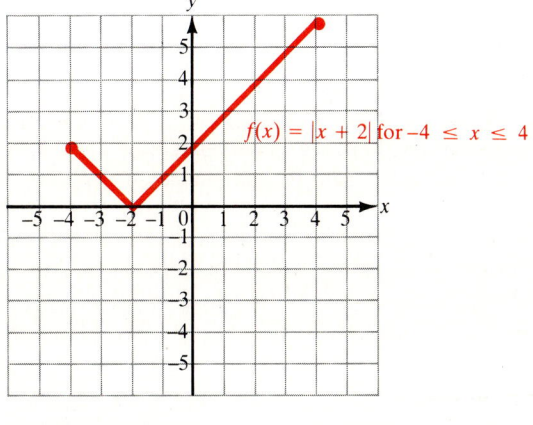

Make Sure

Graph each function by plotting points.

See Example 1 ▶

1. $f(x) = x^2 + 2x + 1$, $-4 \leq x \leq 2$

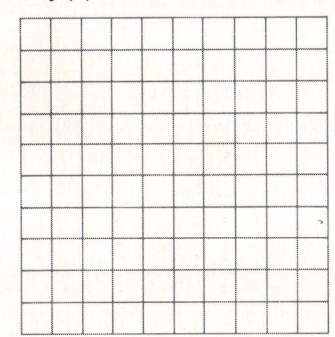

2. $f(x) = x^3 + 2$, $-2 \leq x \leq 2$

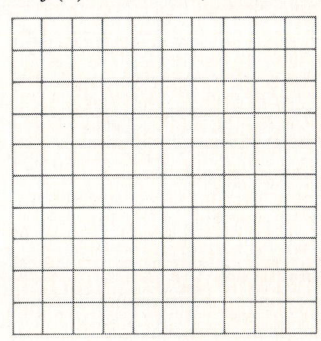

MAKE SURE ANSWERS: See Appendix Selected Answers.

452 Chapter 11 Functions

To graph a function given by an equation containing two or more parts, you graph each part over the indicated domain.

EXAMPLE 2: Graph a function given by an equation containing two or more parts.

Problem ▶ Graph $h(x) = \begin{cases} x & \text{if } x < 0 \\ x^2 & \text{if } 0 \leq x \leq 2 \\ 1 & \text{if } 2 < x \end{cases}$

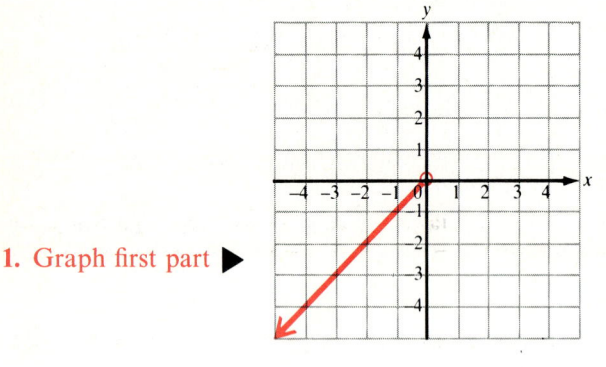

1. Graph first part ▶

$h(x) = \begin{cases} x & \text{if } x < 0 \\ x^2 & \text{if } 0 \leq x \leq 2 \\ 1 & \text{if } 2 < x \end{cases}$

This part of the graph consists of that portion of the identity function $h(x) = x$ for $x < 0$.

2. Graph second part ▶

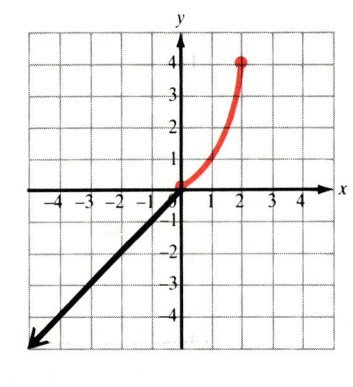

$h(x) = \begin{cases} x & \text{if } x < 0 \\ x^2 & \text{if } 0 \leq x \leq 2 \\ 1 & \text{if } 2 < x \end{cases}$

This part of the graph consists of the portion of $h(x) = x^2$ for $0 \leq x \leq 2$.

3. Graph third part ▶

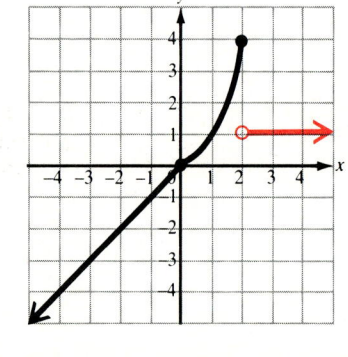

$h(x) = \begin{cases} x & \text{if } x < 0 \\ x^2 & \text{if } 0 \leq x \leq 2 \\ 1 & \text{if } 2 < x \end{cases}$

This part of the graph consists of that portion of the constant function $h(x) = 1$ for $x > 2$.

Solution ▶

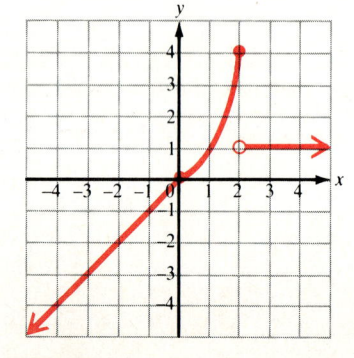

The graph of $h(x) = \begin{cases} x & \text{if } x < 0 \\ x^2 & \text{if } 0 \leq x \leq 2 \\ 1 & \text{if } 2 < x \end{cases}$.

11.4 Graph Functions   453

Make Sure

Graph a function given by equations containing two or more parts.

See Example 2 ▶ $f(x) = \begin{cases} x^2 - y = 9 & \text{if } -3 \leq x \leq 3 \\ 5x - 3y = 15 & \text{if } 3 \leq x \leq 8 \\ 5x + 3y = -15 & \text{if } -8 \leq x \leq -3 \end{cases}$

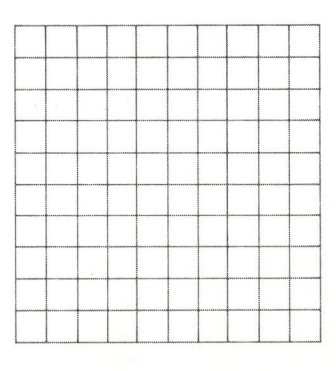

MAKE SURE ANSWER: See Appendix Selected Answers.

Many functions involving absolute value notation can be graphed quickly and easily by renaming the equation as an equation containing two parts.

EXAMPLE 3: Graph a function by renaming its equation.

Problem ▶   Graph $F(x) = |x - 2|$.

1. Rename ▶ $F(x) = |x - 2|$

$F(x) = \begin{cases} x - 2 & \text{if } x - 2 \geq 0 \\ -(x - 2) & \text{if } x - 2 < 0 \end{cases}$ Use the definition of absolute value.

2. Simplify ▶ $F(x) = \begin{cases} x - 2 & \text{if } x \geq 2 \\ -x + 2 & \text{if } x < 2 \end{cases}$ Think: (1) $x - 2 \geq 0$ means $x \geq 2$
(2) $x - 2 < 0$ means $x < 2$

3. Graph first part ▶ $F(x) = \begin{cases} x - 2 & \text{if } x \geq 2 \\ -x + 2 & \text{if } x < 2 \end{cases}$ This part of the graph consists of that portion of the line with slope 1 and y-intercept -2, for $x \geq 2$.

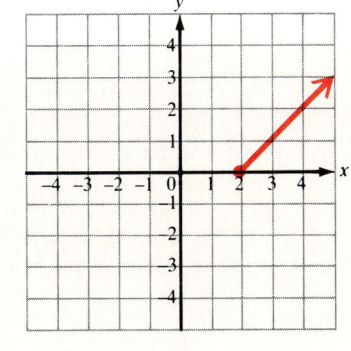

4. Graph second part ▶ $F(x) = \begin{cases} x - 2 & \text{if } x \geq 2 \\ -x + 2 & \text{if } x < 2 \end{cases}$ This part of the graph consists of that portion of the line with slope -1 and y-intercept 2, for $x < 2$.

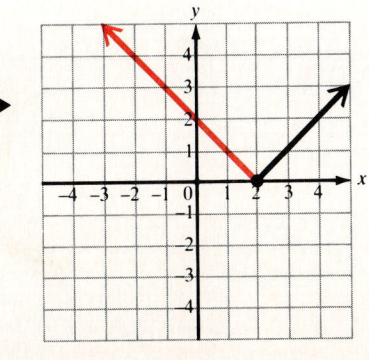

454 Chapter 11 Functions

Solution ▶

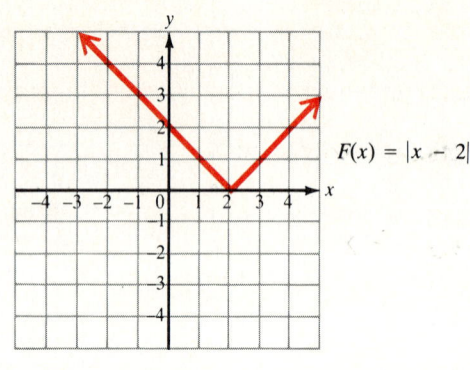

$F(x) = |x - 2|$

Note ▶ You should compare this method with the point-plotting method used in Example 1.

Another Example ▶ Graph $H(x) = |x^2 - 4x|$.

1. Rename ▶ $H(x) = \begin{cases} x^2 - 4x & \text{if } x^2 - 4x \geq 0 \\ -(x^2 - 4x) & \text{if } x^2 - 4x < 0 \end{cases}$

2. Simplify ▶ $H(x) = \begin{cases} x^2 - 4x & \text{if } x \leq 0 \text{ or } x \geq 4 \\ -x^2 + 4x & \text{if } 0 < x < 4 \end{cases}$

Think: (1) if $x^2 - 4x \geq 0$, then $x(x - 4) \geq 0$ and $x \leq 0$ or $x \geq 4$
(2) if $x^2 - 4x < 0$, then $x(x - 4) < 0$ and $0 < x < 4$

3. Graph as before ▶

Solution ▶ $H(x) = |x^2 - 4x|$

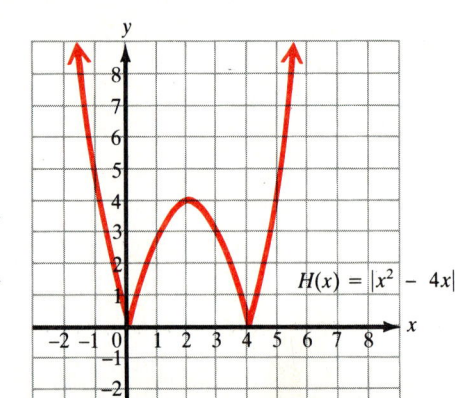

Make Sure

Graph each function by renaming its equation in two or more parts.

See Example 3 ▶ **1.** $f(x) = \left|\dfrac{1}{x}\right|$ **2.** $g(x) = |x - 2| - 3$

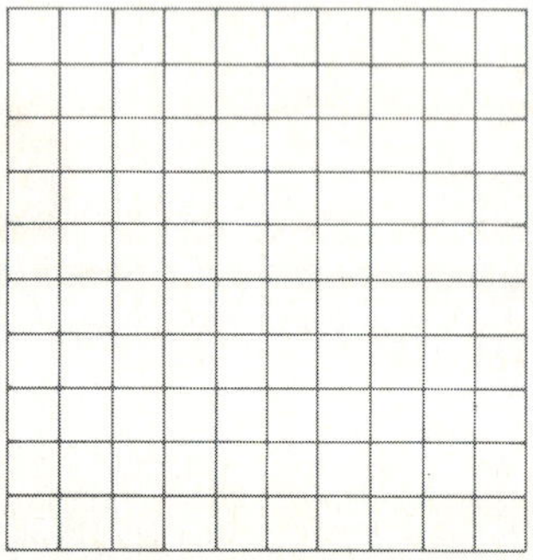

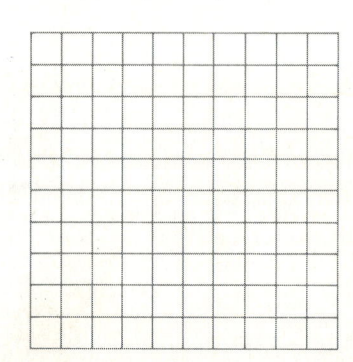

MAKE SURE ANSWERS: See Appendix Selected Answers.

Name _____ Date _____ Class _____

11.4 Practice

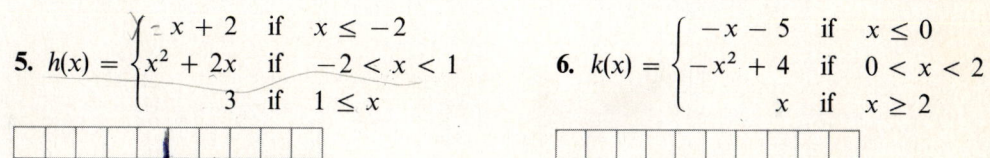

Set 1: Graph each function by plotting points.

1. $f(x) = x^2 + 1$, $-3 \leq x \leq 2$

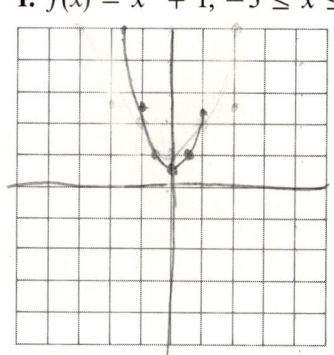

2. $g(x) = x^2 - 4$, $-3 \leq x \leq 3$

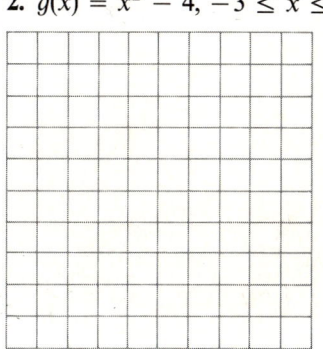

3. $h(x) = x^3 + 1$, $-2 \leq x \leq 2$

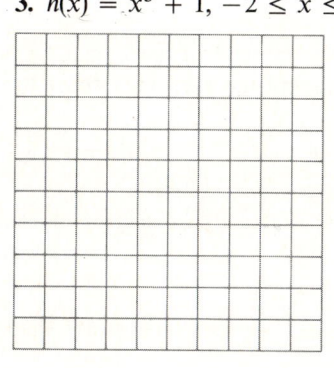

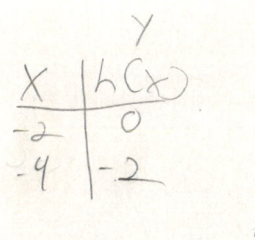

4. $k(x) = x^3 - 2$, $-2 \leq x \leq 2$

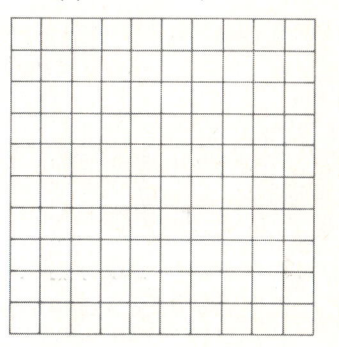

Set 2: Graph functions given by an equation containing two or more parts.

5. $h(x) = \begin{cases} x + 2 & \text{if } x \leq -2 \\ x^2 + 2x & \text{if } -2 < x < 1 \\ 3 & \text{if } 1 \leq x \end{cases}$

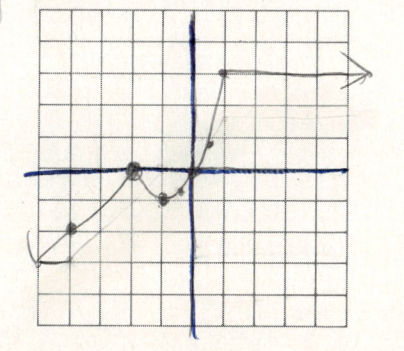

6. $k(x) = \begin{cases} -x - 5 & \text{if } x \leq 0 \\ -x^2 + 4 & \text{if } 0 < x < 2 \\ x & \text{if } x \geq 2 \end{cases}$

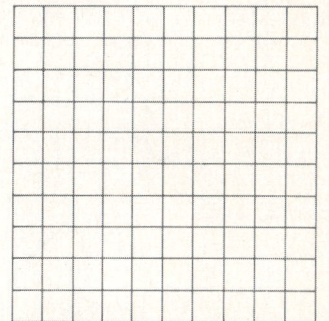

Chapter 11 Functions

Set 3: Graph each function by renaming its equation.

7. $f(x) = \left|\dfrac{1}{x-1}\right|$

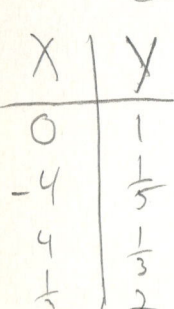

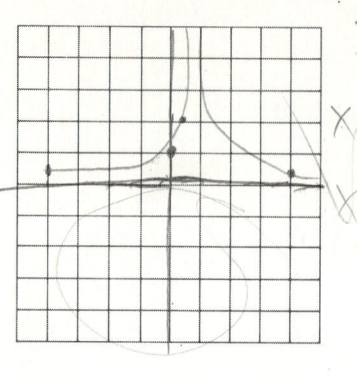

8. $g(x) = \left|\dfrac{1}{x+1}\right|$

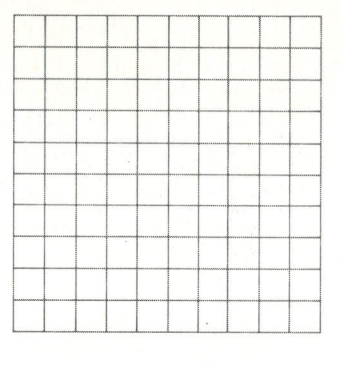

9. $h(x) = |x + 2|$

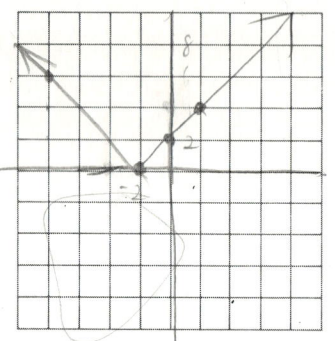

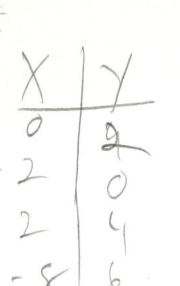

10. $k(x) = |x - 3|$

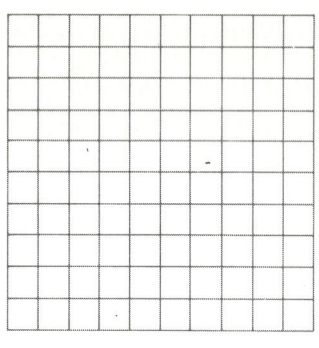

Extra: Graph each equation involving the greatest integer function.

11. $f(x) = 2[x]$

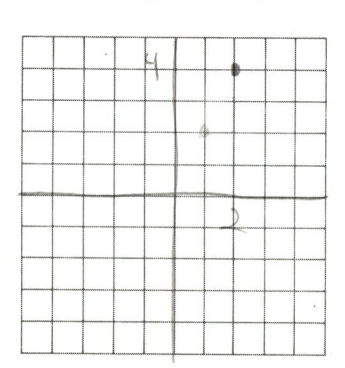

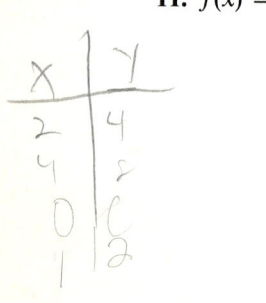

12. $g(x) = -3[x]$

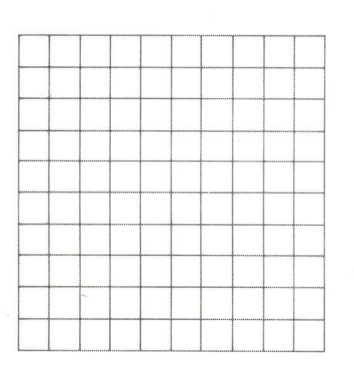

13. $h(x) = [x] - 4$

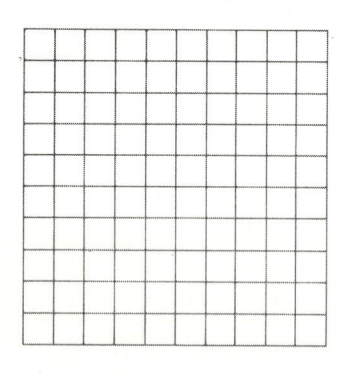

14. $k(x) = [x] + 4$

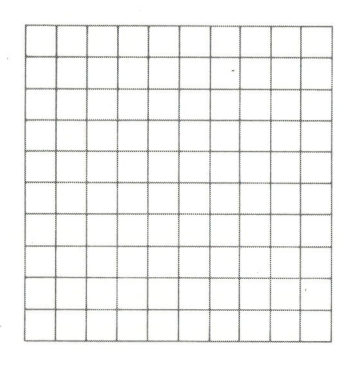

Review: Work these problems on a separate sheet of paper. Attach your work to this page.

Solve each equation for y. (See Lesson 2.6.)
15. $x = y^2 - 3$
16. $x = y^2 + 4$
17. $x = y^2 + 2$
18. $x = y^2 - 4y$
19. $x = y^2 + 2y$
20. $x = y^2 - 3y$
21. $x = y^2 - 4y + 3$
22. $x = y^2 - y + 2$
23. $x = 3y^2 + 2y - 1$

Copyright © 1985 by Harcourt Brace Jovanovich, Inc. All rights reserved.

11.5 Find the Inverse of a Function

Recall ▶ The free-fall experiment on page 431 used t (time) as the independent variable and d (distance) as the dependent variable.

Sometimes it is useful to interchange the independent and the dependent variables in a relation.

Example ▶ Use d as the independent variable and t as the dependent variable in the free-fall experiment.

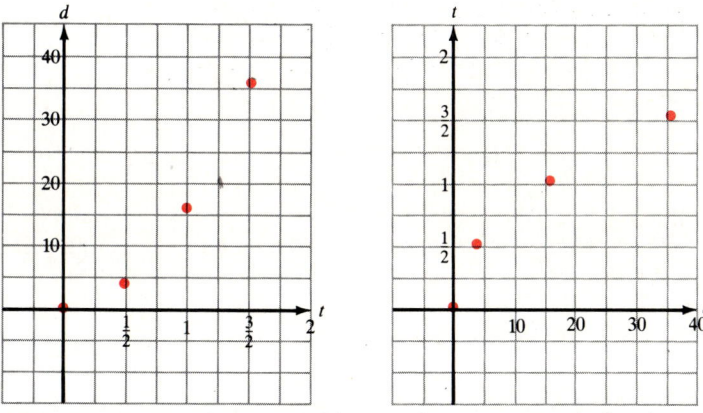

The interchange of the variables produces a new relation between the distance an object falls and the time that it has fallen. This new relation is called an *inverse relation* of the original relation.

> For every relation r, there is an inverse relation denoted by r^{-1}, such that:
> $r^{-1} = \{(x, y) \mid (y, x) \in r\}$

If the relation r is given by a set of ordered pairs, then r^{-1} can be formed by interchanging the order of the components in each ordered pair belonging to r.

> CAUTION: r^{-1} does not mean $\dfrac{1}{r}$.

EXAMPLE 1: Find the inverse of a relation given by a set of ordered pairs.

Problem ▶ Find the inverse relation r^{-1} of the relation r given by $\{(2, 7), (-3, 7), (5, 2)\}$.

Solution ▶ $r = \{(2, 7), (-3, 7), (5, 2)\}$
$r^{-1} = \{(7, 2), (7, -3), (2, 5)\}$ Interchange components of each ordered pair.

Note ▶ Because the inverse relation r^{-1} consists of all the ordered pairs of r with their components interchanged, it must be true that the domains and ranges have also been interchanged.

Example ▶ Write the domain and range for the relation $r = \{(2, 7), (-3, 7), (5, 2)\}$ and its inverse $r^{-1} = \{(7, 2), (7, -3), (2, 5)\}$.

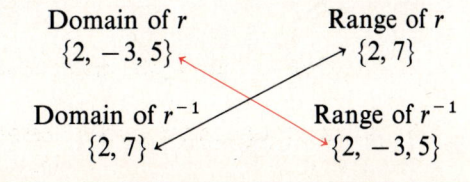

Domain of r Range of r
$\{2, -3, 5\}$ $\{2, 7\}$

Domain of r^{-1} Range of r^{-1}
$\{2, 7\}$ $\{2, -3, 5\}$

> CAUTION: The inverse of a function may not be a function.

458   Chapter 11   Functions

Example ▶ The relation $r = \{(2, 7), (-3, 7), (5, 2)\}$ is a function, but its inverse $r^{-1} = \{(7, 2), (7, -3), (2, 5)\}$ is not a function because the domain member 7 is paired with more than one range member.

Make Sure

Find the inverse of relations given by a set of ordered pairs.

See Example 1 ▶ 1. $\{(2, -2), (3, -3), (-4, 4)\}$ 2. $\{(2, 1), (3, 1), (4, 1), (1, 1)\}$

MAKE SURE ANSWERS: 1. $\{(-2, 2), (-3, 3), (4, -4)\}$ 2. $\{(1, 2), (1, 3), (1, 4), (1, 1)\}$

Because an inverse relation was defined by interchanging the x and y components of the relation, you can form the inverse of an equation containing two variables by interchanging the variables.

EXAMPLE 2: Find the inverse of an equation containing two variables.

Problem ▶ Find the inverse of $y = 3x + 7$.

1. Identify ▶ In $y = 3x + 7$, x is the independent variable and y is the dependent variable.

2. Interchange variables ▶ $x = 3y + 7$

3. Solve for dependent variable ▶ $\dfrac{x - 7}{3} = y$

Solution ▶ $y = \dfrac{1}{3}x - \dfrac{7}{3}$ is the inverse of $y = 3x + 7$.

Note ▶ In this example, both $y = 3x + 7$ and $y = \dfrac{1}{3}x - \dfrac{7}{3}$ are functions.

Another Example ▶ Find f^{-1} if f is specified by $f(x) = x^2 + x - 6$.

$y = x^2 + x - 6$ Think: Write $f(x) = x^2 + x - 6$ as $y = x^2 + x - 6$.

In $y = x^2 + x - 6$: x is the independent variable and y is the dependent variable.

$x = y^2 + y - 6$ Interchange variables

$0 = y^2 + y - 6 - x$ Solve for the dependent variable.

$0 = y^2 + y - (6 + x)$

$y = \dfrac{-1 \pm \sqrt{1^2 - 4 \cdot 1 \cdot [-(6 + x)]}}{2 \cdot 1}$ ⟵ the quadratic formula with $a = 1$, $b = 1$, and $c = -(6 + x)$

$f^{-1}(x) = \dfrac{-1 \pm \sqrt{4x + 25}}{2}$ is the inverse of $f(x) = x^2 + x - 6$.

Note ▶ In the above example, $f(x) = x^2 + x - 6$ is a function, but $f^{-1}(x) = \dfrac{-1 \pm \sqrt{4x + 25}}{2}$ is not a function because each value of $x > -\dfrac{25}{4}$ is paired with two y values.

11.5 Find the Inverse of a Function

Make Sure

Find the inverse of each equation.

See Example 2 ▶ **1.** $y = 3 - 2x$ **2.** $y = x^2 - 6$ **3.** $f(x) = x^2 + 3x + 4$

_____ _____ _____

MAKE SURE ANSWERS: 1. $y = -\dfrac{x}{2} + \dfrac{3}{2}$ **2.** $y = \pm\sqrt{x + 6}$ **3.** $f^{-1}(x) = \dfrac{-3 \pm \sqrt{4x - 7}}{2}$

The concept of symmetry about a line is important in the study of relations and their inverses. It can be shown that the two points (a, b) and (b, a) are symmetrical about the line $y = x$ as shown in Figure 11.7.

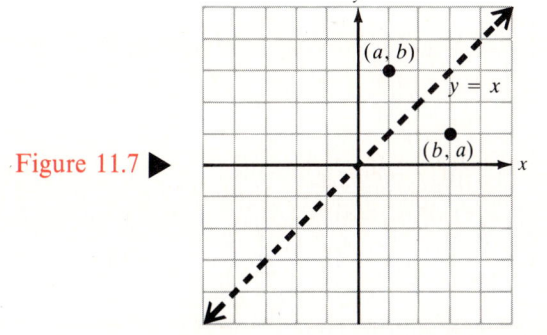

Figure 11.7 ▶

The points (a, b) and (b, a) are sometimes called *reflections* of each other about the line $y = x$.

Since the points (a, b) and (b, a) are symmetrical about the line $y = x$, it follows that the graphs of r and its inverse r^{-1} will be symmetrical about the line $y = x$.

EXAMPLE 3: Graph the inverse of a relation given by its graph.

Problem ▶ Graph the inverse t^{-1} of the relation t that is graphed on Grid C.

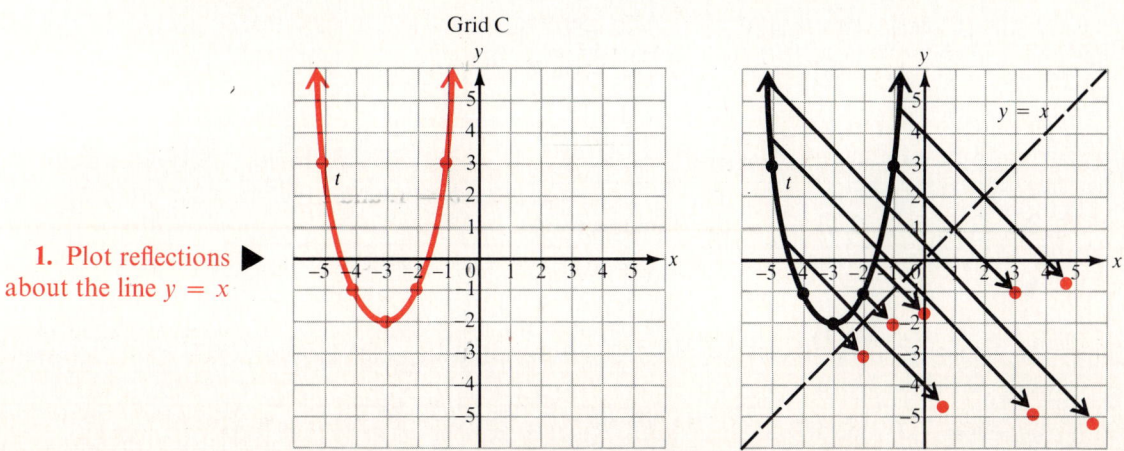

Grid C

1. Plot reflections about the line $y = x$

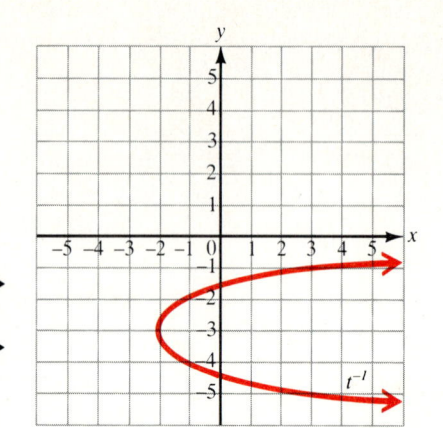

2. Sketch inverse ▶

Solution ▶

Note ▶ The vertical line test shows that the relation t is a function, but its inverse t^{-1} is not a function.

Recall ▶ In Chapter 10 you graphed parabolas that had an equation in $y = ax^2 + bx + c$ form.

Because $x = ay^2 + by + c$ is the inverse of $y = ax^2 + bx + c$, and because the graphs of inverse relations are symmetrical about the line $y = x$, you can graph equations in $x = ay^2 + by + c$ form using the following procedure.

> **Graphing Parabolas**
> To graph a parabola that has an equation of the form $x = ay^2 + by + c$:
> 1. The parabola opens to the right if $a > 0$, and opens to the left if $a < 0$.
> 2. Find the vertex using $(x, y) = \left(\dfrac{4ac - b^2}{4a}, -\dfrac{b}{2a} \right)$.
> 3. Find the x-intercept $(c, 0)$.
> 4. Find the y-intercept(s) by solving $0 = ay^2 + by + c$.
> 5. Find additional solutions of $x = ay^2 + by + c$.
> 6. Draw a smooth curve through the graphs of all of the ordered pairs found in Steps 2–5.

Example ▶ Graph $x = y^2 - 2y - 3$.

1. Identify ▶ In $x = y^2 - 2y - 3$, $a = 1$, $b = -2$, and $c = -3$. Because $a > 0$, the parabola opens to the right.

2. Find vertex ▶
$x = \dfrac{4ac - b^2}{4a}$ | $y = -\dfrac{b}{2a}$

$= \dfrac{4(1)(-3) - (-2)^2}{4(1)}$ | $= -\dfrac{-2}{2(1)}$

$= -4$ | $= 1$

$(-4, 1)$ ⟵ the vertex

3. Find x-intercept ▶ $(c, 0) = (-3, 0)$ ⟵ the x-intercept

4. Find y-intercept(s) ▶ $0 = y^2 - 2y - 3$

$0 = (y + 1)(y - 3)$

$0 = y + 1$ or $0 = y - 3$

$y = -1$ or $y = 3$

$(0, -1), (0, 3)$ ⟵ the y-intercepts

11.5 Find the Inverse of a Function

5. Find additional solutions

y	$x = y^2 - 2y - 3$
-3	12
-2	5
2	-3
4	5
5	12

6. Sketch graph

Solution

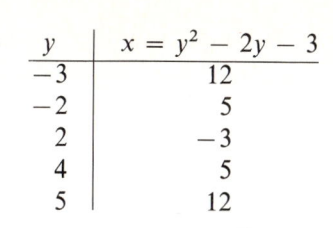

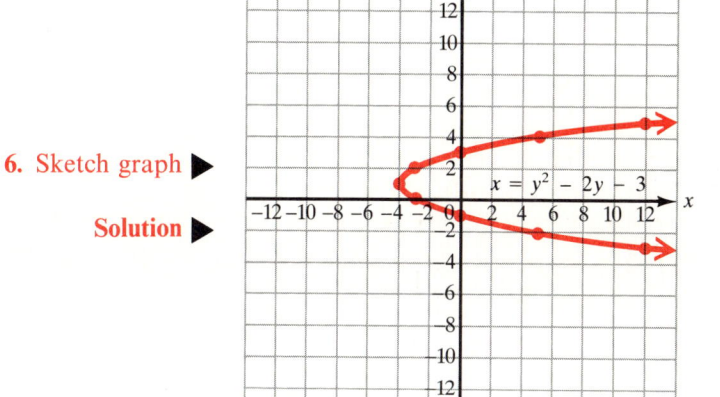

Draw a smooth curve through the graphs of all of the ordered pairs found in Steps 2–5.

The previous examples illustrate the fact that the inverse of a function may or may not be a function. To discover just what is required for a function to have an inverse that is also a function, consider the following functions and their inverses.

G	G^{-1}	H	H^{-1}
D → R	D → R	D → R	D → R
0 → 3	3 → 0	1 → 6	6 → 1
1 → 4	4 → 1	2 → 9	9 → 2
2 → 5	5 → 2	8 → 9	9 → 8

You should observe that G has an inverse G^{-1} that is a function, and that H has an inverse H^{-1} that is not a function. Notice that the function G differs from the function H in that each member of the range of G has only one domain member paired with it.

A function f is said to be a *one-to-one function* if each member of the range is paired with exactly one member of the domain.

> Only one-to-one functions have inverses that are functions.

Make Sure

Sketch the graph of the inverse of each relation given its graph.

See Example 3

1. Sketch r^{-1}

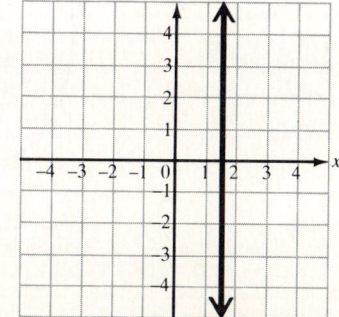

2. Sketch s^{-1}

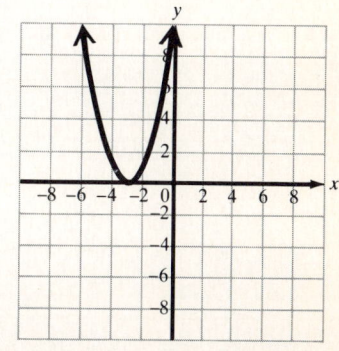

MAKE SURE ANSWERS: See Appendix Selected Answers.

462 Chapter 11 Functions

To identify a one-to-one function from its graph, you can use the *horizontal line test*.

> **The Horizontal Line Test**
> A function is a one-to-one function if each horizontal line in the coordinate system intersects the graph of the function in no more than one point (no points or one point).

EXAMPLE 4: Determine if a function is a one-to-one function.

Problem ▶ Which of the following graphs represent a one-to-one function?

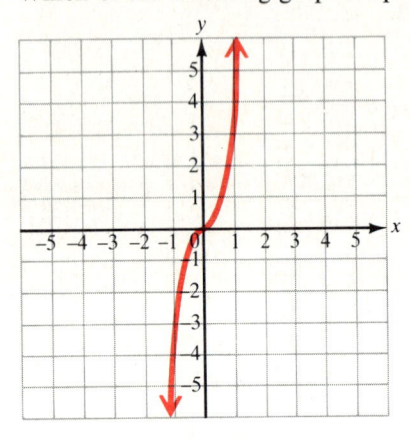

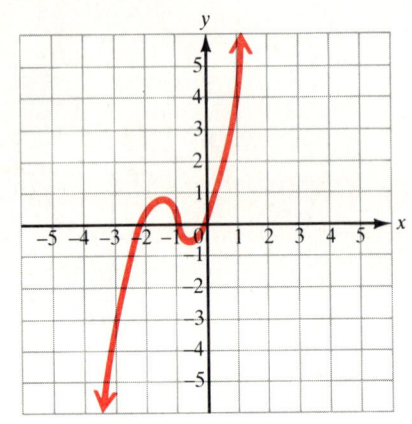

1. Use horizontal line test ▶

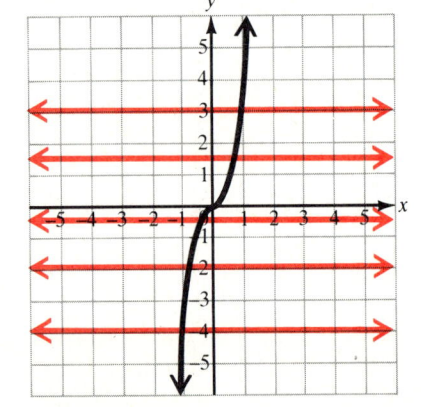

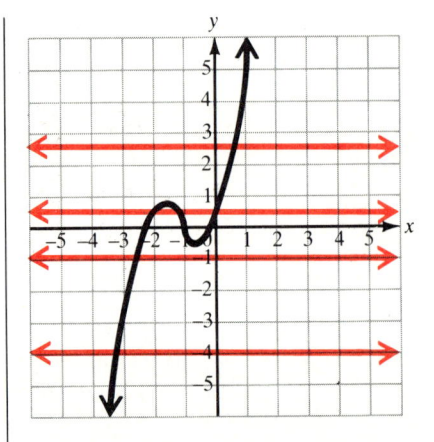

2. Interpret ▶ All horizontal lines intersect the graph at exactly one point.

Some horizontal lines intersect the graph at more than one point.

Solution ▶ This is the graph of a function that is a one-to-one function.

This is the graph of a function that is not a one-to-one function.

Make Sure

Determine if each function has an inverse that is a function.

See Example 4 ▶ 1. 2. 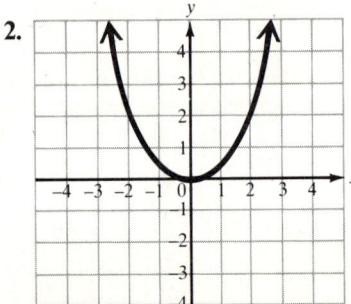 3.

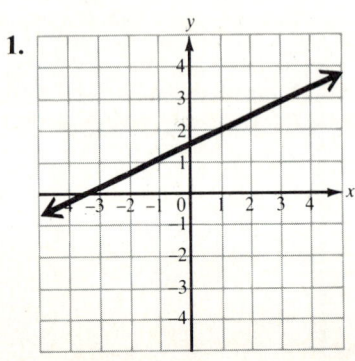

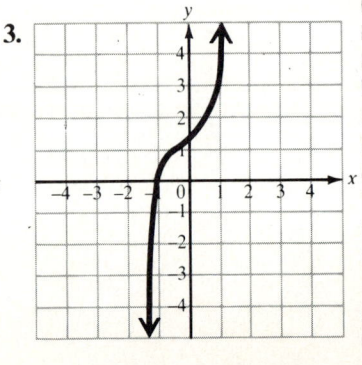

MAKE SURE ANSWERS: 1. one-to-one function **2.** not a one-to-one function **3.** one-to-one function

11.5 Practice

Set 1: Find the inverse of relations given by a set of ordered pairs.

1. $\{(2, 3), (1, 2), (0, 1), (1, 0)\}$
2. $\{(1, 0), (1, 1), (1, 2), (2, 1)\}$

3. $\{(2, 1), (3, 1), (4, 1), (1, 0)\}$
4. $\{(2, 4), (3, 4), (4, 4), (4, 0)\}$

5. $\{(-1, 0), (-1, 1), (1, 0), (0, 1)\}$
6. $\{(-3, 4), (-4, 3), (2, -1), (1, -2)\}$

7. $\{(0, -3), (2, 2), (-3, 0), (2, 3)\}$
8. $\{(5, -2), (-2, 3), (3, -4), (-4, 5)\}$

9. $\{(1, 1), (0, 0), (2, 2), (3, 3)\}$
10. $\{(-1, -1), (0, 0), (-1, 3), (5, 1)\}$

11. $\{(9, 1), (8, 2), (7, 3), (6, 4)\}$
12. $\{(-2, 8), (-3, 7), (4, -6), (5, -5)\}$

Set 2: Find the inverse of each equation.

13. $y = 5 + 3x$
14. $y = 2x - 6$
15. $f(x) = 7 - 6x$

16. $f(x) = 3x + 5$
17. $y = 2x^2 - 3x + 5$
18. $y = 3x^2 - 4x + 1$

19. $f(x) = x^2 - 4x + 4$
20. $f(x) = x^2 - 6x + 6$
21. $y = \dfrac{2x - 3}{3x + 4}$

22. $y = \dfrac{4x + 3}{2 - 5x}$
23. $x^2 - 6x + y = 0$
24. $x^2 + 4x - y = 0$

464 Chapter 11 Functions

Set 3: Sketch the graph of the inverse of each relation given its graph.

25. Sketch f^{-1}

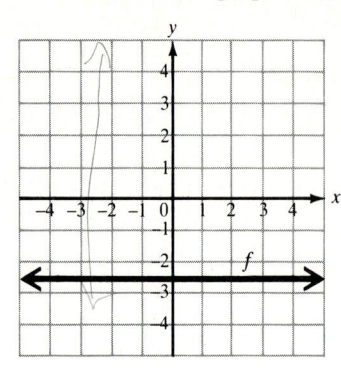

26. Sketch g^{-1}

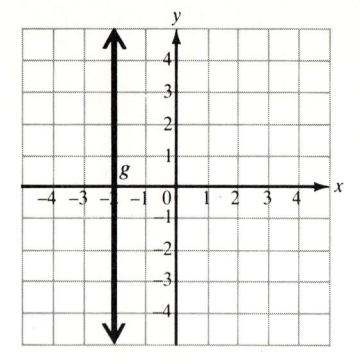

27. Sketch h^{-1}

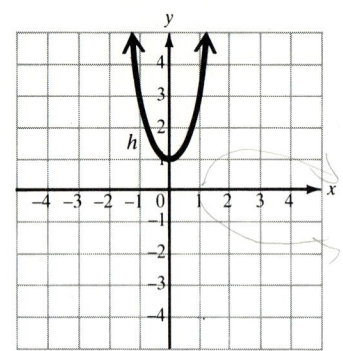

28. Sketch k^{-1}

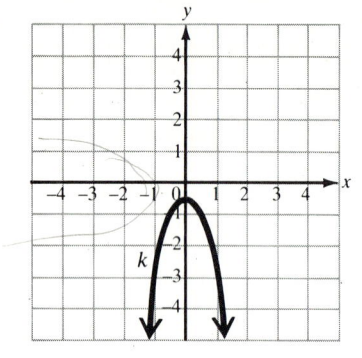

Set 4: Determine if each relation has an inverse that is a function.

29.

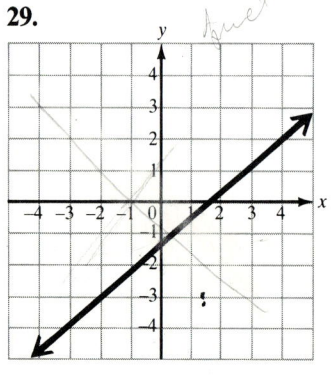

30.

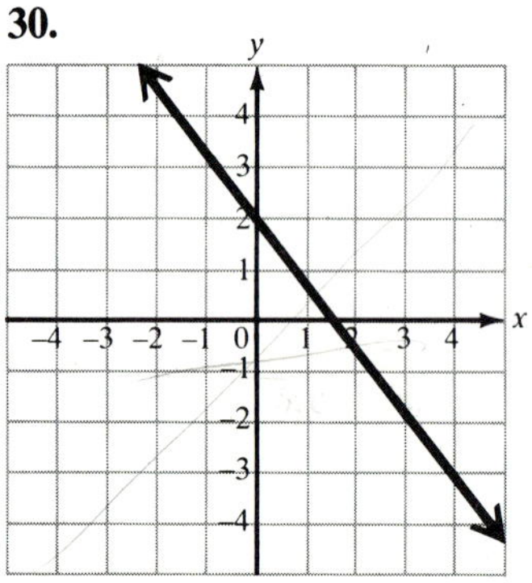

31.

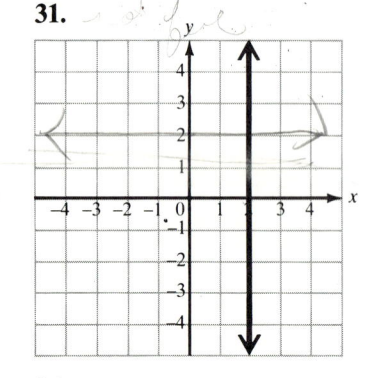

32.

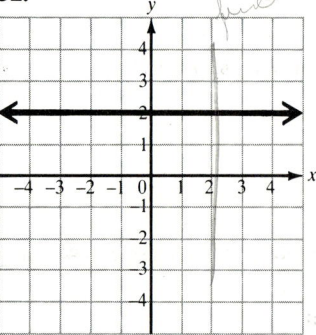

33.

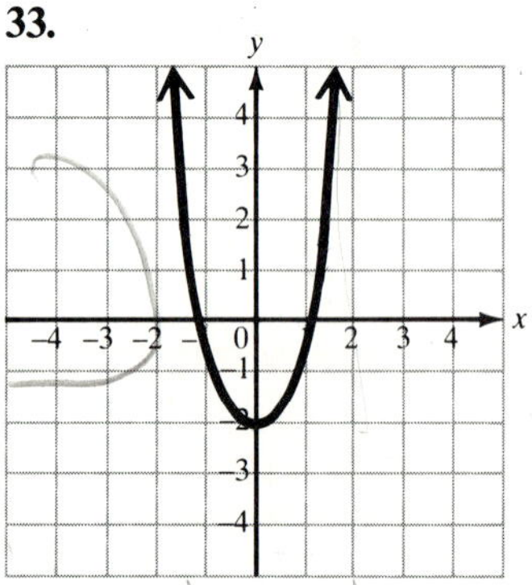

34.

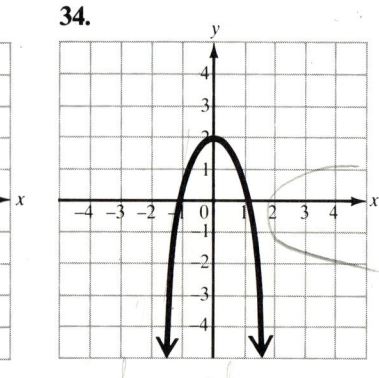

29. _____

30. _____

31. _____

32. _____

33. _____

34. _____

Review: Work these problems on a separate sheet of paper. Attach your work to this page.

Solve each equation for k. (See Lesson 3.1.)

35. $y = k\sqrt{x}$ for $y = 3$ and $x = 4$.

36. $A = k\sqrt{a}$ for $A = 5$ and $a = 3$.

37. $B = \dfrac{k}{\sqrt[3]{b}}$ for $B = \sqrt[3]{2}$ and $b = 4$.

38. $C = \dfrac{k\sqrt[3]{d}}{\sqrt{e}}$ for $C = 1$, $d = 27$, and $e = 8$.

Copyright © 1985 by Harcourt Brace Jovanovich, Inc. All rights reserved.

11.6 Solve Variation Problems

A variation is a special type of function. It relates variables by means of multiplication, division, or both. To use a variation, you must know how to represent it with an equation.

> A variation is called a *direct variation* if it can be represented by an equation of the form $y = kx^n$, where $x > 0$, $n > 0$, and $k \neq 0$. The number k is called the *variation constant* or the *constant of proportionality*.

Examples ▶ (a) The direct variation $d = 2r$ means d varies directly as r. The variation constant is 2.

(b) The direct variation $p = k\sqrt{L}$ means p varies directly as the square root of L. The variation constant is k.

(c) The direct variation $A = \pi r^2$ means A varies directly as the square of r. The variation constant is π.

Note ▶ Direct variations have the property that as one of the variables increases (decreases), the other variable also increases (decreases).

Example ▶ For the direct variation represented by $y = 3x$:

when x equals	0	1	2	3	5	← increasing value of x
y equals	0	3	6	9	15	← increasing value of y

To solve a problem that involves a direct variation, you first translate it to an equation.

EXAMPLE 1: Solve a direct variation.

Problem ▶ If y varies directly as x, and $y = 48$ when $x = 3$, find y when $x = 10$.

1. Translate ▶ y varies directly as x means $y = kx$ for some constant k. Think: y varies directly as x means the exponent $n = 1$.

2. Solve for k ▶ $y = kx$

$y \cdot \dfrac{1}{x} = kx \cdot \dfrac{1}{x}$ Multiply both members by $\dfrac{1}{x}$.

$\dfrac{y}{x} = k$

3. Evaluate for k ▶ $\dfrac{48}{3} = k$ Substitute the given values $y = 48$ and $x = 3$.

$16 = k$ ← variation constant

4. Substitute for k ▶ $y = 16x$ ← original statement of direct variation

5. Evaluate for y ▶ $y = 16 \cdot 10$ Substitute the given value $x = 10$.

$y = 160$

Solution ▶ $y = 160$ when $x = 10$.

466 Chapter 11 Functions

If both (x_1, y_1) and (x_2, y_2) are solutions of the direct variation $y = kx^n$, then both (1) $y_1 = k(x_1)^n$ and (2) $y_2 = k(x_2)^n$ must be true statements. Dividing both members of equation (1) by the respective members of equation (2) produces a proportion as shown below.

$$\frac{y_1}{y_2} = \frac{k(x_1)^n}{k(x_2)^n}$$

$$\frac{y_1}{y_2} = \frac{\cancel{k}}{\cancel{k}} \cdot \frac{(x_1)^n}{(x_2)^n}$$

$$\frac{y_1}{y_2} = \frac{(x_1)^n}{(x_2)^n} \quad \longleftarrow \text{a proportion associated with the direct variation } y = kx^n$$

You can use the above proportion as another way to solve direct variation problems.

Example ▶ If y varies directly as x, and $y = 48$ when $x = 3$, use a proportion to find y when $x = 10$.

1. Write proportion ▶ $\quad \dfrac{y_1}{y_2} = \dfrac{x_1}{x_2} \qquad$ Think: y varies directly as x means the exponent $n = 1$.

2. Substitute ▶ $\quad \dfrac{48}{y_2} = \dfrac{3}{10} \qquad$ Think: $y_1 = 48$, $x_1 = 3$, and $x_2 = 10$.

3. Solve ▶ $\quad 48 \cdot 10 = 3 \cdot y_2$

$$\frac{48 \cdot 10}{3} = y_2$$

$$160 = y_2$$

Solution ▶ $y = 160$ when $x = 10$

Another Example ▶ If y varies directly as the square of x, and $y = 325$ when $x = 5$, find y when $x = 3$.

Method I (Solve for the variation constant)

$$y = kx^2$$

$$y \cdot \frac{1}{x^2} = kx^2 \cdot \frac{1}{x^2}$$

$$\frac{y}{x^2} = k$$

$$\frac{325}{5^2} = k$$

$$13 = k \quad \longleftarrow \text{the variation constant}$$

$$y = kx^2 \quad \longleftarrow \text{original variation}$$

$$y = 13(3)^2$$

$$y = 117$$

Method II (Use the associated proportion)

$$\frac{y_1}{y_2} = \frac{(x_1)^2}{(x_2)^2}$$

$$\frac{325}{y_2} = \frac{5^2}{3^2} \qquad \text{Substitute}$$

$$325 \cdot 9 = 25 \cdot y_2$$

$$\frac{325 \cdot 9}{25} = y_2$$

$$117 = y_2$$

11.6 Solve Variation Problems

Make Sure

Solve each direct variation.

See Example 1 ▶ **1.** If y varies directly as x, and $y = 36$ when $x = 4$, find y when $x = 5$.

2. If H varies directly as the square root of g, and $H = 48$ when $g = 9$, find H when $g = 5$. (Hint: $H = k\sqrt{g}$)

MAKE SURE ANSWERS: 1. 45 2. $16\sqrt{5}$

A variation is called an *inverse variation* if it can be represented by an equation of the form $y = \dfrac{k}{x^n}$, where $x > 0$, $n > 0$, and the variation constant $k \neq 0$.

Examples ▶ (a) The inverse variation $p = \dfrac{12}{a}$ means p varies inversely as a. The variation constant is 12.

(b) The inverse variation $R = \dfrac{1.6 \times 10^{-6}}{d^2}$ means R varies inversely as the square of d.

(c) The inverse variation $c = \dfrac{0.03}{\sqrt{t}}$ means c varies inversely as the square root of t.

Note ▶ Inverse variations have the property that as one of the variables increases (decreases), the other variable decreases (increases).

Example ▶ For the inverse variation represented by $y = \dfrac{16}{x^2}$:

when x equals	1	2	3	4	⟵ increasing values of x
y equals	16	4	$\frac{16}{9}$	1	⟵ decreasing values of y

EXAMPLE 2: Solve an inverse variation.

Problem ▶ If l varies inversely as w^2, and $l = 24$ when $w = 4$, find l when $w = 6$.

1. Translate ▶ l varies inversely as w^2 means $l = \dfrac{k}{w^2}$ for some constant k.

2. Evaluate for k ▶ $lw^2 = k$

$24(4)^2 = k$

$384 = k$ ⟵ variation constant

3. Substitute k ▶ $l = \dfrac{384}{w^2}$

4. Evaluate for l ▶ $l = \dfrac{384}{6^2}$

$l = \dfrac{32}{3}$

Solution ▶ $l = \dfrac{32}{3}$ when $w = 6$.

You can also solve inverse variations by using a proportion. If both (x_1, y_1) and (x_2, y_2) are solutions of the inverse variation $y = \dfrac{k}{x^n}$, then both (1) $y_1 = \dfrac{k}{(x_1)^n}$ and (2) $y_2 = \dfrac{k}{(x_2)^n}$ must be true statements. Dividing both members of equation (1) by the respective members of equation (2) produces a proportion as shown below.

$$\frac{y_1}{y_2} = \frac{\dfrac{k}{(x_1)^n}}{\dfrac{k}{(x_2)^n}}$$

$$\frac{y_1}{y_2} = \frac{k}{(x_1)^n} \div \frac{k}{(x_2)^n}$$

$$\frac{y_1}{y_2} = \frac{k}{(x_1)^n} \cdot \frac{(x_2)^n}{k}$$

$$\frac{y_1}{y_2} = \frac{(x_2)^n}{(x_1)^n} \quad \longleftarrow \text{ a proportion associated with the inverse variation } y = \frac{k}{x^n}$$

It is important to note the difference between the proportion associated with direct variations: $\dfrac{y_1}{y_2} = \dfrac{(x_1)^n}{(x_2)^n}$, and the proportion associated with inverse variations: $\dfrac{y_1}{y_2} = \dfrac{(x_2)^n}{(x_1)^n}$. The following example illustrates how to use a proportion to solve an inverse variation.

Example ▶ If y varies inversely as the square of x, and $y = 9$ when $x = 3$, find y when $x = 15$.

1. Write proportion ▶ $\dfrac{y_1}{y_2} = \dfrac{(x_2)^2}{(x_1)^2}$ Think: y varies inversely as the square of x means the exponent $n = 2$.

2. Substitute ▶ $\dfrac{9}{y_2} = \dfrac{15^2}{3^2}$ Think: $y_1 = 9$, $x_1 = 3$, and $x_2 = 15$.

3. Solve ▶ $81 = 225y_2$

$\dfrac{9}{25} = y_2$

Solution ▶ $y = \dfrac{9}{25}$ when $x = 15$.

Make Sure

Solve each inverse variation.

See Example 2 ▶
1. If V varies inversely as P, and $V = 50$ when $P = 48$, find V when $P = 32$.

2. If Z varies inversely as the cube root of y, and $Z = 54$ when $y = \frac{8}{27}$, find Z when $y = \frac{3}{4}$.

MAKE SURE ANSWERS: 1. 75   2. 12√3/36

11.6 Solve Variation Problems

Joint variations are functions containing more than one independent variable, and the independent variables are related by means of multiplication.

Examples ▶ (a) The joint variation $y = 5wx$ means y varies jointly as w and x.
(b) The joint variation $V = r^2 h$ means V varies jointly as r^2 and h.
(c) The joint variation $V = lwh$ means V varies jointly as l, w, and h.

A joint variation is similar to a direct variation, so it can be solved either by evaluating for the variation constant or by using the associated proportion.

Example ▶ If y varies jointly as u and v^2, and $y = 6480$ when $u = 15$ and $v = 6$, find y when $u = 3$ and $v = 8$.

Method I (evaluate the variation constant)

$$y = kuv^2$$

$$\frac{y}{uv^2} = k$$

$$\frac{6480}{15 \cdot 6^2} = k$$

$$12 = k$$

$$y = 12uv^2$$

$$y = 12 \cdot 3(8)^2$$

$$y = 2304$$

Method II (use the associated proportion)

$$\frac{y_1}{y_2} = \frac{u_1(v_1)^2}{u_2(v_2)^2} \quad \longleftarrow \text{associated proportion}$$

$$\frac{6480}{y_2} = \frac{15 \cdot 6^2}{3 \cdot 8^2}$$

$$\frac{6480 \cdot 3 \cdot 8^2}{15 \cdot 6^2} = y_2$$

$$2304 = y_2$$

Solution ▶ $y = 2304$ when $u = 3$ and $v = 8$.

Make Sure

Solve a joint variation.

See Previous Example ▶

1. If w varies jointly as x and y, and $w = 36$ when $x = 4$ and $y = 5$, find w when $x = 5$ and $y = 7$.

2. If p varies jointly as q and the square root of r and $p = 120$ when $q = 5$ and $r = 9$, find p when $q = 7$ and $r = 4$.

MAKE SURE ANSWERS: 1. 63 2. 112

When different kinds of variations occur together, they are called *combined variations*.

Examples ▶ (a) The combined variation $y = \dfrac{7u}{v}$ means y varies directly as u and inversely as v.

(b) The combined variation $S = \dfrac{2wd^2}{l^3}$ means S varies jointly as w and the square of d, and inversely as the cube of l.

(c) The combined variation $I = \dfrac{kx}{w\sqrt{z}}$ means I varies directly as x, and inversely as the product of w and the square root of z. The variation constant is represented by k.

You generally solve combined variations by evaluating the variation constant.

EXAMPLE 3: Solve a combined variation.

Problem ▶ If y varies jointly as x and w and inversely as the square of z, and $y = 30$ when $x = 3$, $w = 5$, and $z = 2$, find y when $x = 7$, $w = 6$, and $z = 4$.

1. Translate ▶ y varies jointly as x and w and inversely as the square of z means:

$$y = \frac{kxw}{z^2}$$

2. Evaluate for k ▶ $\dfrac{yz^2}{xw} = k$

$\dfrac{30(2)^2}{3 \cdot 5} = k$ Think: $y = 30$ when $x = 3$, $w = 5$, and $z = 2$.

$8 = k$ ⟵ variation constant

3. Substitute k ▶ $y = \dfrac{8xw}{z^2}$ ⟵ original variation

4. Evaluate for y ▶ $y = \dfrac{8 \cdot 7 \cdot 6}{4^2}$

$y = 21$

Solution ▶ $y = 21$ when $x = 7$, $w = 6$, and $z = 4$.

Make Sure

Solve each combined variation.

See Example 3 ▶

1. If y varies directly as x and inversely as the square root of z, and $y = 8$ when $x = 4$ and $z = 9$, find y when $x = 8$ and $z = 16$.

2. If y varies jointly as m and the square of n and inversely as the cube of r, and $y = 12$ when $m = 4$, $n = 3$, and $r = 2$, find y when $m = 6$, $n = 2$, and $r = 3$.

11.6 Practice

Set 1: Solve each direct variation.

1. If a varies directly as b, and $a = 32$ when $b = 4$, find a when $b = 3$.

2. If c varies directly as d, and $c = 24$ when $d = 6$, find c when $d = 4$.

3. If m varies directly as the square of n, and $m = 48$ when $n = 4$, find m when $n = 3$.

4. If p varies directly as the square of q, and $p = 64$ when $q = 2$, find p when $q = 3.5$.

5. If a varies directly as the sum of b and 7, and $a = 19$ when $b = 5$, find a when $b = 20$.

6. If c varies directly as the difference of d and 5, and $c = 13$ when $d = 6$, find c when $d = 9$.

Set 2: Solve each inverse variation.

7. If a varies inversely as b, and $a = 3$ when $b = 4$, find a when $b = 3$.

8. If c varies inversely as d, and $c = 6$ when $d = 4$, find c when $d = 6$.

9. If g varies inversely as the cube of x, and $g = 12$ when $x = \frac{1}{2}$, find g when $x = 2\frac{2}{3}$.

10. If y varies inversely as the cube of z, and $y = 2\frac{1}{2}$ when $z = \frac{1}{5}$, find y when $z = 2$.

11. If a varies inversely as the square root of b, and $a = 4$ when $b = 2$, find a when $b = 3$.

12. If c varies inversely as the square root of d, and $c = \sqrt{6}$ when $d = 2$, find c when $d = 5$.

Set 3: Solve each combined variation.

13. If a varies jointly as b and c, and $a = 54$ when $b = 3$ and $c = 9$, find a when $b = 8$ and $c = 4$.

14. If x varies jointly as y and z, and $x = 80$ when $y = 5$ and $z = 4$, find x when $y = 10$ and $z = 5$.

15. If f varies jointly as g and the square of h, and $f = 64$ when $g = 1$ and $h = 4$, find f when $g = 32$ and $h = 1$.

16. If p varies jointly as q and the square of r, and $p = \frac{3}{16}$ when $q = \frac{2}{3}$ and $r = \frac{3}{4}$, find p when $q = \frac{2}{5}$ and $r = \frac{1}{2}$.

17. If V varies directly as e and inversely as the square root of d, and $V = 10$ when $e = 100$ and $d = 0.01$, find V when $e = 60$ and $d = 0.02$.

18. If V varies directly as t and inversely as the square root of p, and $V = 30$ when $t = 300$ and $p = 5$, find V when $t = 50$ and $p = 4$.

19. If W varies jointly as v and the square of u and inversely as the cube of t, and $W = \frac{9}{4}$ when $u = 3$, $v = 4$, and $t = 2$, find W when $u = 5$, $v = 6$, and $t = 3$.

20. If A varies jointly as c and the square of b and inversely as the cube of d, and $A = \frac{18}{5}$ when $b = \frac{2}{3}$, $c = \frac{3}{5}$, and $d = \frac{1}{3}$, find A when $b = \frac{1}{4}$, $c = \frac{2}{3}$, and $d = \frac{1}{2}$.

21. If H varies directly as the square root of g and inversely as the product of h and the square of f, and $H = 14$ when $g = 0.49$, $f = 0.2$, and $h = 7$, find H when $g = 0.25$, $f = 0.3$, and $h = 5$.

22. If P varies directly as the square root of q and inversely as the product of s and the square of r, and $P = \frac{9}{2}$ when $q = \frac{4}{9}$, $r = \frac{4}{3}$, and $s = \frac{1}{3}$, find P when $q = \frac{1}{2}$, $r = \frac{2}{3}$, and $s = \frac{3}{5}$.

23. If A varies jointly as W and the square of d and inversely as the square root of a, and $A = 800$ when $W = 3$, $d = 4$, and $a = 100$, find A when $W = 4$, $d = 5$, and $a = 500$.

24. If K varies jointly as a and the square of b and inversely as the square root of c, and $K = \frac{9}{14}$ when $a = \frac{6}{7}$, $b = \frac{3}{4}$, and $c = \frac{1}{4}$, find K when $a = \frac{2}{3}$, $b = \frac{2}{5}$, and $c = \frac{1}{5}$.

Review: Work these problems on a separate sheet of paper. Attach your work to this page.

Evaluate expressions containing rational exponents. (See Lesson 8.1.)

25. 2^{-2}
26. 2^{-1}
27. 2^0
28. 2^2
29. 2^3
30. 2^4
31. $\frac{1}{2}^{-2}$
32. $\left(\frac{1}{2}\right)^{-1}$
33. $\left(\frac{1}{2}\right)^0$
34. $\left(\frac{1}{2}\right)^2$
35. $\left(\frac{1}{2}\right)^3$
36. $\left(\frac{1}{2}\right)^4$
37. $\left(\frac{2}{3}\right)^{-2}$
38. $\left(\frac{2}{3}\right)^{-1}$
39. $\left(\frac{2}{3}\right)^0$
40. $\left(\frac{2}{3}\right)^2$
41. $\left(\frac{2}{3}\right)^3$
42. $\left(\frac{2}{3}\right)^4$
43. $\left(-\frac{3}{4}\right)^{-2}$
44. $\left(-\frac{3}{4}\right)^{-1}$
45. $\left(-\frac{3}{4}\right)^0$
46. $\left(-\frac{3}{4}\right)^2$
47. $\left(-\frac{3}{4}\right)^3$
48. $\left(-\frac{3}{4}\right)^4$

Problem Solving 16: Solve Applied Variation Problems

Because applied variation problems come up so often in virtually every field of science and engineering, it is important that you know how to solve them.

EXAMPLE: Solve this applied variation problem.

1. Read and identify ▶ *Electrical Resistance:* At a fixed temperature and chemical composition, the resistance R of a wire varies directly as the length l and inversely as the square of the diameter d. At 20°C, the resistance of 40 feet of copper wire with a diameter of 20 mils (20 thousandths of an inch) is 1 ohm (Ω). What is the resistance of 500 feet of copper wire at 20°C if the diameter is 25 mils?

   Remember, circle the facts and underline the variations.

2. Understand ▶ That resistance R varies directly as the length l and inversely as the square of the diameter d means: $R = \dfrac{kl}{d^2}$ ← general variation formula for any type of wire

3. Decide ▶ To find the resistance R, you **first evaluate for k** and **then evaluate for R**.

4. Evaluate for k ▶ $R = \dfrac{kl}{d^2}$

   $\dfrac{Rd^2}{l} = k$ Solve for k.

   $\dfrac{1(20)^2}{40} = k$ Substitute: $R = 1$ (Ω), $d = 20$ (mils), and $l = 40$ (ft)

   $10 = k$ ← variation constant

5. Evaluate for R ▶ $R = \dfrac{10l}{d^2}$ ← specific variation formula for copper wire at 20°C

   $= \dfrac{10(500)}{(25)^2}$ Substitute: $l = 500$ (ft) and $d = 25$ (mils)

   $= 8$

6. Interpret ▶ $R = 8$ means that at 20°C the resistance of 500 feet of 25-mil copper wire is 8 ohms or 8 Ω.

   Note ▶ Once the specific applied variation formula has been found, you can use it to answer many other practical types of questions.

   Example ▶ At 20°C, how much 40-mil copper wire is needed to create a coil with a resistance of $\frac{1}{2}$ ohm?

   $R = \dfrac{10l}{d^2}$ ← specific variation formula for copper wire at 20°C

   $\dfrac{Rd^2}{10} = l$ Solve for l.

   $\dfrac{\frac{1}{2}(40)^2}{10} = l$ Substitute: $R = \frac{1}{2}$ (Ω) and $d = 40$ (mils).

   $80 = l$ ← 80 feet of 40-mil copper wire is needed to create a resistance of $\frac{1}{2}$ ohm

474 Chapter 11 Functions

Practice: Solve each applied variation problem. Round to the nearest tenth when necessary.

1a. _____

b. _____

c. _____

2a. _____

b. _____

c. _____

3a. _____

b. _____

c. _____

4a. _____

b. _____

c. _____

5a. _____

b. _____

c. _____

6a. _____

b. _____

c. _____

1. *Electrical Resistance* (See the Example): At 20°C, the resistance of $28\frac{1}{2}$ feet of 19-mil silver wire is $\frac{3}{4}$ ohm. a) Find the specific variation formula for silver wire at 20°C. b) What is the resistance of 100 feet of 10-mil silver wire at 20°C? c) At 20°C, how many feet of 1-mil silver wire is needed for a resistance of 1 ohm?

2. *Hooke's Law:* The distance d that a vertical spring stretches varies directly with the weight w hanging from it. A 50 kg weight will stretch a certain spring 5 cm. a) Find the specific variation formula for the given spring. b) How far will the spring stretch if a 20 kg weight is hung from it? c) How heavy a weight must be used to stretch the spring 10 cm?

3. *Weight of an Astronaut:* The weight W of an object varies inversely as the square of the distance d from the object to the center of the earth. A certain astronaut weighs 130 pounds at sea level (4000 miles from the center of the earth). a) Find the specific variation formula for this astronaut. b) How much does the astronaut weigh at an altitude of 22,300 miles in a *geostationary orbit* (hovering over a stationary point on the earth's surface)? c) At what altitude will the astronaut weigh half of her earth weight?

4. *Distance Seen to the Horizon:* The distance d that a person can see to the ocean horizon varies directly as the square root of the person's eye level h above sea level. A person standing on the ocean shore can see 3 miles to the ocean horizon if the eye level is 6 feet. a) Find the specific variation formula for any person with normal eyesight. (Hint: Use $d = k\sqrt{h}$) b) How far can a person see to the ocean horizon if his eye level is 5 feet above sea level? c) How far above sea level must a person's eye level be to see 6 miles to the ocean horizon?

5. *Boyle's Law:* At a constant temperature, the pressure P of a compressed gas varies inversely as the volume V of gas. At 350°K (Kelvin), the pressure on 2 liters of a certain gas is 1.5 atmospheres. a) Find the specific variation formula for the given gas at 350°K. At 350°K, b) what pressure is needed to reduce the volume of the gas to 1.25 liters? c) What is the volume of the gas when 2 atmospheres of pressure is applied?

6. *Ohm's Law:* An electrical current I in amperes (A) varies directly as the electromotive force E in volts (V) and inversely as the resistance R in ohms (Ω). The current flowing when a 6 Ω wire is connected to a 12 V battery is 2 A. a) Find the specific variation formula for any given current. b) What is the resistance of a toaster that is designed to draw 8 A when used at 110 V? c) How much current is flowing through a lamp if the light bulb filament has a resistance of 220 Ω and the lamp is used on a 110 V line?

Copyright © 1985 by Harcourt Brace Jovanovich, Inc. All rights reserved.

7a. _____

b. _____

8a. _____

b. _____

9a. _____

b. _____

c. _____

10a. _____

b. _____

c. _____

11a. _____

b. _____

c. _____

12a. _____

b. _____

c. _____

7. *Coulomb's Law:* The force F between two small electrical charges q and q_1 varies jointly as the charges and inversely as the square of the distance d between them. a) Write the general variation formula. b) If the distance between two particular charges is tripled and one of the charges is doubled, by what factor must the other charge be increased to maintain the original force between them?

8. *Newton's Law of Universal Gravitation:* The force F of attraction between any two masses m and m_1 in the universe varies jointly as masses and inversely as the square of the distance d between them. a) Write the general variation formula. b) If the distance between two particular masses in the universe is halved and one of the masses is tripled, by what factor must the other mass be decreased to maintain the original attractive force between them?

9. *Radio Waves:* The wavelength W of a radio wave varies inversely as its frequency f. A radio wave with a frequency of 720 kilohertz (kHz) has a wavelength of 500 meters (m). a) Find the specific variation formula for this radio wave. b) What radio frequency will have a wavelength of 1000 m? c) How long is the wavelength of a 1000 kHz radio frequency?

10. *Light Intensity:* The intensity I of light varies inversely as the square of the distance d from the source of the light. The intensity of a certain light bulb is 10 lumens at a distance of 10 feet. a) Find the specific variation formula for this light bulb. b) How far from the light bulb will the intensity be 5 lumens? c) What is the intensity of light 5 feet from the same light bulb?

11. *Music Frequency:* The frequency f of a string under constant tension varies inversely as the length l of the string. A certain string has a frequency of 440 vibrations per second (concert A) when its length is 32 inches. a) Find the specific variation formula for this string under the given constant tension. Under the same constant tension, b) what length should the string be to produce 256 vibrations per second (middle C)? c) What would be the frequency of the string if it were cut in half?

12. *Water Pressure:* The water pressure P varies directly as the depth d beneath the surface of water. The water pressure at 20 feet below the water surface is 1248 pounds per square foot (lb/ft^2). a) Find the specific variation formula for water pressure at any depth. b) The world record for scuba diving is 437 feet. What was the water pressure on the scuba diver who set the world record? c) The deepest point in the ocean ever reached by a person was in the Bathyscaphe diving machine. The pressure on the diving machine was 2,234,044.8 lb/ft^2. How deep was this dive?

Copyright © 1985 by Harcourt Brace Jovanovich, Inc. All rights reserved.

13a. _____

b. _____

c. _____

14a. _____

b. _____

c. _____

d. _____

15a. _____

b. _____

c. _____

d. _____

16a. _____

b. _____

c. _____

13. *Period of a Pendulum:* The period T (the time it takes a pendulum to make one complete swing from left to right and back again) of a pendulum varies directly as the square root of the length l and inversely as the square root of the gravitational pull. On earth, the gravitational pull is about 32 feet per second per second (32 ft/s^2), and an 8-foot-long pendulum has a period of π seconds. a) Find the specific variation formula for a pendulum on earth. On earth: b) What is the period of a 2-foot-long pendulum? c) What is the length of a pendulum that beats seconds (has a 2-second period)?

14. *Strength of a Beam:* The safe load L of a horizontal homogeneous rectangular beam (the amount of weight it supports without breaking) varies jointly as the beam's width w and the square of its length l. A certain homogeneous 2″ × 4″ × 6′ (width = 2 inches, thickness = 4 inches, length = 6 feet) beam has a safe load of one ton (2000 pounds). a) Find the specific variation formula for this type of homogeneous beam. For this type of homogeneous beam, what is the safe load for a beam that measures b) 2″ × 4″ × 8′? c) 4″ × 4″ × 6′? d) 1″ × 4″ × 6′?

15. *Amount of Work:* The amount of work W varies jointly as the distance d that an object is moved and the weight w of the object. The work needed to lift 100 pounds 5 feet is 500 foot-pounds (ft-lb). a) Find the specific variation formula for the work needed to move any object. b) How much work is done by pulling 50 pounds 20 feet? c) If it takes 100 ft-lb to lift an object 6 inches, how much does the object weigh? d) If it takes 100 ft-lb to push a 1 ton object, how far was the object pushed?

16. *Elevator Lifting Power:* The time t needed for an elevator to lift a weight varies jointly as the weight w and the distance d that the weight is lifted, and inversely as the horsepower P of the motor. It takes 30 seconds for a certain elevator to lift 5000 pounds a distance of 100 feet with a 500 horsepower (H.P.) motor. a) Find the specific variation formula for this elevator. b) How many horsepower are needed to lift the same weight the same distance in 20 seconds? c) How long would it take any elevator to lift twice the weight half the distance?

According to Guinness

GREATEST HAUL
THE GREATEST LOAD HAULED BY A PAIR OF CLYDESDALE DRAUGHT HORSES WAS 50 PINE LOGS TOTALING 48 TONS HAULED ON A SLEDGE LITTER 275 YARDS ACROSS SNOW ON THE NESTER ESTATE AT EWEN, MICHIGAN, ON FEBRUARY 26, 1893.

FASTEST ELEVATOR
THE FASTEST DOMESTIC PASSENGER ELEVATORS ARE THE EXPRESS ELEVATORS TO THE 60TH FLOOR OF THE 787.4-FT.-TALL "SUNSHINE 60" BUILDING, IKEBUKURO, TOKYO, JAPAN. THEY WERE BUILT BY MITSUBISHI CORP. AND OPERATE AT A SPEED OF 2000 FT. PER MINUTE.

17. _____

18. _____

17. How much work in foot-pounds was done on the greatest load ever hauled? (Use 1 ton = 2000 pounds and 1 yard = 3 feet. See Problem 15.)

18. What is the horsepower of the fastest elevator's motor if speed is based on a load capacity of 5000 pounds? (Assume the elevator is lifted the full height of the building. See Problem 16.)

Copyright © 1985 by Harcourt Brace Jovanovich, Inc. All rights reserved.

Name _____ Date _____ Class _____

Chapter 11 Review

		What to Review if You Have Trouble		
Objectives		Lesson	Example	Page

Find the domain and range of relations ▶

1. Find the **a.** domain and **b.** range of the relation given by $\{(-1, 3), (2, 4), (2, -5), (3, 7)\}$.
   a. _____ b. _____   11.1   1   432

2. Find the **a.** domain and **b.** range of the relation given by the graph on Grid A.
   a. _____ b. _____   11.1   2   433

3. Find the **a.** domain and **b.** range of the relation given by $y = \sqrt{3-x} + 4$.
   a. _____ b. _____   11.1   3   434

Determine if a relation is a function ▶

4. Does $\{(-3, 1), (-3, 7), (2, 5)\}$ represent a function? _____   11.2   1   437

5. Does the graph on Grid B represent a function? _____   11.2   2   438

6. Which equation(s) **a.** $x^2 - y^2 = 4$, **b.** $5x^2 + y = 3$, or **c.** $x^3 - y^3 = 0$ represent functions? _____   11.2   3   439

Find functional values ▶

7. Find **a.** $f(3)$, **b.** $f(-1)$, and **c.** $f(h - 2)$, given $f(x) = 2x^2 - 5x + 7$.   11.3   1   443

   a. _____ b. _____ c. _____

8. Estimate **a.** $g(-5)$, **b.** $g(-1)$, and **c.** $g(4)$, given the graph g on Grid C.   11.3   2   444

   a. _____ b. _____ c. _____

9. Find $f(g(2))$ given $f(x) = 3x - 2$ and $g(x) = x^2 + 1$. _____   11.3   3   445

Graph functions ▶

10. Graph $f(x) = |1 - x|$ for $-4 \leq x \leq 4$ on Grid D.   11.4   1   451

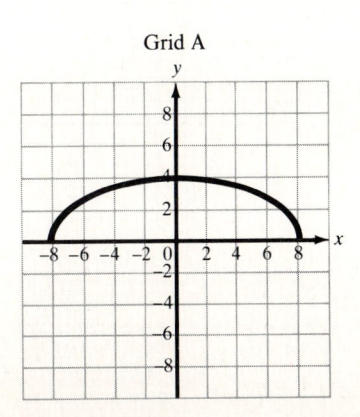

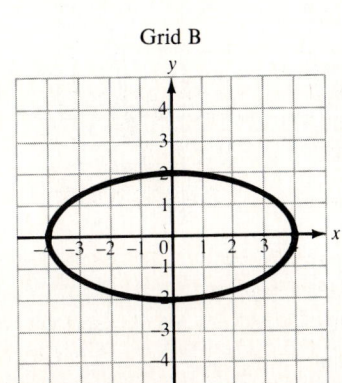

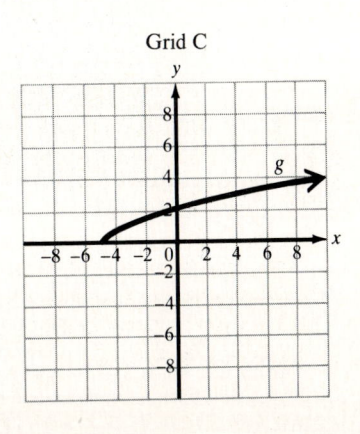

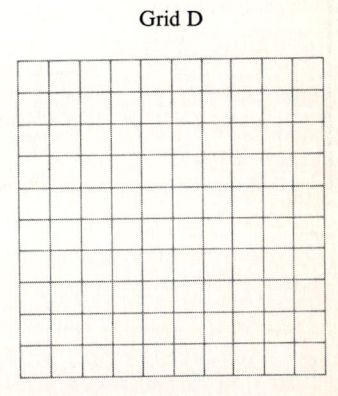

Grid A   Grid B   Grid C   Grid D

Copyright © 1985 by Harcourt Brace Jovanovich, Inc. All rights reserved.

478 Chapter 11 Functions

Find the inverse of a relation

11. Graph $h(x) = \begin{cases} -x & \text{if } x < 0 \\ x^2 - 4 & \text{if } 0 \leq x \leq 2 \\ 3 & \text{if } 2 < x \end{cases}$ on Grid E.   11.4   2   452

12. Graph $F(x) = |2x - 6|$ on Grid F. Use the method of renaming the equation as a two-part equation.   11.4   3   453

13. Find the inverse of the relation given by $\{(-3, 1), (2, 5), (5, -7)\}$.   11.5   1   457

14. Find the inverse of the relation given by $y = 4x - 5$. Write the inverse as an equation solved for y.   11.5   2   458

15. Graph the inverse of the relation given by the graph on Grid G. Put your graph on Grid G.   11.5   3   459

Identify one-to-one functions

16. Does the graph on Grid H specify a function that is a one-to-one function?   11.5   4   462

Solve variation problems

17. Find y when $x = 12$, given that y varies directly as x, and $y = 35$ when $x = 10$.   11.6   1   465

18. Find y when $x = 3$, given that y varies inversely as x^2, and $y = 90$ when $x = 6$.   11.6   2   467

19. If y varies jointly as the product of x and w and inversely as the square of z, and $y = 24$ when $x = 3$, $w = 4$, and $z = 2$, find y when $x = 5$, $w = 3$, and $z = 4$.   11.6   3   470

Solve applied variation problems

20. *Police Speed Traps:* The speed s of a vehicle varies inversely as the time t over a fixed distance. A speed trap is set up so that it takes 8 seconds to cover the *trap distance* between 2 police cars at the speed limit of 25 mph. By how much is a vehicle speeding that covers the trap distance in 5 seconds?   PS 16   —   473

Grid E   Grid F 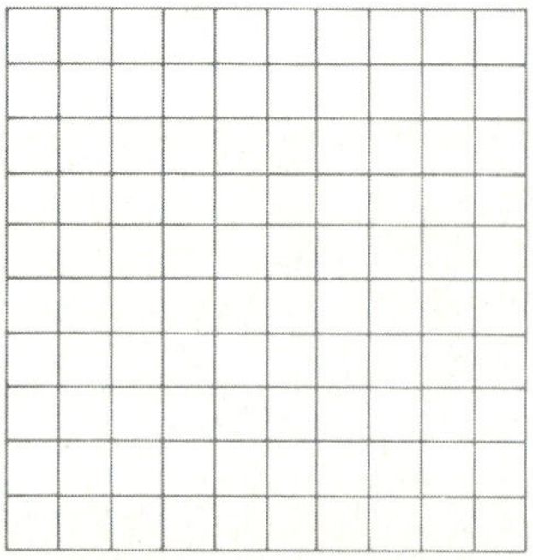   Grid G 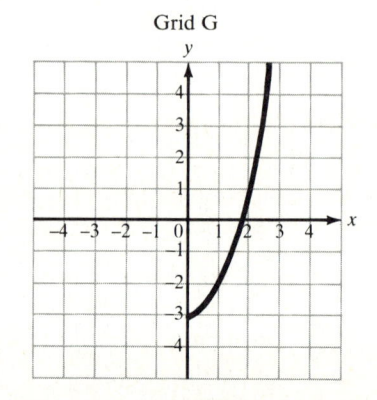   Grid H

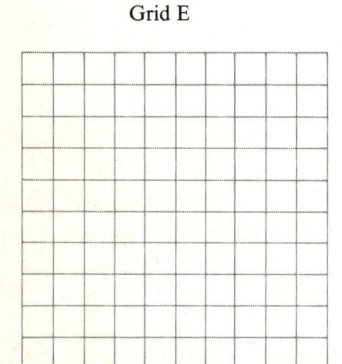

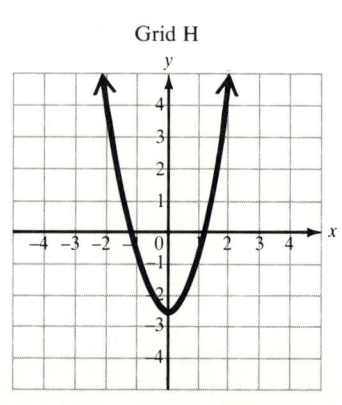

CHAPTER 11 REVIEW ANSWERS: 1a. $\{-1, 2, 3\}$ **b.** $\{-5, 3, 4, 7\}$ **2a.** $\{x \mid -8 \leq x \leq 8\}$ **b.** $\{y \mid 0 \leq y \leq 4\}$ **3a.** $\{x \mid x \leq 3\}$ **b.** $\{y \mid y \geq 4\}$ **4.** It does not represent a function. **5.** It does not represent a function. **6. b** and **c** **7a.** 10 **b.** 14 **c.** $2h^2 - 13h + 25$ **8a.** 0 **b.** 2 **c.** 3 **9.** 13 **10.–12.** See Appendix Selected Answers. **13.** $\{(1, -3), (5, 2), (-7, 5)\}$ **14.** $y = \frac{1}{4}x + \frac{5}{4}$ **15.** See Appendix Selected Answers. **16.** It does not specify a one-to-one function. **17.** 42 **18.** 360 **19.** 7.5 **20.** 15 mph

Copyright © 1985 by Harcourt Brace Jovanovich, Inc. All rights reserved.

FUNCTIONS

12 Exponential and Logarithmic Functions

12.1 Graph Exponential Functions

12.2 Graph Logarithmic Functions

12.3 Use Properties of Logarithmic Functions

12.4 Find Logarithms and Antilogarithms

12.5 Solve Exponential and Logarithmic Equations

PS 17: Solve Problems Using Exponential Formulas

PS 18: Solve Problems Using Logarithmic Formulas

Introduction to Exponential Functions

If the domain of $f(x) = 2^x$ consists only of rational numbers, then for each rational number x, $f(x)$ is the unique real number 2^x as defined in Lesson 8.1.

Examples ▶ (a) If $x = 4$, $f(4) = 2^4$ (b) If $x = -3$, $f(-3) = 2^{-3}$ (c) If $x = \frac{1}{3}$, $f(\frac{1}{3}) = 2^{1/3}$
$= 16$ $= \frac{1}{8}$ $= \sqrt[3]{2}$

If the domain of $f(x) = 2^x$ is to be extended to include irrational numbers, then 2^x must also be defined when x is an irrational number. What meaning can be assigned to the expression $2^{\sqrt{2}}$? A precise definition is beyond the scope of this text; however, an intuitive definition of $2^{\sqrt{2}}$ can be given by using rational approximations of $\sqrt{2}$. The rational number 1.41421356 is an approximation of $\sqrt{2}$.

Because $1.4 < \sqrt{2} < 1.5$, $2^{\sqrt{2}}$ is defined to be a number between $2^{1.4}$ and $2^{1.5}$
Because $1.41 < \sqrt{2} < 1.42$, $2^{\sqrt{2}}$ is defined to be a number between $2^{1.41}$ and $2^{1.42}$
Because $1.414 < \sqrt{2} < 1.415$, $2^{\sqrt{2}}$ is defined to be a number between $2^{1.414}$ and $2^{1.415}$

The above sequence of inequalities indicates that $2^{\sqrt{2}}$ is a number that can be approximated as closely as desired by using closer and closer rational approximations of $\sqrt{2}$.

The above technique can be used to define expressions of the form b^x, where x is any irrational number and $b > 0$.

12.1 Graph Exponential Functions

Many scientific calculators can be used to evaluate expressions of the form b^x by means of a key which is frequently labeled $\boxed{y^x}$ (or on some calculators $\boxed{x^y}$).

EXAMPLE 1: Evaluate expressions of the form b^x using a calculator.

Problems ▶ Evaluate $2^{1.414}$ and $2^{1.415}$ to the nearest thousandth.

1. Enter base ▶ | 2 | | 2 | ← the calculator display

2. Press $\boxed{y^x}$ key ▶ | 2 | | 2 |

3. Enter exponent ▶ | 1.414 | | 1.415 |

4. Press $\boxed{=}$ key ▶ | 2.6647496 ⋯ | | 2.66659735 ⋯ |

Solutions ▶ $2^{1.414} \approx 2.665$ (to the nearest thousandth) and $2^{1.415} \approx 2.667$ (to the nearest thousandth).

Other Examples ▶ (a) $3^{1.34} \approx 4.359$ (b) $5.2^{4.173} \approx 972.483$ (c) $10^{0.23} \approx 1.698$

Make Sure

Evaluate expressions of the form b^x to the nearest thousandth using a calculator.

See Example 1 ▶ 1. $3^{1.7}$ _____   2. $3.1^{2.348}$ _____   3. $0.6^{0.57}$ _____

MAKE SURE ANSWERS: 1. 6.473   2. 14.247   3. 0.747

12.1 Graph Exponential Functions

The function $f(x) = 2^x$ is a member of a special family of functions.

If b is a positive constant ($b \neq 1$), then the function $f(x) = b^x$ is called an *exponential function*. The constant b is called the *base*.

Examples

(a) $f(x) = 2^x$ is an exponential function with base 2.

(b) $f(x) = (\frac{1}{3})^x$ is an exponential function with base $\frac{1}{3}$.

(c) $f(x) = (-4)^x$ is not an exponential function. The base (-4) is not a positive constant.

(d) $f(x) = 1^x$ is not an exponential function. Because $1^x = 1$ for all values of x, this is the constant function $f(x) = 1$.

(e) $f(x) = x^2$ is not an exponential function. The base is not a constant and the exponent is not a variable.

Although a proof is beyond the scope of this text, you need to know that all exponential functions graph to be smooth curves.

EXAMPLE 2: Graph an exponential function.

Problems Graph $f(x) = 2^x$.

1. Make a table

x	-3	-2	-1	0	1	2	3
$f(x) = 2^x$	$\frac{1}{8}$	$\frac{1}{4}$	$\frac{1}{2}$	1	2	4	8

Graph $f(x) = (\frac{1}{3})^x$.

x	-3	-2	-1	0	1	2	3
$f(x) = (\frac{1}{3})^x$	27	9	3	1	$\frac{1}{3}$	$\frac{1}{9}$	$\frac{1}{27}$

2. Graph as before

Solutions

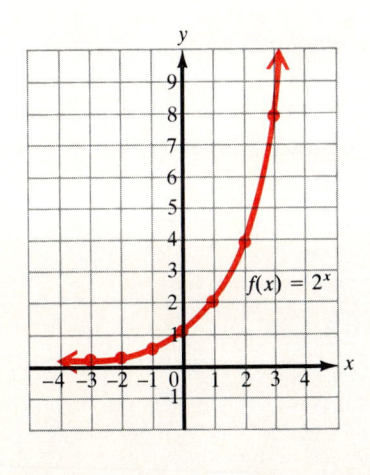

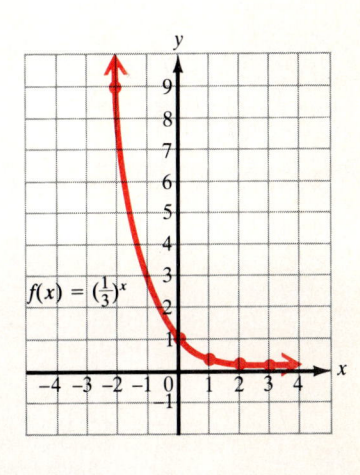

The graphs in Example 2 are typical of the graphs of exponential functions. In fact, the graphs of all exponential functions specified by $f(x) = b^x$ share the following characteristics:

(1) They intersect the y-axis at $(0, 1)$. $f(0) = b^0 = 1$ for any $b > 0$.

(2) If $b > 1$, they increase in height as x increases.
    If $0 < b < 1$, they decrease in height as x increases.

(3) If $b > 1$, they approach the x-axis as x decreases.
    If $0 < b < 1$, they approach the x-axis as x increases.
    The x-axis is a horizontal asymptote for the graphs.

(4) They are one-to-one functions.

482　Chapter 12　Exponential and Logarithmic Functions

Just as the irrational number $\pi(\approx 3.14159)$ arises naturally in some applications involving circles, the irrational number $e(\approx 2.71828)$ arises naturally in the discussion of many physical phenomena. This is why the function specified by $f(x) = e^x$ is called the *natural exponential function*. To graph $f(x) = e^x$, you use a calculator to evaluate e^x for convenient values of x.

Example ▶ Graph $f(x) = e^x$.

1. Make a table ▶

x	-3	-2	-1	0	1	2	3
$f(x) = e^x$	0.050	0.135	0.368	1	2.718	7.389	20.086

Evaluate e^x using a calculator with an $\boxed{e^x}$ key, or use the $\boxed{y^x}$ key with $y \approx 2.71828$.
If a calculator is not available, use Appendix Table 5.

2. Graph as before ▶

Solution ▶

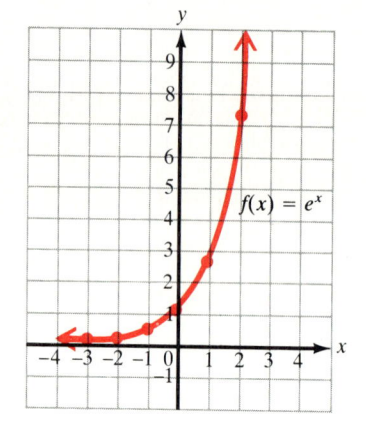

[Graph of $f(x) = e^x$]

Make Sure

Graph each exponential function.

See Example 2 ▶　**1.** $f(x) = 3^x$　　　　　　　　**2.** $g(x) = (0.5)^x$

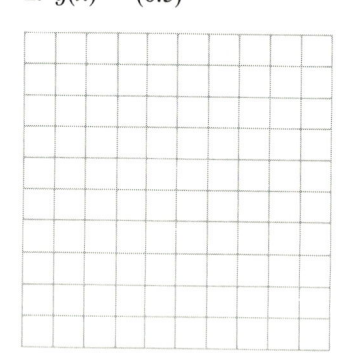

MAKE SURE ANSWERS: See Appendix Selected Answers.

Name _____ Date _____ Class _____

12.1 Practice

Set 1: Evaluate expressions of the form x^y to the nearest thousandth using a calculator.

1. $3^{1.4}$
2. $3^{1.41}$
3. $2^{6.5}$
4. $2^{7.5}$

5. $5^{0.34}$
6. $5^{1.34}$
7. $100^{0.01}$
8. $100^{1.01}$

9. $0.5^{1.414}$
10. $0.5^{1.415}$
11. $0.25^{1.7}$
12. $0.25^{1.8}$

13. $36^{0.4}$
14. $36^{0.3}$
15. $2.71828^{3.14159}$
16. $3.14159^{2.71828}$

Set 2: Graph each exponential function.

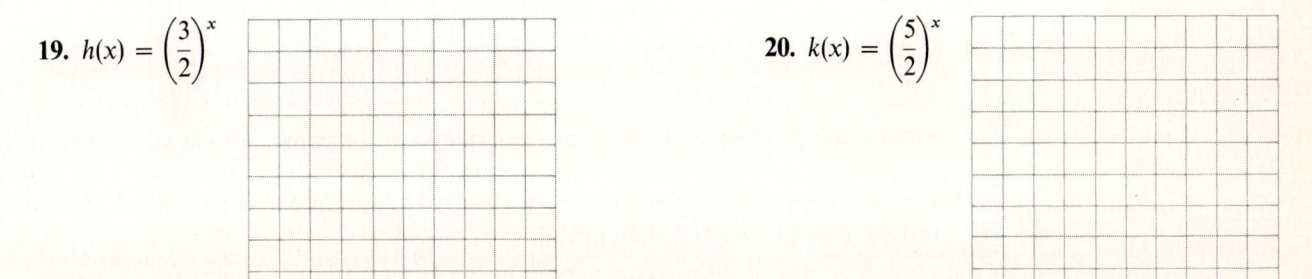

17. $f(x) = 4^x$

18. $g(x) = 5^x$

19. $h(x) = \left(\dfrac{3}{2}\right)^x$

20. $k(x) = \left(\dfrac{5}{2}\right)^x$

484 Chapter 12 Exponential and Logarithmic Functions

21. $f(x) = \left(\dfrac{3}{4}\right)^x$

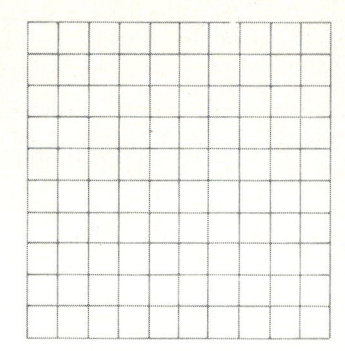

22. $g(x) = \left(\dfrac{1}{4}\right)^x$

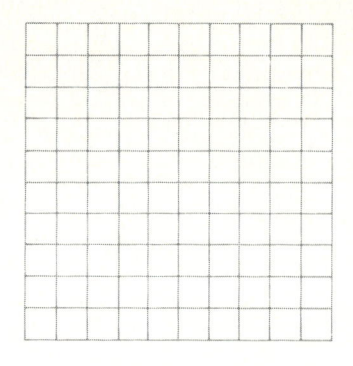

23. $h(x) = \left(\dfrac{2}{5}\right)^x$

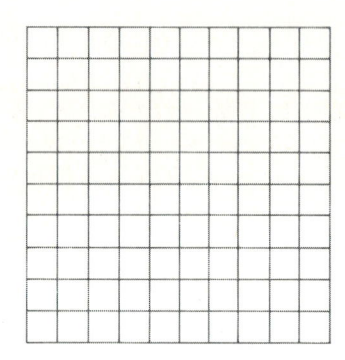

24. $k(x) = \left(\dfrac{3}{5}\right)^x$

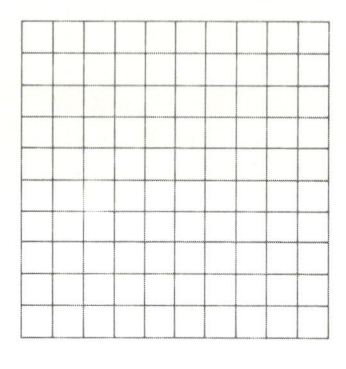

25. $p(x) = \left(\dfrac{2}{3}\right)^x$

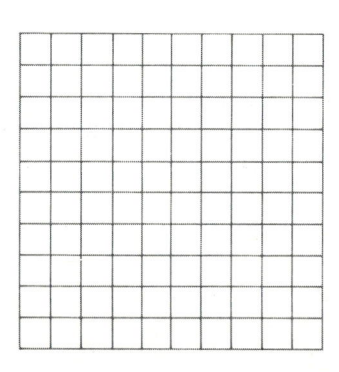

26. $q(x) = \left(\dfrac{4}{5}\right)^x$

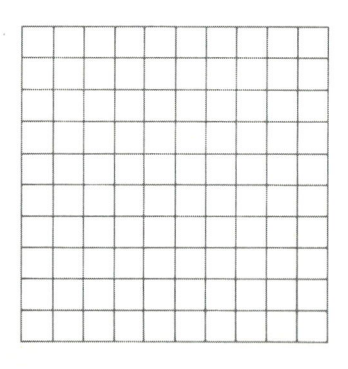

27. $r(x) = (1.4)^x$

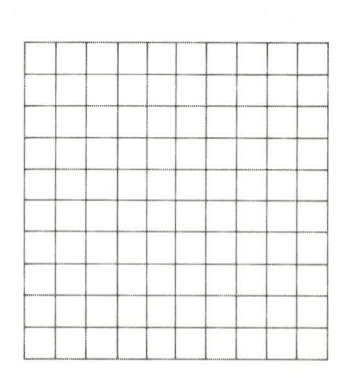

28. $s(x) = (2.6)^x$

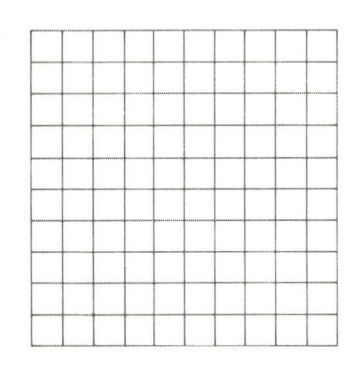

Extra: Work these problems on a separate sheet of paper. Attach your work to this page.

Find the inverse of each equation (see Lesson 11.5, Example 2), and sketch the graph of each inverse. (See Lesson 11.5, Example 3.)

29. $y = 2^x$
30. $y = 3^x$
31. $y = x^3$
32. $y = x^{-4}$
33. $x = y^2$
34. $x = y^3$
35. $x = y^{1/2}$
36. $x = y^4$
37. $y = x^{-1/2}$

Copyright © 1985 by Harcourt Brace Jovanovich, Inc. All rights reserved.

12.2 Graph Logarithmic Functions

Since every exponential function is a one-to-one function, every exponential function will have an inverse which is also a function. The inverse of an exponential function is called a *logarithmic function*.

Recall ▶ To form the inverse of a function specified by an equation, you interchange the variables.

Example ▶ $y = 2^x$ ⟵ an exponential function

$x = 2^y$ ⟵ the inverse of the above exponential function

To solve $x = 2^y$ for the variable y, you will need the following definition of a logarithm.

> If $x > 0$ and b is a positive constant ($b \neq 1$), then the function
> $$b^y = x \text{ means the same as } \log_b x = y.$$
> In the function represented by $y = \log_b x$, b is called the *base*, x is called the *argument*, and y is called the *logarithm*.

Examples ▶ (a) $3^2 = 9$ means the same as $\log_3 9 = 2$. (b) $2^5 = 32$ means the same as $\log_2 32 = 5$.
(c) $(\frac{1}{2})^{-3} = 8$ means the same as $\log_{1/2} 8 = -3$.

You read "$\log_b x$" as "the logarithm (or log) of x base b."

Examples ▶ (a) $\log_3 9 = 2$ is read "the log of 9 base 3 is 2." It means $3^2 = 9$.
(b) $\log_2 32 = 5$ is read "the log of 32 base 2 is 5." It means $2^5 = 32$.
(c) $\log_{1/2} 8 = -3$ is read "the log of 8 base $\frac{1}{2}$ is -3." It means $(\frac{1}{2})^{-3} = 8$.
(d) $\log_{10} \frac{1}{10} = -1$ is read "the log of $\frac{1}{10}$ base 10 is -1." It means $10^{-1} = \frac{1}{10}$.
(e) $\log_5 1 = 0$ is read "the log of 1 base 5 is 0." It means $5^0 = 1$.

It is important to realize that a logarithm is an exponent. If $\log_b x = y$, then by the definition of a logarithm, y is the exponent such that $b^y = x$.

To write logarithmic equations in exponential form, you use the definition of a logarithm.

EXAMPLE 1: Write a logarithmic equation in exponential form.

Problem ▶ Write $\log_2 16 = 4$ in exponential form.

1. Identify parts ▶ $\log_2 16 = 4$ ⟵ the logarithm

   the base

2. Use definition ▶ ⌐the logarithm⌐
   $\log_2 16 = 4$ means $2^4 = 16$ Think: $\log_b x = y$ means the same as $b^y = x$.
   ⌐the base⌐

Solution ▶ $\log_2 16 = 4$ is written as $2^4 = 16$ in exponential form.

Other Examples ▶ (a) $\log_5 125 = 3$ is written as $5^3 = 125$ in exponential form.
(b) $\log_{10} 100 = 2$ is written as $10^2 = 100$ in exponential form.

486 Chapter 12 Exponential and Logarithmic Functions

Make Sure

Write each logarithmic equation in exponential form.

See Example 1 ▶ **1.** $\log_3 243 = 5$ _____   **2.** $\log_{1/3} 81 = -4$ _____

MAKE SURE ANSWERS: **1.** $3^5 = 243$ **2.** $\left(\frac{1}{3}\right)^{-4} = 81$

The definition of logarithms was created so that exponential equations of the form $b^y = x$ could be written in the logarithmic form $\log_b x = y$.

EXAMPLE 2: Write an exponential equation in logarithmic form.

Problem ▶ Write $6^2 = 36$ in logarithmic form.

1. Identify parts ▶
$$\underbrace{6^{\overbrace{2}^{\text{the exponent}}}}_{\text{the base}} = 36$$

2. Use definition ▶ $6^2 = 36$ means $\log_6 36 = 2$ Think: $b^y = x$ means the same as $\log_b x = y$.

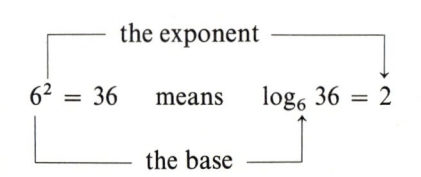

Solution ▶ $6^2 = 36$ is written as $\log_6 36 = 2$ in logarithmic form.

Other Examples ▶ (a) $3^{-2} = \frac{1}{9}$ is written as $\log_3 \frac{1}{9} = -2$ in logarithmic form.
(b) $4^3 = 64$ is written as $\log_4 64 = 3$ in logarithmic form.

Make Sure

Write each exponential equation in logarithmic form.

See Example 2 ▶ **1.** $2^6 = 64$ _____   **2.** $27^{1/3} = 3$ _____

MAKE SURE ANSWERS: **1.** $\log_2 64 = 6$ **2.** $\log_{27} 3 = \frac{1}{3}$

To solve some logarithmic equations that involve a variable, you can write the equation in exponential form.

EXAMPLE 3: Solve logarithmic equations.

Problems ▶ Solve $\log_4 x = -3$, $\log_x 8 = 3$, and $\log_3 81 = x$.

1. Write exponential form ▶

| $\log_4 x = -3$ means $4^{-3} = x$ | $\log_x 8 = 3$ means $x^3 = 8$ | $\log_3 81 = x$ means $3^x = 81$ |

2. Solve ▶

| $\dfrac{1}{4^3} = x$ | $x = \sqrt[3]{8}$ | $3^x = 3^4$ |
| $\dfrac{1}{64} = x$ | $x = 2$ | $x = 4$ |

Solutions ▶ The solution of: $\log_4 x = -3$ is $\tfrac{1}{64}$, $\log_x 8 = 3$ is 2, and $\log_3 81 = x$ is 4.

Note ▶ Because exponential functions are one-to-one functions, you can solve equations of the form $b^x = b^y$ by equating x and y.

Make Sure

Solve each logarithmic equation.

See Example 3 ▶ 1. $\log_3 x = 5$ _____ 2. $\log_x 625 = 4$ _____ 3. $\log_{1/5} 5 = x$ _____

MAKE SURE ANSWERS: 1. 243 2. 5 3. −1

To graph a logarithmic equation, you can prepare a table of solutions using exponential form.

EXAMPLE 4: Graph a logarithmic equation.

Problem ▶ Graph $y = \log_3 x$.

1. Write exponential form ▶ $y = \log_3 x$ means $3^y = x$.

2. Make a table ▶

y	-2	-1	0	1	2	3
$x = 3^y$	$3^{-2} = \tfrac{1}{9}$	$3^{-1} = \tfrac{1}{3}$	$3^0 = 1$	$3^1 = 3$	$3^2 = 9$	$3^3 = 27$

Think: It is convenient to choose y values and solve for the corresponding x values.

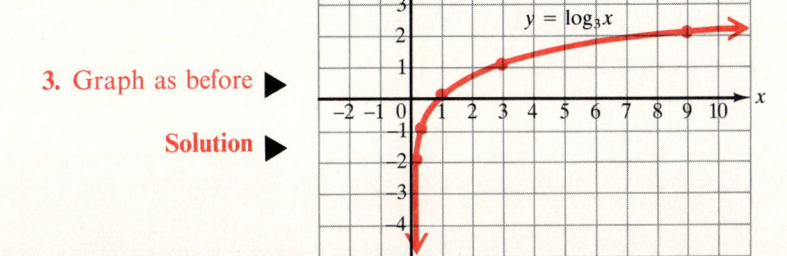

3. Graph as before ▶

Solution ▶

488 Chapter 12 Exponential and Logarithmic Functions

It is important to notice that the logarithmic equation $y = \log_b x$ (or $b^y = x$ in exponential form) specifies a function that is the inverse of $y = b^x$. Therefore another method of graphing $y = \log_b x$ would be to graph its inverse $y = b^x$ and then reflect the graph about the line $y = x$.

Example ▶ Graph $y = \log_2 x$.

1. Write inverse ▶ $y = \log_2 x$ means $2^y = x$.

$2^x = y$ ⟵ the inverse of $y = \log_2 x$

2. Graph inverse and reflect about the line $y = x$

Solution ▶

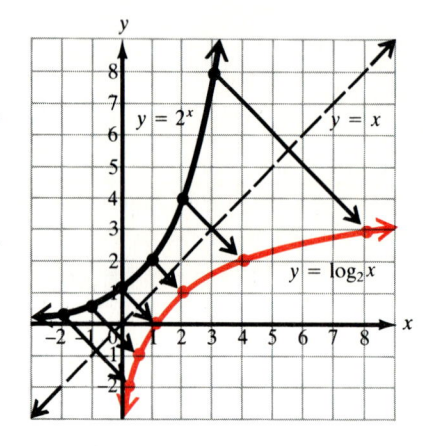

x	$y = 2^x$
-2	$2^{-2} = \frac{1}{4}$
-1	$2^{-1} = \frac{1}{2}$
0	$2^0 = 1$
1	$2^1 = 2$
2	$2^2 = 4$
3	$2^3 = 8$
4	$2^4 = 16$

The above method of graphing a logarithmic equation is more involved than the method used in Example 4; however, it visually illustrates the important concept that the logarithmic equation $y = \log_b x$ specifies a function that is the inverse of the exponential function specified by $y = b^x$.

Make Sure

Graph each logarithmic equation.

See Example 4 ▶ **1.** $y = \log_5 x$

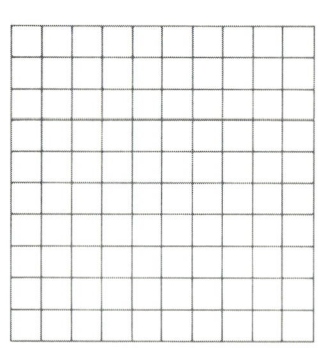

2. $y = \log_{1/3} x$

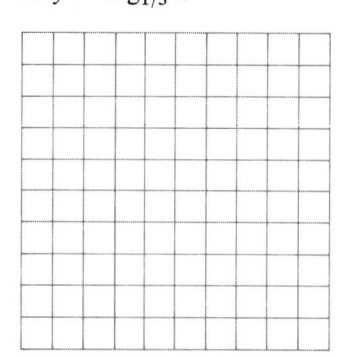

MAKE SURE ANSWERS: See Appendix Selected Answers.

12.2 Practice

Set 1: Write each logarithmic equation in exponential form.

1. $\log_{10} 100 = 2$
2. $\log_{10} 10 = 1$
3. $\log_2 8 = 3$

4. $\log_3 81 = 4$
5. $\log_{1/2} 16 = -4$
6. $\log_{1/5} 625 = -4$

7. $\log_{2/3} \frac{4}{9} = 2$
8. $\log_{4/5} \frac{25}{16} = -2$
9. $\log_2 \sqrt{2} = \frac{1}{2}$

10. $\log_{\sqrt{3}} 3 = 2$
11. $\log_e e = 1$
12. $\log_e 1 = 0$

Set 2: Write each exponential equation in logarithmic form.

13. $10^3 = 1000$
14. $10^4 = 10000$
15. $3^4 = 81$

16. $4^3 = 64$
17. $5^4 = 625$
18. $2^7 = 128$

19. $4^{1/2} = 2$
20. $8^{1/2} = 2\sqrt{2}$
21. $4^{-2} = \frac{1}{16}$

22. $3^{-3} = \frac{1}{27}$
23. $e^1 = e$
24. $e^0 = 1$

Set 3: Solve each logarithmic equation.

25. $\log_2 x = 4$ **26.** $\log_3 x = 3$ **27.** $\log_5 x = -4$

28. $\log_3 x = -2$ **29.** $\log_2 x = \dfrac{1}{2}$ **30.** $\log_{27} x = \dfrac{1}{3}$

31. $\log_x 8 = 3$ **32.** $\log_x 64 = 6$ **33.** $\log_x 27 = 3$

34. $\log_x 6 = \dfrac{1}{2}$ **35.** $\log_{1/9} \dfrac{1}{3} = x$ **36.** $\log_{1/8} 4 = x$

Set 4: Graph each logarithmic equation.

37. $y = \log_2 x$

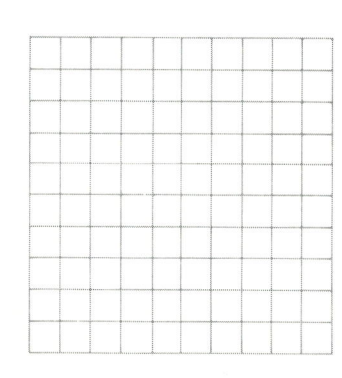

38. $y = \log_4 x$

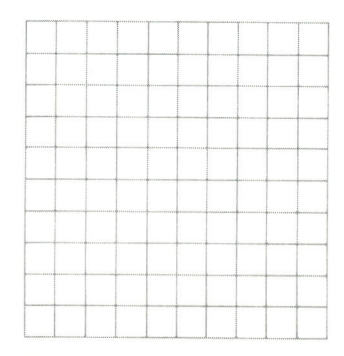

39. $y + \log_{1/3} x = 0$

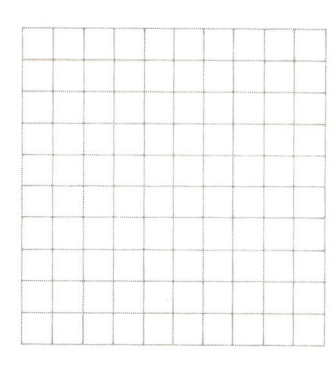

40. $y = \log_e x$

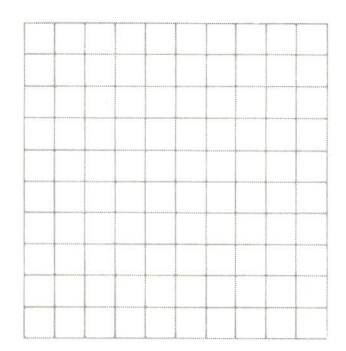

Review: Work these problems on a separate sheet of paper. Attach your work to this page.

Write each radical expression in rational exponential notation. (See Lesson 8.1.)

41. $\sqrt{2}$ **42.** $\sqrt[3]{3}$ **43.** $\sqrt[4]{3}$

44. $\sqrt[3]{9}$ **45.** $\sqrt[3]{4}$ **46.** $\sqrt[4]{8}$

47. $\sqrt[4]{27}$ **48.** $\sqrt[3]{16}$ **49.** $\sqrt[3]{256}$

12.3 Use Properties of Logarithmic Functions

Logarithmic functions are important because of their many problem-solving applications, and because they have special properties that can be used to simplify logarithmic expressions and solve equations. The following identities can help you simplify expressions involving logarithms. Each identity can be established directly from the definition of a logarithm.

> **Logarithmic Identities**
>
> 1. $\log_b (b^p) = p$ The logarithm of the base number to the pth power is p, because $b^p = b^p$.
>
> 2. $\log_b b = 1$ The logarithm of the base number is 1, because $b^1 = b$.
>
> 3. $\log_b 1 = 0$ The logarithm of 1 to any base b is 0, because $b^0 = 1$.

Examples ▶ (a) $\log_{10} (10^2) = 2$ (b) $\log_7 (7^3) = 3$ (c) $\log_e (e^4) = 4$ ⟵ Identity one

(d) $\log_{10} 10 = 1$ (e) $\log_7 7 = 1$ (f) $\log_e e = 1$ ⟵ Identity two

(g) $\log_{10} 1 = 0$ (h) $\log_7 1 = 0$ (i) $\log_e 1 = 0$ ⟵ Identity three

Because logarithms are exponents, the rules for exponents can be stated in terms of logarithms.

> **Properties of Logarithms**
>
> In the following properties, b, M, and N are positive real numbers ($b \neq 1$), and p is a real number.
>
> **Product Property:** $\log_b (MN) = \log_b M + \log_b N$ The log of a product is the sum of the logs.
>
> **Quotient Property:** $\log_b \dfrac{M}{N} = \log_b M - \log_b N$ The log of a quotient is the difference of the logs.
>
> **Power Property:** $\log_b (M^p) = p \log_b M$ The log of the pth power of M is p times the log of M.

To prove the properties of logarithms, you can use the definition of a logarithm and the property that if two positive numbers A and B are equal, then their logarithms are also equal.

Example ▶ Prove $\log_b (MN) = \log_b M + \log_b N$.

Let $\log_b M = x$ and $\log_b N = y$.

Then $M = b^x$ and $N = b^y$.	Change each equation to its exponential form.
$MN = b^x b^y$	Equate the products of the left and right members.
$MN = b^{x+y}$	Use the product rule for exponents.
$\log_b (MN) = \log_b (b^{x+y})$	Equate the logs of both members.
$\log_b (MN) = x + y$	Use $\log_b (b^p) = p$.
$\log_b (MN) = \log_b M + \log_b N$	Substitute $\log_b M$ for x and $\log_b N$ for y.

Make Sure

Prove each property.

See previous Example ▶ **1.** $\log_b \dfrac{M}{N} = \log_b M - \log_b N$

Let $\log_b M = x$ *and* $\log_b N = y$

Then $M = \blacksquare$ and $N = \blacksquare$

$$\dfrac{M}{N} = \dfrac{\blacksquare}{\blacksquare}$$

$$\dfrac{M}{N} = \blacksquare$$

$$\log_b \dfrac{M}{N} = \log_b \blacksquare$$

$$\log_b \dfrac{M}{N} = \blacksquare - \blacksquare$$

$$\log_b \dfrac{M}{N} = \blacksquare - \blacksquare$$

2. $\log_b (M^p) = p \log_b M$

Let $\log_b M = x$

Then $M = \blacksquare$

$M^p = (\blacksquare)^{\blacksquare}$

$= \blacksquare$

$\log_b (M^p) = \log_b \blacksquare$

$\log_b (M^p) = \blacksquare$

$\log_b (M^p) = \blacksquare \log_b \blacksquare$

MAKE SURE ANSWERS: See Appendix Selected Answers.

To write a logarithm in terms of simpler logarithms, you use the properties of logarithms.

EXAMPLE 1: Write a logarithm in terms of simpler logarithms.

Problem ▶ Write $\log_b \sqrt[3]{\dfrac{x^2 y}{z}}$ in terms of $\log_b x$, $\log_b y$, and $\log_b z$.

1. Use exponential notation ▶ $\log_b \sqrt[3]{\dfrac{x^2 y}{z}} = \log_b \left(\dfrac{x^2 y}{z}\right)^{1/3}$

2. Use properties ▶ $= \dfrac{1}{3} \log_b \dfrac{x^2 y}{z}$ ⟵ the Power Property

$= \dfrac{1}{3} [\log_b (x^2 y) - \log_b z]$ ⟵ the Quotient Property

$= \dfrac{1}{3} [\log_b (x^2) + \log_b y - \log_b z]$ ⟵ the Product Property

$= \dfrac{1}{3} [2 \log_b x + \log_b y - \log_b z]$ ⟵ the Power Property

3. Distribute ▶ $= \dfrac{2}{3} \log_b x + \dfrac{1}{3} \log_b y - \dfrac{1}{3} \log_b z$ ⟵ an expression written in terms of $\log_b x$, $\log_b y$, and $\log_b z$

Solution ▶ $\log_b \sqrt[3]{\dfrac{x^2 y}{z}} = \dfrac{2}{3} \log_b x + \dfrac{1}{3} \log_b y - \dfrac{1}{3} \log_b z$

12.3 Use Properties of Logarithmic Functions

Make Sure

Write each logarithm in terms of simpler logarithms.

See Example 1 ▶ **1.** $\log_b (3x^4 y^2 z)$ _____ **2.** $\log_b \sqrt[4]{\dfrac{xy^2}{3z^3}}$ _____

MAKE SURE ANSWERS: 1. $\log_b 3 + 4 \log_b x + 2 \log_b y + \log_b z$ **2.** $\frac{1}{4} \log_b x + \frac{2}{4} \log_b y - \frac{1}{4} \log_b 3 - \frac{3}{4} \log_b z$

To express the sum and/or difference of logarithmic terms as a single logarithm, you use the properties of logarithms.

EXAMPLE 2: Write a logarithmic expression as a single logarithm.

Problem ▶ Write $\dfrac{1}{2} \log_b x + \log_b y - 2 \log_b (yz)$ as a single logarithm with a coefficient of 1.

1. Use properties ▶ $\dfrac{1}{2} \log_b x + \log_b y - 2 \log_b (yz) = \log_b (x^{1/2}) + \log_b y - \log_b (yz)^2$ ⟵ the Power Property

$= \log_b (x^{1/2} y) - \log_b (yz)^2$ ⟵ the Product Property

$= \log_b \dfrac{x^{1/2} y}{(yz)^2}$ ⟵ the Quotient Property

2. Simplify ▶ $= \log_b \dfrac{x^{1/2} y}{y^2 z^2}$

$= \log_b \dfrac{x^{1/2}}{yz^2}$ ⟵ a single logarithm with a coefficient of 1 in simplest form

Solution ▶ $\dfrac{1}{2} \log_b x + \log_b y - 2 \log_b (yz) = \log_b \dfrac{x^{1/2}}{yz^2}$ or $\log_b \dfrac{\sqrt{x}}{yz^2}$

Make Sure

Write each logarithmic expression as a single logarithm.

See Example 2 ▶ **1.** $\dfrac{1}{3} \log_b x + \dfrac{2}{3} \log_b y - \log_b z$ **2.** $\dfrac{3}{4} \log_b x - \dfrac{1}{2} \log_b y - \dfrac{1}{4} \log_b z$

_____ _____

MAKE SURE ANSWERS: 1. $\log_b \dfrac{x^{1/3} y^{2/3}}{z}$ **2.** $\log_b \sqrt[4]{\dfrac{x^3}{y^2 z}}$

You can sometimes use known logarithmic values and the properties of logarithms to approximate other logarithms.

EXAMPLE 3: Find a logarithm using given logarithmic values.

Problem ▶ Find $\log_{10} 120$ given $\log_{10} 2 \approx 0.3010$ and $\log_{10} 3 \approx 0.4771$.

1. Factor argument ▶ $\log_{10} 120 = \log_{10} (2^2 \cdot 3 \cdot 10)$ Think: Write the argument using only powers of 2, 3, and/or 10 (the base).

2. Use properties ▶ $\quad = \log_{10} (2^2) + \log_{10} 3 + \log_{10} 10$ The Product Property.

$\quad = 2(\log_{10} 2) + \log_{10} 3 + 1$ The Power Property and the Identity two.

3. Evaluate ▶ $\quad \approx 2(0.3010) + 0.4771 + 1$

Solution ▶ $\log_{10} 120 \approx 2.0791$

Other Examples ▶ Find a) $\log_{10} 18$, b) $\log_{10} (3\sqrt{2})$, and c) $\log_{10} 15$ given $\log_{10} 2 \approx 0.3010$ and $\log_{10} 3 \approx 0.4771$.

a) $\log_{10} 18 = \log_{10} (2 \cdot 3^2)$

$\quad = \log_{10} 2 + \log_{10} (3^2)$

$\quad = \log_{10} 2 + 2(\log_{10} 3)$

$\quad \approx 0.3010 + 2(0.4771)$

$\quad \approx 1.2552$

b) $\log_{10} (3\sqrt{2}) = \log_{10} (3 \cdot 2^{1/2})$

$\quad = \log_{10} 3 + \log_{10} (2^{1/2})$

$\quad = \log_{10} 3 + \tfrac{1}{2}(\log_{10} 2)$

$\quad \approx 0.4771 + \tfrac{1}{2}(0.3010)$

$\quad \approx 0.6276$

c) $\log_{10} 15 = \log_{10} \dfrac{3 \cdot 10}{2}$ Think: Write the argument using only powers of 2, 3, and/or 10 (the base).

$\quad = \log_{10} (3 \cdot 10) - \log_{10} 2$

$\quad = \log_{10} 3 + \log_{10} 10 - \log_{10} 2$

$\quad \approx 0.4771 + 1 - 0.3010$

$\quad \approx 1.1761$

Make Sure

Find each logarithm using the given logarithmic values.

See Example 3 ▶
1. Find $\log_{10} 108$, given $\log_{10} 2 \approx 0.3010$ and $\log_{10} 3 \approx 0.4771$.

2. Find $\log_{10} (4\sqrt{5})$, given $\log_{10} 2 \approx 0.3010$ and $\log_{10} 5 \approx 0.6990$.

MAKE SURE ANSWERS: 1. 2.0333 2. 0.9515

12.3 Practice

Set 1: Write each logarithm in terms of simpler logarithms.

1. $\log_b (x^3 y^2)$
2. $\log_b (xy^4)$
3. $\log_b \dfrac{x^3}{y^4}$
4. $\log_b \dfrac{x}{y^3}$
5. $\log_b \dfrac{x^3 y}{z^2}$
6. $\log_b \dfrac{x^2 y^3}{z^4}$
7. $\log_b \dfrac{x^2 \sqrt{y}}{z}$
8. $\log_b \dfrac{x \sqrt{y}}{z^3}$
9. $\log_b \dfrac{x \sqrt{y}}{\sqrt[3]{z}}$
10. $\log_b \dfrac{x^2 \sqrt{y}}{\sqrt[3]{z}}$
11. $\log_b \sqrt[4]{\dfrac{x^3 y^2}{z^2 w}}$
12. $\log_b \sqrt[4]{\dfrac{xy^3}{z^2 w^3}}$

Set 2: Write each logarithmic expression as a single logarithm.

13. $\log_b x + \log_b y - \log_b z$
14. $\log_b x - \log_b y + \log_b z$
15. $3 \log_b x + 2 \log_b y - \log_b z$
16. $4 \log_b x - 3 \log_b y + 2 \log_b z$
17. $\dfrac{1}{2} \log_b x - \dfrac{1}{2} \log_b y - \dfrac{1}{2} \log_b z$
18. $\dfrac{1}{3} \log_b x - \dfrac{1}{3} \log_b y - \dfrac{1}{3} \log_b z$
19. $\dfrac{1}{2} \log_b x + 3 \log_b y - 2 \log_b z$
20. $2 \log_b x + \dfrac{1}{3} \log_b y - \log_b z$
21. $2 \log_b x + 2 \log_b y - \dfrac{1}{2} \log_b z$
22. $3 \log_b x + \dfrac{1}{3} \log_b y - \dfrac{2}{3} \log_b z$
23. $\log_b (x^2 - y^2) - \log_b (x + y)$
24. $\log_b (x^3 + y^3) - \log_b (x + y)$

25. _____

26. _____

27. _____

28. _____

29. _____

30. _____

31. _____

32. _____

33. _____

34. _____

35. _____

36. _____

37. _____

38. _____

39. _____

40. _____

41. _____

42. _____

Set 3: Find each logarithm using the following logarithmic values: $\log_{10} 2 \approx 0.3010$, $\log_{10} 3 \approx 0.4771$, $\log_{10} 5 \approx 0.6990$, and $\log_{10} 7 \approx 0.8451$.

25. Find $\log_{10} 4$ **26.** Find $\log_{10} 6$ **27.** Find $\log_{10} 10$

28. Find $\log_{10} 14$ **29.** Find $\log_{10} 12$ **30.** Find $\log_{10} 18$

31. Find $\log_{10} 72$ **32.** Find $\log_{10} 216$ **33.** Find $\log_{10} 1440$

34. Find $\log_{10} 448$ **35.** Find $\log_{10} 3.5$ **36.** Find $\log_{10} 2.5$

37. Find $\log_{10} 17.5$ **38.** Find $\log_{10} 7.5$ **39.** Find $\log_{10} \sqrt{2}$

40. Find $\log_{10} \sqrt{7}$ **41.** Find $\log_{10} \sqrt[3]{16}$ **42.** Find $\log_{10} \sqrt[4]{27}$

Review: Work these problems on a separate sheet of paper. Attach your work to this page.

Write each number in scientific notation. (See Problem Solving 1.)
43. 4350   **44.** 35900   **45.** 0.0743   **46.** 0.00455   **47.** 0.4720

Write each scientific notation in standard form. (See Problem Solving 1.)
48. 2.35×10^2 **49.** 6.44×10^3 **50.** 2.85×10^{-1} **51.** 9.43×10^{-2} **52.** 5.33×10^{-3}

Solve each equation. (See Lesson 2.1.)
53. $x = 0.9562 + (-2)$ **54.** $x = 0.8488 + (-1)$ **55.** $x = 0.5514 + (-3)$
56. $-1.6308 = x + (-2)$ **57.** $-2.5884 = x + (-3)$ **58.** $-0.1029 = x + (-1)$

12.4 Find Logarithms and Antilogarithms

Logarithms with a base of "10" are used so often that they are called *common logarithms* and they are denoted by the following notation.

Common Logarithms: $\log M$ means $\log_{10} M$

If a number is an integral power of 10, then you can find its common logarithm mentally.

Examples ▶
(a) $\log 1000 = 3$, because $10^3 = 1000$. (b) $\log 100 = 2$, because $10^2 = 100$.
(c) $\log 10 = 1$, because $10^1 = 10$. (d) $\log 1 = 0$, because $10^0 = 1$.
(e) $\log 0.1 = -1$, because $10^{-1} = 0.1$. (f) $\log 0.01 = -2$, because $10^{-2} = 0.01$.

To find the common logarithm of a positive number that is not an integral power of 10, you use a common logarithmic table or a calculator that has a $\boxed{\log x}$ key. Appendix Table 2 is a table of common logs. If a number is greater than one and less than ten, and has, at most, two decimal places, then you can read its logarithm directly from the table. A portion of Appendix Table 2 is reproduced in the next two examples to illustrate this procedure.

Example ▶ Find $\log 1.27$ using Appendix Table 2, Common Logarithms (base 10).

1. Identify ▶

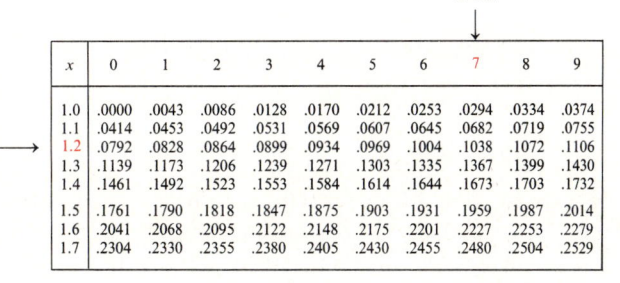

third digit of 1.27

x	0	1	2	3	4	5	6	7	8	9
1.0	.0000	.0043	.0086	.0128	.0170	.0212	.0253	.0294	.0334	.0374
1.1	.0414	.0453	.0492	.0531	.0569	.0607	.0645	.0682	.0719	.0755
1.2	.0792	.0828	.0864	.0899	.0934	.0969	.1004	.1038	.1072	.1106
1.3	.1139	.1173	.1206	.1239	.1271	.1303	.1335	.1367	.1399	.1430
1.4	.1461	.1492	.1523	.1553	.1584	.1614	.1644	.1673	.1703	.1732
1.5	.1761	.1790	.1818	.1847	.1875	.1903	.1931	.1959	.1987	.2014
1.6	.2041	.2068	.2095	.2122	.2148	.2175	.2201	.2227	.2253	.2279
1.7	.2304	.2330	.2355	.2380	.2405	.2430	.2455	.2480	.2504	.2529

first two digits of 1.27

Think: Find the first two digits of the argument in the far left-hand column. Find the third digit in the top row.

2. Locate ▶

x	0	1	2	3	4	5	6	7	8	9
1.0	.0000	.0043	.0086	.0128	.0170	.0212	.0253	.0294	.0334	.0374
1.1	.0414	.0453	.0492	.0531	.0569	.0607	.0645	.0682	.0719	.0755
1.2	.0792	.0828	.0864	.0899	.0934	.0969	.1004	.1038	.1072	.1106
1.3	.1139	.1173	.1206	.1239	.1271	.1303	.1335	.1367	.1399	.1430
1.4	.1461	.1492	.1523	.1553	.1584	.1614	.1644	.1673	.1703	.1732
1.5	.1761	.1790	.1818	.1847	.1875	.1903	.1931	.1959	.1987	.2014
1.6	.2041	.2068	.2095	.2122	.2148	.2175	.2201	.2227	.2253	.2279
1.7	.2304	.2330	.2355	.2380	.2405	.2430	.2455	.2480	.2504	.2529

Think: Find the number at the intersection of the row and column identified in Step 1.

Solution ▶ $\log 1.27 \approx 0.1038$

Note ▶ The rational number 0.1038 is only an approximation of $\log 1.27$, which is an irrational number.

Another Example ▶ Find $\log 9.24$ using Appendix Table 2.

first two digits of 9.24

	0	1	2	3	4	5	6	7	8	9
9.0	.9542	.9547	.9552	.9557	.9562	.9566	.9571	.9576	.9581	.9586
9.1	.9590	.9595	.9600	.9605	.9609	.9614	.9619	.9624	.9628	.9633
9.2	.9638	.9643	.9647	.9652	.9657	.9661	.9666	.9671	.9675	.9680
9.3	.9685	.9689	.9694	.9699	.9703	.9708	.9713	.9717	.9722	.9727
9.4	.9731	.9736	.9741	.9745	.9750	.9754	.9759	.9763	.9768	.9773
9.5	.9777	.9782	.9786	.9791	.9795	.9800	.9805	.9809	.9814	.9818
9.6	.9823	.9827	.9832	.9836	.9841	.9845	.9850	.9854	.9859	.9863
9.7	.9868	.9872	.9877	.9881	.9886	.9890	.9894	.9899	.9903	.9908
9.8	.9912	.9917	.9921	.9926	.9930	.9934	.9939	.9943	.9948	.9952
9.9	.9956	.9961	.9965	.9969	.9974	.9978	.9983	.9987	.9991	.9996
x	0	1	2	3	4	5	6	7	8	9

third digit of 9.24

Think: Find the first two digits of the argument in the far left-hand column. Find the third digit in the top (or bottom) row.

Think: Find the number at the intersection of the row and column.

Appendix Table 2 lists only common logarithms of numbers between 1 and 10; however, you can use scientific notation, the properties of logarithms, and Appendix Table 2 to find the common logarithm of any positive number with, at most, three significant digits.

EXAMPLE 1: Find a common logarithm of a positive number that is not between 1 and 10 using Appendix Table 2.

Problem ▶ Find log 642 using Appendix Table 2.

1. Write scientific notation ▶ $\log 642 = \log(6.42 \times 10^2)$

2. Use properties ▶ $= \log 6.42 + \log(10^2)$ ⟵ the Product Property

$= \log 6.42 + 2$ Think: $\log(10^2) = 2$.

3. Evaluate ▶ $\approx 0.8075 + 2$ Use Appendix Table 2.

Solution ▶ $\log 642 \approx 2.8075$

To find the common logarithm of any positive real number x, you can apply the previous procedures.

Example ▶ If $x = c \times 10^k$ where $1 \le c < 10$ and k is an integer, then:

$\log x = \log(c \times 10^k)$

$ = \log c + \log(10^k)$ ⟵ the Product Property

$ = \log c + k$ ⟵ Identity one in Lesson 12.3

Note ▶ The last equation tells you that to find log x, it is sufficient to know the common logarithms of numbers between 1 and 10.

> If $1 \le c < 10$ and k is an integer such that $\log x = \log c + k$ then:
> the number log c is called the *mantissa*, and
> the integer k is called the *characteristic* of log x.

Examples ▶ (a) $\log 42 = \log(4.2 \times 10^1)$ 　　　(b) $\log 0.00034 = \log(3.4 \times 10^{-4})$

$ = \log 4.2 + \log(10^1)$ 　　　$ = \log 3.4 + \log(10^{-4})$

$ \approx 0.6232 + 1$ 　　　$ \approx 0.5315 + (-4)$

　　　　　↑　　　↑　　　　　　　　　　　　　　　↑　　　↑
　　　mantissa　characteristic　　　　　　　　mantissa　characteristic

Note 1 ▶ Because $1 \le c < 10$, the mantissa (log c) is a number greater than zero but less than one.

Note 2 ▶ The characteristic is the integral exponent used in writing the argument in scientific notation. It may be positive, negative, or zero.

Another Example ▶ $\log 0.00271 = \log 2.71 + (-3)$

$ \approx 0.4330 + (-3)$ ⟵ the characteristic: an integer

　　　　　　↑
　　　the mantissa: a number between 0 and 1

12.4 Find Logarithms and Antilogarithms

Note ▶ If you add the mantissa and the characteristic in the previous example, you get log 0.00271 ≈ −2.567. The mantissa and the characteristic are not visible in this answer. This method of writing a logarithm is inconvenient to use with logarithmic tables because logarithmic tables contain only positive logarithms. When the characteristic of a logarithm is negative, it is customary to leave the logarithm in a form in which the mantissa is visible.

Examples ▶ (a) log 0.00272 ≈ 0.4346 + (−3)

(b) log 0.00272 ≈ 5.4346 + (−3 − 5) = 5.4346 − 8

(c) log 0.00272 ≈ 7.4346 + (−3 − 7) = 7.4346 − 10

Note 1 ▶ The value of the logarithm remains the same if you add a natural number to the mantissa and subtract the same number from the characteristic.

Note 2 ▶ The mantissa 0.4346 is visible in each of the above forms; however, log 0.00272 = 7.4346 − 10 is the most commonly used form.

Another Example ▶ Find log 0.000451.

1. Write scientific notation ▶ log 0.000451 = log (4.51 × 10^{-4})

2. Use properties ▶
$$= \log 4.51 + \log (10^{-4})$$
$$= \log 4.51 + (-4) \quad \text{Think: } \log(10^{-4}) = -4.$$

3. Evaluate ▶
$$\approx 0.6542 - 4 \quad \text{Use Appendix Table 2.}$$
$$\boxed{\approx 6.6542 - 10} \quad \text{Think: Add 6 and subtract 6 to produce the most commonly used form.}$$

Solution ▶ log 0.000451 ≈ 6.6542 − 10

Make Sure

Find each common logarithm using Appendix Table 2.

See Example 1 ▶ **1.** log 452 _____   **2.** log 72 _____   **3.** log 0.00423 _____

MAKE SURE ANSWERS: 1. 2.6551 2. 1.8573 3. 7.6263 − 10

Given the logarithm of a number x, you can find the number x, (called the *antilogarithm*), by using Appendix Table 2 in reverse.

To find the antilogarithm of a number, you use its mantissa and characteristic.

EXAMPLE 2: Find the antilogarithm of a number using Appendix Table 2.

Problem ▶ Find x if $\log x = 3.2455$.

1. Read digits ▶

x	0	1	2	3	4	5	6	7	8	9
1.0	.0000	.0043	.0086	.0128	.0170	.0212	.0253	.0294	.0334	.0374
1.1	.0414	.0453	.0492	.0531	.0569	.0607	.0645	.0682	.0719	.0755
1.2	.0792	.0828	.0864	.0899	.0934	.0969	.1004	.1038	.1072	.1106
1.3	.1139	.1173	.1206	.1239	.1271	.1303	.1335	.1367	.1399	.1430
1.4	.1461	.1492	.1523	.1553	.1584	.1614	.1644	.1675	.1703	.1732
1.5	.1761	.1790	.1818	.1847	.1875	.1903	.1931	.1959	.1987	.2014
1.6	.2041	.2068	.2095	.2122	.2148	.2175	.2201	.2227	.2253	.2279
1.7	.2304	.2330	.2355	.2380	.2405	.2430	.2455	.2480	.2504	.2529
1.8	.2553	.2577	.2601	.2625	.2648	.2672	.2695	.2718	.2742	.2765
1.9	.2788	.2810	.2833	.2856	.2878	.2900	.2923	.2945	.2967	.2989

Think: The mantissa of 3.2455 is 0.2455.

Think: Read the first two digits of the antilog in the far left-hand column. Read the third digit in the top row.

$1{\scriptstyle\wedge}76 \longleftarrow$ the significant digits of x

Position decimal point ▶ $x \approx 1{\scriptstyle\wedge}760.$

Think: Because the characteristic of 3.2455 is 3, the decimal point is 3 places to the right of the caret ($\wedge$).

Solution ▶ The common antilogarithm of $3.2455 \approx 1760$. That is, if $\log x = 3.2455$, then $x \approx 1760$.

Note ▶ It is often convenient to write antilogarithms using scientific notation.

Examples ▶ (a) The common antilogarithm of $3.2455 \approx 1.76 \times 10^3$.
(b) The common antilogarithm of $12.7760 \approx 5.97 \times 10^{12}$.
(c) The common antilogarithm of $(2.6484 - 10) \approx 4.45 \times 10^{-8}$.

Note ▶ In Example (c), it is helpful to first write $2.6484 - 10$ as $0.6484 - 8$.

Make Sure

Solve each equation for N using Appendix Table 2.

See Example 2 ▶ **1.** $\log N = 0.2878$ **2.** $\log N = 3.0374$ **3.** $\log N = 8.8338 - 10$

MAKE SURE ANSWERS: 1. 1.94 2. 1090 3. 0.0682

Logarithms with a base of e (≈ 2.71828) are often used in advanced mathematics. They are called *natural logarithms* and they are denoted by the following notation.

Natural Logarithms: $\ln M$ means $\log_e M$

If a number is an integral power of e, then you can find its natural logarithm mentally.

Examples ▶ (a) $\ln e = 1$, because: $\ln e = \log_e e = 1$. (b) $\ln e^2 = 2$, because: $\ln (e^2) = \log_e (e^2) = 2$.
(c) $\ln 1 = 0$, because: $\ln 1 = \log_e (e^0) = 0$.

12.4 Find Logarithms and Antilogarithms

It is often convenient to find common and natural logarithms with the aid of a calculator.

EXAMPLE 3: Find common and natural logarithms using a calculator.

Problems ▶ Find log 0.026 to the nearest ten-thousandth. | Find ln 57.1 to the nearest ten-thousandth.

1. Enter argument ▶ | 0.026 | | 57.1 ← the calculator display

2. Press logarithmic key ▶ | $-1.5850266\cdots$ Press $\boxed{\text{log}}$ for the common logarithm | $4.0448041\cdots$ Press $\boxed{\text{ln}}$ for the natural logarithm.

Solutions ▶ log 0.026 ≈ −1.5850 | ln 57.1 ≈ 4.0448

Other Examples ▶ (a) log 481 ≈ 2.6821 (b) ln 0.0054 ≈ −5.2214

Make Sure

Find each common and natural logarithm using a calculator. Round your answer to the nearest ten-thousandth.

See Example 3 ▶ 1. log 4.3, ln 4.3   2. log 0.00234, ln 0.00234

————, ————   ————, ————

MAKE SURE ANSWERS: 1. 0.6335, 1.4586 **2.** −2.6308, −6.0576

It is often convenient to find antilogarithms with the aid of a calculator. The following example illustrates the method used on many calculators.

EXAMPLE 4: Find antilogarithms using a calculator.

Problems ▶ Find N to the nearest hundredth if log N = 2.8882. | Find N to the nearest hundredth if ln N = 2.1017.

1. Enter logarithm ▶ | 2.8882 | 2.1017

2. Press inverse key ▶ | $773.03649\cdots$ Press $\boxed{\text{INV}}$ and $\boxed{\text{log}}$, or press $\boxed{10^x}$. | $8.1800642\cdots$ Press $\boxed{\text{INV}}$ and $\boxed{\text{ln}}$, or press $\boxed{e^x}$.

Solutions ▶ $N \approx 773.04$ | $N \approx 8.18$

Other Examples ▶ (a) If log M = 6, $M = 1000000 = 1.0 \times 10^6$. (b) If ln N = 4.5, $N \approx 90.02$.

Make Sure

Solve each equation for N to the nearest hundredth using a calculator.

See Example 4 ▶ 1. log N = 2.2450, ln N = 2.2450   2. log N = 4.2556, ln N = 4.2556

————, ————   ————, ————

MAKE SURE ANSWERS: 1. 175.79, 9.44 **2.** 18,013.58, 70.50

Sometimes it is convenient to evaluate a logarithm using logarithms with a different base. This can be accomplished by the following formula.

> **Change-of-Base Formula:** If $a > 0$ ($a \neq 1$), then: $\log_N M = \dfrac{\log_a M}{\log_a N}$.

Proof ▶ Let $\log_N M = y$

$N^y = M$ Think: $\log_b x = y$ means $b^y = x$.

$\log_a (N^y) = \log_a M$ Think: If $A = B$, then $\log_a A = \log_a B$.

$y \log_a N = \log_a M$ ⟵ the Power Property

$y = \dfrac{\log_a M}{\log_a N}$ Think: Solve for y by dividing both members by $\log_a N$.

$\log_N M = \dfrac{\log_a M}{\log_a N}$ Think: Since $y = \log_N M$, you can substitute $\log_N M$ for y.

EXAMPLE 5: Find logarithms using the Change-of-Base Formula.

Problem ▶ Find $\log_5 20$. Round your answer to the nearest ten-thousandth.

1. Use formula ▶ $\log_5 20 = \dfrac{\log 20}{\log 5}$ Think: $\log_N M = \dfrac{\log_a M}{\log_a N}$

2. Evaluate ▶ $\approx \dfrac{1.3010}{0.6989}$ Use a calculator or Appendix Table 2.

Solution ▶ $\log_5 20 \approx 1.8615$

The Change-of-Base Formula makes it possible to find a natural logarithm using common logarithms and to find a common logarithm using natural logarithms.

Example ▶ Find $\ln 3.86$ using common logarithms, and find $\log 6.2$ using natural logarithms. Give your answers to the nearest ten-thousandth.

1. Use formula ▶ $\ln 3.86 = \dfrac{\log 3.86}{\log e}$ | $\log 6.2 = \dfrac{\ln 6.2}{\ln 10}$

2. Evaluate ▶ $\approx \dfrac{0.5866}{0.4343}$ | $\approx \dfrac{1.8245}{2.3026}$

Solution ▶ $\ln 3.86 \approx 1.3507$ and $\log 6.2 \approx 0.7924$

Make Sure

Find each logarithm to the nearest ten-thousandth using the Change-of-Base Formula.

See Example 5 ▶ **1.** $\log_3 455$ _____   **2.** $\log_2 0.234$ _____   **3.** $\ln 254$ _____

MAKE SURE ANSWERS: 1. 5.5712 **2.** 7.9046 − 10 **3.** 5.5373

12.4 Practice

Set 1: Find each common logarithm using Appendix Table 2.

1. log 24
2. log 35
3. log 525

4. log 672
5. log 89100
6. log 32400

7. log 0.573
8. log 0.234
9. log 0.0321

10. log 0.0835
11. log 0.000431
12. log 0.000763

Set 2: Solve each equation for N using Appendix Table 2.

13. $\log N = 0.7356$
14. $\log N = 0.6010$
15. $\log N = 1.5391$

16. $\log N = 1.6031$
17. $\log N = 2.2878$
18. $\log N = 3.4698$

19. $\log N = 4.9335$
20. $\log N = 5.8414$
21. $\log N = 8.5092 - 10$

Set 3: Find each common and natural logarithm using a calculator. Round your answer to the nearest ten-thousandth.

22. log 5.3
23. log 4.1
24. ln 52
25. ln 63

26. log 98
27. log 2.34
28. ln 5.43
29. ln 4.82

30. log 0.632
31. log 0.0445
32. ln 6450
33. ln 8760

504 Chapter 12 Exponential and Logarithmic Functions

Set 4: Solve each equation for N using a calculator. Round your answers to the nearest hundredth.

34. $\log N = 0.7380$ **35.** $\log N = 0.5514$ **36.** $\ln N = 2.2$ **37.** $\ln N = 3.5$

38. $\log N = 3.1903$ **39.** $\log N = 4.3999$ **40.** $\ln N = 2.71838$ **41.** $\ln N = 7.6211$

42. $\log N = -0.2388$ **43.** $\log N = -0.5952$ **44.** $\ln N = -0.6$ **45.** $\ln N = -0.8$

46. $\log N = 1.9459$ **47.** $\log N = 2.1972$ **48.** $\ln N = 1$ **49.** $\ln N = 5$

Set 5: Find each logarithm to the nearest ten-thousandth using the Change-of-Base Formula.

50. $\log_2 20$ **51.** $\log_2 30$ **52.** $\log_5 2$

53. $\log_5 10$ **54.** $\log_7 0.5$ **55.** $\log_7 0.0016$

56. $\ln 4.38$ **57.** $\ln 1.59$ **58.** $\ln 20.4$

59. $\ln 10.3$ **60.** $\ln 1.09$ **61.** $\ln 4.51$

Extra: Find each logarithm to the nearest ten-thousandth using natural logarithms and the Change-of-Base Formula.

62. $\log 2.3$ **63.** $\log 4.5$ **64.** $\log 3.54$

65. $\log 24.3$ **66.** $\log 52.4$ **67.** $\log 859$

Review: Work these problems on a separate sheet of paper. Attach your work to this page.

Solve each quadratic equation. (See Lesson 9.3.)
68. $x^2 - 5x - 24 = 0$ **69.** $x^2 + 3x - 18 = 0$ **70.** $6x^2 - x - 1 = 0$
71. $x^2 + x - 1 = 0$ **72.** $3x^2 - 5x + 2 = 0$ **73.** $2x^2 - 3x + 1 = 0$
74. $2x^2 - 4x - 3 = 0$ **75.** $3x^2 + 2x - 3 = 0$ **76.** $5x^2 - 5x + 1 = 0$

12.5 Solve Exponential and Logarithmic Equations

Some exponential equations may be solved by writing each member of the equation in terms of a common base and then equating their exponents.

Recall ▶ Because exponential functions are one-to-one functions, you can solve equations of the form $b^x = b^y$ by equating x and y.

EXAMPLE 1: Solve an exponential equation by equating exponents of like bases.

Solve ▶ $2^{3x-7} = 32$

1. Write both members as powers of a common base ▶ $2^{3x-7} = 2^5$ Think: $32 = 2^5$

2. Equate their exponents ▶ $3x - 7 = 5$ Think: $b^x = b^y$ means $x = y$

3. Solve ▶ $3x = 12$

           $x = 4$ ⟵ a proposed solution

4. Check ▶ $2^{3x-7} = 32$ ⟵ original equation

$$\begin{array}{c|c} 2^{3(4)-7} & 32 \\ 2^{12-7} & 32 \\ 2^5 & 32 \\ 32 & 32 \end{array}$$ ⟵ 4 checks

Solution ▶ $x = 4$

Make Sure

Solve each exponential equation by equating exponents of like bases.

See Example 1 ▶   **1.** $2^{z+3} = 16$ **2.** $3^{2y-3} = 243$ **3.** $2^{x^2-3} = 128$

MAKE SURE ANSWERS: 1. 1   2. 4   3. $\pm\sqrt{10}$

Chapter 12 Exponential and Logarithmic Functions

You can solve many exponential equations using logarithms. This is because for any positive numbers M and N, if $M = N$, then: $\log_b M = \log_b N$.

EXAMPLE 2: Solve an exponential equation by equating the logarithms of both members.

Solve ▶ $4^x = 50$

1. Equate logarithms of both members ▶ $\log(4^x) = \log 50$ Think: $A = B$ means $\log_b A = \log_b B$.

2. Use Power Property ▶ $x \log 4 = \log 50$

3. Solve ▶ $x = \dfrac{\log 50}{\log 4}$ ⟵ exact answer

     $\approx \dfrac{1.6990}{0.6021}$ Use a calculator or Appendix Table 2.

     ≈ 2.8218 ⟵ approximation

Solution ▶ $x = \dfrac{\log 50}{\log 4}$ or $x \approx 2.8218$

CAUTION: $\dfrac{\log M}{\log N} \neq \log M - \log N$

Examples ▶   (a) *Correct Method*   (b) *Wrong Method*

$\dfrac{\log 100}{\log 10} = \dfrac{2}{1}$ $\dfrac{\log 100}{\log 10} = \log 100 - \log 10$

$= 2$ $= 2 - 1$

     $= 1$ **No!** The correct answer is 2.

To solve exponential equations that involve the constant e, you often use natural logarithms.

Example ▶   Solve $e^{3x} = 36$.

1. Equate logarithms ▶ $\ln(e^{3x}) = \ln 36$ Think: If the members are equal, then their natural logarithms are also equal.

2. Simplify ▶ $3x = \ln 36$ Think: $\ln(e^a) = a$. Identity one in Lesson 12.3.

3. Solve ▶ $x = \dfrac{\ln 36}{3}$ ⟵ exact answer

     $\approx \dfrac{3.5835}{3}$

     ≈ 1.1945 ⟵ approximation

Solution ▶ $x = \dfrac{\ln 36}{3}$ or $x \approx 1.1945$

12.5 Solve Exponential and Logarithmic Equations

Make Sure

Solve each exponential equation by equating the logarithms of both members.

See Example 2 ▶ **1.** $5^z = 75$ _____   **2.** $2^{y^2} = 256$ _____   **3.** $e^{2x-1} = 20$ _____

MAKE SURE ANSWERS: **1.** $\dfrac{\log 75}{\log 5}$ **2.** $\pm 2\sqrt{2}$ **3.** $\dfrac{\ln 20 + 1}{2}$

You can solve some logarithmic equations by equating the antilogarithms of both members. For equations of the form $\log_b M = \log_b N$, this means M must equal N.

EXAMPLE 3: Solve a logarithmic equation.

Solve ▶ $\log(x^2 + 4) = \log(2x - 4) + \log(x + 1)$

1. Use Product Property ▶ $\log(x^2 + 4) = \log[(2x - 4)(x + 1)]$ Think: $\log M + \log N = \log(MN)$

2. Equate antilogarithms ▶ $x^2 + 4 = (2x - 4)(x + 1)$ Think: $\log_b M = \log_b N$ means $M = N$.

3. Solve ▶
$x^2 + 4 = 2x^2 + 2x - 4x - 4$
$x^2 + 4 = 2x^2 - 2x - 4$
$0 = x^2 - 2x - 8$
$0 = (x + 2)(x - 4)$
$x + 2 = 0 \text{ or } x - 4 = 0$
$x = -2 \text{ or } x = 4$ ⟵ proposed solutions

4. Check ▶ -2 is not a solution because the arguments in the right member are negative when $x = -2$.

$\log(x^2 + 4) = \log(2x - 4) + \log(x + 1)$ ⟵ original equation

$\log(4^2 + 4)$	$\log(2 \cdot 4 - 4) + \log(4 + 1)$
$\log 20$	$\log 4 + \log 5$
$\log 20$	$\log(4 \cdot 5)$
$\log 20$	$\log 20$ ⟵ 4 checks

Solution ▶ $x = 4$

Make Sure

Solve each logarithmic equation.

See Example 3 ▶

1. $\log(x^2) = \log(x+2)$

2. $\log_2(y^2) - \log_2(4y-2) = \log_2 10$

MAKE SURE ANSWERS: 1. 2, −1   2. $20 \pm 2\sqrt{95}$

According to Guinness

SUPERSONIC FLIGHT
THE FIRST SUPERSONIC FLIGHT WAS ACHIEVED OCT. 14, 1947 BY CAPT. (LATER BRIG-GEN.) CHARLES ELWOOD YEAGER, U.S.A.F. RETIRED, OVER EDWARDS AIR FORCE BASE, MUNROC, CALIF. IN A U.S. BELL XS-1 ROCKET PLANE, WITH MACH 1.015 (670 M.P.H.) AT AN ALTITUDE OF 42,000 FT.

3. According to Guinness, how fast is Mach 1, to the nearest whole mile per hour?

4. To the nearest tenth kilometer, at what altitude was the first supersonic flight achieved? (Use 1 mi ≈ 1.609 km.)

On the outside of the U.S. Bell XS-1 rocket plane at 42,000 feet:

5. What was the air temperature using: $a = 0.16(15 - t)$ where a is the altitude in kilometers and t is the temperature in degrees Celsius?

6. What was the atmospheric pressure using the answer to Problem 9 in PS 17?

12.5 Practice

Set 1: Solve each exponential equation by equating exponents of like bases.

1. $2^x = 8$
2. $3^x = 81$
3. $5^x = 625$

4. $4^x = 256$
5. $3^{x+2} = 243$
6. $2^{x+5} = 256$

7. $5^{x-3} = 3125$
8. $6^{x-4} = 36$
9. $3^{2x-3} = 243$

10. $2^{4x+3} = 256$
11. $2^{x^2} = 32$
12. $3^{x^2} = 243$

Set 2: Solve each exponential equation.

13. $3^x = 12$
14. $3^x = 30$
15. $2^x = 40$

16. $2^x = 50$
17. $5^x = 24$
18. $5^x = 35$

19. $e^x = 35$
20. $e^x = 27$
21. $2^{3x-7} = 20$

22. $3^{4x+1} = 42$
23. $3^{x^2} = 15$
24. $2^{x^2} = 12$

Set 3: Solve each logarithmic equation.

25. $\log x = \frac{2}{3} \log 9$

26. $\log x = \frac{1}{2} \log 16$

27. $\log x = \frac{2}{3} \log 27 + 2 \log 2$

28. $\log x = \frac{1}{2} \log 25 - \log 20$

29. $\log x + \log (x + 7) = -1$

30. $\log x + \log (x - 3) = 1$

31. $\ln (x - 2) - \ln (x - 3) = \ln 2$

32. $\ln (3 - x) - \ln (x - 3) = \ln 1$

33. $\log_2 (2x - 3) - \log_2 (1 - 2x) = \log_2 2$

34. $\log_2 (x - 3) - \log_2 (2x + 1) = \log_2 2$

35. $\log (x - 4) + \log (3x - 4) = 1$

36. $\log (x - 3) + \log (5x + 2) = 1$

37. $\log (\log x) = 8$

38. $\ln (\ln x) = 1$

Review: Work these problems on a separate sheet of paper. Attach your work to this page.

Evaluate each expression for the given variable. (See Lesson 1.4.)

39. $5n$ for $n = 1, 3, 5$
40. $-3n$ for $n = 2, 4, 5$
41. n^3 for $n = 1, 3, 4$
42. $(-1)^n$ for $n = 1, 4, 5$
43. $(-1)^{n-1}$ for $n = 1, 4, 5$
44. -1^{n+1} for $n = 1, 3, 4$
45. $\frac{n^3}{n-1}$ for $n = 1, 3, 4$
46. $(n-1)(n-2)$ for $n = 1, 2, 3, 6$
47. $n(n+1)(n+2)$ for $n = 1, 2, 4, 5$
48. $\frac{n}{2}(3n - 2)$ for $n = 1, 2, 3, 4$

Problem Solving 17: Solve Problems Using Exponential Formulas

The accumulated amount A that a principal P will be worth at the end of t years, when invested at an interest rate r and compounded n times a year, is given by the *compound-interest formula*:

$$A = P\left(1 + \frac{r}{n}\right)^{nt}$$

EXAMPLE: Solve this problem using the compound interest formula.

1. Read and identify ▶ On May 6, 1626, the Dutch purchased Manhattan Island from the Manhattan Indians for $24. If the $24 had been invested in a savings account on May 6, 1626, paying 6% interest compounded annually, how much would be in the account on May 5, 1990, 364 years later?

Remember, circle the facts and underline the question.

2. Understand ▶ The question asks you to evaluate $A = P\left(1 + \frac{r}{n}\right)^{nt}$ for A given $P = 24$, $r = 0.06$, $n = 1$, and $t = 364$.

3. Substitute ▶ $A = 24\left(1 + \frac{0.06}{1}\right)^{1 \cdot 364}$ Think: "compounded annually" means $n = 1$.

4. Compute ▶

Paper-and-Pencil Method

$\log A = \log [24(1.06)^{364}]$

$ = \log 24 + \log (1.06^{364})$

$ = \log 24 + 364 \log 1.06$

$ \approx 0.3802 + 1 + 364(0.0253)$

$ = 10.584$

$\log A \approx 10.5894$ means $A \approx 3.89 \times 10^{10}$

Calculator Method

$A = 24(1.06)^{364}$

$ \approx 24 \cdot 1626802968$

$ \approx 3.9043271 \times 10^{10}$

Note: The Calculator Method is shorter and more accurate than the Paper-and-Pencil Method.

5. Interpret ▶ $A \approx 3.89 \times 10^{10}$ (paper-and-pencil method) or $A \approx 3.9043271 \times 10^{10}$ (calculator method) means that on May 5, 1990, the original $24 would be worth about $38,900,000,000 or about $39,043,271,000.

Another Example ▶ How long would it take to double your money at 8% interest compounded quarterly?

The question asks you to evaluate $A = P\left(1 + \frac{r}{n}\right)^{nt}$ for t given $A = 2P$, $r = 0.08$, and $n = 4$.

$2P = P\left(1 + \dfrac{0.08}{4}\right)^{4t}$ Think: "compounded quarterly" means $n = 4$.

$2P = P(1.02)^{4t}$

$2 = 1.02^{4t}$ Divide both members by P.

$\log 2 = \log (1.02^{4t})$ Take the logarithm of both members.

$\log 2 = 4t \log 1.02$ Use the properties of logarithms.

$t = \dfrac{\log 2}{4 \log 1.02}$ Solve for t.

$t = 8.75$ or about 8 years 9 months

512 Chapter 12 Exponential and Logarithmic Functions

Practice: Round to the nearest tenth of a unit when necessary. Solve Problems 1–4 using the compound-interest formula.

1. At $5\frac{3}{4}\%$ interest compounded semi-annually, how much will $1000 amount to at the end of 5 years? (Hint: $n = 2$.)

2. How long will it take to triple your money at 12% interest compounded monthly? (Hint: $n = 12$.)

3. How much should be invested at 6% interest, compounded daily, to amount to $1000 in 5 years? (Hint: $n = 365$.)

4. How long will it take a dollar to lose half its purchasing power at a 4% annual inflation rate?

The general formula for *exponential growth* is $y = y_0 e^{kx}$ where x is the independent variable, y is the dependent variable, and both y_0 and k are positive constants.

Use the exponential growth formula to answer questions 5–8.

5. *Bacteria Growth:* The number y of bacteria present in a culture at time $x = 0$ is 1000. At the end of 2 hours there are 4000 bacteria in the culture. Write the specific exponential growth formula for this particular bacteria. (Hint: Find y_0 and k.)

6. For Problem 5: a) How many bacteria will be in the culture in 24 hours? b) How much time must pass before there are 1,000,000 bacteria in the culture?

7. *Population Growth:* The United States population was 204 million in 1970. By 1980, the United States population had grown to 226.5 million. Write the specific exponential growth formula for the United States.

8. For Problem 7: a) Project the United States population for the year 2000. b) Estimate the year that the United States' population will double that of 1980.

The general formula for *exponential decrease* is $y = y_0 e^{-kx}$ where x is the independent variable, y is the dependent variable, and both y_0 and k are positive constants.

Use the exponential growth decrease formula to answer questions 9–12.

9. *Atmospheric Pressure:* The atmospheric pressure y at sea level is 14.7 pounds per square inch (psi). At altitude 2000 feet, the atmospheric pressure is 13.5 psi. Write the specific exponential decrease formula for atmospheric pressure.

10. For Problem 9: a) What is the atmospheric pressure in mile-high Denver, Colorado? b) A person's blood will boil if the atmospheric pressure drops below 0.9 psi. If you are in an unpressurized airplane, at what altitude would your blood boil?

11. *Carbon 14 Dating:* The amount of carbon 14 y in any given object at time $x = 0$ is 100%. The *half-life* of carbon is 5750 years, which means that only 50% of the original carbon 14 remains at time $x = 5750$. Write the specific exponential decrease formula for carbon 14 dating.

12. For Problem 11: a) The oldest living thing on earth is a plant called the "King Clone," which was estimated to be 11,700 years old using carbon 14 dating. What percent of the original carbon 14 is found in the oldest living thing? b) How much time must pass before there is only one-fourth of the original carbon 14 remaining in an object?

Copyright © 1985 by Harcourt Brace Jovanovich, Inc. All rights reserved.

Problem Solving 18: Solve Problems Using Logarithmic Formulas

Because the range of normal human hearing is so enormous, a special scale in powers of 10 was devised to measure and compare audible sounds.

Sound Intensity: The *intensity level* N in *decibels* (db) of nearby *sound intensity* I in watts/square meter (W/m²) is given by:

$$I = I_0 10^{N/10}$$

where I_0 is the sound intensity at the threshold of normal hearing.

To find the intensity level N of a given sound intensity I, you will find the logarithmic form of the sound intensity formula easier to use.

Example ▶ Find the logarithmic form of $I = I_0 10^{N/10}$.

$\log I = \log (I_0 10^{N/10})$	Take the logarithm of each member.
$\log I = \log I_0 + \log (10^{N/10})$	Use the properties of logarithms.
$\log I = \log I_0 + \dfrac{N}{10} \log 10$	
$\log I = \log I_0 + \dfrac{N}{10}$	Think: $\log 10 = \log_{10} 10 = 1$
$10 \log I = 10 \log I_0 + N$	Multiply both members by 10.
$10 \log I - 10 \log I_0 = N$	Use the Addition Rule.
$10(\log I - \log I_0) = N$	Factor out the GCF 10.
$10 \log \left(\dfrac{I}{I_0}\right) = N$	◀── logarithmic form of the sound intensity formula

EXAMPLE: Solve this problem using the intensity level formula.

1. Read and identify ▶ By international agreement, the sound intensity at the threshold (beginning) of hearing I_0 is 10^{-12} W/m². The sound intensity of normal conversation is (10⁻⁶ W/m²). What is the international intensity level formula by agreement? What is the decibel rating of normal conversation?

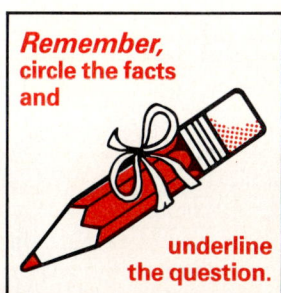

Remember, circle the facts and underline the question.

2. Substitute ▶ $N = 10 \log \left(\dfrac{I}{10^{-12}}\right)$ ◀── international intensity level formula (Answer to first question.)

3. Evaluate ▶ $N = 10 \log \left(\dfrac{10^{-6}}{10^{-12}}\right)$ Think: $I = 10^{-6}$ (W/m²)

$= 10 \log 10^6$ Think: $\dfrac{10^{-6}}{10^{-12}} = 10^{-6-(-12)} = 10^6$

$= 10(6)$

$= 60$

4. Interpret ▶ $N = 60$ means the decibel rating for normal conversation is 60 db. (Answer to second question.)

5. Check ▶ Is the sound intensity 10^{-6} W/m² when $I_0 = 10^{-12}$ W/m² and $N = 60$ db? Yes:
$I = I_0 10^{W/10} = (10^{-12}) 10^{60/10} = (10^{-12}) 10^6 = 10^{-12+6} = 10^{-6}$ ◀── $N = 60$ checks

Copyright © 1985 by Harcourt Brace Jovanovich, Inc. All rights reserved.

Chapter 12 Exponential and Logarithmic Functions

Practice: Round to the nearest tenth of a unit when necessary.

1. The sound intensity for average rock-and-roll music is 0.3 W/m². The sound intensity that causes pain is 1 W/m². The sound intensity that causes hearing damage is 10 W/m² or more. By how many decibels must the average rock-and-roll music be increased to a) cause pain? b) cause hearing damage?

2. The distance d in meters at which sound can be heard when no other sound is present is given by:

$$I = 4\pi d^2 I_0.$$

When no other sound is present, from what distance can you still hear a) rock-and-roll music? b) pain-causing noise? c) the minimum noise that can cause physical damage? d) normal conversation? e) barely audible sounds (threshold of hearing)?

Earthquake Intensity: The *amplitude a* of an earthquake wave is measured on a *seismograph* in millimeters (mm). The *magnitude M* of an earthquake is measured by the *Richter scale* given by the following formula:

$$a = a_0 \, 10^M,$$

where a_0 is the measure of a *zero-level earthquake* on a seismograph 100 km from the *epicenter* (middle) of the earthquake.

3. The logarithmic form of the earthquake-intensity formula is called the earthquake magnitude formula. Find the earthquake magnitude formula.

4. By international agreement, the measure of a zero-level earthquake a_0 is 10^{-3} mm. Write the international earthquake magnitude formula.

5. The *intensity of an earthquake* is the ratio a/a_0. Find the intensity of the San Francisco earthquake of 1906, which measured 8.3 on the Richter scale.

6. How many times as intense was the Alaska Good Friday earthquake of March 28, 1964, with magnitude 8.5 than the Coalinga, California, earthquake of May 2, 1983, with magnitude 6.5?

7. If a seismograph is located 100 km from the epicenter of an earthquake, find the magnitude of the earthquake if the greatest amplitude of the seismic waves is a) 1 micron (0.001 mm) b) 1 mm c) 1 cm d) 1 inch (2.54 cm).

8. The Japan earthquake of 1933 had a magnitude of 8.9. a) Find the amplitude of the largest seismic wave 100 km from the epicenter. b) Find the intensity. c) How many times more intense was the Japan earthquake of 1933 than the San Francisco earthquake of 1906?

According to Guinness

SHOUTING RECORD — JOANNE BROWN, 14, OF DRONFIELD WOODHOUSE, SHEFFIELD, ENGLAND, ATTAINED 113 DECIBELS ON JUNE 15, 1980, AT THE YORKSHIRE TELEVISION CENTRE, SHEFFIELD.

EARTHQUAKE — THE WORLD'S STRONGEST ASSESSABLE EARTHQUAKE HAS BEEN THE LEBU SHOCK NEAR CONCEPCION, CHILE, ON MAY 22, 1960, ASSESSED AT 9.5 ON THE RICHTER SCALE.

9. What was the sound intensity for the shouting record?

10. What was the intensity of the world's strongest assessable earthquake?

Name _____ Date _____ Class _____

Chapter 12 Review

What to Review if You Have Trouble

Objectives			Lesson	Example	Page
Evaluate expressions of the form b^x using a calculator ▶	**1.** Evaluate **a.** $3^{2.6}$ and **b.** $3^{2.7}$ to the nearest thousandth.	**a.** _____ **b.** _____	12.1	1	480
Graph exponential functions ▶	**2a.** Graph $f(x) = 3^x$ on Grid A. **b.** Graph $f(x) = \left(\dfrac{2}{3}\right)^x$ on Grid B.		12.1	2	481
Write logarithmic equations in exponential form ▶	**3.** Write $\log_3 81 = 4$ in exponential form.	_____	12.2	1	485
Write exponential equations in logarithmic form ▶	**4.** Write $5^3 = 125$ in logarithmic form.	_____	12.2	2	486
Solve logarithmic equations ▶	**5.** $\log_6 x = -2$	_____	12.2	3	487
Graph logarithmic equations ▶	**6.** Graph $y = \log_4 x$ on Grid C.		12.2	4	487

Grid A   Grid B   Grid C

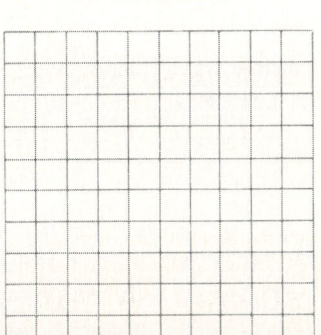

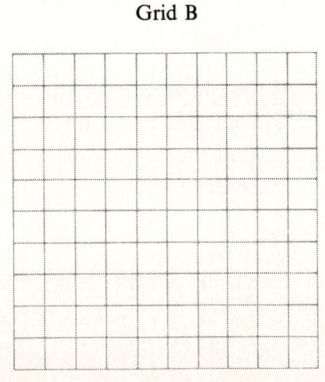

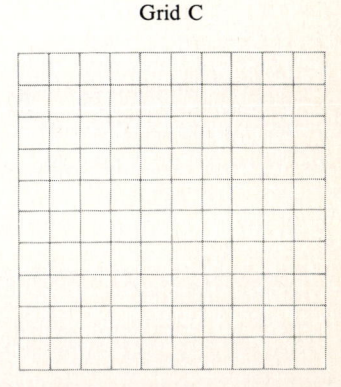

516 Chapter 12 Exponential and Logarithmic Functions

Use properties of logarithms ▶

7. Write $\log_b \left(\dfrac{\sqrt[3]{x}}{yz^2} \right)^2$ in terms of $\log_b x$, $\log_b y$, and $\log_b z$. —— 12.3   1   492

8. Write $\dfrac{1}{3}\log_b x - 2\log_b y + 3\log_b z$ as a single logarithm with a coefficient of 1. —— 12.3   2   493

9. Find $\log_{10} 36$ given $\log_{10} 2 \approx 0.3010$ and $\log_{10} 3 \approx 0.4771$. —— 12.3   3   494

Find logarithms and antilogarithms ▶

10. Find log 937 using Appendix Table 2. —— 12.4   1   498

11. Find the antilogarithm of 1.7612 using Appendix Table 2. —— 12.4   2   500

12. Find **a.** log 23.1 and **b.** ln 14.8 using a calculator. Round each result to the nearest ten-thousandth.   **a.** ——   **b.** ——   12.4   3   501

13. Find **a.** M and **b.** N if log M = 1.9315 and ln N = 2.7758. Use a calculator and round each result to the nearest hundredth.   **a.** ——   **b.** ——   12.4   4   501

14. Find $\log_7 40$ using the Change-of-Base Formula. —— 12.4   5   502

Solve exponential equations ▶

15. $3^{2x-7} = 81$ —— 12.5   1   505

16. $3^x = 40$ —— 12.5   2   506

Solve logarithmic equations ▶

17. $\log 2 + \log(x+2) + \log(x-3) = \log(x^2+3)$ —— 12.5   3   507

Solve problems using exponential formulas ▶

18. The formula for interest that is compounded continuously is

$$A = Pe^{rt}$$

where $e \approx 2.718282$. Which earns more interest a) $r = 7\frac{1}{2}\%$ interest compounded annually or b) $r = 7\frac{1}{4}\%$ interest computed continuously? (Hint: Let $P = 1$ (dollar) and $t = 1$ (year).) —— PS 17   —   511

Solve problems using logarithmic formulas ▶

19. If the amplitude of the largest seismic wave is doubled, by exactly how much is the **a.** magnitude **b.** intensity of the earthquake increased?   **a.** ——   **b.** ——   PS 18   —   513

CHAPTER 12 REVIEW ANSWERS: 1a. 17,399 **b.** 19,419 **2.** See Appendix Selected Answers. **3.** $3^4 = 81$ **4.** $\log_5 125 = 3$ **5.** $\dfrac{1}{36}$ **6.** See Appendix Selected Answers. **7.** $\dfrac{2}{3}\log_b x - 2\log_b y - 4\log_b z$ **8.** $\log_b \dfrac{x^{1/3}z^3}{y^2}$ **9.** 1.556 **10.** 2.9717 **11.** 57.7 **12a.** 1.3636 **b.** 2.6946 **13a.** 85.41 **b.** 16.05 **14.** 1.8957 **15.** 5.5 **16.** 3.3578 **17.** 5 **18. b)** $r = 7\frac{1}{4}\%$ compounded continuously **19a.** log 2 **b.** doubled

FUNCTIONS

13 Sequences and Series

13.1 Use Sequences and Series

13.2 Use Arithmetic Sequences and Series

PS 19: Solve Problems Using Arithmetic Sequences and Series

13.3 Use Geometric Sequences and Series

PS 20: Solve Problems Using Geometric Sequences and Series

13.4 Use the Binomial Expansion

13.1 Use Sequences and Series

A *sequence* is a function whose domain is the set of natural numbers or the set of the first n natural numbers. If the domain is the set of all natural numbers, then the sequence is called an *infinite sequence*. If the domain is the set of the first n natural numbers, then the sequence is called a *finite sequence*.

Examples ▶ (a) $f(n) = 5n + 7$ for $n = 1, 2, 3$, and 4, is a finite sequence. Its domain is $\{1, 2, 3, 4\}$, and its range is found by evaluating $f(n) = 5n + 7$ for $n = 1, 2, 3,$ and 4:

$$f(1) = 5(1) + 7 = 12, f(2) = 5(2) + 7 = 17, f(3) = 5(3) + 7 = 22, f(4) = 5(4) + 7 = 27.$$

In this sequence, the range is $\{12, 17, 22, 27\}$.

(b) $g(n) = \dfrac{1}{n}$ for $n \in N$, is an infinite sequence. Its domain is the set N of all natural numbers.

The kth range member can be found by evaluating $g(n) = \dfrac{1}{n}$ for $n = k$:

$$g(1) = \frac{1}{1} = 1, g(2) = \frac{1}{2}, g(3) = \frac{1}{3}, g(4) = \frac{1}{4}, \cdots, g(k) = \frac{1}{k}, \cdots.$$

In this sequence, the range is the infinite set of rational numbers of the form $\dfrac{1}{k}$ for $k = 1, 2, 3, \cdots$.

The range members of a sequence are called the *terms of the sequence*. The terms of the sequence a are denoted by using the subscript notation $a_1, a_2, a_3, \cdots, a_n, \cdots$ with $n \in N$ and

$$a_1 = a(1), a_2 = a(2), a_3 = a(3), \cdots, a_n = a(n), \cdots.$$

The nth term a_n is called the *general term*.

Example ▶ For the sequence a with terms $3, 8, 15, 24, 35, \cdots, n^2 + 2n, \cdots$

$a_1 = 3$ a_1 is the first term of the sequence.

$a_2 = 8$ a_2 is the second term of the sequence.

$a_3 = 15$ a_3 is the third term of the sequence.

⋮   ⋮   ⋮   ⋮

$a_n = n^2 + 2n$ a_n is the nth term of the sequence.

If the general term a_n of a sequence is defined by a formula involving n, then you can use the formula to find the terms of the sequence.

EXAMPLE 1: Find terms of a sequence using the general term.

Problem ▶ Find the first three terms and the tenth term of the sequence given by $a_n = 3n - 2$.

1. Write formula ▶ $a_n = 3n - 2$ | $a_n = 3n - 2$ | $a_n = 3n - 2$ | $a_n = 3n - 2$

2. Evaluate formula ▶ $a_1 = 3(1) - 2$ | $a_2 = 3(2) - 2$ | $a_3 = 3(3) - 2$ | $a_{10} = 3(10) - 2$

$\qquad\qquad\qquad\quad = 1 \qquad\qquad\qquad = 4 \qquad\qquad\qquad = 7 \qquad\qquad\qquad = 28$

Solution ▶ The first, second, third, and tenth terms of the sequence given by $a_n = 3n - 2$ are 1, 4, 7, and 28, respectively.

Make Sure

Find the first three terms and the tenth term of each sequence using the general term.

See Example 1 ▶ **1.** $a_n = n^3 - 2n + 1$ _____ **2.** $a_n = (-2)^n$ _____

MAKE SURE ANSWERS: 1. 0, 5, 22, 981 **2.** $-2, 4, -8, 1024$

The indicated sum of the terms of a sequence is called a *series*.

Examples ▶
(a) $1 + 3 + 6 + 10 + 15$ is a series associated with the sequence 1, 3, 6, 10, 15.
(b) $1 + \frac{1}{2} + \frac{1}{3} + \frac{1}{4} + \frac{1}{5} + \frac{1}{6}$ is a series associated with the sequence $1, \frac{1}{2}, \frac{1}{3}, \frac{1}{4}, \frac{1}{5}, \frac{1}{6}$.
(c) $1 - 2 + 4 - 8 + 16$ is a series associated with the sequence $1, -2, 4, -8, 16$.

A series is often represented by the following special notation.

Summation Notation

Given a sequence of terms $a_1, a_2, a_3, \cdots, a_n, \cdots$ the symbol $\sum_{k=1}^{n} a_k$ represents the sum of the first n terms of the sequence.

$$\underset{\text{summation notation form}}{\sum_{k=1}^{n} a_k} = \underbrace{a_1 + a_2 + a_3 + \cdots + a_n}_{\text{expanded form}}$$

In the summation notation $\sum_{k=1}^{n} a_k$, the k is called the *index of summation*. The set of natural numbers from 1 to n is called the *range of summation*.

EXAMPLE 2: Evaluate a series written in summation notation.

Problem ▶ Evaluate $\sum_{k=1}^{4} (k + 5)$.

1. Write expanded form ▶ $\sum_{k=1}^{4} (k + 5) = (1 + 5) + (2 + 5) + (3 + 5) + (4 + 5)$ Write the sum of the terms formed by replacing k in $(k + 5)$ with 1, 2, 3, and 4, respectively.

2. Simplify ▶ $= 6 + 7 + 8 + 9$

Solution ▶ $\sum_{k=1}^{4} (k + 5) = 30$

Any letter may be used to denote the index of summation.

Examples ▶ (a) $\sum_{i=1}^{2} \frac{i + 4}{i} = \frac{1 + 4}{1} + \frac{2 + 4}{2}$

$= \frac{5}{1} + \frac{6}{2}$

$= 8$

(b) $\sum_{j=1}^{3} (-1)^j (3j + 5) = (-1)^1(3 \cdot 1 + 5) + (-1)^2(3 \cdot 2 + 5) + (-1)^3(3 \cdot 3 + 5)$

$= (-1)(8) + 1(11) + (-1)(14)$

$= -11$

If every term of a series is the same constant, then the series is called a *constant series*.

Examples ▶ (a) $4 + 4 + 4 + 4 + 4 + 4$ ⟵ a constant series with all 6 terms equal to 4

(b) $\frac{3}{2} + \frac{3}{2} + \frac{3}{2} + \frac{3}{2} + \frac{3}{2} + \frac{3}{2} + \frac{3}{2}$ ⟵ a constant series with all 7 terms equal to $\frac{3}{2}$

To find the sum of a finite constant series, you multiply the number of terms n by the constant term c.

Examples ▶ (a) $4 + 4 + 4 + 4 + 4 + 4 = 6 \cdot 4 = 24$ (b) $\frac{3}{2} + \frac{3}{2} + \frac{3}{2} + \frac{3}{2} + \frac{3}{2} + \frac{3}{2} + \frac{3}{2} = 7 \cdot \frac{3}{2} = \frac{21}{2}$

If all n terms of a constant series equal the constant c, then you can write the series using summation notation as $\sum_{k=1}^{n} c$.

Examples ▶ (a) $4 + 4 + 4 + 4 + 4 + 4 = \sum_{k=1}^{6} 4$ (b) $\frac{3}{2} + \frac{3}{2} + \frac{3}{2} + \frac{3}{2} + \frac{3}{2} + \frac{3}{2} + \frac{3}{2} = \sum_{k=1}^{7} \frac{3}{2}$

Since the sum of a constant series with n terms can be computed by multiplying n times c (the value of each of the constant terms), it follows that you can evaluate the summation notation $\sum_{k=1}^{n} c$ by just multiplying n and c.

If c is a constant and n is a natural number then: $\sum_{k=1}^{n} c = nc$.

Examples ▶ (a) $\sum_{k=1}^{6} 4 = 6 \cdot 4 = 24$ (b) $\sum_{k=1}^{70} 3 = 70 \cdot 3 = 210$ (c) $\sum_{k=1}^{200} (-\frac{5}{4}) = 200(-\frac{5}{4}) = -250$

Make Sure

Evaluate series written in summation notation.

See Example 2 ▶ 1. $\sum_{k=1}^{5} (k^2 + 1)$ _____   2. $\sum_{k=1}^{6} (-\frac{1}{2})^{k-1}$ _____

MAKE SURE ANSWERS: 1. 60 2. $\frac{21}{32}$

Name _____ Date _____ Class _____

13.1 Practice

Set 1: Find the first three terms and the tenth term of such sequence using the general term.

1. $a_n = 2n + 3$
2. $a_n = 3n - 1$
3. $a_n = n^3$

4. $a_n = 2n^2$
5. $a_n = n^2 + 2n - 1$
6. $a_n = n^2 - 5n + 4$

7. $a_n = (-1)^n$
8. $a_n = (-1)^{n-1}$
9. $a_n = \dfrac{(-1)^{n-1}}{n}$

10. $a_n = \dfrac{(-1)^{1-n}}{n}$
11. $a_n = 3(-2)^{n-1}$
12. $a_n = 3\left(-\dfrac{1}{2}\right)^{n-1}$

1. _____
2. _____
3. _____
4. _____
5. _____
6. _____
7. _____
8. _____
9. _____
10. _____
11. _____
12. _____

522   Chapter 13   Sequences and Series

Set 2: Evaluate series written in summation notation.

13. $\sum_{k=1}^{5} 2k$ 14. $\sum_{k=1}^{5} 3k$ 15. $\sum_{k=1}^{6} \left(\frac{1}{2}k - 1\right)$ 16. $\sum_{k=1}^{4} \left(\frac{1}{3}k - 2\right)$

17. $\sum_{k=1}^{6} \frac{1}{4k}$ 18. $\sum_{k=2}^{5} \frac{2}{3k}$ 19. $\sum_{k=2}^{6} 2$ 20. $\sum_{k=1}^{4} 3$

21. $\sum_{k=1}^{6} (-2)^k$ 22. $\sum_{k=1}^{6} (-1)^k$ 23. $\sum_{k=1}^{4} (3k^2 + 4)$ 24. $\sum_{k=1}^{4} (-2k^2 + 5k)$

13. _____
14. _____
15. _____
16. _____
17. _____
18. _____
19. _____
20. _____
21. _____
22. _____
23. _____
24. _____

Review: Work these problems on a separate sheet of paper. Attach your work to this page.

Evaluate each expression for the given values of the variables. (See Lesson 1.4.)

25. $a_1 + (n - 1)d$ if $a_1 = 3, n = 9, d = 5$
26. $a_1 + (n - 1)d$ if $a_1 = 5, n = 11, d = -2$
27. $a_1 + (n - 1)d$ if $a_1 = 81, n = 9, d = -3$
28. $a_1 + (n - 1)d$ if $a_1 = 4, n = 11, d = -\frac{1}{2}$

29. $n\left(\dfrac{a_1 + a_n}{2}\right)$ if $a_1 = 6, n = 17, a_n = 96$

30. $n\left(\dfrac{a_1 + a_n}{2}\right)$ if $a_1 = 14, a_n = -43, n = 20$

Copyright © 1985 by Harcourt Brace Jovanovich, Inc. All rights reserved.

13.2 Use Arithmetic Sequences and Series

Some sequences have the property that every term after the first term can be found by adding the same fixed number to the preceding term.

> A sequence $a_1, a_2, a_3, \cdots, a_n, \cdots$ is called an *arithmetic sequence* (*arithmetic progression*) if there is a constant d, called the *common difference*, such that
> $$a_n = a_{n-1} + d \qquad \text{for every } n \geq 2.$$

EXAMPLE 1: Find terms of an arithmetic sequence.

Problem ▶ Write the first five terms of the arithmetic sequence with $a_1 = 11$ and $d = 20$.

1. Write a_1 ▶ $\quad a_1 = 11$

2. Use definition ▶ $\quad a_2 = a_1 + d = 11 + 20 = 31$

$\quad a_3 = a_2 + d = 31 + 20 = 51$

$\quad a_4 = a_3 + d = 51 + 20 = 71$

$\quad a_5 = a_4 + d = 71 + 20 = 91$

Solution ▶ The first five terms of the arithmetic sequence with $a_1 = 11$ and $d = 20$ are 11, 31, 51, 71, and 91.

Make Sure

Find the first five terms of each arithmetic sequence given the first term and the common difference.

See Example 1 ▶
1. $a_1 = 5, d = 6$
2. $a_1 = 1, d = \dfrac{1}{3}$
3. $a_1 = 5, d = -3$

MAKE SURE ANSWERS: 1. 5, 11, 17, 23, 29   2. 1, $1\frac{1}{3}$, $1\frac{2}{3}$, 2, $2\frac{1}{3}$ 3. 5, 2, −1, −4, −7

Every term of an arithmetic sequence can be written in terms of the first term a_1 and the common difference d.

Example ▶ $\quad a_1 = 3, \qquad a_2 = 10, \qquad a_3 = 17, \qquad a_4 = 24, \qquad a_5 = 31$

can be written as

$a_1 = 3 + \boxed{0} \cdot 7, \; a_2 = 3 + \boxed{1} \cdot 7, \; a_3 = 3 + \boxed{2} \cdot 7, \; a_4 = 3 + \boxed{3} \cdot 7, \; a_5 = 3 + \boxed{4} \cdot 7.$

▶ In each case, the subscript is one more than the coefficient of the common difference. The following nth-term formula is a generalization of the above result.

Chapter 13 Sequences and Series

> The general term of an arithmetic sequence with first term a_1 and common difference d is given by the *nth-term formula*:
>
> $$a_n = a_1 + (n - 1)d$$

EXAMPLE 2: Find the specified unknown of an arithmetic sequence using the *n*th term formula.

Problem ▶ Given an arithmetic sequence with $a_1 = 6$ and $d = 8$, find a_{30}.

1. Write formula ▶ $a_n = a_1 + (n - 1)d$ ⟵ *n*th-term formula for an arithmetic sequence

2. Evaluate ▶ $a_{30} = 6 + (30 - 1)8$ Think: To compute a_{30}, use $n = 30$.

$= 6 + (29)8$

$= 238$

Solution ▶ Given an arithmetic sequence with $a_1 = 6$ and $d = 8$, then: $a_{30} = 238$.

Given any three of the four unknowns in $a_n = a_1 + (n - 1)d$, you can find the remaining unknown.

Examples ▶ Given an arithmetic sequence with:

(a) $a_1 = 20$ and $a_9 = 140$, find d.

(b) $a_{15} = 100$ and $d = -5$, find a_1.

(c) $a_1 = -60$, $a_n = 200$, and $d = 4$, find n.

1. Write formula ▶

$a_n = a_1 + (n - 1)d$ $a_n = a_1 + (n - 1)d$ $a_n = a_1 + (n - 1)d$

2. Solve ▶

$140 = 20 + (9 - 1)d$ $100 = a_1 + (15 - 1)(-5)$ $200 = -60 + (n - 1)(4)$

$140 = 20 + 8d$ $100 = a_1 + (14)(-5)$ $260 = (n - 1)(4)$

$120 = 8d$ $100 = a_1 + (-70)$ $65 = n - 1$

$15 = d$ $170 = a_1$ $66 = n$

Solutions ▶

If $a_1 = 20$ and $a_9 = 140$, then $d = 15$.

If $a_{15} = 100$ and $d = -5$, then $a_1 = 170$.

If $a_1 = -60$, $a_n = 200$, and $d = 4$, then $n = 66$.

Make Sure

Find each specified unknown of an arithmetic sequence using the *n*th-term formula.

See Example 2 ▶   **1.** a_{20} if $a_1 = 2$, $d = 5$ **2.** n if $a_1 = 21$, $a_n = -6$, $d = -3$

MAKE SURE ANSWERS: 1. 97 2. 10

The indicated sum of the terms of an arithmetic sequence is called an *arithmetic series*.

Examples ▶ (a) $1 + 3 + 5 + 7 + 9$ is an arithmetic series associated with the arithmetic sequence 1, 3, 5, 7, 9.

(b) $11 + 8 + 5 + 2 - 1$ is an arithmetic series associated with the arithmetic sequence 11, 8, 5, 2, −1.

(c) $1 + 2 + 4 + 8 + 16 + 32$ is not an arithmetic series because 1, 2, 4, 8, 16, 32 is not an arithmetic sequence.

To find the sum of a finite arithmetic series, you can simply add the terms. However, easier methods can sometimes be applied. For example, when the German mathematician Karl Friedrich Gauss (1777–1855), was a student in elementary school, his teacher assigned the problem of finding the sum of the first one-hundred natural numbers. Most students proceeded to add the numbers in their regular order, but Gauss quickly found the sum using a method similar to the one in the following example.

Example ▶ Find the sum S_{100} of the first one-hundred natural numbers.

1. Write series ▶ $S_{100} = 1 + 2 + 3 + \cdots + 98 + 99 + 100$

2. Write reverse order ▶ $S_{100} = \underline{100 + 99 + 98 + \cdots + 3 + 2 + 1}$

3. Add ▶ $2S_{100} = \underbrace{101 + 101 + 101 + \cdots + 101 + 101 + 101}_{100 \text{ terms}}$

4. Simplify ▶ $2S_{100} = 100 \cdot 101$ Think: Use $\sum_{k=1}^{n} c = nc$ with $n = 100$ and $c = 101$.

   $2S_{100} = 10100$

5. Solve for S_{100} ▶ $S_{100} = \dfrac{10100}{2}$

   $S_{100} = 5050$

Solution ▶ The sum of the first one-hundred natural numbers is 5050.

The method used in the previous problem can be generalized as a formula.

Problem ▶ Develop a formula for the sum S_n of the first n terms of an arithmetic series.

1. Write the series ▶ $S_n = a_1 + (a_1 + d) + (a_1 + 2d) + \cdots + [a_1 + (n-2)d] + [a_1 + (n-1)d]$

2. Write reverse order ▶ $S_n = \underline{a_n + (a_n - d) + (a_n - 2d) + \cdots + [a_n - (n-2)d] + [a_n - (n-1)d]}$

3. Add ▶ $2S_n = \underbrace{(a_1 + a_n) + (a_1 + a_n) + (a_1 + a_n) + \cdots + (a_1 + a_n) + (a_1 + a_n)}_{n \text{ terms}}$

4. Solve for S_n ▶ $2S_n = n(a_1 + a_n)$ Think: Use $\sum_{k=1}^{n} = nc$ with $c = (a_1 + a_n)$.

   $S_n = \dfrac{n(a_1 + a_n)}{2}$

Solution ▶ The sum of the first n terms of an arithmetic series is given by $S_n = \dfrac{n(a_1 + a_n)}{2}$.

You can write the formula $S_n = \dfrac{n(a_1 + a_n)}{2}$ as $S_n = n\left(\dfrac{a_1 + a_n}{2}\right)$. This form of the formula is easy to remember because it states that the sum of the first n terms of an arithmetic series is equal to n times the average of the first term and the nth term.

EXAMPLE 3: Find the sum of the first n terms of an arithmetic series using $S_n = n\left(\dfrac{a_1 + a_n}{2}\right)$.

Problem ▶ Find the sum of the first 51 terms of the arithmetic series with $a_1 = 12$ and $d = 7$.

1. Identify a_1 and n ▶ $a_1 = 12, n = 51, d = 7$ ◀—— given information

2. Find a_n ▶ $a_n = a_1 + (n - 1)d$ Think: Use the nth-term formula to compute a_n.

$a_{51} = 12 + (51 - 1)7$

$= 362$

3. Evaluate formula ▶ $S_n = n\left(\dfrac{a_1 + a_n}{2}\right)$

$S_{51} = 51\left(\dfrac{12 + 362}{2}\right)$

$= 9537$

Solution ▶ The sum of the first 51 terms of the arithmetic series with $a_1 = 12$ and $d = 7$ is 9537.

Another Example ▶ Find the sum of all the even numbers from 80 to 300, inclusive.

1. Identify a_1 and a_n ▶ $a_1 = 80, a_n = 300$ ◀—— given information

2. Find n ▶ $a_n = a_1 + (n - 1)d$ Think: Use the nth-term formula to compute n.

$300 = 80 + (n - 1)2$ Think: Successive even numbers have a common difference of 2.

$300 = 78 + 2n$

$111 = n$

3. Evaluate formula ▶ $S_{111} = 111\left(\dfrac{80 + 300}{2}\right)$ Think: $S_n = n\left(\dfrac{a_1 + a_n}{2}\right)$

$= 21{,}090$

Solution ▶ The sum of all the even numbers from 80 to 300, inclusive, is 21,090.

Make Sure

Find the sum of the first n terms of each arithmetic series using $S_n = n\left(\dfrac{a_1 + a_n}{2}\right)$.

See Example 3 ▶ **1.** Find S_{20} if $a_1 = 2, d = 3$ **2.** Find S_{45} if $a_1 = 7, d = -0.3$

Name _____ Date _____ Class _____

13.2 Practice

Set 1: Find the first five terms of each arithmetic sequence given the first term and common difference.

1. $a_1 = 1, d = 2$
2. $a_1 = 3, d = 5$
3. $a_1 = 4, d = -2$

4. $a_1 = 5, d = -4$
5. $a_1 = -7, d = 7$
6. $a_1 = -5, d = 3$

7. $a_1 = -22, d = -4$
8. $a_1 = -17, d = -5$
9. $a_1 = \frac{1}{2}, d = \frac{2}{3}$

10. $a_1 = \frac{3}{4}, d = \frac{2}{5}$
11. $a_1 = 0.3, d = 1.4$
12. $a_1 = 2.5, d = -1.2$

Set 2: Find each specified unknown of an arithmetic sequence using the nth-term formula.

13. a_{13} if $a_1 = 1, d = 3$
14. a_{21} if $a_1 = 3, d = 5$
15. a_{27} if $a_1 = 3, d = -2$

16. a_{17} if $a_1 = 15, d = -3$
17. a_{15} if $a_1 = -\frac{1}{3}, d = \frac{2}{3}$
18. a_{16} if $a_1 = -\frac{2}{5}, d = \frac{1}{2}$

19. n if $a_1 = \frac{4}{5}, a_n = -\frac{43}{15}, d = -\frac{1}{3}$
20. n if $a_1 = \frac{2}{3}, a_n = -\frac{43}{3}, d = -\frac{3}{4}$

21. a_1 if $a_{25} = 56.6, d = 2.3$
22. a_1 if $a_{18} = 17.4, d = 1.3$

23. d if $a_1 = 81, a_9 = 17$
24. d if $a_1 = 27, a_{19} = -99$

Copyright © 1985 by Harcourt Brace Jovanovich, Inc. All rights reserved.

528 Chapter 13 Sequences and Series

25. _____

26. _____

27. _____

28. _____

29. _____

30. _____

31. _____

32. _____

33. _____

34. _____

35. _____

36. _____

37. _____

38. _____

39. _____

40. _____

41. _____

42. _____

Set 3: Find the sum of the first n terms of each arithmetic series using $S_n = n\left(\dfrac{a_1 + a_n}{2}\right)$.

25. S_{50} if $a_1 = 1, d = 2$ 26. S_{50} if $a_1 = 2, d = 3$ 27. S_{100} if $a_1 = 1, d = 1$

28. S_{100} if $a_1 = 100, d = -1$ 29. S_{30} if $a_1 = 1, d = \dfrac{1}{2}$ 30. S_{30} if $a_1 = 1, d = \dfrac{2}{3}$

31. S_{25} if $a_1 = \dfrac{2}{5}, d = -\dfrac{1}{2}$ 32. S_{28} if $a_1 = \dfrac{4}{5}, d = -\dfrac{1}{3}$ 33. S_{45} if $a_1 = 0.1, d = 0.3$

34. S_{26} if $a_1 = 2.5, d = 1.3$ 35. S_{30} if $a_1 = \dfrac{1}{2}, d = -\dfrac{1}{4}$ 36. S_{18} if $a_1 = \dfrac{3}{4}, d = -\dfrac{2}{3}$

Extra: Find each specified term of an arithmetic sequence given two terms of the sequence.

37. a_2 if $a_1 = 5, a_3 = 11$ 38. a_2 if $a_1 = 3, a_3 = 12$

39. a_3 if $a_2 = \dfrac{2}{3}, a_4 = \dfrac{1}{3}$ 40. a_4 if $a_3 = \dfrac{1}{2}, a_5 = \dfrac{1}{3}$

41. a_8 if $a_7 = 4.3, a_9 = 5.5$ 42. a_{10} if $a_1 = 1.55, a_{11} = 1.54$

Review: Work these problems on a separate sheet of paper. Attach your work to this page.

Evaluate each expression for the given values of the variables. (See Lesson 1.4.)

43. $a_1 r^{n-1}$ if $a_1 = 5, r = 2, n = 7$ 44. $a_1 r^{n-1}$ if $a_1 = 7, r = -1, n = 10$

45. $\dfrac{a_1 - a_1 r^n}{1 - r}$ if $a_1 = 4, r = \dfrac{1}{2}, n = 6$ 46. $\dfrac{a_1 - a_1 r^n}{1 - r}$ if $a_1 = 1, r = -1, n = 9$

47. $\dfrac{a_1}{1 - r}$ if $a_1 = 1, r = \dfrac{1}{2}$ 48. $\dfrac{a_1}{1 - r}$ if $a_1 = 3, r = -\dfrac{1}{3}$

Copyright © 1985 by Harcourt Brace Jovanovich, Inc. All rights reserved.

Problem Solving 19: Solve Problems Using Arithmetic Sequences and Series

Many important applications can be solved using sequences and series.

EXAMPLE: Solve this problem using an arithmetic sequence and series.

1. Read and identify ▶ A free-falling object near the earth's surface will fall from rest (16 feet in the first second), (48 feet during the second second), and (80 feet during the third second), and so on. How far will the object fall during the (tenth second)? How far will the object fall in (10 seconds)? How long will it take for the object to drop (1 mile), to the nearest tenth second?

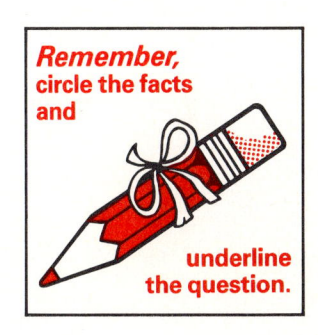

Remember, circle the facts and underline the question.

2. Understand ▶ The terms $a_1 = 16, a_2 = 48, a_3 = 80, \cdots$ form an arithmetic sequence because the common difference is 32.

3. Decide ▶ To solve a problem that involves an arithmetic sequence or series, you **evaluate**

$$a_n = a_1 + (n-1)d \text{ or } S_n = n\left(\frac{a_1 + a_n}{2}\right),$$

respectively.

4. Evaluate a_n ▶ $a_{10} = 16 + (10 - 1)32$ Think: $a_n = a_1 + (n-1)d$
$\qquad = 304$

5. Interpret ▶ $a_{10} = 304$ means that during the tenth second the object will fall **304 feet**.

6. Evaluate S_n ▶ $S_{10} = 10\left(\dfrac{16 + 304}{2}\right)$ Think: $S_n = n\left(\dfrac{a_1 + a_2}{2}\right)$
$\qquad = 1600$

7. Interpret ▶ $S_{10} = 1600$ means that during the first 10 seconds the object will fall **1600 feet**.

8. Solve for n ▶ $S_n = n\left(\dfrac{a_1 + a_n}{2}\right)$

$\qquad S_n = n\left(\dfrac{a_1 + a_1 + (n-1)d}{2}\right)$ Think: $a_n = a_1 + (n-1)d$

$\qquad 5280 = n\left(\dfrac{16 + 16 + (n-1)32}{2}\right)$ Substitute: 1 mile = 5280 feet

$\qquad 5280 = 16n^2$ Simplify.

$\qquad 330 = n^2$

$\qquad \pm\sqrt{330} = n$ Use the Square-Root Rule.

$\qquad 18.2 \approx n$ Find the positive square root and round to the nearest tenth.

9. Interpret ▶ $n \approx 18.2$ means that for the object to fall 1 mile takes about **18.2 seconds**.

Practice: Round to the nearest tenth when necessary.

1. Dry air cools at the rate of $5\frac{1}{2}°$F for each 1000-foot rise. If the ground level temperature is 72°F, a) how high above ground level is the temperature 39°F? b) What is the temperature at 30,000 feet above ground level? c) Write a formula for the temperature T_n in terms of the height above ground level n, in thousands of feet.

2. A certain job pays you a $20,000 starting salary, with a $1500 increase at the end of each year worked. How much can you earn during the a) fifth year? b) first 5 years? c) How long will it take you to earn $30,000 or more per year? d) Would you earn more in the fifth year if the starting salary were $25,000, with annual increases of $1000?

3. In a certain city, the fine for a first parking offense is $12. For each additional parking offense, a fine of $18 is added. How much would you pay for a) the ninth parking offense? b) all 9 parking offenses? c) How many parking offenses does a person who has to pay $210 for the latest parking offense have?

4. The top row of a stack of pipes contains 3 pipes, the next row down contains 4 pipes, the next row down contains 5 pipes, and so on. If there are 40 rows, then how many pipes are in the a) bottom row? b) whole stack? c) How many rows are needed for the stack to contain 196 pipes?

5. A certain object will roll down an inclined plane 2 feet in the first second, 5 feet during the second second, 8 feet during the third second, and so on. How far will the object roll a) during the twelfth second? b) in all 12 seconds? c) How long will it take the object to roll 100 yards?

6. During a free-fall, a certain parachutist fell 12 feet in the first second, 37 feet during the second second, 62 feet during the third second, and so on. How far did she fall a) during the fourteenth second? b) in all 14 seconds? c) How long will it take her to fall 3 miles?

According to Guinness

PENNY PYRAMID
PAUL JAY COHEN OF PHILADELPHIA, PA., BUILT A PYRAMID OF 71,825 PENNIES, WITHOUT ANY ADHESIVE, OVER A PERIOD OF MONTHS.

A stack of 13 pennies is 1 inch high. Assume each stack of pennies in every layer of the pyramid contains 13 pennies.

7. For a pyramid to contain 71,825 pennies, there need to be more than 20 layers. Find the number of layers using

$$p \sum_{k=1}^{n} k^2 = 71{,}825$$

where p is the number of pennies in each stack, n is the number of layers, and k^2 is the number of stacks in each layer (1, 4, 9, 16, 25, $\cdots$ or $1^2, 2^2, 3^2, 4^2, \cdots$).

8. For Problem 7, find the number of pennies that show a) on each side of the pyramid b) on all four sides of the pyramid. (Hint: Each row on each side of the pyramid contains one more stack of pennies than the previous row. Remember, there are 13 pennies to an inch.)

13.3 Use Geometric Sequences and Series

Some sequences have the property that every term after the first can be determined by multiplying the preceding term by the same fixed number.

> A sequence $a_1, a_2, a_3, \ldots, a_n, \ldots$ $(a_1 \neq 0)$ is called a *geometric sequence* (*geometric progression*) if there is a nonzero constant r, called the *common ratio*, such that
> $$a_n = r a_{n-1} \quad \text{for every } n \geq 2.$$

Example ▶ Write the first four terms of the geometric sequence with $a_1 = 3$ and $r = 4$.

$a_1 = 3 \qquad a_2 = ra_1 \qquad a_3 = ra_2 \qquad a_4 = ra_3$
$ \qquad = 4 \cdot 3 \qquad = 4 \cdot 12 \qquad = 4 \cdot 48$
$ \qquad = 12 \qquad = 48 \qquad = 192$

Multiply the preceding term by the common ratio 4.

Every term of a geometric sequence can be written in terms of the first term a_1 and the common ratio r. For instance, the geometric squence with

$$a_1 = 3, \quad a_2 = 6, \quad a_3 = 12, \quad a_4 = 24, \quad a_5 = 48,$$

can be written as

$$a_1 = 3 \cdot 2^0, \quad a_2 = 3 \cdot 2^1, \quad a_3 = 3 \cdot 2^2, \quad a_4 = 3 \cdot 2^3, \quad a_5 = 3 \cdot 2^4.$$

Note ▶ In each case, the subscript is one more than the exponent of the common ratio 2. The following nth-term formula is a generalization of the above result.

> The general term of a geometric sequence with first term a_1 and common ratio r is given by the *nth-term formula*:
> $$a_n = a_1 r^{n-1}$$

EXAMPLE 1: Find a specific term of a geometric sequence using the nth-term formula.

Problem ▶ Given a geometric sequence with $a_1 = 5$ and $r = -2$, find a_{10}.

1. Write formula ▶ $a_n = a_1 r^{n-1}$

2. Evaluate ▶ $a_{10} = 5(-2)^{10-1}$ Think: To compute a_{10}, use $n = 10$.
$\phantom{a_{10}} = 5(-512)$

Solution ▶ Given a geometric sequence with $a_1 = 5$ and $r = -2$, $a_{10} = -2560$.

Because $a_2 = a_1 \cdot r$ in any geometric series, you can compute r using $r = \dfrac{a_2}{a_1}$.

Another Example ▶ Find the ninth term in the geometric series with $a_1 = 512$ and $a_2 = 256$.

1. Find r ▶ $r = \dfrac{a_2}{a_1} = \dfrac{256}{512} = \dfrac{1}{2}$ Think: In any geometric series $r = \dfrac{a_2}{a_1}$.

2. Write formula ▶ $a_n = a_1 r^{n-1}$

3. Evaluate ▶ $a_9 = 512 \left(\dfrac{1}{2}\right)^{9-1} = 512 \left(\dfrac{1}{2}\right)^8 = 512 \left(\dfrac{1}{256}\right) = 2$

Solution ▶ The ninth term in the geometric series with $a_1 = 512$ and $a_2 = 256$ is 2.

Make Sure

Find each specified term of a geometric sequence using the *n*th-term formula.

See Example 1 ▶ 1. Find a_{10} if $a_1 = 1$, $r = 3$. _____ 2. Find a_8 if $a_1 = 2$, r = -5. _____

MAKE SURE ANSWERS: 1. 19683 2. −156250

The indicated sum of the terms of a geometric sequence is called a *geometric series*.

Examples ▶ (a) $2 + 4 + 8 + 16 + 32$ is a geometric series associated with the geometric sequence 2, 4, 8, 16, 32.

(b) $9 + 3 + 1 + \frac{1}{3} + \frac{1}{9} + \frac{1}{27} + \cdots + 9(\frac{1}{3})^{n-1} + \cdots$ is an infinite geometric series because it consists of terms with a common ratio of $\frac{1}{3}$.

To find the sum of a finite geometric series, you can add the terms. However, it is often more convenient to use a summation formula.

Example ▶ Find a summation formula for the first *n* terms of a finite geometric series.

1. Write series ▶ $S_n = a_1 + a_1 r + a_1 r^2 + \cdots + a_1 r^{n-2} + a_1 r^{n-1}$

2. Multiply by *r* ▶ $rS_n = a_1 r + a_1 r^2 + a_1 r^3 + \cdots + a_1 r^{n-1} + a_1 r^n$

3. Subtract rS_n from S_n ▶
$$S_n = a_1 + a_1 r + a_1 r^2 + \cdots + a_1 r^{n-2} + a_1 r^{n-1}$$
$$-rS_n = \phantom{a_1 + {}} a_1 r + a_1 r^2 + \cdots + a_1 r^{n-2} + a_1 r^{n-1} + a_1 r^n \quad \text{Align like terms.}$$
$$S_n - rS_n = a_1 + 0 + 0 + \cdots + 0 + 0 - a_1 r^n \longleftarrow \text{difference}$$

4. Solve for S_n ▶ $S_n - rS_n = a_1 - a_1 r^n$

$S_n(1 - r) = a_1 - a_1 r^n$

Solution ▶ $S_n = \dfrac{a_1 - a_1 r^n}{1 - r}$ provided $r \neq 1$

The sum of the first *n* terms of a geometric series with ratio *r* is given by:
$$S_n = \frac{a_1(1 - r^n)}{1 - r} \quad (r \neq 1).$$

EXAMPLE 2: Find the sum of the first n terms of a geometric series using $S_n = \dfrac{a_1(1-r^n)}{1-r}$.

Problem ▶ Find the sum of the first 12 terms of the geometric series with $a_1 = 8$ and $r = \dfrac{1}{2}$.

1. Write formula ▶ $S_n = \dfrac{a_1(1-r^n)}{1-r}$

2. Evaluate ▶ $S_{12} = \dfrac{8\left[1 - \left(\frac{1}{2}\right)^{12}\right]}{1 - \frac{1}{2}}$ Substitute.

$= \dfrac{8\left[1 - \frac{1}{4096}\right]}{\frac{1}{2}}$ Compute.

$= 8 \cdot \dfrac{4095}{4096} \div \dfrac{1}{2}$

$= \dfrac{4095}{256}$

Solution ▶ The sum of the first 12 terms of the geometric series with $a_1 = 8$ and $r = \dfrac{1}{2}$ is $\dfrac{4095}{256}$.

CAUTION: The formula $S_n = \dfrac{a_1(1-r^n)}{1-r}$ does not apply if $r = 1$. A finite series with a common ratio of 1 would be a constant series, however, and its sum could be found using $\sum_{k=1}^{n} c = nc$.

Make Sure

Find the sum of the first n terms of each geometric series using $S_n = \dfrac{a_1(1-r^n)}{1-r}$.

See Example 2 ▶ 1. S_5 if $a_1 = 2, r = 3$ _____   2. S_7 if $a_1 = \dfrac{1}{2}, r = -\dfrac{1}{2}$ _____

MAKE SURE ANSWERS: 1. 242   2. $\dfrac{43}{128}$

534 Chapter 13 Sequences and Series

Consider the sum of the geometric series $S = \frac{1}{2} + \frac{1}{4} + \frac{1}{8} + \frac{1}{16} + \cdots + \left(\frac{1}{2}\right)^n + \cdots$, which has an infinite number of terms.

For $S = \frac{1}{2} + \frac{1}{4} + \frac{1}{8} + \frac{1}{16} + \cdots + \left(\frac{1}{2}\right)^n + \cdots$

$S_2 = \underbrace{\phantom{\frac{1}{2}+\frac{1}{4}}}_{\frac{3}{4}}$

$S_3 = \underbrace{\phantom{\frac{1}{2}+\frac{1}{4}+\frac{1}{8}}}_{\frac{7}{8}}$

$S_4 = \underbrace{\phantom{\frac{1}{2}+\frac{1}{4}+\frac{1}{8}+\frac{1}{16}}}_{\frac{15}{16}}$

$\vdots$

$S_n = \underbrace{\phantom{\frac{1}{2}+\frac{1}{4}+\frac{1}{8}+\frac{1}{16}+\cdots}}_{\frac{\frac{1}{2}(1-(\frac{1}{2})^n)}{1-\frac{1}{2}}}$

Observe that the power of the ratio $(\frac{1}{2})^n$ can be made as small as desired by making n large enough. Also, because $(\frac{1}{2})^n$ approaches 0 as n increases without bound, the sum

$$S_n = \frac{\frac{1}{2}[1 - (\frac{1}{2})^n]}{1 - \frac{1}{2}} \text{ will approach } \frac{\frac{1}{2}[1 - 0]}{1 - \frac{1}{2}} = \frac{\frac{1}{2}}{1 - \frac{1}{2}} = 1$$

as n increases without bound. The number that S_n is approaching (1 in this case) is called a *limit* and is defined to be the sum of the infinite series. If S_n does not have a limit, then the series does not have a sum.

Consider an infinite geometric series with a ratio of r, such that $|r| < 1$ (that is, $-1 < r < 1$). If $|r| < 1$, then r^n will approach 0 as n increases without bound. The sum

$$S_n = \frac{a_1(1 - r^n)}{1 - r} \text{ must then approach } \frac{a_1(1 - 0)}{1 - r} = \frac{a_1}{1 - r}$$

as n increases without bound.

> The *sum of an infinite geometric series* with a ratio r such that $|r| < 1$, is given by:
>
> $$S = \frac{a_1}{1 - r}.$$

Note ▶ If $|r| \geq 1$, the infinite geometric series does not have a sum.

Example ▶ $1 + 2 + 4 + 8 + 16 + 32 + \cdots$ is an infinite geometric series with a ratio $r = 2$. It does not have a sum because it increases without limit.

> CAUTION: Before you use the formula $S = \frac{a_1}{1 - r}$, it is important to check that the series is an infinite geometric series with a ratio r such that $|r| < 1$.

EXAMPLE 3: Find the sum of an infinite geometric series.

Problem ▶ Find the sum of the infinite geometric series $27 + 9 + 3 + 1 + \frac{1}{3} + \frac{1}{9} + \cdots$.

1. Find r ▶ $r = \frac{a_2}{a_1} = \frac{9}{27} = \frac{1}{3}$

2. Interpret ▶ Because $|r| = \frac{1}{3} < 1$, S can be evaluated by using $S = \frac{a_1}{1-r}$.

3. Evaluate ▶ $S = \frac{a_1}{1-r}$

$= \frac{27}{1 - \frac{1}{3}}$

$= 27 \cdot \frac{3}{2}$

Solution ▶ $S = \frac{81}{2}$

Other Examples ▶ Evaluate the following infinite geometric series (provided they have a sum).

(a) $S = 80 - 40 + 20 - 10 + 5 - \cdots$

$r = \frac{a_2}{a_1} = \frac{-40}{80} = -\frac{1}{2}$

Because $|r| = \left|-\frac{1}{2}\right| = \frac{1}{2} < 1$, S can be evaluated using $S = \frac{a_1}{1-r}$.

$S = \frac{a_1}{1-r}$

$= \frac{80}{1 - (-\frac{1}{2})}$

$= \frac{160}{3}$ or $53\frac{1}{3}$

(b) $S = 3 + 3\sqrt{2} + 6 + 6\sqrt{2} + 12 + \cdots$

$r = \frac{a_2}{a_1} = \frac{3\sqrt{2}}{3} = \sqrt{2}$

Because $|r| = \sqrt{2} > 1$, the sum does not exist.

Make Sure

Find the sum of each infinite geometric series.

See Example 3 ▶
1. $\frac{1}{2} + \frac{1}{4} + \frac{1}{8} + \frac{1}{16} + \cdots$

2. $27 - 9 + 3 - 1 + \frac{1}{3} - \frac{1}{9} + \cdots$

To write a repeating decimal as the quotient of two integers, you can first write the repeating decimal as an infinite geometric series.

EXAMPLE 4: Write a repeating decimal as the quotient of two integers.

Problems ▶ Write both $0.\overline{3}$ and $0.\overline{45}$ as the quotient of two integers.

1. Rewrite ▶
$0.\overline{3} = 0.333\cdots$
$= 0.3 + 0.03 + 0.003 + \cdots$

$0.\overline{45} = 0.454545\cdots$
$= 0.45 + 0.0045 + 0.000045 + \cdots$

2. Find r ▶
$r = \dfrac{a_2}{a_1} = \dfrac{0.03}{0.3} = 0.1$

$r = \dfrac{a_2}{a_1} = \dfrac{0.0045}{0.45} = 0.01$

3. Interpret ▶ Because $|r| = 0.1 < 1$, the sum can be evaluated using $S = \dfrac{a_1}{1-r}$.

Because $|r| = 0.01 < 1$, the sum can be evaluated using $S = \dfrac{a_1}{1-r}$.

4. Evaluate ▶
$S = \dfrac{a_1}{1-r}$
$= \dfrac{0.3}{1-0.1}$
$= \dfrac{0.3}{0.9}$
$= \dfrac{3}{9}$

$S = \dfrac{a_1}{1-r}$.
$= \dfrac{0.45}{1-0.01}$
$= \dfrac{0.45}{0.99}$
$= \dfrac{45}{99}$

Solutions ▶ $0.\overline{3} = \dfrac{1}{3}$

$0.\overline{45} = \dfrac{5}{11}$

Make Sure

Write each repeating decimal as the quotient of two integers.

See Example 4 ▶ 1. $0.\overline{4}$ _____   2. $0.\overline{375}$ _____   3. $0.0\overline{54}$ _____

MAKE SURE ANSWERS: 1. $\dfrac{4}{9}$ 2. $\dfrac{375}{999}$ 3. $\dfrac{3}{55}$

13.3 Practice

Set 1: Find each specified term of a geometric sequence using the nth-term formula.

1. a_8 if $a_1 = 4, r = 2$
2. a_5 if $a_1 = 5, r = 3$
3. a_8 if $a_1 = 15, r = -1$

4. a_6 if $a_1 = 22, r = -2$
5. a_7 if $a_1 = \frac{1}{8}, r = \frac{1}{2}$
6. a_5 if $a_1 = \frac{2}{3}, r = \frac{3}{2}$

7. a_8 if $a_1 = \frac{1}{8}, r = -\frac{1}{2}$
8. a_7 if $a_1 = \frac{2}{3}, r = -\frac{3}{2}$
9. a_5 if $a_1 = 16, r = -\frac{3}{4}$

Set 2: Find the sum of the first n terms of each geometric series using $S_n = \dfrac{a_1(1-r^n)}{1-r}$.

10. S_6 if $a_1 = 3, r = 2$
11. S_9 if $a_1 = 4, r = 2$
12. S_7 if $a_1 = 5, r = -1$

13. S_{12} if $a_1 = 9, r = -1$
14. S_7 if $a_1 = 4, r = -\frac{1}{2}$
15. S_6 if $a_1 = 64, r = -\frac{3}{2}$

16. S_7 if $a_1 = \frac{1}{8}, r = -\frac{2}{3}$
17. S_8 if $a_1 = \frac{1}{27}, r = -\frac{3}{2}$
18. S_7 if $a_1 = -\frac{8}{9}, r = -\frac{3}{2}$

538   Chapter 13   Sequences and Series

Set 3: Find the sum, if it exists, of each infinite geometric series.

19. $16 + 8 + 4 + 2 + \cdots$ 20. $81 + 27 + 9 + 3 + \cdots$ 21. $64 - 32 + 16 - 8 + \cdots$

22. $64 - 16 + 4 - 1 + \cdots$ 23. $\dfrac{1}{2} + \dfrac{1}{6} + \dfrac{1}{18} + \dfrac{1}{54} + \cdots$ 24. $\dfrac{1}{3} + \dfrac{1}{6} + \dfrac{1}{12} + \dfrac{1}{24} + \cdots$

25. $\dfrac{1}{3} - \dfrac{1}{9} + \dfrac{1}{27} - \dfrac{1}{81} + \cdots$ 26. $\dfrac{1}{4} - \dfrac{1}{16} + \dfrac{1}{64} - \dfrac{1}{256} + \cdots$ 27. $\dfrac{1}{2} - \dfrac{1}{3} + \dfrac{2}{9} - \dfrac{4}{27} + \cdots$

Set 4: Write each repeating decimal as the quotient of two integers.

28. $0.\overline{2}$ 29. $0.\overline{1}$ 30. $0.\overline{5}$ 31. $0.\overline{6}$

32. $0.\overline{14}$ 33. $0.\overline{57}$ 34. $1.\overline{31}$ 35. $5.\overline{47}$

36. $4.4\overline{3}$ 37. $5.3\overline{7}$ 38. $0.\overline{142857}$ 39. $0.\overline{428571}$

Extra: Find each specified term of a geometric sequence given two terms of the sequence.

40. a_2 if $a_1 = 1$, $a_3 = \dfrac{4}{9}$ 41. a_3 if $a_1 = 8$, $a_4 = \dfrac{1}{2}$ 42. a_8 if $a_7 = 64$, $a_9 = 16$

43. a_7 if $a_5 = 81$, $a_8 = 3$ 44. a_9 if $a_7 = 16$, $a_{10} = -2$ 45. a_6 if $a_8 = -1$, $a_{11} = \dfrac{1}{27}$

Review: Work these problems on a separate sheet of paper. Attach your work to this page.

Evaluate each expression using the Order of Operations Rule. (See Lesson 1.4.)
46. $7 \cdot 6 \div (1 + 1)$ 47. $21 \cdot 5 \div (2 + 1)$ 48. $35 \cdot 4 \div (3 + 1)$ 49. $21 \cdot 2 \div (5 + 1)$
50. $56 \cdot 5 \div (4 + 1)$ 51. $28 \cdot 2 \div (6 + 1)$ 52. $8 \cdot 7 \div (1 + 1)$ 53. $70 \cdot 4 \div (4 + 1)$

Simplify each term raised to a power. (See Lesson 1.6.)
54. $(2x)^6$ 55. $(3x)^4$ 56. $(4x)^3$ 57. $(2x)^5$
58. $(-3x)^5$ 59. $(-2x)^6$ 60. $(-4x)^4$ 61. $(-x)^7$

Copyright © 1985 by Harcourt Brace Jovanovich, Inc. All rights reserved.

Problem Solving 20: Solve Problems Using Geometric Sequences and Series

Many important applications can be solved using geometric sequences and series.

EXAMPLE: Solve this problem using a geometric sequence and series.

1. Read and identify ▶ A ball dropped from 5 feet up rebounds 80% (0.8) its previous height on each bounce. To the nearest tenth foot, how far does it rebound on the fifth bounce? How far does the ball travel by the time it hits the ground for the fifth time? How far does the ball travel before it comes to rest?

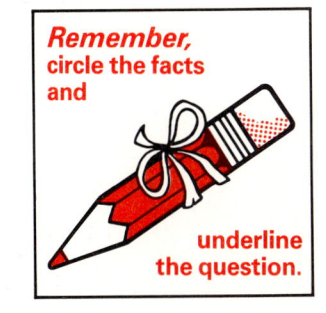

   Remember, circle the facts and underline the question.

2. Understand ▶ The terms $a_1 = 0.8(5) = 4$, $a_2 = 0.8(4) = 3.2$, $a_3 = 0.8(3.2) = 2.56, \cdots$ form a geometric sequence with common ratio 0.8.

3. Decide ▶ To solve a problem that involves a geometric sequence, series, or infinite series, you

   evaluate $a_n = a_1 r^{n-1}$, $S_n = \dfrac{a_1(1 - r^n)}{1 - r}$,

   or $S = \dfrac{a_1}{1 - r}$, respectively.

4. Evaluate a_n ▶ $a_n = a_1 r^{n-1}$
   $a_5 = 4(0.8)^{5-1}$
   $ = 4(0.8)^4$
   $ = 1.6384$
   $ \approx 1.6$ Round to the nearest tenth.

5. Interpret ▶ $a_5 \approx 1.6$ means that on the fifth bounce the ball rebounds about 1.6 feet.

6. Evaluate S_n ▶ $S_n = \dfrac{a_1(1 - r^n)}{1 - r}$

   $S_4 = \dfrac{4(1 - 0.8^4)}{1 - 0.8}$

   $ = \dfrac{4(0.5904)}{0.2}$

   $ = 11.808$ ⟵ the sum of the four rebounds shown above

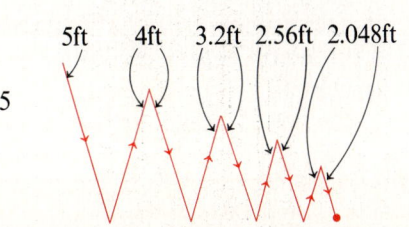

7. Find $2S_n + 5$ ▶ $2S_4 + 5 = 2(11.808) + 5$
   $ = 28.616$
   $ \approx 28.6$

8. Interpret ▶ $2S_4 + 5 \approx 28.6$ means that by the time the ball hits the ground for the fifth time it has traveled about 28.6 feet.

540 Chapter 13 Sequences and Series

9. **Evaluate S** ▶ $S = \dfrac{a_1}{1-r}$

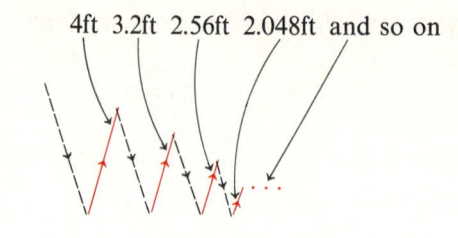

$= \dfrac{4}{1 - 0.8}$

4ft 3.2ft 2.56ft 2.048ft and so on

$= 20$ ◀── the sum of the infinite number of rebounds shown above

5ft 4ft 3.2ft 2.56ft 2.048ft and so on

10. **Find $2S + 5$** ▶ $2S + 5 = 2(20) + 5$

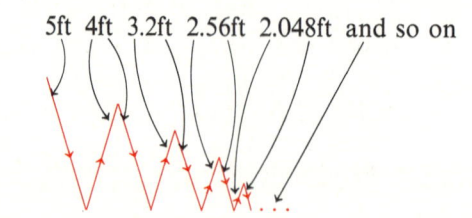

$= 40 + 5$

$= 45$

11. **Interpret** ▶ $2S + 5 = 45$ means that the total distance the ball travels before coming to rest is 45 feet.

Practice: Round to the nearest tenth when necessary.

1a. _____
 b. _____
 c. _____

2. _____

3a. _____
 b. _____

4a. _____
 b. _____
 c. _____

5a. _____
 b. _____

6a. _____
 b. _____

7a. _____
 b. _____

8a. _____
 b. _____

1. A ball dropped from 20 m rebounds $\frac{3}{4}$ its previous height on each bounce. a) How far does it rebound on the eighth bounce? How far does the ball travel by the time it b) hits the ground for the eighth time? c) comes to rest?

2. One penny is put on the first square of a 64-square checkerboard, two pennies on the second square, four pennies on the third square, and so on. How much money will be on the sixty-fourth square?

3. A certain color television set costs $480. The payments for 24 months are $20 per month plus 1% interest, each month on the unpaid balance. Including interest, how much is a) the twelfth payment? b) the total cost of the television set?

4. The value of the average home in California increases 1% each month. How much will the average $100,000 California home be worth in a) 6 months? b) 5 years? c) How long will it take the average California home to double in value?

5. Your direct ancestors are your parents, grandparents, great-grandparents, and so on. Assuming no duplication of ancestors, how many direct ancestors has each of us had over the last a) 7 generations (about 200 years)? b) 10 generations (about 300 years)?

6. Musical notes in cycles per second (cps) form a geometric progression. Concert A is 440 cps and A, an octave (12 notes) higher, is 880 cps. a) What is the constant ratio r? b) How many cycles per second are there for middle C, 9 notes below Concert A?

7. If the mid-points of an equilateral triangle with side 2 ft are joined by straight-line segments, the new figure will be an equilateral triangle. Continuing in this way will form an infinite series of nested equilateral triangles. (see Figure 13.1.) What is the total a) perimeter? b) area of all the equilateral triangles?

8. If the mid-points of a square with side 2 ft are joined by straight-line segments, the new figure will be a square. Continuing in this way will form an infinite series of nested squares. (See Figure 13.2.) What is the total a) perimeter? b) area of all the squares?

Figure 13.1

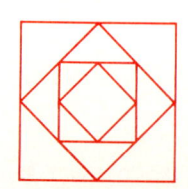

Figure 13.2

Copyright © 1985 by Harcourt Brace Jovanovich, Inc. All rights reserved.

13.4 Use the Binomial Expansion

Every binomial expression of the form $(a + b)^n$ for $n \geq 1$ can be written as a series.

Example ▶ For $n = 1$, $(a + b)^1 = a + b$

$n = 2$, $(a + b)^2 = a^2 + 2ab + b^2$

$n = 3$, $(a + b)^3 = a^3 + 3a^2b + 3ab^2 + b^3$

$n = 4$, $(a + b)^4 = a^4 + 4a^3b + 6a^2b^2 + 4ab^3 + b^4$

$n = 5$, $(a + b)^5 = a^5 + 5a^4b + 10a^3b^2 + 10a^2b^3 + 5ab^4 + b^5$

You can expand $(a + b)^n$ by using repeated multiplication. However, a careful examination of the above expansions will help you discover patterns that can be used to write the expansion of $(a + b)^n$ directly.

Patterns

1. The expansion of $(a + b)^n$ has $n + 1$ terms.

Example ▶ $(a + b)^5 = a^5 + 5a^4b + 10a^3b^2 + 10a^2b^3 + 5ab^4 + b^5$, which has $5 + 1 = 6$ terms.

2. The power of a is n in the first term, and it decreases by 1 for each successive term.

Example ▶ $(a + b)^5 = a^5 + 5a^4b + 10a^3b^2 + 10a^2b^3 + 5a^1b^4 + a^0b^5$

3. The power of b is 0 in the first term, and it increases by 1 for each successive term.

Example ▶ $(a + b)^5 = a^5b^0 + 5a^4b^1 + 10a^3b^2 + 10a^2b^3 + 5a^1b^4 + a^0b^5$

4. The sum of the exponents in each term is n.

$5 + 0 = 5 \quad 4 + 1 = 5 \quad 3 + 2 = 5 \quad 2 + 3 = 5 \quad 1 + 4 = 5 \quad 0 + 5 = 5$

Example ▶ $(a + b)^5 = a^5b^0 + 5a^4b^1 + 10a^3b^2 + 10a^2b^3 + 5a^1b^4 + a^0b^5$

5. The coefficient of the kth term $(k > 1)$ can be obtained from the preceding term by multiplying its coefficient and its exponent of a and dividing this by its exponent of b increased by 1.

Example ▶ $(a + b)^5 = a^5 + 5a^4b^1 + 10a^3b^2 + 10a^2b^3 + 5ab^4 + b^5$

$5 \cdot 4 \div (1 + 1) = 10$

6. The first coefficient is 1. The second coefficient is n. The coefficients form a symmetric pattern. If you start at the left of the expansion and examine successive coefficients, you will observe that they are the same as those obtained by starting at the right and proceeding to the left.

Example ▶ $(a + b)^5 = 1a^5 + 5a^4b + 10a^3b^2 + 10a^2b^3 + 5ab^4 + 1b^5$

same
same
same

542 Chapter 13 Sequences and Series

> **Summary** ▶ **The Binomial Expansion**
>
> To expand $(a + b)^n$ for some natural number n:
>
> 1. Write the literal part of each of the $(n + 1)$ terms. Start the first term with $a^n b^0$ or a^n. For each successive term, decrease the exponent on a by 1 and increase the exponent on b by 1 until you write the last term $a^0 b^n$ or b^n.
> 2. Write n as the coefficient of the second term.
> 3. Starting with the third term, you calculate the kth coefficient from the preceding term by multiplying its coefficient and its exponent of a and dividing this by its exponent of b increased by 1.
> 4. After at least one-half of the coefficients have been determined, write the duplicate coefficients using the symmetry of the coefficients.

Example ▶ Expand $(a + b)^7$.

1. **Write literal parts** ▶ $(a + b)^7 = a^7 + ?a^6 b^1 + ?a^5 b^2 + ?a^4 b^3 + ?a^3 b^4 + ?a^2 b^5 + ?ab^6 + b^7$

2. **Write 2nd coefficient** ▶ $(a + b)^7 = a^7 + 7a^6 b^1 + ?a^5 b^2 + ?a^4 b^3 + ?a^3 b^4 + ?a^2 b^5 + ?ab^6 + b^7$

3. **Calculate coefficient(s) using the preceding term** ▶ $(a + b)^7 = a^7 + 7a^6 b^1 + 21a^5 b^2 + ?a^4 b^3 + ?a^3 b^4 + ?a^2 b^5 + ?ab^6 + b^7$

   $7 \cdot 6 \div (1 + 1)$ See Step 3.

   $(a + b)^7 = a^7 + 7a^6 b^1 + 21a^5 b^2 + 35a^4 b^3 + ?a^3 b^4 + ?a^2 b^5 + ?ab^6 + b^7$

   $21 \cdot 5 \div (2 + 1)$ Stop using this procedure because you have now written at least one-half of the coefficients.

4. **Write duplicate coefficients** ▶ $(a + b)^7 = 1a^7 + 7a^6 b^1 + 21a^5 b^2 + 35a^4 b^3 + 35a^3 b^4 + 21a^2 b^5 + 7ab^6 + 1b^7$

   same
   same
   same
   same

Solution ▶ $(a + b)^7 = a^7 + 7a^6 b + 21a^5 b^2 + 35a^4 b^3 + 35a^3 b^4 + 21a^2 b^5 + 7ab^6 + b^7$

To expand a binomial such as $(5x - 2y)^4$, you use the expansion method with $a = 5x$ and $b = -2y$.

EXAMPLE 1: Find the expansion of a binomial.

Problem ▶ Expand $(5x - 2y)^4$.

1. **Write literal parts** ▶ $(5x - 2y)^4 = (5x)^4 + ?(5x)^3(-2y)^1 + ?(5x)^2(-2y)^2 + ?(5x)(-2y)^3 + (-2y)^4$ Let $a = 5x$ and $b = -2y$.

2. **Write 2nd coefficient** ▶ $(5x - 2y)^4 = (5x)^4 + 4(5x)^3(-2y)^1 + ?(5x)^2(-2y)^2 + ?(5x)(-2y)^3 + (-2y)^4$

3. **Find 3rd coefficient** ▶ $(5x - 2y)^4 = (5x)^4 + 4(5x)^3(-2y)^1 + 6(5x)^2(-2y)^2 + ?(5x)(-2y)^3 + (-2y)^4$

   $(4 \cdot 3) \div (1 + 1)$

4. **Write duplicate coefficients** ▶ $(5x - 2y)^4 = 1(5x)^4 + 4(5x)^3(-2y)^1 + 6(5x)^2(-2y)^2 + 4(5x)(-2y)^3 + 1(-2y)^4$

5. **Simplify** ▶ $= 5^4 x^4 + 4 \cdot 5^3 x^3 (-2)y + 6 \cdot 5^2 x^2 (-2)^2 y^2 + 4 \cdot 5 \cdot x(-2)^3 y^3 + (-2)^4 y^4$

Solution ▶ $(5x - 2y)^4 = 625x^4 - 1000x^3 y + 600x^2 y^2 - 160xy^3 + 16y^4$

13.4 Use the Binomial Expansion   543

Note ▶ The expansion of $(a - b)^n$ is the same as the expansion of $(a + b)^n$ except the even-numbered terms in the expansion of $(a - b)^n$ are preceded by a minus sign.

Make Sure

Find the expansion of each binomial.

See Example 1 ▶ **1.** $(2x + y)^5$ _____

2. $(p - 3q)^6$ _____

MAKE SURE ANSWERS: **1.** $32x^5 + 80x^4y + 80x^3y^2 + 40x^2y^3 + 10xy^4 + y^5$ **2.** $p^6 - 18p^5q + 135p^4q^2 - 540p^3q^3 + 1215p^2q^4 - 1458pq^5 + 729q^6$

To develop a general-term formula for the expansion of $(a + b)^n$, it will be helpful to first define a new function.

The *factorial function* denoted by $n!$ is given by:

$$n! = \begin{cases} n(n-1)\cdots(2)(1) & \text{for } n \in N \text{ and } n > 1 \\ 1 \text{ for } n = 1 \\ 1 \text{ for } n = 0 \end{cases}$$

Note ▶ If n is a natural number greater than 1, then $n!$ is the product of the first n natural numbers.

Examples ▶ (a) $5! = 5 \cdot 4 \cdot 3 \cdot 2 \cdot 1 = 120$ (b) $6! = 6 \cdot 5 \cdot 4 \cdot 3 \cdot 2 \cdot 1 = 720$ (c) $0! = 1$ (by definition)

The factorial function has the important property that $n! = n(n-1)!$.

Examples ▶ (a) $\begin{array}{c|c} 4! & = 4 \cdot (4-1)! \\ \hline 24 & 4 \cdot 3! \\ & 4 \cdot 6 \\ & 24 \end{array}$ (b) $\begin{array}{c|c} 6! & = 6 \cdot (6-1)! \\ \hline 720 & 6 \cdot 5! \\ & 6 \cdot 120 \\ & 720 \end{array}$

To evaluate the quotient of two factorials, you use the property $n! = n(n-1)!$.

Example ▶ $\dfrac{8!}{6!} = \dfrac{8 \cdot 7 \cdot 6!}{6!}$ Think: Write the numerator as a product that contains the factorial that is in the denominator.

$= \dfrac{8 \cdot 7 \cdot 6!}{6!}$ Simplify.

$= 56$

The expression $\binom{n}{k}$, where $0 \leq k \leq n$, and n and $k \in W$, is defined by: $\binom{n}{k} = \dfrac{n!}{k!(n-k)!}$.

To evaluate expressions of the form $\binom{n}{k}$, you use the previous definition and simplify as in the previous example.

EXAMPLE 2: Evaluate expressions of the form $\binom{n}{k}$ for some k such that $0 \le k \le n$, and n and $k \in W$.

Problem ▶ Evaluate $\binom{8}{3}$.

1. Rename $\binom{n}{k}$ ▶ $\binom{8}{3} = \dfrac{8!}{3!(8-3)!}$

2. Simplify denominator ▶ $= \dfrac{8!}{3!\,5!}$

3. Rename numerator ▶ $= \dfrac{8 \cdot 7 \cdot 6 \cdot 5!}{3!\,5!}$ Think: Write the numerator as a product that contains the largest factorial that is in the denominator.

4. Reduce ▶ $= \dfrac{8 \cdot 7 \cdot 6}{3!}$

5. Rename $n!$ ▶ $= \dfrac{8 \cdot 7 \cdot 6}{3 \cdot 2 \cdot 1}$

6. Simplify ▶ $= 56$

Solution ▶ $\binom{8}{3} = 56$

Make Sure

Evaluate each expression of the form $\binom{n}{k}$ for some k such that $0 \le k \le n$, n and $k \in W$.

See Example 2 ▶   **1.** $\binom{7}{4}$ _____   **2.** $\binom{5}{2}$ _____   **3.** $\binom{9}{0}$ _____

MAKE SURE ANSWERS: 1. 35 2. 10 3. 1

To develop a general-term formula for the expansion of $(a + b)^n$, you apply the patterns you used in Example 1. You know that the first term of $(a + b)^n$ is a^n, and that the second term is $na^{n-1}b^1$. The literal part of the third term will be $a^{n-2}b^2$, and you compute its coefficient by using the coefficient and the exponents of the preceding term.

Example ▶ 2nd term ⟶ $na^{n-1}b^1$, $\dfrac{n(n-1)}{2}a^{n-2}b^2$ ⟵ 3rd term

$n(n-1) \div (1+1)$

If there is a fourth term, you compute its coefficient by using the coefficient and the exponents of the third term.

Example ▶ 3rd term ⟶ $\dfrac{n(n-1)}{2}a^{n-2}b^2$, $\dfrac{n(n-1)(n-2)}{3\cdot 2}a^{n-3}b^3$ ⟵ 4th term

$\dfrac{n(n-1)}{2} \cdot (n-2) \div (2+1)$

If there is a fifth term, you compute its coefficient by using the coefficient and the exponents of the fourth term.

Example ▶ 4th term ⟶ $\dfrac{n(n-1)(n-2)}{3\cdot 2}a^{n-3}b^3$, $\dfrac{n(n-1)(n-2)(n-3)}{4\cdot 3\cdot 2}a^{n-4}b^4$ ⟵ 5th term

$\dfrac{n(n-1)(n-2)}{3\cdot 2} \cdot (n-3) \div (3+1)$

Summary ▶ In the general-term formula for $(a+b)^n$:

1. In every term, the exponent of b is one less than the number of the term. The exponent of b in the $(k+1)$th term is k.
2. In every term, the exponent of a is n less the exponent of b. The exponent of a in the $(k+1)$th term is $(n-k)$.
3. The numerator of the coefficient of the $(k+1)$th term consists of the product of k consecutive natural-number factors. The first factor is n. Each successive factor is one less than the preceding factor. The last factor is one more than the exponent of a. Specifically, the numerator of the coefficient of the $(k+1)$th term is: $n(n-1)\cdots(n-k+1)$.
4. The denominator of the coefficient of the $(k+1)$th term is $k!$.

Combining the above observations produces the following general-term formula.

> The $(k+1)$th term in the expansion of $(a+b)^n$ is: $\dfrac{n(n-1)\cdots(n-k+1)}{k!}a^{n-k}b^k$

Note ▶ The coefficient in the above formula can be written more compactly as $\binom{n}{k}$.

Example ▶ Show that $\binom{n}{k} = \dfrac{n(n-1)\cdots(n-k+1)}{k!}$.

1. Rename $\binom{n}{k}$ ▶ $\binom{n}{k} = \dfrac{n!}{k!(n-k)!}$

2. Rename numerator and simplify ▶ $= \dfrac{n(n-1)\cdots(n-k+1)(n-k)!}{k!(n-k)!}$ Think: Write the numerator as a product that involves $(n-k)!$.

Solution ▶ $\binom{n}{k} = \dfrac{n(n-1)\cdots(n-k+1)}{k!}$

13.4 Use the Binomial Expansion   545

> **The General-Term Formula for the Expansion of $(a + b)^n$**
>
> The $(k + 1)$th term in the expansion of $(a + b)^n$ is given by $\binom{n}{k}a^{n-k}b^k$.

To find a specific term in the expansion of $(2c - d)^9$, you use the general-term formula with $a = 2c$ and $b = -d$.

EXAMPLE 3: Find a specific term of a binomial expansion using the general-term formula.

Problem ▶ Find the 7th term in the expansion of $(2c - d)^9$.

1. Write formula ▶ The $(k + 1)$th term of $(a + b)^n = \binom{n}{k}a^{n-k}b^k$.

2. Substitute ▶ The 7th term of $(2c - d)^9 = \binom{9}{6}(2c)^{9-6}(-d)^6$ Since $k + 1 = 7$, $k = 6$. $a = 2c$ and $b = -d$.

3. Simplify ▶
$$= \frac{9!}{6!3!}(2c)^3(-d)^6$$
$$= \frac{9 \cdot 8 \cdot 7 \cdot 6!}{6!3!}(2c)^3(-d)^6$$
$$= \frac{9 \cdot 8 \cdot 7}{3!}(2c)^3(-d)^6$$
$$= \frac{9 \cdot 8 \cdot 7}{3 \cdot 2 \cdot 1}(2c)^3(-d)^6$$
$$= 84(8c^3)d^6$$

Solution ▶ The 7th term of $(2c - d)^9 = 672c^3d^6$.

Make Sure

Find each specific term of a binomial expansion.

See Example 3 ▶ **1.** 4th, $(a + b)^7$ **2.** 5th, $(2x + 3y)^7$ **3.** 4th, $(m - 2n)^6$

MAKE SURE ANSWERS: **1.** $35a^4b^3$ **2.** $22680x^3y^4$ **3.** $-160m^3n^3$

13.4 Practice

Set 1: Write the expansion of each binomial.

1. $(y + z)^3$ _____

2. $(w + x)^6$ _____

3. $(2u - v)^4$ _____

4. $(s - 3t)^5$ _____

5. $(4q - 3r)^4$ _____

6. $(3n - 2p)^5$ _____

Set 2: Evaluate each expression of the form $\binom{n}{k}$ for some k such that $0 \leq k \leq n$, and n and $k \in W$.

7. $\binom{5}{3}$ 8. $\binom{5}{4}$ 9. $\binom{9}{4}$

10. $\binom{9}{5}$ 11. $\binom{8}{1}$ 12. $\binom{13}{0}$

548 Chapter 13 Sequences and Series

Set 3: Find each specific term of a binomial expansion.

13. 5th, $(a + b)^7$ 14. 3rd, $(a + b)^7$ 15. 8th, $(c - d)^9$

16. 4th, $(c - d)^9$ 17. 5th, $(2e + 5d)^9$ 18. 8th, $(2e + 5d)^9$

19. 3rd, $(3f - g)^5$ 20. 5th, $(f - 3g)^5$ 21. 2rd, $(3h - 4k)^{10}$

22. 9th, $(4h - 3k)^{10}$ 23. 8th, $(5m + 3n)^9$ 24. 5th, $(3p + 7q)^8$

13. _____
14. _____
15. _____
16. _____
17. _____
18. _____
19. _____
20. _____
21. _____
22. _____
23. _____
24. _____

Copyright © 1985 by Harcourt Brace Jovanovich, Inc. All rights reserved.

Name _____ Date _____ Class _____

Chapter 13 Review

What to Review if You Have Trouble

Objectives		Lesson	Example	Page
Find terms of a sequence ▶	1. Write the first three terms and the tenth term of the sequence given by $a_n = 5n - 7$. _____ , _____ , _____ , · · · , _____	13.1	1	518
Find the sum of a series ▶	2. Evaluate $\sum_{k=1}^{4} (2k - 3)$. _____	13.1	2	519
Find terms of an arithmetic sequence ▶	3. Write the first five terms of the arithmetic sequence with $a_1 = 8$ and $d = -13$. _____ , _____ , _____ , _____ , _____	13.2	1	523
Find the specified unknown of an arithmetic sequence using the *n*th-term formula ▶	4. Find a_{25} in the arithmetic sequence with $a_1 = 7$ and $d = 11$. _____	13.2	2	524
Find the sum of a finite arithmetic series ▶	5. Find the sum of the first 86 terms of the arithmetic series with $a_1 = 3$ and $d = 4$. _____	13.2	3	526
Find the specified term of a geometric sequence ▶	6. Find a_8 in the geometric sequence with $a_1 = 64$ and $r = \frac{1}{2}$. _____	13.3	1	531
Find the sum of a geometric series ▶	7. Find the sum of the first 10 terms of the geometric series with $a_1 = 12$ and $r = \frac{1}{2}$. _____	13.3	2	533

Copyright © 1985 by Harcourt Brace Jovanovich, Inc. All rights reserved.

550 Chapter 13 Sequences and Series

	8. Find the sum of the infinite geometric series $125 + 25 + 5 + 1 + \frac{1}{5} + \cdots$.	_____	13.3 3	535
Write a repeating decimal as the quotient of two integers ▶	9. Write $0.\overline{27}$ as the quotient of two integers.	_____	13.3 4	536
Expand a binomial ▶	10. Write the expansion of $(2w - 3z)^4$.		13.4 1	542
Evaluate $\binom{n}{k}$ ▶	11. Evaluate $\binom{9}{5}$.	_____	13.4 2	544
Find the indicated term of an expansion ▶	12. Find the 8th term in the expansion of $(2m - n)^{10}$.	_____	13.4 3	546
Solve problems using arithmetic sequences and series ▶	13. The number of times a chime clock strikes each hour is the number of the hour. How many times does a chime clock strike between 9 A.M. and 8 P.M., inclusive?	_____	PS 19 —	529
Solve problems using geometric sequences and series ▶	14. A certain chain letter requires the recipient to send out a copy of the same letter to 10 other people. How many copies of this chain letter will be in circulation after the sixth mailing, assuming all copies are sent by each recipient each time?	_____	PS 20 —	539

CHAPTER 13 REVIEW ANSWERS: 1. $-2, 3, 8, \ldots$ **2.** 8 **3.** 8, $-5, -18, -31, -44$ **4.** 271 **5.** 14878 **6.** $\frac{1}{2}$ **7.** $\frac{3069}{128}$ **8.** $\frac{625}{4}$ **9.** $\frac{3}{11}$ **10.** $16w^4 - 96w^3z + 216w^2z^2 - 216wz^3 + 81z^4$ **11.** 126 **12.** $-960m^3n^7$ **13.** 78 times **14.** 1, 111, 110 copies

Copyright © 1985 by Harcourt Brace Jovanovich, Inc. All rights reserved.

Final Review

To review the entire text, you should complete the following 66 practice problems. The answers are in the Appendix Selected Answers. Try to get all 66 problems correct before taking the final examination. If you have trouble with a particular problem, go back and review the indicated lesson(s).

Use computational rules. (See Lesson 1.4.)

1. Evaluate $-3 - \{4 - [2(5 - 2) - (3 - 9)] - 5\}$.

Use properties. (See Lesson 1.5.)

2. Simplify $4\{-2a + 3[2 - (a - b)] - 3(b - a)\}$ using the Order of Operations Rule.

Use integral exponents. (See Lesson 1.6.)

3. Simplify $(2a^3b^{-2})^3$.

Compute using scientific notation. (See PS 1.)

4. The Sun's mass is 1.99×10^{30} kg. The Moon's mass is 7.37×10^{22} kg. How many times greater is the Sun's mass than the Moon's mass?

Solve equations containing like terms and parentheses. (See Lesson 2.2.)

5. Solve $3(a - 2) - 2(2 - a) = 3(a - 1)$.

Solve equations containing fractions and decimals. (See Lesson 2.3.)

6. Solve $\frac{3}{4}(b - 2) = \frac{3}{5}(2b - 3) + \frac{1}{5}$.

Solve inequalities. (See Lesson 2.4.)

7. Solve $3c + 7 > 4 + 5c$.

Solve absolute value equations. (See Lesson 2.5.)

8. Solve $|d - 3| + 4 = 13$.

Solve literal equations. (See Lesson 2.6.)

9. Solve $ax + by = c$ for y.

Solve problems using equations. (See PS 2.)

10. The perimeter of a rectangle is 26 feet. The length is 2 feet longer than the width. What is the area?

Find linear equations. (See Lesson 3.4.)

11. Find an equation of the line that contains the points $(1, -4)$ and $(-2, 2)$. Write your answer in slope-intercept form.

12. Find an equation of a line perpendicular to $3x = 4y - 12$ that contains the point $(-1, 3)$. Write your answer in slope-intercept form.

552 Final Review

Graph linear inequalities. (See Lesson 3.5.)

13. Graph $2x - 5y \geq 10$.

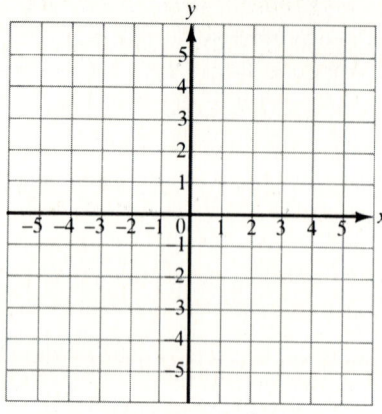

Use linear relations. (See PS 3.)

14. The first session of the 1st United States Congress was in 1789. From then to now, a newly elected Congress has its first session every two years. **a.** Find the linear equation describing the relationship between the number (n) of the U.S. Congress and the year (y) for the first session of that Congress. **b.** Find the number of the U.S. Congress that held its first session in 1985. **c.** In what year was the first session of the 50th U.S. Congress?

Solve second-order systems. (See Lesson 4.2.)

15. Solve $\begin{cases} 3x - 2y = 9 \\ 5x + y = 2 \end{cases}$ using the substitution method.

Solve third-order systems. (See Lesson 4.3.)

16. Solve $\begin{cases} 2x - 3y + z = 5 \\ x + 2y - 2z = 4 \\ 3x - y + 3z = 1 \end{cases}$ using the addition method.

Solve systems using Cramer's Rule. (See Lesson 4.5.)

17. Solve $\begin{cases} 4x - 3y = 3 \\ 3x + 2y = -2 \end{cases}$ using Cramer's Rule.

Solve problems using systems. (See PS 4.)

18. A nurse wants to dilute 500 mL of a 25%-iodine solution (25% iodine and 75% alcohol) to a 10%-iodine solution. How much pure alcohol must be added?

Add and subtract polynomials. (See Lesson 5.1.)

19. $(4p^3 - 8p^2 - 3p + 7) - (2p^3 - 5p^2 + 4p - 3) + (p^2 - 9)$

Multiply polynomials. (See Lesson 5.2.)

20. $(2q - 3)(q^2 - 2q + 4)$

Divide polynomials. (See Lesson 5.4.)

21. $(12x^4y^3 - 6x^3y - 9x^2y^3 + 3x^5y^2) \div (3x^2y)$

Copyright © 1985 by Harcourt Brace Jovanovich, Inc. All rights reserved.

Final Review 553

Name Date Class

22. _____

Simplify expressions involving polynomials. (See Lesson 5.4.)

22. Perform the indicated operations: $(r^3 + 8) \div (r + 2) - (r - 3)(r + 1)$.

Divide using synthetic division. (See Lesson 5.5.)

23. _____

23. $(s^4 - 3s^2 - 5) \div (s - 3)$

Evaluate polynomials. (See PS 6.)

24. _____

24. Principal of $100 is invested in an account paying 10% interest compounded annually. Assuming that there are no further deposits or withdrawals, how much money will be in the account at the end of 4 years?

Factor special polynomials. (See Lesson 6.4.)

25. _____

25. $8u^3 - 27v^3$

26. $t^3 + 14t^2 - 9t - 126$

26. _____

Factor completely. (See Lesson 6.5.)

27. $m^3 + 14m^2 + 48m$

Solve equations by factoring. (See Lesson 6.6.)

27. _____

28. $6n^2 - 13n + 6 = 0$

Factor to solve problems. (See PS 7.)

28. _____

29. The length of a rectangle is $4\frac{1}{2}$ feet longer than the width. The area of the rectangle is 28 square feet. Find the perimeter.

Multiply and divide rational expressions. (See Lesson 7.2.)

29. _____

30. $\dfrac{h^2 - 9}{h + 3} \div \dfrac{3 - h}{h^2 + 9} \cdot \dfrac{h + 3}{h^4 - 81}$

Add and subtract rational expressions. (See Lesson 7.3.)

30. _____

31. $\dfrac{k^2 + k - 6}{k^2 - k + 6} - \dfrac{6k - 12}{k^2 - 4} + \dfrac{k^2 - 3k}{k^2 - 5k + 6}$

Simplify complex fractions. (See Lesson 7.4.)

31. _____

32. $\dfrac{\dfrac{1}{g^2} + \dfrac{1}{g}}{\dfrac{1}{g} - \dfrac{1}{g^2}}$

32. _____

Solve equations containing rational expressions. (See Lesson 7.5.)

33. $z + 3 = \dfrac{z + 3}{z - 3}$

33. _____

Solve problems using rational expressions. (See PS 8.)

34. One number is 5 larger than another number. The sum of their reciprocals is equal to the reciprocal of their product. What are the numbers?

34. _____

Copyright © 1985 by Harcourt Brace Jovanovich, Inc. All rights reserved.

554 Final Review

Compute with rational exponents. (See Lesson 8.1.)

35. Evaluate $(-27)^{2/3}$.

Simplify radicals. (See Lesson 8.2.)

36. Write $\sqrt{\dfrac{8r^6s^2}{10t^3}}$ in simplest radical form. Assume all variables are positive.

Solve radical equations. (See Lesson 8.4.)

37. Solve $\sqrt{4y + 13} - \sqrt{7 - y} = 3$.

Compute with complex numbers. (See Lesson 8.6.)

38. Simplify $(3 - 2i)^2 + \dfrac{4 - i}{3 + 2i}$. Write answer in $a \pm bi$ form.

Solve formulas containing radicals. (See PS 10.)

39. Solve $r = \dfrac{590}{\sqrt{h}}$ (pulse rate formula) for h.

Solve by the quadratic formula. (See Lesson 9.3.)

40. Solve $2x^2 + 4x = 7$ using the quadratic formula.

Rename and solve equations. (See Lesson 9.4.)

41. Solve $w^4 + w^2 - 12 = 0$.

Solve inequalities. (See Lesson 9.5.)

42. Solve $1 \leq \dfrac{x^2 + 2x}{x^2 + 3x - 4}$.

Solve problems using quadratic equations. (See PS 12.)

43. A car traveled 330 miles at a constant rate. A bus traveled the same distance 5 mph faster than the car in ½-hour less time. Find the constant rate for the bus.

Graph parabolas. (See Lesson 10.1.)

44. Graph $y = x^2 - 2x - 3$.

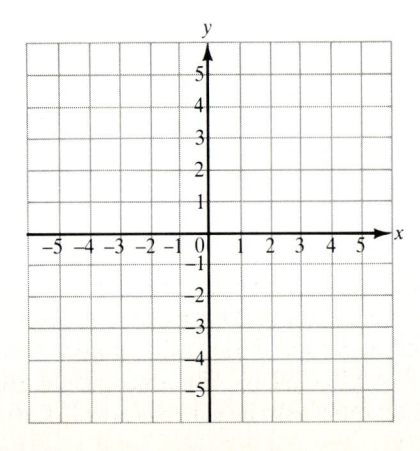

Name _____ Date _____ Class _____

Find the center and radius of circles. (See Lesson 10.2.)

45. Find the center and radius of the circle given by $x^2 + y^2 - 12x + 4y - 24 = 0$.

Graph ellipses and hyperbolas. (See Lessons 10.3 and 10.4.)

46. Graph $4x^2 + 25y^2 = 100$.

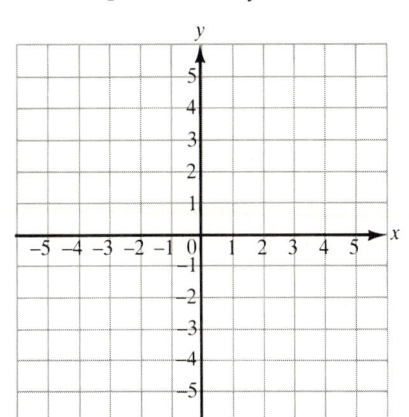

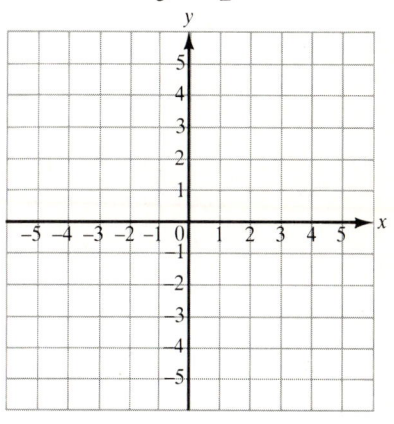

47. Graph $-\dfrac{x^2}{3^2} + \dfrac{y^2}{2^2} = 1$.

Solve quadratic systems. (See Lesson 10.5.)

48. Solve $\begin{cases} x^2 + y^2 = 10 \\ 5x - 2y = 11 \end{cases}$ using the substitution method.

Solve problems using quadratic systems. (See PS 15.)

49. The area of a rectangle is 8 m. The perimeter of the rectangle is 11.4 m. Find the length of the diagonal to the nearest centimeter.

Find the domain and range of relations. (See Lesson 11.1.)

50. Find the **a.** domain and **b.** range of the relation given by $y = \sqrt{x + 2} + 3$.

Identify functions. (See Lesson 11.2.)

51. Which equation(s) (a) $x^2 + y^2 = 6$, (b) $3x^2 - y = 5$, or (c) $y = |x - 3|$ represents a function?

Evaluate functions. (See Lesson 11.3.)

52. Find $g(f(3))$ given $f(x) = 2x - 4$ and $g(x) = x^2 - 11$.

Find the inverse of a function. (See Lesson 11.5.)

53. Find the inverse of the function given by $y = \dfrac{2}{x + 3}$. Write the inverse as an equation solved for y.

Solve applied variation problems. (See PS 16.)

54. *Free-Falling Objects:* The distance d an object falls from rest (disregarding air resistance) varies directly as the square of the time t it falls. Near the earth's surface, a free-falling object will fall 100 feet in $2\frac{1}{2}$ seconds. **a.** Find the specific variation formula for a free-falling object near the earth's surface. For a free-falling object near the earth's surface **b.** how far will it fall in 5 seconds? **c.** how long will it take to fall 1 mile (5280 feet) to the nearest tenth second?

556 Final Review

Graph logarithmic functions. (See Lesson 12.2.)

55. Graph $y = \log_2 x$.

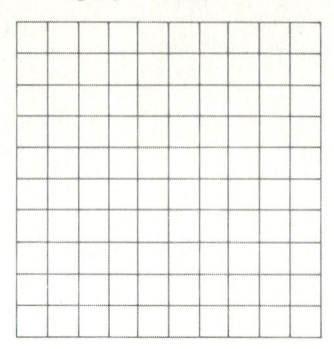

Use properties of logarithmic functions. (See Lesson 12.3.)

56. Write $\frac{1}{2} \log_b x - 3 \log_b y + \frac{3}{4} \log_b z$ as a single logarithm with a coefficient of 1.

Find logarithms and antilogarithms. (See Lesson 12.4.)

57. Find log 1540 using Appendix Table 2.

58. Find the antilogarithm of $(8.5403 - 10)$ using Appendix Table 2.

Solve exponential and logarithmic equations. (See Lesson 12.5.)

59. Solve $2^{4x-5} - 16 = 0$.

60. Solve $\log (x + 3) + \log (x - 2) - \log 3 = \log (x^2 - 17)$

Solve problems using exponential formulas. (See PS 17.)

61. In 1803, the United States doubled its area by purchasing Louisiana from France for about 2.83 cents per acre. Assuming an annual inflation rate of 5% per year, how much would one of these acres be worth in 1993, some 190 years later?

Evaluate given summation notation. (See Lesson 13.1.)

62. Evaluate $\sum_{k=1}^{4} (3k - 7)$.

Find the sum of an arithmetic series. (See Lesson 13.2.)

63. Find the sum of the first 101 terms of the arithmetic series with $a_1 = 7$ and $d = 6$.

Find the sum of an infinite geometric series. (See Lesson 13.3.)

64. Find the sum of the infinite geometric series $64 + 16 + 4 + 1 + \frac{1}{4} + \frac{1}{16} + \cdots$.

Find terms in a binomial expansion. (See Lesson 13.4.)

65. Find the 4th term in the expansion of $(a - 2b)^7$.

Solve problems using geometric sequences and series. (See PS 20.)

66. A ball dropped from 6 feet up rebounds 75% of its previous height on each bounce. To the nearest tenth foot: **a.** how far does it rebound up on the fourth bounce? **b.** How far does the ball travel by the time it hits the ground for the fourth time? **c.** How far does the ball travel before it comes to rest?

FINAL REVIEW ANSWERS: See Appendix Selected Answers.

Copyright © 1985 by Harcourt Brace Jovanovich, Inc. All rights reserved.

Appendix

Table 1: Powers and Roots (from 1 to 100)

Table 2: Common Logarithms (base 10)

Table 3: Geometry Formulas

Table 4: Conversion Factors

Table 5: Exponential Functions e^x and e^{-x}

Table 6: Natural Logarithms (base e)

Selected Answers: Answers to all odd-numbered exercises and problems and certain Make Sure and Review exercises.

TABLE 1 Powers and Roots (from 1 to 100)

n	n^2	$\sqrt{n}$	n^3	$\sqrt[3]{n}$	n	n^2	$\sqrt{n}$	n^3	$\sqrt[3]{n}$
1	1	1.000	1	1.000	51	2,601	7.141	132,651	3.708
2	4	1.414	8	1.260	52	2,704	7.211	140,608	3.733
3	9	1.732	27	1.442	53	2,809	7.280	148,877	3.756
4	16	2.000	64	1.587	54	2,916	7.348	157,464	3.780
5	25	2.236	125	1.710	55	3,025	7.416	166,375	3.803
6	36	2.449	216	1.817	56	3,136	7.483	175,616	3.826
7	49	2.646	343	1.913	57	3,249	7.550	185,193	3.849
8	64	2.828	512	2.000	58	3,364	7.616	195,112	3.871
9	81	3.000	729	2.080	59	3,481	7.681	205,379	3.893
10	100	3.162	1,000	2.154	60	3,600	7.746	216,000	3.915
11	121	3.317	1,331	2.224	61	3,721	7.810	226,981	3.936
12	144	3.464	1,728	2.289	62	3,844	7.874	238,328	3.958
13	169	3.606	2,197	2.351	63	3,969	7.937	250,047	3.979
14	196	3.742	2,744	2.410	64	4,096	8.000	262,144	4.000
15	225	3.873	3,375	2.466	65	4,225	8.062	274,625	4.021
16	256	4.000	4,096	2.520	66	4,356	8.124	287,496	4.041
17	289	4.123	4,913	2.571	67	4,489	8.185	300,763	4.062
18	324	4.243	5,832	2.621	68	4,624	8.246	314,432	4.082
19	361	4.359	6,859	2.668	69	4,761	8.307	328,509	4.102
20	400	4.472	8,000	2.714	70	4,900	8.367	343,000	4.121
21	441	4.583	9,261	2.759	71	5,041	8.426	357,911	4.141
22	484	4.690	10,648	2.802	72	5,184	8.485	373,248	4.160
23	529	4.796	12,167	2.844	73	5,329	8.544	389,017	4.179
24	576	4.899	13,824	2.884	74	5,476	8.602	405,224	4.198
25	625	5.000	15,625	2.924	75	5,625	8.660	421,875	4.217
26	676	5.099	17,576	2.962	76	5,776	8.718	438,976	4.236
27	729	5.196	19,683	3.000	77	5,929	8.775	456,533	4.254
28	784	5.292	21,952	3.037	78	6,084	8.832	474,552	4.273
29	841	5.385	24,389	3.072	79	6,241	8.888	493,039	4.291
30	900	5.477	27,000	3.107	80	6,400	8.944	512,000	4.309
31	961	5.568	29,791	3.141	81	6,561	9.000	531,441	4.327
32	1,024	5.657	32,768	3.175	82	6,724	9.055	551,368	4.344
33	1,089	5.745	35,937	3.208	83	6,889	9.110	571,787	4.362
34	1,156	5.831	39,304	3.240	84	7,056	9.165	592,704	4.380
35	1,225	5.916	42,875	3.271	85	7,225	9.220	614,125	4.397
36	1,296	6.000	46,656	3.302	86	7,396	9.274	636,056	4.414
37	1,369	6.083	50,653	3.332	87	7,569	9.327	658,503	4.431
38	1,444	6.164	54,872	3.362	88	7,744	9.381	681,472	4.448
39	1,521	6.245	59,319	3.391	89	7,921	9.434	704,969	4.465
40	1,600	6.325	64,000	3.420	90	8,100	9.487	729,000	4.481
41	1,681	6.403	68,921	3.448	91	8,281	9.539	753,571	4.498
42	1,764	6.481	74,088	3.476	92	8,464	9.592	778,688	4.514
43	1,849	6.557	79,507	3.503	93	8,649	9.644	804,357	4.531
44	1,936	6.633	85,184	3.530	94	8,836	9.695	830,584	4.547
45	2,025	6.708	91,125	3.557	95	9,025	9.747	857,375	4.563
46	2,116	6.782	97,336	3.583	96	9,216	9.798	884,736	4.579
47	2,209	6.856	103,823	3.609	97	9,409	9.849	912,673	4.595
48	2,304	6.928	110,592	3.634	98	9,604	9.899	941,192	4.610
49	2,401	7.000	117,649	3.659	99	9,801	9.950	970,299	4.626
50	2,500	7.071	125,000	3.684	100	10,000	10.000	1,000,000	4.642

TABLE 2 Common Logarithms (base 10)

x	0	1	2	3	4	5	6	7	8	9
1.0	.0000	.0043	.0086	.0128	.0170	.0212	.0253	.0294	.0334	.0374
1.1	.0414	.0453	.0492	.0531	.0569	.0607	.0645	.0682	.0719	.0755
1.2	.0792	.0828	.0864	.0899	.0934	.0969	.1004	.1038	.1072	.1106
1.3	.1139	.1173	.1206	.1239	.1271	.1303	.1335	.1367	.1399	.1430
1.4	.1461	.1492	.1523	.1553	.1584	.1614	.1644	.1673	.1703	.1732
1.5	.1761	.1790	.1818	.1847	.1875	.1903	.1931	.1959	.1987	.2014
1.6	.2041	.2068	.2095	.2122	.2148	.2175	.2201	.2227	.2253	.2279
1.7	.2304	.2330	.2355	.2380	.2405	.2430	.2455	.2480	.2504	.2529
1.8	.2553	.2577	.2601	.2625	.2648	.2672	.2695	.2718	.2742	.2765
1.9	.2788	.2810	.2833	.2856	.2878	.2900	.2923	.2945	.2967	.2989
2.0	.3010	.3032	.3054	.3075	.3096	.3118	.3139	.3160	.3181	.3201
2.1	.3222	.3243	.3263	.3284	.3304	.3324	.3345	.3365	.3385	.3404
2.2	.3424	.3444	.3464	.3483	.3502	.3522	.3541	.3560	.3579	.3598
2.3	.3617	.3636	.3655	.3674	.3692	.3711	.3729	.3747	.3766	.3784
2.4	.3802	.3820	.3838	.3856	.3874	.3892	.3909	.3927	.3945	.3962
2.5	.3979	.3997	.4014	.4031	.4048	.4065	.4082	.4099	.4116	.4133
2.6	.4150	.4166	.4183	.4200	.4216	.4232	.4249	.4265	.4281	.4298
2.7	.4314	.4330	.4346	.4362	.4378	.4393	.4409	.4425	.4440	.4456
2.8	.4472	.4487	.4502	.4518	.4533	.4548	.4564	.4579	.4594	.4609
2.9	.4624	.4639	.4654	.4669	.4683	.4698	.4713	.4728	.4742	.4757
3.0	.4771	.4786	.4800	.4814	.4829	.4843	.4857	.4871	.4886	.4900
3.1	.4914	.4928	.4942	.4955	.4969	.4983	.4997	.5011	.5024	.5038
3.2	.5051	.5065	.5079	.5092	.5105	.5119	.5132	.5145	.5159	.5172
3.3	.5185	.5198	.5211	.5224	.5237	.5250	.5263	.5276	.5289	.5302
3.4	.5315	.5328	.5340	.5353	.5366	.5378	.5391	.5403	.5416	.5428
3.5	.5441	.5453	.5465	.5478	.5490	.5502	.5514	.5527	.5539	.5551
3.6	.5563	.5575	.5587	.5599	.5611	.5623	.5635	.5647	.5658	.5670
3.7	.5682	.5694	.5705	.5717	.5729	.5740	.5752	.5763	.5775	.5786
3.8	.5798	.5809	.5821	.5832	.5843	.5855	.5866	.5877	.5888	.5899
3.9	.5911	.5922	.5933	.5944	.5955	.5966	.5977	.5988	.5999	.6010
4.0	.6021	.6031	.6042	.6053	.6064	.6075	.6085	.6096	.6107	.6117
4.1	.6128	.6138	.6149	.6160	.6170	.6180	.6191	.6201	.6212	.6222
4.2	.6232	.6243	.6253	.6263	.6274	.6284	.6294	.6304	.6314	.6325
4.3	.6335	.6345	.6355	.6365	.6375	.6385	.6395	.6405	.6415	.6425
4.4	.6435	.6444	.6454	.6464	.6474	.6484	.6493	.6503	.6513	.6522
4.5	.6532	.6542	.6551	.6561	.6571	.6580	.6590	.6599	.6609	.6618
4.6	.6628	.6637	.6646	.6656	.6665	.6675	.6684	.6693	.6702	.6712
4.7	.6721	.6730	.6739	.6749	.6758	.6767	.6776	.6785	.6794	.6803
4.8	.6812	.6821	.6830	.6839	.6848	.6857	.6866	.6875	.6884	.6893
4.9	.6902	.6911	.6920	.6928	.6937	.6946	.6955	.6964	.6972	.6981
5.0	.6990	.6998	.7007	.7016	.7024	.7033	.7042	.7050	.7059	.7067
5.1	.7076	.7084	.7093	.7101	.7110	.7118	.7126	.7135	.7143	.7152
5.2	.7160	.7168	.7177	.7185	.7193	.7202	.7210	.7218	.7226	.7235
5.3	.7243	.7251	.7259	.7267	.7275	.7284	.7292	.7300	.7308	.7316
5.4	.7324	.7332	.7340	.7348	.7356	.7364	.7372	.7380	.7388	.7396
x	0	1	2	3	4	5	6	7	8	9

TABLE 2 (*Continued*)

x	0	1	2	3	4	5	6	7	8	9
5.5	.7404	.7412	.7419	.7427	.7435	.7443	.7451	.7459	.7466	.7474
5.6	.7482	.7490	.7497	.7505	.7513	.7520	.7528	.7536	.7543	.7551
5.7	.7559	.7566	.7574	.7582	.7589	.7597	.7604	.7612	.7619	.7627
5.8	.7634	.7642	.7649	.7657	.7664	.7672	.7679	.7686	.7694	.7701
5.9	.7709	.7716	.7723	.7731	.7738	.7745	.7752	.7760	.7767	.7774
6.0	.7782	.7789	.7796	.7803	.7810	.7818	.7825	.7832	.7839	.7846
6.1	.7853	.7860	.7868	.7875	.7882	.7889	.7896	.7903	.7910	.7917
6.2	.7924	.7931	.7938	.7945	.7952	.7959	.7966	.7973	.7980	.7987
6.3	.7993	.8000	.8007	.8014	.8021	.8028	.8035	.8041	.8048	.8055
6.4	.8062	.8069	.8075	.8082	.8089	.8096	.8102	.8109	.8116	.8122
6.5	.8129	.8136	.8142	.8149	.8156	.8162	.8169	.8176	.8182	.8189
6.6	.8195	.8202	.8209	.8215	.8222	.8228	.8235	.8241	.8248	.8254
6.7	.8261	.8267	.8274	.8280	.8287	.8293	.8299	.8306	.8312	.8319
6.8	.8325	.8331	.8338	.8344	.8351	.8357	.8363	.8370	.8376	.8382
6.9	.8388	.8395	.8401	.8407	.8414	.8420	.8426	.8432	.8439	.8445
7.0	.8451	.8457	.8463	.8470	.8476	.8482	.8488	.8494	.8500	.8506
7.1	.8513	.8519	.8525	.8531	.8537	.8543	.8549	.8555	.8561	.8567
7.2	.8573	.8579	.8585	.8591	.8597	.8603	.8609	.8615	.8621	.8627
7.3	.8633	.8639	.8645	.8651	.8657	.8663	.8669	.8675	.8681	.8686
7.4	.8692	.8698	.8704	.8710	.8716	.8722	.8727	.8733	.8739	.8745
7.5	.8751	.8756	.8762	.8768	.8774	.8779	.8785	.8791	.8797	.8802
7.6	.8808	.8814	.8820	.8825	.8831	.8837	.8842	.8848	.8854	.8859
7.7	.8865	.8871	.8876	.8882	.8887	.8893	.8899	.8904	.8910	.8915
7.8	.8921	.8927	.8932	.8938	.8943	.8949	.8954	.8960	.8965	.8971
7.9	.8976	.8982	.8987	.8993	.8998	.9004	.9009	.9015	.9020	.9025
8.0	.9031	.9036	.9042	.9047	.9053	.9058	.9063	.9069	.9074	.9079
8.1	.9085	.9090	.9096	.9101	.9106	.9112	.9117	.9122	.9128	.9133
8.2	.9138	.9143	.9149	.9154	.9159	.9165	.9170	.9175	.9180	.9186
8.3	.9191	.9196	.9201	.9206	.9212	.9217	.9222	.9227	.9232	.9238
8.4	.9243	.9248	.9253	.9258	.9263	.9269	.9274	.9279	.9284	.9289
8.5	.9294	.9299	.9304	.9309	.9315	.9320	.9325	.9330	.9335	.9340
8.6	.9345	.9350	.9355	.9360	.9365	.9370	.9375	.9380	.9385	.9390
8.7	.9395	.9400	.9405	.9410	.9415	.9420	.9425	.9430	.9435	.9440
8.8	.9445	.9450	.9455	.9460	.9465	.9469	.9474	.9479	.9484	.9489
8.9	.9494	.9499	.9504	.9509	.9513	.9518	.9523	.9528	.9533	.9538
9.0	.9542	.9547	.9552	.9557	.9562	.9566	.9571	.9576	.9581	.9586
9.1	.9590	.9595	.9600	.9605	.9609	.9614	.9619	.9624	.9628	.9633
9.2	.9638	.9643	.9647	.9652	.9657	.9661	.9666	.9671	.9675	.9680
9.3	.9685	.9689	.9694	.9699	.9703	.9708	.9713	.9717	.9722	.9727
9.4	.9731	.9736	.9741	.9745	.9750	.9754	.9759	.9763	.9768	.9773
9.5	.9777	.9782	.9786	.9791	.9795	.9800	.9805	.9809	.9814	.9818
9.6	.9823	.9827	.9832	.9836	.9841	.9845	.9850	.9854	.9859	.9863
9.7	.9868	.9872	.9877	.9881	.9886	.9890	.9894	.9899	.9903	.9908
9.8	.9912	.9917	.9921	.9926	.9930	.9934	.9939	.9943	.9948	.9952
9.9	.9956	.9961	.9965	.9969	.9974	.9978	.9983	.9987	.9991	.9996
x	0	1	2	3	4	5	6	7	8	9

Appendix Table 3   **A-5**

TABLE 3 Geometry Formulas

Figure		Perimeter (P)	Area (A)
Square		$P = 4s$	$A = s^2$
Rectangle		$P = 2(l + w)$	$A = lw$
Parallelogram		$P = 2(a + b)$	$A = bh$
Triangle		$P = a + b + c$	$A = \frac{1}{2}bh$

		Circumference (C)	Area (A)
Circle		$C = \pi d$ $C = 2\pi r$	$A = \pi r^2$

		Volume (V)	Surface Area (SA)
Cube		$V = e^3$	$SA = 6e^2$
Rectangular Prism (box)		$V = lwh$	$SA = 2(lw + lh + wh)$
Cylinder		$V = \pi r^2 h$	$SA = 2\pi r(r + h)$
Sphere		$V = \frac{4}{3}\pi r^3$	$SA = 4\pi r^2$

TABLE 4 Conversion Factors

	U.S. Customary/Metric From	To	Paper-and-Pencil Conversion Factors Multiply By	Calculator Conversion Factors Multiply By
Length	inches (in.)	millimeters (mm)	25	**25.4**
	inches	centimeters (cm)	2.5	**2.54**
	feet (ft)	meters (m)	0.3	**0.3048**
	yards (yd)	meters	0.9	**0.9144**
	miles (mi)	kilometers (km)	1.6	1.609
Capacity	drops (gtt)	milliliters (mL)	16	16.23
	teaspoons (tsp)	milliliters	5	4.929
	tablespoons (tbsp)	milliliters	15	14.79
	fluid ounces (fl oz)	milliliters	30	29.57
	cups (c)	liters (L)	0.24	0.2366
	pints (pt)	liters	0.47	0.4732
	quarts (qt)	liters	0.95	0.9464
	gallons (gal)	liters	3.8	3.785
Weight (Mass)	ounces (oz)	grams (g)	28	28.35
	pounds (lb)	kilograms (kg)	0.45	0.4536
	tons (T)	tonnes (t)	0.9	0.9072
Area	square inches (in.^{2})	square centimeters (cm^2)	6.5	6.452
	square feet (ft^2)	square meters (m^2)	0.09	0.09290
	square yards (yd^2)	square meters	0.8	0.8361
	square miles (mi^2)	square kilometers (km^2)	2.6	2.590
	acres (A)	hectares (ha)	0.4	0.4047
Volume	cubic inches (in.^{3})	cubic centimeters (cm^3 or cc)	16	16.39
	cubic feet (ft^3)	cubic meters (m^3)	0.03	0.02832
	cubic yards (yd^3)	cubic meters	0.8	0.7646
Temperature	degrees Fahrenheit (°F)	degrees Celsius (°C)	$\frac{5}{9}$ (after subtracting 32)	0.5556 (after subtracting 32)

Note: All conversion factors in bold type are exact. All others are rounded.

TABLE 4 (Continued)

	Metric/U.S. Customary *From*	*To*	*Paper-and-Pencil Conversion Factors* Multiply By	*Calculator Conversion Factors* Multiply By
Length	millimeters (mm)	inches (in.)	0.04	0.03937
	centimeters (cm)	inches	0.4	0.3937
	meters (m)	feet (ft)	3.3	3.280
	meters	yards (yd)	1.1	1.094
	kilometers (km)	miles (mi)	0.6	0.6214
Capacity	milliliters (mL)	drops (gtt)	0.06	0.06161
	milliliters	teaspoons (tsp)	0.2	0.2029
	milliliters	tablespoons (tbsp)	0.07	0.06763
	milliliters	fluid ounces (fl oz)	0.03	0.03381
	liters (L)	cups (c)	4.2	4.227
	liters	pints (pt)	2.1	2.113
	liters	quarts (qt)	1.1	1.057
	liters	gallons (gal)	0.26	0.2642
Mass (Weight)	grams (g)	ounces (oz)	0.035	0.03527
	kilograms (kg)	pounds (lb)	2.2	2.205
	tonnes (t)	tons (T)	1.1	1.102
Area	square centimeters (cm^2)	square inches (in.^{2})	0.16	0.1550
	square meters (m^2)	square feet (ft^2)	11	10.76
	square meters	square yards (yd^2)	1.2	1.196
	square kilometers (km^2)	square miles (mi^2)	0.4	0.3861
	hectares (ha)	acres (A)	2.5	2.471
Volume	cubic centimeters (cm^3)	cubic inches (in.^{3})	0.06	0.06102
	cubic meters (m^3)	cubic feet (ft^3)	35	35.31
	cubic meters	cubic yards (yd^3)	1.3	1.308
Temperature	degrees Celsius (°C)	degrees Fahrenheit (°F)	$\frac{9}{5}$ (then add 32)	**1.8** (then add 32)

Note: All conversion factors in bold type are exact. All others are rounded.

TABLE 5 Exponential Functions e^x and e^{-x}

x	e^x	e^{-x}	x	e^x	e^{-x}
0.00	1.0000	1.0000	**1.5**	4.4817	0.2231
0.01	1.0101	0.9901	**1.6**	4.9530	0.2019
0.02	1.0202	0.9802	**1.7**	5.4739	0.1827
0.03	1.0305	0.9702	**1.8**	6.0496	0.1653
0.04	1.0408	0.9608	**1.9**	6.6859	0.1496
0.05	1.0513	0.9512	**2.0**	7.3891	0.1353
0.06	1.0618	0.9418	**2.1**	8.1662	0.1225
0.07	1.0725	0.9324	**2.2**	9.0250	0.1108
0.08	1.0833	0.9231	**2.3**	9.9742	0.1003
0.09	1.0942	0.9139	**2.4**	11.023	0.0907
0.10	1.1052	0.9048	**2.5**	12.182	0.0821
0.11	1.1163	0.8958	**2.6**	13.464	0.0743
0.12	1.1275	0.8869	**2.7**	14.880	0.0672
0.13	1.1388	0.8781	**2.8**	16.445	0.0608
0.14	1.1503	0.8694	**2.9**	18.174	0.0550
0.15	1.1618	0.8607	**3.0**	20.086	0.0498
0.16	1.1735	0.8521	**3.1**	22.198	0.0450
0.17	1.1853	0.8437	**3.2**	24.533	0.0408
0.18	1.1972	0.8353	**3.3**	27.113	0.0369
0.19	1.2092	0.8270	**3.4**	29.964	0.0334
0.20	1.2214	0.8187	**3.5**	33.115	0.0302
0.21	1.2337	0.8106	**3.6**	36.598	0.0273
0.22	1.2461	0.8025	**3.7**	40.447	0.0247
0.23	1.2586	0.7945	**3.8**	44.701	0.0224
0.24	1.2712	0.7866	**3.9**	49.402	0.0202
0.25	1.2840	0.7788	**4.0**	54.598	0.0183
0.30	1.3499	0.7408	**4.1**	60.340	0.0166
0.35	1.4191	0.7047	**4.2**	66.686	0.0150
0.40	1.4918	0.6703	**4.3**	73.700	0.0136
0.45	1.5683	0.6376	**4.4**	81.451	0.0123
0.50	1.6487	0.6065	**4.5**	90.017	0.0111
0.55	1.7333	0.5769	**4.6**	99.484	0.0101
0.60	1.8221	0.5488	**4.7**	109.95	0.0091
0.65	1.9155	0.5220	**4.8**	121.51	0.0082
0.70	2.0138	0.4966	**4.9**	134.29	0.0074
0.75	2.1170	0.4724	**5.0**	148.41	0.0067
0.80	2.2255	0.4493	**5.5**	244.69	0.0041
0.85	2.3396	0.4274	**6.0**	403.43	0.0025
0.90	2.4596	0.4066	**6.5**	665.14	0.0015
0.95	2.5857	0.3867	**7.0**	1096.6	0.0009
1.0	2.7183	0.3679	**7.5**	1808.0	0.0006
1.1	3.0042	0.3329	**8.0**	2981.0	0.0003
1.2	3.3201	0.3012	**8.5**	4914.8	0.0002
1.3	3.6693	0.2725	**9.0**	8103.1	0.0001
1.4	4.0552	0.2466	**10.0**	22,026	0.00005

TABLE 6 Natural Logarithms (base e)

n	$\log_e n$	n	$\log_e n$	n	$\log_e n$
		4.5	1.5041	9.0	2.1972
0.1	−2.3026	4.6	1.5261	9.1	2.2083
0.2	−1.6094	4.7	1.5476	9.2	2.2192
0.3	−1.2040	4.8	1.5686	9.3	2.2300
0.4	−0.9163	4.9	1.5892	9.4	2.2407
0.5	−0.6931	5.0	1.6094	9.5	2.2513
0.6	−0.5108	5.1	1.6292	9.6	2.2618
0.7	−0.3567	5.2	1.6487	9.7	2.2721
0.8	−0.2231	5.3	1.6677	9.8	2.2824
0.9	−0.1054	5.4	1.6864	9.9	2.2925
1.0	0.0000	5.5	1.7047	10	2.3026
1.1	0.0953	5.6	1.7228	11	2.3979
1.2	0.1823	5.7	1.7405	12	2.4849
1.3	0.2624	5.8	1.7579	13	2.5649
1.4	0.3365	5.9	1.7750	14	2.6391
1.5	0.4055	6.0	1.7918	15	2.7081
1.6	0.4700	6.1	1.8083	16	2.7726
1.7	0.5306	6.2	1.8245	17	2.8332
1.8	0.5878	6.3	1.8405	18	2.8904
1.9	0.6419	6.4	1.8563	19	2.9444
2.0	0.6931	6.5	1.8718	20	2.9957
2.1	0.7419	6.6	1.8871	25	3.2189
2.2	0.7885	6.7	1.9021	30	3.4012
2.3	0.8329	6.8	1.9169	35	3.5553
2.4	0.8755	6.9	1.9315	40	3.6889
2.5	0.9163	7.0	1.9459	45	3.8067
2.6	0.9555	7.1	1.9601	50	3.9120
2.7	0.9933	7.2	1.9741	55	4.0073
2.8	1.0296	7.3	1.9879	60	4.0943
2.9	1.0647	7.4	2.0015	65	4.1744
3.0	1.0986	7.5	2.0149	70	4.2485
3.1	1.1314	7.6	2.0281	75	4.3175
3.2	1.1632	7.7	2.0412	80	4.3820
3.3	1.1939	7.8	2.0541	85	4.4427
3.4	1.2238	7.9	2.0669	90	4.4998
3.5	1.2528	8.0	2.0794	100	4.6052
3.6	1.2809	8.1	2.0919	110	4.7005
3.7	1.3083	8.2	2.1041	120	4.7875
3.8	1.3350	8.3	2.1163	130	4.8676
3.9	1.3610	8.4	2.1282	140	4.9416
4.0	1.3863	8.5	2.1401	150	5.0106
4.1	1.4110	8.6	2.1518	160	5.0752
4.2	1.4351	8.7	2.1633	170	5.1358
4.3	1.4586	8.8	2.1748	180	5.1930
4.4	1.4816	8.9	2.1861	190	5.2470

APPENDIX SELECTED ANSWERS

Chapter 1 Answers

▶ Practice 1.1, pp. 7–8

1. $\{2\}$ 3. $\{a\}$ 5. $\{0\}$ 7. $\{-2\}$ 9. $\{1, 2\}$ 11. $\{a, b\}$ 13. $\{1, 2, 3\}$ 15. $\{1, 2, 3\}$ 17. $\{a, b, c\}$ 19. $\{3\}$
21. $\{1, 2, 3, 4, 5, 6\}$ 23. W 25. $\emptyset$ 27. $\emptyset$ 29. W 31. $\emptyset$ 33. I 35. Q 37. $\emptyset, \{2\}$ 39. $\emptyset, \{0\}$
41. $\emptyset, \{2\}, \{4\}, \{2, 4\}$ 43. $\emptyset, \{r\}, \{s\}, \{r, s\}$ 45. $\emptyset, \{0\}, \{3\}, \{5\}, \{0, 3\}, \{0, 5\}, \{3, 5\}, \{0, 3, 5\}$
47. $\emptyset, \{a\}, \{b\}, \{d\}, \{e\}, \{a, b\}, \{a, d\}, \{a, e\}, \{b, d\}, \{b, e\}, \{d, e\}, \{a, b, d\}, \{a, b, e\}, \{a, d, e\}, \{b, d, e\}, \{a, b, d, e\}$ 49. true
51. false 53. false 55. true 57. false 59. false

▶ Make Sure Answers for Lesson 1.2, Example 1, p. 9

1. [number line with points at $-3\frac{1}{2}$, -1, 0, $2\frac{1}{2}$]
2. [number line with points at -4, $-1\frac{1}{2}$, 1, $2\frac{1}{2}$]

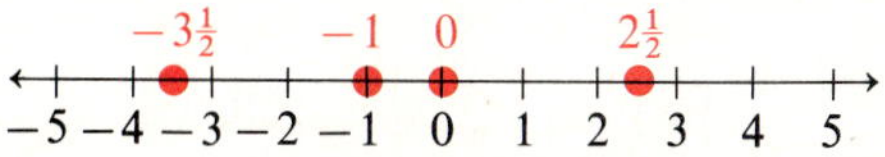

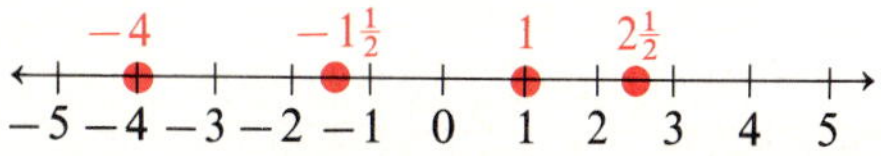

▶ Make Sure Answers for Lesson 1.2, Example 3, p. 11

1. [number line showing $z > -2$]
2. [number line showing $y \leq 2$]

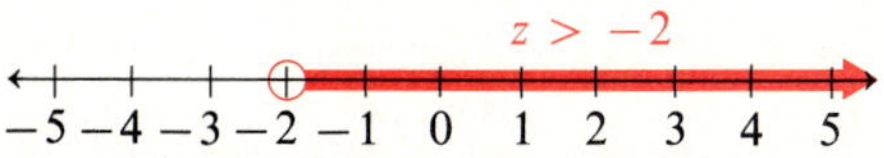

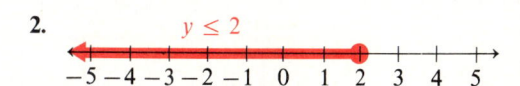

▶ Make Sure Answers for Lesson 1.2, Example 4, p. 12

1. [number line showing $z > 1$ or $z \leq -2$]
2. [number line showing $-3 < y \leq 2$]

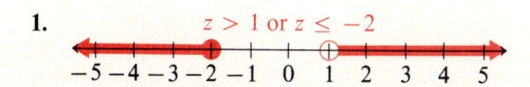

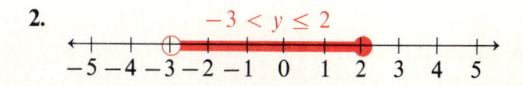

▶ Practice 1.2, pp. 13–14

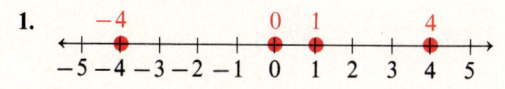

1. [number line with points at -4, 0, 1, 4]
3. [number line with points at -2, 1, 3, 5]
5. [number line with points at $-2\frac{1}{2}$, -1, $\frac{1}{2}$]
7. 7 9. 2 11. $2\frac{1}{2}$ 13. $3\frac{1}{3}$ 15. 4.23 17. $\sqrt{2}$
19. [number line showing $z > 1$]
21. [number line showing $x < 4$]
23. [number line showing $v \geq -2\frac{1}{2}$]
25. [number line showing $t \leq 3\frac{1}{2}$]
27. [number line showing $r > 2.3$]
29. [number line showing $p \not> 2$]

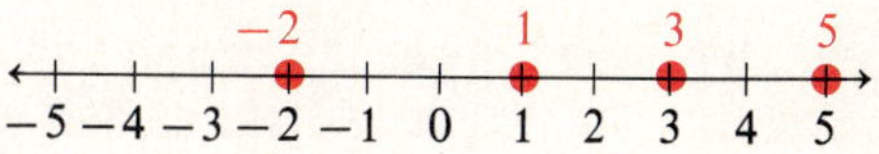

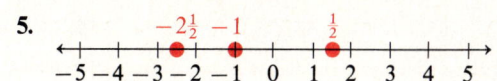

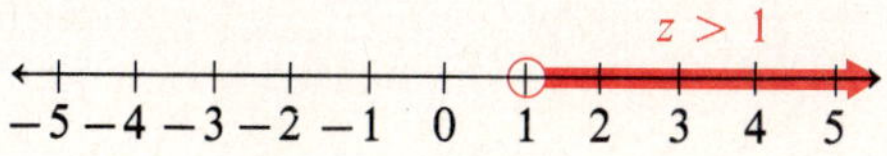

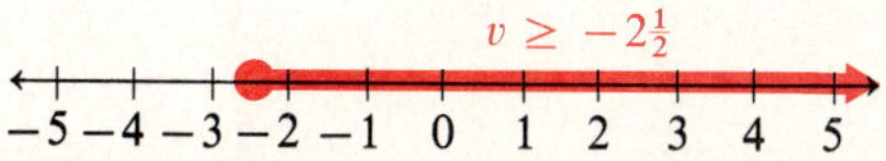

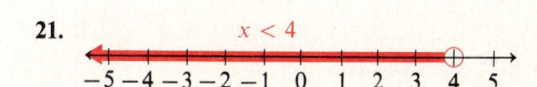

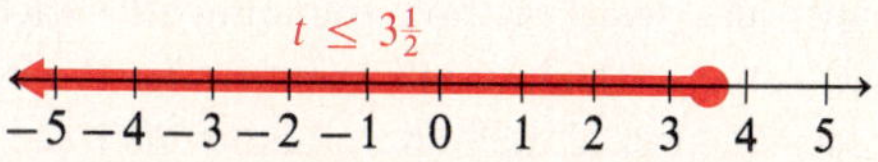

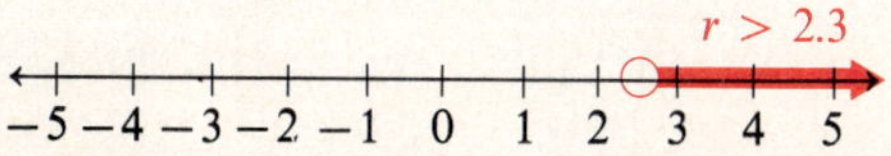

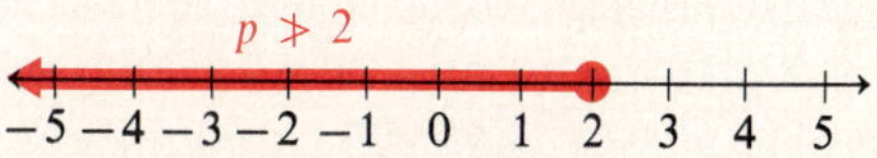

31. 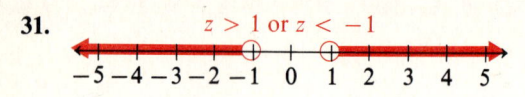 $z > 1$ or $z < -1$

33. 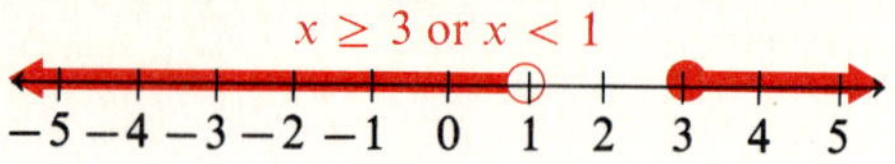 $x \geq 3$ or $x < 1$

35. 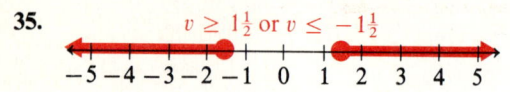 $v \geq 1\frac{1}{2}$ or $v \leq -1\frac{1}{2}$

37. 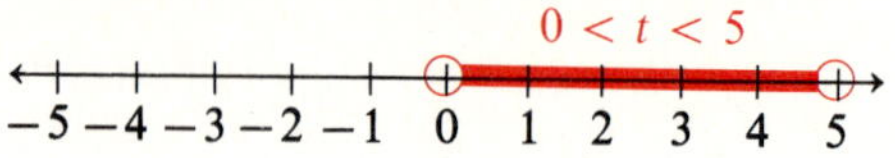 $0 < t < 5$

39. 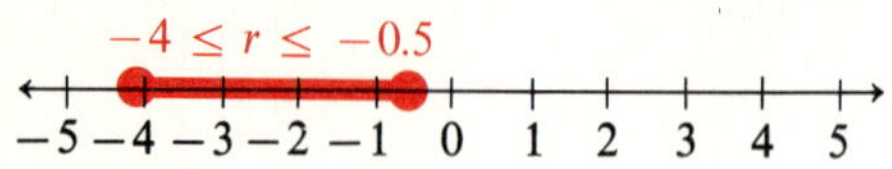 $-4 \leq r \leq -0.5$

41. 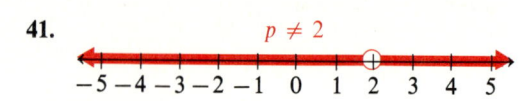 $p \neq 2$

43. true **45.** false **47.** true **49.** true **51.** false **53.** $\frac{9}{7}$ **55.** $\frac{22}{15}$ **57.** $\frac{19}{12}$ **59.** $\frac{59}{36}$ **61.** $\frac{2}{3}$ **63.** $\frac{7}{15}$ **65.** $\frac{10}{7}$ **67.** $\frac{17}{6}$ **69.** $\frac{15}{28}$ **71.** $\frac{3}{4}$ **73.** $\frac{10}{21}$ **75.** $\frac{3}{2}$ **77.** $\frac{10}{9}$ **79.** $\frac{9}{10}$ **81.** $\frac{3}{8}$ **83.** 10

▶ **Practice 1.3, pp. 19–20**

1. 4 **3.** 7 **5.** -16 **7.** -3.4 **9.** -1 **11.** $\frac{5}{28}$ **13.** -3 **15.** -14 **17.** 9 **19.** 1.4 **21.** 1 **23.** $\frac{17}{40}$ **25.** -12 **27.** -30 **29.** 20 **31.** -9.43 **33.** $\frac{12}{25}$ **35.** $-\frac{2}{5}$ **37.** -2 **39.** -4 **41.** 9 **43.** 4 **45.** $-\frac{3}{2}$ **47.** $-\frac{3}{8}$ **49.** -10 **51.** -47 **53.** -8 **55.** -8.2 **57.** 21.5 **59.** $-\frac{12}{7}$ **61.** $\frac{2}{3}$ **63.** -0.3 **65.** -4 **67.** 5 **69.** 5 **71.** -7 **73.** -6 **75.** 0 **77.** -210 **79.** 60 **81.** -60 **83.** $\frac{2}{3}$ **85.** $\frac{3}{4}$ **87.** $\frac{8}{3}$ **89.** $\frac{4}{3}$ **91.** $\frac{16}{81}$ **93.** $\frac{12}{35}$

▶ **Practice 1.4, pp. 27–30**

1. 16 **3.** 9 **5.** -8 **7.** 16 **9.** 1.728 **11.** $-\frac{8}{27}$ **13.** 5^3 **15.** $(-3)^4$ **17.** $3^5 x^4$ **19.** $(-3)^3 x^4 y$ **21.** $(\frac{2}{3})^6 w^3$ **23.** $(a+b)^3$ **25.** 9 **27.** 12 **29.** 2 **31.** 101 **33.** 26 **35.** 497 **37.** -1 **39.** 3 **41.** 8 **43.** -11 **45.** -14 **47.** 18 **49.** -5 **51.** -21 **53.** -1 **55.** 10 **57.** $\frac{5}{3}$ **59.** $-\frac{15}{2}$ **61.** 3 **63.** -2 **65.** 20 **67.** -8 **69.** 46 **71.** -24 **73.** -7 **75.** -19 **77.** 13 **79.** -61 **81.** 29 **83.** 3 **85.** -8 **87.** -21 **89.** 18

▶ **Practice 1.5, pp. 35–36**

1. Commutative property of addition **3.** Commutative property of addition **5.** Commutative property of addition **7.** Commutative property of addition **9.** Commutative property of multiplication **11.** Associative property of multiplication **13.** Commutative property of multiplication **15.** Commutative property of addition **17.** Commutative property of addition **19.** $5z + 15$ **21.** $-3x + 15$ **23.** $-3v + 5u$ **25.** $4q - 6r + 8s$ **27.** $-9a + 3b - 6c$ **29.** $5w - 10x + 15y - 20$ **31.** $7z$ **33.** $5x$ **35.** $-4v$ **37.** $-t$ **39.** $2r - 4$ **41.** $-2de + 4df$ **43.** $-27a - 12$ **45.** $28c - 56$ **47.** $4x + y$ **49.** $-2m + 4n$ **51.** $-20a + 24b$ **53.** $-3a + 18b - 36c$ **55.** $\frac{3}{4}$ **57.** $-\frac{2}{3}$ **59.** 1 **61.** -3 **63.** 3 **65.** -2 **67.** -6 **69.** 4 **71.** 8 **73.** 81

▶ **Practice 1.6, pp. 41–42**

1. a^3 **3.** $\frac{1}{c^5}$ **5.** $\frac{g}{e^3 f^2 h^2}$ **7.** $\frac{3r^2 t^2}{4su}$ **9.** $\frac{6b^2 df^4}{7ac^3 e}$ **11.** $\frac{4r}{9q^2}$ **13.** a^9 **15.** c^3 **17.** ef^2 **19.** $\frac{3}{4n^8}$ **21.** $\frac{8s^5 t^4}{117}$ **23.** $\frac{3x^3 y^3 z}{8}$ **25.** a^{15} **27.** $81c^{12}$ **29.** $\frac{9e^8}{f^6}$ **31.** $\frac{16p^{28}}{81m^{12} n^4}$ **33.** $\frac{-32v^{25}}{243t^5 u^{10}}$ **35.** $\frac{4a^{10} c^8}{9b^{10}}$ **37.** $\frac{4b^6}{a^{10}}$ **39.** $\frac{f^{18}}{27e^9}$ **41.** $\frac{p^7}{108m^7 n^8}$ **43.** $\frac{144u^8 v^{14}}{w^{10}}$ **45.** $2a^7 b^8$ **47.** $\frac{81g^{12} h^4}{256k^4}$ **49.** -3 **51.** 12 **53.** $-3a$ **55.** $13c$ **57.** a **59.** c **61.** $3e$ **63.** $4g$ **65.** $\frac{1}{4}$ **67.** $-\frac{1}{5}$ **69.** $\frac{3}{2}$ **71.** -4 **73.** a **75.** c **77.** e **79.** g

▶ **Problem Solving 1, pp. 44–46**

1. 2.00×10^{20} N (paper-and-pencil method; rounding after each computation)
   1.99×10^{20} N (calculator method; rounding only after the final computation)
3. 4.42×10^{20} N (paper-and-pencil method; rounding after each computation)
   4.41×10^{20} N (calculator method; rounding only after the final computation)
5. 3.33×10^5 times **7.** 1.97×10^8 mi² **9.** 2×10^{16} cubic light-years **11.** 5.54×10^3 kg/m³ **13.** 1.41×10^3 kg/m³ **15.** 1.11×10^7 mi² **17.** 1×10^{22} kg **19.** $-460°$F

Chapter 1 Review, p. 47–48

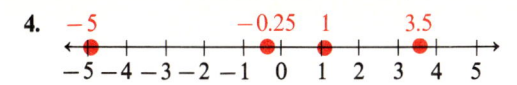

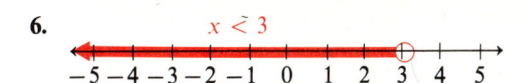

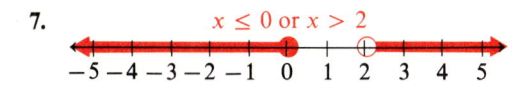

4. Number line showing points at -5, -0.25, 1, 3.5 on scale -5 to 5.

6. Number line showing $x < 3$ on scale -5 to 5.

7. Number line showing $x \leq 0$ or $x > 2$ on scale -5 to 5.

Chapter 2 Answers

Practice 2.1, pp. 53–54

1. 2   3. 8   5. -8 7. 8   9. -5 11. 8   13. 5   15. $\frac{8}{3}$ 17. $-\frac{1}{2}$ 19. -6 21. 7   23. -4 25. $\frac{7}{2}$
27. -1 29. 2   31. 0   33. $-\frac{4}{5}$ 35. $\frac{1}{2}$ 37. $2z - 6$ 39. $-3x + 15$ 41. $-6uv + 2tu + 50u$ 43. $5y$
45. $-2p$ 47. $3r + 2s$

According to Guinness, p. 58

5. $-58.9°C$ 6. $7.3°F$

Practice 2.2, pp. 59–60

1. 2   3. -3 5. -4 7. $-\frac{1}{2}$ 9. $\frac{5}{2}$ 11. 0.55   13. -12 15. -3 17. -9 19. $\frac{11}{9}$ 21. 0   23. 3
25. 7   27. 6   29. -6 31. 1   33. $-\frac{2}{3}$ 35. 5   37. 2, 3   39. $-2, -1$ 41. $-2, 3$ 43. $-1, 0$ 45. $-\frac{2}{3}, \frac{3}{2}$
47. $-\frac{1}{4}, \frac{1}{5}$ 49. 2   51. $\frac{1}{3}$ 53. $\frac{14}{5}$ 55. $\frac{1}{2}$ 57. $\frac{13}{17}$ 59. $-\frac{9}{4}$ 61. no solution   63. all real numbers
65. no solution   67. 0   69. no solution   71. all real numbers   73. no solution   75. -4 77. -1 79. 0.25
81. 0.05   83. $0.\overline{3}$ 85. $0.1\overline{6}$ 87. 12   89. 60   91. 4   93. 9   95. $4x - 24$ 97. $16z - 15$

Practice 2.3, pp. 65–66

1. $\frac{4}{3}$ 3. 1   5. 7   7. 6   9. 0.7   11. 1.625   13. 40   15. 8   17. 40   19. $\frac{2}{11}$ 21. -2 23. $-\frac{1}{10}$
25. $-\frac{3}{10}$ 27. 12   29. 16   31. 40   33. -3 35. 7   37. 8   39. $-\frac{5}{2}$ 41. 3   43. 2   45. $\frac{5}{3}$ 47. $-\frac{6}{5}$

Problem Solving 2, p. 68

1. 6 ft, 24 ft   3. 26m, 13m, 78m   5. $\frac{91}{4}$ ft^2 7. 2808 ft^2 9. 11 years old (Dick), 22 years old (Alan), 33 years old (Peter)
11. 53.2 ft   13. 85,612.1 ft^3 15. 341,333.3 ft^3

Practice 2.4, pp. 73–74

1. $\{z | z > 2\}$ 3. $\{x | x < 7\}$ 5. $\{v | v \geq 1\}$ 7. $\{t | t < 3\}$ 9. $\{r | r \geq -\frac{3}{4}\}$ 11. $\{n | n > \frac{15}{2}\}$ 13. $\{q | q > 5\}$
15. $\{s | s > -\frac{3}{5}\}$ 17. $\{u | u \leq -3\}$ 19. $\{z | z \leq -1 \text{ or } z \geq 3\}$ 21. $\{x | x < 0 \text{ or } x > \frac{1}{2}\}$ 23. $\{v | v \leq 2 \text{ or } v > 5\}$
25. $\{z | -\frac{3}{2} < z < \frac{5}{2}\}$ 27. $\{n | 9 < n < 15\}$ 29. $\{q | 1 \leq q \leq 3\}$ 31. $\{s | 2 \leq s \leq 3\}$ 33. $\{u | -\frac{5}{4} \leq u \leq \frac{3}{4}\}$ 35. 3
37. $1\frac{2}{3}$ 39. Number line showing $x > -2$.
41. Number line showing $x > 0$.
43. Number line showing $x > 3$ or $x < -2$.
45. Number line showing $0 < x < 3$.

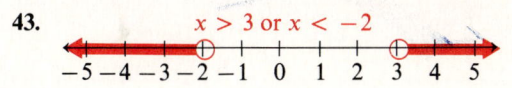

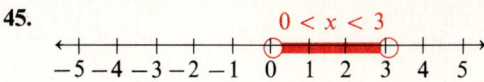

Make Sure Answers for Lesson 2.5, Example 2, p. 77

1. $\{a | a \leq 2 \text{ or } a \geq 4\}$: Number line showing $|a - 3| \geq 1$.

2. $\{b | b > 4 \text{ or } b < 1\}$: Number line showing $3 < |2b - 5|$.

A-14 Appendix Selected Answers

▶ **Make Sure Answers for Lesson 2.5, Example 3, p. 78**

1. $\{c \mid -4 \leq c \leq -2\}$: [number line with $|c+3| \leq 1$, closed circles at -4 and -2]

2. $\{d \mid -\frac{1}{2} < d < \frac{7}{2}\}$: [number line with $4 > |3-2d|$, open circles at $-\frac{1}{2}$ and $\frac{7}{2}$]

▶ **Practice 2.5, pp. 79–80**

1. $-8, 2$ 3. $-\frac{1}{3}, \frac{5}{3}$ 5. $-3, 7$ 7. $\{q \mid q \leq 0 \text{ or } q \geq 4\}$: [number line with $|q-2| \geq 2$, closed at 0 and 4]

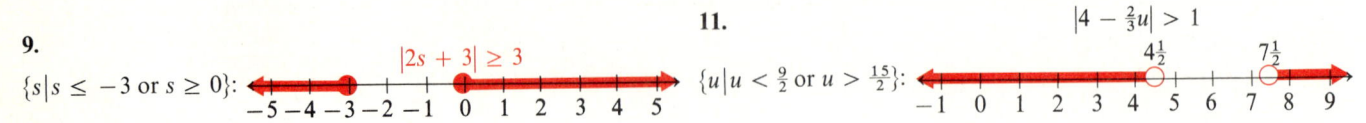

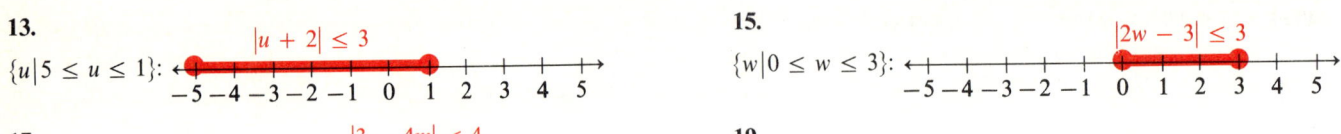

9. $\{s \mid s \leq -3 \text{ or } s \geq 0\}$: [number line with $|2s+3| \geq 3$, closed at -3 and 0]

11. $\{u \mid u < \frac{9}{2} \text{ or } u > \frac{15}{2}\}$: [number line with $|4 - \frac{2}{3}u| > 1$, open at $4\frac{1}{2}$ and $7\frac{1}{2}$]

13. $\{u \mid 5 \leq u \leq 1\}$: [number line with $|u+2| \leq 3$, closed at -5 and 1]

15. $\{w \mid 0 \leq w \leq 3\}$: [number line with $|2w-3| \leq 3$, closed at 0 and 3]

17. $\{m \mid -\frac{1}{4} < m < \frac{7}{4}\}$: [number line with $|3-4m| < 4$, open at $-\frac{1}{4}$ and $\frac{7}{4}$]

19. $\{z \mid -1 \leq z \leq 1\}$: [number line with $|z|-3 \leq -2$, closed at -1 and 1]

21. $\{t \mid t < -3 \text{ or } t > 3\}$: [number line with $-2|t| < -6$, open at -3 and 3]

23. $\{p \mid p \leq -\frac{4}{3} \text{ or } p \leq \frac{10}{3}\}$: [number line with $3|p-1|-4 \geq 3$, closed at $-1\frac{1}{3}$ and $3\frac{1}{3}$]

25. $\frac{5}{8}$ 27. $-\frac{3}{7}$ 29. $\frac{6}{5}$ 31. $\frac{8}{3}$ 33. 1 35. $\frac{20}{13}$

▶ **Practice 2.6, pp. 83–84**

1. $y = \dfrac{2x+5}{3}$ 3. $z = \dfrac{c-by}{a}$ 5. $z = by + b$ 7. $y = \dfrac{6z-5}{2}$ 9. $z = \dfrac{y}{2}$ 11. $y = z$ 13. $z = \dfrac{4y-6}{3}$

15. $x = w$ 17. $x = \dfrac{a^2}{a-b}$ 19. $z = \dfrac{ax+by}{c}$ 21. $x = \dfrac{cy}{a-b}$ 23. $z = \dfrac{cy-ab}{a}$ 25. $x = 0$ 27. $c = \dfrac{a-b}{a}$

29. $c = \dfrac{b-a}{b}$ 31. $c = \dfrac{b}{ax+a}$ 33. $y = \dfrac{ax-1}{a-1}$ 35. $b = \dfrac{-ac}{b-c}$ or $\dfrac{ac}{c-b}$ 37. 1 39. 3 41. 0 43. -18

Chapter 3 Answers

▶ **Make Sure Answers for Lesson 3.1, Example 1, p. 89**

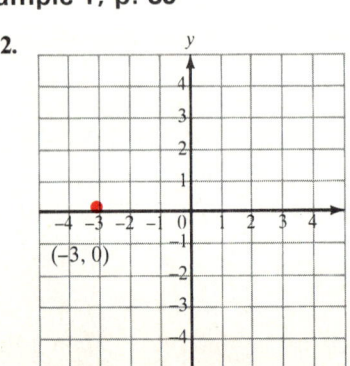

1. [graph showing point $(4, -1)$]

2. [graph showing point $(-3, 0)$]

Chapter 3 Answers A-15

▶ **Practice 3.1, pp. 93–94**

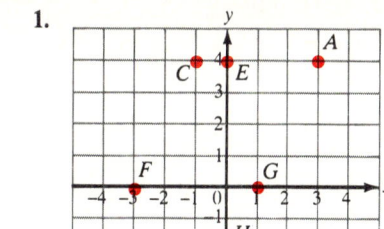

1. [graph with points A(3,4), B(2,-3), C(-2,4), D(-3,-4), E(0,4), F(-3,0), G(1,0), H(0,-2)]

3. $A(5, 3); B(-4, 4); C(-1, -5); D(5, -2); E(3, 0); F(0, 3); G(-4, 0); H(0, -1)$
5. yes 7. no 9. yes 11. no 13. yes 15. yes
17. $(-4, 15), (0, 7), (2, 3)$ 19. $(-23, -5), (-3, 0), (17, 5)$
21. $(-3, -4), (0, -2), (3, 0)$ 23. $(2.5, -1.5), (4, 0), (8.3, 4.3)$
25. $(-4, 24), (-2, 21), (4, 12)$ 27. $(-\frac{99}{8}, -3), (-\frac{45}{8}, 3), (-\frac{9}{2}, 4)$
29. $(-2, -1), (1, 8), (3, 14)$ 31. $-3, -1$ 33. $-4, -7$
35. $-\frac{1}{6}, -\frac{13}{6}$ 37. $4, -2$ 39. $-\frac{5}{6}, -\frac{7}{2}$ 41. $y = -\frac{2}{3}x$
43. $y = \dfrac{-4x + 6}{3}$ 45. $y = \dfrac{-x + 4}{3}$ 47. $y = \dfrac{-5x + 4}{3}$

▶ **Make Sure Answers for Lesson 3.2, Example 1, p. 96**

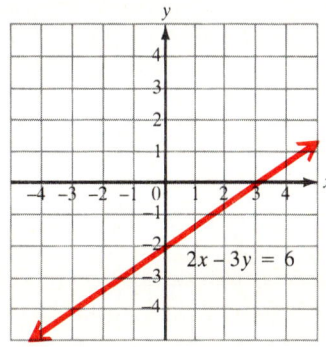

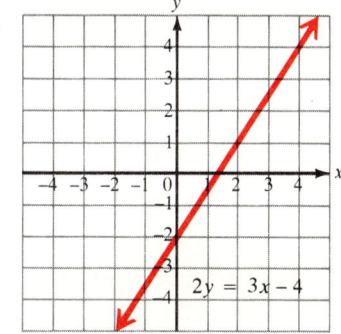

1. [graph of $2x - 3y = 6$]
2. [graph of $2y = 3x - 4$]

▶ **Make Sure Answers for Lesson 3.2, Example 2, p. 97**

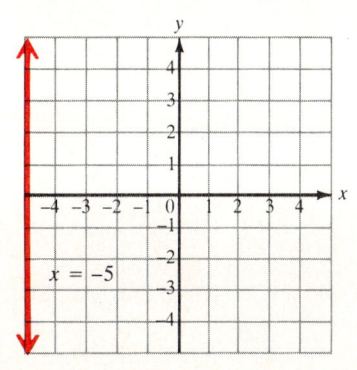

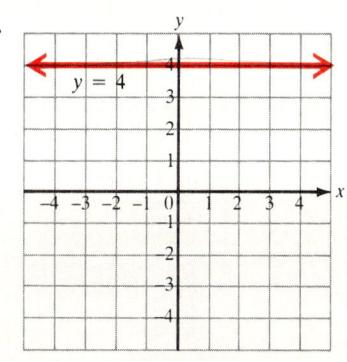

1. [graph of $x = -5$]
2. [graph of $y = 4$]

▶ **Make Sure Answers for Lesson 3.2, Example 3, p. 98**

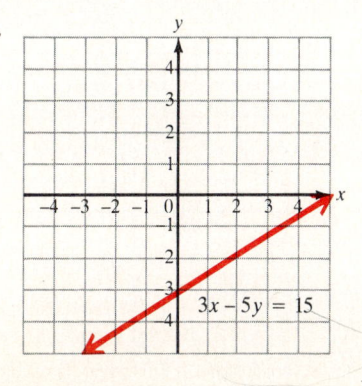

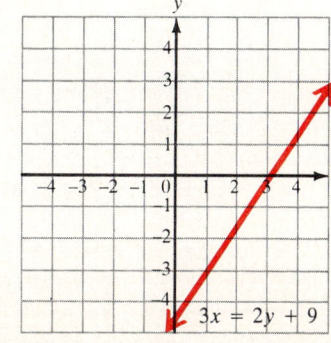

1. [graph of $3x - 5y = 15$]
2. [graph of $3x = 2y + 9$]

A-16 Appendix Selected Answers

▶ **Practice 3.2, pp. 99–100**

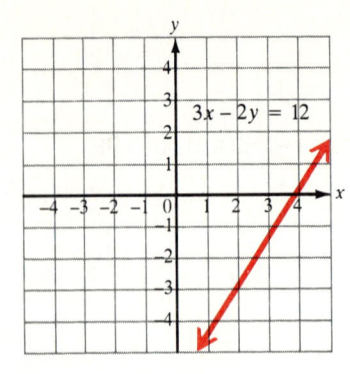

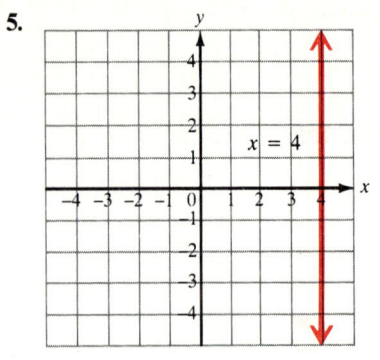

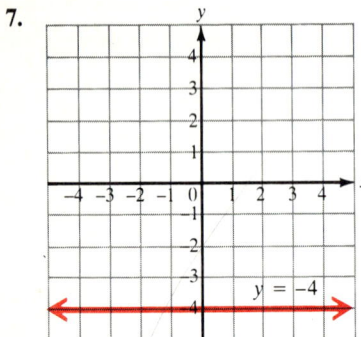

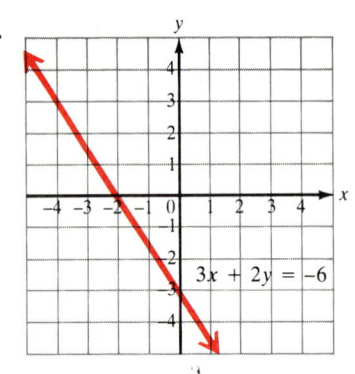

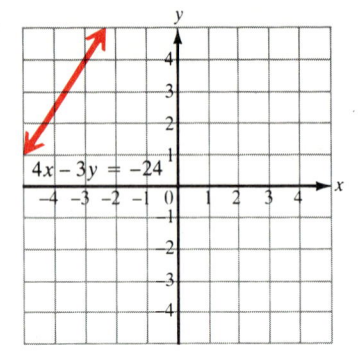

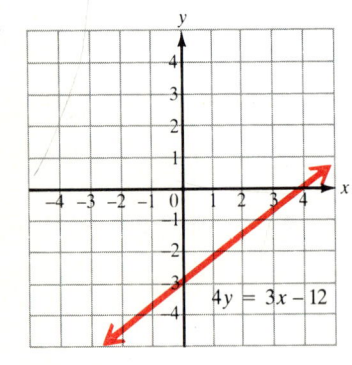

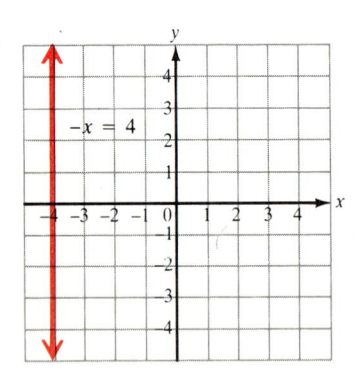

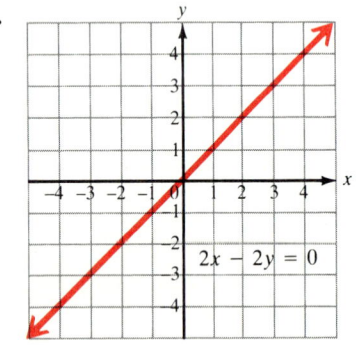

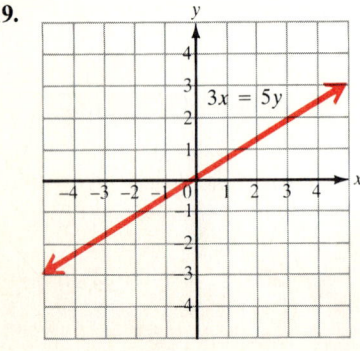

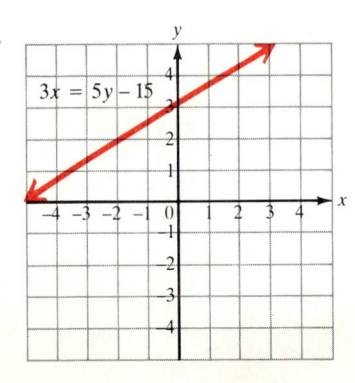

1. Graph of $x + 3y = 4$
3. Graph of $3x - 2y = 12$
5. Graph of $x = 4$
7. Graph of $y = -4$
9. Graph of $3x + 2y = -6$
11. Graph of $4x - 3y = -24$
13. Graph of $4y = 3x - 12$
15. Graph of $-x = 4$
17. Graph of $2x - 2y = 0$
19. Graph of $3x = 5y$
21. Graph of $3x = 5y - 15$
23. Graph of $\frac{2}{3}x + \frac{3}{4}y = 3$

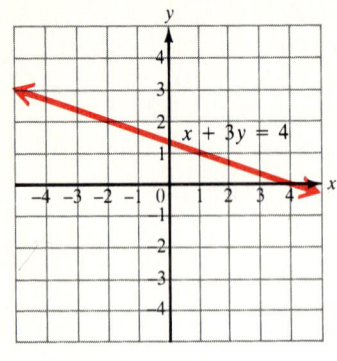

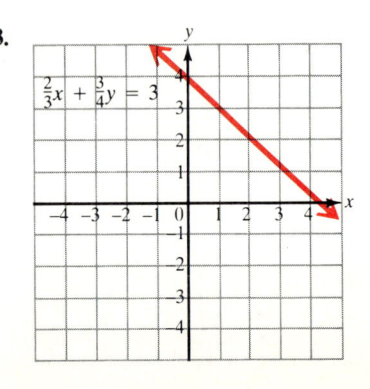

25. $\frac{3}{4}$ 27. $-\frac{1}{3}$ 29. -2 31. $\frac{2}{3}$ 33. 2 35. $-\frac{1}{2}$ 37. -1 39. 0 41. $y = 2x - 3$ 43. $y = -x - 5$
45. $y = -\frac{3}{5}x + \frac{28}{5}$

▶ **Make Sure Answers for Lesson 3.3, Example 3, p. 106**

1. [graph of $y = \frac{2}{3}x + 4$]

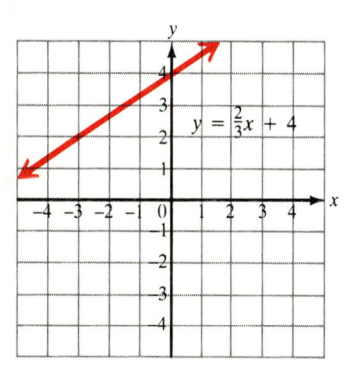

2. [graph of $3x + 2y = 6$]

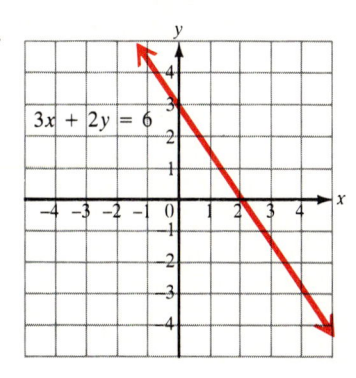

▶ **Practice 3.3, pp. 107–108**

1. 1 3. -3 5. $\frac{4}{7}$ 7. $-\frac{7}{5}$ 9. no slope 11. 0 13. $y = 3x + 2$ 15. $y = -4x + 4$ 17. $y = -3x - 2$
19. $y = 0x - 2$ 21. $y = x + 2$ 23. $y = 0x + 0$ 25. $y = \frac{3}{4}x + 4$ 27. $y = -\frac{5}{3}x + 1$ 29. $y = -\frac{4}{7}x + \frac{2}{3}$

31. [graph of $x + y = 7$]

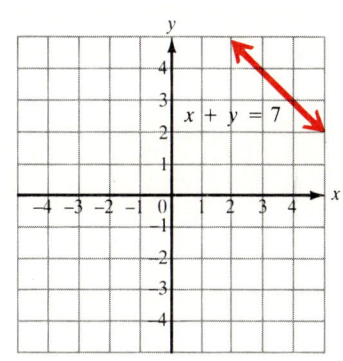

33. [graph of $2x + 3y = 9$]

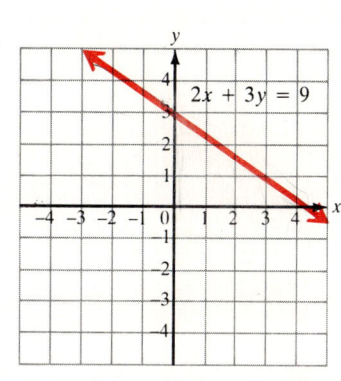

35. [graph of $4y - 3x = 2$]

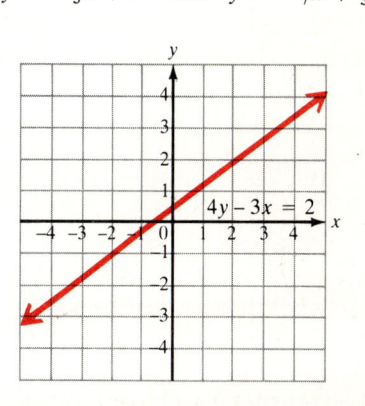

37. $\frac{1}{4}$

39. $-\frac{1}{3}$ 41. 3 43. $-\frac{3}{2}$ 45. $\frac{1}{3}$ 47. $-\frac{5}{2}$ 49. $\frac{4}{3}$ 51. $-\frac{3}{2}$ 53. $y = 2x - 1$ 55. $x = \frac{3}{2}y + \frac{9}{2}$ 57. $y = 3$

▶ **Practice 3.4, pp. 115–16**

1. $y = 2x - 1$ 3. $y = -x - 1$ 5. $y = \frac{4}{3}x + \frac{4}{3}$ 7. $y = \frac{4}{3}x - 2$ 9. $y = -\frac{2}{3}x + \frac{8}{3}$ 11. $y = -\frac{1}{2}x + \frac{1}{2}$
13. $y = 2x - 4$ 15. $y = 0x + 3$ 17. $x = 0$ 19. $y = 4x - 16$ 21. $y = -\frac{4}{3}x + 0$ 23. $y = 0x - 5$
25. $y = -\frac{1}{2}x + 4$ 27. $y = \frac{2}{3}x - \frac{13}{3}$ 29. $x = -3$ 31. $y = -\frac{1}{2}x + 2$ 33. $y = \frac{2}{3}x + \frac{5}{3}$ 35. $x = 2$ 37. no
39. no 41. yes 43. no 45. yes

47. [graph of $x - y = 0$]

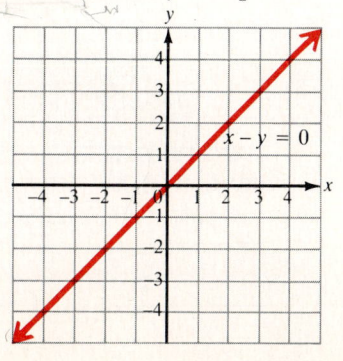

49. [graph of $4x - 3y = 12$]

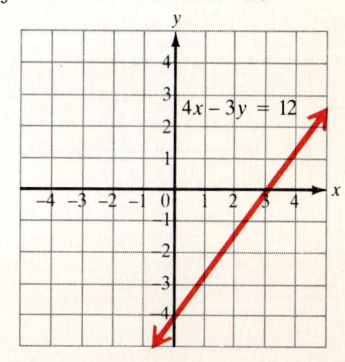

51. [graph of $3x + y = -3$]

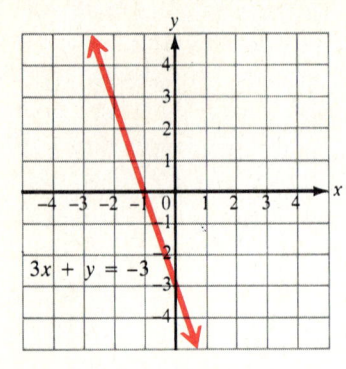

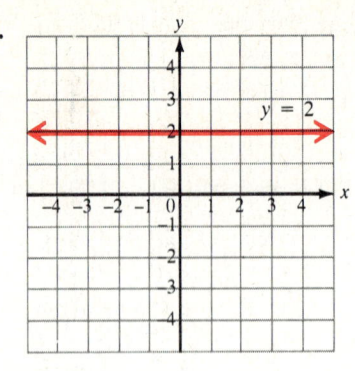

53. [graph of $y = 2$]

▶ **Problem Solving 3, pp. 119–20**

1a. $h = 4p + 100$ or $p = \frac{1}{4}h - 25$ **b.** $250   **c.** 100   **3a.** $t = \frac{1}{4}n + 40$ or $n = 4t - 140$ **b.** 65°F   **c.** 180 times
5a. $n = 3l - 25$ or $l = \frac{1}{3}n + \frac{25}{3}$ **b.** $6\frac{1}{2}$ size   **c.** $11\frac{1}{3}$ in.
7a. $y = 2n + 1787$ or $n = \frac{1}{2}y - \frac{1787}{2}$ **b.** 99th Congress   **c.** 1887 A.D.
9a. $y = 0.436p + 1881.2$ or $p = 2.295y - 4317.5$ **b.** 272.5m   **c.** 2012 A.D.
11a. $c = -\frac{600}{47}a + \frac{142400}{47}$ or $a = -\frac{47}{600}c + \frac{712}{3}$ **b.** 2455 calories   **c.** $80\frac{2}{3}$ years

▶ **Make Sure Answers for Lesson 3.5, Example 1, p. 122**

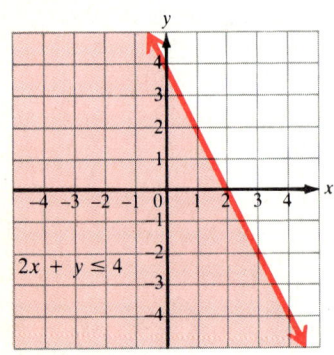

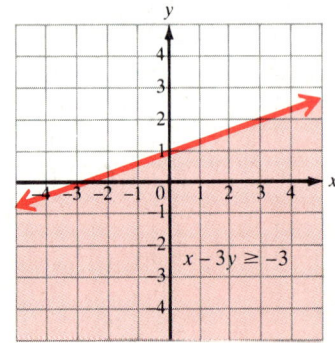

1. [graph of $2x + y \leq 4$]

2. [graph of $x - 3y \geq -3$]

▶ **Make Sure Answers for Lesson 3.5, Example 2, p. 123**

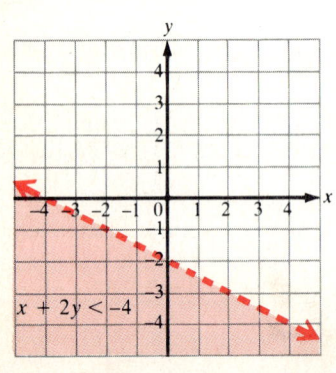

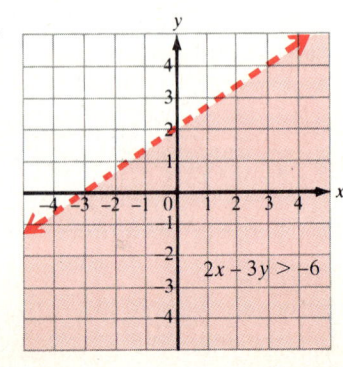

1. [graph of $x + 2y < -4$]

2. [graph of $2x - 3y > -6$]

Chapter 3 Answers A-19

▶ **Make Sure Answers for Lesson 3.5, Example 3, p. 125**

1. [graph: $x > -2$] 2. [graph: $y > 0$]

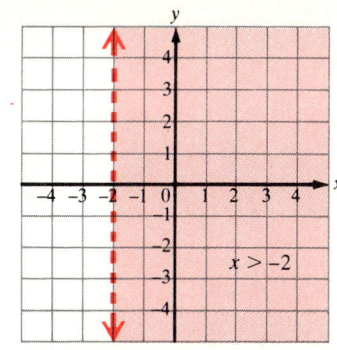

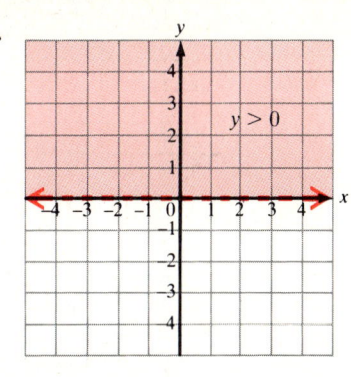

▶ **Make Sure Answers for Lesson 3.5, Example 4, p. 126**

1. [graph: $y \geq -2x$] 2. [graph: $y \leq \frac{2}{3}x$]

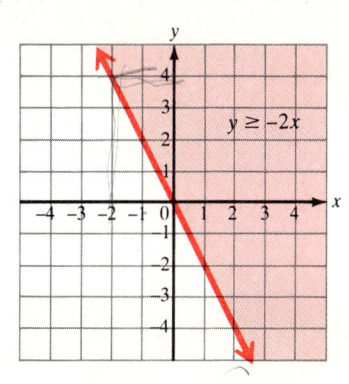

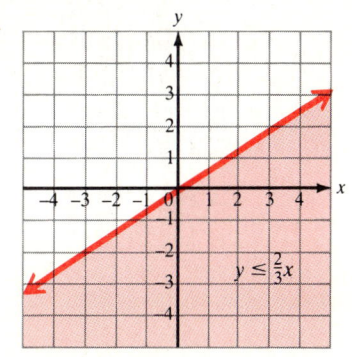

▶ **According to Guinness, p. 126**

3. $t = \frac{1}{100}d + \frac{36}{5}$ or $d = 100t - 720$ 5. 535.2°F

▶ **Practice 3.5, pp. 127–28**

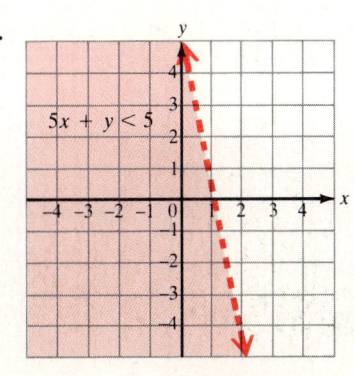

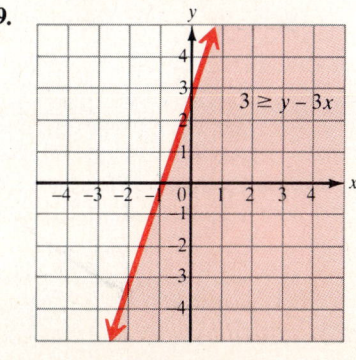

1. [graph: $x + y \leq 3$]   3. [graph: $5x + y < 5$]   5. [graph: $x > 0$]

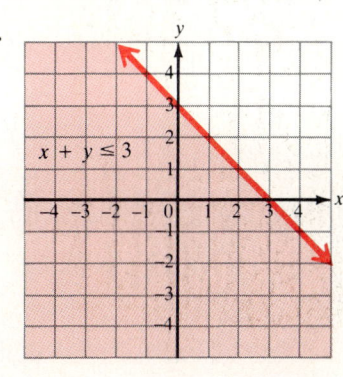

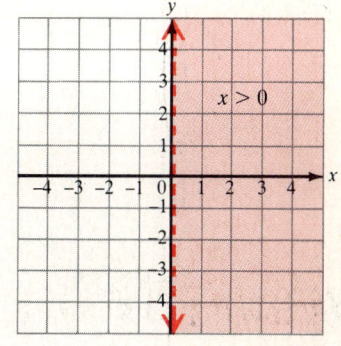

7. [graph: $y \leq 2x$]   9. [graph: $3 \geq y - 3x$]   11. [graph: $-2 < x - 2y$]

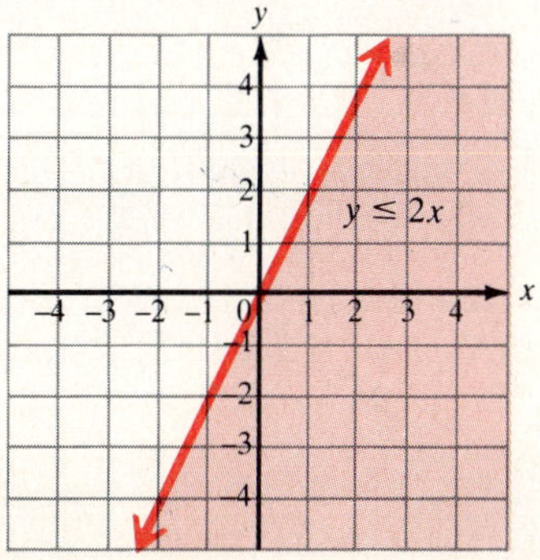

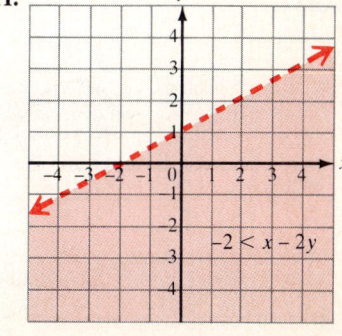

A-20 Appendix Selected Answers

13. [graph: $-3 \geq x$]

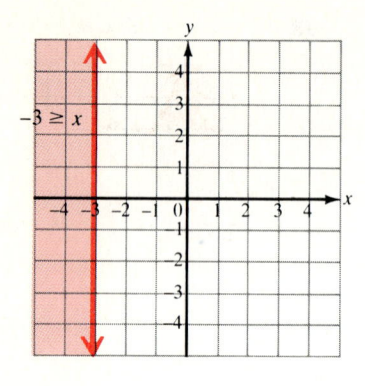

15. [graph: $2y < 3x$]

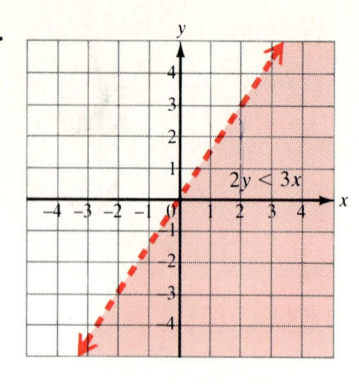

17. yes   19. no   21. no

23. [graph: $x = 3$]

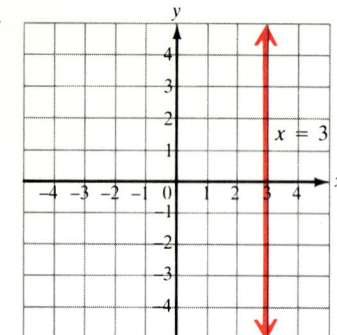

25. [graph: $3x + 2y = 10$]

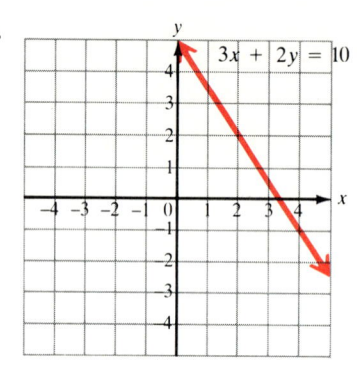

27. [graph: $5x + 3y = 18$]

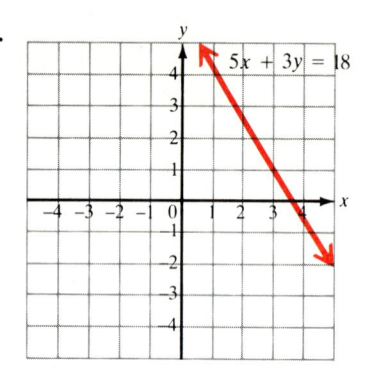

▶ Chapter 3 Review, pp. 129–30

1. Grid A [graph: point $(-2, 3)$]

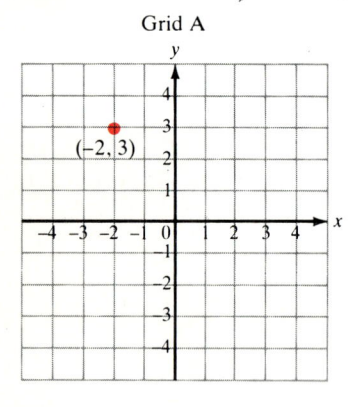

5. Grid B [graph: $2x - y = 3$]

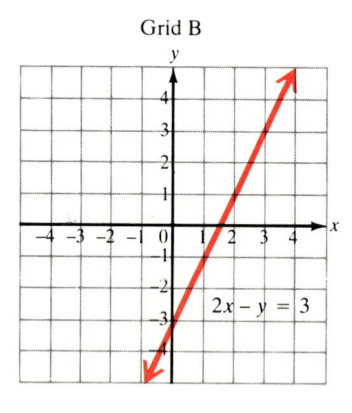

6. Grid C [graph: $y = -3$]

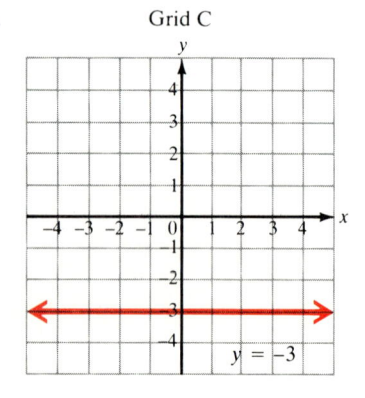

7. and 10. Grid D [graph: $3x - 2y = 4$ and $3x - 2y = 6$]

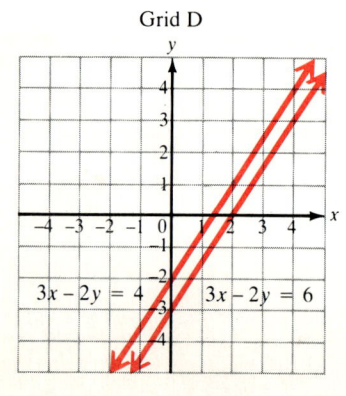

15. Grid E [graph: $2x - y > 4$]

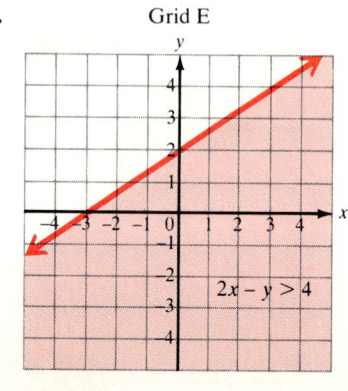

Chapter 4 Answers   A-21

16. Grid F

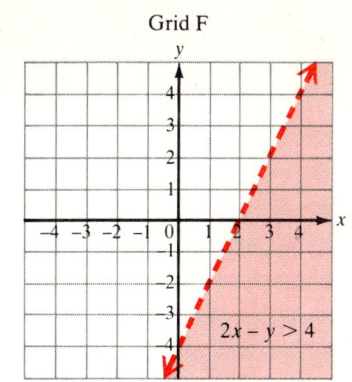

17. Grid G — $y > -1$

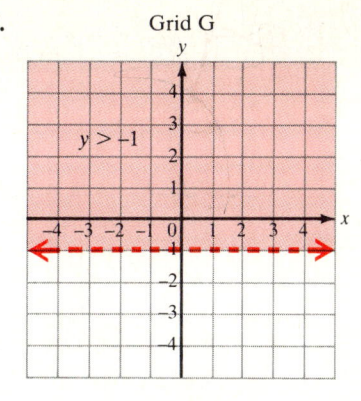

18. Grid H — $y \le -2x$

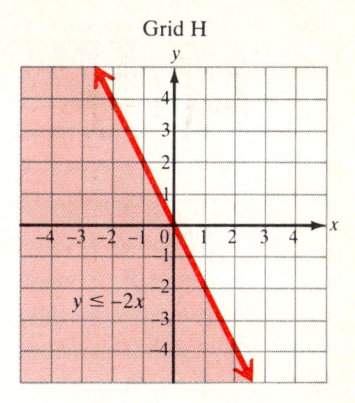

Chapter 4 Answers

▶ **Make Sure Answers for Lesson 4.1, Example 2, p. 134**

1. $3x + 2y = 5$; $2x + 5y = -4$; solution: $(3, -2)$

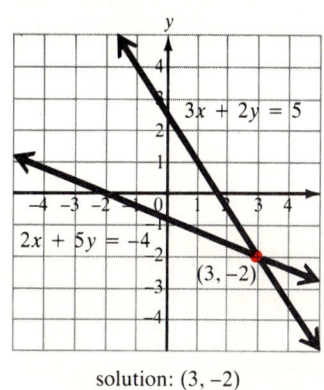

2. $5x + 2y = 12$; $x + y = 3$; solution: $(2, 1)$

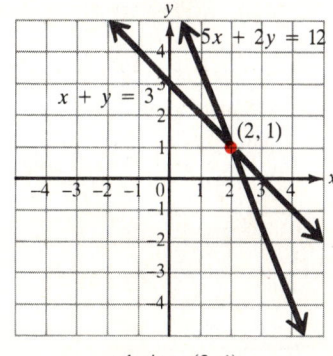

▶ **Practice 4.1, pp. 135–36**

1. yes 3. no 5. yes

7. $x + y = 3$; $x - y = 3$; solution: $(3, 0)$

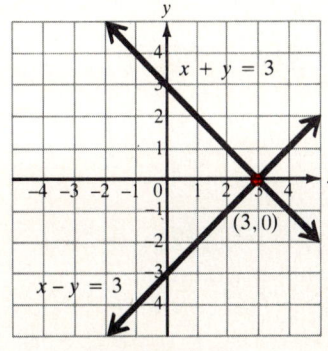

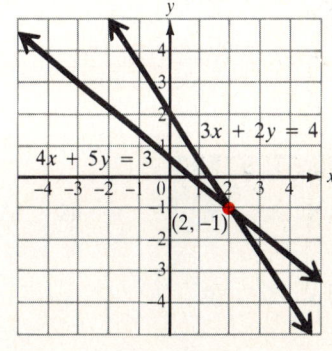

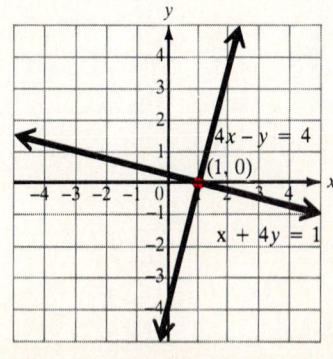

9. $3x + 2y = 4$; $4x + 5y = 3$; solution: $(2, -1)$

11. $4x - y = 4$; $x + 4y = 1$; solution: $(1, 0)$

13. $2x = 3y + 1$; $4x = 6y + 2$; infinitely many solutions, dependent system

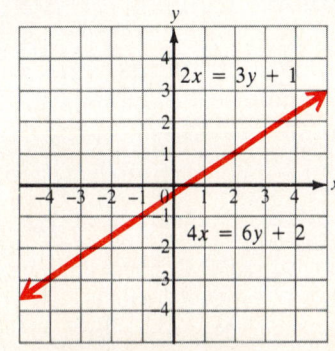

15.

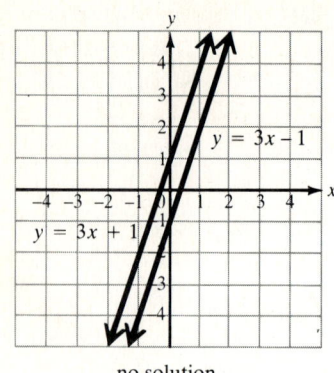

no solution, inconsistent system

17.

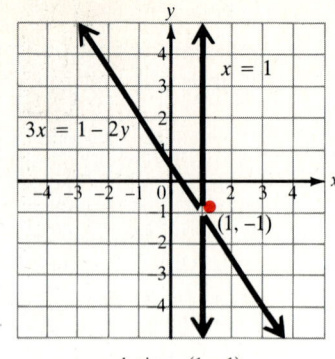

solution: $(1, -1)$

19. $-2x + 3$ **21.** $\dfrac{-2x + 3}{4}$ **23.** $\dfrac{12x - 9}{2}$ **25.** $\dfrac{-12 - 24y}{2}$ **27.** $y = 2x - 4$ **29.** $y = \dfrac{2x - 2}{3}$ **31.** $x = \dfrac{4 - 2y}{3}$

▶ **Practice 4.2, pp. 141–42**

1. $(4, 1)$ **3.** $(3, 2)$ **5.** $(\frac{1}{2}, \frac{2}{3})$ **7.** $(4, 2)$ **9.** infinitely many solutions, dependent system
11. no solution, inconsistent system **13.** $(2, 3)$ **15.** $(3, 2)$ **17.** $(2, 3)$ **19.** $(0, 2)$ **21.** no solution, inconsistent system
23. $(-\frac{1}{2}, 0)$ **25.** $6x - 8y - 4z$ **27.** $-4x + 6y - 8z + 10$ **29.** $-2x + 3y - z - 4$ **31.** yes **33.** no **35.** yes

▶ **Problem Solving 4, p. 144**

1a. 4 gal **b.** 4.8 gal **3a.** 9 kg of \$1.20/kg, 3 kg of \$0.40/kg **b.** 75% **c.** $\frac{1}{4}$ **5a.** 9%-saline **b.** $\frac{91}{100}$
7a. 0.625 gal **b.** 5.625 gal **9a.** $3\frac{1}{3}$ gal **b.** 4 gal

▶ **Practice 4.3, pp. 149–50**

1. yes **3.** no **5.** $(3, 2, 1)$ **7.** $(-1, -2, 2)$ **9.** $(2, 0, -3)$ **11.** $(1, -1, 2)$ **13.** $(0, 2, 4)$ **15.** $(\frac{5}{6}, \frac{1}{9}, \frac{1}{18})$
17. no solution, inconsistent system **19.** infinitely many solutions, dependent system **21.** -3 **23.** -2 **25.** 6
27. 3 **29.** -8 **31.** 8 **33.** -10

▶ **Problem Solving 5, p. 152**

1. 95 **3.** 82 **5.** 95 **7.** 92 **9.** 48 **11.** 85 or 58

▶ **Practice 4.4, pp. 157–58**

1. 5 **3.** 8 **5.** 3 **7.** -8 **9.** 9 **11.** 1 **13.** $1\begin{vmatrix} 0 & 2 \\ 3 & 1 \end{vmatrix} - (2)\begin{vmatrix} -1 & 2 \\ 2 & 1 \end{vmatrix} + 3\begin{vmatrix} -1 & 0 \\ 2 & 3 \end{vmatrix}$

15. $2\begin{vmatrix} 4 & 2 \\ 2 & 3 \end{vmatrix} - (-3)\begin{vmatrix} -1 & -2 \\ 2 & 3 \end{vmatrix} + 0\begin{vmatrix} -1 & -2 \\ 4 & 2 \end{vmatrix}$ **17.** 30 **19.** 18 **21.** 5 **23.** -6 **25.** 1 **27.** -3 **29.** -6
31. 6 **33.** $-\frac{8}{3}$ **35.** 9 **37.** -14 **39.** 9

▶ **Practice 4.5, pp. 163–64**

1. $(3, 2)$ **3.** $(5, -1)$ **5.** $(\frac{1}{2}, \frac{2}{3})$ **7.** $(\frac{3}{2}, \frac{5}{3})$ **9.** infinitely many solutions, dependent system
11. no solution, inconsistent system **13.** $(3, 2, 1)$ **15.** $(2, -1, 1)$ **17.** $(2, -1, 0)$ **19.** $(\frac{1}{2}, \frac{1}{2}, \frac{1}{2})$ **21.** $(-3, 1, 2)$
23. $(0, 1, 0)$ **25.** $-x - 3$ **27.** $-x^2 + 3x - 2$ **29.** $-3 + 2x + 3x^3$ **31.** $-1 + 3x^4 - 2x$ **33.** $6x + 2$
35. $2x + 2$ **37.** $-x^2 - 2x - 2$

Chapter 4 Review, p. 165–66

2.

solution: (3, 1)

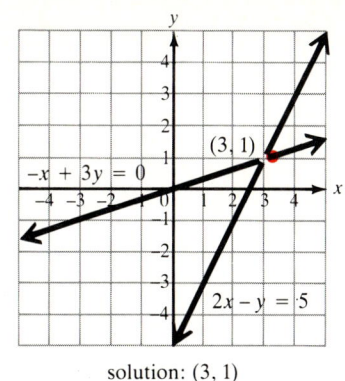

Chapter 5 Answers

Practice 5.1, pp. 173–74

1. $5a + 11$ 3. $8c^2 + 5c + 10$ 5. $4e^3 + 2e^2 - 2e - 4$ 7. $2g^3 + 2$ 9. $2x^3 + 5$ 11. $x^2 + 2$ 13. $2y^2 + 12$
15. $-2k^2 + 2$ 17. $-b^2 + b + 2$ 19. $-3ab - 4b$ 21. $-9c^2d + 5c$ 23. $-5a^2 - 3ab + b^2$ 25. $-abc - 7b^2c - b^2$
27. $-2c^2 - 2c + 9$ 29. $-q^5 - 3p^3q + 5pq^3$ 31. $x^2 - z^2$ 33. $b - 3c$ 35. $a - b + 2c - 2d$ 37. $10x - 7y$
39. $4x^2 - 3y$ 41. $2c^2 - 5d^2$ 43. -24 45. -40 47. 24 49. 6 51. $12z - 20$ 53. $-15x - 12$
55. $6v^2 - 4v$ 57. $2t^4 - 2t^3 - 2t^2$ 59. z^7 61. x^4 63. $2v^7$ 65. $6t^7$

Problem Solving 6, pp. 175–76

1. $110   3. $133.10   5. $1191.02   7. $110   9. $364.10   11. $3374.62

13.

TABLE 5.1 Height/Weight Table for an Average Person in the United States					
Height	Weight	Height	Weight	Height	Weight
2 ft 6 in.	14 lbs.	4 ft 6 in.	79 lbs.	5 ft 8 in.	157 lbs.
3 ft	23 lbs.	5 ft 2 in.	119 lbs.	5 ft 10 in.	172 lbs.
3 ft 6 in.	37 lbs.	5 ft 4 in.	131 lbs.	6 ft	187 lbs.
4 ft	55 lbs.	5 ft 6 in.	144 lbs.	6 ft 2 in.	203 lbs.

15a. 1.6% b. 4.6%

Practice 5.2, pp. 181–82

1. $6a^3$ 3. $-12a^6b^9$ 5. $12x^5y^7z^7$ 7. $12x^4 - 20x^3 + 8x^2$ 9. $-12u^2v^4 + 15u^6v^5 - 9u^4v^7$
11. $-3q^5r + 9q^4r^2 - 6q^3r^5$ 13. $12x^2 + 29x + 15$ 15. $z^5 - 2z^4 - 7z^2 - 2z + 6$
17. $3v^5 + 2v^4w + 4v^3w - 11v^3w^2 + 8v^2w^2 + 4v^2w^3 - 4vw^3$ 19. $45a^5b^{11}c^2$ 21. $10g^4 - 23g^2h + 12h^2$ 23. $p^3 - q^3$
25. $8x^3 - 27y^3$ 27. $a^4 - 2a^2b^2 + b^4$ 29. $2a^2 + 2ab$ 31. $10e^2 - 13f^2$ 33. $8z$ 35. $-x$ 37. v 39. $-6t$
41. $-5r$ 43. $-4p$ 45. $9z^2$ 47. $16x^2$ 49. $9v^2$ 51. $-3t^2$ 53. $-18r^2$ 55. $30p^2$

Practice 5.3, pp. 187–88

1. $a^2 + 8a + 15$ 3. $10c^2 - c - 21$ 5. $20e^2 - 47ef + 21f^2$ 7. $a^2 + 6a + 9$ 9. $16c^2 + 24cd + 9d^2$
11. $4e^4 - 12e^2f + 9f^2$ 13. $z^2 - 9$ 15. $9 - 4x^2$ 17. $9u^2 - 4v^2$ 19. $6x^2 + 19x + 10$ 21. $64 + 80z + 25z^2$
23. $16v^2 - 9w^2$ 25. $81q^2 - 16r^2$ 27. $h^4 - k^4$ 29. $9c^4 - 12c^2d^2 + 4d^4$ 31. $10ef - 5f^2$ 33. $16m^2 - mn - 6n^2$
35. -1 37. -2 39. 3 41. -5 43. z 45. x^3 47. v^6 49. t^2 51. $-2x - 1$ 53. $8z - 7$
55. $v^2 + v - 3$

Practice 5.4, pp. 195–96

1. a^3 3. $-c^3$ 5. w^2z^3 7. $3a^2c$ 9. $-\dfrac{1}{7p^3q^2r^2}$ 11. $\dfrac{1}{2x^3y^4z^4}$ 13. $2a^2 - 3$ 15. $-c - 2$ 17. $2a - 1$
19. $4a - 2b + 3a^2b^3$ 21. $2w^3z - 6z^2 + \dfrac{5w}{z^2}$ 23. $-2xy + 3y^2 - \dfrac{4y^2}{x^2}$ 25. $z - 2$ 27. $2x + 1$ 29. $u + v$
31. $3r^3 - 5r + 4 + \dfrac{r}{2r^2 + 6r - 1}$ 33. $13z^2 - 24z + 13$ 35. $4v^2 - 21v + 22$ 37. $10t^2 - 31t + 19$ 39. 2
41. 1 43. -10 45. -11 47. -6 49. -4 51. -3 53. 6 55. $-2z + 3$ 57. $x^2 + 3x - 2$
59. $-v^3 + 4v^2 + 5v - 3$ 61. $-7t^3 - t + 3$ 63. $-r^3 + 1$

Practice 5.5, pp. 201–202

1. $z + 3$ 3. $x + 3$ 5. $v - 2$ 7. $t^2 - 4$ 9. $r^3 - 2r^2 + 9r - 18$ 11. $9p + 36 + \dfrac{80}{p - 4}$ 13. $m^2 - 3m + 9$
15. $6p + 2$ 17. $-r^3 + r^2 - r + 1$ 19. yes   21. yes   23. no   25. yes   27. no   29. yes   31. no   33. no
35. yes   37. $2 \cdot 3$ 39. 2^4 41. $2^2 \cdot 3^2$ 43. $2 \cdot 3^3$ 45. $2 \cdot 5 \cdot 7$ 47. $2^5 \cdot 3$ 49. z^5 51. x^5 53. $2v$
55. $-3t^4$ 57. $-7m^3n^4$ 59. $4s^2$

Chapter 6 Answers

Practice 6.1, pp. 209–10

1. $2x$ 3. z^2 5. v^3 7. $2qr^2$ 9. $-hk^5$ 11. $-4d^2e^3$ 13. $2a^3(a^2 + 2 + 3a)$ 15. $3c^2(4c^2 + 5c - 3)$
17. $2e^2f(3ef - 4e^2f^4 + 6)$ 19. $m^2n^3(5m^2n^4 - 15mn^2 + 12)$ 21. $rs^2(4s^2t^3 - 9r^2s^3 + 12r^3t)$ 23. $4(6x^3y^5 - 9x^4z^3 + 10y^3z^2)$
25. $(3x + 1)(4x + 5)$ 27. $(9x + 1)(4x + 1)$ 29. $(e^2 + 1)(5e + 7)$ 31. $(2g + 1)(4g + 7)$ 33. $(3k^2 + 2)(k + 1)$
35. $(3n - 2)(4n + 5)$ 37. $(3q^2 + 2)(3q - 7)$ 39. $(5s - 4)(2s - 3)$ 41. $(8u - 5)(u + 3)$ 43. $(4w^2 + 3)(2w - 5)$
45. $(2y + 3)(6y - 5)$ 47. $(5 - 3x)(1 + 4x)$ 49. $z^2 + 2z + 1$ 51. $x^2 + 2x - 8$ 53. $v^2 + 2v - 15$
55. $t^2 + 2t - 15$ 57. $r^2 - 6r + 8$ 59. $p^2 + 8p + 15$ 61. $m^2 - 9m + 14$ 63. $h^2 + 4h - 32$
65. $f^2 - f - 2$ 67. $d^2 - 5d + 6$ 69. $y^2 - 6y + 5$ 71. $w^2 - 6w - 7$

Practice 6.2, pp. 215–16

1. $(x + 2)(x + 1)$ 3. $(z + 4)(z + 2)$ 5. $(u + 6)^2$ 7. $(x - 2)^2$ 9. $(x - 3)(x - 5)$ 11. $(e - 9)(e - 4)$
13. $(p + 2)(p - 1)$ 15. $(r + 3)(r - 2)$ 17. $(t + 9)(t - 3)$ 19. $(z + 1)(z - 5)$ 21. $(x + 2)(x - 8)$
23. $(v + 4)(v - 12)$ 25. $(x + 8)(x + 9)$ 27. $(24 - x)(3 - x)$ 29. $(e + 10f)(e - 7f)$ 31. $(m + 6n)(m - 14n)$
33. $(r^2 + 8)(r^2 + 9)$ 35. $(t^2 + 4)(t^2 - 2)$ 37. $(z + 4)^2$ 39. $(x + 7)(x + 2)$ 41. $(v - 8)(v - 3)$ 43. $6z^2 + 17z + 12$
45. $15x^2 - 26x + 8$ 47. $8v^2 - 6v - 9$ 49. $12t^2 + t - 6$ 51. $15r^2 - 4r - 3$ 53. $8p^2 + 2p - 3$
55. $4m^2 - 12m + 9$ 57. $4h^2 + 20h + 25$ 59. $15f^2 + 14f - 8$ 61. $6y^2 - 13y - 5$ 63. $12w^2 + 28w + 15$

Practice 6.3, pp. 223–26

1. $(3x + 2)(x + 1)$ 3. $(2y + 3)(y + 3)$ 5. $(3z + 1)(4z + 5)$ 7. $(2p - 1)(p - 5)$ 9. $(7r - 9)(r - 1)$
11. $(4t - 1)(9t - 11)$ 13. $(5v + 7)(v - 1)$ 15. $(3x - 4)(x + 2)$ 17. $(z + 1)(6z - 5)$ 19. $(2x + 1)(x - 7)$
21. $(5y + 2)(y - 3)$ 23. $(2z + 1)(4z - 7)$ 25. $(3z + 2)(z + 1)$ 27. $(3x - 2)(4x + 5)$ 29. $(3v - 7)(3v + 2)$
31. $(4t - 3)(t + 2)$ 33. $(2r - 3)(5r - 4)$ 35. $(p + 3)(8p - 5)$ 37. $(4x + 3)(2x - 5)$ 39. $(2h + 3)(6h - 5)$
41. $(3f - 1)(4f - 1)$ 43. $(2 + 9d)(5 + 3d)$ 45. $(1 + 4x)(5 - 3x)$ 47. $(3 + 8z)(4 - 3z)$ 49. $(3x + 4y)(4x - 9y)$
51. $(3u + 2v)(9u - 4v)$ 53. $(9q - 4p)(3q - 2p)$ 55. $(2z^2 + 1)(z^2 - 7)$ 57. $(5x^2 - 2)(x^2 + 3)$ 59. $(4v^2 + 5)(3v^2 - 2)$
61. $(3s^2 + 2t^2)(3s^2 - 7t^2)$ 63. $(2x^2 + 3y^2)(6x^2 - 5y^2)$ 65. $(3e^2f^2 + 4)(4e^2f^2 + 1)$ 67. $(2x^3 + y^3)(x^3 - 7y^3)$
69. $(2x - 13)(x - 5)$ 71. $(4v^2 - 8v + 3)(v^2 - 2v + 5)$ 73. $z^2 + 6z + 9$ 75. $9x^2 + 24x + 16$ 77. $16v^2 - 24v + 9$
79. $9t^2 - 4$ 81. $16r^2 - 1$ 83. $(p + 3)(4p - 5)$ 85. $(a + 4)(a - 3)$ 87. $(2c + 3)(c - 3)$ 89. $(3 - 2e)(5e - 2)$

Practice 6.4, pp. 231–32

1. $(r + 6)(r - 6)$ 3. $(2p + 7)(2p - 7)$ 5. $(9k^2 + 8m^2)(9k^2 - 8m^2)$ 7. $(z + 3)^2$ 9. $(2x - 5)^2$ 11. $(2u - 11v)^2$
13. $(f + 3)(f^2 - 3f + 9)$ 15. $(d - 1)(d^2 + d + 1)$ 17. $(2a^2 - 3b^2)(4a^4 + 6a^2b^2 + 9b^4)$ 19. $(z^2 + 2)(z + 1)$
21. $(w + x + v)(w + x - v)$ 23. $(2p - 3)(p - 2q)(p^2 + 2pq + 4q^2)$ 25. $(a - b + c)(a - b - c)$ 27. $(2g + 3)^2(2g - 3)^2$

29. $(4k^2 + 9)(2k + 3)(2k - 3)$ **31.** $(2n^2 + 1)(3n^2 + 7)$ **33.** $(3q - 4r)(9q^2 + 12qr + 16r^2)$
35. $(2u + 1)(2u - 1)(16u^4 + 4u^2 + 1)$ **37.** $(w + x + 2)(w - x - 2)$ **39.** $(a - b)(a^2 + ab + b^2 - 1)$
41. $(x + y)(x + y + ax - ay)$ **43.** $5z(z^2 - 2z + 5)$ **45.** $3wx(w + 3x - 4x^2)$ **47.** $6s^2t^2(4s^2 - 6st - 3t^2)$ **49.** $(p - 2)^2$
51. $(m + 4)(m + 3)$ **53.** $(3h + 4)(2h - 3)$ **55.** $(3f + 4)(4f - 3)$ **57.** $(3d + 4)(4d + 3)$

▶ **Practice 6.5, pp. 237–38**

1. $4(2a + 5b)(2a - 5b)$ **3.** $3f(4e + 3f)(4e - 3f)$ **5.** $9(p^2 + 4q^2)(p + 2q)(p - 2q)$ **7.** $z(y + 3)(y - 9)$
9. $(v^2 - 3)(v + 2)(v - 2)$ **11.** $(t^2 + 3u^2)(t + u)(t - u)$ **13.** $(2a - 5b)(c + 4d)$ **15.** $3k(k + 4)(k + 1)(k - 1)$
17. $3pq(p + q)(p^2 - 3q^2)$ **19.** $(a + b)^2(a - b)$ **21.** $5e^2f(e^3 + 4e^2f^2 + 3f^4)$ **23.** $3km(4k^2 + 9m^2)(2k + 3m)(2k - 3m)$
25. $e(e + f)(e^2 - ef + f^2 - 5)$ **27.** $(m + n + 2)(m + n - 2)$ **29.** $r(3r + 2s + 4t)(3r + 2s - 4t)$
31. irreducible over integers **33.** $a(ab^2 + 8)(ab^2 - 2)$ **35.** $(e + 2f)(e^2 - 2ef + 4f^2)(e - 2f)(e^2 + 2ef + 4f^2)$
37. $-2, 3$ **39.** $3, 5$ **41.** $\frac{3}{4}, \frac{4}{3}$ **43.** $-\frac{1}{2}, \frac{4}{3}, 3$ **45.** $-\frac{3}{4}, 1, \frac{3}{2}$ **47.** $2p(2p - 3)$ **49.** $4m(3m^2 - 2m + 4)$
51. $5e^2f(2eg^2 + 3f^2 + 4e^2f^2g)$

▶ **Practice 6.6, pp. 243–46**

1. $-2, 0$ **3.** $0, \frac{1}{2}$ **5.** $0, 5$ **7.** $0, \frac{9}{4}$ **9.** $0, 36$ **11.** $-3, 3$ **13.** $-7, 7$ **15.** $-\frac{3}{4}, \frac{3}{4}$ **17.** $-4, 4$ **19.** $-8, -2$
21. $-9, 4$ **23.** $\frac{4}{3}, 2$ **25.** $-\frac{4}{3}, 2$ **27.** $-3, -2, 2$ **29.** $-4, 4, 7$ **31.** $-5, -2, 2$ **33.** $-3, 3$ **35.** $0, 36$ **37.** $0, \frac{4}{5}$
39. $-\frac{2}{5}, \frac{2}{5}$ **41.** $-6, -3$ **43.** $-\frac{9}{4}, \frac{3}{4}$ **45.** $-\frac{3}{4}, \frac{3}{2}$ **47.** $-3, -\frac{2}{3}, \frac{2}{3}$ **49.** $-25, -\frac{7}{4}, \frac{7}{4}$ **51.** $-\frac{7}{4}, \frac{7}{4}$ **53.** $1, 5$
55. $0, 6$ **57.** $-3, 0, 3$ **59.** $2, 3, 4$ **61.** $-4, 0, \frac{3}{2}$ **63.** $-5, -2, 0, 2$ **65.** 2 **67.** 3 **69.** $\frac{1}{2}$
71. irreducible over integers **73.** $(2x - 3)(3x + 4)$ **75.** $(6v + 5)(4v - 3)$ **77.** $(t + 5)(t - 5)$ **79.** $(4r + 1)(4r - 1)$
81. $(6p + 5)(6p - 5)$

▶ **Problem Solving 7, p. 248**

1. 5 mi over **3.** 3 **5.** 8 **7.** 14 **9.** 50 **11.** 5 **13.** 2 **15.** 12 in. (h) × 8 in. (b)
17a. $2 \times 2 \times 2$ **b.** $4 \times 3 \times 2$

Chapter 7 Answers

▶ **Practice 7.1, pp. 257–58**

1. 0 **3.** 2 **5.** $-3, 0$ **7.** $0, 3$ **9.** $-2, 2$ **11.** $-3, 3$ **13.** $-\frac{3}{2}, 1$ **15.** $-\frac{1}{2}$ **17.** $-2, 0, 3$ **19.** yes **21.** no
23. no **25.** yes **27.** no **29.** yes **31.** $\dfrac{2}{z - 3}$ **33.** $\dfrac{2}{x - 4}$ **35.** $\dfrac{1}{v - 2}$ **37.** $-\dfrac{3}{2t - 1}$ **39.** $-\dfrac{1}{p^2 + pr + r^2}$
41. $\dfrac{2m + 3}{3 - 2m}$ **43.** $\dfrac{h + 3}{2h(4h - 3)}$ **45.** $-\dfrac{e + 2f}{4f^2 + 2ef + e^2}$ **47.** $\dfrac{3 + b}{1 + 3b}$ **49.** $a + b$ **51.** 1 **53.** $(a - b)^2$
55. $(z + 2)^2$ **57.** $(3x + 1)(4x + 5)$ **59.** $(6v - 1)(v + 5)$ **61.** $(2t + 1)(4t - 7)$ **63.** $(2a - 3)(2a + 3)$
65. $(c + 7d)(c - 7d)$ **67.** $(g - 2)(g^2 + 2g + 4)$ **69.** $(3m - 2n)(9m^2 + 6mn + 4n^2)$

▶ **Practice 7.2, pp. 263–64**

1. $\dfrac{4z - 3}{4 - 3z}$ **3.** $-\dfrac{2 - 3x}{3x + 2}$ **5.** $\dfrac{v - 2}{v + 2}$ **7.** $\dfrac{t + 3}{t^2 + 2t + 4}$ **9.** $-\dfrac{1}{a + 2}$ **11.** $-\dfrac{(3f - e)(f + 2e)}{2e^2 - ef + 3f^2}$ **13.** -1 **15.** $\dfrac{4}{c + 3}$
17. $\dfrac{e^2 + 3e + 9}{e - 3}$ **19.** $\dfrac{(m - 5)(m + 7)}{(m + 4)(m - 3)}$ **21.** $\dfrac{a^2 + 2a}{a^2 + 5a + 6}$ **23.** $\dfrac{c^2 + 10c + 25}{c^2 - 25}$ **25.** $\dfrac{6e^2 + 13e + 6}{9e^2 + 12e + 4}$
27. $-\dfrac{8g^3 + 12g^2 - 18g + 27}{12g - 9 - 4g^2}$ **29.** $z^2 + 2z - 1$ **31.** $x^2 - 1$ **33.** $z^2 + 1$ **35.** $5x - 7$ **37.** $a^2 - ab + b^2$
39. $\dfrac{d + 2}{d + 3}$

A-26 Appendix Selected Answers

▶ **Practice 7.3, pp. 269–70**

1. $\dfrac{2a^2-a}{2a+3}$ 3. $\dfrac{c}{2c-3}$ 5. $\dfrac{43}{21e}$ 7. $\dfrac{-3g}{g^2-9}$ 9. $\dfrac{5k^2-6k-1}{(3k-2)(2k+3)(2k-3)}$ 11. $-\dfrac{b+2}{3b^2}$ 13. $\dfrac{1}{12d}$ 15. $\dfrac{-2}{3f-7}$

17. $\dfrac{2h^3+3h^2-17h-29}{(h+3)(h-3)}$ 19. 1 21. $\dfrac{x+1}{x-1}$ 23. $-\dfrac{v^2-2v+2}{(v+3)(v-3)(v+2)}$ 25. $\dfrac{3t^2-7t+4}{(t-1)(t-3)(t+2)}$

27. $\dfrac{2(3a^2+6a+5)}{3(a+2)}$ 29. $\dfrac{2(2c+3)}{2c-3}$ 31. $\dfrac{1+\frac{1}{e}}{1-\frac{1}{e}}$ or $\dfrac{e+1}{e-1}$ 33. $z+y$ 35. u^2-v^2 37. p^2+q^2 39. h^2-k

41. c^3-d^2

▶ **Practice 7.4, pp. 277–78**

1. $\dfrac{a+1}{a-1}$ 3. $\dfrac{c+1}{c-1}$ 5. $-\dfrac{p-1}{p+1}$ 7. $\dfrac{1-x^2}{x^2+1}$ 9. $\dfrac{v+1}{v-1}$ 11. -1 13. $\dfrac{3(z-1)}{z+1}$ 15. $\dfrac{2(x+4)}{x-5}$

17. $\dfrac{(m+7n)(2m+n)}{(m+n)^2}$ 19. $\dfrac{(c^3-1)(c-2)}{c(c^3-2c^2-c+1)}$ 21. $\dfrac{g^2}{g^2-1}$ 23. $\dfrac{k^2-k-1}{-k^2+k-1}$ 25. $\tfrac{1}{5}$ 27. $\tfrac{38}{11}$

29. 38 31. 0 33. $-1, 6$

▶ **Practice 7.5, pp. 283–84**

1. -17 3. 0 5. -1 7. -6 9. no solution 11. 0 13. $\{x\mid x\in \mathscr{R} \text{ and } x\neq 0\}$ 15. $-\tfrac{4}{3}, 1$ 17. $-\tfrac{11}{5}$

19. $-\tfrac{4}{7}$ 21. 0 23. $-3, 8$ 25. $-1, -\tfrac{1}{5}$ 27. no solution 29. $2, -11$ 31. z^7 33. t 35. $\dfrac{1}{p^3}$ 37. k

39. a^6 41. $\dfrac{1}{c^6}$ 43. $\dfrac{1}{e^3}$ 45. g^2

▶ **Problem Solving 8, pp. 285–88**

1. 64.2 hours 3. $34\tfrac{1}{2}$ minutes 5. 10 mph 7. 3 mph 9. 6 minutes 11. 3 hr 20 minutes 13. 175 mph
15. 62.5 mph 17. 2 hours 19. $4\tfrac{5}{7}$ days 21. $1303\tfrac{1}{3}$ mi 23. $4, 6$ 25. 4 27. -3 29. $\tfrac{2}{3}$ 31. 27
33. 6 minutes

▶ **Problem Solving 9, p. 290**

1. $E = I(R_1 + R_2)$ 3. $a = s(1-r)$ 5. $f = \dfrac{ab}{a+b}$ 7. $t = \dfrac{1+fm-f}{mf}$ 9. $R = \dfrac{E - I\cdot \frac{r}{n}}{I}$ 11. $c^2 - \dfrac{vv_1v_2}{v_1+v_2-v}$

13. $B = \dfrac{24I}{T(n+1)}$ 15. $F = RS + RP - S - P$ 17. $R = \dfrac{E}{I}$ 19. $R = 400$ ohms

Chapter 8 Answers

▶ **Practice 8.1, pp. 299–300**

1. $\sqrt[3]{4}$ 3. $\sqrt[3]{\tfrac{1}{16}}$ 5. $\sqrt[6]{a}$ 7. $\sqrt[5]{\dfrac{1}{c^4}}$ 9. $\sqrt[5]{e^2f^2}$ 11. $m\sqrt[4]{n^3}$ 13. 2 15. 25 17. $\tfrac{1}{4}$ 19. 625 21. -36

23. $\tfrac{1}{16}$ 25. $a^{3/2}$ 27. $c^{2/5}$ 29. $\dfrac{1}{e^{2/3}f^2}$ 31. $\dfrac{-3}{mn}$ 33. $r^{5/4} + 2r^{43/15} + 5r^{15/4}$ 35. $u^{253/180}v^{167/120}$ 37. $6a$

39. $3c^2$ 41. e 43. $\dfrac{1}{g^{1/12}}$ 45. $m^{14/3}n^{10/3}$ 47. $\dfrac{s^{7/6}}{r^{7/6}}$ 49. $2^3\cdot 3$ 51. $2^5\cdot 3$ 53. $2^3\cdot 3^2\cdot 5$ 55. 2^{10} 57. 2^4

59. $2^3\cdot 3^2$ 61. $2^2\cdot 3^3\cdot 5^2$ 63. $5\cdot 7^3\cdot 11^2$ 65. $2\cdot 3^3\cdot 5^2\cdot 7^2$

Chapter 8 Answers   **A-27**

▶ **Practice 8.2, pp. 305–306**

1. $4a^2\sqrt{2a}$ 3. $2c^3d^2\sqrt[3]{9c}$ 5. $-3g^2h^2\sqrt[3]{h}$ 7. $-2p^3q\sqrt[3]{pq^2}$ 9. $(t-3)\sqrt[3]{t-3}$ 11. $|v-2|$ 13. $\sqrt{2}$ 15. $\sqrt{a}$
17. $2c^2d^3$ 19. $\dfrac{\sqrt{21}}{3}$ 21. $\dfrac{\sqrt{2a}}{a}$ 23. $2\sqrt[3]{2}$ 25. $\dfrac{3\sqrt[3]{2c^2}}{2c}$ 27. $\dfrac{\sqrt[3]{4e^2}}{2}$ 29. $3g\sqrt[3]{g}$ 31. $\dfrac{5y\sqrt{y}}{8z}$ 33. $\dfrac{z\sqrt[3]{z}}{3v}$
35. $\dfrac{\sqrt{6b}}{2b}$ 37. $\dfrac{\sqrt[3]{36d}}{6d}$ 39. $\dfrac{\sqrt{10gh}}{8h^2}$ 41. $\dfrac{p^3\sqrt{7q}}{2q^2}$ 43. $\dfrac{3u\sqrt[3]{49v}}{14v}$ 45. $\dfrac{2\sqrt[4]{686yz^2}}{21z}$ 47. $(x+1)\sqrt[4]{x+1}$
49. $(2x-1)^2\sqrt[8]{2x-1}$ 51. $x+3$ 53. $|x^2-1|$ 55. z^2-3z 57. x^2-5x 59. u^2-v^2 61. $6p^2+pq-12q^2$
63. $6gh+9gk+8hk+12k^2$ 65. $12c^2-7cd+d^2$

▶ **Practice 8.3, pp. 311–12**

1. $12\sqrt{3}$ 3. $18a^3\sqrt{5a}$ 5. $6c\sqrt[3]{3c^2}$ 7. $7\sqrt[3]{2e^2}$ 9. $6g\sqrt{gh}$ 11. $6\sqrt{2}$ 13. $5\sqrt{7}-7$ 15. $5b+15b^2$
17. $18e+9\sqrt{6ef}-12f$ 19. $50h+20\sqrt{6hk}+12k$ 21. $4p-12q$ 23. $\dfrac{6-2\sqrt{3}}{3}$ 25. $-5\sqrt{2}+5\sqrt{3}$
27. $\dfrac{5\sqrt{b}-5\sqrt{3}}{b-3}$ 29. $\dfrac{\sqrt{ef}-f}{e-f}$ 31. $\dfrac{16\sqrt{15}-75}{-34}$ 33. $\dfrac{2m+5\sqrt{mn}+2n}{4m-n}$ 35. $\dfrac{12r+17\sqrt{rs}+6s}{9r-4s}$
37. $\dfrac{10v\sqrt{15}+95\sqrt{vw}+10w\sqrt{15}}{20v-75w}$ 39. $\dfrac{x-\sqrt{x^2-y^2}}{-y}$ 41. y^2-6y+9 43. $w+7+6\sqrt{w-2}$
45. $3u-7-6\sqrt{3u-2}$ 47. $-1, 2$ 49. $-1, \tfrac{3}{2}$ 51. $\tfrac{1}{2}$ 53. $-4, -3$

▶ **Practice 8.4, pp. 317–18**

1. 17 3. $-4, 4$ 5. $-7, 7$ 7. 2 9. $\tfrac{3}{2}$ 11. 4 13. 7 15. no solution 17. 2 19. -3 21. 2 23. 3
25. $3, 4$ 27. $-3, -2$ 29. $-5, -4$ 31. $(3, 1)$ 33. $(-2, 1)$ 35. $5+4y$ 37. $1-7u$ 39. $12-7w$

▶ **Problem Solving 10, pp. 319–20**

1. $P=\dfrac{E^2}{R}$ 3. $P=I^2R$ 5. $Z^2=X^2+R^2$ 7. $d^2=l^2-5.3s^2$ 9. $L=\dfrac{T^2g}{4\pi^2}$ 11. $L=\dfrac{1}{4\pi Cf^2}$ 13. $m=\dfrac{Tk}{4\pi^2}$
15. $g=\dfrac{v^2(r+h)}{r^2}$ 17. $m=M\sqrt{1-\dfrac{v^2}{c^2}}$ 19. $\theta=X\sqrt{1+(2\pi ft)^2}$ 21. $E=I\sqrt{R^2+(\omega L)^2}$ 23. $L=\dfrac{\dfrac{1}{C_1}+\dfrac{1}{C_2}}{4\pi^2 f^2}$
25. $a=s-\dfrac{A^2}{s(s-b)(s-c)}$

▶ **According to Guinness, p. 326**

3. 2648 ft

▶ **Problem Solving 11, pp. 321–22**

1. 3.1 sec. 3. 3.3 ft 5. increased by 9.75 ft 7. 244.9 mi 9. 3 mi 11. 68,651.2 mph 13. 22.9 mi

▶ **Practice 8.5, pp. 327–28**

1. $i\sqrt{2}$ 3. $4i\sqrt{2}$ 5. $4i$ 7. $-i$ 9. $ab^3i\sqrt{a}$ 11. $6ei\sqrt{2e}$ 13. $7+7i$ 15. $8-3i$ 17. $1-5i$ 19. $-1+9i$
21. $8-8i$ 23. $2-2i$ 25. $9-3i$ 27. $-7+8i$ 29. $10-10i$ 31. $4-2i$ 33. $5-5i$ 35. $a=2, b=5$
37. $a=-4, b=-5$ 39. $a=1, b=4$ 41. $a=12, b=-7$ 43. $a=-1, b=-1$ 45. $a=0, b=-4$
47. $a=\tfrac{1}{2}, b=0$ 49. $a=1, b=10$ 51. $x=2, y=4$ 53. yes 55. $-20+11\sqrt{5}$ 57. -38 59. $\dfrac{3+\sqrt{5}}{2}$
61. $\dfrac{10-2\sqrt{3}}{11}$ 63. $\dfrac{19-8\sqrt{5}}{-41}$

Practice 8.6, pp. 333–34

1. $12 + 15i$ 3. $-6 - 10i$ 5. $-7 + 22i$ 7. $26 - 7i$ 9. $13 + 0i$ 11. $-21 + 20i$ 13. $-1 - 2i$
15. $-\frac{12}{13} + \frac{8}{13}i$ 17. $0 - 1i$ 19. $\frac{4}{13} - \frac{7}{13}i$ 21. $\frac{7}{2} - \frac{1}{2}i$ 23. $-\frac{11}{13} - \frac{16}{13}i$ 25. $\frac{5}{13} + \frac{12}{13}i$ 27. $\frac{20}{29} + \frac{21}{29}i$ 29. $1 - 1i$
31. $2 + 3i$ 33. $-\frac{7}{25} - \frac{24}{25}i$ 35. $-\frac{1}{2} - \frac{11}{4}i$ 37. $-\frac{13}{5} + \frac{9}{5}i$ 39. $\frac{5}{4} - \frac{1}{4}i$ 41. $-\frac{7}{625} + \frac{24}{625}i$ 43. $\frac{7}{11} - \frac{6\sqrt{2}}{11}i$
45. $-1 - 1i$ 47. yes 49. yes 51. yes 53. $6y(y + 2)$ 55. $3w(2w - 1)$ 57. $3u(2u + 3)$ 59. $6s(2s - 3)$
61. $-1, 0$ 63. $0, \frac{1}{2}$ 65. $-\frac{3}{2}, 0$ 67. $0, \frac{3}{2}$

Chapter 9 Answers

Practice 9.1, pp. 343–44

1. $2z^2 - 4z - 5 = 0, a = 2, b = -4, c = -5$ 3. $3x^2 - 4x + 6 = 0, a = 3, b = -4, c = 6$
5. $5v^2 - 7v - 3 = 0, a = 5, b = -7, c = -3$ 7. $3t^2 - 4t + 5 = 0, a = 3, b = -4, c = 5$
9. $5r^2 - 4r - 7 = 0, a = 5, b = -4, c = -7$ 11. $4p^2 - 5p + 6, a = 4, b = -5, c = 6$ 13. $0, 2$ 15. $-2, 0$
17. $0, \frac{3}{4}$ 19. ± 3 21. $\pm 2i\sqrt{2}$ 23. $\pm \frac{4}{3}i$ 25. $\pm \frac{\sqrt{6}}{2}i$ 27. $0, \frac{3}{5}$ 29. $\pm \frac{5\sqrt{6}}{8}$ 31. $-\frac{6}{7}, 0$ 33. $0, \frac{3}{2}$
35. $\pm \frac{\sqrt{15}}{2}i$ 37. $\frac{3}{8}$ 39. $\frac{5}{12}$ 41. $\frac{1}{3}$ 43. $\frac{3}{2}$ 45. $(z - 2)^2$ 47. $(x - 4)^2$ 49. $(v + \frac{5}{2})^2$ 51. $(s - \frac{5}{8})^2$

Practice 9.2, pp. 349–50

1. $-2, 0$ 3. $-5, -1$ 5. $-9, -3$ 7. $2, 12$ 9. $-6, -2$ 11. $\frac{3 \pm \sqrt{2}}{2}$ 13. $-3, 1$ 15. $-1, 7$ 17. $-3 \pm \sqrt{5}$
19. $-2 \pm i$ 21. $\frac{-5 \pm \sqrt{37}}{2}$ 23. $-1, -\frac{1}{2}$ 25. $1, 5$ 27. $\frac{5 \pm \sqrt{13}}{6}$ 29. $\frac{3 \pm i\sqrt{11}}{10}$ 31. $z^2 - z - 6$ 33. $x^2 + 9$
35. $v^2 - 2v + 2$ 37. $4\sqrt{3}$ 39. $2i\sqrt{6}$ 41. $2i\sqrt{2}$ 43. $2i\sqrt{5}$ 45. $2z^2 - 4z + 1 = 0, a = 2, b = -4, c = 1$
47. $3x^2 - 5x = 0, a = 3, b = -5, c = 0$ 49. $v^2 - 3v + 2 = 0, a = 1, b = -3, c = 2$

Practice 9.3, pp. 357–58

1. -2 3. $-\frac{1}{2}, \frac{3}{2}$ 5. $\frac{2}{3}, \frac{3}{4}$ 7. $\frac{1 \pm \sqrt{5}}{2}$ 9. $-1 + \sqrt{2}$ 11. $\frac{3 \pm i\sqrt{7}}{4}$ 13. $-\sqrt{2}$ 15. one real solution
17. two real solutions 19. two real solutions 21. two real solutions 23. two nonreal number solutions
25. two nonreal number solutions 27. $x^2 + x - 6 = 0$ 29. $12x^2 - 5x - 2 = 0$ 31. $x^2 + 9 = 0$ 33. $x^2 - 3 = 0$
35. $x^2 + 10x + 25 = 0$ 37. $x^2 - 4x + 13 = 0$ 39. -27 41. $\frac{5 \pm \sqrt{33}}{2}$ 43. 4

Practice 9.4, pp. 363–64

1. $2, 3$ 3. $2 \pm \sqrt{5}$ 5. $\frac{3 \pm i\sqrt{11}}{2}$ 7. $\pm \frac{1}{2}i$ 9. $-1, 3$ 11. $\frac{-2 - \sqrt{19}}{6}$ 13. $\pm \frac{\sqrt{6}}{2}, \pm \frac{\sqrt{6}}{3}i$ 15. 49
17. $64, -8$ 19. $4 \pm \sqrt{10}$ 21. $\frac{-1 \pm \sqrt{29}}{8}$ 23. 4 25. $-3, 3$ 27. $-\frac{5}{3}, \frac{3}{2}$ 29. $-\sqrt{2}, \sqrt{3}$ 31. 3
33. $-4, -3, 0, 3$ 35. $2, 3, 4$

Problem Solving 12, pp. 367–68

1. 17.8 cm 3. 90 mph, 120 mph 5. 5 ft by 5 ft. 7. 23 cm, 13 cm 9. 8 hours 11. 945 mi 13. 50 mph
15. 5 by 10 17. 5 mph 19a. 6 km 21. 132 ft

Chapter 10 Answers   A-29

▶ **Practice 9.5, pp. 373–74**

1. $\{z \mid -3 < z < 0 \text{ or } z > 3\}$ 3. $\{x \mid -1 < x \leq 0\}$ 5. $\{v \mid -2 < v < 2 \text{ or } v \geq 4\}$ 7. $\{z \mid z \leq 2 \text{ or } z \geq 3\}$
9. $\left\{x \mid \dfrac{-1-\sqrt{5}}{2} < x < \dfrac{-1+\sqrt{5}}{2}\right\}$ 11. $\{v \mid -\tfrac{1}{2} \leq v \leq 2\}$ 13. $\{z \mid -1 < z < 0 \text{ or } 1 < z\}$
15. $\{w \mid w \leq -\tfrac{4}{3} \text{ or } \tfrac{3}{2} \leq w \leq 2\}$ 17. $\{z \mid z < \tfrac{1}{2} \text{ or } 1 < z < 3 \text{ or } 3 < z\}$ 19. $\{x \mid x \in \mathcal{R}\}$ 21. $\{v \mid -2 < v < 2 \text{ or } v > 4\}$
23. $-\tfrac{1}{3}$ 25. $\tfrac{1}{5}$ 27. $\tfrac{1}{2}$ 29. $\tfrac{2}{3}, \tfrac{3}{4}$ 31. $\tfrac{1}{4}, \tfrac{1}{3}$ 33. $\dfrac{1 \pm i\sqrt{3}}{2}$ 35. $-\tfrac{1}{2}$ or 2 37. $\dfrac{3 \pm i\sqrt{7}}{4}$

Chapter 10 Answers

▶ **Make Sure Answers for Lesson 10.1, Example 3, p. 384**

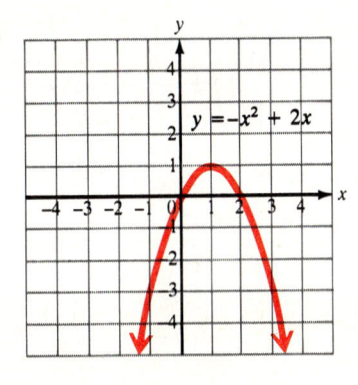

1. [Graph of $y = x^2 - 3$]

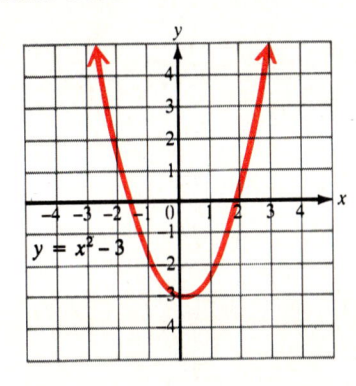

2. [Graph of $y = -x^2 + 2x$]

3. [Graph of $y = x^2 - 4x + 4$]

4. [Graph of $y = -x^2 - 3x + 4$]

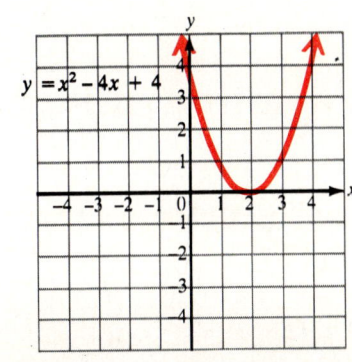

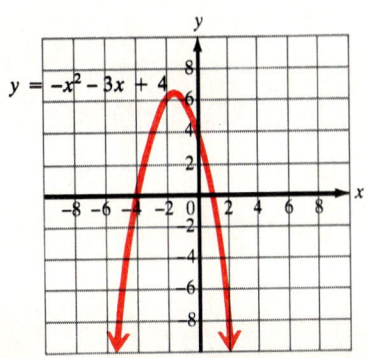

▶ **According to Guinness, p. 384**

5. $y = -\dfrac{5}{2809}x^2$

▶ **Practice 10.1, pp. 385–86**

1. $(0, 0), (2, 0)$ 3. $(-\tfrac{3}{2}, 0), (\tfrac{3}{2}, 0)$ 5. $(-\tfrac{1}{2}, 0), (2, 0)$ 7. $(0, 3), x = 0$ 9. $(2, -3), x = 2$ 11. $(1, 0), x = 1$

13. [Graph of $y = 2x^2 + 8$]

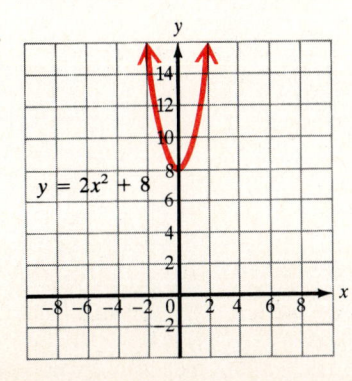

15. [Graph of $y = x^2 - 4x - 5$]

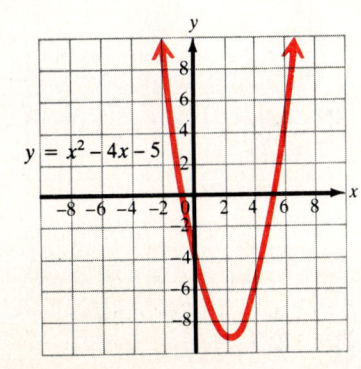

17. [Graph of $y = 2x^2$]

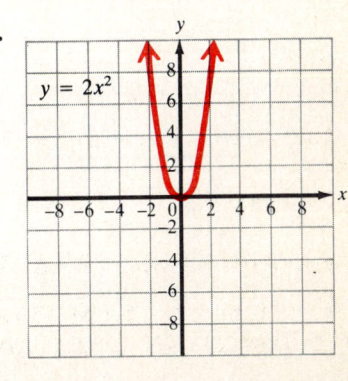

A-30 Appendix Selected Answers

19. $y = -3x^2$

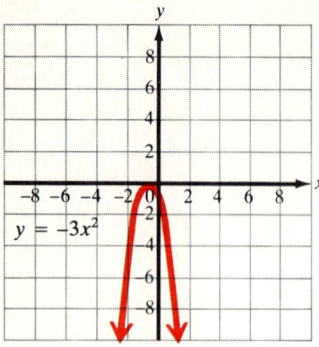

21. $y = 3x^2 - 12x$

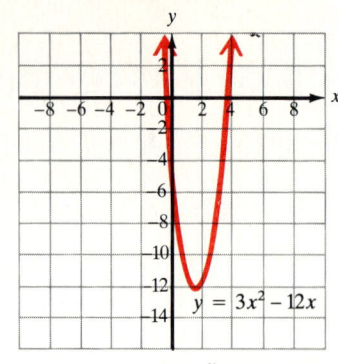

23. $y = -x^2 + 4$

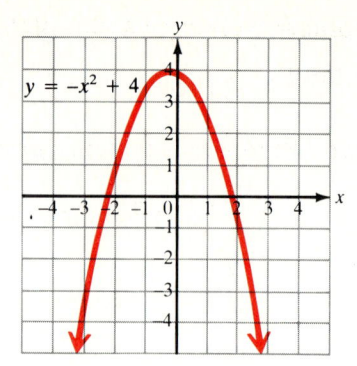

25. $y = x^2 + 4x + 6$

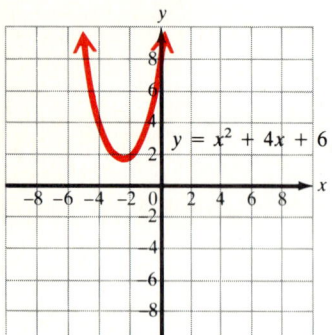

27. $y = 3x^2 + 2x - 8$

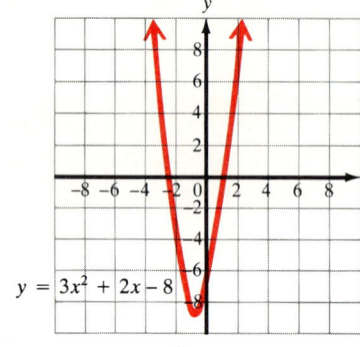

29. $y = 2x^2 + 2x + 1$

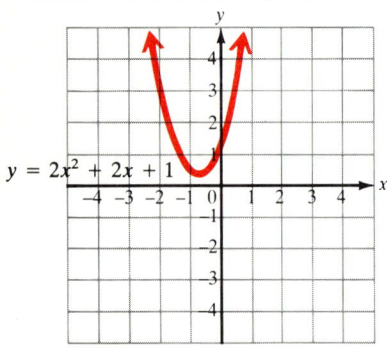

31. $y = 3x^2 - x - 4$

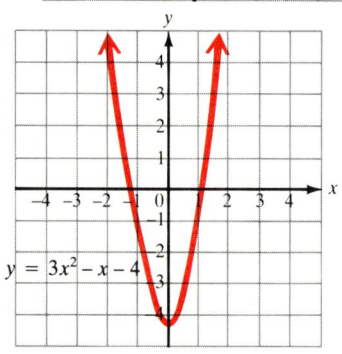

33. $y = 4 + x - 3x^2$

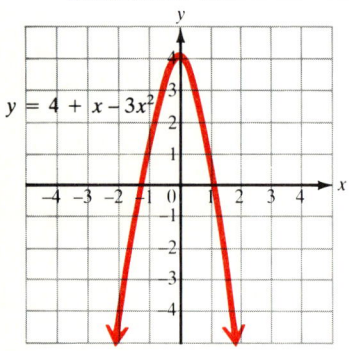

35. $x^2 - 8x + 2y + 4 = 0$

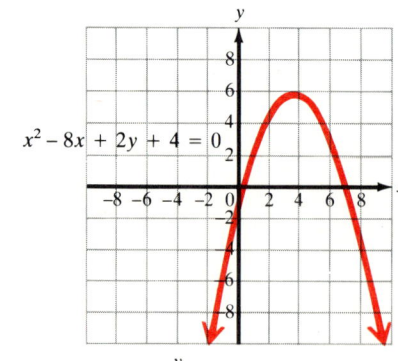

37. $x^2 - 4x + 2y + 2 = 0$

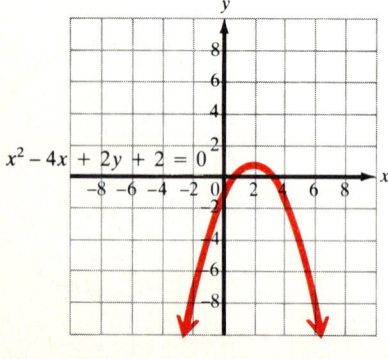

39. $2x^2 + 4x - y + 3 = 0$

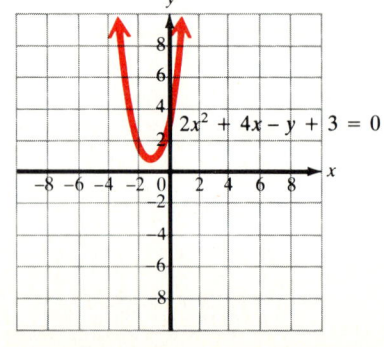

41. 10   **43.** 29   **45.** 53   **47.** $-5, 5$ **49.** $-2\sqrt{2}, 2\sqrt{2}$

▶ **Problem Solving 13, pp. 388**

1. 256 ft² **3.** 21 ft **5.** $\frac{1}{6}$ mi² **7.** 16 m² **9.** $-\frac{1}{2}, \frac{1}{2}$ **11.** $\sqrt{2}$ mi **13.** 57,600 ft²

▶ **Make Sure Answers for Lesson 10.2, Example 3, p. 393**

1. $(x + 1)^2 + (y - 2)^2 = 9$

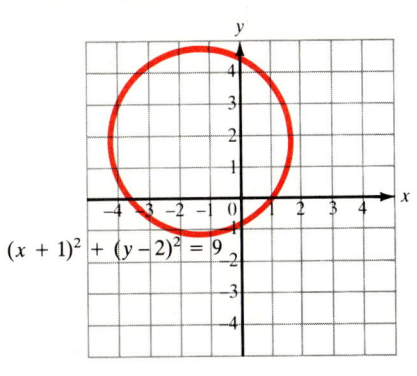

2. $(x - 3)^2 + (y + 2)^2 = 4$

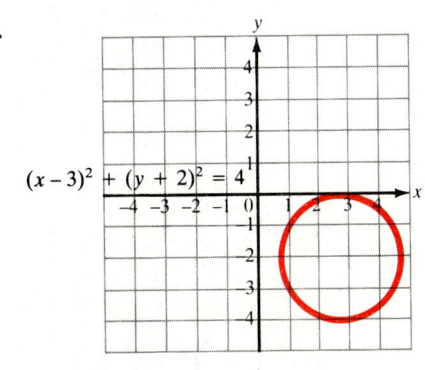

▶ **Practice 10.2, pp. 395–96**

1. 5 **3.** 13 **5.** 7 **7.** $x^2 + y^2 = 3^2$ **9.** $(x - 2)^2 + (y - 3)^2 = 2^2$ **11.** $[x - (-5)]^2 + (y - 3)^2 = 1^2$

13. $x^2 + y^2 = 36$

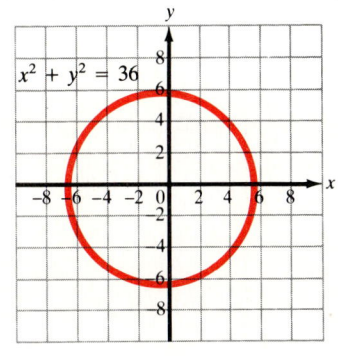

15. $(x - 3)^2 + y^2 = 4$

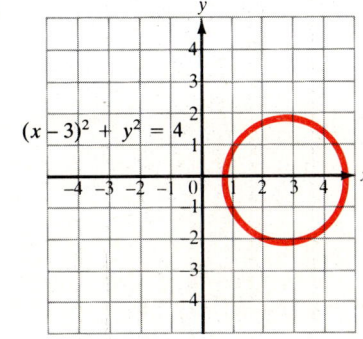

17. $x^2 + (y - 1)^2 = 1$

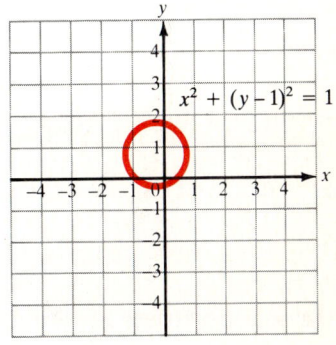

19. $(x - 2)^2 + (y - 4)^2 = 4$

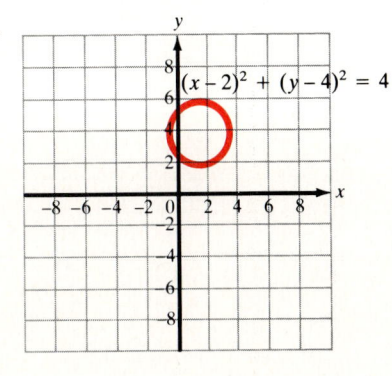

21. $(x - 2)^2 + (y + 1)^2 = 9$

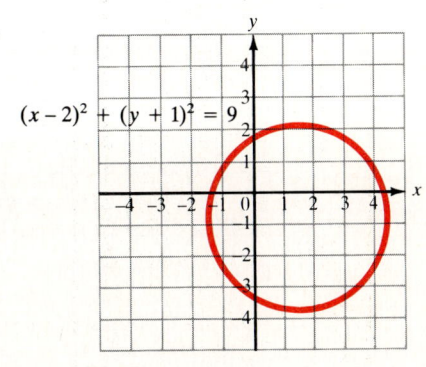

23. $(x - 1)^2 + (y + 2)^2 = 4$

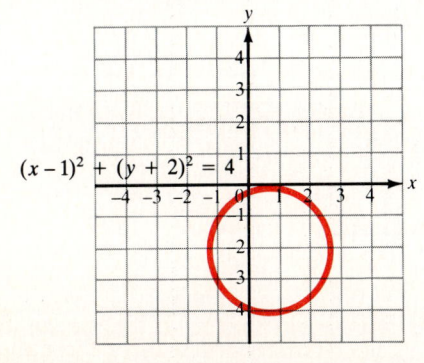

25. $(x + 2)^2 + (y - 1)^2 = 9$

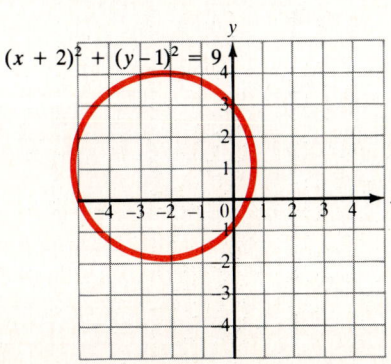

27. $(x - 1)^2 + y^2 = 16$

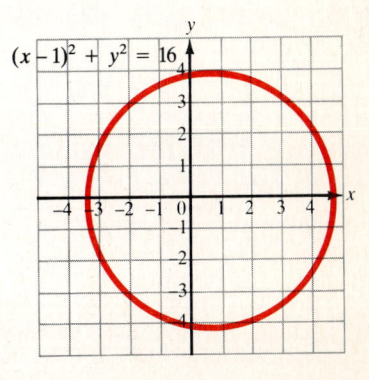

29. $(0, 0), 2$ **31.** $(3, 0), 3$ **33.** $(4, 3), 5$ **35.** $(-1, -2), 5$ **37.** $(-2, 4), 5$ **39.** $(\frac{2}{3}, \frac{1}{3}), \sqrt{5}$

41. (graph of $x + 4y = -4$)

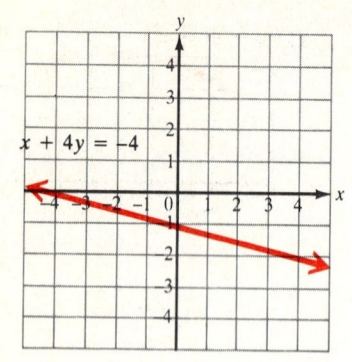

43. (graph of $2x + 3y = 6$)

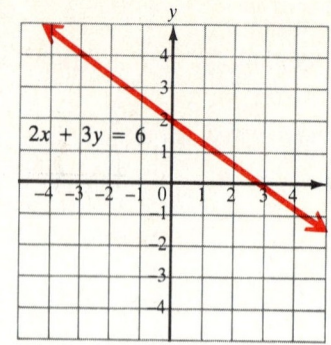

45. (graph of $5x + 2y = 10$)

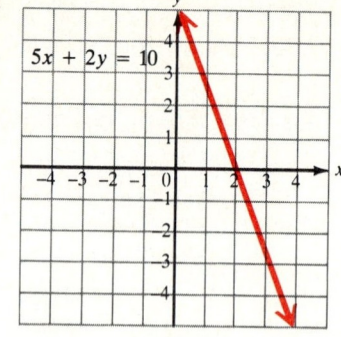

47. (graph of $3x + 5y = 0$)

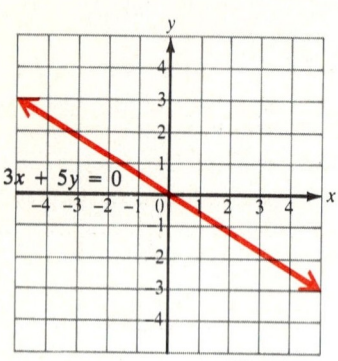

49. (graph of $x = 5$)

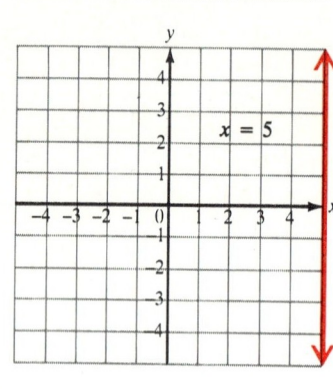

51. (graph of $y = 3$)

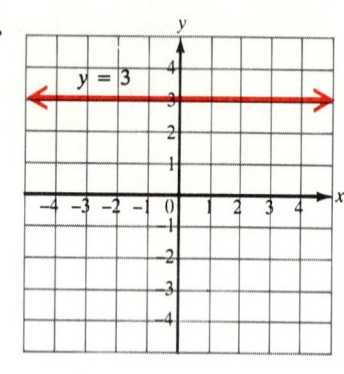

▶ **Problem Solving 14, pp. 397–98**

1. 15 ft   3. 17 m   5. 8 m, 15 m   7. $d = s\sqrt{2}$, $s = \dfrac{d\sqrt{2}}{2}$ 9. $d = e\sqrt{3}$ 11. 12 ft   13. $8\sqrt{2}$ cm   15. $1\frac{1}{2}$ ft

17. $\sqrt{29}$ mi   19. 15,400 ft

▶ **Make Sure Answers for Lesson 10.3, Example 2, p. 402**

1. $\dfrac{x^2}{4} + \dfrac{y^2}{25} = 1$ (graph)

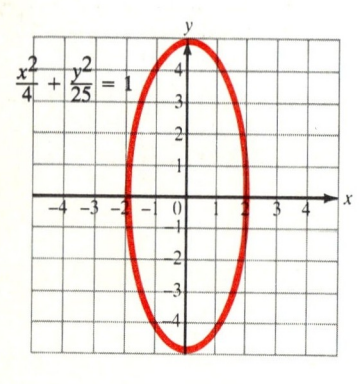

2. $\dfrac{x^2}{16} + \dfrac{y^2}{9} = 1$ (graph)

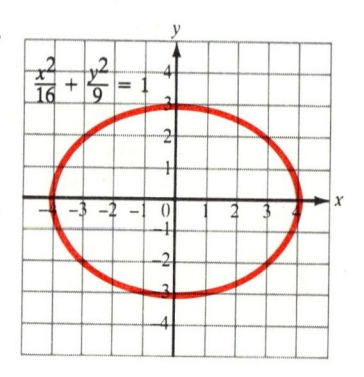

▶ **Practice 10.3, pp. 403–404**

1. $\dfrac{x^2}{1^2} + \dfrac{y^2}{2^2} = 1$ 3. $\dfrac{x^2}{3^2} + \dfrac{y^2}{8^2} = 1$ 5. $\dfrac{x^2}{\sqrt{2}^2} + \dfrac{y^2}{2^2} = 1$ 7. $\dfrac{x^2}{1} + \dfrac{y^2}{16} = 1$ (graph)

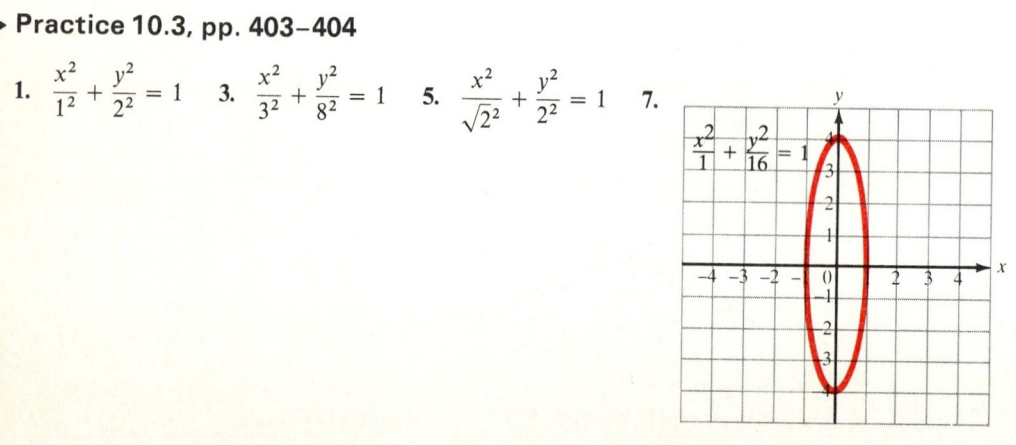

Chapter 10 Answers   A-33

9. [graph of $\frac{x^2}{16} + \frac{y^2}{4} = 1$]

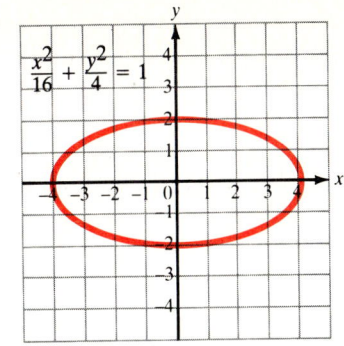

11. [graph of $4x^2 + 9y^2 = 36$]

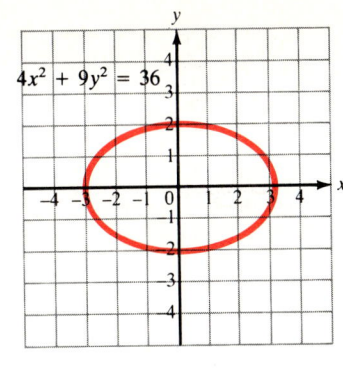

13. [graph of $16x^2 + 25y^2 = 400$]

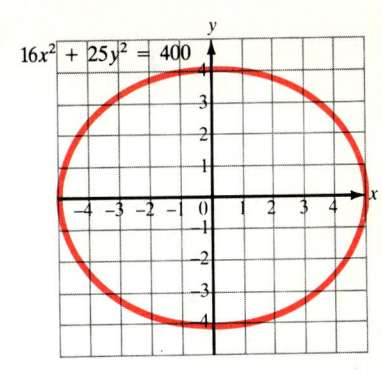

15. [graph of $\frac{x^2}{16} + \frac{y^2}{1} = 1$]

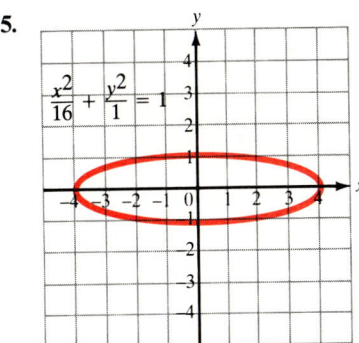

17. [graph of $\frac{x^2}{16} + \frac{y^2}{25} = 1$]

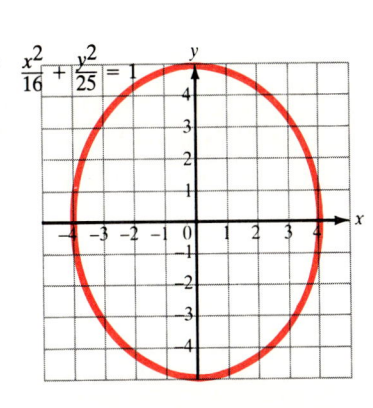

19. [graph of $\frac{x^2}{4} + \frac{y^2}{9} = 1$]

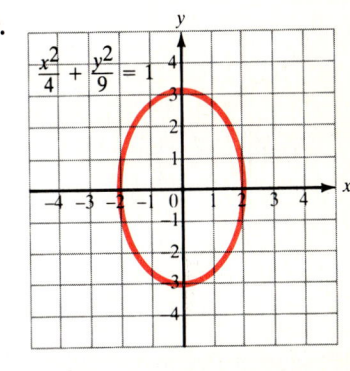

21. [graph of $4x^2 + y^2 = 4$]

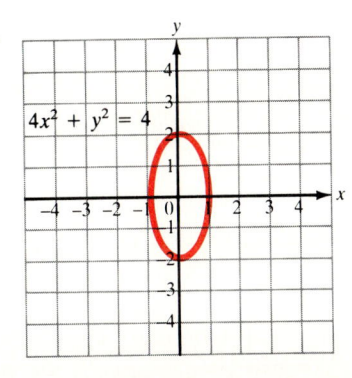

23. [graph of $16x^2 + 9y^2 = 144$]

25. [graph of $x^2 + 25y^2 = 25$]

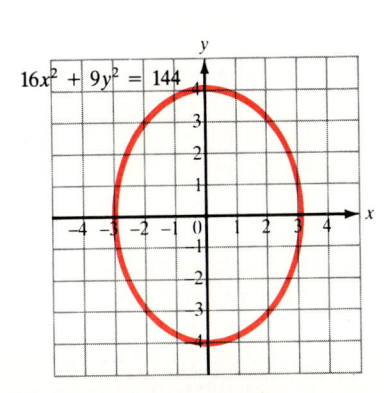

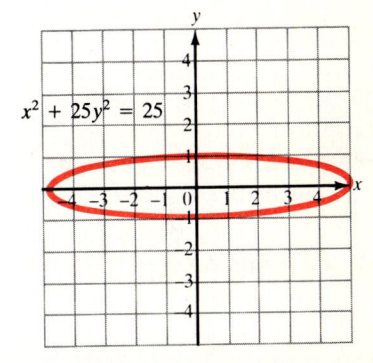

27. 1, 2, 4, 8   29. 1, 2, 4, 8   31. 1, 2, 4, −1

▶ **Make Sure Answers for Lesson 10.4, Example 1, p. 406**

1. [graph of $\frac{x^2}{16} - \frac{y^2}{25} = 1$]

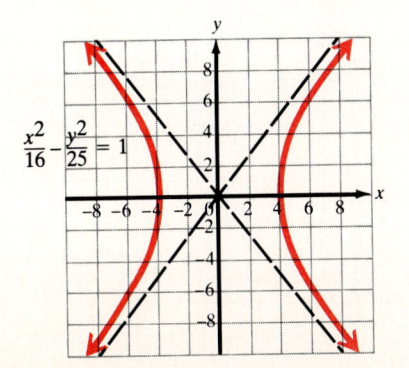

2. [graph of $\frac{x^2}{16} - \frac{y^2}{4} = 1$]

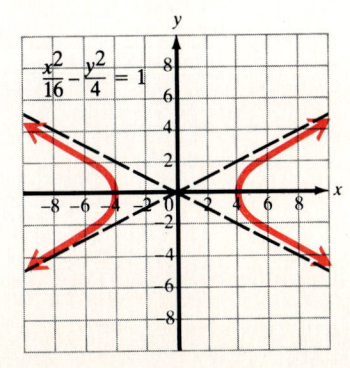

A-34 Appendix Selected Answers

▶ **Make Sure Answers for Lesson 10.4, Example 2, p. 407**

1. $-\dfrac{x^2}{9} + \dfrac{y^2}{25} = 1$

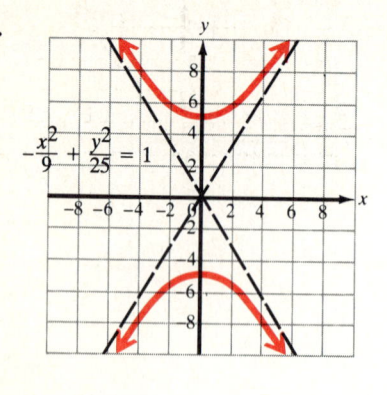

2. $-\dfrac{x^2}{16} + \dfrac{y^2}{1} = 1$

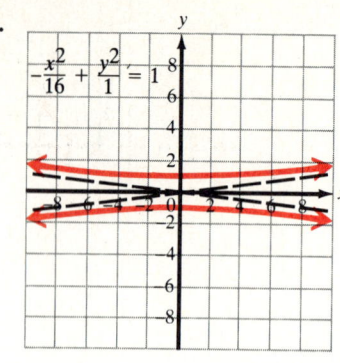

▶ **Make Sure Answers for Lesson 10.4, Example 3, p. 408**

1. $xy = 6$

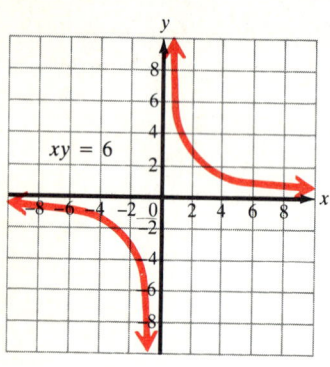

2. $xy = -6$

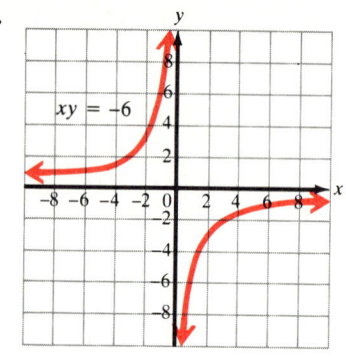

▶ **Practice 10.4, pp. 409–10**

1. $\dfrac{x^2}{4} - \dfrac{y^2}{9} = 1$

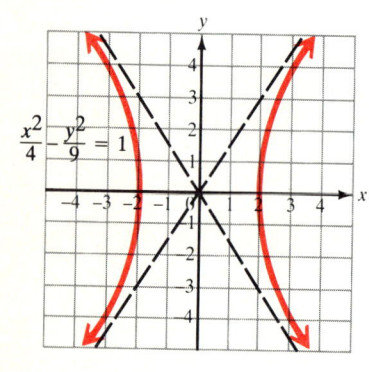

3. $\dfrac{x^2}{9} - \dfrac{y^2}{4} = 1$

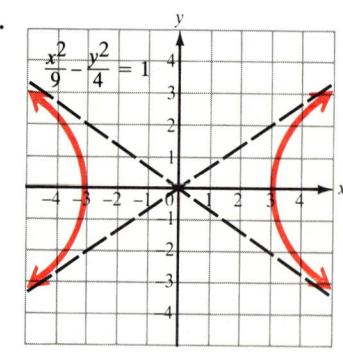

5. $-\dfrac{x^2}{4} + \dfrac{y^2}{16} = 1$

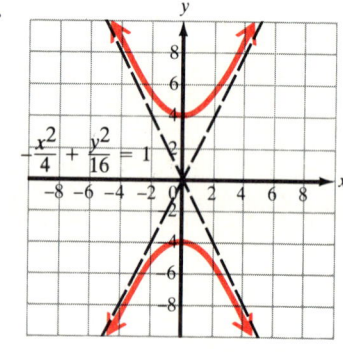

7. $-\dfrac{x^2}{9} + \dfrac{y^2}{4} = 1$

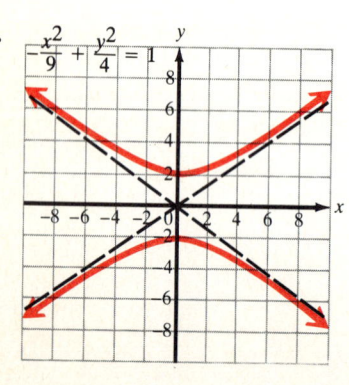

9. $xy = 4$

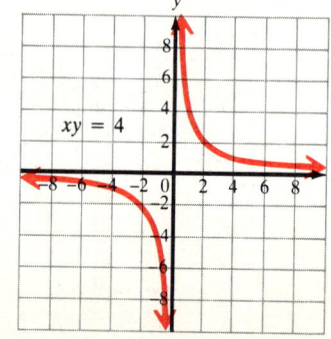

11. $xy = -12$

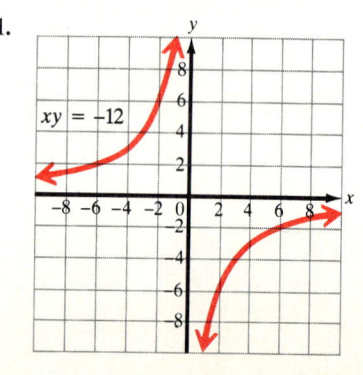

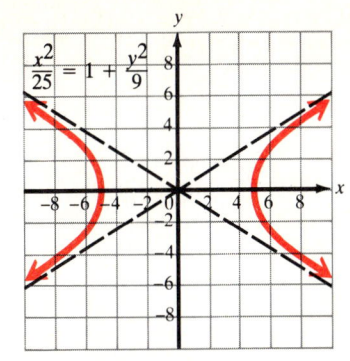

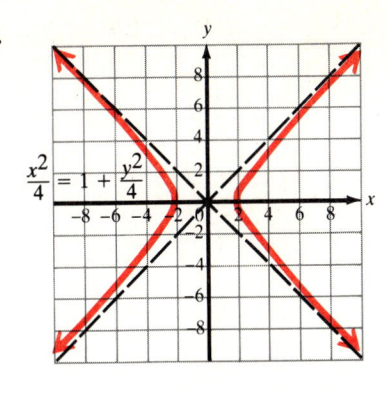

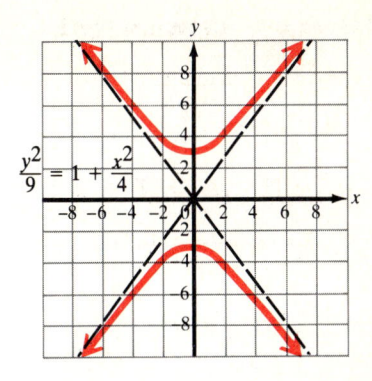

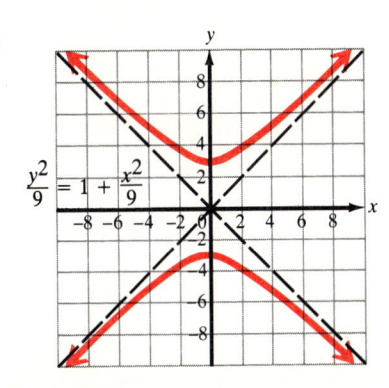

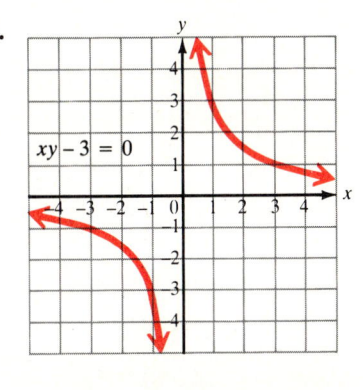

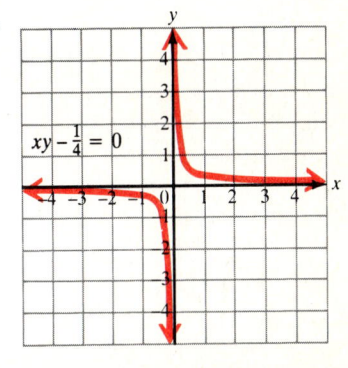

Chapter 10 Answers A-35

13. $\frac{x^2}{25} = 1 + \frac{y^2}{9}$

15. $\frac{x^2}{4} = 1 + \frac{y^2}{4}$

17. $\frac{y^2}{9} = 1 + \frac{x^2}{4}$

19. $\frac{y^2}{9} = 1 + \frac{x^2}{9}$

21. $xy - 3 = 0$

23. $xy - \frac{1}{4} = 0$

25. $3x - 4y \geq 12$

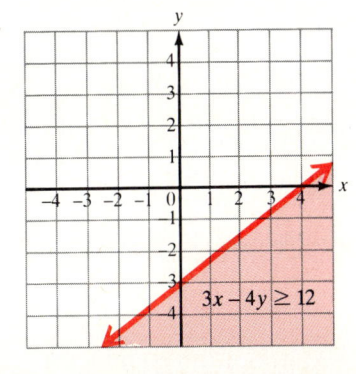

27. $1 < y \leq 4$

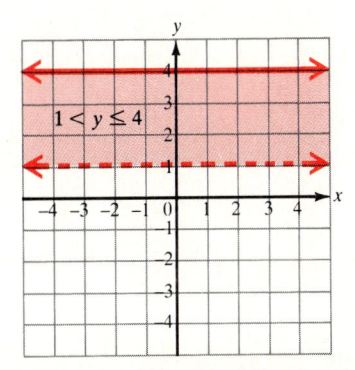

29. $(3, 2)$ 31. $(2, 3)$ 33. $(3, -2)$

▶ **Make Sure Answers for Lesson 10.5, Example 3, p. 418**

1. $\begin{cases} xy \geq 3 \\ x + y \leq 4 \end{cases}$

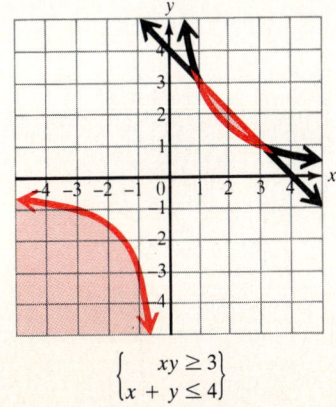

2. $\begin{cases} x^2 - 4y^2 \geq 4 \\ x^2 - y < 9 \end{cases}$

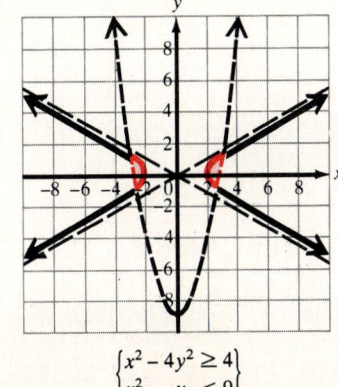

Practice 10.5, pp. 419–22

1. $(4, -3), (-4, 3)$ 3. $(0, 9), (3, 0)$ 5. $(4, -4), (\frac{48}{13}, -\frac{32}{13})$ 7. No solution 9. $(2, 2)$ 11. $(-2, 0), (2, 0)$
13. $(-2, 2), (2, 2), (-\sqrt{7}, -1), (\sqrt{7}, -1)$ 15. $(0, -3), (0, 3)$ 17. $(0, 3)(-\frac{\sqrt{15}}{2}, -\frac{3}{4}), (\frac{\sqrt{15}}{2}, -\frac{3}{4})$ 19. $(0, 0), (\frac{72}{25}, \frac{54}{25})$
21. $(0, 1)$ 23. $(-15, 135), (0, 0)$ 25. 27.

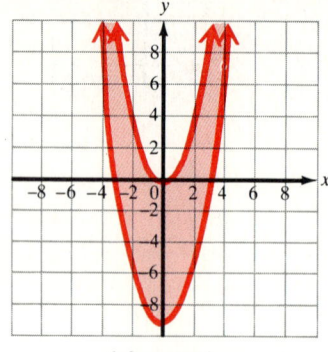

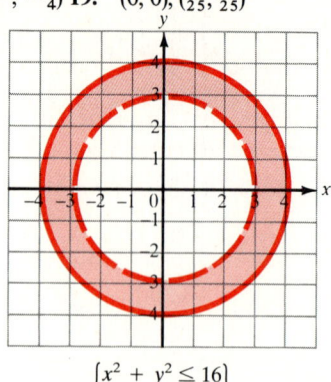

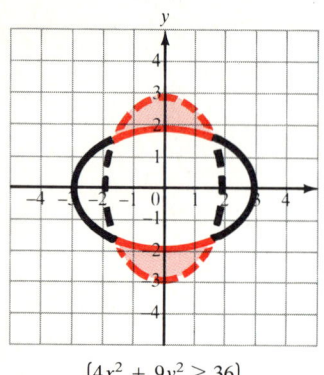

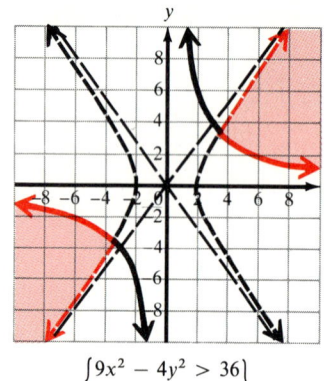

29. 31. 33.

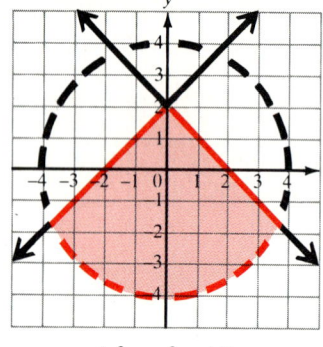

35.

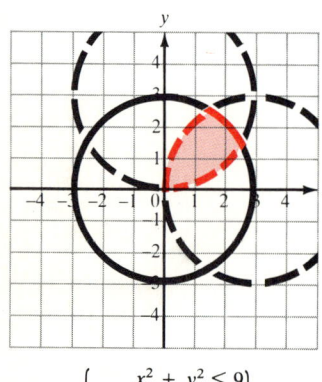

37. $21, 7, 1$ 39. $-1, \frac{7}{2}, \frac{13}{2}$ 41. $\frac{7}{2}, \frac{11}{4}, -\frac{1}{4}$

Problem Solving 15, pp. 425–26

1. $\frac{5 + \sqrt{5}}{2}$ and $\frac{5 - \sqrt{5}}{2}$ 3. $1 + \sqrt{5}$ and $-1 + \sqrt{5}$ or $1 - \sqrt{5}$ and $-1 - \sqrt{5}$ 5. 18 ft by 12 ft. 7. $(16 + 6\sqrt{2})$ cm
9. 58 and 85 11. 48 yds 13. 48 ft² 15. 96 ft and 94 ft. 17. $(9 + \sqrt{53})$ yds 19. 17 mi or 16 mi
21. $\frac{-3 + 3\sqrt{5}}{2}$ and $\frac{3 + 3\sqrt{5}}{2}$ or $\frac{-3 - 3\sqrt{5}}{2}$ and $\frac{3 - 3\sqrt{5}}{2}$ 23. 60 cm 25. $(2 + 2\sqrt{2})$ ft
27. 50 km/h 29a. 5 hr b. 50 mph 31. 8 cm and 3 cm 33. 1575 ft by 246 ft

Chapter 10 Review, pp. 305–306

3. Grid A: $y = x^2 - 6x + 8$

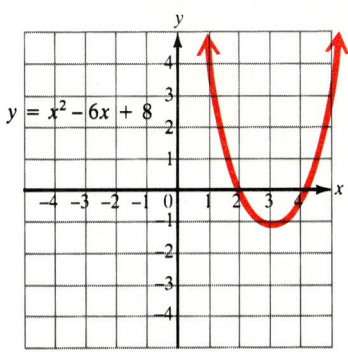

6. Grid B: $(x+2)^2 + (y-3)^2 = 9$

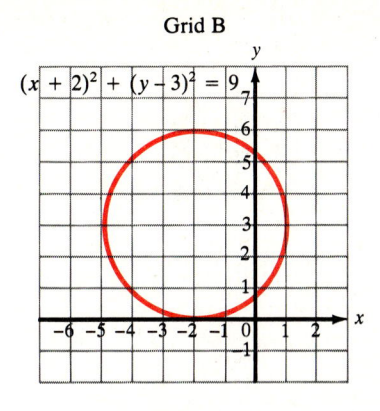

9a. Grid C: $\dfrac{x^2}{81} + \dfrac{y^2}{36} = 1$

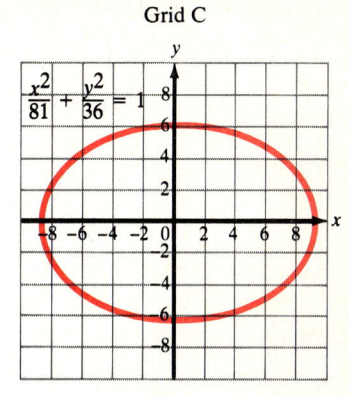

9b. Grid D: $\dfrac{x^2}{25} + \dfrac{y^2}{49} = 1$

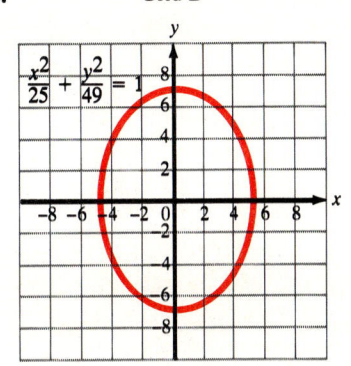

10. Grid E: $\dfrac{x^2}{5^2} - \dfrac{y^2}{3^2} = 1$

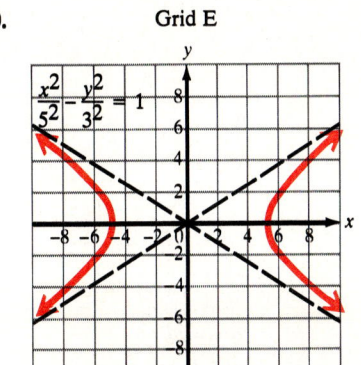

11. Grid F: $-\dfrac{x^2}{3^2} + \dfrac{y^2}{2^2} = 1$

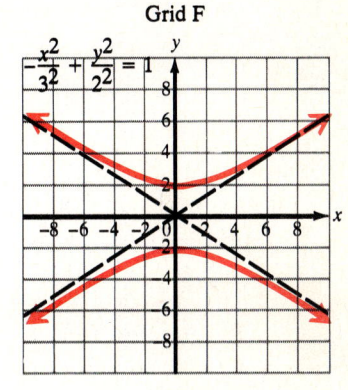

12. Grid G: $xy = 4$

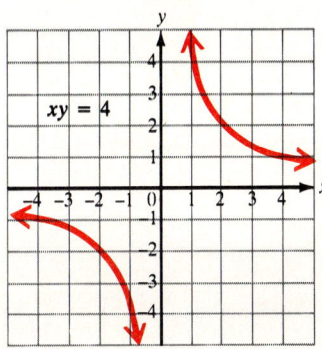

15. Grid H: $\begin{cases} x^2 + y^2 < 20 \\ y \leq x^2 \end{cases}$

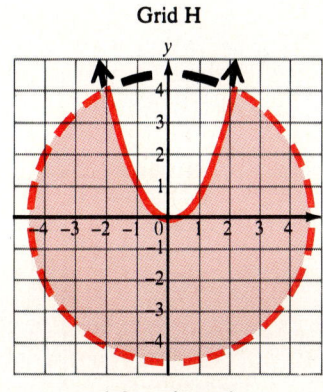

Chapter 11 Answers

Practice 11.1, pp. 435–36

1. $D = \{1, 2, 4, 5\}$, $R = \{2, 3, 5\}$ 3. $D = \{-4, -3, -2, -1\}$, $R = \{1, 2, 3, 4\}$ 5. $D = \{1\}$, $R = \{2, 3, 4, 5\}$
7. $D = \{1, 2, 3, 4\}$, $R = \{1\}$ 9. $D = \{1, 2\}$, $R = \{1, 2, 3, 4\}$ 11. $D = \{1, 2\}$, $R = \{1, 2\}$
13. $D = \{-3, 0, 2, 4\}$, $R = \{-3, 0, 1, 3\}$ 15. $D = \{x \mid -2 \leq x \leq 4\}$, $R = \{y \mid 0 \leq y \leq 5\}$
17. $D = \{x \mid -4 \leq x \leq 4\}$, $R = \{y \mid -2 \leq y \leq 2\}$ 19. $D = \{x \mid x \in \mathcal{R}\}$, $R = \{y \mid y \in \mathcal{R}\}$ 21. $D = \{x \mid x \in \mathcal{R}\}$, $R = \{y \mid y \geq 0\}$
23. $D = \{x \mid x \in \mathcal{R}\}$, $R = \{y \mid y \geq 4\}$ 25. $D = \{x \mid x \in \mathcal{R}\}$, $R = \{y \mid y = 3\}$ 27. $D = \{x \mid x \in \mathcal{R}\}$, $R = \{y \mid y \geq 0\}$
29. $D = \{x \mid x \geq 0\}$, $R = \{y \mid y \in \mathcal{R}\}$ 31. $D = \{x \mid x \geq 0\}$, $R = \{y \mid y \geq 0\}$ 33. $D = \{x \mid x \geq 3\}$, $R = \{y \mid y \geq 0\}$
35. $D = \{x \mid -2 \leq x \leq 2\}$, $R = \{y \mid 0 \leq y \leq 2\}$ 37. $D = \{x \mid x \in \mathcal{R} \text{ and } x \neq 1\}$, $R = \{y \mid y \in \mathcal{R} \text{ and } y \neq 0\}$
39. $D = \{x \mid x \in \mathcal{R} \text{ and } x \neq 2 \text{ or } -2\}$, $R = \{y \mid y > 1 \text{ or } y \leq 0\}$ 41. $D = \{x \mid x \in \mathcal{R} \text{ and } x \neq -1\}$, $R = \{y \mid y \in \mathcal{R} \text{ and } y \neq 1\}$

A-38 Appendix Selected Answers

43. $y = \dfrac{6 - 3x}{2}$ **45.** $y = 6x - x^2$ **47.** $y = \dfrac{1}{x - 3}$ **49.** $y = \dfrac{5x - 160}{9}$ **51.** $y = \pm\sqrt{25 - x^2}$ **53.** $y = \pm\tfrac{2}{3}\sqrt{9 - x^2}$

▶ According to Guinness, p. 440

5. yes, because for each domain value r there is a unique range value A **7.** 5034 ft^2 **9.** $\tfrac{1}{8}$ ft^2

▶ Practice 11.2, pp. 441–42

1. function   **3.** not a function   **5.** function   **7.** function   **9.** function   **11.** function   **13.** function   **15.** function
17. function   **19.** function   **21.** function   **23.** not a function   **25.** not a function   **27.** function   **29.** not a function
31. not a function   **33.** function   **35.** function   **37.** $-4, 5, 11, 17$ **39.** $5, -1, -3, 15$ **41.** $19, 1, 4, 6, 4$

▶ Practice 11.3, pp. 447–48

1. $12, 0, 27$ **3.** $-17, 1, -7$ **5.** $11, 1, 1$ **7.** $-1, -1, y^3 - 2y^2 - 1$ **9.** $7, x^2 + 2x + 4, x^2 - 2x + 4$
11. $G(3) \approx 1, G(0) \approx -\tfrac{1}{2}, G(-2) \approx -\tfrac{3}{2}$ **13.** $g(-2) \approx 5, g(0) \approx -3, g(3) \approx 0$ **15.** $-5, 3$ **17.** $99, 39$ **19.** $73, -19$
21. $-8, 8$ **23.** $9, 9$ **25.** $4, 1$ **27.** $a^2 + 2a + 3, a^2 + 6a + 9$ **29.** $x^2 - 4x + 4, x^2 - 2x$ **31.** $4x^2 - 1, 4x^2 + 1$

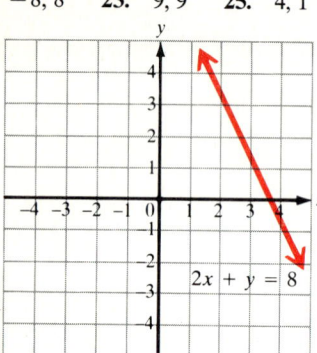

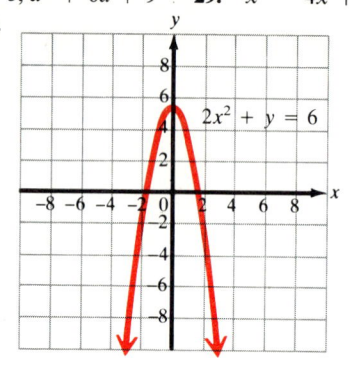

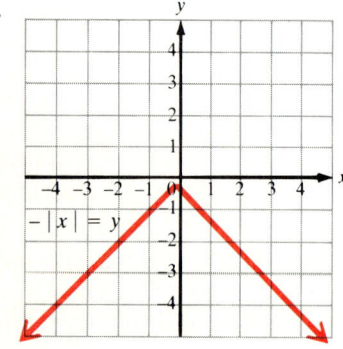

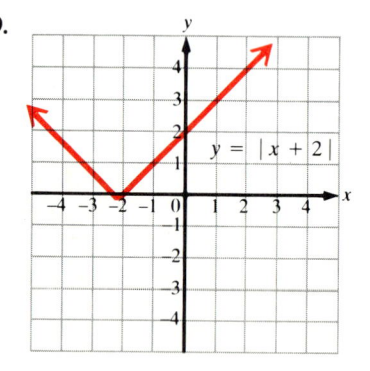

33. [Graph of $2x + y = 8$]

35. [Graph of $2x^2 + y = 6$]

37. [Graph of $-|x| = y$]

39. [Graph of $y = |x + 2|$]

▶ Make Sure Answers for Lesson 11.4, Example 1, p. 451

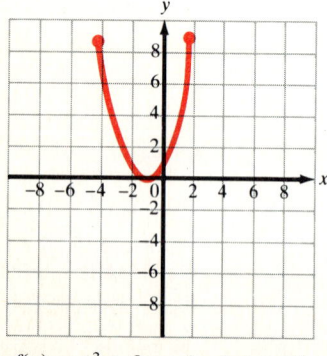

1. [Graph of $f(x) = x^2 + 2x + 1, -4 \le x \le 2$]

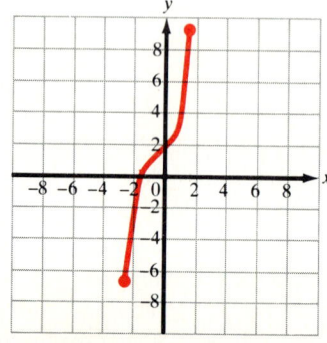

2. [Graph of $f(x) = x^3 + 2, -2 \le x \le 2$]

Chapter 11 Answers   A-39

▶ **Make Sure Answers for Lesson 11.4, Example 2, p. 452**

1.

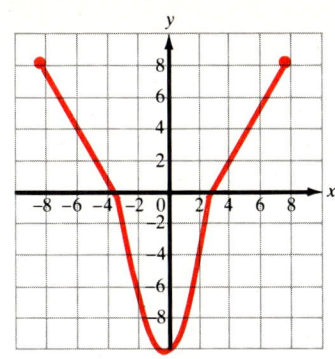

$$f(x) = \begin{cases} x^2 - y = 9 \text{ if } -3 \leq x \leq 3 \\ 5x - 3y = 15 \text{ if } 3 \leq x \leq 8 \\ 5y + 3y = -15 \text{ if } -8 \leq x \leq -3 \end{cases}$$

▶ **Make Sure Answers for Lesson 11.4, Example 3, p. 454**

1. $f(x) = \left|\frac{1}{x}\right|$

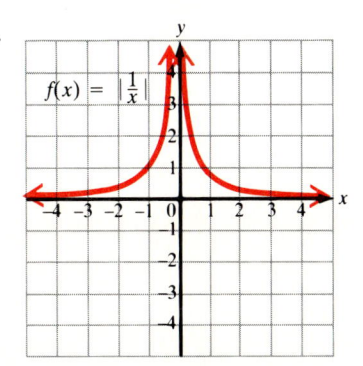

2. $g(x) = |x-2| - 3$

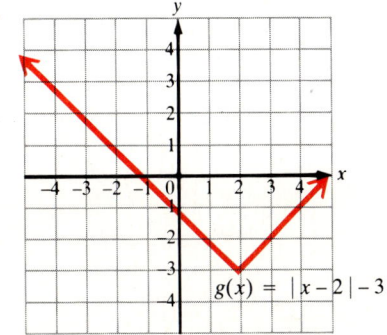

▶ **Practice 11.4, pp. 455–56**

1. $f(x) = x^2 + 1, -3 \leq x \leq 2$

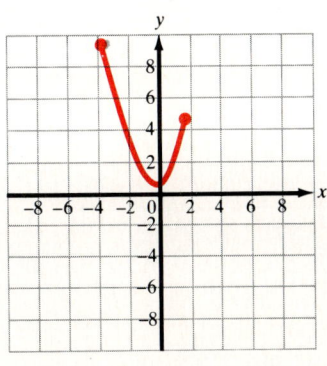

3. $h(x) = x^3 + 1, -2 \leq x \leq 2$

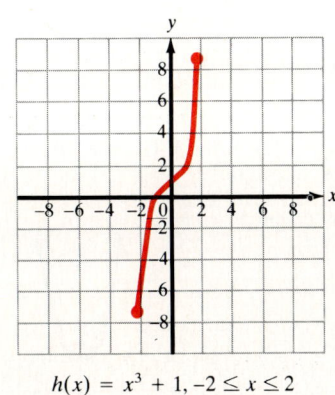

5. $h(x) = \begin{cases} x + 2 \text{ if } x \leq -2 \\ x^2 + 2x \text{ if } -2 < x < 1 \\ 3 \text{ if } 1 \leq x \end{cases}$

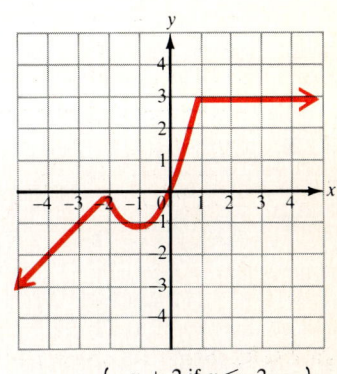

7. $f(x) = \left|\frac{1}{x-1}\right|$

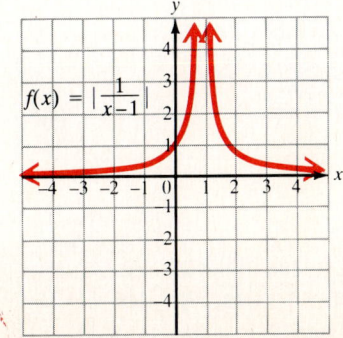

9. $h(x) = |x+2|$

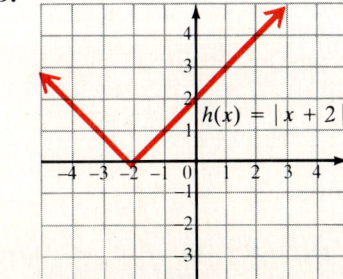

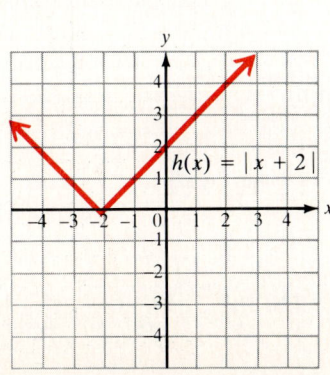

A-40 Appendix Selected Answers

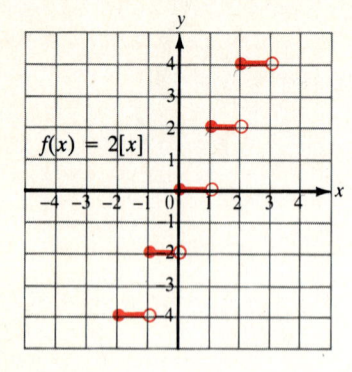

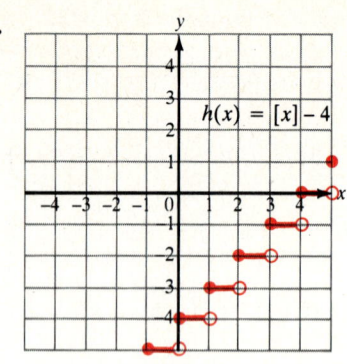

11. [graph of $f(x) = 2[x]$]

13. [graph of $h(x) = [x] - 4$]

15. $y = \pm\sqrt{x+3}$ **17.** $y = \pm\sqrt{x-2}$ **19.** $y = -1 \pm \sqrt{1+x}$ **21.** $y = 2 \pm \sqrt{1+x}$ **23.** $y = \dfrac{-1 \pm \sqrt{4+3x}}{3}$

▶ **Make Sure Answers for Lesson 11.5, Example 3, p. 461**

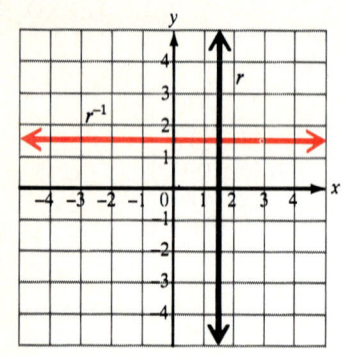

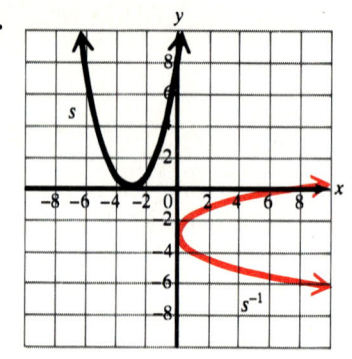

1. [graph of r and r^{-1}]

2. [graph of s and s^{-1}]

▶ **Practice 11.5, pp. 463–64**

1. {(3, 2), (2, 1), (1, 0), (0, 1)} **3.** {(1, 2), (1, 3), (1, 4), (0, 1)} **5.** {(0, −1), (1, −1), (0, 1), (1, 0)} **7.** {(−3, 0), (2, 2), (0, −3), (3, 2)}
9. {(1, 1), (0, 0), (2, 2), (3, 3)} **11.** {(1, 9), (2, 8), (3, 7), (4, 6)} **13.** $y = \dfrac{x-5}{3}$ **15.** $y = \dfrac{7-x}{6}$ **17.** $y = \dfrac{3 \pm \sqrt{8x-31}}{4}$
19. $f^{-1}(x) = 2 \pm \sqrt{x}$ **21.** $y = -\dfrac{4x+3}{3x-2}$ **23.** $y = 3 \pm \sqrt{9-x}$

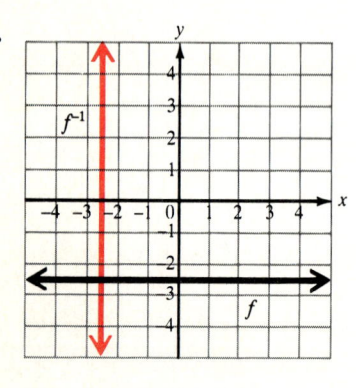

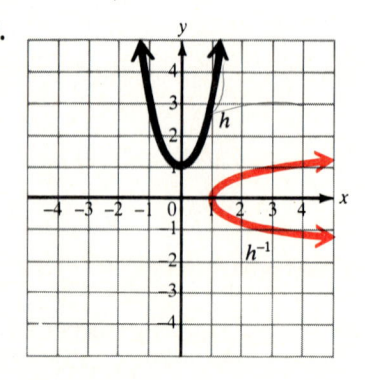

25. [graph of f and f^{-1}]

27. [graph of h and h^{-1}]

29. yes **31.** yes **33.** no **35.** $\frac{3}{2}$ **37.** 2 **39.** $3\sqrt[3]{2}$

▶ **Practice 11.6. pp. 471–72**

1. 24 **3.** 27 **5.** $\frac{171}{4}$ **7.** 4 **9.** $\frac{81}{1024}$ **11.** $\frac{4\sqrt{6}}{3}$ **13.** 64 **15.** 128 **17.** $3\sqrt{2}$ **19.** $\frac{25}{9}$ **21.** $\frac{56}{9}$
23. $\frac{1000\sqrt{5}}{3}$ **25.** $\frac{1}{4}$ **27.** 1 **29.** 8 **31.** 4 **33.** 1 **35.** $\frac{1}{8}$ **37.** $\frac{9}{4}$ **39.** 1 **41.** $\frac{8}{27}$ **43.** $\frac{16}{9}$ **45.** 1
47. $-\frac{64}{27}$

Chapter 12 Answers A-41

▶ **Problem Solving 16, pp. 474–76**

1a. $R = \dfrac{9.5l}{d^2}$ **b.** 9.5 ohms   **c.** 10.5 ft   **3a.** $w = \dfrac{2{,}080{,}000{,}000}{d^2}$ **b.** 4.2 cm   **c.** 5656.9 kg

5a. $P = \dfrac{3}{V}$ **b.** 2.4 atmospheres   **c.** $\tfrac{3}{2}$ liters   **7a.** $F = \dfrac{kgg_1}{d^2}$ **b.** by a factor of 4

9a. $w = \dfrac{360000}{f}$ **b.** 360 kHz   **c.** 360 m   **11a.** $f = \dfrac{12800}{l}$ **b.** 50 inches   **c.** 800 vps

13a. $T = \dfrac{l}{2}\sqrt{\dfrac{2}{g}}$ **b.** $\tfrac{1}{4}$ sec.   **c.** 16 ft   **15a** $W = dw$ **b.** 1000 ft lbs   **c.** 200 lbs   **d.** $\tfrac{1}{20}$ ft   **17.** 79,200,000 ft lbs

▶ **Chapter 11 Review, p. 477**

10. Grid D

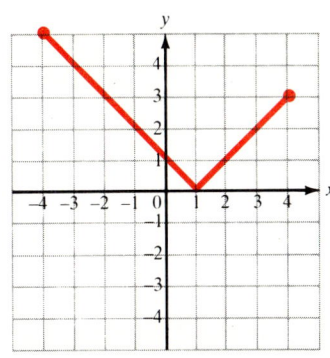

$f(x) = |1 - x|, \; -4 \le x \le 4$

11. Grid E

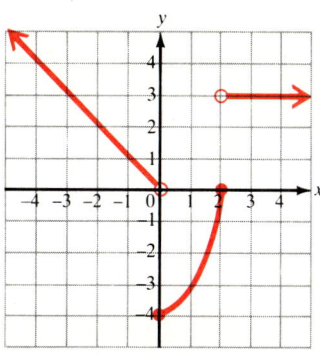

$h(x) = \begin{cases} -x & \text{if } x < 0 \\ x^2 - 4 & \text{if } 0 \le x \le 2 \\ 3 & \text{if } 2 < x \end{cases}$

12. Grid F

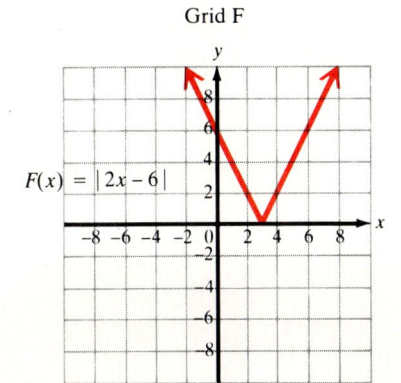

$F(x) = |2x - 6|$

15. Grid G

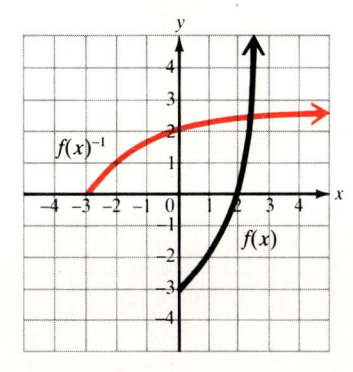

$f(x)^{-1}$, $f(x)$

Chapter 12 Answers

▶ **Make Sure Answers for Lesson 12.1, Example 2, p. 482**

1. $f(x) = 3^x$

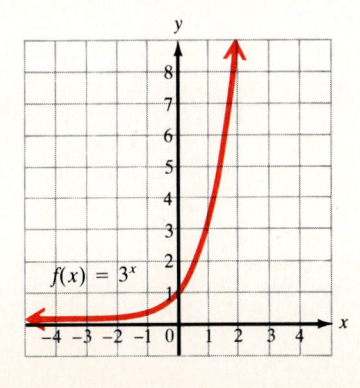

3. $g(x) = (0.5)^x$

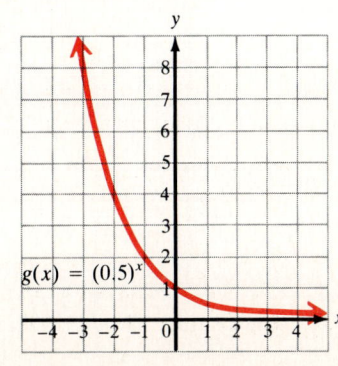

A-42 Appendix Selected Answers

▶ **Practice 12.1, pp. 483–84**

1. ≈ 4.656 3. ≈ 90.510 5. ≈ 1.728 7. ≈ 1.047 9. ≈ 0.375 11. ≈ 0.095 13. ≈ 0.082 15. ≈ 2.930

17. $f(x) = 4^x$

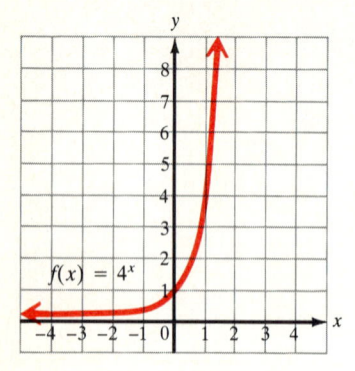

19. $h(x) = \left(\frac{3}{2}\right)^x$

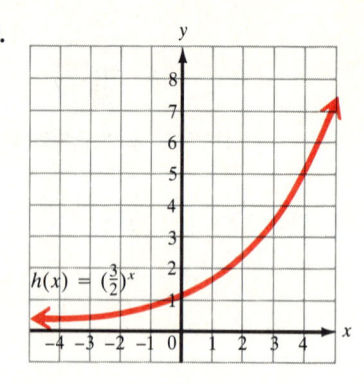

21. $f(x) = \left(\frac{3}{4}\right)^x$

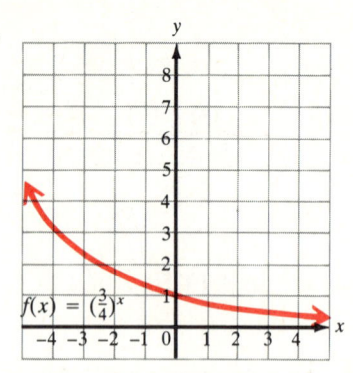

23. $h(x) = \left(\frac{2}{5}\right)^x$

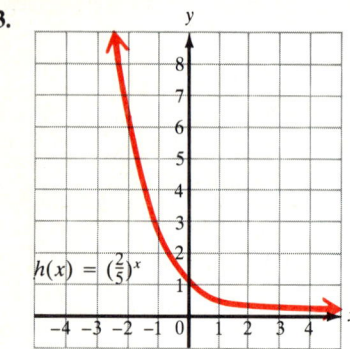

25. $p(x) = \left(\frac{2}{3}\right)^x$

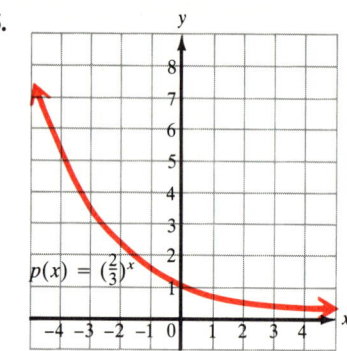

27. $r(x) = (1.4)^x$

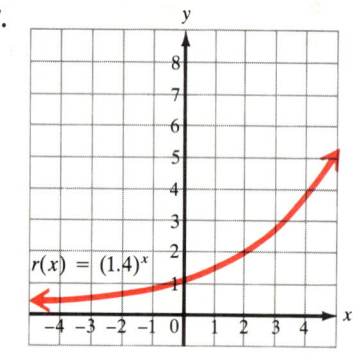

29. $x = 2^y$

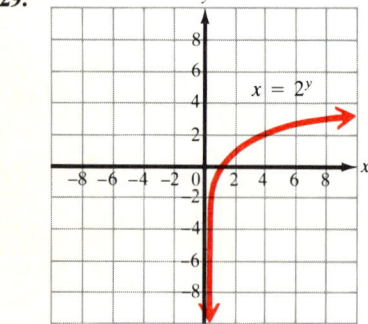

31. $x = y^3$

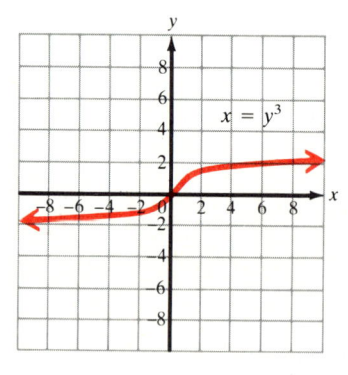

33. $y = x^2$

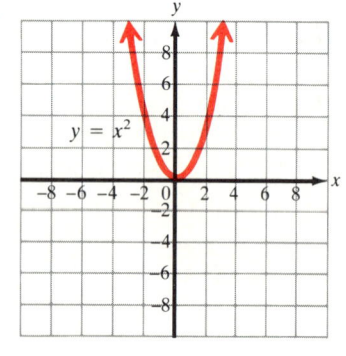

35. $y = x^{\frac{1}{2}}$

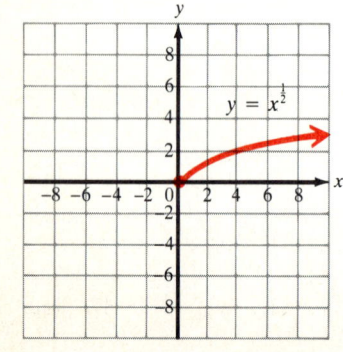

37. $x = y^{-\frac{1}{2}}$

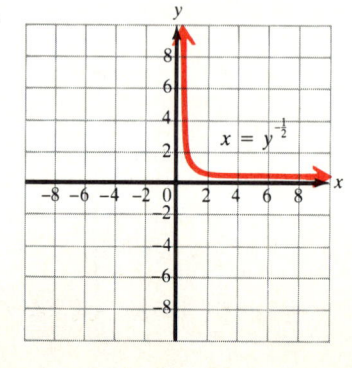

Chapter 12 Answers   A-43

▶ **Make Sure Answers for Lesson 12.1, Example 4, p. 488**

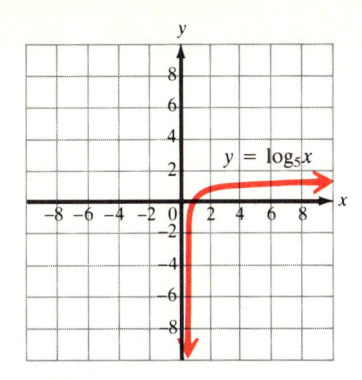

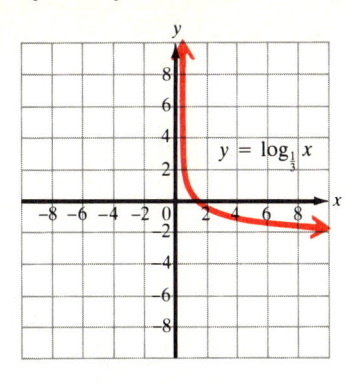

1. [graph of $y = \log_5 x$]
2. [graph of $y = \log_{\frac{1}{3}} x$]

▶ **Practice 12.2, pp. 489–90**

1. $10^2 = 100$ 3. $2^3 = 8$ 5. $(\frac{1}{2})^{-4} = 16$ 7. $(\frac{2}{3})^2 = \frac{4}{9}$ 9. $2^{1/2} = \sqrt{2}$ 11. $e^1 = e$ 13. $\log_{10} 1000 = 3$
15. $\log_3 81 = 4$ 17. $\log_5 625 = 4$ 19. $\log_4 2 = \frac{1}{2}$ 21. $\log_4 \frac{1}{16} = -2$ 23. $\log_e e = 1$ 25. 16   27. $\frac{1}{625}$ 29. $\sqrt{2}$
31. 2   33. 3   35. $\frac{1}{2}$ 37. [graph of $y = \log_2 x$]   39. [graph of $y + \log_{\frac{1}{3}} x = 0$]

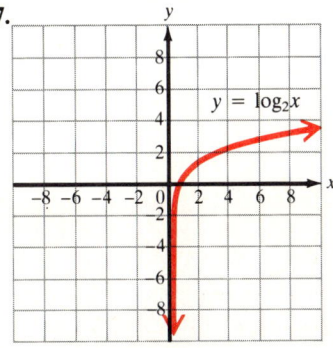

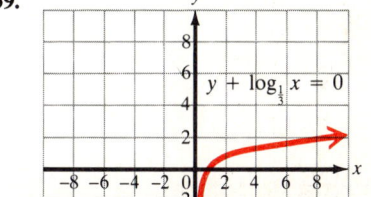

41. $2^{1/2}$ 43. $3^{1/4}$ 45. $2^{2/3}$ 47. $3^{3/4}$ 49. $2^{8/3}$

▶ **Make Sure Answers for Example p. 492**

1. $\log_b \dfrac{M}{N} = \log_b M - \log_b N$

   Let $\log_b M = x$ and $\log_b N = y$

   Then $M = b^x$ and $N = b^y$

   $\dfrac{M}{N} = \dfrac{b^x}{b^y}$

   $\dfrac{M}{N} = b^{x-y}$

   $\log_b \dfrac{M}{N} = \log_b b^{x-y}$

   $= x - y$

   $= \log_b M - \log_b N$

2. $\log_b M^p = p \log_b M$

   Let $\log_b M = x$

   Then $M = b^x$

   $M^p = (b^x)^p$

   $= b^{xp}$

   $\log_b M^p = \log_b b^{xp}$

   $= xp$

   $= p \log_b M$

▶ **Practice 12.3, pp. 495–96**

1. $3 \log_b x + 2 \log_b y$ 3. $3 \log_b x - 4 \log_b y$ 5. $3 \log_b x + \log_b y - 2 \log_b z$ 7. $2 \log_b x + \frac{1}{2} \log_b y - \log_b z$
9. $\log_b x + \frac{1}{2} \log_b y - \frac{1}{3} \log_b z$ 11. $\frac{3}{4} \log_b x + \frac{1}{2} \log_b y - \frac{1}{2} \log_b z - \frac{1}{4} \log_b w$ 13. $\log_b \dfrac{xy}{z}$ 15. $\log_b \dfrac{x^3 y^2}{z}$ 17. $\log_b \sqrt{\dfrac{x}{yz}}$
19. $\log_b \dfrac{y \sqrt[3]{x}}{z^2}$ 21. $\log_b \dfrac{x^2 y^2}{\sqrt{z}}$ 23. $\log_b (x - y)$ 25. 0.6020   27. 1.0000   29. 1.0791   31. 1.8572   33. 3.1582

A-44   Appendix Selected Answers

35. 0.5441 **37.** 1.2431 **39.** 0.1505 **41.** 0.4013 **43.** 4.35×10^3 **45.** 7.43×10^{-2} **47.** 4.72×10^{-1} **49.** 6440
51. 0.0943 **53.** -1.0438 **55.** -2.4486 **57.** 0.4116

▶ **Practice 12.4, pp. 503–504**

1. 1.3802 **3.** 2.7202 **5.** 4.9499 **7.** $9.7582 - 10$ **9.** $8.5065 - 10$ **11.** $6.6345 - 10$ **13.** 5.44 **15.** 34.6
17. 194 **19.** 85,800 **21.** 0.0323 **23.** ≈ 0.6128 **25.** ≈ 4.1431 **27.** ≈ 0.3692 **29.** ≈ 1.5728 **31.** ≈ -1.3516
33. ≈ 9.0780 **35.** ≈ 3.56 **37.** ≈ 33.12 **39.** ≈ 25113.08 **41.** ≈ 2040.81 **43.** ≈ 0.25 **45.** ≈ 0.45
47. ≈ 157.47 **49.** ≈ 148.41 **51.** ≈ 4.9069 **53.** ≈ 1.4307 **55.** ≈ -3.3083 **57.** ≈ 0.4637 **59.** ≈ 2.3321
61. ≈ 1.5063 **63.** ≈ 0.6532 **65.** ≈ 1.3856 **67.** ≈ 2.9340 **69.** $-6, 3$ **71.** $\dfrac{-1 \pm \sqrt{5}}{2}$ **73.** $\tfrac{1}{2}, 1$ **75.** $\dfrac{-1 \pm \sqrt{10}}{3}$

▶ **According to Guinness, p. 508**

3. 660 mph   **5.** $-65°C$

▶ **Practice 12.5, pp. 509–10**

1. 3 **3.** 4 **5.** 3 **7.** 8 **9.** 4 **11.** $\pm\sqrt{5}$ **13.** $\dfrac{\log 12}{\log 3} \approx 2.2619$ **15.** $\dfrac{\log 40}{\log 2} \approx 5.3219$ **17.** $\dfrac{\log 24}{\log 5} \approx 1.9746$

19. $\ln 35 \approx 3.5553$ **21.** $\left[\dfrac{\log 20}{\log 2} + 7\right] \div 3 \approx 3.7740$ **23.** $\pm\dfrac{\log 15}{\log 3} \approx \pm 1.5700$ **25.** $3\sqrt[3]{3}$ **27.** 36 **29.** $\dfrac{-7 + \sqrt{53}}{2}$

31. 4 **33.** no solution **35.** $\dfrac{8 \pm \sqrt{46}}{3}$ **37.** $10^{(10^8)}$ **39.** 5, 15, 25 **41.** 1, 27, 64 **43.** 1, -1, 1 **45.** undefined, $\tfrac{27}{2}, \tfrac{64}{3}$

47. 6, 24, 120, 210

▶ **Problem Solving 17, p. 512**

1. $1327.70 **3.** $740.81 **5.** $y = 1000e^{0.69315x}$ **7.** $y = 204e^{0.0105x}$ **9.** $y = 14.7e^{-0.000426k}$ **11.** $y = e^{-0.00001205x}$

▶ **Problem Solving 18, p. 514**

1a. 5.2dB **b.** 15.2dB **3.** $M = \log a - \log a_0$ **5.** $10^{8.3} = 1.995 \times 10^8$ **7a.** 0 **b.** 3 **c.** 4 **d.** 4.405
9. $10^{-0.7}$ W/m²

▶ **Chapter 12 Review, pp. 515–16**

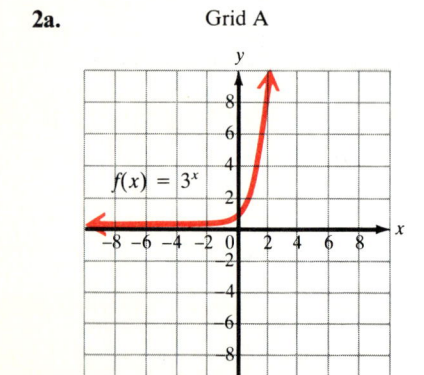

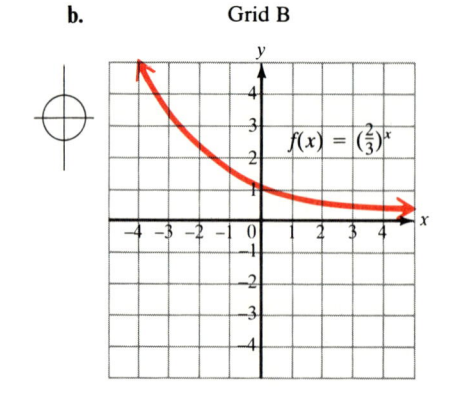

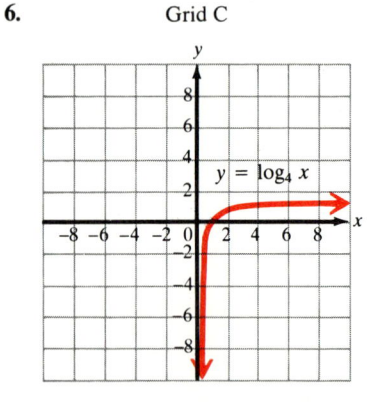

2a. Grid A — $f(x) = 3^x$
b. Grid B — $f(x) = \left(\tfrac{2}{3}\right)^x$
6. Grid C — $y = \log_4 x$

Chapter 13 Answers

▶ **Practice 13.1, pp. 521–22**

1. 5, 7, 9, 23 **3.** 1, 8, 27, 1000 **5.** 2, 7, 14, 119 **7.** $-1, 1, -1, 1$ **9.** $1, -\tfrac{1}{2}, \tfrac{1}{3}, -\tfrac{1}{10}$ **11.** $3, -6, 12, -1536$ **13.** 30
15. $\tfrac{9}{2}$ **17.** $\tfrac{49}{80}$ **19.** 10 **21.** 42 **23.** 106 **25.** 43 **27.** 57 **29.** 867

Final Review Answers   A-45

▶ **Practice 13.2, pp. 527–28**

1. 1, 3, 5, 7, 9   **3.** 4, 2, 0, −2, −4   **5.** −7, 0, 7, 14, 21   **7.** −22, −26, −30, −34, −38   **9.** $\frac{1}{2}, \frac{7}{6}, \frac{11}{6}, \frac{5}{2}, \frac{19}{6}$
11. 0.3, 1.7, 3.1, 4.5, 5.9   **13.** 37   **15.** −49   **17.** 9   **19.** 12   **21.** 1.4   **23.** −8   **25.** 2500   **27.** 5050   **29.** $\frac{495}{2}$
31. −140   **33.** 301.5   **35.** $-\frac{375}{4}$ **37.** 8   **39.** $\frac{1}{2}$ **41.** 4.9   **43.** 320   **45.** $\frac{63}{8}$ **47.** 2

▶ **Problem Solving 19, p. 530**

1a. 6000 ft   **b.** −93°F   **c.** $T_n = [72 - \frac{1}{2}n]°F$ **3a.** $156   **b.** $756   **c.** 12
5a. 35 ft   **b.** 222 ft   **c.** 14 sec to the nearest second   **7.** 25 layers

▶ **Practice 13.3, pp. 537–38**

1. 512   **3.** −15   **5.** $\frac{1}{512}$ **7.** $-\frac{1}{1024}$ **9.** $\frac{81}{16}$ **11.** 2044   **13.** 0   **15.** −266   **17.** $-\frac{1261}{3456}$ **19.** 32   **21.** $\frac{128}{3}$
23. $\frac{3}{4}$ **25.** $\frac{1}{4}$ **27.** $\frac{3}{10}$ **29.** $\frac{1}{9}$ **31.** $\frac{2}{3}$ **33.** $\frac{19}{33}$ **35.** $\frac{542}{99}$ **37.** $\frac{242}{45}$ **39.** $\frac{3}{7}$ **41.** $\sqrt[3]{2}$ **43.** 9   **45.** −9
47. 35   **49.** 7   **51.** 8   **53.** 56   **55.** $81x^4$ **57.** $32x^5$ **59.** $64x^6$ **61.** $-x^7$

▶ **Problem Solving 20, p. 540**

1a. 2.7   **b.** 120 m   **c.** 140 m   **3a.** 10,000   **b.** 10^{2048} **c.** 10^n **5a.** $22.20, $510   **7a.** 10.3 cm   **b.** 88.2 cm
c. 160 cm   **9a.** 128   **b.** 1024   **11a.** 12   **b.** $\frac{4\sqrt{3}}{3}$

▶ **Practice 13.4, pp. 547–48**

1. $y^3 + 3y^2z + 3yz^2 + z^3$ **3.** $16u^4 - 32u^3v + 24u^2v^2 - 8uv^3 + v^4$ **5.** $256q^4 - 768q^3r + 864q^2r^2 - 432qr^3 + 81r^4$
7. 10   **9.** 126   **11.** 8   **13.** $35a^3b^4$ **15.** $-36c^2d^7$ **17.** $2{,}520{,}000e^5d^4$ **19.** $270f^3g^2$ **21.** $-787{,}320h^9k$
23. $1{,}968{,}300m^2n^7$

Final Review, pp. 551–56

1. 10   **2.** $24 - 8a$ **3.** $8a^9b^{-6}$ **4.** 27,000,000 times   **5.** $\frac{7}{2}$ **6.** $\frac{2}{9}$ **7.** $\{c \mid c < \frac{3}{2}\}$ **8.** −6, 12
9. $y = \frac{c}{b} - \frac{a}{b}x$ **10.** $41\frac{1}{4}$ square feet   **11.** $y = -2x - 2$ **12.** $y = \frac{3}{4}x + \frac{15}{4}$ **13.**

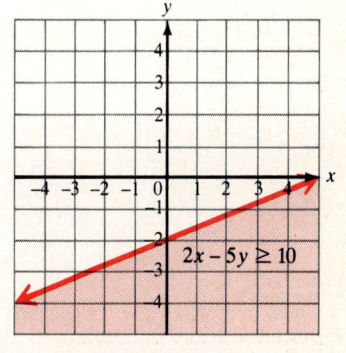

$2x - 5y \geq 10$

14a. $y = 2n + 1787$ **b.** 99th   **c.** 1887   **15.** (1, −3)   **16.** (2, −1, −2)   **17.** (0, −1)   **18.** 750 mL
19. $2p^3 - 2p^2 - 7p + 1$ **20.** $2q^3 - 7q^2 + 14q - 12$ **21.** $4x^2y^2 - 2x - 3y^2 + x^3y$ **22.** 7
23. $s^3 + 3s^2 + 6s + 18 + \frac{49}{s-3}$ **24.** $1464.10   **25.** $(2u - 3v)(4u^2 + 6uv + 9v^2)$ **26.** $(t + 14)(t - 3)(t + 3)$
27. $m(m + 6)(m + 8)$ **28.** $\frac{3}{2}, \frac{2}{3}$ **29.** 23 ft   **30.** $\frac{1}{h-3}$ **31.** $\frac{3k^3 - 8k^2 + 18k - 24}{k^3 - 3k^2 - 4k + 12}$ **32.** $\frac{1+g}{g-1}$
33. −2, 3   **34.** 3 and −2   **35.** 9   **36.** $\frac{2r^3s\sqrt{5t}}{5t^2}$ **37.** 3   **38.** $\frac{75}{13} + (-\frac{167}{13})i$
39. $h = \frac{348{,}100}{r^2}$; $h \neq 0$ and $r \neq 0$ **40.** $\frac{-2 \pm 3\sqrt{2}}{2}$ **41.** $\pm 2i, \pm\sqrt{3}$

42. $\{x | x < -4 \text{ or } 1 < x \leq 4\}$ **43.** 60 mph   **44.** [graph of $y = x^2 - 2x - 3$]   **45.** center $(6, -2)$, radius 8

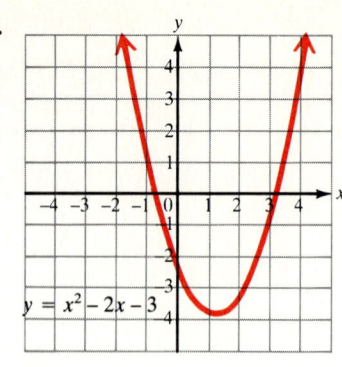

46. [graph of $4x^2 + 25y^2 = 100$]   **47.** [graph of $-\dfrac{x^2}{3^2} + \dfrac{y^2}{2^2} = 1$]

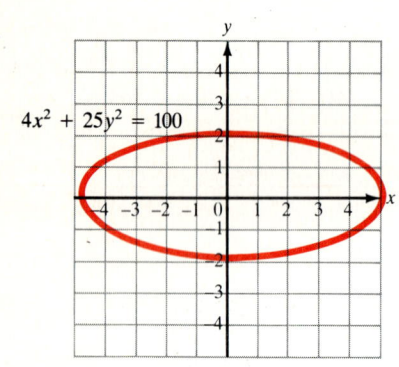

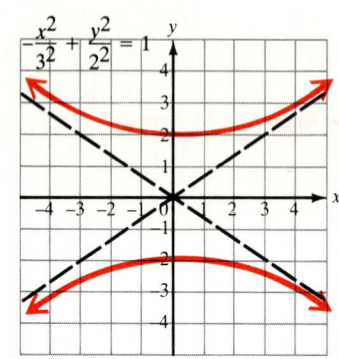

48. $\left(\dfrac{81}{29}, \dfrac{43}{29}\right), (1, -3)$ **49.** 406 cm   **50a.** $\{x | x \geq -2\}$ **b.** $\{y | y \geq 3\}$ **51.** b and c **52.** -7

53. $y = \dfrac{2}{x} - 3$ **54a.** $d = 16t^2$ **b.** 400 ft   **c.** 18.2 sec   **55.** [graph of $y = \log_2 x$]

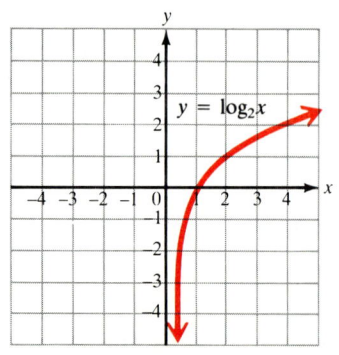

56. $\log \dfrac{x^{1/2} z^{3/4}}{y^3}$ **57.** 3.1875   **58.** 0.0347   **59.** $\dfrac{9}{4}$ **60.** 5   **61.** \$30,200   **62.** 2   **63.** 31,007   **64.** $\dfrac{256}{3}$

65. $-280a^4 b^3$ **66a.** 1.9 ft   **b.** 26.8 ft   **c.** 42 ft

Index

abscissa, 88
absolute value, 10, 75
absolute value equations, 75–76
absolute value functions, 450
absolute value inequalities, 76–78
Absolute Value Rules for Equations and Inequalities, 75–78
ac method of factoring, 220–22
addend, 15
addition
    associative property of, 31–32
    commutative property of, 31–32
    of complex numbers, 325
    distributive property of multiplication over, 32
    of like radicals, 307–308
    of polynomials, 170–71
    of rational expressions, 265–68
    of signed numbers, 15
addition method
    for solving quadratic systems, 414–15
    for solving second-order systems, 139
    for solving third-order systems, 146
Addition Rule for Equations, 50–51
Addition Rule for Inequalities, 69
additive inverse, 2
    of polynomials, 171–72
algebraic methods, 137–40
antilogarithms, 497, 500–501
argument, 485
arithmetic sequence/arithmetic progression, 523–25
arrow diagrams, 430
associative property of addition, 31–32
associative property of multiplication, 31–32
asymptotes, 405
axes, 88
axis of symmetry, 378–79

bars, 23, 25
base, 21
    of exponential functions, 481
    of logarithmic functions, 485
binomial expansion, 541–46
binomials, 168
    completing the square of, 346
    multiplying, 183–86
    squaring, 184–85, 227, 316
boundary value, 11
braces, 2, 23
brackets, 23
branch (of hyperbola), 405
break-even point, 119
broken curves, 416

carbon 14 dating, 512
Cartesian coordinate system, 95
    see also rectangular coordinate system
center
    of a circle, 391
    of an ellipse, 400
    of a hyperbola, 405
change in x, 101
change in y, 101
Change-of-Base Formula, 502
characteristic, logarithmic, 498
chords, 400
circles, 378
    graphing, 389–94
closed regions, 416
coefficient determinant, 159
coinciding lines, 112
combined variations, 469–70
combining like terms, 34
common denominators, rational expressions containing, 265–68
    see also LCD
common differences, 523

common logarithms, 497
common ratio, 531
commutative property of addition, 31–32
commutative property of multiplication, 31–32
compact form, 72
completing the square, 346–48
complex fractions, 271–76
complex numbers, 323–26, 329–32
complex solutions, 411
composite numbers, 206
compound inequalities, 12
compound-interest, 511
computational rules, 21–26
    for scientific notation, 43
conditional equalities, 56
conic sections, 378
conjugates of complex numbers, 330
consistent system, 132
constant functions, 449
constant of proportionality, 465
constant series, 520
constants, 3
    variation, 465
coordinates, 9, 88
correspondences, 430
Cramer, Gabriel, 159
Cramer's Rule, 159–62
critical numbers, 369
critical point method, 369
    for solving quadratic inequalities, 371–72
    for solving rational inequalities, 369–70
cube roots, 228
cubes, 228–29

decibels, 513
decimals, 4–5
    solving equations containing, 64

degree
  of a monomial, 169
  of a polynomial, 169
denominator, rationalizing, 303
denominator of complex fractions, 271
density, 46
dependent system, 132, 138
dependent variables, 432
Descartes, René, 95
determinants, 153–56
difference, 16
  common, 523
difference of two squares, 227
direct variation, 465–67
discriminants, 354–55
distance-between-two-points formula, 390
distinct parallel lines, 113
distributive property of multiplication
      over addition, 32–33
dividend, 17
division
  of complex numbers, 330–32
  of polynomials, 189–94, 197–200
  by radical factors, 310
  of rational expressions, 260
  of signed numbers, 17–18
  synthetic, 197–200
  using scientific notation, 43
Division Method, 273–76
divisor, 17
domain, 430, 432–34

e, 482
element of a set, 2
ellipses, 378
  graphing, 399–402
ellipsis symbol, 2
empty set, 3
equality value, 369
equations
  absolute value, 75–76
  conditional, 56
  containing rational expressions,
      279–82
  exponential, 480–82
  first-degree, 50–52, 56
  graphs of, 95
  linear, 95–98, 145
  literal, 81–82
  logarithmic, 485–88
  Power Principle for, 313
  quadratic, 338–72
  radical, 313–16, 360–61
  rational, 359–60
  solving, 50–68, 75–76, 81–82
  solving by factoring, 239–42
  specifying relations by, 431
  system of, 132
equivalent inequalities, 69
excluded value, 252–53

exponential equations, 505–508
exponential functions, 480–82
  base of, 481
  natural, 482
exponential notation, 21–23
exponents, 21
  integral, 37–40
  Product Rule for, 38–39
  Quotient Rule for, 39
  rational, 294–98
expressions, 23
Extended Power Rule, 40
Extended Zero-Product Property for
      Polynomials, 241–42
extraneous solution, 313

factorial function, 543
factoring, 206–48
  ac method, 220–22
  general strategy, 233–36
  simplifying radicals by, 301–302
  to solve quadratic equations, 340
factors, 16
  rationalizing, 309
finite sequences, 518
first-degree equations, 50–52
  in one variable, 50
FOIL method, 183–84, 309, 329
fractions, 4
  complex, 271–76
  solving equations containing, 63
functional notation, 443–46
Fundamental Rule for Rational
      Expressions, 255
Fundamental Rule Method, 271–72
Fundamental Theorem of Arithmetic, 206
functions, 437–40
  absolute value, 450
  constant, 449
  evaluating, 443–46
  exponential, 480–82
  factorial, 543
  graphing, 451–54
  greatest integer, 450
  identity, 449
  logarithmic, 485–88, 491–94
  one-to-one, 461
  quadratic, 449
  step, 450

Gauss, Karl Friedrich, 525
GCF (greatest common factor), 206–208
general factoring strategy, 233–36
general term, 518
geometric sequence/geometric
      progression, 531
geometric series, 532–36
  infinite, 534
Golden Rectangle, 368

graphing, 9
  circles, 389–94
  compound inequalities, 12
  conic sections, 378–408
  domain, 433
  ellipses, 399–402
  to estimate functional values, 444
  exponential functions, 480–82
  functions, 438–40, 449–54
  hyperbolas, 405–408
  inverses of relations, 459–62
  linear equations, 95–98
  linear inequalities, 121–26
  logarithmic, 487–88
  ordered pairs, 88–91
  parabolas, 378–84
  quadratic systems, 411, 417–18
  range, 433
  second-order systems, 133–34
  simple inequalities, 11
  slope, 103–106
  specifying relations by, 431, 432
  third-order systems, 146
greatest common factor (GCF), 206–208
greatest integer function, 450
grouping symbols, 23–25
  nested, 24–25

half-life, 512
half-planes, 121
horizontal line test, 462
hyperbolas, 378
  graphing, 405–408
hypotenuse, 389

i, 323
identities, logarithmic, 491
identity, 56
identity functions, 449
imaginary numbers, 324
imaginary part, 324
imaginary units, 323
incomplete quadratic equations, 339
inconsistent system, 132, 138
independent system, 132, 138
independent variables, 432
index, 294, 308
index of summation, 519
inequalities
  absolute value, 76–78
  Addition Rule for, 69
  compound, 12
  equivalent, 69
  involving a product of linear
      factors, 372
  linear, 121–26
  Multiplication Rule for, 70
  quadratic, 371–72
  quadratic systems of, 416–18

rational, 369–70
simple, 11
solutions of, 69
solving, 69–72, 76–78
inequality symbols, 69
infinite geometric series, 534–35
infinite sequences, 518
integers, 2, 6
integral exponents, 37–40
integral polynomials, 206
intercept method, 97–98
intersection of sets, 3
interval, 369
inverse operations, 17
inverse relations, 457
inverse variation, 467–68
irrational numbers, 5, 6

joint variations, 469

kth term, 541

LCD (least common denominator), 63
  of rational expressions, 265–68
least common denominator (LCD), 63
left members, 50
legs of a right triangle, 389
like radicals, 307–308
  combining, 307
like terms, 55
limits, 534
linear equations, 95–98, 109–14, 145
linear inequalities, 121–26
linear relationships, 117–19
linear terms, 338
literal equations, 81–82
literal part, 25
logarithmic equations, 505–508
logarithmic functions, 485–88, 491–94
logarithmic identities, 491
logarithms, 485
  common, 497
  natural, 500
  properties of, 491

major axis, 400
mantissa, 498
maximum values, 387
members, 2
minimum values, 387
minor axis, 400
minors, 154
minuend, 16
monomials, 168
  degree of, 169
  dividing, 189–90
  factoring, 206

multiplying, 177
squaring, 227
multiplication
  associative property of, 31–32
  commutative property of, 31–32
  of complex numbers, 329–30
  distributive property of, over addition, 32
  distributive property of, over subtraction, 33
  of polynomials, 177–80
  of radical expressions, 308–309
  of rational expressions, 259
  of signed numbers, 16–17
  using scientific notation, 43
Multiplication Rule for Equations, 51–52
Multiplication Rule for Inequalities, 70

nappes, 378
natural exponential function, 482
natural logarithms, 500–502
natural numbers, 2, 6
negative numbers, 2
nested grouping symbols, 24–25
Newton's Constant, 44
nonterminating nonrepeating decimals, 4–5
nth-term formula, 523–24, 531
null set, 3
number line, 2
  see also real-number line
numbers
  complex, 323–26, 329–32
  composite, 206
  critical, 369
  imaginary, 324
  irrational, 5, 6
  natural, 2, 6
  rational, 4–5, 6
  real, 5–46, 22, 323
  signed, 2, 15–18
  whole, 2, 6
numerator, 271
numerical coefficients, 25

one-to-one functions, 461
open regions, 416
opposites, 2
Order of Operations Rule, 23–24
order of the radical, 294
order relations, 9–10
ordered pairs, 88–91
  specifying relations with, 430, 431
ordered triple, 145
ordinate, 88
origin, 2, 88

parabolas, 378
  graphing, 378–84, 460

parallel lines, 112
  distinct, 113
parentheses, 23, 33
  solving equations containing, 56–57
percents, 64
perfect square trinomial (PST), 227–28
perpendicular lines, 114
point-slope formula, 109
plane, 145–46
plotting a point, 89
polynomials, 168–200
  additive inverse of, 171
  degree of, 169
  Extended Zero-Product Property for, 241–42
  factoring, 206–48
  integral, 206
  irreducible over the integers, 214
  special, 227–30
  standard form, 170
  Zero-Product Property for, 239–41
positive numbers, 2
Power Principle for Equations, 313, 361
Power Property, 491
Power Rule, 40
powers, 43
prime factorization, 206
prime numbers, 206
principal diagonal, 153
principal square roots, 294
product, 16
Product Property, 491
Product Rule for Exponents, 38–39
Product Rule for Radicals, 301, 308, 329
proportion, 468
PST (perfect square trinomial), 227–28
Pythagoras, 5, 389
Pythagorean Theorem, 389, 397–98

quadrants, 88
quadratic equations, 338–72
  incomplete, 339–42
  renaming as, 359–62
quadratic formula, 351–56
quadratic function, 449
quadratic inequalities, 371–72
quadratic systems of equations, 411–18
quadratic systems of inequalities, 416–18
quadratic term, 338
quotient, 17
Quotient Property, 491
Quotient Rule for Exponents, 39
Quotient Rule for Radicals, 301

radical equations, 313–16, 360–61
radical expressions, 294
radical form, simplest, 303
radical notation, 294
radicals, 23
  computing with, 307–10

radicals (*continued*)
   like, 307–308
   order of, 294
   reducing the order of, 302
   rules of, 301
   simplifying, 301–304
radicand, 294
radius, 391
range, 430, 432–34
range of summation, 519
rational equations, renaming as quadratic equations, 359–60
rational exponents, 294–98
   simplifying expressions containing, 297–98
rational expressions, 252–90
   adding, 265–68
   dividing, 260
   equations containing, 279–82
   multiplying, 259
   reducing to lowest terms, 255–56
   renaming in higher terms, 262
   simplifying, 252–56
   subtracting, 265–68
   undefined, 253
rational inequalities, 369–70
rational numbers, 4–5, 6
rationalizing factors, 309
real numbers, 5–46, 323
   squaring, 22
real-number line, 9
real part, 324
real solutions, 411
reciprocal, 51
rectangular coordinate system, 88, 95
reducing the order of the radical, 302
reflections, 459
relations, 430–32
   as functions, 437–40
repeating decimals, 4–5
Richter scale, 514
right members, 50
right triangle, 389–90
roots, 50
   see also cube roots; square roots
roster method, 2
rules of radicals, 301

scientific notation, 43–44
second-order determinant, 153–54
second-order systems, 132–34, 137–40
   Cramer's Rule for, 159–60
secondary diagonal, 153
sequences, 518–19
   arithmetic, 523–24
   geometric, 531–32
series, 519–20
   arithmetic, 525–26
   constant, 520
   geometric, 532–36
set-builder notation, 56

sets, 2–6
   intersection of, 3
   null, 3
   solution, 11, 50, 69
   union of, 4
sign array, 154
signed numbers, 2
   computing with, 15–18
significant digits, 43
simple inequalities, 11
simplest radical form, 303–304
simplifying terms, 39
slope, 101–106
slope-intercept form, 104–106
solid curves, 416
solution of the system, 132
solution sets, 11, 50, 69
solutions, 11, 50
   complex, 411
   proposed, 51
   real, 411
special polynomials, 227–30
special products, 183–86
Square Root Rule, 341–42, 345
square roots, 22–23
   principal, 294
squares
   of binomials, 184–85
   perfect, 227
standard equation
   of a circle, 392
   of an ellipse, 400
   of a hyperbola, 405
standard form of a polynomial, 239
step function, 450
Straight line depreciation, 117
subsets, 5–6
substitution method, 137, 412–14
subtraction
   of complex numbers, 325
   distributive property of multiplication over, 33
   of like radicals, 307–308
   of polynomials, 172
   of rational expressions, 265–68
   of signed numbers, 16
subtrahend, 16
summation, 519
summation notation, 519–20
symmetry, 459
   axis of, 378–79
synthetic division, 197–200
system of equations, 132–34

tables, specifying relations with, 431
terminating decimals, 4–5
terms, 25
terms of the sequence, 518–19
third-order determinant, 154–56
third-order systems, 145–48

   Cramer's Rule for, 161–62
transverse axis, 405
triangle, right, 389–90
trinomials, 168
   factoring, 211–14, 217–22

union of sets, 4

value of the determinant, 153
variable expressions, 26
variables, 3, 432
variation constant, 465
variations, 465–70
vertex of a parabola, 378–79
vertices of an ellipse, 400
vertices of a hyperbola, 405

whole numbers, 2, 6
word problems
   applied variation problems, 473–76
   digit problems, 151–52
   evaluating formulas containing radicals, 321–22
   evaluating polynomials, 175–76
   finding maximum and minimum values, 387–88
   linear relationships, 117–20
   mixture problems, 143–44
   scientific notation, 44–46
   using arithmetic sequences and series, 529–30
   using equations, 67–68
   using exponential formulas, 511–12
   using factoring, 247–48
   using geometric sequences and series, 539–40
   using logarithmic formulas, 513–14
   using Pythagorean Theorem, 397–98
   using quadratic systems, 423–26
   using rational expressions, 285–90

x-axis, 88
x-coordinate, 88
x-intercept, 97

y-axis, 88
y-coordinate, 88
y-intercept, 97

zero
   absolute value of, 10
   division by, 18
   as origin, 2
   as signed number, 2
zero-product property, 57–58
Zero-Product Property for Polynomials, 239–41
   Extended, 241–42